GREEN
Communications

Theoretical Fundamentals, Algorithms and Applications

Edited by
Jinsong Wu • Sundeep Rangan • Honggang Zhang

CRC Press
Taylor & Francis Group
Boca Raton London New York

CRC Press is an imprint of the
Taylor & Francis Group, an **informa** business

CRC Press
Taylor & Francis Group
6000 Broken Sound Parkway NW, Suite 300
Boca Raton, FL 33487-2742

First issued in paperback 2016

Version Date: 20120801

ISBN 13: 978-1-138-19980-4 (pbk)
ISBN 13: 978-1-4665-0107-2 (hbk)

Visit the Taylor & Francis Web site at
http://www.taylorandfrancis.com

and the CRC Press Web site at
http://www.crcpress.com

Dedicated to the promotion for the research and development in green communications and computing

Contents

3   Energy Saving Solutions and Practices of China Mobile   51

*Sen Bian, Huabin Tang, Xidong Wang, Jinsong Wu, and Zhihong
Zhang, Chih-Lin I, and Bill Huang*

4   Standard Methodologies for Energy Efficiency Assessment   83

Raffaele Bolla, Roberto Bruschi, and Chiara Lombardo

5   Reciprocal Learning for Energy-Efficient Opportunistic Spectrum Access in Cognitive Radio Networks   103

Xianfu Chen, Zhifeng Zhao, Honggang Zhang, Jinsong Wu, and Tao Chen

6   Green Communications for Carbon Emission Reductions: Architectures and Standards   125

Charles Despins, Fabrice Labeau, Richard Labelle, Mohamed Chériet, Alberto Leon-Garcia, and Omar Cherkaoui

10 Energy-Efficient Management of Campus PCs   **247**

Luca Chiaraviglio and Marco Mellia

II Focus on Wireless Communications   **277**

13 Green Wireless Communications under Quality of Service Constraints   321

Deli Qiao, Mustafa Cenk Gursoy, and Senem Velipasalar

14 On the Energy Efficiency-Spectral Efficiency Trade-off in Cellular Systems   353

Fabien Héliot, Efstathios Katranaras, Oluwakayode Onireti, and Muhammad Ali Imran

xiv

18 Energy Efficient Communications in MIMO Wireless Channels 519

Vineeth S. Varma, E. Veronica Belmega, Samson Lasaulce, and Mérouane Debbah

19 Minimising Power Consumption to Achieve More Efficient Green Cellular Radio Base Station Designs   555

John S. Thompson, Peter M. Grant, Simon Fletcher, and Tim O'Farrell

20 Energy Conservation of Mobile Terminals in Multi-cell TDMA Networks   579

Liqun Fu, Hongseok Kim, Jianwei Huang, Soung Chang Liew, and Mung Chiang

21 Energy Efficiency for Wireless Relay Systems   615

Jinho Choi, Duc To, Weixi Xing, Ye Wu, and Shugong Xu

III Focus on Wireline Communications   691

Preface

The telecommunications and information community today is not only facing huge challenges but also embracing great opportunities. The enormous demands for ubiquitous wireless and Internet services have been on the rise in the past and are ever-growing with the proliferation of multimedia-rich mobile communication devices (e.g., smart phones). The volume of transmitted data continues to increase by a factor of approximately 10 every five years. The unprecedented thirst for increased data rates is resulting in the need for ever-increasing capacity-density in broadband telecommunications networks. Even though this has been hugely beneficial for the development of contemporary society, this has also resulted in an unsustainable increase in system complexity, energy consumption, and burgeoning environmental footprint. It is coming to a consensus that the information and communication technologies (ICT) industry has emerged as one of the major contributors to the world power consumption and greenhouse gas (GHG) emission. Currently, 3% of the world-wide energy is consumed by the ICT infrastructures which generate about 2% of the world-wide CO_2 emissions, comparable to the CO_2 emissions by all commercial airplanes or one quarter of the CO_2 emissions by all vehicles around the world. The ICT sector's carbon footprint is expected to quickly grow to 1.4 Gigaton CO_2 equivalents by 2020, nearly 2.7% of the overall carbon footprint from all human activities. Accordingly, increasing awareness of the potential environmental impact induced by the greenhouse gas emissions and exhaustion of non-renewable energy resources spur the critical need to improve the energy efficiency of the telecommunication systems and devices. Moreover, the European Council has set forth a great target as the "20 20 by 2020" initiative, which calls for a reduction of 20% in greenhouse gases and a 20% share of renewable energies in European Union (EU) energy consumption by 2020. In addition to the environmental concerns, there are also economical benefits for telecommunication network operators to reduce the power consumption of their networks. Recently, it has been reported that energy costs can account for as much as half of the annual operating expenses of a mobile service provider. Therefore, a significant proportion of the overall costs in capital expenditure (CapEx) and operational expenditure (OpEx) could be saved through the improvement of energy efficiency. In this context, energy efficient information and communication technologies is expected to play a major active role in the reduction of the world-wide energy requirements in optimizing energy generation, transportation, and consumption. Doubtlessly, it has already been a world-wide concern that if the aggregated energy consumption of all networking systems and devices would follow the growth trajectories of Internet traffics (i.e., about 50% per year), the environmental and economical consequences would lead to a daunting nightmare for all human beings. Thus, making ICT equipments and applications "greener" in terms of energy consumption could not only have a tangible positive impact on environment, but also help telecommunications operators to attain long-term profitability. Moreover, energy-efficient communications and networks could help the world reduce dependence on fossil fuel, enable efficient energy resources distribution, and ultimately achieve sustainable prosperity around the world. Recently, a myriad of communications and information technologies have already been exploited by the global Smart Grid initiatives to empower conventional power grid to support two-way energy and information flow. To meet the fundamental challenges of increasing the energy efficiency in telecommunication systems and devices, it is necessary to resort to a plethora of paradigm-shifting novel technologies in complying with the philosophies and principles of environmental-friendly green communications and computing, such as energy-efficient network architecture & protocols, energy-efficient wireless and wireline transmission techniques, energy-aware backbone wired infrastructures, energy-efficient data centers and cloud computing, green smart home and Smart Grid, green cognitive networks, opportunistic spectrum sharing without causing harmful interference pollution, and so on.

Accordingly, this book is completely devoted to the energy efficiency improvement within the framework of green communications, which presents a holistic view of the relevant fundamental scientific challenges and essential technical approaches for increasing the whole-scale energy efficiency in communications and computing networks. Although there have been a number of publications, conferences, tutorials, short courses and international projects in this area, there does not exist a single comprehensive reference book introducing the state-of-the-art of the technological development and advancement in green communications to the potential readers in a structured manner. Therefore, there is a profound need for a systematic reference book that thoroughly covers a wide range of topics within green communications, including definitions of energy efficiency, energy consumption metrics, theoretical fundamentals and algorithms, basic implementation functionalities, current applications and future research issues for green communications.

The foremost objective of this book is to encourage the relevant academic researchers (including graduate students), industrial engineers and designers, and regulatory practitioners to take up this exciting emerging research field as quickly as possible. Meanwhile, it is our hope that this book will also provide informative advices to motivate the existing research, industrial and regulatory communities to take forward the state-of-the-art in new ways. According to different focuses on green topics, this book is divided in three parts: general topics (Part I), wireless communications (Part II), and wireline communications (Part III). The first part of the book provides detailed instructions to various representative killer applications and case studies of green communications, such as Smart Grid, cloud computing, carbon emission reduction, cognitive radio, FPGA power consumption, and energy efficiency assessment standards. Afterwards, various key wireless aspects of green communications are thoroughly addressed in Part II, ranging from traffic load management in cellular networks, energy-efficient base stations and mobile terminals, MIMO and relaying systems, video steaming, radio access networks (RAN), dynamic spectrum accessing, to sensor and ad hoc networks. Part III is focused on the area of wireline communications, covering optical networks, IP-over-WDM backbone networks, virtual infrastructures planning, and energy-aware networks management & contents distribution.

This book is made possible by the extensive support of numerous individuals, which comprises solid contributions from many prominent researchers working in the green communications area around the world. Firstly, we are deeply indebted to our active authors, who share in our vision on the energy-efficient and environmental-friendly green communications and have produced well-written and high-quality chapters. Definitely, it would not have been possible to publish a book of this quality and breadth without the active authors sacrificing their days and nights to put together these excellent chapters in a good shape. Secondly, we owe our special thanks to the support from the Technical Sub-Committee on Green Communications and Computing (TSCGCC) of IEEE Communications Society. The editors and quite a few key authors of this book are the founding members of TSCGCC. Thirdly, we would like to thank the CRC Press editorial team, especially Ruijun He and Amber Donley, for their patience, hard-work, guidance, support, and encouragement during the whole period of the creation of this book. Finally, we would like to thank our families for their warm support.

Jinsong Wu
Bell Laboratories, China

Sundeep Rangan
Polytechnic Institute of New York University, United States

Honggang Zhang
Zhejiang University, China

Editor Biographies

Jinsong Wu

Jinsong Wu is the Founding Chair of Technical Subcommittee on Green Communications and Computing (TSCGCC), IEEE Communications Society, which was officially approved and established in December 2011. He is the Vice-Chair, Track on Green Communication Systems and Networks, the Selected Areas in Communications Symposium, IEEE GLOBECOM 2012. He is one of Technical Program Committee Chairs, the 2012 IEEE Online Conference on Green Communications. He is the proposer and the Moderator/Chair of the Technical Panels on Green Communications and Computing in the IEEE INFOCOM 2012, IEEE ICC 2012, and IEEE GLOBECOM 2012. He obtained Ph.D. degree in electrical engineering from Queen's University, Kingston, Canada. Since 2010, he has worked as Research Scientist in Bell Laboratories, Shanghai, China. He has experienced research and development positions relevant to communications engineering in Nortel Networks Canada, Philips Research USA, and Sprint-Nextel USA. His recent research interests lie in green communications and computing, communications theory and signal processing, cognitive networks, space-time-frequency processing and coding, cooperative communications, quality of service, iterative processing, and communication optimization. He has served as technical program committee members in more than 26 leading international telecommunications relevant conferences or workshops, such as IEEE GLOBECOM, IEEE ICC (main symposia and workshops), IEEE VTC (main tracks and workshops), IEEE ISCIT, iCOST, WAC, FutureTech, and so on. He currently is an IEEE Senior Member.

Sundeep Rangan

Sundeep Rangan received the B.A.Sc. at the University of Waterloo, Canada and the M.Sc. and Ph.D. at the University of California, Berkeley, all in Electrical Engineering. He has held postdoctoral appointments at the University of Michigan, Ann Arbor and Bell Labs. In 2000, he co-founded (with four others)

Flarion Technologies, a spin off of Bell Labs, that developed Flash OFDM, one of the first cellular OFDM data systems. Flarion grew to over 150 employees with trials worldwide. In 2006, Flarion was acquired by Qualcomm Technologies where Dr. Rangan was a Director of Engineering involved in OFDM infrastructure products. He joined the Electrical and Computer Engineering Department at Polytechnic Institute of New York University, USA, in 2010. His research interests are in wireless communications, signal processing, information theory and control theory.

Honggang Zhang

Honggang Zhang is a Full Professor of Department of Information Science and Electronic Engineering as well as the Co-Director of York-Zhejiang Lab for Cognitive Radio and Green Communications at the Zhejiang University, China. He is an Honorary Visiting Professor of the University of York, UK. He received the Ph.D. degree in Electrical Engineering from Kagoshima University, Japan, in March 1999. From October 1999 to March 2002, he was with the Telecommunications Advancement Organization (TAO) of Japan, as a TAO Research Fellow. From April 2002 to November 2002, he joined the TOYOTA IT Center. From December 2002 to August 2004, he has been with the UWB Research Consortium, the Communications Research Laboratory (CRL) and the National Institute of Information and Communications Technology (NICT) of Japan. He was the principle author and contributor for proposing DS-UWB in IEEE 802.15 WPAN standardization task group. From September 2004 to February 2008, he has been with CREATE-NET (Italy), where he leaded its wireless teams in exploring Cognitive Radio (CR) and UWB technologies while participated the European FP6/FP7 projects (EUWB, PULSERS 2). Dr. Honggang Zhang serves as the Chair of Technical Committee on Cognitive Networks (TCCN) of the IEEE Communications Society (ComSoc). He was the Co-Chair of IEEE Globecom 2008 Symposium. He was the founding TPC Co-Chair of CrownCom 2006 as well as the Steering Committee Member of CrownCom 2006-2009. In the area of green communications, Dr. Honggang Zhang was the Lead Guest Editor of the IEEE Communications Magazine special issues on "Green Communications". He was the General Chair of IEEE/ACM GreenCom 2010 (2010 IEEE/ACM International Conference on Green Computing and Communications) and the Co-Chair of the IEEE International Workshop on Green Communications (GreenComm 2010-2011) in conjunction with IEEE ICC/Globecom. He is the co-author/editor of the book "Cognitive Communications - Distributed Artificial Intelligence (DAI), Regulatory Policy & Economics, Implementation", published by Wiley Press.

Part I

General Topics

1

Runtime-Controlled Energy Reduction Techniques for FPGAs

Assem A. M. Bsoul

University of British Columbia, Canada

Stuart Dueck

University of British Columbia, Canada

Steven J. E. Wilton

University of British Columbia, Canada

CONTENTS

1.1 Introduction

Green communications can only be as efficient as the underlying technology used to implement protocols and algorithms. Over the past 50 years, advances in Integrated Circuit fabrication technology have provided system designers with unprecedented amounts of computing power. This has led to dramatic changes in feature sets of mobile devices along with the infrastructure required to support this new "mobile world". The exponential growth in semiconductor technology, however, has led to a new problem: the cost of designing, verifying, and fabricating an integrated circuit has become so prohibitive, that today, very few enterprises are capable of creating state-of-the-art chips. This has led many designers to employ standard parts such as Field-Programmable Gate Arrays (FPGAs) to leverage advanced process technologies without the cost and risk of designing a custom integrated circuit. FPGAs are pre-designed substrates which can be configured, in seconds, to implement any digital circuit. By amortizing the fabrication cost over all FPGA users, these devices essentially provide large-scale integration without requiring access to a state-of-the-art chip fabrication plant.

When using FPGAs for "green communications" applications, the power dissipation of these devices is critical. A recent study observed that typical FPGA designs can require 11 times more power than a fixed-function chip [1]. This can be problematic, both in hand-held devices, which rely on battery power for operation, and in the base station, where power consumption and heating is increasingly becoming a concern. In order to continue to provide the advantages of configurability to designers of leading edge devices, researchers have invested considerable effort into finding techniques to reduce the power consumption of FPGAs.

In this chapter, we focus on the power-efficiency of integrated circuit technology, focusing on FPGA techniques. Understanding the capabilities and limitations of the underlying technology is essential when designing a "green" device. We first give an overview of FPGA technology, and show how the configurability inherent in FPGAs leads to increased power dissipation. We then outline methods to reduce the power dissipation, both in custom integrated circuits and in FPGA devices. We then describe a new FPGA architecture that has been optimized for applications with long idle times; such a device may be very suitable for the types of applications described elsewhere in this book.

1.2 Island-Style Field-Programmable Gate Arrays

To better understand power dissipation reduction techniques, it is important to understand how an FPGA achieves its flexibility. FPGAs contain large numbers of pre-fabricated tracks connected using programmable switches; these tracks and switches are responsible for the majority of the power dissipation in an FPGA. In this section, we describe the architecture (internal structure) of an FPGA, as well as the Computer-Aided Design (CAD) algorithms that map circuits to an FPGA. The architecture we describe is a simplified model that is representative of commercial FPGAs from vendors.

1.2.1 Architecture

Figure 1.1 shows the internal structure of an FPGA. This architecture is referred to as a *tile-based* architecture because the same tile is replicated many times through the chip. Each tile (or slice) consists of a logic cluster (also known as a logic block or configurable logic block) and the routing resources that provide connectivity between the cluster and the routing channels, and between the individual wires in the different routing channels.

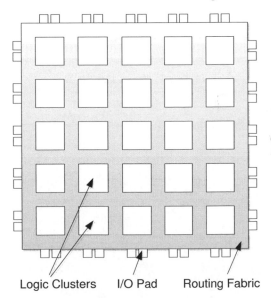

Logic Clusters   I/O Pad   Routing Fabric

FIGURE 1.1 General architecture for tile-based FPGAs.

Each logic cluster consists of a number of function generators that can implement sequential or combinational functions. These function generators are called basic logic elements (BLE). Figure 1.2 shows a logic cluster (LC) of N BLEs. An internal switch matrix in the cluster provides connectivity between the inputs of the cluster and the outputs of the BLEs to the inputs of the BLEs. Each BLE (Figure 1.3) has a K-input lookup table (K-LUT), which consists of a multiplexer and 2^K configuration bits; each LUT can implement any function of K inputs by configuring the function in the 2^K configuration bits, and providing the K inputs to the select lines of the multiplexer. The output of a K-LUT can be registered or unregistered by appropriately configuring the 2:1 multiplexer shown in the figure.

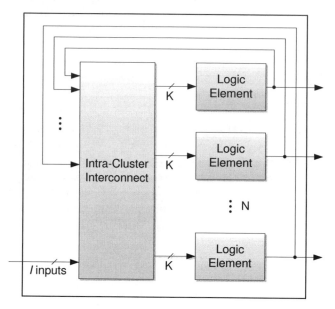

FIGURE 1.2 A cluster of *N* BLEs.

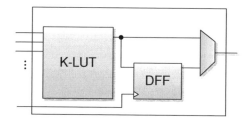

FIGURE 1.3 Basic logic element.

The logic clusters are surrounded by a sea of routing resources. Figure 1.4 shows the routing architecture for island-style FPGAs. The routing resources consist of fixed pre-fabricated wires that can be connected using programmable switches. The pre-fabricated wires lie in horizontal and vertical channels, and each wire typically spans one or more logic blocks. At the intersection of each horizontal and vertical channel is a *switch block*; each switch block contains programmable switches that provide connectivity between the incident wires. Programmable switch blocks also exist in the *connection blocks* to provide connectivity between the fixed wires and the neighboring logic blocks. The actual topology of the wires and switches varies between devices, and has been the subject of intense academic and industrial optimization studies. The selection of a suitable topology balances the routability and speed of the device with the area overhead of the programmable switches.

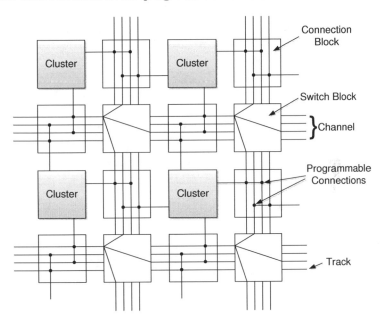

FIGURE 1.4 Routing architecture for island-style FPGA.

In actual FPGAs, other resources typically exist that provide additional functionality, such as adders carry chains, embedded memory blocks, DSP blocks, and embedded processor cores. These blocks provide area, performance, and power benefits compared to implementing the corresponding circuitry in the general-purpose logic blocks.

1.2.2   CAD Flow for FPGAs

In order to implement an application on an FPGA, each configuration SRAM cell need to be set appropriately. Since there are typically millions of these SRAM cells, it is only feasible to generate the values of these cells using appropriate computer-aided design (CAD) tools. FPGA CAD tools have evolved significantly since their appearance in the 1980s [2]. Design entry is done usually using high-level hardware languages such as VHDL and Verilog, or higher-level languages such as System-C. The CAD tools then transform the high-level description of a circuit to the appropriate SRAM cells values; the task is decomposed into multiple steps to make the process more tractable. Figure 1.5 shows the steps in an FPGA CAD flow, with the output being a programming file that is used to configure the programmable resources in an FPGA device.

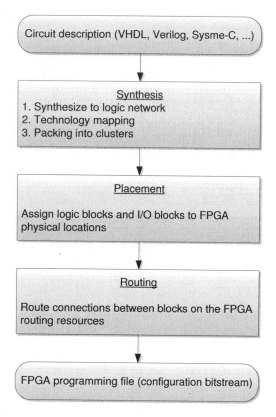

Circuit description (VHDL, Verilog, Sysme-C, ...)

Synthesis
1. Synthesize to logic network
2. Technology mapping
3. Packing into clusters

Placement

Assign logic blocks and I/O blocks to FPGA physical locations

Routing

Route connections between blocks on the FPGA routing resources

FPGA programming file (configuration bitstream)

FIGURE 1.5 General FPGA CAD flow.

The high level description of a circuit is transformed into a network of logic gates in the *synthesis* step. During synthesis, technology-independent logic optimization is performed to simplify the logic and remove redundancies whenever possible. After that, the resulting logic is mapped to LUTs and flip-flops (FFs) that correspond to the components in the target FPGA device; this is called technology-dependent mapping. These LUTs and FFs are packed into logic blocks (or logic cells, LCs) in the packing step. Thus, the final output of this step is a network of logic blocks that will be mapped to the physical locations on an FPGA device.

Following the synthesis step, the *placement* step assigns each block (LC or IO block) to a physical location on an FPGA device. The main objective during placement is to produce a routable circuit, i.e., a circuit for which the router can route the nets that connect the different blocks on the available FPGA routing resources. Other objectives include optimizing the timing performance of the circuit by minimizing the resulting critical path delay.

The final step in the CAD flow is *routing*. In this step, the nets that connect the different blocks in a circuit are mapped to the available routing tracks in an FPGA device, and the switches that connect the routing tracks are set appropriately. In addition to producing a legal routing by using the available routing resources, routers strive to find solutions with good timing and power performance by keeping critical wires as short as possible.

Industrial CAD tools also provide platforms for developing embedded system-on-a-chip applications, such as the SOPC builder from Altera and the XPS from Xilinx. These tools facilitate designing large applications for FPGAs by providing the ability to integrate IP cores into larger systems.

1.3 Power Dissipation in VLSI

In this section, we describe the sources of power dissipation in CMOS integrated circuits. Power can be dissipated in two ways: as a result of gates switching logic state (this is termed *dynamic power*) and as a result of static current that occurs when gates are not switching (this is termed *static power*). Each of these will be described separately.

In the following, it is important to differentiate between *power* and *energy*. Power is measured in Watts while energy is expressed in Joules, where $1\text{ W} = 1\text{ J/s}$. The instantaneous power through a device is computed by taking the product of the voltage across it and the current through it.

$$P(t) = I(t)V(t) \tag{1.1}$$

Energy can be computed as the integral of the instantaneous power over a time interval T, which is basically the area under the power curve.

$$E = \int_0^T P(t)\,\mathrm{d}t \tag{1.2}$$

The average power is then

$$P_{ave} = \frac{E}{T} = \frac{1}{T}\int_0^T P(t)\,\mathrm{d}t \tag{1.3}$$

In Figure 1.6 the height denotes the power and the area under the curve is the energy. Approach 1 uses more power for a shorter period of time, but both approaches use the same amount of energy.

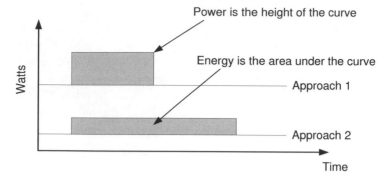

FIGURE 1.6 Power vs. energy.

1.3.1   Dynamic Power

Dynamic power dissipation is due to switching and short-circuit current:

$$P_{dynamic} = P_{switching} + P_{sc} \qquad (1.4)$$

Switching power is due to the charging and discharging of parasitic load capacitances, as illustrated in Figure 1.7. The amount of dynamic power dissipated is given by

$$P_{switching} = C_{Load}V_{DD}^2 f_{switch} \qquad (1.5)$$

where C_{Load} is the capacitive load and f_{switch} is the frequency of the input switching. Power is only dissipated when the load capacitance is charged and then discharged, so the frequency of this will typically not be the same as the clock frequency of a circuit.

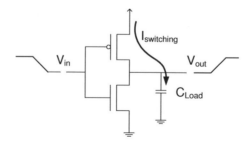

FIGURE 1.7 Switching current charges the load capacitance.

Short-circuit current, sometimes referred to as crowbar current, occurs when both pMOS and nMOS networks are partially on (Figure 1.8). The power dissipated is proportional to the duration of time the crowbar current occurs, t_{sc}, the supply voltage, V_{DD}, the total switching current, I_{peak}, and the switching frequency, $f_{0 \rightarrow 1}$. The short-circuit power can be written as:

$$P_{sc} = t_{sc}V_{DD}I_{peak}f_{0 \rightarrow 1} \qquad (1.6)$$

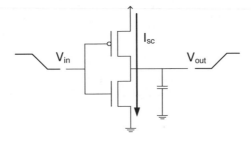

FIGURE 1.8 Short circuit current.

1.3.2 Static Power

Static power dissipation is due to three main sources of leakage current in a CMOS device, the first two of which are dominant.

- Subthreshold Leakage (I_{sub}): the current that flows from the drain to the source region when the transistor is in the off state.

- Gate Leakage (I_{gate}): the current that flows from the gate to the substrate through the dielectric.

- Junction Leakage (I_{junct}): the current that flows across reverse-biased junctions between the diffusion regions of a transistor and the substrate.

As the size of transistors continues to decrease, the contribution of static power compared to dynamic power is increasing. It was recently shown that even with power-optimized process technology, static and dynamic power can be within the same range for FPGAs based on 28 nm technology node [3]. There are various methods to reduce static and dynamic power, some of which will be discussed in Section 1.4.

1.4 Energy Reduction Techniques for FPGAs

This section describes techniques that are commonly used to reduce energy dissipation. These techniques include process, circuit, architecture, system, and CAD enhancements. Although these techniques will be described in the context of FPGA energy reduction, most of the techniques also apply to non-configurable integrated circuits as well.

1.4.1 Clock Gating

The power dissipated in clock networks contributes up to 22% of the power in an FPGA tile [4]. The clock signal to idle functional units can be gated, when not needed, to reduce power consumption. Clock gating is found in ASICs [5,6] and microprocessors [7]. Figure 1.9 shows a clock signal routed to a group of registers. The clock and enable signals are inputs to an *and* gate and the output connects to the clock ports of the registers. The gated clock signal can be connected to as many registers as desired to gate a region. If the enable signal is logic low the registers will not change value, thus eliminating dynamic power dissipation.

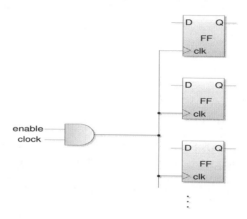

FIGURE 1.9 Clock gating for a group of registers

This technique is utilized by state-of-the-art FPGAs to reduce clock tree power consumption. Fine-grained clock gating is available through the use of clock-enable signals on registers in FPGAs from Altera, Xilinx, Microsemi and SiliconBlue [8–11]. Altera, Xilinx, and Lattice allow control of the clock tree using static or dynamic methods [12–14]. However, this feature is only available at the clock region granularity and often requires extra logic.

Zhang et al. found that RTL-level clock gating is less efficient in FP-GAs than in ASICs [15]. This is expected due to the large area overhead of RTL-level gating in FPGAs. Another study showed that the enable signals for gating the clock network have a direct relation to the efficiency of clock gating [16].

1.4.2   Controlled Body Biasing

Body biasing is used to alter the threshold voltage (V_{th}) of a MOSFET by adjusting the voltage potential between the source and body. The relationship can be seen in the following equation:

$$V_{th} = V_{TO} + \gamma \left(\sqrt{V_{SB} + 2\phi} - \sqrt{2\phi} \right) \qquad (1.7)$$

where V_{SB} is the voltage difference between the source and the body, V_{TO} is the threshold voltage with a zero-V_{SB}, γ is the body effect parameter, and

2ϕ is the surface potential parameter. Higher thresholds reduce static leakage power, while lower thresholds enable faster switching times. Altera employs a software-controlled back-biasing technique which can configure a tile to low-power or high-speed mode depending on performance requirements [17]. Devices which are not on the critical path can be biased to switch slower and to dissipate less leakage power.

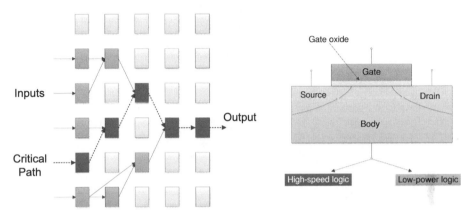

FIGURE 1.10 An example design with the critical path showing. The body voltage can be adjusted depending on if the device lies on the critical path.

Figure 1.10 shows an example FPGA design. Certain logic blocks are on the critical path and will benefit from high-speed logic. The rest of the FPGA can be configured with low-power logic as long as this does not change the critical path.

1.4.3 Dynamic Reconfiguration

Dynamic reconfiguration (DR) can be used to overwrite an FPGA region at run-time with a new configuration bitstream [18]. This does not affect the rest of the design, but incurs time and power overhead. Current FPGA compile tools allow DR at a coarse granularity.

Figure 1.11 shows an example of a system employing DR. Areas of the FPGA can be designated as Reconfigurable Partitions (RP), which act as containers for different design configurations. A Reconfigurable Module (RM) is the logic that is loaded into an RP. In this example, two regions are designated as RPs, and each region has three associated RMs. All RMs associated with a given RP must have the same port list, so that they can attach to the static logic in a uniform manner. An RP can support as many RMs as desired, but only one RM associated with an RP can be active at a single time. At run-time, active RM can be loaded into each RP without affecting the rest of the design. This "time-multiplexing" allows multiple hardware implementations in a smaller area.

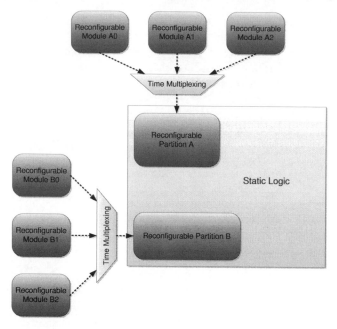

FIGURE 1.11 Dynamic reconfiguration set up using: static logic, reconfigurable partitions, and reconfigurable modules.

DR can be used to dynamically scale the clock frequency of an on-chip digital clock manager (DCM) [19]. Liu et al. proposed using DR to turn off modules in a design which are not being used, by loading blank designs [20]. This resulted in a reduction in the dynamic power dissipation of arithmetic modules. However, the blank region still experiences the same static power dissipation.

1.4.4   Multiple Supply Voltages

Reducing supply voltage has a quadratic effect on dynamic power. Typically SoCs use voltage islands to take advantage of this relationship. The supply rail voltage for functional blocks is allowed to vary depending on power and performance requirements [21]. Most FPGA-related works use only low (V_{DDL}) and high (V_{DDH}) voltage islands due to the complexity and hardware required to connect two voltage domains.

Li et al. proposed an architecture which supports a dual-V_{DD}/dual-V_{th} FPGA fabric [22]. During chip fabrication, logic blocks are selected to be powered by (V_{DDH}) or (V_{DDL}) while SRAM configuration cells use high-V_{th} to reduce leakage power. This work was extended by the same authors to allow voltage level configuration of some logic blocks at compile time [23]. Li et al. again followed up this work by enabling their architecture to employ programmable dual-V_{DD} and power gating for routing resources [24]. Voltage level shifters are required to connect the output of (V_{DDL}) with the input of (V_{DDH}) logic blocks [25].

1.4.5 CAD Techniques

Other power reduction techniques for FPGAs include enhancing the computer-aided design (CAD) algorithms so that they seek low-power implementation of user circuits. These techniques typically take advantage of detailed power models [26–28] to direct decision making in the compilation stage. Included in this group are clock-aware optimizations [13, 16, 29], power-aware placement and routing [30–32], and power-aware technology mapping [30, 33, 34]. There are other CAD techniques for specific low-power architectures. These include dual V_{DD}-aware voltage assignment [35] and mapping [36], synthesis of a power state controller using the data flow graph (DFG) of an application [37], and sleep mode-aware region-constrained placement [38].

1.4.6 Other Techniques

There exist many other static and dynamic power dissipation reduction techniques at the device- and circuit-level for FPGAs.

Altera and Xilinx both use different oxide thickness (triple gate oxide) transistors in their FPGAs. This process enables a trade-off between performance and static power. They can reduce dynamic power by reducing capacitance through the use of low-K dielectrics. Additionally they also employ low-core voltage, larger transistor lengths, and higher V_{th} [17, 39].

Other published techniques include configuration-bit flipping [40], low-power circuits for routing resources [41, 42], and reduction of glitch power dissipation [43, 44].

1.5 Runtime-Controlled Power-Gated FPGAs

In this section, we describe a new FPGA architecture that employs dynamic power gating to significantly reduce the static power dissipation of FPGA applications with long idle times.

Previous techniques to reduce the power dissipation of FPGAs have focused on reducing both the dynamic and static (leakage) power of these devices. Static power constitutes a major component of the power consumption in sub-90 nm FPGAs [3, 45]. In hand-held devices, such as mobile phones that employ communication ICs and other types of ICs, it is conceivable that the leakage power will be even more significant, since these devices are often used in an "always on" state, remaining idle except for short bursts of activity.

Thus, it is essential to have low-leakage FPGAs in order to be used for these types of applications.

Power gating [46] has been used for ASIC designs as an effective technique to reduce static power during idle periods. Figure 1.12 (a) illustrates the basic architecture idea. By connecting the supply voltage or the ground of a circuit component through a power switch, also called a *sleep transistor*, the circuit component can be turned on or off by turning the corresponding power switch on or off. Turning off the power switch limits the leakage current of the whole circuit to that of the switch, thus significantly reducing leakage power.

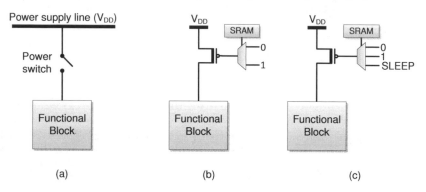

(a)   (b)   (c)

FIGURE 1.12 Illustration of the basic idea of power gating.

The power switch can be an nMOS (connects to GND) or a pMOS (connects to V_{DD}) transistor. In this discussion we focus on the pMOS transistor because it is more commonly used in power gating designs. Designing the power switch involves balancing different tradeoffs, such as performance, power, and area overhead [47]. Investigating these trade-offs is beyond the scope of this chapter.

Figure 1.13 shows an illustration of the effect of power gating. The shaded area in the power diagrams is the energy consumption. As can be seen in the figure, applying power gating by turning off functional blocks when they are idle reduces the energy dissipation of the system significantly.

Statically-controlled power gating for FPGAs was proposed in previous works [37,38,48,49]. The basic idea is illustrated in Figure 1.12 (b). The power switch can be turned on or off when the FPGA is configured, and cannot be changed until a new configuration is loaded. This is achieved by appropriately setting the SRAM memory that drives the 2:1 multiplexer select lines. This scheme allows turning off blocks that are not used in a given application. Since designers often tend to select the smallest FPGA device that fits their application, the benefits from this scheme might be limited.

In this section, *dynamically-controlled power gating* FPGA architecture is introduced as a technique to allow run-time control of the power state of a functional block. This enables turning off the functional blocks in an application during their idle periods, thus limiting their leakage power when

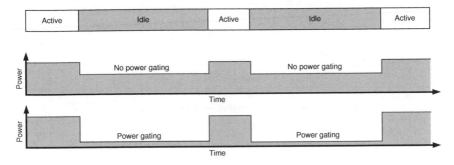

FIGURE 1.13 Power gating reduces energy consumption. Energy is the shaded area under the power curve.

no useful task is being performed. The basic idea is shown in Figure 1.12 (c). The select lines of the 3:1 multiplexer can be set (through proper configuration for the SRAM memory cells) to select the SLEEP signal as the control input for the power switch. This SLEEP signal can be controlled by a control logic to turn off the functional block when it is idle in order to enter a low-power mode.

1.5.1 Architecture Description

Figure 1.14 shows an example system-on-a-chip (SoC) that can be implemented on an FPGA or on an ASIC. The discussion in this section will focus on FPGA-based implementations. This SoC is composed of a microprocessor core and several IP cores connected on the system's bus. These IP cores perform functionality specific to the target application. For example, in a communications application, such as a network router, there will be an Ethernet core to manage the Ethernet interfaces, a memory controller core to interface with off-chip memory modules, a serial interface core to configure the router, and many other cores.

At run-time, there might be time periods when there is no network traffic or when the router is not being configured by a network administrator, for example. In these periods, it is desirable to turn off the IP cores that are idle to reduce the power consumption of the system. This can be done by using power gating that is dynamically controlled based on predefined conditions.

In order to enable such capability on an FPGA, the architecture of typical FPGAs need to be altered. The main requirements are to provide power to FPGA blocks through power switches, and to enable routing the power control signals to these switches using the general purpose routing fabric of the FPGA. Figure 1.15 illustrates this. The figure shows how the control signals that control the power state of modules M1 and M2 are routed to the power gating regions in an FPGA.

The power gating architecture is for one logic cluster, or a region of logic

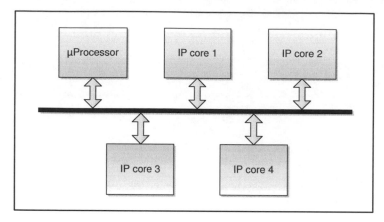

FIGURE 1.14 An example system-on-a-chip that can be implemented on an FPGA. Some of the IP cores can be turned off (using power gating) during their idle periods.

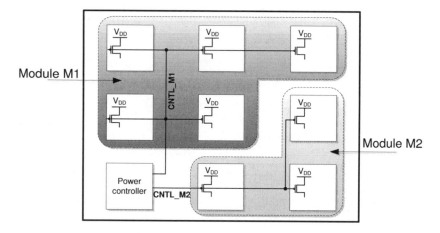

FIGURE 1.15 Using a power controller in an FPGA to control the power state of the different modules in an application. Control signals are routed on the general purpose routing fabric.

clusters, of an FPGA and its bordering routing channels. Figure 1.16 shows the architecture for a region of size two (region size R means $R \times R$ logic clusters) and its bordering routing channels. The gray-coloured blocks are the ones that are connected to power lines through power switches that are controlled by the signal PG_CNTL. In the figure, the power switch is shown only for the internal part of the power gating region. Note that switch blocks are not power-gated because they may be used to route signals that are unrelated to the power-gated blocks.

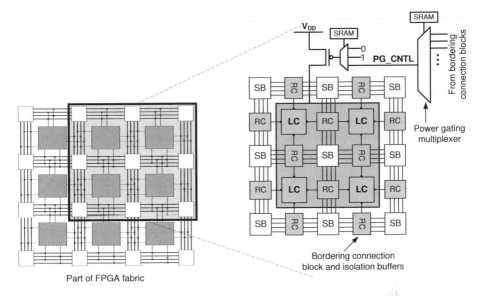

FIGURE 1.16 Dynamic power gating architecture and the zoom-in for one region.

The power gating multiplexer in Figure 1.16 selects (using configuration SRAM bits) which of the bordering connection blocks outputs will be used as the control signal. This scheme enables routing control signals from anywhere in the chip to the power switches of the individual FPGA power gating regions; this is done in a similar way other signals would be routed to the inputs of logic clusters.

The power switches for the gray-colored components in Figure 1.16 are similar to the one shown in Figure 1.12 (c) with the 3:1 multiplexer. This enables turning on/off the block statically, or dynamically by controlling it by a control signal coming from elsewhere in the chip. Note that there are more details related to the design of the power switches for the bordering routing channels, because these channels have some circuit components that are shared between the two neighboring regions. The basic idea is to turn off the routing channel only when both of the neighboring regions are turned off or if the output of the connection blocks in that channel are not used by any logic cluster. Further details can be found in [50].

There is a trade-off between the region size and the area/power overheads, and the flexibility of mapping applications to the architecture. As the region size increases, the area overhead decreases and the power reduction increases. At the same time, as the region size increases, the flexibility decreases since larger group of related logic clusters need to be mapped to the same physical power gating region in order to achieve energy saving during idle periods. Some of these trade-offs are experimentally investigated in Section 1.5.2.

1.5.2   Evaluation

In this subsection, the power gating architecture presented in Subsection 1.5.1 is evaluated. Firstly, the architecture is evaluated in terms of power reduction and area overhead, compared to an architecture that does not support power gating. A circuit model of the architecture was built and simulated using HSPICE circuit simulation tool. Secondly, we evaluate the efficiency of the proposed architecture for an example application using randomly-generated execution traces.

1.5.2.1   Architecture Evaluation

Different region sizes were simulated using HSPICE assuming fixed basic FPGA architecture parameters. A 45 nm technology node was used [51] with $V_{DD} = 1$ V. Power consumption was measured assuming the worst case temperature of 85°C. The sizing for the power switches (sleep transistors) was iteratively performed until a constraint of 10% speed degradation is met. The reported results in this section are for the components that are power-gated, i.e., we do not include results for switch blocks.

As can be seen in Figure 1.17, as region size increases, power increases for both the proposed architecture and the ungated architecture. At the same time, the power reduction increases. Power reduction ranges from 81% for the smallest region of size $R = 1$, to 95% for a region of size $R = 7$.

Also, Figure 1.18 shows that the incurred area overhead from adding additional circuit components to support power gating decreases as the region size increases. The area overhead is relatively small for large region sizes, i.e., less than 2% for $R = 4$. Note that larger region sizes also result in smaller wakeup current per tile, which is desirable as it reduces the overhead required to handle the wakeup current, and improves wakeup time as has been discussed in [52].

In addition to investigating the region size effect on the power reduction and area overhead, the effect of other FPGA architecture parameters, such as routing channel width and cluster size, was investigated in [50].

1.5.2.2   Example Application

In order to evaluate the effectiveness of the proposed architecture in reducing energy of FPGA systems, we evaluated the energy for a collection of cores that represent what might be included in a communication application, such as a network router. Table 1.1 lists the hardware cores that have been used in this section. All of these cores are from Altera QUIP benchmark library [53].

The same architecture parameters in the previous subsection were used for synthesis and power analysis. In order to simulate real application behavior, we generated a random trace for each core that represents the state of the core (active or idle) during the execution time. In the active state, the core is assumed to consume dynamic power that was obtained from power analysis

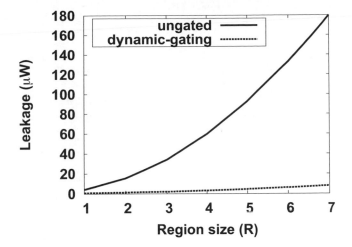

FIGURE 1.17 Leakage power for increased region sizes.

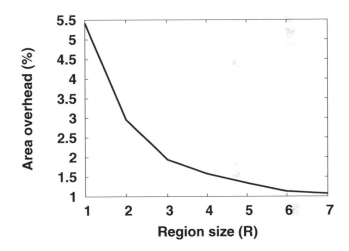

FIGURE 1.18 Area overhead for increased region sizes.

using the FPGA Power Model from [28]. In the idle state, the core is assumed to consume leakage power, which was obtained from HSPICE simulation results.

Figure 1.19 shows the amount of energy reduction for the application as a function of the assumed activity. An execution time of 5 hours was assumed, with a step size of 100 ms that is used to generate the execution trace. As can be expected, energy reduction increases for smaller activities. The results show that even for a system with activity as high as 30%, energy saving could be achieved using the power gating architecture described in this chapter.

The results in this section assumed only power gating for logic clusters.

TABLE 1.1

IP cores used for evaluating the FPGA power gating architecture.

IP core	Logic Clusters
Memory Controller	800
MIPS Processor	905
DMA Core	920
Ethernet Controller	400
I²C Interface	58
DES Encryption	276

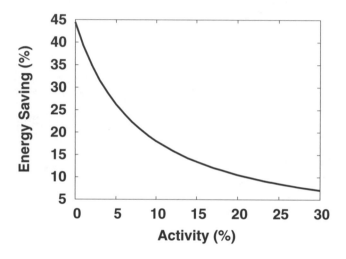

FIGURE 1.19 Energy reduction as a function the system's activity.

Switch blocks leakage power is more than 50% of a tile's leakage power in an FPGA; thus, power gating for switch blocks, in addition to logic clusters, would result in much larger energy savings.

1.6 Conclusion

The power dissipation of FPGAs is critical especially when they are used for "green communications" applications. Therefore, it is important to find new techniques to reduce the power consumption of FPGAs.

In this chapter, we provided an overview of methods used to reduce the power dissipation in integrated circuits, focusing on FPGAs. We then described a new FPGA architecture optimized for applications that experience long idle times. The described architecture is based on run-time controlled power gating. The results show that this architecture is very efficient in reducing idle-period power consumption.

Future directions include extending the power gating architecture described in this chapter to include more FPGA components, such as switch blocks, that consume large amount of leakage power during idle periods. Furthermore, it is important to design CAD tools that automate the design of applications for the power gating architecture described in this chapter; this would significantly simplify system design.

Bibliography

[1] I. Kuon and J. Rose, "Measuring the Gap Between FPGAs And ASICs," in *Proceedings of the 14th ACM/SIGDA International Symposium on Field-Programmable Gate Arrays*, 2006, pp. 21–30.

[2] L. Scheffer, L. Lavagno, and G. Martin, Eds., *EDA for IC Implementation, Circuit Design, and Process Technology*, 1st ed. CRC, 2006.

[3] J. Hussein, M. Klein, and M. Hart. (2011, June) Lowering Power at 28 nm with Xilinx 7 Series FPGAs. Xilinx, Inc. white paper WP389 (v1.1). Xilinx, Inc. [Online]. Available: http://www.xilinx.com/support/documentation/white_papers/wp389_Lowering_Power_at_28nm.pdf

[4] L. Shang, A. S. Kaviani, and K. Bathala, "Dynamic Power Consumption in Virtex^{TM}-II FPGA Family," in *Proceedings of the Tenth ACM/SIGDA International Symposium on Field-Programmable Gate Arrays*, 2002, pp. 157–164.

[5] M. Donno, A. Ivaldi, L. Benini, and E. Macii, "Clock-Tree Power Optimization Based on RTL Clock-Gating," in *Proceedings of the 40th Design Automation Conference*, 2003, pp. 622–627.

[6] Q. Wang and S. Roy, "Power Minimization by Clock Root Gating," in *Proceedings of the 2003 Asia and South Pacific Design Automation Conference*, 2003, pp. 249–254.

[7] V. Venkatachalam and M. Franz, "Power Reduction Techniques For Microprocessor Systems," *ACM Comput. Surv.*, vol. 37, no. 3, pp. 195–237, 2005.

[8] F. Rivoallon, "Reducing Switching Power with Intelligent Clock Gating," Xilinx, Inc., White paper WP370, May 2010. [Online]. Available: http://www.xilinx.com/support/documentation/white_papers/wp370_Intelligent_Clock_Gating.pdf

[9] *iCE65 Ultra Low-Power mobileFPGA Family Datasheet*, http://www.siliconbluetech.com/media/downloads/iCE65Datasheet.pdf,

SiliconBlue, May 2010. [Online]. Available: http://www.siliconbluetech.com/media/downloads/iCE65Datasheet.pdf

[10] *Stratix Device Handbook*, Altera Corporation, July 2005. [Online]. Available: http://www.altera.com/literature/hb/stx/stratix_handbook.pdf

[11] *IGLOO Low-Power Flash FPGAs Handbook*, Actel Corporation, November 2009. [Online]. Available: http://www.actel.com/documents/IGLOO_HB.pdf

[12] *Stratix III Device Handbook*, http://www.altera.com/literature/hb/stx3/stx3_siii5v1.pdf, Altera Corporation, July 2010. [Online]. Available: http://www.altera.com/literature/hb/stx3/stx3_siii5v1.pdf

[13] Q. Wang, S. Gupta, and J. H. Anderson, "Clock Power Reduction for Virtex-5 FPGAs," in *Proceeding of the 17th ACM/SIGDA International Symposium on Field-Programmable Gate Arrays*, 2009, pp. 13–22.

[14] *LatticeECP3 Family Handbook*, Lattice Semiconductor Corp., June 2010. [Online]. Available: http://www.latticesemi.com/documents/hb1009.zip

[15] Y. Zhang, J. Roivainen, and A. Mammela, "Clock-Gating in FPGAs: A Novel and Comparative Evaluation," in *Proceedings of the 9th EUROMICRO Conference on Digital System Design*, 2006, pp. 584–590.

[16] S. Huda, M. Mallick, and J. Anderson, "Clock Gating Architectures for FPGA Power Reduction," in *Proceedings of the International Conference on Field Programmable Logic and Applications*, 2009, pp. 112–118.

[17] (2008, August) 40-nm FPGA Power Management and Advantages. Altera, Corp. white paper WP-01059-1.2 (v1.2). Altera, Corp. [Online]. Available: http://www.altera.com/literature/wp/wp-01059-stratix-iv-40nm-power-management.pdf

[18] *Virtex-5 FPGA Configuration User Guide*, Xilinx, Inc., August 2009.

[19] K. Paulsson, M. Hübner, S. Bayar, and J. Becker, "Exploitation of Run-Time Partial Reconfiguration for Dynamic Power Management in Xilinx Spartan III-based Systems," in *ReCoSoC Workshop*, 2007.

[20] S. Liu, R. N. Pittman, A. Forin, and J.-L. Gaudiot, "On Energy Efficiency of Reconfigurable Systems with Run-Time Partial Reconfiguration," in *Proceedings of the 21st IEEE International Conference on Application-Specific Systems, Architectures and Processors*, 2010.

[21] D. E. Lackey, P. S. Zuchowski, T. R. Bednar, D. W. Stout, S. W. Gould, and J. M. Cohn, "Managing Power and Performance for System-on-Chip Designs using Voltage Islands," in *Proceedings of the 2002 IEEE/ACM international conference on Computer-Aided Design*, 2002, pp. 195–202.

[22] F. Li, Y. Lin, L. He, and J. Cong, "Low-Power FPGA using Pre-Defined Dual-Vdd/Dual-Vt Fabrics," in *Proceedings of the 12th ACM/SIGDA*

International Symposium on Field-Programmable Gate Arrays, 2004, pp. 42–50.

[23] F. Li, Y. Lin, and L. He, "FPGA Power Reduction Using Configurable Dual-Vdd," in *Proceedings of the 41st Design Automation Conference*, 2004, pp. 735–740.

[24] ——, "Vdd Programmability to Reduce FPGA Interconnect Power," in *Proceedings of the 2004 IEEE/ACM International Conference on Computer-Aided Design*, 2004, pp. 760–765.

[25] R. Puri, L. Stok, J. Cohn, D. Kung, D. Pan, D. Sylvester, A. Srivastava, and S. Kulkarni, "Pushing ASIC Performance in a Power Envelope," in *Proceedings of the 40th Design Automation Conference*, 2003, pp. 788–793.

[26] F. Li, D. Chen, L. He, and J. Cong, "Architecture Evaluation for Power-Efficient FPGAs," in *Proceedings of the 11th ACM/SIGDA International Symposium on Field-Programmable Gate Arrays*, 2003, pp. 175–184.

[27] J. H. Anderson and F. N. Najm, "Power Estimation Techniques for FP-GAs," *IEEE Trans. Very Large Scale Integr. Syst.*, vol. 12, no. 10, pp. 1015–1027, 2004.

[28] K. K. W. Poon, S. J. E. Wilton, and A. Yan, "A Detailed Power Model for Field-Programmable Gate Arrays," *ACM Trans. Des. Autom. Electron. Syst.*, vol. 10, no. 2, pp. 279–302, 2005.

[29] J. Lamoureux and S. Wilton, "Clock-Aware Placement for FPGAs," in *Proceedings of the International Conference on Field Programmable Logic and Applications*, 2007, pp. 124–131.

[30] J. Lamoureux and S. J. E. Wilton, "On the Interaction Between Power-Aware FPGA CAD Algorithms," in *Proceedings of the 2003 IEEE/ACM International Conference on Computer-Aided Design*, 2003, p. 701.

[31] K. Vorwerk, M. Raman, J. Dunoyer, Y. chung Hsu, A. Kundu, and A. Kennings, "A Technique for Minimizing Power During FPGA Placement," in *Proceedings of the 2008 International Conference on Field Programmable Logic and Applications*, 8–10 2008, pp. 233–238.

[32] Q. Dinh, D. Chen, and M. Wong, "A Routing Approach to Reduce Glitches in Low Power FPGAs," *IEEE Transactions on Computer-Aided Design of Integrated Circuits and Systems*, vol. 29, no. 2, pp. 235–240, February 2010.

[33] J. Anderson and F. Najm, "Power-Aware Technology Mapping for LUT-Based FPGAs," in *Proceedings of the 2002 IEEEInternational Conference on Field-Programmable Technology*, 2002, pp. 211–218.

[34] L. Cheng, D. Chen, and M. Wong, "GlitchMap: An FPGA Technology Mapper for Low Power Considering Glitches," in *Proceedings of the 44th ACM/IEEE Design Automation Conference*, 2007, pp. 318–323.

[35] Y. Lin and L. He, "Leakage Efficient Chip-Level Dual-Vdd Assignment with Time Slack Allocation for FPGA Power Reduction," in *Proceedings of the 42nd Design Automation Conference*, 2005, pp. 720–725.

[36] D. Chen and J. Cong, "Delay Optimal Low-Power Circuit Clustering for FPGAs with Dual Supply Voltages," in *Proceedings of the 2004 International Symposium on Low Power Electronics and Design*, 2004, pp. 70–73.

[37] R. P. Bharadwaj, R. Konar, P. T. Balsara, and D. Bhatia, "Exploiting Temporal Idleness to Reduce Leakage Power in Programmable Architectures," in *Proceedings of the 2005 Asia and South Pacific Design Automation Conference*, 2005, pp. 651–656.

[38] A. Gayasen, Y. Tsai, N. Vijaykrishnan, M. Kandemir, M. J. Irwin, and T. Tuan, "Reducing Leakage Energy in FPGAs Using Region-Constrained Placement," in *Proceedings of the 12th ACM/SIGDA International Symposium on Field-Programmable Gate Arrays*, 2004, pp. 51–58.

[39] M. Klein. (2009, April) Power Consumption at 40 and 45 nm. Xilinx, Inc. white paper WP298 (v1.0). Xilinx, Inc. [Online]. Available: http://www.xilinx.com/support/documentation/white_papers/wp298.pdf

[40] J. H. Anderson, F. N. Najm, and T. Tuan, "Active Leakage Power Optimization for FPGAs," in *Proceedings of the 12th ACM/SIGDA International Symposium on Field-Programmable Gate Arrays*, 2004, pp. 33–41.

[41] J. H. Anderson and F. N. Najm, "Low-Power Programmable Routing Circuitry for FPGAs," in *Proceedings of the 2004 IEEE/ACM International Conference on Computer-Aided Design*, 2004, pp. 602–609.

[42] M. Lin and A. El Gamal, "A Low-Power Field-Programmable Gate Array Routing Fabric," *IEEE Trans. Very Large Scale Integr. Syst.*, vol. 17, no. 10, pp. 1481–1494, 2009.

[43] J. Lamoureux, G. G. Lemieux, and S. J. E. Wilton, "Glitchless: An Active Glitch Minimization Technique for FPGAs," in *Proceedings of the 15th ACM/SIGDA International Symposium on Field-Programmable Gate Arrays*, 2007, pp. 156–165.

[44] T. S. Czajkowski and S. D. Brown, "Using Negative Edge Triggered FFs to Reduce Glitching Power in FPGACircuits," in *Proceedings of the 44th Design Automation Conference*, 2007, pp. 324–329.

[45] (2005, August) Stratix II vs. Virtex-4 Power Comparison & Estimation Accuracy. Altera Corp. white paper WP-S20805-01 (v1.0). Altera Corp. [Online]. Available: www.altera.com/literature/wp/wp_s2v4_pwr_acc.pdf

[46] S. Henzler, *Power Management of Digital Circuits in Deep Sub-Micron CMOS Technologies (Springer Series in Advanced Microelectronics)*. Secaucus, NJ, USA: Springer-Verlag New York, Inc., 2007.

[47] K. Shi and D. Howard, "Challenges in Sleep Transistor Design and Implementation in Low-Power Designs," in *DAC '06: Proceedings of the 43rd annual Design Automation Conference.* New York, NY, USA: ACM, 2006, pp. 113–116.

[48] Y. Lin, F. Li, and L. He, "Routing Track Duplication with Fine-Grained Power-Gating for FPGA Interconnect Power Reduction," in *Proceedings of the 2005 Asia and South Pacific Design Automation Conference*, 2005, pp. 645–650.

[49] T. Tuan, S. Kao, A. Rahman, S. Das, and S. Trimberger, "A 90 nm Low-Power FPGA for Battery-Powered Applications," in *Proceedings of the 14th ACM/SIGDA International Symposium on Field-Programmable Gate Arrays*, 2006, pp. 3–11.

[50] A. A. M. Bsoul and S. J. E. Wilton, "An FPGA Architecture Supporting Dynamically Controlled Power Gating," in *Field-Programmable Technology, 2010. (FPT). Proceedings. 2010 IEEE International Conference on*, December 2010, pp. 1–8.

[51] W. Zhao and Y. Cao, "New Generation of Predictive Technology Model for Sub-45nm Design Exploration," *Quality Electronic Design, International Symposium on*, vol. 0, pp. 585–590, 2006.

[52] A. A. M. Bsoul and S. J. E. Wilton, "A Configurable Architecture to Limit Wakeup Current in Dynamically-Controlled Power-Gated FPGAs," in *Proceedings of the 20th ACM/SIGDA International Symposium on Field-Programmable Gate Arrays*, 2012.

[53] "Quartus II University Interface Program," http://www.altera.com/education/univ/research/quip/unv-quip.html.

Author contact information

Assem A. M. Bsoul, Stuart Dueck, and Steven J. E. Wilton are with University of British Columbia, Canada, Email: absoul@ece.ubc.ca, stevew@ece.ubc.ca.

2

Smart Grid and ICT's Role in Its Evolution

Hassan Farhangi

British Columbia Institute of Technology, Vancouver, Canada

CONTENTS

2.1   Climate Change

While there is debate around the real causes of Climate Change, Green House Gas (GHG) emissions as a result of widespread use of fossil-based fuels by major economies around the world has been thought of playing a significant role in perpetuating the negative impacts of the phenomenon known as Climate Change. Regardless of whether GHG emissions is the sole culprit of the unusual, and often devastating changes in the climate patterns around the world, the global understanding has been sought over mitigating further dependence on fossil fuels by the developed countries. What further accentuates that desire, is not only the political and social instability of the regions which have traditionally supplied such fuels, but the fact that such fuel are finite in nature and due to be substantially exhausted in the not too distant future. It is interesting to note that the political and social turmoil associated with traditional sources of fossil fuels has given rise to the justification for many special interest groups in the developed world to call for "drilling closer to home". This view often ignores the fact that fossil fuel in the developed world often lies in "difficult to reach" and technologically challenging areas, which do not lend themselves to relatively risk-free exploration and exploitation. Recent environmental disasters, such as the oil spill in the Gulf of Mexico is a clear and undisputable indication of the dangers associated with "drilling closer to home". Consequently, to get out of our energy conundrum, it seems that our societies have no choice but to review and question the way our economies generate and utilize energy. Most studies of this nature reveal the wasteful and unsustainable processes and approaches which we have so far used in energy production and use. Conversion of one form of energy into another, transmission of energy from one place to another, distribution of energy through our urban and rural communities, and management of energy resources have all been imperfect, to say the least. Such wasteful approaches to energy use have been the hallmark of the last century, which has now come back to haunt us in terms of devastating consequences associated with Climate Change. It is in that light that Smart Grid has been inadvertently positioned as the silver bullet to address the Climate Change and Energy Independence issues. Smart Grid is expected to enable unprecedented degrees of conservation, efficiencies and utilization of alternative sources of energy, thus substantially reducing this century's dependence on fossil fuels. It is notable that regardless of which development category they belong to, the developed countries, as well as the developing countries, have put together ambitious plans for the development of next generation electric grid, also called Smart Grid, as the main engine for the development of their economies and the well-being of their population. However, the fact remains that Smart Grid is still a collection of concepts and ideas, whose full impact cannot be realized until a rich portfolio of innovative technologies, system architectures, integration solutions and social-economic components are available cost-effectively and in concert to address the energy supply and demand issues which individual countries across the world are grappling with. And as such, energy independence should be perceived by the world community as a global problem longing for global solutions. As will be demonstrated in the rest of this chapter, Information and Communication Technologies are poised to play a critical role in bringing about the full spectrum of functionalities which Smart Grid promises. After all, Smart Grid is all about pervasive monitoring and control, which could not be realized without a comprehensive blanket of communication technologies, encompassing all utility assets, and enabling the intelligence implanted in each node to contribute to the overall system capabilities and functionalities which Smart Grid is expected to provide.

2.2 Conventional Electricity Grid

The existing electricity grid is a product of rapid urbanization and infrastructure developments in various parts of the world in the past century. Although in different geographies, the utility companies have adopted similar
technologies, the growth of the electrical power system has been influenced by
economic, political and geographical factors, unique to each utility company.

Nevertheless, despite such differences, the topology of the existing electric
power system has remained virtually unchanged. Since its inception, the power
industry has created clear demarcations between Generation, Transmission
and Distribution subsystems, and as such has influenced different levels of
automation, evolution and transformation in each silo.

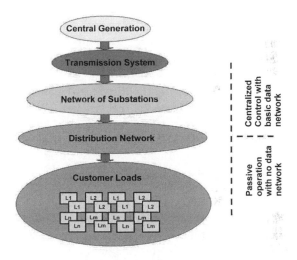

FIGURE 2.1 Hierarchical structure of existing electricity grid [1] © 2010 IEEE

As Fig. 2.1 demonstrates, the existing electricity grid is strictly hierarchical
in nature, with power plants at the top of the chain, ensuring power delivery to
customers' loads at the bottom of the chain. The system as such is essentially
a one-way pipeline where the source has no real-time information about the
service parameters of the termination points. The system is therefore intrinsically over-engineered to withstand maximum anticipated peak demand across
its aggregated load. And since the peak demand is an infrequent occurrence
except for certain periods of time, the system is inherently inefficient.

Moreover, unprecedented rise in demand for electric power, coupled with
lagging investments in the electric power infrastructure, has made the system

more unstable. Having exhausted the safe margins, any unforeseen surge in demand, or anomalies across the distribution network, causing component failures, may trigger catastrophic black-outs.

Nevertheless, to facilitate troubleshooting and upkeep of the expensive upstream assets, the utility companies have introduced various levels of command and control functions. A typical example is the widely deployed system known as SCADA (Supervisory Control And Data Acquisition). Although, such systems enable Utility companies to have a limited control over their upstream functions, the Distribution Network, remains outside the real-time control sphere of the utility companies. And the picture is almost the same across the world. For instance, in North America, which has established one of the world's most advanced electric power systems, less than a quarter of the distribution network is equipped with information and communications systems, and the distribution automation penetration at the system feeder level is estimated at only 15-20 percent.

The utility industry across the world is thus trying to address numerous challenges such as generation diversification, optimal deployment of expensive assets, demand response, energy conservation, reduction of industry's overall carbon footprint, etc. It is evident that such critical issues cannot be addressed within the confines of the existing electricity grid.

The existing electricity grid is unidirectional in nature. It converts only 1/3 of fuel energy into electricity without recovering the waste heat. Almost 8% of its output is lost along its transmission lines, while 20% of its generation capacity exists to meet peak demand only (i.e. 5% of time). In addition to that, due to hierarchical topology of its assets, the existing electricity grid suffers from domino effect failures.

The next generation electricity grid, known as Smart Grid (or Intelligent Grid) is expected to address the major shortcoming of the existing grid. In essence, the Smart Grid needs to provide the utility companies with full visibility and pervasive control over their assets and services. Smart Grid is required to be self-healing and resilient to system anomalies. And last but not least, Smart Grid needs to empower its stakeholders to define and realize new ways of engagements and energy transactions across the system.

2.3   Overview of Smart Grid

As the backbone of the power industry, the electricity grid is now the focus of assorted technological innovations. Utilities in North America and across the world are taking solid steps towards incorporating new technologies in many aspects of their operations and infrastructure. At the core of this transformation, is the need to make more efficient use of current assets. As such, Asset Management is at the base of Smart Grid Development for many utility companies in the world. It is upon this base, that utilities build the foundation of Smart grid through careful overhaul of their IT, Communication and Circuit infrastructure. As Fig 2.2 depicts, one may envisage the move towards Smart Grid as a well planned overlay of Communication and Information Technologies over the existing and future utility assets and infrastructure.

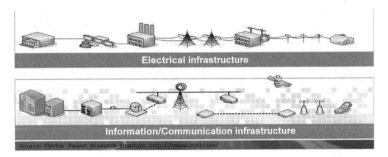

FIGURE 2.2 Overlay of communication and information technologies over electricity infrastructure [2]

By default, Smart Grid is the amalgamation of all technologies, concepts, solutions and methodologies which allow the silo hierarchies of generation, transmission and distribution to be replaced with an end-to-end organically intelligent and fully integrated environment where business processes, objectives and needs of all stake holders are supported through efficient exchange of data, services and transactions. Smart Grid is therefore defined as a grid which accommodates a wide variety of generation options, e.g. central, distributed, intermittent and mobile. It empowers the consumer to interact with the energy management systems to manage their energy use and reduce their energy costs.

Smart Grid is also a self-healing system. It predicts looming failures and takes corrective actions to avoid or mitigate system problems. Smart Grid uses information technology to continually optimize the use of its capital assets while minimizing operations and maintenance costs. To allow pervasive control and monitoring, Smart Grid is emerging as a convergence of Information Technology and Communication Technology with Power System Engineering. Fig. 2.3 depicts the salient features of Smart Grid in comparison to the existing grid.

2.4 Smart Grid Applications

Within the context of Smart Grid capabilities, communication and data management play an important role. These basic ingredients enable the utilities

Existing Grid	Intelligent Grid
Electromechanical	Digital
One–Way communication	Two–Way communication
Centralized Generation	Distributed Generation
Hierarchical	Network
Few Sensors	Sensors throughout
Blind	Self–monitoring
Manual Restoration	Self–Healing
Failures & Blackouts	Adaptive & Islanding
Manual Check/Test	Remote Check/Test
Limited Control	Pervasive Control
Few customer choices	Many custormer choices

FIGURE 2.3 Comparisons between existing grids and intelligent grids [1] ©
2010 IEEE

to place a layer of intelligence over their current and future infrastructure,
thereby allowing the introduction of new applications and processes in their
business. As discussed earlier, the convergence of Communication Technology
and Information Technology with Power System, assisted by an array of new
approaches, technologies and applications, allows the existing grid to traverse
the complex, yet staged, trajectory of architecture, protocols and standards
towards the Smart Grid.

Moreover, the organic growth of this well-designed layer of intelligence over
utility assets, enables Smart Grid's fundamental applications to emerge. It is
interesting to note that although the foundation of Smart Grid is built upon
a lateral integration of these basic ingredients, Smart Grid capabilities will be
built upon vertical integration of the upper layer applications. For instance, a
critical capability such as Demand Response and Dynamic Pricing, depicted
in Fig. 2.4, may not be feasible without tight integration between a variety of
middleware, field assets and termination devices placed in various management
layers, which provide inputs, as well as get influenced by the real-time load
status of the system.

This is further accentuated in Fig 2.5, where various utility corporate functions are built on top of an integrated communication network, which in turn
is placed over utility's field assets, termination devices and customer controlled
loads and generators. It is interesting to note that the level of sophistication
of a given utility corporate function hinges upon its linkages with, connections
to, and use of data generated by other management layers of the organization.
Moreover, such functions do not need to have all those ties to other layers in

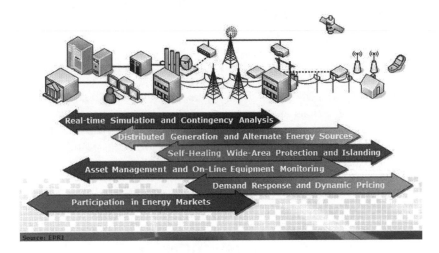

FIGURE 2.4 Smart Grid capabilities built upon vertical integration [3]

place before they could function. Rather, their internal decision making processes would be considerably enriched when that level of integration and access to data mines are built into the system. As such, the evolutionary process of utility functions should be built such that the architecture could support further integration between applications, allowing the applications to learn and become more sophisticated over time, and enabling the system to support application's decision making processes with increasing access to system-wide real-time data.

As an example, in the early stages of Smart Grid roll out, a utility may choose to have their corporate Billing & Accounting function interface directly with a vendor's MDMS (Metering Data Management System), and by doing that minimize the possible disruptions to its business processes and functions. Such an approach, will allow the utility to have an early exposure to Advanced Metering Infrastructure and reap the benefits of automated billing, revenue management and streamlined operations.

However, utilities well know that Smart Meters are capable of much more than capturing consumption data only. By design, most Smart Meters are also capable of producing instantaneous and real-time sample values related to line parameters (Voltage, Current, Power Factor, etc). Such information is of great value to many other Smart Grid applications and capabilities which the utility may be planning to develop in the future. As an example, as the level of Smart Grid roll out progresses, the utility may want to integrate the aforementioned function with other middleware such as Energy Management System (for dynamic energy pricing), Distribution Management System (for load curtailment, load shedding and load management), Geographic Information System (for fault location), Outage Management System (for automatic restoration), Volt/VAR & CVR Optimization (for real-time adaptive feeder voltage/VAR optimization), etc.

The aforementioned applications could only function if accurate real-time system-wide data, such as those generated by Smart Meters are made available to other Smart Grid applications on-demand. And that would only be possible if the utility ICT architecture is built to facilitate gradual and phased integration of all applications and functions within different management layers of the system. Fig. 2.6 demonstrates the phased introduction of Smart Grid applications and the fundamental technologies which are required to support them.

2.5   Command and Control Strategies

As discussed earlier, what distinguishes Smart Grid from the existing grid is the pervasiveness of real-time monitoring, command and control across "all" nodes of the system. Note the emphasis placed on the word "all", as this is an acknowledgement of the fact that the existing grid is reasonably smart as well. There are a lot of sophisticated command and control functions built into the existing grid, to make it one of the engineering wonders of the past century.

However, the fundamental difficulty with the existing grid is that such intelligence, monitoring and control is limited to the upper layers of the system hierarchy, ie primarily in Generation and Transmission only. The Distribution side of the system, which is spread over much larger geographic areas, and incorporating much larger number of nodes and components, has virtually no monitoring and automation built into it. And considering the common belief that 90% of system anomalies, disturbances and faults have their root in the distribution system, one can appreciate the importance of extending the reach of upstream intelligence into downstream systems.

Smart Grid system thus requires all its nodes to have the capability of generating real-time data and communicate those with other nodes across the system. Such data would be required to enable enterprise applications to have access to, and apply appropriate controls, to all the nodes of the system, regardless of where these nodes are or what their functions may be. In other words, monitoring and control within Smart Grid system has to be an end-to-end exercise, and encompassing all nodes within generation, transmission and distribution areas of the network.

The next question which arises is what are the command and control strategies which Smart Grid systems should implement? Should Smart Grid continue with the existing grid's centralized command and control hierarchy and open-loop decision making process, where all data points need to be captured and brought into a central command and control center to enable human operators to take system control actions, including but not limited to the required adjustments and/or configurations?

The difficulty with extending pervasive monitoring and command and control onto a centrally controlled smart grid system is the cost and the feasibility associated with the task of data management and end-to-end communication. The fact of the matter is that compared to the existing grid, smart grid is expected to have substantially more nodes and components to monitor and control. The task of data capture, communication and control is no longer limited to generation plants or major transmission or receiving substations, but to every single component and asset that utility has in their entire network from generation plants to customer meters.

That requirement makes a centralized command and control strategy unworkable and inefficient. Many events in Smart Grid network are local in nature, and as such have to be dealt with locally and through built-in localized intelligence. Higher level systems and applications may need to be informed of the occurrence of the events and the local actions taken to attend to them, but they certainly should not be required to assume the task of time-sensitive and critical control, prescribing and/or implementing remedies for local system issues while sitting miles away in other domains and networks. An Energy Management System (EMS) in a feeder substation should not have to wait for a command from a central Control Room to deal with a looming overload situation which is limited to that substation only. The EMS should first deal with the event and then report the situation and the action taken in due time.

Moreover, the technologies required to realize inherently centralized command and control at such scale are either unavailable commercially or not cost-effective for mass deployment, given the necessary throughput and the required response time which such systems require. That means that Smart Grid command and control have to be "distributed" by nature, operating on the basis of local decision making using a set of global as well as local attributes and parameters. By default, this notion points to the need for the constituent components of Smart Grid system to have the inherent ability to generate, receive, process and analyze real-time data, and make informed decisions and/or attend to their pre-assigned tasks based on the set of local and global attributes available to them.

Fig. 2.7 shows a distributed command and control architecture for Smart Grid in which physical separation of networks and domains are identified, with clear demarcations between networks and their constituent components. Moreover, and despite clear boundaries between different networks (which may have their own purpose-built cyber-security and data-access strategies and regimes as a function of their constituency, functions and associations), the figure emphasizes the need for end-to-end communication and data exchange protocols to ensure seamless communication between nodes across the entire network.

This approach also advocates the distributed command and control system to be built over a system of intelligent agents which may exist as stationary implants (associated with individual nodes) or mobile agents (associated with networks). These agents are capable of generating, capturing, processing and analyzing information and data from their immediate environment and make decisions related to their functions using a set of local and global parameters and attributes. The agents are empowered to negotiate with other agents, take actions, and broadcast their status to higher level applications if and when required. Stationary agents' sphere of influence is limited to the node within which they operate (e.g. a protection agent implanted within a relay), whereas mobile agents traverse the length of a network and are in charge of global tasks (e.g. intrusion detection agents operating within the boundaries of a particular network, implementing its access control protocol and regime).

2.6   Smart Grid Building Blocks

Utility companies are fully cognizant of the difficulties involved in transitioning their infrastructure, organizations and processes towards an uncertain future. The fact of the matter is that irrespective of all the capabilities which Smart Grid promises to yield, as a provider of a critical service, utilities' primary concern remains in keeping the lights on. Mapping the strategies discussed so far to a practical architecture, one can readily see that Smart Grid cannot and should not be a replacement for the existing electricity grid, but a complement to it. In other words, Smart Grid would and should co-exist with the exiting electricity grid, adding to its capabilities, functionalities and capacities through an evolutionary path. This therefore necessitates a topology for the Smart Grid that allows for organic growth, inclusion of forward looking technologies and full backward compatibility with the existing legacy systems.

At the core, the Smart Grid will therefore materialize through an ad-hoc integration of complementary components, subsystems and functions under the pervasive control of a highly intelligent and distributed command and control system. Furthermore, organic growth and evolution of the Smart Grid is envisaged to necessitate plug & play integration of certain basic structures which are called Intelligent (or Smart) Microgrids. Microgrids are defined as an interconnected network of distributed energy systems (loads & resources) that can function connected to or separate from the electricity grid.

In other words, one may consider Microgrid as a collection of loads, generation plants and command and control systems which can function in isolation from the electricity grid, or in concert with it. The smartness of Microgrid is thus determined by the level of closed-loop intelligence which allows the desired level of automation, energy management and protection to be built into the system. This definition is further extended to encompass scaled-down subsets of Microgrid, such as a building, a home or even an electric car as a Nanogrid. Regardless of such differentiations, the basic characteristics of Microgrids are their ability to implement desired Smart Grid capabilities and functions in a well-controlled environment, minimizing the risks associated with the level of maturity, or lack thereof, of the required technologies used in realizing the planned Smart Grid capabilities and functionalities.

As such, it is expected that the quest for transition to Smart grid would see the emergence of a variety of Smart Microgrids with different capabilities, functions and applications. Increasingly, these Microgrids will assume a much greater role in providing services to their associated loads, thereby reducing the overall load on the electricity grid. As such, the electricity grid is poised to transform itself into a collection of integrated Smart Microgrids, capable of engaging each other with various degrees of power transactions. It goes without saying that Information and Communication Technologies will provide the required plumbing for that transformation, determining not only the level of sophistication and intelligence within each Microgrid, but also their level of integration within the larger Smart Grid system.

In essence, a Microgrid incorporates not only the three major components of Generation, Loads and Smart Controls, but also a flexible and highly programmable command/control overlay which enables the microgrid to operate as an independent entity within a much larger system in both grid-tied as well as islanded modes, typically integrating the following 7 components:

1. Power plants, capable of meeting the local demand, as well as feeding the unused energy back to the electricity grid. Such power plants are known as co-generators that often use renewable sources of energy, such as wind, sun and biomass. Some Microgrids also recover and distribute the heat byproducts of these power plants in the form of hot water for local use (CHP: Combined Heat & Power).

2. Variety of loads, including residential, office, commercial and industrial loads.

3. Local and distributed power storage capability to smooth out the intermittent performance of renewable energy sources.

4. Smart Meters and Sensors, capable of measuring multitudes of consumption parameters (e.g. active power, reactive power, voltage, current, demand, etc) with acceptable precision and accuracy. Smart Meters should be tamper-proof and capable of soft connect/disconnect for load/service control.

5. Communication infrastructure, enabling system components to exchange information and commands securely and reliably.

6. Smart Terminations, Loads and Appliances, capable of communicating their status and accepting commands to adjust/control their performance and service level based on user and/or utility requirements.

7. Core-Intelligence, comprising of integrated networking, computing and communication infrastructure, appearing as Energy Management applications which allow command and control on all nodes of the network. These should be capable of identifying each termination, querying them, exchanging data and command with them and managing the collected data for scheduled and/or on-demand transfer to the higher level intelligence, residing in the Smart Grid.

As Fig. 2.8 depicts, the Smart Grid is therefore expected to emerge as a well-planned plug & play integration of Smart Microgrids, which will be interconnected through dedicated highways for command/data and power exchange. The emergence of these Smart Microgrids, and the degree of their interplay and integration will be a function of rapidly escalating Smart Grid capabilities and requirements. It is also expected that not all Microgrids will be equal. Depending on their diversity of load, mix of primary energy sources, geography, economics, etc, Microgrids will be built with different capabilities, assets and structures. As noted earlier, Information and Communication Technologies shall play an important role in providing Microgrids with their planned capabilities and functionalities.

2.7   Smart Grid Network Topologies

The end-to-end command and control architecture depicted in Fig. 2.7 has in its core the basic premise that service and information connectivity has to be provided between where the power is generated, and where it is consumed. Moreover, Smart Grid strives to empower the consumers to be "prosumers", ie the consumers should have the ability to not only consume power but also inject power back into the system if they so choose to. That means that Smart Grid in its more advanced mode should be viewed as a network of networks which facilitates energy transactions between its nodes.

Presently, however, the grid is in a transitional mode, in which monitoring and control is gradually trickling down from upstream utility domain towards downstream consumer domain. And given the fact that roles of different actors in the system, and therefore their needs and requirements for command and control, will remain essentially the same for the foreseeable future, the actors need to be grouped together in different networks based on their associations and interactions with the larger system.

It is in that light that the command and control architecture depicted in Fig. 2.7 is further clarified as the network topology presented in Fig. 2.8. Here the actors which collectively define the system load are depicted as Home Area Network (with variations based on the nature of the load such as Commercial Area Network, Industrial Area Network, Neighborhood Area Network, etc), while utility assets which constitute the bulk of power system network (primarily the distribution side) is called Local Area Network (with variations based on actual assets such as Substation Area Network, Field Area Network, etc) and Wide Area Network, representing the collection of utility's back-office tools and applications covering the rest of upstream assets. It should be noted that at the present time there are no universally accepted terminologies and conventions on the naming of such networks. Nevertheless, the fact remains that Smart Grid continues to emerge as an integrated system of different groupings of actors and/or functions with specific requirements for command, control and data access.

Therefore, the topology of the emerging smart grid will resemble a hybrid system, the core-intelligence of which grows as a function of its maturity and extent. Fig. 2.8 shows a highly abstracted topology of Smart Grid in transition. As discussed earlier, Smart Grid Network Topology will follow the path of refinement as set by the smart grid roadmap which each utility will choose and the resulting developments which will be set in motion.

2.8   Smart Grid Communication Systems

As discussed earlier, smart grid communication systems need to enable end-to-end connectivity and data-exchange capability between different nodes in the system. Fig. 2.10 further clarifies the network topology depicted in Fig. 2.9 as a hybrid communication overlay with clear demarcations as Home Area Network (HAN, serving nodes within consumer premises), Local Area Network (LAN, serving Smart Meters, Sensors, Capacitor Banks, Voltage Regulators), and in general all intelligent components which reside downstream of substation (and therefore outside substation fence) but outside HAN. Realistically, at the present point in time, most LAN's may contain Smart Meters and nothing else. For simplicity, Fig. 2.10 groups Substation nodes (Distribution as well as Transmission), upstream utility assets as well as Utility's core network as WAN (Wide Area Network).

The reason for such classification is attributed in part to the security, access, bandwidth and latency requirements of the prevailing functions. For instance, Fig. 2.9 suggests that applications such as Energy Management System and Demand Response to be the predominant applications in the HAN environment. Note that currently in most North American jurisdictions, Utilities are not permitted to go beyond the meter, and as such the ultimate reach of their Smart Grid applications is limited to LAN. The assumption in that case is that the Utility would provide energy pricing signals and tariff information to customer's EMS systems (directly through a secondary communication link provided by the meter such as a Zigbee radio link or narrow-band PLC, or possibly through internet), and allow them to take predetermined actions (based on user preferences) if they so choose to. As far as HAN is concerned, the industry is converging on Zigbee (2.4 GHz ISM Band) or various derivatives of PLC (Power Line Communication) as the predominant communication technologies and Smart Energy Profile 2.0 as the messaging protocol for connectivity between HAN nodes.

Most utilities in the developed and the developing world envisage the roll-out of Smart Meters as the first step towards Smart Grid. Together with Smart Meters, communication systems, data aggregation nodes and Metering Data Management Systems are also required, the collection of which is commonly known as the AMI (Advanced Metering Infrastructure) sub-system.

Smart Meters ordinarily incorporate a modular communication interface, which enable the meter to communicate with its corresponding DAU (Data Aggregation Unit) through a variety of wired and wireless communication technologies. Utilities are aware of the fact that an optimum communication system is required to deal with massive bursts of data (on polled or event basis) which Smart Meters are expected to produce. Such information may vary considerably in size and latency requirements. They include consumption data (active and reactive energy, max demand and average demand) which may need typical bandwidths of 256 Kbps with e2e latencies in the order of tens of seconds, all the way to other types of time-sensitive information, such as alarms, sample values or over-the-air firmware upgrades which may require bursts of 1 Mbps with e2e latencies in the order of a few milliseconds.

Fig. 2.9 lists some of the existing communication medium and technologies for LAN, which include Radio Frequency (Zigbee, proprietary and standard ISM Band RF, WiFi and WiMax), Powerline Career (including narrowband and broadband PLC) and fiber technologies, in addition to communication protocols such as ANSI C12.22 (North America) or DLMS/COSEM (Europe and Rest Of the World) as well as a host of other protocols which various international standardization committees are currently developing.

Each of these technologies will meet certain bandwidth, security and latency requirements, and as such they will be optimal to carry certain throughput and meet certain latency targets. Consequently, utilities should base their choice of LAN communication system and protocol on their smart grid roadmap, schedule and desired ROI (return on investments). Any oversight in such choices will result in either an inability to seamlessly implement future functionalities, or incur inordinate costs in supporting an over-engineered communication system which the utility does not particularly need for its operations.

In contrast to novelties associated with HAN and LAN technologies, Utility companies already have extensive Wide Area Networks in place, which support their enterprise functions, corporate applications, expensive assets and workforce. A combination of extensive fiber network, supported by point-to-point microwave radios, and even high throughput Power Line Communication over transmission lines are in use in most utilities in support of their legacy functions. Recently, WiMax (developed from IEEE 802.16e) has been touted as the main contender for backbone utility communication, including support of mobile workforce.

2.9   Smart Grid ICT Future Developments

As discussed in previous sections, the emergence of many applications and capabilities of Smart Grid hinges upon the availability and/or feasibility of the Information and Communication Technologies which they require for implementation and operation. General consensus in the community of practitioners is that although a rich portfolio of ICT technologies do exists, not each and every one of those technologies could readily meet Smart Grid requirements in terms of performance, cost or integration. Efforts may be required to re-purpose or re-engineer some of those technologies for application within a Smart Grid environment, as well as work would be required to develop new ICT technologies capable of being incorporated into the up and coming Smart grid system.

Consequently, Smart Grid ICT Researchers in academia, public and private research labs are working towards:

1. technology-agnostic topologies for HAN, LAN and WAN networks,

2. protocols to minimize traffic,

3. service discovery, dynamic routing and channel access,

4. security, authentication, redundancy and reliability,

5. quality of service for different information types,

6. data mining for consumer and utility data,

7. visualization techniques for customer and utility information and interaction portals.

The above list is only a sample of what is happening in the research scene today and by no means is meant to be comprehensive or exhaustive. Suffice to say that intensive research is being carried out into protocols and security of critical infrastructure, and in the development of robust, secure and reliable communications and networking infrastructure, and last but not least efforts are being spent across the board to design new data structures for efficient visualization, storage and mining of data.

In summary, future developments of ICT is envisaged to focus on:

- Media-agnostic topology for HAN, LAN & WAN networks in Smart Grid & their associated protocols to support asymmetric communication (real-time, event & polled) among termination points,

- Seamless exchange of data & command through hybrid technologies within each network (Zigbee, Narrowband PLC, Broadband PLC, Narrowband ISM Radios, WiFi and WiMax),

- Robust authentication methods associated with various access functionality based on user security level and efficient encryption of command & data with minimal overhead.

2.10   Suggested Readings

As the final remark, we draw the attention of the readers to further research in this area by pointing to a list of state-of-art research works in these relevant topics: [1,4–15].

Bibliography

[1] H. Farhangi, "The path of smart grid," *IEEE Power & Energy*, vol. 8, no. 1, Jan. 2010.

[2] "http://www.oe.energy.gov/SmartGridIntroduction.htm."

[3] G. Gilchrist, "An overview of smart grid standards," Nov. 2008, http://www.ieso.ca/imoweb/pubs/smart_grid/Gilchrist-EnerNex.pdf.

[4] H. Farhangi, "Campus based smart microgrid," in *Proc. Cigre International Conference in Smart Grid, Bologna*, Sept. 2011.

[5] D. Moore and D. McDonnell, "Smart grid vision meets distribution utility reality," *Electric Light & Power Opinion Editorial*, Mar. 2007.

[6] H. Farhangi, "Intelligent micro grid research at BCIT," *EnergyBiz Smart Grid sup*, July 2008.

[7] ——, "Intelligent micro grid research at BCIT," in *Proc. IEEE Electrical Power & Energy Conference (EPEC)*, Oct. 2008.

[8] A. Palizban and H. Farhangi, "Low voltage distribution substation integration in smart microgrid," in *Proc. IEEE International Conference on Power Electronics*, June 2011.

[9] A. Vojdani, "Integration challenges of the smart grid enterprise integration of DR & meter data," in *Proc. IEEE lectrical Power & Energy Conference EPEC*, Oct. 2008.

[10] M. Smith, "Overview of federal r&d on microgrid technologies," in *Proc. Kythonos Symposium on Microgrids 2008*, June 2008.

[11] K. Moslehi, "Intelligent infrastructure for coordinated control of a self-healing power grid," in *Proc. IEEE Electrical Power & Energy Conference (EPEC)*, Oct. 2008.

[12] K. Mauch and A. Foss, "Smart grid technology overview," *Natural Resources Canada*, Sept. 2005.

[13] S.J.Anders, *The emerging Smart Grid.* Energy Policy Initiative Center, University of San Diego School of Law, Oct. 2007.

[14] S. Amin and B.F.Wollenberg, "Toward a smart grid: power delivery for the 21st century," *IEEE Power and Energy Magazine*, Sept. 2005.

[15] G. Stanciulescu, H. Farhangi, A. Palizban, and N. Stanchev, "Com tech for smart microgrid," in *Proc. IEEE PES Innovative Smart Grid Technologies (ISGT)*, Jan. 2012.

Author Contact Information

Hassan Farhangi is with British Columbia Institute of Technology, Vancouver, Canada, Email: Hassan_Farhangi@bcit.ca.

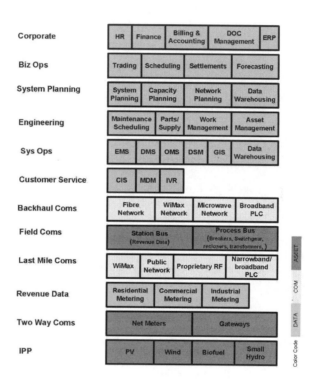

FIGURE 2.5 Smart Grid ICT architecture

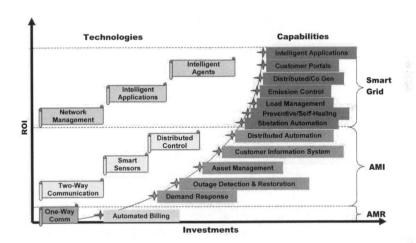

FIGURE 2.6 Phased introduction of Smart Grid applications and the fundamental technologies [1] © 2010 IEEE

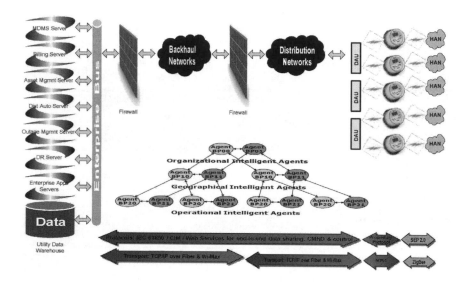

FIGURE 2.7 Distributed command and control architecture for Smart Grid

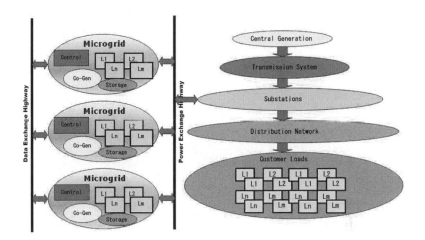

FIGURE 2.8 Smart Grid: well-planned plug & play integration of Smart Microgrids [1] © 2010 IEEE

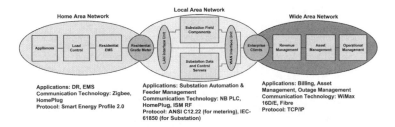

FIGURE 2.9 Hybrid communication overlay among HAN,LAN, and WAN [4] © 2011 Cigre

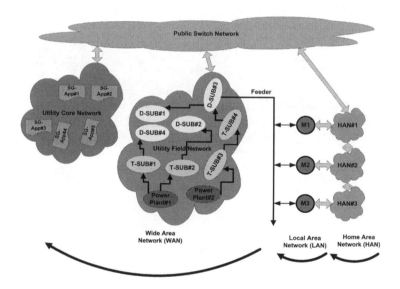

FIGURE 2.10 Smart Grid network topology

3

Energy Saving Solutions and Practices of China Mobile

Sen Bian, Huabin Tang, Xidong Wang

China Mobile Research Institute, China

Jinsong Wu

Bell Laboratories, China

Zhihong Zhang, Chih-Lin I, and Bill Huang

China Mobile Research Institute, China

CONTENTS

3.1   Introduction

3.1.1   Energy Consumption Status and Objective of CMCC

China has treated energy conservation and environmental protection as a national strategic policy. Energy conservation is not only an indispensable part of business activities in China Mobile Communications Corporation (CMCC), but also one of the main corporate social responsibilities. CMCC focuses on six green aspects, including the conservation of energy, materials, lands, and costs, as well as waste recycling and society informationization. CMCC has established a green-dedicated unit, "Green action plan working group", to promote energy conservation. Compared with the energy consumption per unit volume of business in 2005, in 2010, CMCC has realized the challenge goal of 40% reduction, and in the same time, signed a voluntary energy-saving and emission reduction agreement with the China National Ministry of Industry and Information Technology, which has stated that "Compared with in 2008, in the end of December 2012, the energy consumption per unit of volume consumption per year will be decreased by 20% , achieving the electricity saving of 11.8 billion degrees". Energy-saving and emission reduction may enable the business to enhance the soft-power, save operating expenditure, realize low-cost, and highly-efficient operations, improve the efficiency and value of the enterprise to obtain the rapid, healthy and sustainable development.

3.1.2 Views on Energy Saving of Wireless Communication Devices and Systems from CMCC

Through strengthening the statistical analysis and sharing relevant implementation experiences and information, CMCC has effectively set up energy saving strategies to make notable contributions in the areas of communications energy-saving and emission reduction. CMCC has successfully cooperated with the WWF (World Wild Life), IDG (International Data Group), Bell Laboratories, and other relevant institutions to generate solution reports of a low-carbon communications to significantly improve network efficiency and reduce network energy consumption, and make effective efforts in energy saving standardization for the communications industry by the year of 2016.

Currently, CMCC has started the implementation of green actions, including the determination of the operational objectives, the extensive investigations in industries, the development of relevant technical standards and specifications, and the evaluation of energy saving in the processes of centralized purchasing the equipments of machine rooms. Through the above efforts, CMCC has achieved significant results in energy-saving, including creating 54 items of relevant technological innovations, exhibiting scale and concentration efficiency, improving the energy efficiency of system components and devices, increasing the cooling energy efficiency, and supporting device automatic dormancy.

CMCC has also achieved significant achievements in the aspects of renewable energy utilization. CMCC has established more than 7000 base stations relying on renewable energy, among which there are more than 1 thousand base stations using renewable energy located over the altitude of more than 4 thousand meters in the Qinghai-Tibet Plateau, China, according to the best knowledge of us, the largest solar base station group in the world. As the global leader achieving green communications, CMCC has established possibly the largest green networks. CMCC is actively promoting the use of virtualization technologies based cloud computing, building the green marketing model, and supporting electronic commercial distribution channels.

3.2 Energy Saving Solutions for Wireless Networks

Compared with the core networks, wireless access networks are located at the bottom of the network layers, consuming 64% of total energy consumption, and thus, energy-saving technologies for wireless base stations have a significant importance. To enhance the energy efficiency of base stations, one of main solutions is to improve RF (radio frequency) efficiency and reduce antenna losses, maximize the antenna input power so as to increase energy efficiency of the wireless base station sites.

3.2.1   The High-Energy-Efficient Hardware Platform

Traditional centralized macro base station equipment includes baseband processing boards, the master, transmission, monitor, clock, transceiver, power amplifier modules and other components. Each distributed base station partitioned the function components existed in the centralized macro base station into two relatively independent parts connected via fiber-optic communication channel: one part is the centralized base station enclosure, while the other part including RF power amplifiers and some other components except antennas is called a remote radio head (RRH) [1]which is also called remote radio unit (RRU) [1]. Distributed base station architecture was originated from the third generation (3G) system, which mainly utilized only remote radio over optical fiber, but, when GSM (Global System for Mobile Communications [2]) systems adopted distributed base stations, two kinds of RRH modes, remote radio over optical fiber and remote baseband-signal over optical fiber, were introduced, mainly because the key resources to GSM systems and 3G systems are different: 3G systems using code division multiple access approaches, and thus the key resources are baseband signals. The key resources to GSM systems are the carrier frequencies and the baseband processing functions are relatively easy to be implemented, and thus placing baseband processing functions into either BBU (baseband unit) or RRU will not cause a significant impact to system performance. In GSM/GPRS(General Packet Radio Service)/EDGE(Enhanced Data Rate for GSM Evolution) systems, BBU communicates with the BSC via the Abis interface traffic [2], while RRU communicates with mobile terminals through the Um interface [2]. The two RRH modes have no significant differences in the network organization, operation, maintenance, future upgrade and expansion, evolution and so on. According to the current situation of the main equipment vendors, the mostly used implementation ways of two RRH modes are described as follows [3]:

1. Remote radio over optical fiber [4]: in this case, the BBU may include the baseband processing board, master, transmission, monitoring, and clock component modules, while the RRU may include the transceiver, and power amplifier component modules. The BBU is installed indoors, while the RRU is installed outdoor near antennas.

2. Remote baseband-signal over optical fiber: in this case, the BBU may include the master, transmission, monitoring, and clock component modules, while the RRU may include the baseband processing board, transceiver, and power amplifier component modules. The BBU is installed indoors, while the RRU is installed outdoor near the antennas.

3.2.1.1   RRUs in Distributed Base Stations

RRU in distributed base station can be classified as dual-density carrier-frequency RRU [5] and multi-carrier RRU. Dual-density carrier-frequency RRU, as shown in Fig. 3.1, adopts SCPA (single-carrier power amplifier) technology that is narrowband PA (power amplifier), and a carrier-frequency board only outputs an RF signal frequency. In the case of SCPA, multiple-carrier signals have to be combined and bring some signal attenuation, reducing the efficiency of the whole machine by about 5% - 6% (total amplifier output power / total power consumption of the base station). Dual-density carrier-frequency RRU is with small capacity due to using a single carrier frequency, module mainly used in the scene with low requirements, such as suburban signal blindspot, rural-farms coverage, high-speed rail, highways, tunnels, and so on.

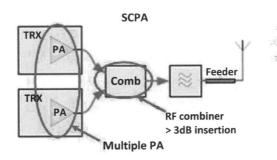

FIGURE 3.1 Dual density carrier-frequency RRU schematic

Multi-carrier RRU, as shown in Fig. 3.2, supports multi-carrier signals in addition to the distributed characteristics. Multi-carrier transceiver module has high integration density, and single such a module may support up to 6 carrier-frequencies. Within the same cell, multiple carrier-frequencies are directly combined through the digital IF signals, and are amplified simultaneously using single a PA module, MCPA (multi-carrier power amplifier). A large base station can run without the need of an external multi-carrier combiner so as to avoid the combination loss and be able to expand without additional hardware. The whole machine efficiency using MCPA is much, approximately 13% -15%, higher than that using SCPA so as to meet the requirements of high-traffic scenarios such as in CBD / dense urban areas, residential areas, and indoor-coverage areas.

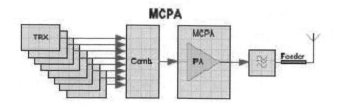

FIGURE 3.2 Schematic diagram of multi-carrier RRU

3.2.1.2   Distributed RRU Networking

There are three networking modes among RRUs and BBU: star, chain and ring modes. Devices supporting ring networking is not mature, and thus ring modes only has limited applications. Currently, star networking is mostly recommended.

Networking Mode 1: star

As shown in Fig. 3.3, each individual RRU pulls fiber to the BBU, and different RRUs has different optical docking interfaces at the BBU. The advantage of this network is scalable, while the disadvantage is high consumption of fiber resources, so this mode is suitable for fiber-rich areas. Star networking is mainly used in the same site for multi-sectors applications, usually used in the case that the BBU and the RRU are closely located.

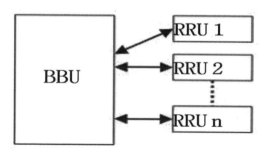

FIGURE 3.3 Star network diagram

Networking Mode 2: chain

Chain network formation is composed of multiple chain-connected RRUs before a RRU and BBU connection, as shown in Fig. 3.4. The advantage of chain networking is that the optical port resource utilization is high, but because of many chain-connected links, any RRU failure will lead to the whole chain failure. Chain network formation is suitable for trunk-road and railway coverage.

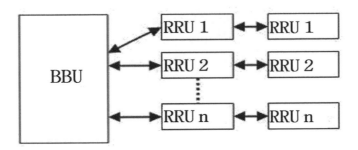

FIGURE 3.4 Chain network diagram

Networking Mode 3: ring

Ring network formation, as shown in Fig. 3.5, is composed of multiple-ring-connected RRUs, and two pairs of fiber between a RRU and a BBU use different physical routes, thus ensuring at the same time two pairs of fiber will not be in failure status simultaneously. This network is used to improve transmission system reliability over chain networking.

This approach will tap each end of the chain RRU on one fiber port of the BBU, respectively. When the middle one port fails, the ring can automatically switch network into the two-chain form continue to provide services. It does not occur that a large area of RRUs do not work when the ring-type network formation is used. Although this method greatly improves reliability, it requires to use two optical ports, thus using more optical port resources. So the practice should be based on the actual importance of the environment and the number of optical ports to determine what modes of networking should be used.

For the coverage in large venues, in order to guarantee to cover the whole place and ensure that the business will not be interrupted in the case of a single point of failure, the ring network is more appropriate.

3.2.1.3 Multi-Carrier Distributed Base Station

The traditional design of GSM base station mainly adopts single-carrier power amplifier (SCPA), and single-sector multiple carriers are realized through combining multiple single-carrier power amplifiers. The introduction of passive combiner brings the signal loss (usually through a combiner, the signal is attenuated about 3.5dB; through two combiners, the signal is attenuated about 7dB), reducing the machine efficiency of GSM base station based on

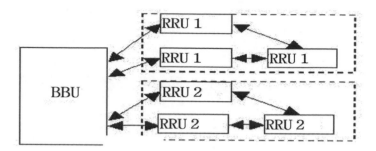

FIGURE 3.5 Ring network diagram

traditional single-carrier techniques. However, in multi-carrier base stations, multi-baseband signals experience digital-conversion process, pass the digital IF combiner into a broadband multi-carrier signal, and then go through a broadband multi-carrier power amplifier (MCPA) for signal amplification directly into the antenna system. In the receiver, the antenna receives the multi-channel RF signals, and after IF(intermediate frequency) down-conversion and digital down conversion, digitally filtered baseband signal is then digitally processed.

Compared with traditional SCPA based GSM distributed base station, MCPA technology-based multi-carrier distributed base stations, shown in Fig. 3.6, have the following features:

1. Large capacity and high density of integration: A single BBU supports at least 36 TRX (Transmiter/Receiver), supports at least six RRUs. Each RRU supports 4-8 TRX, higher density of integration enables smaller equipment size and area.

2. High efficiency: A BBU of 2G mode consumes less than 2W per TRX. Under the GMSK modulation mode, RRU machine power efficiency (4 TRX) minimum 22

3. Broadband: In 2G base station, the instantaneous operating bandwidth is 20MHz bandwidth for the GSM 900MHz band, and 25MHz bandwidth for the GSM 1800MHz band;

4. High efficiency: Multi-carrier power amplifiers are supported so that output power is shared between multiple carriers. Resource-sharing between the carriers and the sectors is supported so as to improve resource utilization;

5. High performance: 2G interference cancelation technology is supported in hardware. 2G 2-antenna transmit diversity and 4-antenna receive diversity are supported;

6. Capabilities of continued evolution: Support for EDGE Evolution (RED-HOT A / B, HUGE A / B), Support for LTE evolution (LTE bandwidths within the same frequency band are not less than 10MHz);

7. Flexible interface specification: the unified standard optical fiber interfaces between BBUs and RRUs, support for daisy-chained fiber connection of at least six logic cells, Abis interface supports the latest GSM Abis existing processes and algorithms.

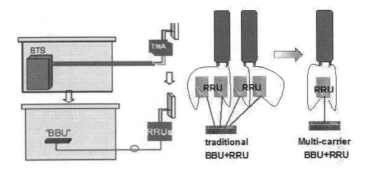

FIGURE 3.6 Schematic diagram of multi-carrier distributed base station

3.2.1.4 Energy Saving Advantages

The GSM distributed base stations mainly has the following advantages:

1. Installing RRUs near the antennas can eliminate 3dB feeder loss existed in the traditional base station, enabling lower power amplifier output power requirements, thereby improving the coverage performance or reducing machine power consumption.

2. Reduce the workload in construction, since BBU and RRU are smaller and lighter, which will enable a single person to carry;

3. Compared with the traditional base stations, the equipment power consumption and the operation costs of the base station can be partly reduced;

4. The baseband part is placed at the centralized location, while the RF part is placed near the antennas. This way which may help save rental costs due to smaller room or space requirements and reduce the requirements for supporting facilities;

5. RRU remote pull-away separates the functions from the relevant power consumptions, and a RRU exhausts low power consumption than a traditional base station does, which may enable to design some renewable energy (wind, solar, etc.) based RRU;

6. With the maturity of MCPA technologies, MCPA based distributed base stations may help satisfy the needs of customers under a vast majority of current network scenarios and applications. With low requirements of the resource of station locations, MCPA based distributed base stations support the flexible construction of cellular networks. In addition, MCPA based distributed base stations not only own soft capacities of quick and easy remote expansion, but also support flexible configurations of the number of carrier frequencies through sharing remote licenses in order to improve the utilization of online carrier frequencies in the network. The multi-carrier technologies help the broadbandization of the GSM systems, which enables GSM base stations to smoothly evolve into LTE(long term evolution) and effectively protect existing investments.

3.2.2   Software Power-Saving Features

3.2.2.1   Intelligent Power-Saving Technology

Under the frequent changes of carrier-frequency load in the traffic channels, the power consumption of the base station system is quite volatile. One of main energy saving technologies in wireless communications is to adopt software-controlled dynamic power control. For instance, in the GSM systems, the priority of primary BCCH(Broadcast Control Channel [2]) carrier-frequency can be set to the highest, that is to say, the TCH(Traffic Channel [2]) is assigned to the primary BCCH carrier frequency at the highest priority when voice traffic comes. To improve the utilization of the primary BCCH carrier frequency, higher carrier-frequencies are with higher priorities, and the other carrier-frequencies can be cut-off. The cut-off may be performed in ways: either at the pre-determined time segment according to idle time statistics or in accordance with the changes of communication volume. The above-mentioned second way is based on some intelligent algorithms to open or close some carrier frequencies, called "Load based TRX / PA cut-off". Further, the more precise cut-off control can be performed in finer granularity level, such as time slots, called "Time slot based on the PA cut-off".

Currently, main GSM equipment vendors are at least able to support one of the two techniques, " Time slot based on the PA cut-off" and "Load based TRX / PA cut-off". The following Fig. 3.7 is the sketch diagram of energy-saving through smart cut-off technology to optimize the power allocation .

1. In the evening, the carrier frequencies without carrying traffic can be closed using "Load based TRX / PA cut-off" according to the statistics and prediction of communication volumes.

2. In the case of low traffic, through granular management, "Time slot based on the PA cut-off" can be performed to improve the energy efficiency in the RF communication channels.

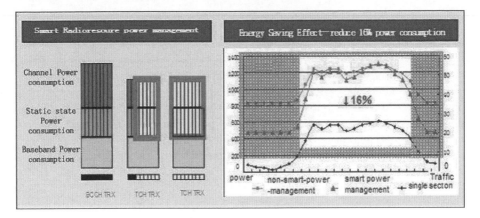

FIGURE 3.7 Sketch diagram of the smart turn-off technology

Either load or time slot based TRX / PA cut-off is implemented through the software-controlled adjustment to either the switching regulator of TRX or PA offset based on the network communication volume. The adjustment to TRX switch is mainly to open or close the PA work voltage, while the adjustment to PA offset switch is mainly to open or close PA bias voltage to control the PA bias current.

Load based TRX / PA cut-off

1. Load based TRX cut-off According to the traffic situation of each carrier-frequency module, BSC (Base Station Controller) makes judgement to shutdown the the PA work voltage corresponding to the voice carrier-frequency module whose idle time is longer than some predetermined threshold time. Then, when the monitored traffic load increases over a certain configurable threshold, BSC will immediately activate the cut-off TRX to meet the traffic demand. The activation time for carrier frequency is up to minute-level.

2. Load based on the PA cut-off According to the traffic situation of each carrier-frequency module, BSC makes judgement to shutdown the the PA bias voltage corresponding to of the voice carrier-frequency module whose idle time is longer than some predetermined threshold time. Then, when the monitored traffic load increases over a certain configurable threshold, it takes the millisecond-level time to re-open the amplifier bias voltage to meet the traffic demand.

The power saving effects of Load based TRX / PA cut-off is pretty strongly proportional to the number of carrier frequency configured in the cell, and is inversely proportional to the length of time segment of voice traffic and utilization of wireless systems.

Time slot based on the PA cut-off

The power-amplifier-tube of carrier-frequency amplifier is the main device that the carrier frequency transmits power to antennas. Because of the limitation of the linearity of the device, even if it does not transmit power, it also require to impose a fixed bias voltage, making the amplifier work in the linear region, whose power consumption is called static power consumption. Another part of power consumption also includes dynamic power consumption, which occurs when voice traffic comes. The more dynamic power consumption is, the higher the carrier-frequency output power; in the case of no traffic, the dynamic power consumption is 0. The technique of "Time slot based on the PA cut-off" is that, when no traffic is in the time slot, that dynamic power is 0 in this case, the static power consumption is further cut off, the power consumption in the time slot is 0. Time slot based on the PA cut-off enable PA to switch between on and off per time slot basis.

The power saving effects of time slot based TRX / PA cut-off is pretty strongly inversely proportional to average amount of voice traffic and utilization of wireless systems.

3.2.3 GSM Intelligent Power-Saving System Based on Base Station Sector Switching

3.2.3.1 System Overview

Through OMC(operation & maintenance center) software, the base station energy-saving control station system intelligently controls the increase and decrease of the power, and turning-on and turning-off of the sector, and BEC realizes the base station energy-saving through intelligent power switching based on the power detection. Base station energy-saving system mainly consists of two parts, BEC hardware and OMC software. BEC(Base station energy-conservation) is mainly responsible for switching and the distribution and transmission of RF signals, while system OMC software is mainly responsible for the intelligent control of the base station power and the integrated management of BEC. The whole system architecture is shown in Fig. 3.8.

BEC hardware components

BEC mainly consists of RF switches, power allocation unit, power supply unit, monitor board, and modem. The functions of BEC hardware units are introduced as follows:

1. RF switch: The RF switching realizes omni-directional coverage for a single cell, which is realized through switching to different RF transmission branches. After finishing RF switching, the system OMC software sends **OMC-R** commands to the base stations so as to minimize or soft-close the power switches for the other two sectors **B** and **C** or to minimize

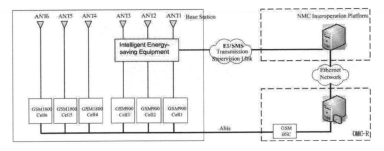

FIGURE 3.8 Energy saving system architecture of base station

the power soft-off area, and the main sector **A** becomes omni-directional coverage through RF switching mentioned above.

2. Power allocation unit: The signals from the primary-sector of the cell are uniformly powered for all directions in the energy-saving state.

3. power supply unit: The power supply supports 48V or 24V DC(direct current) mode.

4. Monitoring board and the modem: The equipment operation status is monitored in real-time, and abnormal states are automatically reported.

Software Components

The system software architecture, shown in Fig. 3.9, uses a distributed three-tier server architecture, including communications servers, application servers, and the clients. Communications servers are responsible for the communications among hardware equipments of the energy-saving systems and alarm SMS(Short Message Service) forwarding. Application servers are responsible for the application and business logic processing, including the base station power control, BEC control, data synchronization among base stations and energy saving systems, and the analysis of voice traffic data, and so on. The client ends are responsible for the operation and maintenance functions of customers, consisted of some client-end business logic components.

System workflow

OMC is connected with the OMCR(Operation & Maintenance Center Radio) database for the base station, which provides the analysis of traffic data on a regular basis for different base stations, and determines the configuration and on-off modes of power saving functions of the base stations. After power saving functions of the base stations are turned on, OMC either lowers or turns off the power for the secondary-sector, an then BEC switches the operation mode for the primary-sector to omni-directional coverage mode so as to save

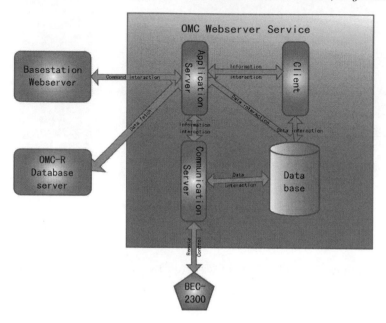

FIGURE 3.9 System software architecture diagram

the energy consumption of the base station. When power saving functions of the base station are turned off, OMC either increases or turns on the power for the secondary-sector, and then BEC switches back to normal three-sector coverage mode.

3.2.3.2   Typical Power-Saving Proposals

GSM power-saving system can meet the application requirements of a single GSM network or a GSM / DCS dual-network, and achieve different levels of power-saving modes through different RF power levels.

Application proposal for single GSM/DCS dual-network

Single-network energy-saving proposals, as shown in Figure 3.10, mainly support the base stations providing services for independent either GSM900 or DCS1800 networks. The energy-saving system takes full advantage of the tidal characteristics of voice traffic in communication networks to maximally save power consumption of the base stations during network idle states. Since the volume of voice traffic in daytime is high, the equipments are configured to the non-energy-saving state, and the base station is configured to per-sector independent coverage mode for a three-sector cell so as to meet the needs of users. Since the volume of voice traffic in nighttime is low, OMC software will minimize the output power of two secondary sectors, BEC switches to the

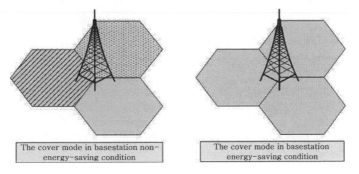

FIGURE 3.10 Single network (GSM or DCS) application scenario

energy-saving state, the signals from the primary sector is uniformly powered through a power allocation unit to three sectors to realize omni-directional coverage.

Application proposal for a dual-GSM/DCS network

Dual-network energy-saving proposals, as shown in Figure 3.11, mainly support the base stations providing joint services for both GSM900 and DCS1800 networks. In urban areas, the motivation for joint station construction of DCS and GSM900 systems is mainly to increase the network capacity. When the capacity of a 900MHz cell can meet the needs of the whole site traffic, it is desirable to only maintain a 900MHz system for coverage. Compared with a single GSM/DCS network, dual-network with co-location base station site may significantly save the energy consumption by the unit of 5 cells.

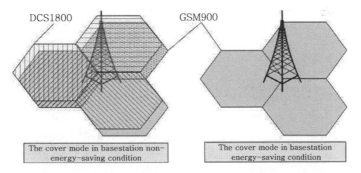

FIGURE 3.11 GSM/DCS dual-network application scenario

3.2.4   Highly Energy-Efficient RAN Network Architecture

With the rapidly growing mobile Internet traffic, the expenditure for the construction, operation, and upgrade of wireless access network is continually increasing, and traditional high capital expenditure, operation, and maintenance costs of wireless access networks make the mobile operators lose the competitiveness in the mobile Internet market. C-RAN, suitable for high-performance low-cost mobile Internet, will become the future trends of radio access network. C-RAN are consisted of the centralized baseband processing pools, collaborative wireless network using remote radio-frequency units and antennas, and real-time cloud-based infrastructure based on open platforms. Centralized baseband processing can greatly reduce the number of base stations to cover the same area. Collaborative remote radio-frequency units and antennas can improve the system spectral efficiency. Real-time cloud-based infrastructure based on open platforms and base station virtualization technology can help reduce costs, share processing resources, reduce energy consumption, and increase infrastructure utilization. The detailed discussions of C-RAN are provided in chapter 11.

3.3   Energy Saving Proposals for Core Networks and Data Centers

3.3.1   Energy Saving of Data Center Facilities

3.3.1.1   Data Center Energy Efficiency Assessment and Optimization

A considerable number of data centers currently used in CMCC have been in operation for quite a while. At the time of the construction of those earlier data centers, the levels of business requirements and technology development were much lower than the current ones, and thus the power densities of the data centers are notably lower, since it had not paid enough attention to reducing energy consumption. As the results, the existing data centers have no enough spaces, and have low power and cooling efficiency, which seriously hamper the further development of IT systems and increase the operating costs of data centers.

One of important tasks in energy saving for data centers is to comprehensively assess energy efficiency, and to perform some targeted solutions to the data centers with serious problems in high energy consumptions.

Energy efficiency assessment to existing data centers

CMCC has raised its own energy efficiency assessment criteria for overall data centers in order to perform a reasonable assessment of energy efficiency of existing data centers, promote the measurement and management of energy consumptions of data centers, and actively take various techniques and measures to improve energy efficiency.

In the chosen evaluation criteria, PUE (Power Usage Effectiveness, energy efficiency), generally accepted by international data centers as the energy efficiency measure, is calculated as follows [6]: PUE = Total power consumption of the data center / the power consumption of IT equipments, where

1. the total power consumption of the data centers is the power consumption of the whole data center to maintain the normal operation (if the building where a data center is located is also used as an office building and other purposes, the power consumption for office purposes should not be included in the power consumption for overall facilities of the data center), including IT equipments, air conditioning and power supply systems.

2. the power consumption of IT equipment includes the power consumption of various types of servers, digital storage and network equipments.

It can be known according to the definition that the physical significance of PUE is how much is really consumed in the IT equipment among the power consumption of the total data center. The main way to reduce the PUE of the data center is to reduce the power consumption of air conditioning and power supply systems.

Since some of the data centers is currently even unable to provide the PUE based on actual measurement values, the evaluation of overall energy efficiency of data centers requires to install the necessary electricity meters and other measuring devices to capture, record, analyze, and evaluate the electricity consumption of various types of equipment in data centers. This will lead that the data centers treat the measurement of power consumption as the routine steps of the relevant operation management and maintenance. The measurement of power consumption will gradually become more thorough, delicate, and automatic.

3.3.1.2 The Analysis of the Main Factors Affecting Energy Efficiency

In a recent investigation of the existing status on the energy efficiency of data centers, a comprehensive analysis and comparison of PUE was obtained for more than a hundred of data centers all over the country. From this investigation, main factors impacting the energy efficiency of data centers are summarized as follows:

1. Utilization of data centers: When the utilization of the data center is lower than that of its design capacity, the power supply equipment, such as UPS (uninterruptible power supply), of the data center has lower conversion efficiency, and the relevant waste of resources of air conditioners becomes more serious, which often leads to a higher PUE value.

For example, the designed power densities of existing data centers usually around from 3KW(kilowatts) to 5KW per machine rack. The average PUE for the data center with the measured power density, 3KW per machine rack, is about 2.19. With the decrease of the measured power density of data center per machine rack [7], the average PUE is gradually increased, such as the average PUE 2.38 for the measured power density 2KW per machine rack about 2.38, and the average PUE 2.65 for 1 KW per machine rack. If the measured power density of the data center is less than 0.5KW per machine rack, the PUE is likely to exceed 4.0.

2. Separation of hot and cold channels Whether or not to clearly separate the hot and cold channels so as to isolate the hot air and cold air in the data center is one of the most important factors affecting the cooling efficiency of the data center and the power consumption of the air conditioning cooling system. Good isolation of hot and cold channels within the data center ensures a reasonable air flow distribution, so that cold air is effectively conveyed to cool IT equipments. Through the analysis of measured data, we know that, especially for high single-rack data center with higher power, the isolation of hot and cold channels may reduce up to 30% power consumption of air conditioning systems.

3. The types of air conditioning systems: There are three types of air-conditioning systems used in existing data centers: air-cooled air conditioning, water-cooled air conditioning [8], or other natural cooling-source based (air-side or water-side economizers) air-conditioning systems, which are with different cooling efficiencies, that ia to say, to provide the same cooling capacity, the electricity consumptions for different air-conditioning systems are quite different, and thus the relevant PUE values are also quite different.

   According to rough statistics, the average PUE for the data center with air-cooled air-conditioning is larger than 2.5. Compared with the data center with air-cooled air-conditioning, the data center with water-cooled air conditioning may reduce energy consumption by about 25%. In a part of the operation time, with the aid of natural cooling-source air conditioning, the energy consumption can be reduced to a half of the case of air-cooled air-conditioning.

4. The ways of air distribution:

   The ways of air distribution, such as upper air distribution and underfloor air distribution [9–11], as well as whether or not the amount of air distribution can be automatically adjusted, may notably impact the PUE value of the data center.

   For the data center using air duct or air hood for air distribution, the cold air is easy to mix with the rising hot air causes so as to reduce the cooling efficiency. In contrast, the adoption of underfloor air distribution methods may help reduce the air conditioning system power consumption by 10%. Further, if the air distribution of the data center is precisely controlled according to the amount of heat produced by IT equipments, the power consumption of air conditioning systems may be possibly reduced by 50%.

3.3.1.3 Technologies and Measures for Optimizing Energy Efficiency

The purpose to collect and analyze the data of energy efficiency of the existing data centers is to propose targeted measures to improve energy efficiency. The relevant proposals are mainly summarized as follows:

(a) Optimize the cooling efficiency of the data center: The core approach is in accordance with the principle of "separation of hot and cold channels" to organize and optimize return air-flow of air conditioning systems in the data center, minimize or eliminate the mixture of cold and hot air in the data center, and reduce the PUE through the effective use of cooling-air resources.

(b) Improve the energy efficiency of power supply and air-conditioning system: This may be achieved through reducing power loss in the process of power transmission and conversion and choosing more energy-efficient air conditioning systems.

(c) Adjust the management of the operation and maintenance of the data center to support higher energy efficiency: The relevant measures to enhance the overall energy efficiency of the data center include monitoring energy consumption and relevant environmental parameters, performing energy-saving operation modes of equipments, and relaxing the constraints of temperature and humidity within an appropriate scope, and so on.

In the recent assessment and optimization of energy efficiency of existing data centers, through separating the cold and hot channels, turning off redundant air conditioning equipments, performing accurate control of air distribution, and increasing the amount of installed computing-machine capacity of the data center, the PUE of some current data centers would be reduced from larger than 2.5 to lower than 2.0.

3.3.1.4 The Construction of the New Generation of Green Data Center

In order to meet the needs of continual business development and the rapidly growing demand of IT systems, CMCC is also accelerating the pace of the construction of the next-generation data centers with the overall goal of "large-scale, low-cost, large-storage, and energy-saving."

Energy efficiency is not only undoubtedly one of the most important features for the new generation of data centers but also the key to reduce the operating costs of data centers. To achieve this goal, the research and application of modular data centers, natural cooling systems, high-voltage direct-current electricity distribution, and other relevant technologies have received extensive attentions.

Modular data center

The so-called modular data center refers to the data center whose IT equipment, air conditioning and power supply systems are designed using relative standard industrialization modules, which can be quickly transported to the designated location for rapid deployment or expansion of the data center according to the needs of the business [12, 13].

Containerized data center (CDC) [12, 13] is a form of modular data centers. CDC usually chooses the ISO (International Organization for Standardization) standard shipping container for containing IT(information technology), air conditioning, or power equipments. CDC can be pre-assembled in the factory, quickly transported, and deployed. There are also some modular data centers more flexible than the CDC, which can be assembled in-demand on-site and now receive more and more applications.

Modular data centers own the following advantages:

1. Acceleration of deployment: Compared with the traditional data centers, container data centers have shorter cycles of civil construction and hardware deployment, thereby accelerating the deployment of data center and IT systems. The construction cycle of a traditional data center, about 2 years, can now be shortened to from 3 to 6 months.

2. High density and energy-efficiency: Container data centers are typically designed to accommodate IT equipments with high power density, and for example, the power density per rack is four times more than that of a traditional data center. In addition, since the cooling source is closer to the equipment and a fully enclosed design is chosen, cooling efficiency is greatly improved, and thus space and energy efficiency are also improved, while operating costs are reduced.

3. Easy expansion: Available container data centers can be used to extend existing data centers, which enhances the flexibility in choosing locations of new data centers. The modular expansion of existing data centers can be done gradually according to the needs of the business in order to reduce the scale of initial investment to avoid the low efficiency and waste of resources for the case of very limited initial installed capacity.

4. Professional production and technical services. The design, manufacture and operation and maintenance of data center infrastructure are provided by professional solution providers, which greatly help the customers simplify the preparation and construction outsourcing of the data centers.

At present, CMCC is trying to use the containerized or modular approach to build new data centers with the power density as high as 15KW(kilowatts) per rack, greatly improving the space utilization of data centers, and through specialized solutions of modular data center to achieve higher energy efficiency.

Natural cooling

The so-called natural cooling or free cooling [14] is to utilize low temperature water or air from the nature to cool the data center instead of the traditional air-cooled air-conditioning compressor or water-cooled air-conditioning unit, thereby minimizing the power consumption of air conditioning and refrigeration systems.

According to the specific implementations of natural cooling, they can be classified into two ways, direct and indirect natural cooling: Direct cooling is to directly use outdoors cold water as cooling water or directly use outdoor cold air to cool the equipment, but there are potential concerns in cleanliness and other issues to be solved for cold air or cold water. An indirect cooling system is realized through the addition of a plate type of heat exchanger and cold condenser in parallel, so that, when the outdoor temperature is low, the heat generated by the data center is removed through the heat exchanger.

To take advantage of natural cooling, air conditioning systems are usually designed to be able to switch to the use of different cooling sources. When the outdoor temperature is higher than a certain threshold value, water-cooled air conditioning or other non-natural-cooling air-conditioning systems are in operation. When the outdoor temperature falls below the threshold value, the air conditioning systems switch to the use of natural cooling. The availability of natural cooling depends on the time, location, climate, and natural conditions for the data center. For example, in northern China, most of the time each year natural cooling is available to use for data centers. The reasonable design of data centers may further extend the time availability of natural cooling.

One of the most important factors in the location selection of the the green data center is whether or not to effectively use natural cooling sources. For example, the new data centers of Microsoft in Ireland, Google in Finland and Facebook in Sweden, all based on the considerations of local cold climatic conditions, can almost completely use natural cooling for air conditioning systems.

At present, the construction of newly established data centers of CMCC will be based on the principles of effectively utilizing natural cooling and reducing the operating costs. CMCC has planned to construct large-scale cloud computing centers in the provinces of Heilongjiang and Inner Mongolia, China, which will fully take advantage of local abundant power resources, low ambient temperature, and other favorable natural conditions.

HVDC

Traditional data centers usually use UPS power supply systems, but, due to the need of multiple AC(alternate current)-DC conversions, the power conversion efficiency of power supplies is relatively low.

Particularly, to improve the reliability of UPS, it is required to use redundant backup systems, resulting in the increase of the system complexity and the decrease of the energy efficiency. UPS power supply systems usually operate at the conditions of lower load (40% of the full load). The conversion efficiency of a UPS system is 80% or less system efficiency of a single UPS.

If the HVDC(high-voltage direct-current) is used to power the system [15], the side of the switching power supply system will reduce a DC/AC conversion and the side of the IT equipment will reduce a AC/DC conversion, which can greatly improve the system reliability and conversion efficiency. Typically, the power conversion efficiency of single-system HVDC can be increased to 76%. Since the reliability of HVDC power supply system has greatly been improved, it is not necessary to use redundant and parallel UPS systems, and thus the actual conversion efficiency of HVDC power supply system is increased by of 18% or more.

CMCC has already had some trial studies and tests in HVDC power supply for data centers, and then will substitute UPS with HVDC systems to improve the energy efficiency of the power supply system in the next-generation green data centers.

3.3.2   Energy Saving of IT Equipment

3.3.2.1   Improve Equipment Efficiency Requirements

The requirements of energy-saving and emission reduction for IT equipments in CMCC, have been changed from simply concerning the configurations and performance of the servers to using per-watt performance to measure the energy efficiency of the equipments.

Testing And Grading Of The Energy Efficiency Of Equipment

All needed servers to buy are required to perform energy-related tests. Using some performance test method for a specific server (such as SPEC performance test), the overall power consumption is measured and recorded in the case of no load, half load, and full load operation, respectively. The energy consumption of servers with different models from different manufacturers may be thus compared.

It has been shown through practical tests that, under the the same configuration and load, the power difference of two servers was even more than 50%. Under normal circumstances, compared with a blade server [16], a rack server [17] can save about 15% of the power consumption, which may be used to calculate the power consumption of the server and provide an assessing reference for procurement.

The grading standards of energy saving for the servers are obtained based on the testing results of the server power consumption and server parameters of the weight, volume, area, and so on. There are four classes adopted in CMCC, A,B, C,and D, in descending order to measure the energy-saving efficiency.

It is required to purchase equipments with energy saving Class C or above, and equipments with the Class A or Class B are preferred in procurement, which incentives manufacturers to develop and sell more energy-efficient servers.

Customized Server

Facebook recently announced Open Compute project, claiming that, under the same processing capacity, their customized servers consume 38%. less power than traditional servers. Inspired by the energy-efficient customized servers of Facebook and some other Internet companies, CMCC have started considering customized servers.

The core objective of ongoing research and pilot studies of customized servers is to lower TCO(total cost of ownership), whose crucial point is to reduce the power consumption of servers. According to the needs of business, the CPU (central processing unit), memory or hard drive of the server is customized through removing unnecessary components, adopting low-power components and more efficient power supply, and performing precise control of the power consumption of the server, and so on, which may reduce the power consumption of the server by from 15% to 20% from the current value.

3.3.2.2 Applications of Server Virtualization

Integration of existing IT systems using server virtualization [18] may not only effectively improve resource utilization and dynamic resource allocation but also help reduce the energy consumption of IT systems.

Server virtualization is to either partition the physical resources into multiple independent virtual resources or quickly and dynamically allocate the same resources to different applications during different time durations. It is not necessary for each application to purchase a physically independent server using virtualization. With the increased applications of virtualization, the number of required severs may be significantly reduced, while the utilization of servers may be improved,

Utilization of IT equipment has important impact to its own power consumption. For example, it has been shown that the lower utilization of servers, the lower the device efficiency. Even if the CPU utilization of the server is only 10%, its electricity consumption is almost equal to 60% of that at full utilization. Unfortunately, before the use of virtualization, most of the average server utilization of existing data centers and IT systems is between 10% and 30%, and thus about 80% of the total consumption of electricity is exhausted in the standby operation of the servers.

Currently, server virtualization has been used in many IT systems within CMCC. According to a recent survey, the integration ratio of about 1 over 5 after virtualization. In this case, the systems reduce the number of servers by 80%. Even if considering increased energy consumption after virtualization, there is at least 60% energy consumption reduction.

Moreover, with the aid of server virtualization, it is possible to shut down some servers without the interruption of the business in order to further reduce the energy consumption.

At present most of the virtualization support "live migration (Live Migration)" function, the virtual machine can be almost without interruption, the migration from a physical device to another physical device. Based on this virtualization technology, under appropriate circumstances, the virtual machines in the servers with very low utilization can be centralized into a few servers with higher utilization, while other servers will be shut down to save power.

3.3.2.3   Power Management for Cloud Computing

From the experiments conducted in CMCC, the most effective energy saving measure is turning off the servers during idle time, through which nearly 50% energy saving is achieved. When turning off the power switch, the server has but still the access to PDU, which may still consume some power, about from 6 to 20W. Thus, the server which is not in operation for a long time should be powered down completely;

To solve this problem, CMCC includes a power management module in a cloud computing platform [19]. The following list the relevant energy saving measures in CMCC:

1. Closing idle resources in the resource pool: The preferred measure used in the power management module is to close idle resources in the resource pool in time.With the increase of the demand of resources, the scheduling system can dynamically start or close the resources through the management interface to meet the needs of resources of the business.

2. Power-aware load scheduling: In order to support the idle shutdown measure, the scheduling system should be power-aware. Most of the loads should be centrally scheduled to a small amount of servers through dynamic "Live Migration" , so that more idle servers can be shut down to save power.

3. Properly adjusting the CPU frequency to control the power consumption of the server. To deal with the task which is required to be always on-line even if the server utilization is low, the CPU frequency can be dynamically adjusted to achieve power control. When the server is relatively idle, the CPU frequency can be reduced to reduce power consumption. On the

contrary, as the server load increases, the CPU frequency can be increased to obtain better performance.

3.4 Energy-Saving Solutions for Accessory Equipments

3.4.1 Energy-Saving Solutions for the Refrigeration System

The current main energy-saving solution for the refrigeration system of base stations is based on the natural cooling source, i.e., through the use of natural cooling source, the operation time of non-cooling-source based air conditioning systems is reduced to achieve energy savings. There are two mostly used natural-cooling-source energy-saving solutions for base station in promotion and trial, intelligent ventilation and heat pipe heat exchanger unit.

3.4.1.1 Smart Ventilation System

The smart ventilation equipment, shown in Fig. 3.12, in a communication base station is the ventilation unit which provides the dust-filtered outdoor air to the indoor base station. In order to cool the internal air of the base station, this equipment directly introduces the external cold air into the base station and discharge the internal hot air to outdoors without the use of any refrigeration components [20].

Under the premise of environmental requirements for the operation of the base station, according to monitored indoor and outdoor temperature and humidity of the base station, the intelligent ventilation system determines the smart ventilation logic of the base station. When certain conditions are satisfied, the intelligent ventilation system stops the non-cooling-source air conditioning equipment and directly brings the outdoor cold air into the base station for natural cooling to effectively reduce the running time of the air conditioning equipment or substitute the air conditioning equipment so as to reduce base station power consumption.

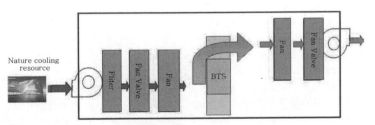

FIGURE 3.12 Intelligent ventilation system structure

3.4.1.2   Heat-Pipe Heat-Exchanger Unit

A heat pipe heat exchanger unit used in a communication base station may transfer from indoor thermal energy to outdoor one through a heat-pipe based environmental control utilizing the difference between indoor and outdoor air temperature, thereby reducing the indoor temperature [21]. Without the use of compressors and other refrigeration components this equipment may be able to exploit the difference between indoor and outdoor air temperature and utilize the phase transition of working fluid refrigerants to achieve the natural cycle of of refrigerants, thus efficiently using the natural cooling sources.

When the difference between indoor and outdoor air temperature exceeds a certain pre-configured threshold, through the dynamic balance generated by the evaporation and condensation cycle of working fluid refrigerants enclosed in the pipe, the heat pipe heat exchanger unit exploits the relative small difference between indoor and outdoor air temperature to transfer the indoor heat intensively and quickly to to the outside, reducing indoor temperatures. After the indoor blower drives air flow inside the data center room, the heat in the air is transferred into the evaporator, which reduces the indoor air temperature, and forces the working refrigerant in the evaporator evaporate into a gaseous state. Then, after the working refrigerant vapor flows from the main gas tube into the outdoor condenser, which is cooled by the colder outdoor air, the refrigerant vapor releases heat in the condenser and is condensed into a liquid state. The liquid refrigerant liquid then flows back to the evaporator. The above process cycle runs continually to transfer the interior heat into the external environment. The above cycle is completed through the heat equilibrium of the working medium, which is initiated upon natural forces generated by the difference between indoor and outdoor natural temperatures. This enables high efficient energy-saving using the outdoor cooling source to discharge the heat of a indoor engine room.

3.4.2   The Energy Saving Solutions for Backup Power Systems

As a backup power system, the traditional lead-acid batteries need to maintain the temperature at 25 degrees Celsius, which requires to maintain a constant temperature inside the data center room of the communication base station through the cooling system, and thus brings great energy cost. Therefore, how to reduce the power consumption of the refrigeration system supporting backup power systems has been one of the key aspects in energy-saving emission reduction. In addition to the energy-saving measures for cooling as discussed in Section 3.4.1.2, the application of new high-performance rechargeable batteries (such as lithium iron-phosphate batteries) and thermostat for rechargeable batteries is the effective means to improve the energy efficiency of the cooling system.

3.4.2.1 Lithium Iron-Phosphate Battery

Lithium iron phosphate battery [22] is a kind of lithium ion battery using iron phosphate as cathode materials. The charging external environmental temperature range of lithium iron-phosphate is from 0 to 45 C°, while the discharging external environmental temperature range is from -20 to 60 C°. Lithium iron-phosphate works well for high external environmental temperature:

1. For external environmental temperature 65 C°, the internal temperature may be as high as 95 C°.

2. When the battery discharge ends, the structural safety of the battery in good condition even for the battery internal temperature 160 C°.

However, lithium iron phosphate cathode materials are with poor electrical conductivity, and thus it is required for lithium iron phosphate cathode materials to be mixed with conductive carbon materials or conductive metal particles. Note that the charging temperature for lithium iron phosphate battery is generally from 0 C° to 55 C°:. the battery charge for the temperature of lower than 0 C° will introduce irreversible damage, while the battery charge for the temperature of higher than 55 C° may appear dangerously delithiated phenomena.

The actual capacity of a lithium iron phosphate battery will be affected by the ambient temperature, discharge rate, and the discharge termination voltage. Currently, due to the lack of experimental data, the formula of capacity of lithium iron phosphate batteries still can not be obtained. For the calculation of the capacity of a lithium iron phosphate battery, it is required for manufacturers to provide the data tables for constant-current discharge, used as look-up tables for selection.

3.4.2.2 The Thermostat for Rechargeable Battery

The thermostat of a rechargeable battery is a separate space to install the rechargeable battery in order to perform a separate temperature control to the battery, which may improve the operating temperature of the main equipment and power equipment to reduce the energy consumption. The system using this approach can provide a working environment of from 15 C° to 25 C° to a rechargeable battery, while the operating temperature of the base station equipment is improved from 25 C° to the range between 30 C° to 40 C° to reduce the site energy consumption of the temperature control equipment. At present, the regulation measures of local temperature for rechargeable batteries are: buried incubator, compressor thermostat.and semiconductor thermostat. The required overall designed life time of semiconductor / compressor thermostats should be not less than 10 years, while the designed life time of the buried incubator system should not be less than 5 years. The system should meet the carrier-grade requirements with the continuous uninterrupted operation of 7×24 hours.

3.4.3   Energy Saving Solutions Using Renewable Energy

3.4.3.1   Solar Energy Utilization for Base Stations

A complete photovoltaic system includes solar cell matrix (also called photovoltaic array), charge controller, battery pack, stands, functional circuit elements, transmission and distribution cables, and other ancillary systems. A solar cell matrix (also called photovoltaic array) consists of a number of blocks of photovoltaic modules through series or parallel connections to achieve different voltage levels, current values, and power outputs. The electricity obtained through solar photovoltaic power generation is controlled by charge controller to directly to provide electricity to appropriate circuits or loads, while the excessive energy is stored in rechargeable batteries which may be used at night or in the period of insufficient power supply.

The solar energy powered base stations have recently attracted many attentions from research and academic communiites [23]. The desired annual average sunshine time for solar power supply is larger than 1800 hours. If the newly established base stations are far away from power system stations of the city and the construction cost of the electricity transmission line is larger than (or exceed) 70% of the total investment in solar power system, the solar power systems are recommended. The base station using the actual power level less than 1000W, for an example that the distance of electricity transmission line from the power system station is not less than 5km, recommended the use of solar power systems. For repeaters, if the distance of electricity transmission line from the power system station is longer than 2km(kilometers), the solar power systems are recommended. The above recommended distance of electricity transmission line from the power system station may be changed due to actual different construction costs in different locations.

3.4.3.2   Base Stations Powered by Wind Energy Technologies

Modern wind turbines are mostly based on horizontal axes, while only very few vertical axis wind turbine manufacturers are available [24]. A typical modern horizontal axis wind turbine consists of blades, wheel hubs (which together with the blades are called turbine wheels), nacelle, gearbox, generator, tower, base, control system, brake system, yaw system, hydraulic devices, and so on. The working principle is: When wind blows the blades, the turbine wheel rotates due to aerodynamic effects, and the rotation of turbine wheel is accelerated through the gear box (or the growth machine) to drive generators to generate electricity. Currently some manufacturers launch no gear box units, which may reduce vibration, noise, and improve power generation efficiency at the higher cost. In the mobile base stations, no gear box units are recommended.

The complementary wind-solar power generation system [25], shown in Fig. 3.13, consists of solar cells, small wind turbine, the system controller, rechargeable battery pack, reversing module, and so on. Solar energy and wind energy are transformed into electrical energy through wind generators and solar cells, respectively, and then the battery is charged with the aid of complementary wind and solar controller.

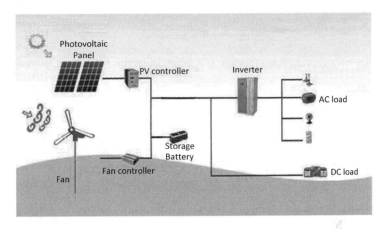

FIGURE 3.13 Main components and diagram of complementary wind-solar power generation system

Bibliography

[1] *Common Public Radio Interface (CPRI) Specification V4.1*, Feb. 2009.

[2] 3GPP, *TS 44.060 V8.14.0 Technical Specification - 3rd Generation Partnership Project; Technical Specification Group GSM/EDGE Radio Access Network; General Packet Radio Service (GPRS); Mobile Station (MS) - Base Station System (BSS) interface; Radio Link Control/Medium Access Control (RLC/MAC) protocol; (Release 8).*, Dec. 2011.

[3] Y. Yang, C. Lim, and A. Nirmalathas, "Comparison of energy consumption of integrated optical-wireless access networks," in *Proc. Optical Fiber Communication Conference and Exposition / National Fiber Optic Engineers Conference (OFC/NFOEC)*, Mar. 2011.

[4] S. Iezekiel, *Microwave Photonics: Devices and Applications.* Wiley-IEEE Press, Oct. 2009.

[5] ZTE, "Reliance communications: Built to succeed," *ZTE Technologies*, Sept. 2011.

[6] C. Belady, *White paper: Metrics & Measurements - Green grid data center power efficiency metrics, PUE and DCIE.* Green Grid, Oct. 2007.

[7] R. Schmidt and M. Iyengar, "Effect of data center layout on rack inlet air temperatures," in *Proc. ASME InterPACK 2005*, July 2005.

[8] V. Mulay, D. Agonafer, and R. Schmidt, "Liquid cooling in data centers," in *Proc. International Mechanical Engineering Congress and Exposition*, Oct. 2008.

[9] F. Bauman and T. Webster, "Outlook for underfloor air distribution," *ASHRAE Journal*, June 2001.

[10] T. Webster, F. Bauman, and J. Reese, "Underfloor air distribution: Thermal stratification," *ASHRAE Journal*, May 2002.

[11] K.S.Lee, Z.Jiang, and Q.Chen, "Air distribution effectiveness with stratified air distribution," *ASHRAE Transactions*, Feb. 2009.

[12] J. Hamilton, "Architecture for modular data centers," in *Proc. 3rd Biennial Conference on Innovative Data Systems Research (CIDR)*, Jan. 2007.

[13] K. V. Vishwanath, A. Greenberg, and D. A.Reed, "Modular data centers: how to design them?" in *Proc. the International Symposium on High Performance Distributed Computing (HPDC 2009) the ACM workshop on Large-Scale system and application performance (LSAP)*, June 2009.

[14] E. Shaviv, A. Yezioro, and I. G.Capeluto, "Thermal mass and night ventilation as passive cooling design strategy," *Renewable Energy*, vol. 24, no. 34, pp. 445–452, Nov. 2001.

[15] V. Agelidis, G. Demetriades, and N. Flourentzou, "Recent advances in high-voltage direct-current power transmission systems," in *Proc. IEEE International Conference on Industrial Technology (ICIT)*, Dec. 2006.

[16] K. Mani and B. Jee, *On the Edge: A Comprehensive Guide to Blade Server Technology.* Wiley, Dec. 2007.

[17] J. Choi, Y. Kim, and et al., "A CFD-based tool for studying temperature in rack-mounted servers," *IEEE Trans. Computers*, vol. 57, no. 8, pp. 1129–1142, Aug. 2008.

[18] J. Daniels, "Server virtualization architecture and implementation," *ACM Crossroads*, vol. 16, no. 1, Sept. 2009.

[19] Q. Zhang, L. Cheng, and R. Boutaba, "Cloud computing: state-of-the-art and research challenges," *Journal of Internet Services and Applications*, vol. 1, no. 1, pp. 7–18, June 2010.

[20] J. Axley, S. Emmerich, S. Dols, and G. Walton, "An approach to the design of natural and hybrid ventilation systems for cooling buildings," in *Proc. International Conference on Indoor Air Quality and Climate in Monterey*, June 2002.

[21] R. Simons, "Estimating temper-atures in an air-cooled closed box electronics enclosure," *Electronics Cooling*, vol. 11, no. 1, Feb. 2005.

[22] H. Xie and Z. Zhou, "Physical and electrochemical properties of mix-doped lithium iron phosphate as cathode material for lithium ion battery," *Electrochimica Acta*, vol. 51, no. 10, Feb. 2010.

[23] J. Louhi, "Energy efficiency of modern cellular base stations," in *Proc. International Telecommunications Energy Conference*, Sept. 2009, pp. 475–476.

[24] J. F. Manwell, J. G. McGowan, and A. L. Rogers, *Wind Energy Explained: Theory, Design and Application.* Wiley, Feb. 2010.

[25] F. Giraud and Z. Salameh, "Steady-state performance of a grid-connected rooftop hybrid wind-photovoltaic power system with battery storage," *IEEE Trans. Energy Conversion*, vol. 16, no. 1, pp. 1–7, Mar. 2001.

Author Contact Information

Sen Bian, Huabin Tang, Xidong Wang, Zhihong Zhang, Chih-Lin I, and Bill Huang are with China Mobile Research Institute, China Mobile Communications Corporation, China, Email: biansen@chinamobile.com, tanghuabin@chinamobile.com, wangxidong@chinamobile.com, zhangzhihong@chinamobile.com, icl@chinamobile.com, bill.huang@chinamobile.com. Jinsong Wu is with Bell Laboratories, Alcatel-Lucent, China, Email: wujs@ieee.org.

4

Standard Methodologies for Energy Efficiency Assessment

Raffaele Bolla

DIST - University of Genoa, Italy, and CNIT, University of Genoa Research Unit, Italy

Roberto Bruschi

CNIT, University of Genoa Research Unit, Italy

Chiara Lombardo

DIST - University of Genoa, Italy, and CNIT, University of Genoa Research Unit, Italy

CONTENTS

4.1 Introduction

For many years, the main concern of the ICT industries was network performance, so that the new services being offered could work properly and be available to a larger and larger number of customers. The solution adopted to achieve this goal has been just feeding the networks with more bandwidth: as a result, today networks are constantly working at rates that are hardly reached at rush hours. Although the expansion of the Internet is not such a recent phenomenon, the research for more efficient solutions has been disregarded until recently. But this problem can not be postponed anymore. The main cause that brought many Telecom vendors to reconsider their strategies is the constant increase of energy costs, but the impact on the environment is also alarming. In fact, the ICT industries produce 2% of the global CO_2 emissions, overcoming the carbon footprint due to aviation [1]. In this context, ground-breaking work on energy consumption in the Internet was conducted by Gupta et al. [2] in 2003, and by Christensen et al. [3] in 2004, showing the importance of energy efficiency in communication networks. Today, researchers, service providers and manufacturers are constantly occupied in this field, with the creation of numerous projects in order to obtain and standardize effective solutions. Among others, a partnership among universities, device manufacturers and Telcos has been established for the creation of an initiative funded by the European Community, called ECONET [4]. The purpose of this initiative consists in developing innovative solutions and device prototypes for wired network infrastructures within 2013. It is also worth mentioning the Home Gateway Initiative (HGI) [5]: its task consists in defining a set of requirements and specifications to make possible the detection of inconsistencies among future standards.

The strongest effort in the research community is currently the development of power management capabilities to be introduced in future network devices, allowing the system to reduce its energy consumption without affecting the service capabilities. In this respect, it becomes crucial to provide a correct evaluation of the trade-offs between power saving and performance degradation, in order to clearly evaluate and compare such capabilities. The existing methodologies work fine as long as the device software configuration is concerned, even though they are not completely able to characterize different hardware and architectural choices. Future propositions will be required to provide consistency with the evolution of Internet devices, especially when power management capabilities will be applied at large scale.

This chapter will introduce some benchmarking methodologies that have been proposed for this purpose, and will provide a critical analysis of their effectiveness through an accurate description of their characteristics, assessing their efficiency through comparisons and test results.

4.2 An Overview on Novel Green Network Technologies

4.2.1 Internet Evolution and Concern over Energy Wastes

Many ISPs and public organizations have collected statistic data on the power requirements of network infrastructures in the last few years. These data show an alarming growing trend of energy consumption: in 2006, for example, Telecom Italia has increased its energy demand by 7.95% with respect to 2005 and by 12.08% to 2004 [6] [7].

Since both the number of customers and of new services being offered are constantly increasing, and so data traffic volume doubles every 18 months according to Moore's law [8], power consumption is becoming an issue that can no longer be neglected. In fact, according to the perspective presented by Tucker et al. in [9] [10], which is derived from data sheets of state-of-the-art commercial devices and projections of future technologies, energy consumption is going to be the future bottleneck for Internet evolution. In order to overcome this huge problem, researchers and ICT companies are trying to develop optimization strategies able to reduce the carbon footprint of network devices without penalizing performance. Of course, the solution is not easy or fast to achieve, and as a matter of fact it can not be absolute: different kinds of devices bring different issues. It is easy to see that the number of home gateways grows with the customers, but also core routers increase consequently their energy demands. Although the data plane is the most critical element for all kinds of devices, network interfaces are responsible for the highest energy consumption at the access level, while the biggest constraint in core networks is represented by lookup tables.

4.2.2 Current Approaches and Concepts for Low-Energy Networking

Since network devices are obtained from specialized computer hardware, the power management methodologies that have been proposed so far are derived by optimization strategies already existing in processors. These methodologies can be classified as re-engineering, dynamic adaptation and smart standby:

1. **Re-Engineering** approaches consist in introducing new energy-efficient technologies and optimally exploit them inside the network architecture. These technologies can be modern silicons, memories and interfaces.

2. **Dynamic Adaptation** approaches allow to modulate the service capacity according to the current traffic load. This is usually performed through two techniques, already used in processors, called Adaptive Rate (AR) and Low Power Idle (LPI). Adaptive Rate allows a dynamic reduction of the device working rate, by tuning the processing engine frequency or voltage, or throttling (i.e gating or disabling) the CPU clock. The result is

a reduction of the power consumption, at the price of lower performance. Low Power Idle, instead, allows reducing power consumption by turning off sub- components when not employed and waking them up, as rapidly as possible, when the system needs their activity.

Both Adaptive Rate and Low Power Idle are implemented by pre-selecting a set of HW configurations that can provide different trade-offs between power and performance. Of course, Adaptive Rate and Low Power Idle can be jointly exploited to better fit system requirements. It is worth noting that the effectiveness of these methods strictly depends on the traffic and network characteristics. For example, the application of Low Power Idle capabilities shows better results when incoming traffic presents a high burstiness, so that the system has enough time to spend in an idle state before a new burst is received.

Finally, dynamic adaptation capabilities require an optimization policy to manage their application. Such policies already exist in off-the-shelf systems and are developed as SW applications called governors. Their application in commercial products is not easy because of the computational complexity, mostly due to the estimation of the current workload.

3. **Sleeping and standby** approaches can be considered as a deeper idle state: using power management primitives, devices can be turned almost completely off, providing higher power savings than idle states at the price of longer wake up times. The main issue with this approach is represented by the loss of network connectivity. In fact, when a device is sleeping it cannot be seen on the network, thus a wake up time also includes a re-connection phase, in which the device has to send signaling traffic to communicate its presence on the network and allow updating the forwarding tables. This is probably the predominant reason why many PCs and servers are left fully powered even when not employed. Recent solutions propose the introduction of a proxy to take charge of the host network presence during sleeping times.

4.3   Benchmarking Methodologies

As previously mentioned, the main reasons that brought interest to the greening of Internet technologies are both the environmental concern and the growth of energy costs. In fact, the first research initiatives were conceived in Europe, where energy costs are historically higher, while interest rose later in the United States. Focusing on benchmarking methodologies and efficiency metrics, the aim is to standardize a set of measurements, including the testbed, and a feasible representation of results, in order to compare in an objective way the energy-efficiency level even among different device types. From the vendors' point of view, it will give a chance to show their leadership in both performance and power efficiency.

Taking this widespread interest into account, it is not surprising that different proposals on benchmarking methodologies are coming from vendors as well as from research communities and standardization bodies. Therefore, the next subsections will give a detailed description of some solutions coming from these backgrounds, describing their test procedures before critically comparing them. RFC 2544 [11] will be also introduced: although it is not a standard conceived for power measurements, it is very accurate for what concerns the characterization of network performance, and some of its tests are used by other methodologies as a qualification phase.

4.3.1 ECR 3.0.1

ECR 3.0.1 [12] is a proprietary standard owned by Juniper and IXIA, presented in the document "Network and Telecom Equipment-Energy and Performance Assessment". This document introduces tests for the determination of energy efficiency for packet-based networks and telecom equipment. Its tests are best suited for packet-based and medium- to large-scale network and telecom systems, as they can exhibit performance numbers lower than the face value of connected ports, hence showing a dependence between power consumption and throughput.

Four tests are defined, of which one is mandatory and the others are optional: energy consumption in relation to variable load is the mandatory test, then there are energy consumption in relation to static load change, component level energy footprint and embedded energy monitoring capabilities. There is also a fifth test that is a stub. Tests are performed in order to evaluate energy efficiency, defined as the energy consumption normalized to the effective throughput. After setting up the system, the first thing to do is to measure the maximum load that can be sustained for the given configuration at zero packet loss. For this purpose, it is possible to exploit RFC 2544 Throughput Test, that will be introduced later in this section:

Energy consumption in relation to variable load: Traffic is sent at 100%, 50%, 30%, 10% of predetermined throughput for 1200 seconds measuring power consumption, then there is an additional measure with no traffic transmitted. Average power consumption is calculated for each traffic percentage. These values are employed to obtain the Energy Consumption Rating (ECR), the ECR over Variable Load cycle (ECR-VL) and the ECR over EXtended idle load cycle (ECR-EX):

1. ECR is the ratio between power consumption measured at 100% and measured maximum effective throughput:

$$ECR = \frac{E_{100}}{T_f} [\frac{W}{Gbps}].$$ (4.1)

2. ECR-VL represents the ratio between the sum of the weighted powers measured varying the throughput (E_x) and the weighted throughputs themselves:

$$ECR - VL = \frac{\alpha * E_{100} + \beta * E_{50} + \gamma * E_{30} + \delta * E_{10} + \epsilon * E_i}{\alpha * T_f + \beta * T_{50} + \gamma * T_{30} + \delta * T_{10}} [\frac{W}{Gbps}].$$

(4.2)

3. ECR-EX represents the ratio between the sum of the weighted powers measured varying the number of active ports (P_x) and the weighted corresponding throughputs:

$$ECR - EX = \frac{\alpha * E_{100} + \beta * P_{50} + \gamma * P_{30} + \delta * P_{10} + \epsilon * P_i}{\alpha * T_f + \beta * T_{50} + \gamma * T_{30} + \delta * T_{10}} [\frac{W}{Gbps}]$$

(4.3)

in the previous expressions, $\alpha + \beta + \gamma + \delta + \epsilon = 1$.

The average energy consumption here determined can be also used to estimate the system energy consumption over a projected lifetime:

1. **Energy consumption in relation to static load change:** This time, traffic is always sent at full rate, again for 1200 seconds, but the number of active ports is reduced at each step.

2. **Component level energy footprint:** This optional test is feasible only for redundant systems, as it calculates the energy footprint of each component. Traffic is first sent to the System Under Test (SUT) at full throughput for 1200 seconds, then the Component Under Test (CUT) is removed and the new maximum throughput is calculated. Finally the system receives traffic at this new rate. The component footprint is the difference between the power measured before and after the component removal. Using this procedure, it is possible to create a library containing the energy footprint of each redundant component.

3. **Embedded energy monitoring capabilities:** This test provides real-time energy consumption estimates thanks to embedded sensors. As it exploits the results of the first and second tests, it does not add any cost.

4.3.2   ATIS 060015.03.2009

ATIS 060015.03.2009 [13] is an ANSI standard presented in the document "Energy Efficiency for Telecommunication Equipment: Methodology for Measurement and Reporting for Router and Ethernet Switch Products". It provides a methodology to calculate an energy efficiency ratio conceived for equipment categorized as Enterprise, Service Provider and Branch office routers, and Ethernet switch products. The method takes into account systems modularity, which consists of a chassis (shelf) with multiple slots that can be equipped with a variety of cards and/or service modules.

Aside from the value produced by this standard test, called Telecommunication Energy Efficiency Ratio (TEER), and the methodology to obtain it, the standard also specifies a document called application description, which is a textual description of the required system and a diagram showing the correct topology. The application description is fundamental for a correct system set up. It allows to convert a requested system into an actual configuration, as it includes the exact topology and protocols that each required interface must support. Maximum data rate for each different interface is also reported, at the speeds declared in appendix of the document.

The modularity of the devices that are taken into account by this standard makes it hard to produce a single set of results, since there could be as many TEERs as possible configurations: clearly it is not feasible to test all possibilities. To overcome this limitation, the equipment supplier maintains a database containing module-level power consumption data, so that it is possible to obtain a TEER for any requested configuration, called declared TEER ($TEER_{DEC}$). There is also a subset of systems certified and whose reports are available on request. Those measure reports are called certified TEER ($TEER_{CERT}$).

In order to obtain both $TEER_{DEC}$ and $TEER_{CERT}$, the first thing to do is to review the application description. Since it also contains the data rates for each interface, there is no need for a qualification phase. After configuring the system following the application description directions, it is possible to start power measurements, the total data rate being provided by the application description. To allow a good level of consistency with respect to a real world behavior, Internet MIX (IMIX) traffic profile [14] is used on Ethernet ports, which means that frames of different length are generated by the test equipment according to a statistic distribution and transmitted to the system under test. Each module is measured independently, hence it is possible even for certified TEER to provide results of different configurations performing a single test.

TEER is defined as a ratio of maximum demonstrated throughput (T_d) to weighted power (energy consumption rate) P_w:

$$TEER = \frac{T_d}{P_w}. \tag{4.4}$$

Weighted energy consumption is calculated with the following formula:

$$P_w = a * P_{u1} + b * P_{u2} + c * P_{u3}[\frac{Mbps}{W}]. \tag{4.5}$$

In the previous expression, $a+b+c = 1$. These weights are chosen according to the device type. P_{u1}, P_{u2} and P_{u3} represent power consumed at 100%, 30% or 10% according to the device type, and 0% of the maximum throughput.

For the modular case, the ratio in presence of n components becomes the following:

$$TEER = \frac{\sum_{i=1}^{n} T_i}{\sum_{i=1}^{n} P_{wi}}[\frac{Mbps}{W}]. \tag{4.6}$$

4.3.3 RFC 2544

The purpose of the tests discussed in RFC 2544 is to introduce a method that allows comparison among performance of heterogeneous SUTs. Moreover, the document specifies the reporting format for each particular test, enhancing ease of comparison among tested devices.

The tests take into account all the possible conditions a system can support, having regard for architectural considerations, supported protocols and traffic shape. After choosing an initial benchmarking configuration, tests are run without any further parameters modification, aside from those required by that specific test:

1. **Throughput Test:** the throughput is the fastest rate at which the count of test frames transmitted by the SUT is equal to the number of test frames sent to it by the tester. For each available frame length the maximum throughput is calculated; this provides not only the first test result, but also a binding datum for the next measures.

2. **Latency Test:** the procedure that determines latency uses the previous result as a starting point. Latency represents the interval between the time at which a frame is fully transmitted and the time at which the same frame is received. This test, along with the following ones, has to be repeated at least 20 times to get an average value.

3. **Back-To-Back Frame Test:** as previously sketched, only one test is performed generating bursty traffic. In the back-to-back frame test, bursty traffic characterized by the maximum burst length allowed is generated by the tester. The back-to-back value is the number of frames in the longest burst that the SUT can handle without losing any data.

4.3.4 Other Upcoming Propositions

This subsection introduces three methodologies that are currently drafts but particularly interesting for the sake of this overview, since they are directly derived from the standards introduced so far.

The research effort of IETF on benchmarking methodologies [15] defines suggestions for measuring power usage of live networks under different traffic loads and various switch and router configuration settings. It provides a suite which can be deployed on any networking device.

The factors affecting power consumption are divided into device and traffic factors. The first group includes the chassis power requirement, the number of line cards, active ports and their settings and utilization, the presence of Ternary Content Access Memories (TCAMs) and the firmware version.

Traffic factors are packet size, inter-packet delay and the percentage of CPU traffic. However, this second group is basically neglected as it is considered less relevant for the global power consumption of a device.

Traffic used to perform tests follows the definitions of RFC 2544. The NECR (Network Energy Consumption Rate) defines the power usage increase in milliWatts per Mbps of data at the physical layer. The NECR will depend on the line card, the port and the other factors defined earlier. For the effective use of the NECR the base power of the chassis, a line card and a port need to be specified when there is no load. The measurements must take into consideration power optimization techniques when there is no traffic on any port of a line card.

ITU-T is currently working on two recommendations, one concerning benchmarking methodologies [16], the other one about metrics [17]. ITU-T is working in collaboration with ATIS and ECR initiatives, so the upcoming standards will be related to the methodologies presented in the former subsections. The main innovation is represented by an exhaustive characterization of possible systems, device types and topologies.

4.4 Comparison among the Standards

The main characteristics of the three standards discussed so far have been summarized in Table 4.1. First, it is worth noting again that RFC 2544 is

TABLE 4.1 Comparison of the Analyzed Benchmarking Standards

Standard	Benchmarking scenario	Main results obtained
ECR 3.0.1	1. CBR traffic 2. Traffic at different loads 3. Maximum packet size	1. ECR [W/Gbps] 2. ECR-VL [W/Gbps], ECR-EX [W/Gbps] 3. Energy Bill Estimates [$]
ATIS 060015.03.2009	1. CBR traffic 2. Traffic at different loads 3. IMIX traffic	1. DTEER [Mbps/W] 2. CTEER [Mbps/W]
RFC 2544	1. CBR traffic 2. Bursty traffic 3. Traffic at different loads 4. Range of packet sizes	1. Maximum Throughput [Gbps] 2. Max, average, min latency [us] 3. Maximum burst length [burst/frames]

not a standard for power consumption measures; however its comparison to the other standards turns out to be appropriate if performance evaluation is considered.

If ECR 3.0.1 and ATIS 060015.03.2009 are taken into consideration, some similarities are easy to find. The first common feature can be seen in the presence of a metric composed by some combination of throughput and power consumption. Although each method also presents other indexes, this definition of energy efficiency appears to be globally taken for granted, not only in these studies but as an assumption in the largest part of today's research initiatives. With respect to the first two methodologies, RFC 2544 appears to have a wider application field: the reason is clearly the device type they are addressed to. In fact, while the latter is conceived to embrace all possible network devices and topologies, the others are designed for transport/core routers. As a result, the testbed description and the performed tests are more specific for what concerns ECR and ATIS, while RFC 2544 has a more general approach, in which the strongest emphasis is put on the importance of a thorough description of the tested configurations.

Considering again ECR 3.0.1 and ATIS 060015.03.2009, the first one presents a larger number of tests and results. Although its test procedures are quite longer, they can be fully automated, so their application is fast and almost inexpensive. On the other hand, ECR tests are performed at a fixed packet size, which is neither accurate nor realistic with respect to real Internet traffic. ATIS, instead, uses IMIX traffic, which allows to simulate a real world behavior.

Another important feature of ATIS 060015.03.2009 is the application description: by describing in a complete and also standard way the testbed and the device, tests are easily reproducible guaranteeing comparison among the results.

Taking again ECR 3.0.1 and ATIS 060015.03.2009 into account, there are two main aspects affecting the effectiveness of their representations: the use of CBR (Constant Bit Rate) traffic during tests and the absence of time metrics to represent network performance.

Of course, CBR traffic can be used to get an immediate and simple idea of network performance, but real Internet traffic has a bursty behavior, which means that these tests can of course be accurate, but not realistic. In the results shown in the next section, CBR traffic will be included as a limit case along with the results obtained at varying burst length.

The only parameter employed in all three methodologies to characterize network performance is throughput. It definitely plays an important role in energy efficiency, as it can be directly related to power consumption, but depending on the type of traffic being processed it might not completely characterize a device behavior. In fact, the drawback of the introduction of the power management capabilities previously described can be seen in longer delays more than in traffic loss. In detail, reducing the service capacity produces higher latencies, while all methods based on standby cause wake-up times. The impact of service times is even clearer taking into account the importance of applications calling for some QoS requirements. Without the introduction of time metrics, a system processing real-time traffic could be considered efficient from the energetic point of view even if it introduced long delays.

In the next section some measurements and results from a real testbed will be shown. Research prototypes (SW routers) will be used to confirm the features and critical points previously discussed. The use of SW Routers allows to take into account the perspective of Future Internet devices without going too far from the behavior of commercial network equipment, hence guaranteeing a good level of generality to our discussion.

4.5 Performance Evaluation

In order to confirm the considerations moved in the previous section, results obtained with the ECR 3.0.1 methodology will now be compared with the set of benchmarkings proposed in [18] and shown in the next subsection. In detail, we want to show how metrics including latency can evaluate in a more complete way the energy efficiency of a device with power managent capabilities. To do this, tests have been performed on a Linux-based software router [19], as it includes a power management mechanism performed by means of the ACPI (Advanced Configuration and Power Interface) technology [20].

4.5.1 Extending Current Evaluation Methodologies

As previously sketched, both AR and LPI techniques have the increasing of latency as a drawback. Adaptive Rate causes a stretching of packet service times, while Low Power Idle introduces an additional delay in packet service, due to wake-up times.

Moreover, Adaptive Rate causes larger packet service times, and consequently shorter idle periods. For all these reasons, it is necessary to collect and put data together to quantify what we gain in terms of energy saving and what we lose in terms of service delays.

Energy gain represents the power saving obtained thanks to power management in comparison to a scenario with no such capabilities:

$$\Phi_\% = \frac{\Phi_{max} - \Phi_c}{\Phi_{max}} \tag{4.7}$$

where Φ_c is the current power consumption, Φ_{max} is the maximum consumption reached by the device. The result gets closer to 1 as consumption decreases.

Performance degradation is expressed as a ratio between the values of packet latency in the case of an ideal network device (i.e., with an infinite processing capacity, which allows to neglect forwarding times and just consider transmission times), and the ones measured with the real SUT:

$$L_\% = \frac{\widetilde{L}_i + \frac{1}{2}(L_i^{max} - L_i^{min})}{\widetilde{L}_r + \frac{1}{2}(L_r^{max} - L_r^{min})} \tag{4.8}$$

where parameters with the i index represent the latencies for the ideal device, and those with the r index are measured on the real SUT. L^{min}, $\widetilde{L}$ and L^{max} are the minimum, average and maximum values, respectively, of packet latencies.

The values for minimum, average and maximum latency in the ideal case have to be computed by starting from the packet transmission times on input and output links.

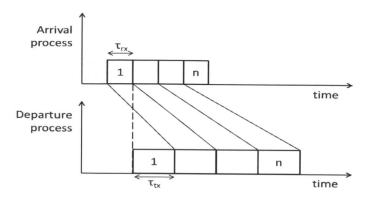

FIGURE 4.1 Packet burst processing in an "ideal" SUT.

First, it is worth noting that if the packet transmission time on the input link is longer or equal to the one of output link, then $L_i^{min} = \widetilde{L}_i = L_i^{max}$. Otherwise, as shown in Fig. 4.1, these latency values have to be estimated by considering traffic burstiness in an explicit way.

We can simply compute the L_i^{min} value as the sum of the packet transmission times on the input and output links (τ_{rx}, τ_{tx}):

$$L_i^{min} = \tau_{rx} + \tau_{tx}. \tag{4.9}$$

With reference again to the simple example in Fig. 4.1, the L_i^{max} parameter can be calculated as follows:

$$L_i^{max} = \tau_{rx} + n\tau_{tx} - (n-1)\tau_{rx} = n\tau_{tx} - (n-2)\tau_{rx}, \tag{4.10}$$

where n is the length of the incoming bursts.

Thus, the $\widetilde{L}_i$ value can be simply expressed as:

$$\widetilde{L}_i = \frac{n+1}{2}\tau_{tx} - \frac{n-3}{2}\tau_{rx}. \tag{4.11}$$

4.5.2   Test Results

Tests have been performed, as previously mentioned, on a software router supporting a different set of working frequencies (P-states) and idle states (C-states), as in the ACPI standard. The frequencies used to perform these tests are $P_0 = 2{,}67$ GHz and $P_8 = 1{,}60$ GHz, idle states are C_1 and C_3 (deepest idle state). For each P- and C-state configuration, tests have been repeated varying the throughput. We used packets of 64 and 1500 B and bursts of 1, 10, 50, 100 and 200 packets.

An Ixia Router Tester [21] is connected to the software router and sends traffic to it; when traffic is sent back the tester can determine outgoing traffic statistics. An Agilent watt-meter [22] is also connected to the software router, in order to register the power consumed during the tests. Both the tester and the watt-meter report their results to a controller.

Both the ECR 3.0.1 methodologies and the one introduced above have been implemented to characterize the software router behavior. Moreover, in order to improve comparison among them, ECR 3.0.1 measurements have been performed also for bursty traffic (bursts of 200 packets), as reported in Table 4.2.

TABLE 4.2 ECR-VL at different energy-aware configurations, packet and burst sizes.

Offered load	P_0, C_1				P_0, C_3				P_8, C_1				P_8, C_3			
	64 B		1500 B		64 B		1500 B		64 B		1500 B		64 B		1500 B	
	1 pkt	200 pkt	1 pkt	200 pkt	1 pkt	200 pkt	1 pkt	200 pkt	1 pkt	200 pkt	1 pkt	200 pkt	1 pkt	200 pkt	1 pkt	200 pkt
100%	258	258	199	199	231	231	182	182	490	490	196	196	450	450	182	182
50%	508	505	398	394	463	460	361	339	490	490	392	392	450	450	363	339
30%	769	762	669	662	708	696	614	510	746	746	662	662	662	662	610	510
10%	2463	2413	1940	1910	2275	1850	1770	1360	2400	2400	1920	1920	2238	1850	1760	1360

In order to fully understand these results, it is important to keep in mind that the ECR index has the physical meaning of power consumption needed to move one Gigabit of data per second, hence smaller values are preferable. In this respect, it is clear that a higher throughput reduces ECR, so the lowest values are reported on the first row of the table. We can also notice that, at full throughput, no configuration shows difference between CBR and bursty traffic: at high speeds, bursts are so close to one another that they tend to behave as a continuous flow. As traffic speed decreases, bursty traffic is still seen as constant for P-state P_8: here, the reason is the lower service capacity. If latency was included in the index computation, variations would definitely be more visible.

Considering in detail the level of energy efficiency provided by the tested configurations, packets of size 1500 Bytes provide the best results for all P- and C-state couples. This is not surprising, as longer packets mean less headers to process. What is particularly interesting, is considering which configuration gives the best results at varying offered load. If we go through the first row of the table again, we can see that the lowest ECR is obtained for P_0, C_1: this state allows to keep a high performance level, compensating a higher power consumption. On the other hand, when in P_8, C_1, power consumption is not discernibly lower, while the working frequency is not enough to avoid packet loss. The other traffic loads present a completely different balance: since no choice of P- and C-states causes losses, and so all throughputs are the same, the most important role in the ECR computation is played by power consumption.

Finally, taking a global look at the results again, it is worth noting that the P-state plays the most important role: while the collected values change visibly as the working frequency varies, differences are smoother entering a deeper idle state, especially for offered loads down to 50%. This result confirms the main limit of this benchmarking methodology: since latency is not taken into consideration, there is no way to discriminate performance in the absence of packet loss. Of course, this may not be a limit when data traffic is transmitted, but in the presence of voice or other traffic that needs specific bounds on QoS this definition of energy efficiency is not effective.

Let us see how energy efficiency representation changes by introducing latency. Figures 4.2, 4.3, 4.4, 4.5, 4.6, 4.7, 4.8, and 4.9 show $L_\%$ and $\Phi_\%$ computed at varying burst size. Each line represents a different offered load, for packet sizes of 64 and 1500 Bytes. Taking a first global glance at all figures, we can see that $L_\%$ decreases as the burst size grows, while energy gain is higher. This general trend guarantees a good characterization between power consumption and network performance.

In more details, the P-state visibly affects power consumption, as we have already seen in the ECR results. But in that case, the influence of C-states seemed to be scarce. Here, we can see that the idle state has a strong impact on latency: in Figure 4.2 results at lower offered loads are comparable for both packet sizes, while in Figure 4.3 they are similar only in the presence of packet loss (full load). This means that a deeper idle state affects more short packets, while for longer ones it is still possible to achieve a good performance level. The same considerations are valid for Figures 4.6 and 4.7.

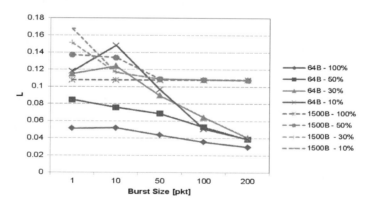

FIGURE 4.2 $L_\%$ values for the P_0, C_1 configuration according to 64 and 1500 B(Bytes) sized datagrams and different traffic load and burstiness levels.

The impact of the P-state on power consumption is stronger for short packets, as they push the system tho the limit more than the long ones. In particular, considering energy gain, we see no big differences between Figures 4.4 and 4.5, while Figures 4.4 and 4.6, in which the working frequency varies, present clear variations for high traffic levels.

Finally, in order to give a global idea on how this trade-off appears, besides

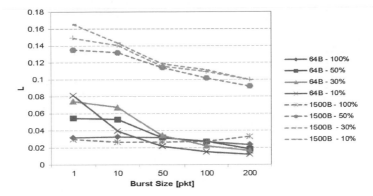

FIGURE 4.3 $L_\%$ values for the P_0, C_3 configuration according to 64 and 1500 B sized datagrams and different traffic load and burstiness levels.

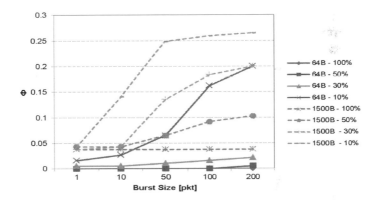

FIGURE 4.4 $\Phi\%$ values for the P_0, C_1 configuration according to 64 and 1500 B sized datagrams and different traffic load and burstiness levels.

the decreasing trend of performance degradations and the growing one of energy gains, it is possible to see that results in Figures 4.2, 4.3, 4.6, and 4.7 cover a wider range of values for short bursts, whereas in Figures 4.4, 4.5, 4.8, and 4.9 this spreading appears for bursts longer than 50 packets.

4.6 Conclusions

This work tried to present the most appealing solutions concerning benchmarking methodologies used to evaluate the energy efficiency level of Internet devices. First, a general overview on the power management capabilities under

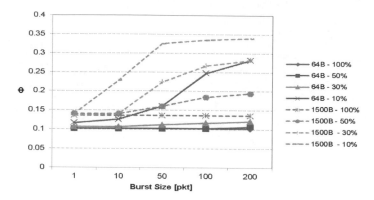

FIGURE 4.5 $\Phi\%$ values for the P_0, C_3 configuration according to 64 and 1500 B sized datagrams and different traffic load and burstiness levels.

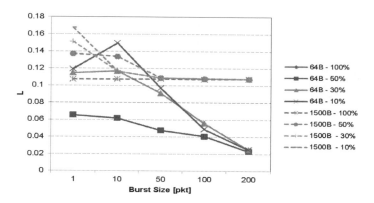

FIGURE 4.6 $L_\%$ values for the P_8, C_1 configuration according to 64 and 1500 B sized datagrams and different traffic load and burstiness levels.

study has been provided, then the focus has been on the existing benchmarking methodologies for the evaluation of energy efficiency. The most important existing standards have been thoroughly described one by one, then comparisons among them have been provided, in order to catch their characteristics and identify room for some possible improvements.

With this respect, tests have been performed to compare the results obtained with the ECR 3.0.1 methodology and a different benchmarking that takes into account latency. The device under test was a software router, which already has power management capabilities. Test results showed how the second set of indexes allows to represent the trade-off between power consumption and network performance in the presence of power management capabilities in a clear and, at the same time, complete way. Furthermore, results in terms of

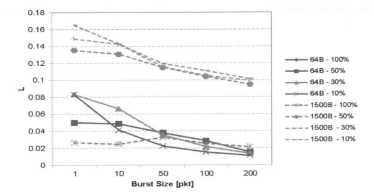

FIGURE 4.7 $L_\%$ values for the P_8, C_3 configuration according to 64 and 1500 B sized datagrams and different traffic load and burstiness levels.

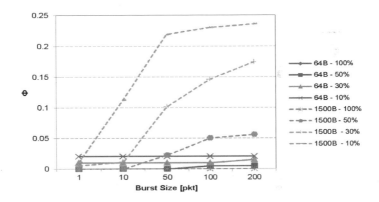

FIGURE 4.8 $\Phi\%$ values for the P_8, C_1 configuration according to 64 and 1500 B sized datagrams and different traffic load and burstiness levels.

energy gain and performance degradation also allowed to capture the different effects of each power management strategy on the device global behavior.

Acknowledgement

This work was supported by the ECONET Integrated Project, funded by the European Commission under the grant agreement no. 258454.

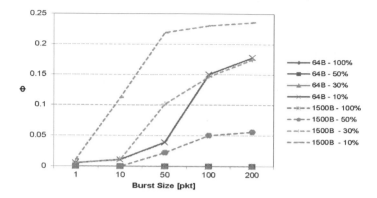

FIGURE 4.9 $\Phi\%$ values for the P_8, C_3 configuration according to 64 and 1500 B sized datagrams and different traffic load and burstiness levels.

Bibliography

[1] "ICT and CO_2 Emissions," Dec. 2008, Parliament Office of Science and Technology. [Online]. Available: http://www.parliament.uk/documents/post/postpn319.pdf

[2] M. Gupta and S. Singh, "Greening of the Internet," in *Proc. ACM SIG-COMM Conf. (SIGCOMM 03)*, Aug. 2003, pp. 19–26.

[3] K. Christensen, B. Nordman, and R. Brown, "Power Management in Networked Devices," *IEEE Computer*, vol. 37, no. 8, pp. 91–93, Aug. 2004.

[4] "The low Energy COnsumption NETworks (ECONET) project," IP project founded by the EC in ICT-Call 5 of the 7th Work-Programme, Grant agreement no. 258454.

[5] "HGI puts Energy Saving on the Agenda," The Home Gateway Initiative (HGI). [Online]. Available: http://www.homegatewayinitiative.org/press/PR010.pdf

[6] C. Bianco, F. Cucchietti, and G. Griffa, "Energy Consumption Trends in the Next Generation Access Network - a Telco Perspective," in *Proc. the 29th Int. Telecommun. Energy Conf., 2007. INTELEC 2007.*, Sept. 2007, pp. 737–742, Rome, Italy.

[7] Telecom Italia, "The Environment." [Online]. Available: http://www.telecomitalia.it/sostenibilita2006/english/B05.html

[8] G.-Q. Zhang, Q.-F. Yang, S.-Q. Cheng, and T. Zhou, "Evolution of the Internet and its Cores," *New Journal of Physics*, vol. 27, no. 3, pp. 1–11, Dec. 2008.

[9] R. S. Tucker, R. Parthiban, J. Baliga, K. Hinton, R. W. A. Aire, and W. V.Sorin, "Evolution of WDM Optical IP Networks: A Cost and Energy Perspective," *IEEE Journal of Lightwave Technology*, vol. 10, no. 12, pp. 243 – 252, Feb. 2009.

[10] R. S. Tucker, "Optical Packet-Switched WDM Networks - a Cost and Energy Perspective," in *Proc. Opt. Fiber Commun. Conf. Expo. (OFC 2008)*, no. 27, Mar. 2008, San Diego, CA, USA.

[11] S. Bradner and J. McQuaid, "Benchmarking methodology for network interconnect devices," http://www.ietf.org/rfc/rfc2544.txt, IETF Request for Comments no. 2544 (RFC 2544).

[12] A. Alimian, B. Nordman, and D. Kharitonov, "Network and Telecom Equipment - Energy and Performance Assessment," http://www.ecrinitiative.org/pdfs/ECR_3_0_1.pdf, Dec. 2010, Draft 3.0.1.

[13] E. Eckert, K. Sievert, L. Rabinovich, and C. Underkoffler, "Energy Efficiency for Telecommunication Equipment: Methodology for Measurement and Reporting for Router and Ethernet Switch Products," July 2009.

[14] "IMIX (Internet MIX)," http://spcprev.spirentcom.com/documents/4079 .pdf, Mar. 2006.

[15] V. Manral, "Benchmarking Power Usage of Networking Devices," http://tools.ietf.org/html/draft-manral-bmwg-power-usage-02.

[16] "Energy Efficiency Measurement for Telecommunication Equipment," http://www.itu.int/ITU-T/workprog/wp_item.aspx?isn=7735.

[17] "Energy Efficiency Metrics for Telecommunication Equipment," http://www.itu.int/ITU-T/workprog/wp_item.aspx?isn=7732.

[18] R. Bolla, R. Bruschi, C. Lombardo, and D. Suino, "Evaluating the Energy-Awareness of Future Internet Devices," in *IEEE 12th Conference on High Performance Switching and Routing*, Jul. 2011, pp. 36–43, Cartagena, Spain.

[19] R. Bolla, R. Bruschi, and A. Ranieri, "Green Support for PC-based Software Router: Performance Evaluation and Modeling," in *Proc. of the 2009 IEEE International Conference on Communications (IEEE ICC 2009)*, Jun. 2009, pp. 1–6, Dresden, Germany.

[20] "The Advanced Configuration & Power Interface (ACPI) Specification," Revision 4.0. [Online]. Available: http://www.acpi.info/

[21] "The Ixia XM2 router tester." [Online]. Available: http://www.ixiacom. com/products/chassis/display?skey=ch$\_$optixia$\_$xm2

[22] "Agilent u2700 modular system." [Online]. Available: http://www.home.agilent.com/agilent/redirector.jspx?ckey= 1537556&action=ref&lc=ita&cc=IT&cname=AGILENT_EDITORIAL

Author Contact Information

Raffaele Bolla and Chiara Lombardo are with DIST - University of Genoa, Italy, and CNIT, University of Genoa Research Unit, Italy, Email: raffaele.bolla@unige.it, chiara@reti.dist.unige.it. Roberto Bruschi is with CNIT, University of Genoa Research Unit, Italy, Email: roberto.bruschi@cnit.it.

5

Reciprocal Learning for Energy-Efficient Opportunistic Spectrum Access in Cognitive Radio Networks

Xianfu Chen, Zhifeng Zhao, Honggang Zhang

Zhejiang University, China

Jinsong Wu

Bell Laboratories, China

Tao Chen

VTT Technical Research Centre of Finland, Finland

CONTENTS

5.1 Introduction

Recently, there has been extensive research on cognitive radio (CR) which bridges the enormous gulf in time and space between the regulation and the potential spectrum efficiency ([1–3] and the references therein). One of the critical challenges is how to realize the coexistence of primary users (PUs) and cognitive users (CUs) accessing the same part of the authorized spectrum. The PUs have priority in accessing the frequency bands while the CUs

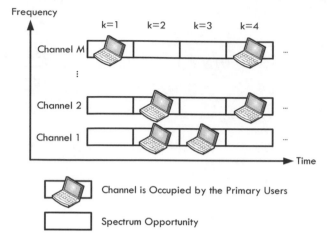

FIGURE 5.1 Channel model

opportunistically transmit when the spectrum is idle, as illustrated in Figure 5.1. To avoid colliding with the licensed network, the CUs must sense before transmission to determine whether there are PUs over the selected channel. However, the PUs' stochastic behavior patterns challenge the selection process.

First, due to the resource and hardware constraints, each CU can only access a part of the available channels simultaneously. Under such scenarios, the problem is modeled as a Partially Observed Markov Decision Process (POMDP) [4, 5]. In this chapter, we assume that the the probability that each channel is free is known a priori. The goal is to design an opportunistic spectrum access (OSA) protocol, in order to fully and efficiently utilize the spectrum opportunities. Then, in the presence of multiple CUs, the protocol must account for the competition among different CUs over the same channel (see Figure 5.2).

In order to avoid significant reduction in the overall network performance caused by possible collisions among the CUs, we model the OSA in CR network as a non-cooperative game [6]. In general, game theoretic approaches have been exploited to determine the communication resources of multiple interacting users. For instance, non-cooperative multi-radio channel allocation for wireless networks is studied in [7], where the authors study the existence of Nash equilibrium (NE) in a static game and show that the channel allocation results in a load-balancing solution. A distributed pricing approach is proposed to enable the CUs to reach a good NE [8].

Game theory bases its solution on equilibrium. A user behaving within an equilibrium is often explained in terms of its beliefs about the strategies of its opponents. How users reach such beliefs through interactions is by learning. The distinction between learning and non-learning users is simply that

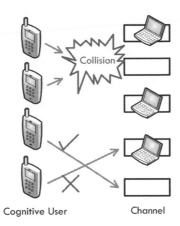

Cognitive User   Channel

FIGURE 5.2 Collision model

the former change their beliefs, whereas the latter's beliefs are static. A reinforcement learning algorithm is developed to design power control in wireless ad-hoc networks [9], where it is shown that the learning dynamics eventually converge to pure NE. In [10], the authors propose two stochastic learning algorithms for wireless users to dynamically and efficiently allocate discrete power levels, and show that pure and mixed equilibria exist given certain conditions.

This chapter is concerned with developing distributed learning OSA protocols for the CUs from not only the game-theoretic, but also the learning perspective. To utilize the spectrum opportunities more efficiently, we adopt the conjectural variations model introduced by Bowley [11] to encourage cooperation among the CUs and enable them to build beliefs about how the others react in response to their own strategy changes. The belief functions of CUs reflect an awareness that there are strategic interaction mechanisms in which they do not correctly perceive how the future choices of their rivals depend on the past. Specifically, by deploying such a behavior model, the CUs will no longer adopt myopically selfish behaviors, but rather they will form conjectural beliefs about how their strategy changes will influence the responses of other CUs and, based on these beliefs, they will try to maximize their own utility functions [12].

5.2   Network Model and Problem Formulation

As shown in Figure 5.3, let $\mathcal{N} = \{1, \ldots, N\}$ be a set number of CUs and $\mathcal{M} = \{1, \ldots, M\}$ be a set of orthogonal channels with equal bandwidth W. Without loss of generality, the bandwidth is normalized to be $W = 1$. All PUs

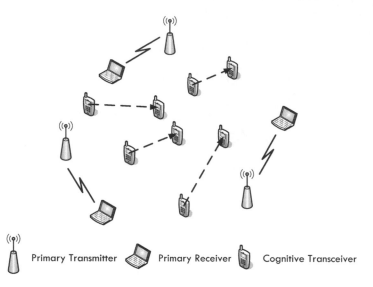

FIGURE 5.3 System model

and CUs in the network are operated in a time-slotted fashion. A CU suggests a CR link consisting of a pair of CR transmitter and CR receiver. We use i to refer to the index of CUs, m to refer to the index of channels, and k to refer to the time-slot index. In each time slot k, the PU transmits over channel m with probability $1 - \mu_m \geq 0$. In other words, let $S_m(k)$ be a random variable indicating the channel state, such that

$$S_m(k) = \begin{cases} 0, & \text{channel } m \text{ occupied in time slot } k; \\ 1, & \text{otherwise.} \end{cases}$$

Thus, given the idle probability μ_m, $S_m(k)$ is a Bernoulli random variable with Probability Density Function (pdf)

$$u\left(s_m(k); \mu_m\right) = \left(\mu_m\right)^{s_m(k)} \left(1 - \mu_m\right)^{1-s_m(k)}, \text{ for } s_m(k) \in \{0, 1\}.$$

Given the mean availability vector $\boldsymbol{\mu} = [\mu_1, \ldots, \mu_M]$, $S_m(k)$ is independent for each m and k. Figure 5.3 shows the system model considered in this chapter.

In our model, all CUs try to exploit the available channels of PUs during each time slot, given that $\boldsymbol{\mu}$ is initially known to all CUs. Our goal is to design efficient protocols for CUs to compete which channel to access. At the beginning of time slot k, each CU $i \in \mathcal{N}$ selects one channel $a_i(k) \in \mathcal{M}$ according to its strategy π_i^k for channel access. A strategy π_i^k is defined to be a probability vector $\pi_i^k = \left(\pi_i^k(1), \ldots, \pi_i^k(M)\right)$, where $\pi_i^k(m)$ means the persistence probability with which CU i accesses channel m at time slot k. We assume that channel sensing is perfect at all CUs.

If the sensing result indicates that the channel $a_i(k)$ is free, i.e., $S_{a_i(k)}(k) =$

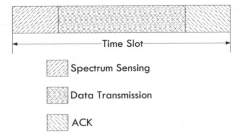

FIGURE 5.4 Time slot structure

1, the CUs selecting this channel compete to transmit. Furthermore, we assume the collision model under which each CU always transmits, and if two or more CUs transmit over the same channel then none of the transmissions are successful (see Figure 5.2). At the end of the same time slot, CU i receives an acknowledgement (ACK) $Z_{i,a_i(k)}(k)$ that equals 1 if the transmission went through and equals 0 otherwise. In other words, $Z_{i,a_i(k)}(k)$ suggests whether there are multiple CUs choosing the idle channel $a_i(k)$. The time slot structure is shown in Figure 5.4.

By incorporating the collision model without avoidance mechanism, the utility that the CU i obtains by accessing channel $a_i(k)$ is the number of bits that it can transmit during time slot k,

$$B_i(k) = S_{a_i(k)}(k)Z_{i,a_i(k)}(k).$$

It is clear that $B_i(k)$ is a random variable that depends on the PUs' traffics and, more importantly for us, the channel access strategies implemented by the CUs. The CU i's expected transmitted bits during time slot k is given as

$$
\begin{aligned}
U_i\left(\pi_i^k, \pi_{-i}^k\right) &= \mathrm{E}\left\{B_i(k)\right\} \\
&= \mathrm{E}\left\{S_{a_i(k)}(k)Z_{i,a_i(k)}(k)\right\} \\
&= \sum_{m \in \mathcal{M}} \mu_m \pi_i^k(m) \prod_{j \in \mathcal{N}\backslash\{i\}}\left(1 - \pi_j^k(m)\right),
\end{aligned}
$$

where $\pi_{-i}^k = \left(\pi_1^k, \ldots, \pi_{i-1}^k, \pi_{i+1}^k, \ldots, \pi_N^k\right)$. From (5.1), we can see that CU i's expected utility at each time slot k depends not only on its own channel access strategy, but also the other CUs' channel access strategies.

The overarching goal in the rest of this chapter is to design the channel access strategy π_i^k for each CU $i \in \mathcal{N}$ such that

$$\max_{\pi_i^k} U_i\left(\pi_i^k, \pi_{-i}^k\right)$$

s.t.

$$\sum_{m \in \mathcal{M}} \pi_i^k(m) = 1, \tag{5.1}$$

which is a non-cooperative game. An important solution for (5.1) is the NE. First, we define the best response strategy of CU i to its opponents' strategies $\boldsymbol{\pi}^k_{-i}$ as π^*_i that achieves the maximum expected utility:

$$U_i\left(\pi^k_i, \boldsymbol{\pi}^k_{-i}\right) \leq U_i\left(\pi^*_i, \boldsymbol{\pi}^k_{-i}\right).$$

An NE is a joint strategy $(\pi^*_1, \ldots, \pi^*_N)$ such that each individual π^*_i is a best response strategy to the others. The NE describes a status quo, where no CU can benefit by changing its strategy as long as all other CUs keep their strategies constant.

In this chapter, we limit to the scenarios where all CUs undertake distributed learning and channel access with very limited information exchange. Hence, in general, the channel access strategy π^k_i employed by CU i at time slot k, is obtained through the previous $k-1$ observations:

$$\begin{aligned}\Theta_i(k) = \{&a_i(1), S_{a_i(1)}(1), Z_{i,a_i(1)}(1), \ldots, a_i(k-1), S_{a_i(k-1)}(k-1), \\ &Z_{i,a_i(k-1)}(k-1)\},\end{aligned}$$

$k \geq 2$, i.e. $a_i(k)$ is drawn according to $\pi^k_i(\Theta_i(k))$. Notice that $a_i(k)$ is the channel being accessed at time slot k, $S_{a_i(k)}(k)$ is the sensing outcome, and $Z_{i,a_i(k)}(k)$ shows whether the transmission is successful. If $k = 1$, there is no accumulated information, thus $\Theta_i(1) = \varnothing$ and $a_i(1)$ could be random, that is, the CU i may randomly select channel $a_i(1)$ from $\mathcal{M}$ with probability $\pi^1_i(a_i(1))$, such that $\sum\limits_{a_i(1) \in \mathcal{M}} \pi^1_i(a_i(1)) = 1$.

5.3 Distributed Learning Algorithms with Dynamic Conjectures

In this section, to promote cooperation, we design a simple and intuitive rule each CU has that links the other CUs' strategies to its own current OSA strategy. Such a rule reflects an awareness that there are strategic interactions. CUs with such beliefs do not correctly perceive how the future OSA strategies of their opponents depend on the past. Thus, we propose a conjecture model which embodies the beliefs. Each CU has beliefs about how its opponents' current OSA strategies relates to its own past channel selection strategy. Furthermore, based on the conjecture model, we propose two distributed OSA algorithms to achieve the optimal transmission configuration.

As discussed before, each CU has beliefs concerning the way in which other CUs react are a dynamic version of conjectures. Each CU thinks any change in its current channel selection strategy will induce other CUs to make well-defined changes in the next time slot. Specifically, we need to express the expected contention measure $b^k_{i,m} = \prod\limits_{j \in \mathcal{N}\setminus\{i\}} \left(1 - \pi^k_j(m)\right)$ through $\tilde{b}_{i,m}$, that is

$$\widetilde{b}_{i,m}^{k} = b_{i,m}^{k-1} - \delta_{i,m}\left(\pi_i^k(m) - \pi_i^{k-1}(m)\right), \tag{5.2}$$

with $\delta_{i,m} > 0$, for $i \in \mathcal{N}$ and $m \in \mathcal{M}$. Particularly, we set $\pi_i^0(m) = \pi_i^1(m)$ when $k = 1$. That is, the CU i believes that a change of $\pi_i^k(m) - \pi_i^{k-1}(m)$ in its channel selection strategy at time slot k will induce a change of $\delta_{i,m}\left(\pi_i^k(m) - \pi_i^{k-1}(m)\right)$ in the expected contention measure exactly corresponding to the channel selection strategies of the other CUs. Although CU i may be aware that other CUs are subject to many influences on their channel selection strategies, when making its own decision, it is only concerned with other CUs' reactions to itself. In another word, CU i does not take into account whether or not CU $j(j \in \mathcal{N} \setminus \{i\})$ might react to changes in channel selection strategy made by CU $l(l \in \mathcal{N} \setminus \{i, j\})$.

Among different possibilities of capturing the expected contention measure $\widetilde{b}_{i,m}^{k}$, the linear model represented in (5.2) is the simplest form based on which a CU can model the impact of its changes in OSA strategy to others. The conjecture models deployed by the CUs are based on the concept of reciprocity, which refers to interaction mechanisms in which the CUs repeatedly interact when accessing the channels. If they realize that their probabilities of interacting with each other in the future is high, they will consider their influence on the OSA strategies of other CUs, which is captured in the conjecture model by the positive parameter $\delta_{i,m}$. Otherwise, they will act myopically, which will lead to terrible performance reduction.

Thus the CU *is* expected transmitted bits during time slot k can be rewritten as

$$U_i\left(\pi_i^k, \widetilde{\mathbf{b}}_i^k\right) = \sum_{m=1}^{M} \mu_m \pi_i^k(m)\left[\widetilde{b}_{i,m}^{k-1} - \delta_{i,m}\left(\pi_i^k(m) - \pi_i^{k-1}(m)\right)\right]. \tag{5.3}$$

where $\widetilde{\mathbf{b}}_i^k = \left(\widetilde{b}_{i,1}^k, \ldots, \widetilde{b}_{i,M}^k\right)$. Each CU i selfishly chooses a strategy π_i^k to maximize its expected utility at each time slot k, which it would achieve if the other CUs reacted to its channel selection strategy change according to $\delta_{i,m}\left(\pi_i^k(m) - \pi_i^{k-1}(m)\right)$. Based on $\widetilde{\mathbf{b}}_i^k$, each CU $i \in \mathcal{N}$ designs its channel access strategy π_i^k in order to

$$\max_{\pi_i^k} U_i\left(\pi_i^k, \widetilde{\mathbf{b}}_i^k\right)$$

s.t.

$$\sum_{m \in \mathcal{M}} \pi_i^k(m) = 1. \tag{5.4}$$

The solution of (5.4) is conjecture equilibrium (CE), which is defined as follows.

Definition: A configuration of conjectures $\left(\widetilde{\mathbf{b}}_1^*, \ldots, \widetilde{\mathbf{b}}_N^*\right)$ and a joint strat-

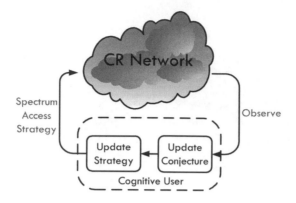

FIGURE 5.5 Conjecture based learning process

egy $(\pi_1^*, \ldots, \pi_N^*)$ constitute a CE if, for each CU $i \in \mathcal{N}$,

$$\widetilde{b}_{i,m}^* = \prod_{\substack{j \in \mathcal{N} \setminus \{i\}}}^{N} \left(1 - \pi_j^*(m)\right),$$

for $\forall m \in \mathcal{M}$, and

$$\pi_i^* = \arg\max_{\pi_i^k} U_i\left(\pi_i^k, \widetilde{\mathbf{b}}_i^*\right).$$

Figure 5.5 shows the learning process of CU i based on the conjecture $\widetilde{\mathbf{b}}_i^k$.

5.3.1 A Best Response Learning Method for OSA

If the CUs operate at a low level of sophistication in forming their beliefs, it seems unreasonable that they solve an infinite horizon dynamic programming problem at each time slot. On the other side, if the CUs employ a high level of sophistication in making their decisions, however, they ascribe a low level of sophistication to the decision makings of others. The reasons why we make the this assumption are two-fold. First, the CUs may unilaterally think that others are not as sophisticated as they are. Second, the CUs may be aware that the others are as sophisticated as they are, but mistakenly take the attitude that forming the beliefs that the others are equally sophisticated will complicate their decision making too much. In either case, the CUs may regard (2) as a good approximation of what OSA strategies others choose.

5.3.1.1 The Best Response Strategies

Along with the previous discussion, we develop the best response channel selection strategy function for each CU. We treat $b_{i,m}^1$ and $\pi_i^1(m)$ as initial parameters and then find an optimal channel selection strategy for CU i that consists of a sequence of single slot policy functions $\pi_i^k(m) = \psi_{i,m}^k\left(\pi_i^{k-1}(m), b_{i,m}^{k-1}\right)$, which gives best response behavior for CU i at any time slot k given its beliefs $\delta_{i,m}$.

Theorem 5.1. *The infinite horizon best response opportunistic spectrum access strategy for CU i is given by*

$$\pi_i^k(m) = \begin{cases} \left\{\dfrac{1}{2}\pi_i^{k-1}(m) + \dfrac{1}{2\delta_{i,m}}b_{i,m}^{k-1} + \dfrac{\lambda_i^k}{2\delta_{i,m}\mu_m}\right\}_0^1, & \text{if } \mu_m > 0, \\ 0, & \text{if } \mu_m = 0, \end{cases} \quad (5.5)$$

where λ_i^k is the constant that satisfies $\sum_{m\in\mathcal{M}} \pi_i^k(m) = 1$. Here, $\{x\}_0^1$ denotes the Euclidean projection of x onto the interval $[0,1]$, i.e., $\{x\}_0^1 = 0$ if $x < 0$, $\{x\}_0^1 = x$ if $0 \le x \le 1$, and $\{x\}_0^1 = 1$ if $x > 1$.

Proof. The best response opportunistic spectrum access strategy π_i^k at time slot k maximizes $U_i\left(\pi_i^k, \widetilde{\mathbf{b}}_i^k\right)$. Now, we should solve the following optimization problem

$$\max_{\pi_i^k} \quad U_i\left(\pi_i^k, \widetilde{\mathbf{b}}_i^k\right)$$

$$\text{s.t.}$$

$$\text{C1:} \qquad \sum_{m\in\mathcal{M}} \pi_i^k(m) = 1$$

$$\text{C2:} \qquad \pi_i^k(m) \ge 0.$$

In our formulation, each individual optimization problem is a convex problem with linear constraints C1-C2. So the Lagrangian function for CU i can be written as

$$\mathcal{L}_i^k = U_i\left(\pi_i^k, \mathbf{b}_i^k\right) + \lambda_i^k\left(\sum_{m=1}^M \pi_i^k(m) - 1\right) + \sum_{m=1}^M \gamma_{i,m}^k \pi_i^k(m), \qquad (5.6)$$

where λ_i^k and $\gamma_{i,m}^k$ are Lagrangian multipliers (non-negative real numbers). The Karush-Kuhn-Tucker (K.K.T.) conditions [13] are given by

$$\frac{\partial \mathcal{L}_i^k}{\partial \pi_i^k(m)} = \mu_m\left[-2\delta_{i,m}\pi_i^k(m) + b_{i,m}^{k-1} + \delta_{i,m}\pi_i^{k-1}(m)\right] + \lambda_i^k + \gamma_{i,m}^k = 0$$

$$\pi_i^k(m) \ge 0$$

$$\gamma_{i,m}^k \pi_i^k(m) = 0$$

$$\sum_{m\in\mathcal{M}} \pi_i^k(m) = 1.$$

It's easy to check that

$$\pi_i^k(m) = \begin{cases} \left\{ \dfrac{1}{2}\pi_i^{k-1}(m) + \dfrac{1}{2\delta_{i,m}}b_{i,m}^{k-1} + \dfrac{\lambda_i^k}{2\delta_{i,m}\mu_m} \right\}_0^1, & \text{if } \mu_m > 0, \\ 0, & \text{if } \mu_m = 0, \end{cases}$$

where λ_i^k is the constant that satisfies $\sum\limits_{m \in \mathcal{M}} \pi_i^k(m) = 1$. This concludes the proof.   □

Remark 1: We can see from Theorem 5.1 that it's not rational for each CU i to follow the channel selection strategy π_i^k obtained at the current time slot in the future. This is because π_i^k is based on the conjectures about the current channel selection strategies of the others while the other CUs' future channel selection strategies, in general, are dynamic. Thus a CU needs to recalculate another infinite horizon strategy in the same way in each subsequent time slot.

The detailed description of the distributed best response learning protocol for opportunistic spectrum access is summarized in Algorithm 5.1. Next, we are concerned with the convergence of this algorithm.

Algorithm 5.1. A Best Response Learning OSA Algorithm

1. ***Initialize:*** *$k = 1$, the channel access strategies π_i^1 and the parameters $\delta_{i,m} > 0$ in CU i's conjecture functions, for $\forall i \in \mathcal{N}$ and $\forall m \in \mathcal{M}$.*

2. ***Learning:***

    a)   *Set $k \leftarrow k + 1$.*
    b)   *For $\forall i \in \{1, \ldots, N\}$, do*

$$\pi_i^k(m) = \begin{cases} \left\{ \dfrac{1}{2}\pi_i^{k-1}(m) + \dfrac{1}{2\delta_{i,m}}b_{i,m}^{k-1} + \dfrac{\lambda_i^k}{2\delta_{i,m}\mu_m} \right\}_0^1, & \text{if } \mu_m > 0; \\ 0, & \text{if } \mu_m = 0. \end{cases}$$

    c)   *CU i decides to access channel m at time slot k with probability $\pi_i^k(m)$.*

3. ***End Learning***

5.3.1.2   Global Convergence of the Dynamic Network

With the best response policies, the CUs should wonder whether their beliefs have any negative effects. Our conjecture model expressed in (5.2) suggests that errors exist in the conjectures and the CUs learn from the previous observations to try to improve the utilities at the next time slot. For this reason, we shall assume that the dynamics of the network will appear reasonably consistent to the CUs if the values of the conjectures stabilize when time passes.

We will show in Theorem 5.2 that the dynamic network is stable if each $\psi_{i,m}^k$ is a contraction mapping.

Theorem 5.2. *Suppose that*

$$\delta_{i,m} \geq N - 1, \ \text{for } \forall i \in \mathcal{N} \text{ and } \forall m \in \mathcal{M}, \tag{5.7}$$

the dynamic network has a unique steady state; that is, regardless of any initial value chosen for $\pi_i^1(m)$, the best response OSA strategy $\pi_i^k(m)$ converges to π_i^* as $k \to \infty$.

Proof. Without loss of generality, we assume that $\mu_m > 0$, for $\forall m \in \mathcal{M}$. At the moment, the best response channel selection strategy in (5.5) can be rewritten as

$$
\begin{aligned}
\pi_i^k(m) &= \frac{1}{2}\pi_i^{k-1}(m) + \frac{1}{2\delta_{i,m}}b_{i,m}^{k-1} + \frac{\lambda_i^k}{2\delta_{i,m}\mu_m} \\
&= \frac{1}{2}\pi_i^{k-1}(m) + \frac{1}{2\delta_{i,m}}\prod_{j\in\mathcal{N}\setminus\{i\}}\left(1 - \pi_j^{k-1}(m)\right) + \frac{\lambda_i^k}{2\delta_{i,m}\mu_m}.
\end{aligned}\tag{5.8}
$$

It's easy to get the sum of the absolute values of the partial derivatives of (8) with respect to $\left(\pi_1^{k-1}(m), \dots, \pi_N^{k-1}(m)\right)$,

$$
\begin{aligned}
q_1 &= \frac{1}{2} + \frac{1}{2\delta_{i,m}} \sum_{j\in\mathcal{N}\setminus\{i\}} \prod_{l\in\mathcal{N}\setminus\{i,j\}}\left(1 - \pi_l^{k-1}(m)\right) \\
&\leq \frac{1}{2} + \sum_{j\in\mathcal{N}\setminus\{i\}} \frac{1}{2\delta_{i,m}} \\
&= \frac{1}{2} + \frac{N-1}{2\delta_{i,m}}.
\end{aligned}
$$

If the condition in (5.7) is met, there exists a positive ε, such that $q_1 = 1 - \varepsilon < 1$. That is, the function $\psi_{i,m}^k$ satisfies Lipschitz condition [14]. Therefore, for $\forall i \in \mathcal{N}$ and $\forall m \in \mathcal{M}$, as $k \to \infty$ the $\pi_i^k(m)$ converges to $\pi_i^*(m)$ by the contraction mapping theorem [15]. $\qquad\square$

Remark 2: We think that our conjecture model and the dynamics it generates are much less appealing if (5.7) does not hold. In such a case, the CUs' conjectures about others are continuously falsified, and yet the CUs do not change the way they form them. We find it hard to believe that the CUs would continue to stick to their belief adjustment rules under such circumstances. On the contrary, if the network converges to a steady state, the CUs' beliefs eventually cease to be falsified and our approach is justified.

5.3.2 OSA by Gradient Ascent Algorithms

How strategies of multiple agents evolve over time while interacting with one another is an important aspect in multi-agent reinforcement learning (RL) [16–20]. To begin with, we suppose that there is only one state in our considered networking environment. The techniques, for tackling RL problems, then match the topic we discuss in this chapter. The action $a_i(k)$ of each CU i at each time slot k is to select one channel $a_i(k) \in \mathcal{M}$ to access, the reward is defined to be the utility that it obtains by accessing channel $a_i(k)$.

During the learning procedure, there is no assumption on the behaviors of other CUs. That is, there is little information exchange among all users. This results in an extra level of learning, the purpose of which is to learn the other CUs' OSA strategies. Accordingly, each CU chooses the best mixed strategy π_i^k rather than the best channel in order to avoid severe collisions. This motivates the development of gradient ascent learners [21–23].

We consider in this chapter the General Infinitesimal Gradient Ascent (GIGA) algorithm [22]. At each time slot, each CU updates its probabilities for channel access gradually in the ascent direction of its conjectured utility defined in (5.6). That is, at time slot k, CU i updates its strategy according to

$$\pi_i^k(m) = \pi_i^{k-1}(m) + \eta \frac{\partial \mathcal{L}_i^k}{\partial \pi_i(m)}\Big|_{\pi_i(m)=\pi_i^{k-1}(m)}. \tag{5.9}$$

Each CU updates its OSA strategy along the gradient direction of its conjecture utility with some step size η. We assume that the step size is same for $\forall i \in \mathcal{N}$ and $\forall m \in \mathcal{M}$. If the step size is small enough, the learning procedure evolves smoothly. So effectively, $\frac{\partial \mathcal{L}_i^k}{\partial \pi_i(m)}\Big|_{\pi_i(m)=\pi_i^{k-1}(m)} > 0$ means the probability of choosing a good channel increases by a rate. Similarly, the probability of choosing a bad channel decreases by a rate. Substituting the utility function (5.6) into (5.9), we have

$$\pi_i^k(m) = \left\{ \pi_i^{k-1}(m) + \eta\big[\mu_m \big(b_{i,m}^{k-1} - \delta_{i,m}\pi_i^{k-1}(m)\big) + \lambda_i^k\big] \right\}_0^1, \tag{5.10}$$

where λ_i^k is chosen such that $\sum_{m\in\mathcal{M}} \pi_i^k(m) = 1$. The detailed description of the reinforcement learning algorithm is summarized in Algorithm 5.2.

Algorithm 5.2. A Gradient Ascent Learning OSA Algorithm

1. ***Initialize:*** $k = 1$, *the channel access strategies* π_i^1 *and the parameters* $\delta_{i,m} > 0$ *in CU i's conjecture functions, for* $\forall i \in \mathcal{N}$ *and* $\forall m \in \mathcal{M}$.

2. ***Learning:***

    a)   *Set* $k \leftarrow k+1$.

    b)   *For* $\forall i \in \{1,\ldots,N\}$, *do*

$$\pi_i^k(m) = \left\{ \pi_i^{k-1}(m) + \eta\big[\mu_m \big(b_{i,m}^{k-1} - \delta_{i,m}\pi_i^{k-1}(m)\big) + \lambda_i^k\big] \right\}_0^1.$$

   c)   *CU i decides to access channel m at time slot k with probability* $\pi_i^k(m)$.

3. End Learning

Theorem 5.3. *Suppose that*

$$\delta_{i,m} \geq N - 1, \ for \ \forall i \in \mathcal{N} \ and \ \forall m \in \mathcal{M},$$

and the step size η *is small enough, the OSA strategy* $\pi_i^k(m)$ *obtained by Algorithm 5.2 converges to* $\pi_i^*(m)$ *as* $k \to \infty$.

Proof. If η is small enough, (5.10) can be rewritten as

$$
\begin{aligned}
\pi_i^k(m) &= \pi_i^{k-1}(m) + \eta \left[\mu_m \left(b_{i,m}^{k-1} - \delta_{i,m} \pi_i^{k-1}(m) \right) + \lambda_i^k \right] \\
&= \left(1 - \eta \mu_m \delta_{i,m} \right) \pi_i^{k-1}(m) + \eta \mu_m \prod_{j \in \mathcal{N} \setminus \{i\}} \left(1 - \pi_j^{k-1}(m) \right) + \eta \lambda_i^k \quad (5.11)
\end{aligned}
$$

for $\forall i \in \mathcal{N}$ and $\forall m \in \mathcal{M}$. Thus, the sum of the absolute values of the partial derivatives of (5.11) with respect to $\left(\pi_1^{k-1}(m), \ldots, \pi_N^{k-1}(m) \right)$ is given by,

$$
\begin{aligned}
q_2 &= 1 - \eta \mu_m \delta_{i,m} + \eta \mu_m \sum_{j \in \mathcal{N} \setminus \{i\}} \prod_{l \in \mathcal{N} \setminus \{i,j\}}^{N} \left(1 - \pi_l^{k-1}(m) \right) \\
&\leq 1 - \eta \mu_m \delta_{i,m} + \sum_{j \in \mathcal{N} \setminus \{i\}} \eta \mu_m \\
&= 1 - \eta \mu_m [\delta_{i,m} - (N - 1)].
\end{aligned}
$$

If the conditions in Theorem 5.3 are met, there exists a positive ε, such that $q_2 = 1 - \varepsilon < 1$. That is, the function $\psi_{i,m}^k$ satisfies Lipschitz condition [14]. Therefore, as $k \to \infty$, the strategy $\pi_i^k(m)$ converges by the contraction mapping theorem [15]. $\square$

5.4   Numerical Results

In this section, we present simulations for evaluating the performance of the algorithms developed earlier in this chapter.

We first consider a relatively simple case in which there are 2 CUs and 2 channels with probabilities of availability 0.6 and 0.8. Denote the probability of CU 1 choosing channel 1 by α and choosing channel 2 by $1 - \alpha$. In the same way, CU 2 chooses channel 1 with probability β and choose channel 2 with probability $1 - \beta$. For each CU, the initial transmission strategies are

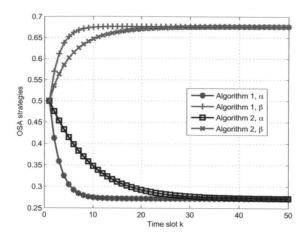

FIGURE 5.6 Strategy dynamics of Algorithm 5.1 and 5.2

set to be $\alpha^1 = 0.5000$ and $\beta^1 = 0.5000$, $\delta_{i,m}$ is uniformly distributed between 2 and 5, and the step size in Algorithm 5.2 $\eta = 0.0300$. Figure 5.6 compares the trajectory of the opportunistic spectrum access strategy updates in both Algorithm 5.1 and Algorithm 5.2, under the assumption that each CU i can perfectly conjecture the $\prod_{j \in \mathcal{N} \setminus \{i\}} \left(1 - \pi_j^k(m)\right)$. The best response method converges in around 8 iterations and the reinforcement learning based method experiences a more smooth trajectory and the same optimal channel access strategies are obtained after about 40 iterations.

The initial strategies α^1 and β^1 do not affect the convergence of our algorithms to the optimal mixed strategies. To show this, we can set (α^1, β^1) to be $(0.4500, 0.5500)$ and $(0.7000, 0.8000)$. It is shown in Figure 5.7 that, as we expect, the trajectory still converges to the same optimal mixed strategies as in Figure 5.6. We can notice that the curves are ultimately attracted by $(\alpha = 0.2726, \beta = 0.6765)$.

Different CUs have different learning abilities, that is, they may have different beliefs. If the CUs have the same beliefs of what their opponents react to their own OSA strategy changes, it will result in the symmetric strategies. The curves in Figure 5.8 indicate that the strategy dynamics converge to $\alpha = \beta = 0.5355$.

Further, for a more general case, we consider that the system has $M = 6$ channels with idle probabilities characterized by Bernoulli distributions with evenly spaced parameters ranging from 0.4 to 0.9. The initial OSA strategy $\pi_i^1(m)$ is set to be $\frac{1}{6}$ for all $i \in \{1, \ldots, 5\}$ and $m \in \{1, \ldots, 6\}$, and the belief $\delta_{i,m}$ is uniformly distributed between 6 and 10. The step size in Algorithm 5.2 is $\eta = 0.0300$.

First, we numerically compare the overall network performance of the two

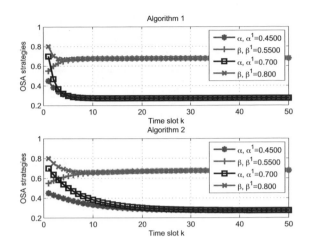

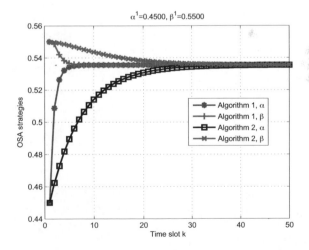

FIGURE 5.7 Strategy dynamics of Algorithm 5.1 and 5.2 with different initial values of α^1 and β^1

FIGURE 5.8 Strategy dynamics of Algorithm 5.1 and 5.2 with the same belief parameter $\delta_{i,m}$

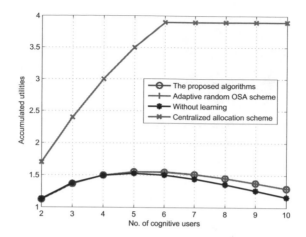

FIGURE 5.9 Comparison of the accumulated utilities corresponding to different OSA schemes

algorithms proposed in this chapter with three existing protocols, i.e. adaptive random OSA scheme, no learning scheme, and centralized allocation scheme: (1) Adaptive random OSA scheme: If each CU target the best channel, then it would result in terrible collisions. When there is no communication among multiple CUs, they need to randomize channel access in order to avoid collisions. However, if the CUs randomize at every time slot, there is a finite probability of collisions in every time slot. The CUs need to adapt to a collision-free configuration to ensure that the collisions are logarithmic [24]. Hence, the OSA strategies converge to the symmetric strategies.

(2) No learning scheme: Without learning capability, every CU i selects channel m according to its OSA strategy π_i, which is given as

$$\pi_i(m) = \frac{\mu_m}{\sum\limits_{l \in \mathcal{M}} \mu_l},$$

for $\forall i \in \mathcal{N}$ and $\forall m \in \mathcal{M}$.

(3) Centralized allocation scheme: In the ideal scenario, a central agent orthogonally allocates the CUs to the N-best channels (i.e., the channels with highest entries in $\boldsymbol{\mu}$), the accumulated expected utilities is given by

$$U = \sum_{i \in \mathcal{N}} \boldsymbol{\mu}(i^*),$$

where i^* is the ith-highest entry in $\boldsymbol{\mu}$. If the number of CUs N is greater than the number of channels M, we assume that the central agent allocates any M out of N CUs to the channels.

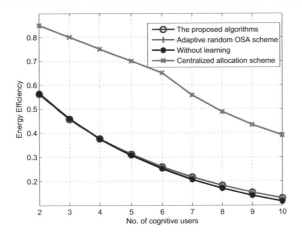

FIGURE 5.10 Comparison of the energy efficiency corresponding to different OSA schemes

As shown in Figure 5.9, the proposed algorithms and the adaptive random OSA scheme achieve significant accumulated utilities compared with no learning scheme. In addition, we can find from Figure 5.9 that unlike the centralized allocation scheme, the accumulated utilities of other OSA solutions increase versus the number of CUs, but decrease when the number of CUs exceeds the number of channels. The reason is obvious: when $N \leq M$, more CUs, the utilization of channels is better exploited; yet, when $N > M$, the collisions among the CUs cannot be avoided, thus cause the reduction in overall network performance.

We assume that CUs are always working with the same transmission power level. In this chapter, we also consider the energy efficiency, which is defined as following,

$$\text{Energy Efficiency} = \frac{\text{Efficient Energy Used for Successful Transmission}}{\text{Overall Energy Consumption}}.$$

The curves in Figure 5.10 show the energy efficiency of different OSA schemes. By applying the learning algorithms proposed in this chapter, severe collisions among the CUs can be alleviated, thus, achieving reduction in the number of re-transmissions and energy saving.

Next, we evaluate the fairness of the algorithms proposed in this chapter using the quantitative fairness index [25],

$$\mathbf{F} = \frac{\varrho(U_i)}{\varrho(U_i) + \sigma(U_i)},$$

where ϱ and σ are, respectively, the mean and the standard deviation of each

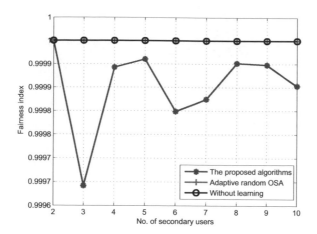

FIGURE 5.11 Comparison of the achieved fairness index of different OSA schemes

CU i's utility U_i over all the data flows. Figure 5.11 evaluates the fairness characteristics of the algorithms we proposed in this chapter, the adaptive random OSA scheme proposed in [24], and the no learning scheme. We can see that they are comparable in their fairness performance and the achieved fairness indexes are nearly the same.

5.5   Conclusion

We have studied opportunistic spectrum access problems within the CR networking framework. In order to encourage cooperation among the CUs, we proposed two distributed algorithms based on the conjectural variation theory to improve their performance. Each CU forms its own belief about the influence of the OSA strategy changes to the other CUs, and thus learns the optimal strategies maximizing the utility. The optimal strategies are obtained from the interaction outcomes among the intelligent CUs. Additionally, we prove the convergence of the dynamic networking environment under the two distributed learning solutions. We have also presented experimental results to illustrate that the proposed algorithms achieve significant accumulated utilities, as well as the energy efficiency performance.

Bibliography

[1] J. Mitola and G. Q. Maguire, "Cognitive radios: Making software radios more personal," *IEEE Personal Communications*, vol. 6, no. 4, pp. 13–18, Aug. 1999.

[2] S. Haykin, "Cognitive radio: Brain-empowered wireless communications," *IEEE journal of Selected Areas in Communications*, vol. 23, no. 2, pp. 201–220, Feb. 2005.

[3] I. F. Akyildiz, W.-Y. Lee, and K. R. Chowdhury, "CRAHNs: Cognitive radio ad hoc networks," *Ad Hoc Networks*, vol. 7, pp. 810–836, July 2009.

[4] J. Unnikrishnan and V. V. Veeravalli, "Algorithms for dynamic spectrum access with learning for cognitive radio," *IEEE Transactions on Signal Processing*, vol. 58, no. 2, pp. 750–760, Feb. 2010.

[5] Q. Zhao, L. Tong, A. Swami, and Y. Chen, "Decentralized cognitive mac for opportunistic spectrum access in ad hoc networks: A pomdp framework," *IEEE journal on Selected Areas in Communicatioin*, vol. 25, no. 3, pp. 589–600, April 2007.

[6] J. Nash, "Non-cooperative games," *The Annals of Mathematics*, vol. 54, no. 2, pp. pp. 286–295, 1951.

[7] M. Felegyhazi, M. Cagalj, S. S. Bidokhti, and J.-P. Hubaux, "Non-cooperative multi-radio channel allocation in wireless networks," in *Proc. of 26th IEEE International Conference on Computer Communications (INFOCOM)*, Anchorage, USA, May 2007.

[8] F. Wang, M. Krunz, and S. Cui, "Price-based spectrum management in cognitive radio networks," *IEEE journal of Selected Topics in Signal Processing*, vol. 2, pp. 74–87, Feb. 2008.

[9] C. Long, Q. Zhang, B. Li, H. Yang, and X. Guan, "Non-cooperative power control for wireless ad hoc networks with repeated games," *IEEE journal on Selected Areas in Communications*, vol. 25, no. 6, pp. 1101–1112, Aug. 2007.

[10] Y. Xing and R. Chandramouli, "Stochastic learning solution for distributed discrete power control game in wireless data networks," *IEEE/ACM Transactions on Networking*, vol. 16, no. 4, pp. 932–944, Aug. 2008.

[11] A. L. Bowley, *The Methematical Groundwork of Economics.* Oxford: Oxford University Press, 1924.

[12] L. Lai, H. E. Gamal, H. Jiang, and H. V. Poor, "Cognitive medium access: Exploration, exploitation and competition," *IEEE Transactions on Mobile Computing*, vol. 10, no. 2, pp. 239–253, Feb. 2011.

[13] S. Boyd and L. Vandenberghe, *Convex Optimization.* Cambridge, UK: Cambridge University Press, 2004.

[14] H. Jeffreys and B. Jeffreys, *Methods of Mathematical Physics*, 3rd ed. Cambridge, UK: Cambridge University Press, 2000.

[15] A. Granas and J. Dugundji, *Fixed Point Theory.* New York: Springer-Verlag, 2003.

[16] J. Hu and M. P. Wellman, "Nash q-learning for general-sum stochastic games," *journal of Machine Learning Research 4*, pp. 1039–1069, 2003.

[17] C. Watkins and P. Dayan, "Q-learning," *Machine learning*, vol. 8, pp. 279–292, 1992.

[18] M. Weinberg and J. S. Rosenschein, "Best-response multiagent learning in non-stationay environments," in *Proc. of the Third International Joint Conference on Autonomous Agents and Multiagent Systems*, New York, USA, July 2004.

[19] R. S. Sutton and A. G. Barto, *Reinforcement Learning: An Introduction.* Cambridge, MA: MIT Press, 1998.

[20] M. P. Wellman and J. Hu, "Conjectural equilibrium in multiagent learning," *Machine Learning*, vol. 33, pp. 179–200, 1998.

[21] S. Singh, M. Kearns, and Y. Mansour, "Nash convergence of gradient dynamics in general-sum games," in *Proc. of the 16th Conference on Uncertainty in Artificial Intelligent*, San Francisco, CA, USA, 2000.

[22] M. Zinkevich, "Online convex programming and generalized infinitesimal gradient ascent," in *Proc. of 20th International Conference on Machine Learning (ICML)*, Washington, USA, Aug. 2003.

[23] S. Abdallah and V. Lesser, "A multiagent reinforcement learning algorithm with non-linear dynamics," *journal of Artificial Intelligence Research*, vol. 33, pp. 521–549, 2008.

[24] A. Anandkumar, N. Michael, A. K. Tang, and A. Swami, "Distributed algorithms for learning and cognitive medium access with logarithmic regret," *IEEE journal on Selected Areas in Communications*, vol. 29, no. 4, pp. 731–745, April 2011.

[25] D. Qiao and K. G. Shin, "Achieving efficient channel utilization and weighted fairness for data communications in ieee 802.11 wlan under the dcf," in *Proc. of 10th International Workshop on Quality of Service*, Miami, USA, May 2002.

Author Contact Information

Xianfu Chen, Zhifeng Zhao, Honggang Zhang are with Zhejiang University, China, Email: zhaozf@zju.edu.cn, honggangzhang@zju.edu.cn. Jinsong Wu is with Bell Laboratories, Alcatel-Lucent, China, Email: wujs@ieee.org. Tao Chen is with VTT Technical Research Centre of Finland, Finland, Email: Tao.Chen@vtt.fi.

6

Green Communications for Carbon Emission Reductions: Architectures and Standards

Charles Despins

Prompt inc. and École de Technologie Supérieure, Canada

Fabrice Labeau

McGill University, Canada

Richard Labelle

The Aylmer Group, Canada

Mohamed Chériet

École de Technologie Supérieure, Canada

Alberto Leon-Garcia

University of Toronto, Canada

Omar Cherkaoui

Université du Québec à Montréal, Canada

CONTENTS

6.1 Introduction

As climate change has emerged as a key global challenge over the last decade, the international community is now increasingly unified in a call to action. It is a challenge that not only jeopardizes the sustainability of our planet; it poses significant, long-term threats to the global economy. According to former UK Government and World Bank Chief Economist Lord Stern [1]: "If no action is taken, the overall costs and risks of climate change will be equivalent to losing at least 5% of global gross domestic product (GDP) each year. Not acting now would incur a wider range of risks and impacts and the estimates of damage could rise to 20% of global GDP or more. The Information and Communications Technologies (ICT) industry is estimated to contribute 2% to 3% of global Greenhouse Gases (GHG) emissions, a share that is quickly rising.

More significantly however, the ICT industry also has the potential to, by 2020, reduce current global GHG emissions by 15% (7.8 Gt CO2e) and generate worldwide energy cost savings of 800 billion (US) dollars annually [1]. In 2008 alone, Green ICT, Green communications and related applications helped to eliminate 376 million metric tons of carbon 800 billion (US) dollars annually. By 2013, this volume could exceed 1.2 billion metric tons. This creates a huge potential for the ICT industry, academia, and governments to demonstrate leadership on the climate change issue while at the same time exploiting this largely untapped economic opportunity. As the ICT sector's products and services are generally productivity and economic development tools (e.g. through the availability of broadband), and as well in view of the mitigated outcomes of recent global Climate Change Summits, Green ICT therefore offers an unparalleled potential to reconcile sustainable development and economic development.

Fully exploiting this opportunity will require a holistic combination of public policy frameworks and technology developments to ensure "Greening of ICT" infrastructures and also "Greening through ICT" various sectors of economic and human activity i.e. the so-called "enabling" effect such as through intelligent transport systems, smart buildings, etc. In order to develop a carbon-focused Green communications research perspective, the relationship between energy efficiency and carbon emission reductions must first be understood. Most Green communications research is generally geared to improving energy efficiency, a theme at which both the academic and industrial communications communities excel. The current broadband data explosion, notably on wireless platforms, is so strong that the industry is actively looking at means, such as improved energy efficiency, to reduce network operating costs.

However, while energy efficiency is an important target in terms of cost savings and can in some cases directly contribute to reducing GHG emissions, it can also, in many other situations, have little or no impact on reducing these GHG emissions. To understand the relationship between energy efficiency and GHG emissions, it is important to note that GHG emissions from ICT are indirect in that they don't directly produce GHG gases through their operation but rather through the generation of electricity needed to power them. In GHG parlance, they are referred to as Scope 2 emissions. Primary emissions from a power plant are classified as Scope 1. In most of the world, over 50% of this electricity comes from coal-powered electrical generating stations, the primary source of the ICT GHG emissions. As most ICT infrastructures operate 24 hours a day, they thus have very little impact on a utility's peak power consumption. Although there are a number of energy efficiency projects underway to reduce power consumption of PCs and consumer devices when they are not in use, this will have little impact on their base load power usage. Base load power is primarily produced by coal, as the cost of power from these plants is very low and as they are most efficient at full power operation.

More expensive gas and hydro-electricity are generally used to provide peak power only. Therefore, although many power utilities may claim a power mix which is relatively balanced between coal, gas, nuclear and hydro, the ICT sector consumes a disproportionate percentage of coal power generated base load electricity. Most power utilities will also not commit to reducing the power from their coal plants commensurate with reductions in energy consumption. Instead, they will resell the power to other users as it is very costly not to have a coal plant operating at 100% capacity. Alternatively, they may reduce operation of the more expensive gas plants and/or reduce imported hydroelectric power or the use of solar and wind-based energy. In such cases, energy efficiency gains do not yield any reductions in carbon emissions. The link between energy efficiency and GHG emission reductions is therefore far from direct. In this sense, the important target is not "how much" but "what type" of energy is consumed by ICT networks and applications. While improving the energy efficiency of ICT products and services remains an important target without which carbon emissions will increase, the concepts described in this chapter therefore emphasize the importance of an ultimate focus on GHG emission reductions for network and access design.

This chapter thus focuses on network architectures and technologies designed with energy efficiency as an important consideration but with the prime purpose of reducing carbon emissions. While the enabling "Greening through ICT" effect is an important opportunity, this chapter deals with greening (in the carbon-neutral sense) ICT infrastructure. Various emerging testbeds are described to validate these concepts. Finally, the chapter concludes with a description of global standards to properly assess the carbon abatement potential of various network architectures and enabling technologies, an essential tool for the ICT industry to fully participate in the future and emerging carbon trading economy.

6.2   Network Architectures and Technologies to Reduce Carbon Emissions

6.2.1   Networks and Protocols

Reducing GHG emissions for communication networks involves dealing with the trade-offs between performance, energy savings and QoS. Migrating to optical networks is the long-term solution to reduce GHG emissions, as in general optical communications devices consume less energy than electronic ones. Network devices can also be designed with features that create an opportunity for energy-saving operation such as turning off network interfaces and throttling of processors. Network protocols could also be optimized, or even be redeveloped in a way that enhances the energy-efficient operation of the network elements. However, an essential part of the solution is to transfer transport services from energy inefficient to "greener" devices. This can be achieved by delegating forwarding operations to routers which are located at green powered data centers. Network devices will simply perform signaling operations which need low bandwidth, therefore reducing the total amount of GHG emissions of the communication network. In such an network model, customer premise flows feed into edge routers which in turn are aggregated in core switches that are interconnected to other core switches including those attached to large data centers. A green network architecture also requires the relocation of physical infrastructure to sites with renewable energy. This is possible through new traffic grooming and virtualization techniques.

The optimum configuration of green networks can be formulated as an optimization of the placement and sizing of the resources to meet traffic demand forecasts. In [2], this model was used to investigate the design of IP over WDM networks to minimize overall energy consumption. The study found that routers account for approximately 90% of the overall power consumed in a range of network scenarios. Consequently, the optimum designs essentially minimize the number of router ports and maximize the use of optical bypass to minimize the number of router hops. In [3], future packet and circuit switch technologies were compared and it was found that the relative power consumption performance between electronic routers and optical circuit-switches will persist. Clearly placement of router functionality at renewable energy sites will be an important element in the design of green networks.

Consider now an architecture where router forwarding functionality is concentrated at sites with renewable energy. As before edge router flows feed into core routers; however core router ports are now virtualized and the edge router flows are aggregated and transported using optical paths to remote sites where the physical resources that provide router functionality are located. Existing large-scale router designs may initially provide router functionality, so the aggregation will lead to better port fill levels. We note however that entirely new physical system designs to provide virtualized router functionality also become possible. We speculate that the economy of scale benefits of cloud computing may also accrue to these green networks.

The lack of sustained availability of wind and solar power require the development of monitoring, prediction, and reallocation mechanisms to ensure uninterrupted operation of green networks. The use of virtualization enables a separation between the IP and underlying substrate and facilitates the replacement of the underlying physical resources with the availability of renewable energy. New network topology design and fault tolerance methodologies will be thus required to ensure that green networks provided uninterrupted service.

6.2.2 Integrated Optical-Wireless Access

Wireless access is rightly considered as a major power consumer. The development of competing wireless access systems has been done to increase capacity and coverage at reduced cost, but the actual GHG impact of these deployments has never been considered in their design. Wireless access requires the presence of an infrastructure, which is mostly transparent to users, but plays a major role in determining the energy consumption profile of the overall access system, and as a consequence its carbon footprint. Beyond the environmental concerns raised by the current wireless access infrastructure, there are more down-to-earth OpEx issues that operators are currently facing in terms of powering this infrastructure.

Among potential avenues, virtual telco access clouds are seen as a potential way to aggressively re-use the existing infrastructure by sharing it among operators [4] as well as to leverage virtualization concepts to reduce carbon emissions; beyond the issues of pricing and security that need to be solved, it is also unclear at this point whether such a solution would yield any net benefits, as traffic peaks experienced by most operators tend to be correlated spatially and in time. This situation would most surely need to be combined with powering down of some portions of the infrastructure in low demand locations and periods.

The realization of virtual access clouds needs to rely on the virtualization of resources that currently eminently physical: base stations in wireless access form, together with the backbone connections linking them to the operator, constitute the main building blocks of an infrastructure that is a major power consumer. From an architectural standpoint, the combination of centralized processing and distributed radiation could lead to large energy efficiency gains.

The development of so-called remote radio heads (RRH), and feeding through Radio over Fibre [5, 6] are the critical enablers to develop such distributed clouds: this architectural revolution relies on a solid and reliable fast backbone network, and allows for the processing that is normally done in the base station to be carried on off-site, while keeping the base station as simple and power efficient as possible. Beyond obvious advantages in terms of maintenance, this architecture also promises to deliver transmission advantages, as processing can now be done based on data collected at different points in the networks, throughout one or several cells. Coordinated Multi Point (CoMP) operation in LTE [7] is just a simple example of the potential architecture shifts: inter-cell centralized beamforming or intra-cell distributed arrays could yield large energy efficiency gains. From a virtualization perspective, this offloading of computational services "to the cloud" also has obvious consolidation advantages, including relaxed availability specifications on individual components [6] due to the inherent redundancy of a cloud-based approach.

For cases where such an aggressively distributed architecture is not feasible, it is also important to reconsider base station efficiency as a way to reduce the carbon footprint of the overall access network, as, according to some operator data, base stations consume on the order of 55% of the total wireless access energy. The distribution of computational power and energy consumption in different infrastructure sites also opens up opportunities for reducing the system's carbon footprint through clean power fed to certain parts of the infrastructure. Along these lines, an architecture using off-grid base stations would also lead to large carbon footprint gains, but would face the common issues of fluctuations of renewable energy (solar/wind) during operation. Through the scheduling of sleep modes, the development of more energy efficient BS chips (analog or digital), and the deployment of energy efficient algorithms for baseband operation, one can hope to dramatically increase the amount of power consumed by a BS that is actually radiated as useful signal (Currently, it is estimated that only 5 to 10% of the energy consumed in a base station actually accounts for useful radiated energy).

The use of cognition (i.e., the capability of a radio system to sense its radio environment and adapt its transmission in some respect to this environment) has been widely touted as a path to "green communications" [8]. Initially developed to enable communication by secondary users in frequency bands occupied intermittently by primary, licensed users, cognitive radio has evolved into an overall approach to communications which is simply intelligently adaptive: sensing the radio environment, choosing a proper action to take, transmitting while continuing to monitor the environment are steps that are followed by cognitive radios to achieve this intelligence. Initial objectives for action would be to transmit as much information as possible while not impeding primary users; in a Green Radio environment, the objectives and criteria for decision and action need to incorporate local or network-wide energy considerations as well as the concomitant GHG footprint metrics; once this is done, the arsenal of tools developed in cognitive radio over the last decade [9, 10] can be deployed to the benefit of greener radio access.

6.2.3   Data Centers

Server consolidation is one of the first steps to achieve energy efficiency in data center. As many enterprise servers do not maximize the utilization of available server resources all of the time, co-locating applications allows for a reduction in the total number of physical servers, minimizes server idle time as well as the total data center space requirements. Consolidation reduces the total power consumed by the applications because existing servers are not energy proportional, i.e., a significant amount of power is consumed even at idle time. Though server features like voltage and frequency scaling modify this curve, there is still substantial power drawn at idle or low utilization. Consolidation thus provides an opportunity to reduce the overall power consumed by operating the servers. For example, if two identical servers each utilizing, say 40% of the resources and drawing 80% of peak power were consolidated onto a single server, the consolidated server would be able to deliver identical performance at significantly less than the 160% (80 + 80) of the peak power. However, the key to effective consolidation is to estimate the resource requirements of applications and to utilize these estimates along with the power profile of the physical servers to determine the consolidation strategy that can provide the best space-power benefits.

There can be three categories of server consolidation, namely static, semi-static and dynamic consolidation. In static consolidation, applications are placed on physical servers for a long time period (e.g. months, years), and not migrated continuously in reaction to load changes. Semi-static refers to the mode of consolidating these applications on a daily or weekly basis. On the other hand, dynamic consolidation spans a couple of hours and requires a runtime placement manager to migrate virtual machines automatically in response to workload variations. Most of data center operators are using static consolidation with supports from vendors [11].

The next step of server consolidation is virtualization. Indeed, the times when a server is made of a single CPU, a single Ethernet card (with a unique MAC address) and a single operating system are long gone. Today servers are built with multiple sockets, each socket contains multiple cores and each core is capable of running multiple threads simultaneously. These servers have also multiple network interfaces to connect to different networks. These network interfaces are now evolving to support server virtualization. Server virtualization is a technique that allows use of all the available cores without modifying/rewriting the applications. VMware ESX, XEN, KVM and Microsoft Hyper-V are widely used virtualization products in the market. They allow multiple virtual machines (VMs) to run on a single physical server through the coordination of a hypervisor.

A VM is an instantiation of a logical server that behaves as a standalone server, but it shares the hardware resources with the other VMs. VM to VM communication is achieved using "software switch" module, which creates a different model compared to standalone servers. Virtualization-based data center is so-called cloud where VMs are basic units to run applications. The GreenStar Network project, to be described later in this chapter, is such an example. A data center includes physical servers linked by Ethernet switches. A logical server running in a VM connects to the software switch module in the hypervisor and this in turn connects to one or more Ethernet switches. This creates a virtual network to provide cloud services.

Virtual machines (VMs) can be moved, copied, created and deleted according to management policies. Although energy efficiency can be achieved by hardware consolidation, server hibernating or hardware design optimization, the degree of GHG reduction based on an energy efficient strategy in data centers is limited. Virtualization allows services not to be managed within a data center site, and to be moved to other sites where they can operate in the greenest possible way. Not only must the aspect of load be considered, but also the data center cooling requirements generated by a service have to be measured before migrating operations. Such management is realized transparently for end users.

Generally, the number of VMs is much larger than the number of physical servers leading to scalability and manageability concerns. In addition, virtualization software can also move the VMs between different physical servers making the management more challenging. In order to manage a large set of VMs, a management by delegation approach is chosen, where each VM is managed by an agent (so called engine) and management tasks will be assigned to engines in a dynamic manner. The main tasks of an engine are to start, stop VM and report VM states, including power-consuming measurements. Engines are designed as intelligent entities which are able to run user applications and report VM errors. Generally, an engine is associated with a host's hypervisor and corresponding VMs.

Power management and its integration with cloud management solutions have been proposed in recent research. In [12], a Distributed Resource Scheduler is proposed to perform automated load balancing in response to CPU and memory pressure. Similarly, in [13], a dynamic consolidation and redistribution of VMs is proposed for managing performance and SL violations. In [14], authors propose Entropy, which uses constraint programming to determine a globally optimal solution for VM scheduling in contrast to the first-fit decreasing heuristic used by [13, 15], which can result in globally sub-optimal placement of VMs. In [16], authors propose a scheduling algorithm for VM allocation taking into account power consumption of VMs and the hosts. In [17], optimal reallocation of VMs in order to reduce their carbon footprint and energy consumption over a virtual cloud is studied using a heuristic algorithm.

Greener data centers based on server virtualization will require the development of new network services and new protocols in order to support new features such as resource pooling and virtualization, resource elasticity and flexibility, self-management, multi-tenancy, security and privacy, performance and reliability, agility and responsiveness, integration with enterprise legacy systems, as well as portability and standard compliance. A typical traditional network architecture for a data center is a hierarchy organized from a layer of servers in racks at the bottom to a layer of core routers at the top with a Top of Rack (ToR) switch to aggregate and access routers. Each TOR use a 1 Gig Ethernet interface to connect a rack of more than 60 physical servers in which we can have more than 10 VM machines. In those cases, we have to support the connectivity between more than 10 thousands servers in one data center. This virtualization adds huge complexity to the connectivity by an explosion of the number of connections between servers. Unfortunately, this conventional design suffers from multiple fundamental limitations: configuration (address allocation), isolation, load balancing, agility and cost effectiveness. The migration of VM between servers requires a high bandwidth and low latency and on-demand connection between servers. Many research projects try to circumvent this inflexibility and insufficient bisection bandwidth by exploring new designs such as VL2 [18], PortLand [19], DCell [20], BCube [21]. Most of the research has focused on providing a more energy-efficient communication channel by optimizing the physical characteristics of the link through energy consumption proportional to the number of handled packets: Folded-Clos [22], Flattened Butterfly [23], ElasticTree [24], MPI run-time system [25], Hyper-X network [26], ALR [27].

Major data center owners have also proposed to use the Openflow protocol [28] as new way to reduce network data center complexity and power consumption. The Openflow specification uses an external controller which sends routing decisions into the network. The Openflow approach gives a high level of flexibility for the design of the data center network in order to provide a reliable, scalable, cost-effective computing infrastructure for massive Internet services. Based on a high level of programmability and isolation, Openflow reduces energy consumption and operational costs while improving efficiency.

The Netvirt Project [29] has developed a stackable 4X100 G switch mainly dedicated for huge data centers [30]. A large scale data center with 10,000 servers can be connected with only eight such switches. Each switch supports 32 ports of 10 Gbps. By using Openflow, this switch offers a high level of programmability where aggregation functions and routing can easily be inserted on the same board. The concept also ensures full isolation of each domain.

6.3 Testbeds

6.3.1   Greenstar

The GreenStar Network (GSN) [31] is a green cyber-infrastructure pilot testbed to share infrastructure and maximize lower-cost power with "follow the wind, follow the sun" networks. Several zero-carbon energy sites have been selected for the location of network and computing resources in a testbed that can be managed using Infrastructure as a Service (IaaS). This project includes international collaboration between Canada, the United States, Spain, Ireland, Australia, and China. GSN, an alliance of Canada's leading ICT companies, universities, and international partners, is led by the École de Technologie Supérieure (ÉTS) in Montreal. GSN will develop the world's first Internet network whose nodes will be powered entirely by hydroelectricity, wind, and solar energy and yet will provide the same reliability to users as the current Internet network does. The GreenStar Network will be applied to two green ICT service provision scenarios:

- The creation and enactment of a green IT protocol based on the utilization of a green data center;

- The development of management and technical policies that leverage virtualization mobility to facilitate use of renewable energy within the GSN.

The GSN Carbon Measurement Protocol will also be the world's first such protocol. Although the International Organization for Standardization (ISO) 14064 standard - upon which the protocol is based - is straightforward; its specialization to ICT will require synergistic solutions relating to power and performance measurement as well as network and system operation. The protocol's success will be measured by its uptake within industry, academia, and government. The GreenStar Network is shown in Figure 6.1 where the Canadian section has the largest deployment of six GSN nodes powered by sun, wind and hydroelectricity. It is connected to the European green nodes in Ireland (HEAnet), Iceland (NORDUnet), Spain (i2CAT), the Netherlands (SURFnet), and some other nodes in other parts of the world such as in China (WiCo), Egypt (Smart Village) and USA (ESNet). The idea behind the GSN project is that a carbon neutral network must consist of data centers built in proximity to clean power sources, and user applications will be moved to be executed in such data centers. Such a network must provide an ability to migrate entire virtual machines (routers and servers) to alternate data center locations with an underlying communication network supported by a high-speed optical layer having up to, e.g., 1,000 Gbps bandwidth capacity. Leveraging clean power sources means a greater dependency on power from renewable energy sources which however can be unreliable and unpredictable. While storage systems and large distributed electrical grids are part of the solution, ICT can play a critical role in developing smart solutions where products and services can adapt to this variable and unpredictable power availability. The GSN is the first example of such thinking, where applications are shuttled around a high-speed network to sites that have available power.

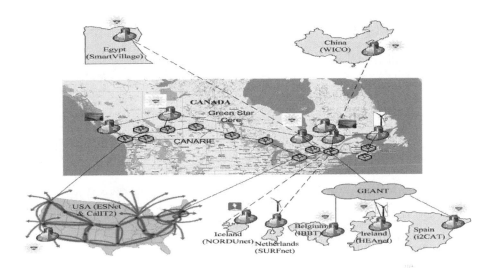

FIGURE 6.1 The GreenStar network

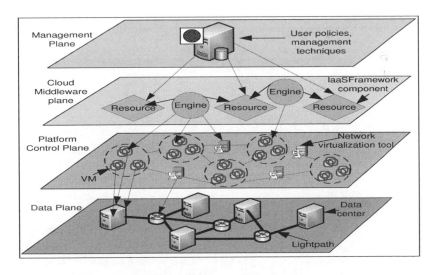

FIGURE 6.2 Layered GSN architecture.

Figure 6.2 illustrates the layered architecture of the GSN. The data plane layer includes massive physical resources, such as storage servers and application servers linked by controlled lightpaths. The platform control plane layer implements the platform-level services that provide cloud computing and networking capabilities to GSN services. The cloud middleware plane layer provides platform as service capabilities based on IaaS framework components. The top management plane layer focuses on applications by making use of services provided by the lower layers. In each GSN node, servers are installed in outdoor enclosures where the climate control is powered by green energy. Therefore, only data centers and core network equipment are considered to evaluate the GSN GHG emission reduction potential. As an example, experiments are performed in GSN data centers using a University of Calgary application, GeoChronos, which enables the Earth observation community to share data and scientific applications. The application runs on a 48-core multiprocessor server system. According to our estimation using [32], this system emits 30 tons of GHG annually, assuming it is powered by the Alberta fossil-based electrical grid. This number does not account for emissions of local switches and routers providing communications to GeoChronos. If the system is moved to a green powered server, 30 tons of GHG emissions may be saved yearly. However, such a migration consumes about 19.62 kW/day, which is equivalent to 1.5 tons of GHG emission annually, assuming that a nine-node core network is composed of IP routers powered entirely by fossil-based energy and that migrations are done twice a day. Consequently in this example, the GSN yields annual GHG emission savings of 28.5 tons. GHG emissions of a core network and data centers are also compared in Figure 6.3, assuming both are powered by fossil energy. For example, for a service with 520 Gbytes data volume, 1357 tons of GHG (the difference between the two curves) may thus be saved if the service is provided by the GSN.

6.3.2   NetVirt

The Netvirt project [29], is investigating how to build virtualized nodes. The virtualization is pushed at the core of the network. The virtualized node approach helps reduce the size of network infrastructure and hence energy consumption. The Netvirt project investigates the main challenges of network virtualization (NV), notably in the following areas: the discovery and "advertisement" of network resources; the creation and management of a sliced node across diverse resources; the extension of virtualization to wireless and optical links; the implementation of virtualization across diverse resources and across layers of a protocol stack; the management of slices; the service offering from the network infrastructure providers to the network providers; the set of capabilities enabled by infrastructure virtualization.

Underlying these research challenges, the Netvirt project is implementing a virtualized node, in order to validate and experiment on the new infrastructure.

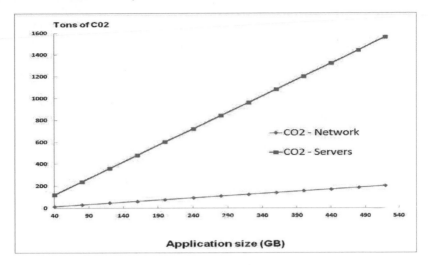

FIGURE 6.3 GHG emission comparison between 48-core data servers and a nine-node core network.

6.3.3 SAVI

The "Smart Applications on Virtual Infrastructures" (SAVI) project [33] is investigating the design of future application platforms and is therefore linked to the enabling effect of the Green ICT opportunity. An application platform consists of the software and infrastructure (personal devices, wireless and wired access networks, Internet, and computing clouds) that are involved in the delivery of applications. The openness of the Internet protocol allows for many application platforms to coexist. Distinct application platforms focused around particularly valuable services can lead to the emergence of distinct vibrant ecosystems. For example, the application platform around the iPhone has led to an explosion in applications that has triggered increased use of wireless access and Internet. It is noteworthy that the new capabilities in the iPhone (location, social network awareness, voice and image recognition) build on services provided by the computing cloud and not just the device itself.

SAVI is based on the premise that advances in all elements of the infrastructure (ubiquitous mobile and fixed very-high-bandwidth access, intelligent latency-sensitive high-bandwidth network-based services, cloud computing on local and remote data centers) will provide an opportunity to develop future application platforms that are cost effective and environmentally sustainable and that will support more powerful, intelligent and disruptive applications in the business, consumer, and public sectors. For example application platforms can support participatory sensing, where individuals and communities can systematically collect and analyze data for use in discovery of patterns and relationships, in contexts such as health and wellness, education, or community sustainable practices. Application platforms will also enable future smart infrastructures that will control the usage of resources and thus address major socioeconomic problems in areas such as energy, transportation, and green house gas emissions.

Consider the following scenario: a Major League team is in the final game of the 2020 World Series and they are mounting a comeback in the last inning. Reporters and pundits report on the game through live audio and multi-angle video streaming feeds that are followed by the public on their 10G iPads and Greenberries. Fans in the stadium and throughout the city share their reactions and commentary using social networking apps and contribute their own audio and video streams from personal devices to shared media repositories. Using speech and video recognition, audio and video streams are tagged in real time with timestamps, location, and by subject. Viewers anywhere create one or more customized views, e.g. from home plate, of the MVP, etc., using the Kaleidoscope app.

As the team wins the game, data traffic peaks. The "smart edge" activates the maximum number of pico-cells to provide capacity to carry and backhaul massive data streaming content wirelessly. Local virtual circuits route data to storage in the edge and beyond. The smart edge must ensure that local security personnel continue to have their own secure and dedicated channels for the purpose of law enforcement.

As the crowd moves to local restaurants and bars to celebrate, monitoring in the smart edge and core platforms detect and predict decreasing volumes of data traffic, and the power of pico-cells is dynamically and progressively turned off. Stadium traffic is reduced to fit in one optical wavelength and power is turned down on lasers and receivers on the local backhaul. In contrast, dormant pico-cells in restaurants and bars in the area are powered up and backhaul capacity to core data centers is increased to accommodate the demand as fans turn to Kaleidoscope on giant screens to relive their favorite moments.

The above scenario illustrates the agility and flexibility required in future application platforms. The SAVI Network will address the design of future application platforms built on flexible, versatile and evolvable infrastructure that can readily deploy, maintain, and retire the large-scale, possibly short-lived, distributed applications that will be typical in the future applications marketplace. The figure below shows the infrastructure elements of an application platform. Future users will typically access the application platform through a mobile device that connects to a ubiquitous very-high-bandwidth, integrated wireless/optical access network. The application platform provides connectivity to services that support the application of interest. Many services will be supported by massive-scale distant data centers located at sites of renewable energy. Other services will require low latency (e.g. alarms in smart grids, safety applications in transportation, monitoring in remote health) or processing of large volumes of local data (e.g. video capture in lecture rooms) provided by converged network and computing resources at the "smart edge" of the network, such as the premises of service providers.

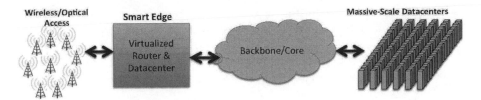

FIGURE 6.4 Scope of SAVI application platforms.

SAVI considers the computing cloud to include the smart edge, and it is investigating the interplay between the smart edge and remote data centers in the delivery of applications. SAVI also adopts the hypothesis that all computing and networking resources, including the integrated wireless/optical access, can be virtualized and managed using infrastructure- and platform-as-a-service principles. A key SAVI objective is to develop application-development methodologies and the intelligent and automated management and control systems that will transform the application platform into a powerful and highly flexible enabler of a broad range of applications.

SAVI is focused on five themes:

- Smart Applications. Investigate and design reusable frameworks for rapid design and deployment of software applications enabled by virtualized application platforms. SAVI is focusing on three classes of applications with high potential impact in future application platforms: a) Data-intensive applications that require high performance for users accessing data; b) Mobile applications, with support for a multitude of "typical" event types, including frequent location updating, P2P communications and services, and on-the-cloud-workflow management; and c) Media applications to support flexible, redundant caching, with transparent and efficient transcoding and streaming services.

- Resource Control and Management. Adaptive resource management to provide effective, efficient and reliable support for applications across all elements of the application platform (encompassing data centers and telecom infrastructure) while attaining economic, sustainability, and other high-level objectives. (In essence, computing and communications infrastructure become "programmable.")

- Smart Converged Edge. Investigate and design smart-edge converged infrastructure (future telecom infrastructure) that uses virtualization and cloud computing principles to dynamically support multiple network protocols and high-bandwidth, latency-sensitive applications. (The traditional infrastructure is completely replaced by an infrastructure that can simultaneously support multiple Internet protocols, each customized for a specific set of requirements.)

- Integrated Wireless Optical Access. Very high speed, energy-proportional adaptive virtual access networks based on dense small cells and dynamic optical backhaul. (Increasing the speed delivered to smart phones requires replacing conventional cellular architectures with energy-efficient small wireless cells with fiber optics connections to the core network.)

- SAVI Application-Platform Testbed. Build an application platform testbed that includes integrated wireless/optical access, smart converged edge, wide-area network connectivity and data centers; Use of the testbed to test future Internet architecture alternatives and smart applications; Develop and provide tools for an ecosystem of open source projects.

6.4   Carbon Standards for Communications Technologies

As described in the previous sections, green cyber-infrastructure research and the testbeds used to validate and demonstrate the results of such research will produce greener ICTs. However, the economic viability of these new technologies in the low carbon economy and carbon credit marketplace of the 21st century will rely on quantifying the "green impact" of these technologies, particularly when applied in non-ICT sectors where ICT has the greatest GHG emission reduction potential. As such, one of the key issues related to the role that ICTs play in abating climate change is the development of a methodology to accurately and scientifically measure the impact of ICTs on climate change specifically and on the environment in general.

The impact of ICTs themselves, as well as the beneficial impact of smart technologies such as smart motor systems, intelligent electrical systems including smart grids, smart buildings, smart transportation systems, smart cities and dematerialization need to be better understood. The role of ICTs in enabling the full use of renewable energy resources also has to be understood.

Understanding and measuring emission sources as well as demonstrating improvements in emission reductions over time, and finally achieving all this through science-based, standardized methods, remains one of the main challenges facing ICT manufacturers and users. The question of embodied carbon also constitutes a critical gap. Life cycle analysis methods are being developed for this purpose.

ISO standard 14064, which has been published by the International Organization for Standardization (ISO), specifies principles and requirements for determining GHG emissions or removals. However, ISO 14064 does not specify how these are to be measured in different sectors or industries. Recognizing the significance of the question, the International Telecommunication Union (ITU) and various other standard-setting organizations have been in the process of developing standard methods to measure ICT emissions as well as the contribution of ICTs to emission reductions. These standards will define how to measure energy consumption and GHG emissions, along with the appropriate metrics. Once the standards-based measures are universally accepted and applied to collect the data, developing a response can then be addressed using a combination of policy and technical tools. Standards are one such policy instrument.

At this time, the ITU and its many members as well as other standard-setting organization are still discussing the way forward and receiving comments from the many members (in the case of ITU these are telecommunication service providers and ICT manufacturers of one type or another) and other interested parties. Much of their work is focused on developing materials and methods to undertake environmental impact assessments (EIAs) that are scientific and gathering feedback not only from the many and various standard-setting organizations at the national as well as international levels, but from the ICT sector as well as businesses in general. Efforts by the World Business Council for Sustainable Development (WBCSD), the World Resources Institute, the International Telecommunication Union and others underscore this concern and the efforts ongoing to achieve this objective.

While the immediate impacts of ICT use on material utilization and energy consumption are relatively straightforward to determine, measuring the environmental benefits, i.e. more efficient energy and material use as a result of "smart" processes such as smart grids, buildings, logistics, transportation and "smart motor efficiencies", dematerialization, etc. that accrue from using ICTs have not been measured accurately or using standard methods and may not be measured easily, especially when one has to take into consideration the business environment and product life cycles.

The problem is compounded by the need to not only measure environmental impact arising from the use of ICTs, but from all phases of the life cycle of an ICT component or service. This requires to adoption of life cycle analysis (LCA) technologies. For ICTs, LCA methodologies are still being developed, and much of the methodological effort is focused on this aspect of EIA for ICT use in organizations and projects.

This has led some researchers to conclude that traditional environmental assessment approaches "are insufficient to accommodate the digital technology revolution and cannot accommodate the challenge of measuring the impacts of ICT on environmental sustainability" [34].

6.4.1 Power and Performance Measurement Challenges

Computers, peripherals and network equipment can exist in different operational states. Invariably, many of these components are always on if they are located in data centres, server rooms or are components of IP transmission networks. Technologies and standards are already being experimented with to move from "always on" devices to "always available" devices and states that can be powered up on demand only and which are otherwise maintained in an off power situation or in a low power consumption mode.

Estimating the power load of devices that are switched on intermittently and on demand presents a challenge when developing power ratings for these devices. The ITU and other standard-setting organizations are researching various options to come up with technical solutions which can limit energy consumption. These organizations are also researching standard energy consumption measurement methods and technologies [35].

While several standards exist that encourage energy conservation (Energy Star, 80 plus, etc.), a universally agreed upon method to measure and report energy consumption in ICT equipment has not yet been defined. The ITU is working on developing a common carbon calculator standard to compare the short and long term benefits of different telecommunications systems as well as developing a methodology to quantify embedded carbon along with several other initiatives.

However, the main priority and challenge is measuring the impact of ICTs on organizations, businesses and human settlements, i.e. cities. We know how to measure energy consumption and carbon emissions even though there has been no agreement on a standard way of doing this. However, we do not have a clear understanding of the energy and material balances related to ICT use by society. This is now becoming the priority for the Green ICT community.

6.4.2   Network and System Operation Challenges

According to research ongoing at the ITU on this question, and given the move away from energy inefficient legacy networks to more energy efficient NGNs including fiber optic transmission systems, it is suggested that telecommunication service providers and manufacturers take the following actions to further reduce energy consumption while improving the efficiency of energy use [36]:

- Maximizing network capacity.

- Improving IP systems so that they are more energy efficient

- Reducing the energy requirements of VoIP services and multimedia applications

- Lessening the number of electronic devices used in order to reduce emissions from the manufacturing and distribution of devices

- Lowering the overall consumption of energy in data centres and developing more energy efficient servers

- NGNs related technology should make use of more tolerant climatic range specification so that they can operate at higher temperatures that would limit the requirement for cooling while at the same time allowing ambient air cooling, a factor that is especially important in temperate and cold climate.

Challenges include developing a better understanding of the energy consumption and related GHG emissions of fixed line vs. wireless/mobile networks in order to further investigate the following questions:

- The extent to which fiber optic technology can provide additional speed and increased range while reducing power consumption

- Research on comparing the power consumption of fixed and mobile networks given that the applications and usage behaviour are different

- The information from this research will inform network providers of impact of fixed vs. mobile networks.

6.4.3   The Need for Carbon Standards Adapted to Communications Technologies

Several clean energy technologies are expected to have a significant impact in reducing GHG emissions. These include smart energy systems and specifically smart grids, smart buildings, smart logistics and transportation system, smart motor systems in general and dematerialization technologies. These clean technologies or smart technologies depend on ICTs. One of the most important technologies will be wireless sensor networks (WSNs). Estimates of the true GHG abatement potential of these technologies is important in order to identify research and development priorities and to make best use of limited investment funds that are now increasingly finding their way into cleantech startups and established companies.

Knowing the environmental impact of ICTs and related appliances and smart technologies is important to consumers and increasingly, to environmental and consumer protection agencies and to power companies and utilities. One of the main advantages of ICTs is that they can be used to also report on GHG emissions and energy use. Knowledge of the energy consumption and GHG emission patterns of consumers can have a very important feedback effect on the consumers and can lead to changes in behaviour that can dramatically reduce emissions. Sensor networks are important components of many of clean energy technologies and they can be used for reporting on energy use as well as be used for controlling energy use. Research on the telecommunications properties of these technologies is ongoing, but increasingly, research on their power consumption and on their potential to abate climate change is also taking centre stage. As energy prices increase and as efforts to reduce and conserve energy grow in the face of growing energy demand and ever increasing prices, these technologies and the standards that will underpin their use are becoming essential.

6.4.4   Standards Development in the International Telecommunications Union (ITU)

The ITU Study Group 5 is preparing recommendations for methodologies to deal with among other things, the environmental impact of ICT [37] of goods, networks and services; in organizations; of projects in cities.

The recommendations that are being developed will deal with both the GHG emissions associated with the use of ICTs and how these can be minimized and at the same time the abatement potential of ICTs and how these can be maximized. The methodology will seek to quantify both aspects. A recommendation providing an overview of the issues and documenting the principles for implementing these recommendations has already been published.

Results from this work are expected in September 2011 in some cases ("Goods, networks and services" as well as "Organizations") and later in 2012 in the other cases. The recommendation on ICT goods, networks and services will focus on the one hand on life cycle analysis and will describe steps to follow in order to assess the environmental impacts of ICTs over their entire life cycle. The recommendation specifies the steps that are to be taken and it is expected that the recommendation will receive consent in Sep. 2011 and then sent to ITU for approval. While the recommendation calls for specific actions, it is subject to improvement and a process has been put into place by the ITU and its members for this purpose.

The second recommendation on ICTs in organizations seeks to allow organizations to assess their GHG emissions and their energy consumption [38]. The recommendation will deal with assessing the life cycle environmental impact of ICT Goods, Networks and Services used by an organization ("ICT in organizations"), assessing the environmental impact of an ICT organization ("ICT organizations"), interpreting these impacts and reporting these impacts in a fair and transparent way.

The recommendation includes information about defining the parts of the organization that are to be included in the measurement of emissions, the identification of emission sources that are to be documented in the assessment using an life cycle analysis approach, emissions related not only to production activities but also to all operational activities including emissions of staff generated in their daily commute and business travel as well as emissions from freight and transport operations, emissions resulting from decommissioning ICT equipment, etc. This means that both direct and indirect emissions are to be accounted for. Calculations are to be made for a base year and this is to be used for comparative purposes to help identify priority concerns and measure progress. The recommendation also includes a methodology for collecting data and calculating emissions and related factors and indicators, identifying uncertainties and reporting.

ITU is not the only organization dealing with these issues. The International Electrotechnical Commission (IEC) has a working group looking at two issues. The "Quantification methodology of greenhouse gas emissions for electrical and electronic products and systems" and a "Quantification Methodology of greenhouse gas emission reductions for electrical and electronic products and systems from the project baseline".

Several bodies are working together to address the issue of measuring and reporting on GHG emissions from a life cycle analysis perspective , including ITU, IEC, the Carbon trust working with the British Standards Institute in the United Kingdom , the International Organization for Standardization (ISO), the Japanese Industrial Standards (JIS) and the World Business Council for Sustainable Development (WBCSD) as well as the World Resources Institute (WRI) .

Several other standards are being developed and much research is ongoing. As ICTs become ever more important as tools for abating climate change, their role will continue to draw attention and research efforts from not only standard-setting organizations but also from ICT companies and associations from around the world.

6.5   Conclusion

Green ICT, when focused not only on energy efficiency but also specifically on carbon emission mitigation, offers a huge opportunity for the ICT industry as well as an unparalleled tool to reconcile sustainable development and economic development. Concepts such as virtualized infrastructure and smart applications, enabling "Greening ICT" and "Greening through ICT", open up research avenues on both core network and access network design. The outputs of such research endeavors will also need to be coalesced with holistic public policy frameworks in order to take full advantage of the opportunity.

Bibliography

[1] Smart 2020, "Enabling the low-carbon economy in the information age," The Climate Group, London, U.K., 2008. [Online]. Available: http://www.smart2020.org

[2] G. Shen and R. Tucker, "Energy-Minimized Design for IP over WDM Networks," *IEEE Journal on Optical Communication Networks*, vol. 20, pp. 176–186, June 2009.

[3] S. Aleksic, "Analysis of Power Consumption in Future High-Capacity Network Nodes," *IEEE Journal on Optical Communication Networks*, vol. 20, pp. 245–258, August 2009.

[4] A. Reaz, V. Ramamurthi, and M. Tornatore, "Cloud-over-WOBAN (CoW): an Offloading-Enabled Access Network Design," in *Proceedings of 2011 IEEE International Conference on Communications (ICC)*, June 2010, pp. 1–5.

[5] F. H. P. Fitzek and M. D. K. (eds.), *Cooperation in Wireless Networks: Principles and Applications.* Springer, 2006.

[6] P. Bosch, A. Duminuco, F. Pianese, and T. Wood, "Telco clouds and virtual telco: Consolidation, convergence, and beyond," in *Proc. 2011 IFIP/IEEE International Symposium on Integrated Network Management (IM)*, May 2011, pp. 982–988.

[7] A. Ghosh, R. Ratasuk, B. Mondal, N. Mangalvedhe, and T. Thomas, "LTE-advanced: next-generation wireless broadband technology [Invited Paper]," *IEEE Wireless Communications*, vol. 17, no. 3, pp. 10–22, June 2010.

[8] G. Gür and F. Alagöz, "Green Wireless Communications via Cognitive Dimension: An Overview," *IEEE Network*, vol. 25, no. 2, p. 51, March-April 2011.

[9] J. Ma, G. Li, and B. Juang, "Signal Processing in Cognitive Radio," *Proceedings of the IEEE*, vol. 97, no. 5, pp. 805–823, May 2009.

[10] S. Haykin, D. Thomson, and J. Reed, "Spectrum sensing for cognitive radio," *Proceedings of the IEEE*, vol. 97, no. 5, pp. 849–877, May 2009.

[11] VMWare Guided Consolidation. [Online]. Available: http://www.vmware.com/products/vi/vc/features.html

[12] VMware Dynamic Resource Scheduler. [Online]. Available: http://www.vmware.com/products/vi/vc/drs.html

[13] N. Bobroff, A. Kochut, and K. Beaty, "Dynamic placement of virtual machines for managing sla violations," in *Proc. International Symposium on Integrated Network Management '07*, 2007.

[14] F. Hermenier, X. Lorca, J. Menaud, G. Muller, and J. Lawall, "Entropy: a consolidation manager for clusters," in *Proceedings of the 2009 ACM SIGPLAN/SIGOPS international conference on Virtual execution environments*, 2009, pp. 41–50.

[15] T. Wood, P. Shenoy, A. Venkataramani, and M. Yousif, "Black-box and gray-box strategies for virtual machine migration," in *Proc. of NSDI*, 2007.

[16] G. Dhiman, G. Marchetti, and T. Rosing, "vGreen: A system for energy efficient computing in virtualized environments," in *Proc. of ISLPED*, 2009, pp. 243–248.

[17] F. Moghaddam, M. Cheriet, and K. Nguyen, "Low Carbon Virtual Private Clouds," in *2011 IEEE International Conference on Cloud Computing (CLOUD)*, 2011, pp. 259–266.

[18] A. Greenberg, J. R. Hamilton, N. Jain, S. Kandula, C. Kim, P. Lahiri, D. A. Maltz, P. Patel, and S. Sengupta, "VL2: a scalable and flexible data center network," *In SIGCOMM '09: Proceedings of the ACM SIGCOMM 2009 Conference on Data Communication*, pp. 51–62, 2009.

[19] R. N. Mysore, A. Pamboris, N. Harrington, N. Huang, P. Miri, S. Radhakrishnan, V. Subramanya, and A. Vahdat, "Portland: a scalable fault-tolerant layer 2 data center network fabric," *ACM SIGCOMM Computer Communication Review*, vol. 39, no. 4, pp. 39–50, 2009.

[20] C. Guo, H. Wu, K. Tan, L. Shi, Y. Zhang, and S. Lu, "Dcell: a scalable and fault-tolerant network structure for data centers," *ACM SIGCOMM Computer Communication Review*, vol. 38, no. 4, pp. 75–86, August 2008.

[21] C. Guo, G. Lu, D. Li, H. Wu, X. Zhang, Y. Shi, C. Tian, Y. Zhang, and S. Lu, "BCube: a high performance, server-centric network architecture for modular data centers," *ACM SIGCOMM Computer Communication Review*, vol. 39, no. 4, pp. 63–74, August 2009.

[22] M. Al-Fares, A. Loukissas, and A. Vahdat, "A scalable, commodity data center network architecture," in *Proceedings of the ACM SIGCOMM 2008 Conference on Data communication*, 2008, pp. 63–74.

[23] J. Kim, W. Dally, and D. Abts, "Flattened butterfly: a cost-efficient topology for high-radix networks," in *Proc. the 34th Annual International Symposium on Computer Architecture*, 2007, pp. 126–137.

[24] B. Heller, S. Seetharaman, P. Mahadevan, Y. Yiakoumis, P. Sharma, S. Banerjee, and N. McKeown, "ElasticTree: Saving energy in data center networks," in *Proceedings of the 7th USENIX Symposium on Networked System Design and Implementation (NSDI)*, 2010, pp. 249–264.

[25] J. Li, L. Zhang, C. Lefurgy, R. Treumann, and W. E. Denzel, "Thrifty interconnection network for hpc systems," in *Proceedings of the 23rd International Conference on Supercomputing (ICS)*, 2009, pp. 505–506.

[26] J. H. Ahn, N. Binkert, A. Davis, M. McLaren, and R. S. Schreiber, "HyperX: topology, routing, and packaging of efficient large-scale networks," in *Proceedings of the Conference on High Performance Computing Networking, Storage and Analysis*, 2009, pp. 1–11.

[27] C. Gunaratne, K. Christensen, B. Nordman, and S. Suen, "Reducing the energy consumption of Ethernet with adaptive link rate (ALR)," *IEEE Trans. Computers*, pp. 448–461, 2008.

[28] M. Brunner, "Programmable Flow-based Networking with OpenFlow," in *Proc. of ETSI Future Network Technologies Workshop*, March 2008, pp. 448–461.

[29] O. E. Ferkouss, I. Snaiki, O. Mounaouar, H. Dahmouni, Y. Lemieux, R. B. Ali, and O. Cherkaoui, "A 100 Gig Network Processor Platform for OpenFlow," in *CNSM 2011*, Paris, France.

[30] O. E. Ferkouss, S. Correia, R. B. Ali, Y. Lemieux, M. Julien, M. Tatipamula, and O. Cherkaoui, "On the Flexibility of MPLS Application over an OpenFlow-enabled Network," in *Proc. IEEE GLOBECOMM 2011*, Houston, Texas, December 2011.

[31] www.greenstarnetwork.com.

[32] LivClean Carbon Offset Solution, "How is this Calculated?" http://www.livclean.ca.

[33] A. Leon-Garcia, "Smart Applications on Virtual Infrastructure," *Canada - EU Future Internet Workshop*, April 2011.

[34] L. Ly and H. R. Thomas, "A review of research on the environmental impact of e-business ad ICT," Environ Int. 2007, pp. 841–849, August 2007.

[35] ITU. 2009. Deliverable 2. Gap Analysis. Focus Group on ICTs and Climate Change (FG ICT&CC). Report to TSAG / Deliverables. [Online]. Available: http://www.itu.int/oth/T3307000004/en

[36] ITU. 2008. NGNs and energy efficiency. ITU-T Technology Watch Report 7. [Online]. Available: http://www.itu.int/oth/T2301000007/en

[37] J.-M. Canet, "Methodologies for assessment of environmental impacts of ICT," PowerPoint presentation, 12 slides. ITU-T Study Group "Environment and Climate Change". ITU Green Standards Week., Sep. 2011.

[38] G. Buty, "ICT in organizations current status of the ITU-T SG5 recommendation," PowerPoint presentation, 12 slides. ITU-T Study Group "Environment and Climate Change". ITU Green Standards Week., Sep. 2011.

Author contact information

Charles Despins is the president of Prompt-Quebec and adjunct professor at INRS-EMT, Email: cdespins@promptinc.org. Fabrice Labeau is an Associate Professor with Department of Electrical and Computer Engineering, McGill University, Canada. Alberto Leon-Garcia is a Professor with Department of Electrical and Computer Engineering, University of Toronto, 10 King's College Circle, Toronto, ON, Canada M5S 3G4, Email: alberto.leongarcia@utoronto.ca.

7

The Role of ICT in the Evolution Towards Smart Grids

Raffaele Bolla

DITEN-University of Genoa, Italy

Flavio Cucchietti

Telecom Italia, TILab., Turin, Italy

Matteo Repetto

CNIT - Research Unit of Genoa, Italy

CONTENTS

Electricity is a convenient and flexible way to provide power to a huge set of appliances. Effective and reliable supply of this kind of energy largely depends on its ubiquitous transmission and distribution. However, the architecture of current power grids was designed forty years ago or even more, when generation by few large oligopolistic utilities was mostly based on fossil fuels.

Today, the awareness of finite availability of fossil resources, the attention to environmental impact and pollution and the trend towards liberalization and deregulation are bringing new models for the energy market and the electrical grid, which are expected to become ever more intelligent and autonomic infrastructures, largely relying on ICT technologies. We try to fill the gap in knowledge about this evolution towards *smart grids*, in order to outline new challenges and opportunities for ICT.

7.1   Introduction

Modern society and economy largely depend upon the massive availability of energy for many different purposes: heating and cooling of buildings, household and work appliances, transportation, industries, and so on. Electricity is a convenient and flexible way to provide power to a huge set of appliances, and its importance is expected to grow with the introduction of electrical vehicles for transportation.

Generation of electrical energy takes place in most countries by means of large plants based on fossil fuels; electricity is then transported across the territory and distributed to the final users. The current grid architecture has been designed, deployed and operated since the mid of the 20th century according to this model and only allows for the presence of few large utilities (often owned by states). Power plants are efficient when they run at their designed generation capacity; however, the system load varies during the hours and the seasons. This leads to waste of fuel during low load conditions and to the use of inefficient backup generators to compensate transient peaks, thus affecting the cost of energy and bringing several issues concerning pollution; indeed, energy management has been traditionally oriented to keep constant the frequency and the voltage of the network, not to reduce fuel consumption in power plants.

As a matter of fact, the use of carbon-based technologies for energy generation will be no more sustainable in the future mainly because it is responsible for large part of greenhouse gas (GHG) emission [1] and because of the finite availability of fossil resources. Renewables are expected to gain more and more importance in the future, insomuch as there are already studies claiming they can meet all energy demand [2]; however, their exploitation requires delocalization of small/medium power plants, but current energy distribution networks have been designed for "unidirectional" flow of power from large plants to users and are not suitable for distributed generation. Further, availability of some renewable sources (wind, sun) is intermittent and only partially predictable and this would require massive energy storage and a large effort to control the final load by actively involving consumers to modify their consumption according to the current production.

As a consequence of this new generation model, electrical grids are expected to radically change their behavior. However, economic policies and market liberalization are pushing for lower energy prices, so utilities will be forced to reduce their costs; hence the above evolution cannot happen with large investment in new infrastructures. Instead, new ways of operating power systems have to be found, especially for what concerns transmission and distribution, bringing more intelligence into the grid to automate and control its functioning. This requires to overcome legacy proprietary and non-interoperable solutions (hardware and operating systems) usually found in the monopolistic and oligopolistic energy markets and to design open, common, flexible and scalable ICT infrastructures to effectively support the new models for "smart" grids.

To fully take advantage of its role in the upcoming energy systems, ICT should be aware of the challenges it is going to face and should be prepared to seize its opportunities by offering suitable products and solutions for the new market. Many stakeholders (standardization bodies, utilities, governments, research institutes) have already undertaken this road by setting up several working groups that aim at depicting use-cases, scenarios and architectures for smart grids. We review and summarize most of the work done so far, focusing on the specific aspects of communication and networking. Our work provides a general understanding about the evolution towards smart grids, current communication architectures for them and hot topics for the time being.

7.2   Legacy Energy Grids

An electrical system is composed of three main components [3]: production points, energy grids and consumer units. Traditionally, production points are large power plants that generate energy from various sources: coal, oil, hydroelectric, nuclear, renewables; they are usually located far away from urban areas and in convenient places for fuel supplying.

Energy grids are made of transmission, sub-transmission and distribution lines (see Fig. 7.1). Transmission lines carry high-voltage electrical energy all over the country to the main consumer zones; they build a meshed interconnection of power plants for redundancy and load sharing. They are also used for international links.

Sub-transmission lines carry high-voltage electricity from transmission networks to the main consumer centers. Distribution lines carry medium and low-voltage electricity to medium and small consumer centers. Sometimes sub-transmission networks are included in distribution networks. Different layouts can be used for these networks, e.g. open loop or radial, with different characteristics of reliability and redundancy.

The connection among the different grids is made in substations. Substations are composed by various electric devices: feeders, transformers, circuit breakers, capacitors, voltage regulators, switches. Their main tasks concern conversion among high, medium and low voltage, switching among different lines in case of faults and forking of electric lines.

Power grids are subjected to several disturbances: shunt faults, equipment failure with subsequent isolation, switching surges and lighting strikes, mechanical damages. Sometimes restoration may be undertaken by grid equipment, but often the breakdown may cause oscillations and voltage/frequency deviations that require the intervention of an Energy Management System (EMS), which monitors the grid state and controls its operation; the latter needs a proper ICT infrastructure made of communication equipment, control software, human interfaces.

An EMS relies on a Supervisory Control And Data Acquisition (SCADA) infrastructure; it is a single control center that gathers data from local controls and substations and displays them in a meaningful way to the control operators. The EMS processes SCADA data in various ways, including topology identification by using the dynamic data from switchgears, isolators and other connectors. These data can also be combined with current and voltage measurements to determine the system state. Notwithstanding ICT infrastructures for SCADA are continuously evolving to provide more reliability and control [4], they are essentially based on centralized solutions [5].

Legacy electricity grids are not ready to fulfill all the requirements for the evolution that is taking place. Key factors of this process are the necessity to reduce GHG emission and the new market models, which require the integra-

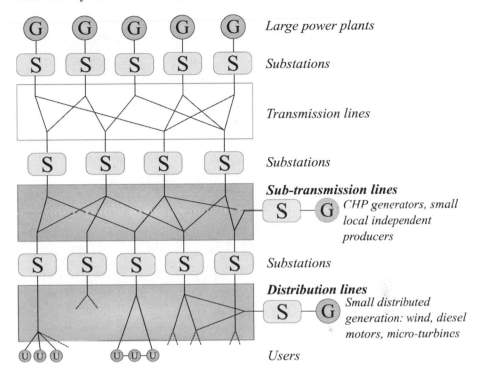

FIGURE 7.1 The layout of energy grids with distributed generation.

tion of microgeneration and renewables into the current system. The reduction in GHG emission will drive to the gradually replacement of fossil fuels, which are finite resources and could become very expensive in the near future, with renewables. Renewables are well-suited to be used in distributed generation, as of low investment and of their availability in large areas. The fight against GHG emission will also foster the widespread adoption of electric vehicles.

Traditionally, energy systems were characterized by three main roles [6]: producers, transmission operators and distributors; in many countries, there have been just few operators for long time, bringing to monopolistic or oligopolistic markets. The general conviction is that this situation leads to economic inefficiencies, discrimination and abuse of the dominant position; the regulatory trend is therefore towards market liberalization. This process allows the decentralization of energy production by small and micro generation plants with very low investment; they need to be connected to different parts of the grid, depending on their capacity (see on the right side of Figure 7.1).

The evolution will change the whole electricity supply chain, from generation, transmission and distribution to the customer and consumption side; Fig. 7.2 depicts a simplified view of the future grid ecosystem. First of all, the presence of a large number of electrical vehicles will bring higher loads

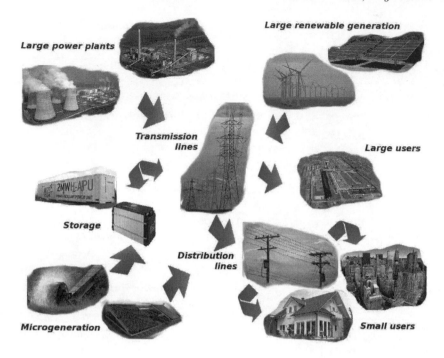

FIGURE 7.2 A simplified view of the future grid ecosystem.

in the network. Second, the current system based on unidirectional flow of energy from few bulk generators far from load will progressively see an increasing number of "prosumers", namely users that are both (micro)producer and consumer. This heterogeneity of sources will make difficult to provide high quality energy with stable voltage and frequency, as of difficulties in maintaining synchronization among generators. Finally, the inconstant and unpredictable availability of renewable power in different hours and seasons will require flexible dynamic loads and large storage capacity to find an optimal balance between availability and request of electrical energy; then, a long term challenge will be the optimization among different energy domains (electricity, gas, heat).

The electrical grid will have a key role in such evolution, as it shall address a number of challenges: high power loading, increasing distance between bulk generation and load, intermittent renewables, new loads (hybrid/electrical vehicles), utility unbundling, energy trading, regulation. It will be required to provide flexible, scalable, secure, reliable, resilient and efficient operation to the whole energy system. As of these requirements, much more intelligence than today is expected in the new electricity networks, thus leading to the definition of "smart grid".

7.3   The Concept of Smart Grid

Smart grids can be defined [7]

> an electricity network that can cost efficiently integrate the behavior and actions of all users connected to it - generators, consumers and those that do both - in order to ensure economically efficient, sustainable power system with low losses and high levels of quality and security of supply and safety.

Beyond this general concept, the actual essence of a smart grid varies from country to country, according to regional regulations, economy, grid status, primary energy resources. However, common objectives for smart grids can be identified as follows [8]:

- *Robustness*: reliability and resilience are essential to provide stability in network operation, avoiding wide-area breakouts. Smart grids must also assure service continuity by redundancy and self-healing in emergency situations as natural disasters, extreme weather conditions and intentional damaging.

- *Security and safety*: continuous supply of electricity is vital in many processes; if an intruder could control even a small set of devices, he could cause the collapse of the whole system (the electrical grid is prone to cascading failures when it operates at its maximum generation capacity). The massive introduction of ICT infrastructure would expose the grid to many remote security threats, exploiting vulnerabilities in intelligent electronic devices (IED), communication protocols and network topology [9]; data confidentiality, access control, authentication and integrity verification are key issues for smart grids. Further, the large amount of customer's data also rise huge concerns about privacy [10, 11].

- *Compatibility*: many heterogeneous power sources have to coexist that have very different characteristics: bulk generation from large power plants provides continuous energy but with low flexibility to follow dynamic load, whilst distributed generation often results in intermittent and unforeseeable energy provisioning.

- *Efficiency and balancing*: adaptive load and energy storage should match the current power production in order to avoid wasting; power dispersion should be limited by localizing production as close as possible to consumption points.

- *Optimization*: optimizing assets and grid management allows reducing costs and improving competitiveness.

- *Business*: diverse actors should trade energy in an open market with proper revenues for their investments.

- *Integration*: information and data for grid operation and management should conform to common, uniform and standard formats and representation models.

- *Sustainability*: energy should come from sources with long availability (if possible, renewable), low environmental impact and high safety for citizens.

7.4   Hot Topics for Smart Grids

The concept of smart grid is quite ambitious and implies a number of different issues; they can be condensed in a set of hot topics of interest for researchers, standardization bodies and regulatory frameworks. Further, to make the concept of smart grid more concrete, a huge number of scenarios and use cases have been proposed and continuously updated [12, 13].

7.4.1   Load Management

The generation from efficient power plants is quite static, while that from renewables is intermittent; however, the load can vary during the hours and this requires the use of dynamic but inefficient generators in traditional grid architectures. The principle of load management is quite simple and is based on adapting the current consumption in the grid to the available power. Usually, these policies foster the use of energy during off-peaks hours, as nighttime and weekends.

Load shedding is a one-sided decision to cut off electricity to some areas; it is the last resort utilities may use to prevent the total blackout of their grid whenever they cannot satisfy the current energy demand. This is a drastic intervention and should not be used regularly over time.

Demand side management (DSM) and demand response (DR) involve programs developed to influence the electricity usage patterns of customers [14, 15]. DSM pursues energy efficiency and conservation by customer education and deployment of low power appliances and systems (lighting retrofits, building automation upgrades, re-commissioning, HVAC improvements, variable frequency drives, etc.); DR are activities designed to change the amount and/or timing of customer's use of electricity, mainly in response to dynamic prices of financial incentives[1].

Energy efficiency and conservation is the utilities' legacy approach to load

---

[1]Unfortunately, there is not a common agreement on these terms: someone uses them in-

management; recently, the interest has been shifting to demand response to price signals. The main background for DR is that the operation of many appliances actually is not affected by small delays: this is especially true for air conditioning, refrigerators, washing machines, dishwashers. DR usually implies automatic mechanisms to adapt the load to the current power supply and can be implemented by two main mechanisms:

- *load curtailment*: that involves rewarding the customer for reducing their load during critical peak periods. Usually, this is done by incentive payment for direct load control of customers' appliances and by curtailable and interruptible rates;

- *dynamic pricing*: that means raising prices during peak periods and lowering them during off-peak periods. Several pricing scheme have been proposed: time of use pricing (TOU), critical peak pricing (CPP), extreme day pricing (EDP), extreme day CPP (ED-CPP), real time pricing (RTP). Real time prices should reflect the current availability of energy in the grid and may be delivered to users to shift their power consumption when energy is cheaper. Several communication channels could be used for this purpose: customer meters, in home displays, automated phone calls, e-mails, pagers, web sites; if the decision is taken by an automated system, it should conform to the customer's preferences and settings (i.e., his profile), in order to avoid the interruption of critical or vital tasks.

Whichever of incentives or dynamic pricing is used, the service can be undertaken by third parties in the market. Despite the concept could appear quite simple in theory, demand response requires to develop a comprehensive commercial and technical framework, mainly concerning equipment installed at the consumer premises: electrical appliances and thermal or electrical energy storage systems.

7.4.2 Wide Area Situation Awareness

Wide Area Situation Awareness (WASA) concerns monitoring and displaying in real time the grid status and its performance. This is used to control the behavior and performance of system components, in order to detect failures and to respond to problems before disruptions can arise. Most issues in this context have traditionally concerned the stability of frequency and voltage.

Legacy control systems are based on rigid and non-interoperable SCADA architectures. Typically these are vertical hierarchical architectures that reflect the classical structure of the power grid: measurements and data flow up from bottom (equipment and metering infrastructures) to higher levels (management centers), while control information is transmitted in the opposite direction. Other common issues of these infrastructures are the presence of

---

terchangeably, someone claims DR is a kind of DSM and someone gives two complementary meanings.

vendor-specific protocols and interfaces (gateways are usually required among different grid areas and for external applications), the monolithic (i.e., not modularized) layout of applications for energy management, the heterogeneity of functionalities, interface protocols and data models, the vulnerabilities due to the complexity and interdependency among systems.

Future grids are expected to integrate a virtually unlimited number of sensors and meters to support demand response and energy-aware applications; this will produce a huge amount of critical information for grid operation to be collected, exchanged and managed in a trustworthy way, requiring bi-directional flows among different layers. Network-centric architectures (e.g., Service-Oriented Architectures (SOA) [16]) and resource aggregation and virtualization (e.g., virtual power plants [17], cell-based virtual utilities [18] and microgrids [19]) are among possible solutions proposed to achieve flexibility and scalability in the grid control and monitoring infrastructure. Communication will likely be built upon standard communication technologies based on wired and wireless local area networks in local domains (home, factory, plant, substation) and wide area networks to provide global connectivity.

7.4.3 Distributed Generation

Smart grids are expected to include a large number of small and micro generation systems, typically owned by small/medium enterprises or private individuals. They usually exploit renewable sources (sun, wind, water) or co-generation (combined heat and power), thus reducing GHG emissions. The main challenge with distributed generation is energy accuracy and quality, which means voltage stability and frequency synchronization. This will likely imply plenty of information to be exchanged in real time, expecting for high bandwidth and low latency inside the communication network.

A second challenge is that the amount of energy produced cannot be accurately predicted; it depends on meteorological conditions for renewables and the industrial processes for co-generation. Yet timely and reliable communication is needed to balance the variability of energy supply with demand response, load shedding and, as a last resort, back-up generators.

7.4.4 Microgrids

The availability of distributed resources gives a real chance to have local generation and consumption of electricity. A group of electrical sources and loads can be organized in a semi-autonomous system called microgrid. It is usually connected and synchronous to the main electrical grid, but if the latter fails it can be isolated and keep on operating.

Microgrids can reduce the bulk of energy transferred by long transmission lines and bring more efficiency due to less power losses. They also increase the local reliability of power systems and provide better efficiency by consuming local resources (wind, water, sun, biomasses).

7.4.5 Energy Storage

Energy storage involves the conversion of electricity into some other form of energy: chemical (batteries, hydrogen), potential (water), kinetic (flywheels), elastic (air pressure), thermal (water, molten salt), magnetic (superconducting coils). Electricity may also be stored into supercapacitors, but their capacity is limited with respect to the solutions above.

Until now, few installations are present in distribution systems, in bulk power plants and at large utilities that need uninterruptible power supply (e.g., telecoms). The massive integration of unpredictable and intermittent renewables, as well as the push for more efficient power generation from fossil fuel, will require the installation of huge distributed and widespread storage capacity. Grid operators will boost the deployment of such infrastructure by offering real time pricing to their customers, thus enabling them to get revenues by purchasing energy when it is cheaper and by reselling it when this operation becomes economically convenient.

Energy storage can also be used to improve the quality of energy: it provides fast voltage sag correction by responding rapidly and automatically during short to medium duration disturbances. This will avoid customer outages and ameliorate power quality problems.

Finally, it is worth noting the relevance plug-in electric vehicle (PEV) are expected to have for energy storage. They will be equipped with high-capacity batteries and can be used as "mobile" energy storage.

7.4.6 AMI Systems

Accounting of energy consumption has always been a big deal for utilities, whose meters are installed at customers' premises. The availability of new communication technologies has allowed *Automatic Meter Reading* (AMR), that is a technology for automatically retrieving of consumption, diagnostic and status data from meters (typically for water and energy). Touch technologies and short radio communications are used for "walk-by" meter reading; wireless networks and power line communications enable to send data directly to accounting servers. The Telegestore[2] system by Enel Energia in Italy is the first AMR system widely available inside a country.

The interaction between utilities and customers for demand response, dynamic pricing, electric storage, PEV management and other services in smart grids (e.g., voltage, frequency and watt control) will require a further evolution, yielding to the deployment of Advanced Metering Infrastructures (AMI). It will build a two way communication channel towards customers to collect data and to carry information; this will allow (near) real time response to network conditions through dynamic pricing and load shedding. Some vendors

---

[2]http://www.enel.com/en-GB/innovation/project_technology/smart_grids/smart_meter.aspx.

are already delivering AMI architectures to collect data from a large number of different sites, such as customer's home and low-voltage stations [20].

7.4.7   Electric Vehicles

Transportation consumes about 20% of the global energy supply (80% of which comes from fossil fuels) and is responsible of about 14% of GHG emissions [1]. Internal combustion engines have very low efficiency for energy conversion with respects to wind, water and gas turbines and electric motors[3]. Although the use of electric motors in vehicles dates back to the beginning of the 20th century, they only won the attention of the market recently. Hybrid Electric Vehicles (HEVs) combine an internal combustion engine with an electric motor; the latter is powered by batteries, which in turn are charged by the combustion engine and sometimes by regeneration systems as brakes and suspensions. Plug-in Electric Vehicles (PEVs) are equipped with electric motors and batteries, which are charged by an external electric power source (typically a standard wall socket). Finally, Plug-in Hybrid Electric Vehicles (PHEVs) have an internal combustion engine, an electric motor, batteries and an external plug for charging.

Widespread adoption of PEVs and PHEVs will bring additional and critical load to the grid. However, due to their large capacity for energy storage, they will have a high potential for power injection into the same grid. Large-scale integration of electrical vehicles will require intelligence programs to manage charging/discharging operations to meet the dynamic network load; this includes turning off or reducing the charge rate (perhaps according to the current level of charge) and even getting back power during critical grid events. Obviously, owners of vehicles should be willing to grant energy operators these possibilities; again, economic incentives and real time prices are feasible ways to cut customers in on the energy deal.

Another typical matter of electric vehicles concerns mobility. Customers may plug their P(H)EV at their home or other places; thus proper public charging infrastructures will have to be deployed that allow flexible and secure transactions, by also taking into account the existence of custom tariffs and the possibility to present the user with a single and unified energy bill. Communication issues concerning identification and security arise for this kind of "mobile nodes" that access the network from different locations. Finally, mobility of electrical vehicles will also bring the potential of having huge energy availability close to the main load, thus avoiding to transfer unnecessary power over the grid.

---

[3]A simple comparison table can be found on Wikipedia (URL: http://en.wikipedia.org/wiki/Energy_conversion_efficiency). Please consider those values as indicative; thorough data should be collected from more specific technical description of each technology.

7.4.8   Energy Gateways

Energy gateways represent the primary aggregators of information coming from private sub-domains (homes, buildings, factories). They monitor and control energy resources inside the sub-domain: plug-in electric vehicle, generation systems, appliances. They should work with smart devices with electric metering and communication capabilities as well as with legacy devices without such capabilities.

Their functions include load and charging of local equipment, profile management, storage of energy usage data, information retrieval and sharing to/from external sources; they may participate in higher-layer aggregations (microgrids) or may directly interact with utilities and third-party service providers.

Energy-effective and interoperable integration of local devices will be the major challenge; in addition, reliable communication to the rest of the grid will be required. The viable road for an energy-effective solution will likely be the use of low-power hardware platforms with multiple network interfaces; modular and versatile software suites will enable the development of custom services and applications [21].

7.4.9   Man-Machine Interfaces

Despite the high level of smartness that will be bring into future energy grids, human control will still be necessary to check that network behavior be congruent to what expected. Given the size and complexity of the system, this task will require suitable abstraction, representation and interaction with status and operating information, as well as proper control tools to control and modify the configuration of single or groups of devices.

Most of data will regard energy consumption, energy storage, power generation, resource and appliance state, real time prices; application should also provide alerting for abnormal states or energy use, as well as critical situations concerning voltage drops or frequency shifts. Information may be collected in homes, premises, plants and substations, both locally or remotely; it will be made accessible anywhere, anytime and anyway, by fixed devices (PCs, (IP)TVs, Internet video phones, in-home displays) and mobile terminals (handhelds, smart phones, tablets).

Flexibility and scalability are required to allow management of equipment in different domains, whose composition may change dynamically: homes, buildings, franchise shops, factory areas, closed residential groups. Such infrastructures may be deployed by private owners, public utilities or third-party service providers. Vendors are already providing simple installations in this context [22].

7.4.10   Network Management

Communication requirements that come out from use-cases can be used to identify different domains and sub-domains, with different networking solutions: public and private, wired and wireless. This variety of environments will make network management rather complex: different applications, services and actors will have different performance requirement, which are quite difficult to map on generic topologies. Network management will face numerous issues concerning performance, priority, quality of service, configuration, robustness, resilience and so on.

7.4.11   Grid Management

Grid management involves different aspects covering regular and emergency situations. It is strictly related with activities falling in other use-cases, e.g., wide area awareness, load management (demand response and energy efficiency) and electric vehicles.

Main issues for grid management are the collection and analysis in real time of large amounts of information from heterogeneous sources: demand response schemes, distributed energy resources (loads and generators), energy storage systems, plug-in electrical vehicles, AMIs. Such information must be filtered, aggregated and fitted into suitable models for successive analysis that aims at detecting abnormal, critical and dangerous situations, as well as deviation from regular operations, which the grid is required to deal with. The outcomes from this process are also made available to the energy management system and to aggregators at a higher layer of abstraction [23, 24]. Specific actions include emergency coordination, load shedding schemes, service restoration (e.g., by reconfiguration of feeders and switches).

Another typical objective in large grid management is the organization in microgrids, which tries to concentrate and balance locally the production and consumption of energy. This is impossible in the current grid structure, but it will be feasible with the massive introduction of distributed energy resources as photovoltaic and wind generators. Main issues in this context are the transfer of the status of distributed energy resources to local management systems (e.g., a building or a group of premises) and mechanisms to share the current power generation capacity among the members of a group, in order to balance the temporary scarceness of some of them.

The smaller form of grid management will be building and home energy management systems, which will manage the local facility and component operation. These systems will make extensive use of sensing and metering devices for context awareness; information about the current status (power consumption, energy prices and availability) will be used to control and manage electrical automation systems, heating, ventilation and air conditioning equipment (HVAC), plug-in electric vehicles and the same sensing and metering devices, which must result in energy-effective solutions.

As of the very different context of these scenarios, ICT is expected to provide a large set of heterogeneous solutions, with different characteristics in terms of communication coverage, scalability, computation and storage requirements.

7.4.12 Market Operations

Deregulation, liberalization, distributed generation, integration of renewables are key factors of the radical change undertaken in the energy system. They will bring new system operators (bulk power producers, micro-producers, transmission and distribution operators) that will give rise to a new market structure, mainly based on energy trading and supply of ancillary services. New figures are expected to provide brokerage among the energy players (producers, distributors and consumers) and auxiliary financial and administrative services.

This market will consist of exchange of information and energy, dynamic pricing, accounting and other services that will rise; data model and representation, communication infrastructures and services, security paradigms and interfaces are therefore required to cope with these different sets of issues.

7.5 Political and Standardization Initiatives

The interest towards smart grids is mainly boosted by worldwide political initiatives pursuing GHG emission reduction, energy independence from oil, and economic growth through innovation. That results in several types of actions: establishment of commissions and task forces, issuing regulations, and mandates to standardization bodies. Most of them came in Europe, North America and China.

7.5.1 Policies and Directives

7.5.1.1 Europe

Europe is jointly planning its energy policy in an integrated approach to combat climate changes, to pursue security and independence of carbon resources, to strengthen its competitiveness [25, 26]. Starting from 2006 the European Commission has developed an European Strategic Energy Technology Plan (SET-Plan) [27], together with industries and the research community. It consists of a "roadmap" towards low-carbon technologies to be achieved through different steps in 2020, 2030 and 2050 [28]; currently, it identifies eight main areas with strong potential: wind, solar, electricity grids, bioenergy, carbon capture and storage, sustainable nuclear, fuel cell and hydrogen and smart cities [29].

Key issues of this strategy are energy efficiency, renewable sources and new energy technologies. The first initiative in 2007 was the definition of the "20-20-20" targets to be met by 2020: a reduction of GHG emission of at least 20% below 1990 levels, a 20% of energy from renewables, a reduction of 20% of primary energy needs compared with projected levels. This strategy complies with the three pillars of the European energy policy: security of supply, sustainability and market efficiency.

The European Commission undertook several initiatives towards this purpose. In 2008 it proposed the climate and energy package [30], a binding legislation to share the effort among associated countries according to their economic possibilities. In 2009, new architectures and infrastructures for energy grid management and communication were required by the directive about renewable energy [31]. Concurrently, it published the Third Energy Package, a further step in the process of liberalization of the electricity and gas markets. Within this package, the Electricity and the Gas Directives [32, 33] demand the member states to deploy smart meters in their networks, with the aim of covering at least 80% of households by 2020. Afterward, the importance of distributed energy generation and smart metering was highlighted again in the directive of energy performance of buildings [34].

In addition to the regulatory framework, the European Commission has also fostered collaboration and joint initiatives among the main stakeholders.

The European Technology Platform[4] for Smart Grids began its work in 2005 to foster the development of European electricity networks looking towards 2020 and beyond [35]. Its mission is to support the deployment of smart grids in Europe by coordinating and advising the different stakeholders fora and associations, and by ensuring consistency and relevance with the EU policy. Smart grid ETP aims at building a shared vision and a strategic overview in Europe, promoting research, development, demonstration and deployment projects. Until now, its main achievements have been three documents. The vision [36] is driven by the combined effects of market liberalization for both transmission and distribution networks, the change in generation technologies to meet environmental targets and the future use of electricity. The strategic research agenda [37] describes the main areas to be investigated, technical and non-technical, in the short/medium term in Europe. The strategic deployment document [38] describes the priorities for the introduction of innovation in the electricity networks and the benefits that such innovation will deliver for all stakeholders. In 2009, the ETP was restructured and the so called SmartGrids ETP Forum has substituted the Advisory Council in taking the lead of the whole platform structure and setting the new programme of activities [39].

In the framework of the SET Plan, the European Electricity Grids Ini-

---

[4]European Technology Platforms (ETPs) are industry-led stakeholder fora charged with defining research priorities and action plans in a broad range of technological areas.

tiative (EEGI) comes from a group of electricity transmission (ENTSO-E[5]) and distribution (EDSO-SG[6]) system operators and proposes a 9-years research, development and demonstration program to accelerate innovation and to address most critical issues to reach the targets on energy and climate for 2020 and beyond. The main outcome of this initiative is the roadmap and implementation plan [40], which was prepared by ENTSO-E and EDSO-SG in close collaboration with the European Commission, ERGEG[7] and other relevant stakeholders.

In 2009 the Commission established the Smart Grid Task Force [41], which was designed to provide a joint picture about different aspects concerning the deployment of smart grids (regulatory, technological and commercial); its main goal is to produce a set of recommendations to guide the design of a consistent, cost-effective, efficient and fair implementation of new energy grids across Europe according to the directives of the Third Energy Package [42]. The Task Force should depict a clear vision about expectations, evolution, technology and deployment, by following the work of other groups in this area: the Smart Grids European Technology Platform, the Smart Grids Forum and the European Electricity Grids Initiative. It should also consider standardization activities undertaken by European organizations (functionality, interoperability, smart meters). The expected policy directions will be focused on the period 2010-2020. The work of the Task Force was organized into four expert groups that already provided their first reports on functionalities for smart grid [43], regulatory recommendation for data safety, data handling and data protection [44], roles and responsibility of actors involved in smart grids deployment [45] and smart grids aspects related to gas [46].

The European Commission has also cared about the convergence towards standard and interoperable solutions and issued three mandates to European standardization organizations. The first concerns open system architectures, communication protocols and interoperability for utility meters [47]. The second requires an European framework that integrates communication technologies, digital computing, electrical architectures and that facilitates the implementation of smart grid services and functionalities [48]. Finally, the third mandate concerns electrical vehicles [49].

7.5.1.2 North America

The electric grid of the United States has changed very little since first commercial deployment dating back to the end of the 19th century [50]; many utilities even do not have real time information about the state of the grid yet. The upgrade of the current electricity grid is considered necessary by the U.S. government to keep America's worldwide leading role, to create new jobs

---

[5]European Network of Transmission System Operators for Electricity.

[6]European Distribution System Operators Association for SmartGrids.

[7]The European Regulators Group for Electricity and Gas is a body of independent national energy regulatory authorities set up by the European Commission to advise the EU executive on energy issues.

and to boost the economic growth. Obama's administration outlined a vision to double the use of clean energy by 2035 and to put one million electric vehicles on the road by 2015, in order to secure the energy future of America [51].

The Energy Independence and Security Act of 2007 (EISA) required policies and initiatives to modernize the national transmission and distribution systems. This act established the Federal Smart Grid Task Force to ensure awareness, coordination and integration of the diverse activities related to smart grid technologies among the different federal agencies, practices and services [52]. The *Office of Electricity Delivery and Energy Reliability* of the Department of Energy is also active with several initiatives [53]: it produced a book introducing smart grids and it is conducting a series of e-forums to discuss various issues including costs, benefits, value proposition to consumers, implementation, and deployment.

The act also assigned the National Institute of Standards and Technology (NIST) the coordination of the development of a framework of standards for smart grids [54]. Gathering experts from utilities and ICT industries, NIST proposed use cases and architectures for smart grid information networks in its report concerning the framework and roadmap for interoperability standards [55]. In addition, NIST set up the public-private partnership Smart Grid Interoperability Panel [56]; it selected a number of Priority Action Programs (PAPs) among its technical activities that are chartered to address areas in which standards require development or revision to complete the NIST framework[8].

Following the push for aggregation and innovation, private initiatives and consortia were set up too. The Green Smart Grid Initiative (GSGI) is a nonprofit partnership focusing on climate change and smart grids [57], especially for what concerns energy efficiency and renewable sources.

To support and to accelerate the development of smart grid technologies, the government allocated $ 4.5 billion in the American Recovery and Reinvestment Act of 2009 for modernizing the electric grid and for implementing demonstration and deployment programs. However, to date, the policies related to smart grid technologies adopted by different states have resulted in a lot of different grids; the federal government has undertaken a policy framework pressing the modernization of electricity networks by taking advantage of latest information, energy and communication technologies [58]. This framework is built upon four pillars: enabling cost-effective smart grid investments, unlocking the potential for innovation in the electric sector, empowering consumers and enabling them to make informed decisions and securing the grid.

---

[8]For example, PAP #1 (Internet) and PAP #2 (Wireless) are cooperating to quantify network requirements for smart grids, and then identifying Internet and radio technologies that meet these requirements.

7.5.1.3 Asia

The most important Asian countries are undertaking different initiatives to reduce GHG emission and to make their energy systems more efficient. In Japan the main objectives are to promote the use of renewable sources, to build an infrastructure for electrical vehicles and to create new services in smart grids. In China, the stimulus plan of 2008 has been making large investments in improving the capacity and reliability of electricity networks; integration of renewables and energy efficiency are expected as secondary effects. The budget allocated by China's government for smart grids is surpassing that of United States. In South Korea, the focus is on monitoring the energy use and increasing the production from green sources.

7.5.2 Research Projects in Europe

Several research projects concerning smart grids have been financed by the European Commission in the last five years. Initially, the main question was to establish how and to what extent ICT would help the evolution of electricity grids towards new market models, distributed generation, integration of renewables and demand management; these topics are faced in CRISP [59] Afterward, other projects have focused on more specific issues.

Efficient energy management by active demand mechanisms and by demand response schemes for domestic appliances is considered in ADDRESS [60] and Smart-A [61], respectively.

EU-DEEP [23] focuses on the importance of distributed energy resources by addressing technical challenges, economic values and business models, and drawing a set of recommendations. Fenix [24] develops the concept of *Virtual Power Plant* to abstract and model the presence of a huge number of distributed energy resources, while More Microgrids [62] deals with the definition of small grids of distributed energy resources and their integration in the whole network.

Monitoring and management of the distribution grid through the development of middleware for secondary substation nodes are the objectives of OpenNode [63]; another initiative targeting the distribution grid is Open-Meter [64], which pursues the development of an open metering standard in EU countries. Providing uniform and homogeneous access to sensors and devices in the environment by a common middleware is the goal of Hydra [65], which tries overcoming the large number of existing heterogeneous machine-to-machine interfaces.

Integration and management of distributed generation is taken into account in ADINE [66]; MetaPV [67] specifically deals with photovoltaic systems, while Night Wind [68] and EWIS [69] consider wind farms.

G4V [70] and MERGE [71] study the integration of electrical vehicles into the grid, GROW-DERS [72] faces different energy storage mechanisms.

Many projects restrict their research field to the home environment. Energy efficiency is the main topic for DEHEMS [73]; AIM [74] focuses on modeling and management of domestic appliances and Beywatch [75] addresses smart management of generation, user profiles, consumes. In some cases, the interest is widened to entire buildings: IntUBE [76] deals with energy efficiency of single and groups of buildings and eDIANA [77] develops middleware and platforms to integrate buildings as nodes in the grid. A further issue concerns making users aware of energy and environmental issues: BeAware [78] proposes an interactive game which monitors the environment, advices users and awards them according to their behavior.

Finally, there have also been some test fields around Europe. In INTE-GRAL [79] and SmartHouse/SmartGrid [80], multi-agent systems and home gateways are used to control local energy production, energy market, demand-side, and load forecast. Several sites are considered, representative of normal, emergency and critical situations.

7.5.3   Standardization Efforts

Standardization bodies are active worldwide to avoid the fragmentation of smart grid activities into a plethora of different independent and non-interoperable solutions; in many cases these organizations operate under specific mandates coming from governments.

The Joint Working Group (JWG) set up by CEN, CENELEC and ETSI was originated by the mandates from the European Union [47–49], whose aims were to promote the use of smart metering systems, to secure interoperability, to protect the customers and to ensure system reliability. This group has already prepared an overview of requirements for European standardization of smart grids, including state-of-the-art of existing standards and other relevant work, together with recommendations and plans for future work [81].

ETSI (*European Telecommunications Standards Institute*) has an active Technical Committee on Machine-to-Machine (TC M2M), which is about communication among machines without (or just limited) human intervention in buildings (home/office automation), production (line automation and management, quality control), healthcare, security (crime/terrorism, disaster/emergency), energy (production, distribution), retail (stores, tourism), transport (road vehicles, supply chain); this initiative also falls within the scope of M/441 [47]. TC M2M has explicitly considered the smart metering use case [82] and a specific Work Item was created to deal with application of M2M architecture for the needs of smart grids[9]; an early draft was already produced by this group [83] and the final publication is scheduled within

---

[9]ETSI Work Item DTR/M2M-00011. Status at http://webapp.etsi.org/WorkProgram/Report_WorkItem.asp?WKI_ID=34662.

few months[10]. Related activities concern powerline communication (ETSI TC PLT[11]) and next-generation network architectures (ETSI TC TISPAN[12]).

IEC (*International Electrotechnical Commission*) addresses smart grid topics by its Strategic Group 3 [84]. The starting point of this group was the analysis of existing standards available within IEC and their applicability to smart grid scenarios, the identification of potential gaps and the definition of a roadmap to help the smart grid vision to become a reality [85]. The SG3 develops the so-called Mapping Tool, a virtual graphical map that allows to position any standard in relation to its role in smart grids. Further standard development will take place starting from the framework of other IEC Technical Committees, especially TC 57 (power system control equipment and systems), TC 8 (electrical energy supply), TC 82 (solar photovoltaic energy systems) and TC 13 (electrical energy measurement, tariff and load control).

ITU (*International Telecommunication Union*) established a Focus Group on Smart Grid (FGSmart) [86]. This group collects documents, ideas and liaisons with other entities to provide recommendations, use cases and architectures on smart grid from a communication perspective. It does not produce new standards: instead, its purpose is to investigate potentials, to select issues, to identifies trends and to follow other initiatives; it will supply ITU with a clear understanding of smart grid technologies, attributes and challenges, to be used as a decision base for future activities. The main outcomes from this effort are a set of live documents about overview [8], use cases [12], architectures [87] and requirements [88].

IEEE (*Institute of Electrical and Electronics Engineers*), on the strength of its multidisciplinary nature and knowledge background (digital library and standards), is carving out a leading role in the field of smart grids [89]. It has several standardization activities on smart grids, concerning interoperability, interconnection and photovoltaic systems. The 2030 Smart Grid series of standards focus on interoperability of the electric power system with end-use applications and loads. They deal with characteristics, functional performance and evaluation criteria; specific working groups were set up for electrical transportation and energy storage. The 1547 series covers interconnection of distributed energy resources, mainly from an electrical perspective, but with specific issues related to communication.

IETF (*Internet Engineering Task Force*) is naturally engaged in smart grids, leading the whole TCP/IP stack used in the Internet. Indeed, IETF tried to organize some of the large set of protocol standards into a structured framework for smart grids [90]. Beside this, it has other activities concerning sensor networks and energy management that are not explicitly targeted to

---

[10]DTR/M2M-00011 Work Item Schedule: `http://webapp.etsi.org/workProgram/Report_Schedule.asp?WKI_ID=34662`.

[11]ETSI Powerline Telecommunications: `http://portal.etsi.org/portal/server.pt/community/PLT/326`.

[12]ETSI Telecoms & Internet converged Services & Protocols for Advanced Networks: `http://portal.etsi.org/portal/server.pt/community/TISPAN/339`.

smart grids, but could be used in AMI or Home Area Networks (HANs). About sensor networks, the focus is on communication among nodes with constrained resources (computation, power, memory, lifetime); the 6LoWPAN working group [91] is facing the use of IPv6 and has already outcomes about requirements and transmission of IPv6 packets over IEEE 802.15 networks, the ROLL working group [92] is studying generic routing issues in ad hoc topologies, and the CoRE working group [93] provides a resource-oriented framework for applications exchanging data on constrained IP networks, where nodes may be up and running occasionally. About energy management, the EMAN working group [94] aims at operating communication networks with a minimal amount of energy, yet providing sufficient performance to meet service level objectives.

NIST (*National Institute of Standards and Technologies*) is in charge of leading and coordinating the development of an interoperability framework for smart grid systems and devices, with special focus on ICT and communication issues. Its effort has mainly been devoted to depict a reference picture of future smart grids by an abstract reference model, to identify a large set of standards relevant in smart grid deployment, to trace a roadmap to guide the development of interoperable standards [55] and to publish guidelines for smart grid cyber security [95]. NIST also promotes collaboration and coordination among different stakeholders and standardization bodies (see, for example, the smart grid interoperability panel [56]), in order to close the gaps in achieving smart grid interoperability in the near future.

The ISO/IEC JTC 1[13] created the Special Working Group on Smart Grid (SWG – Smart Grid) [96] to identify market requirements and standardization gaps, to encourage JTC 1 to address the need for ISO/IEC standards, to promote these standards, to coordinate JTC 1 activities on smart grid among IEC, ISO, ITU-T and other standardization bodies (especially the IEC Strategic Group 3), to periodically report results and recommendations to JTC 1. JTC also has a Working Group on Sensor Networks[14] (WG7), which is working on sensor network architectures and interfaces to support smart grid systems, in order to visualize sensor/device status and data/information flow in large scalable heterogeneous network systems, including the geospatial information ones.

TIA (*Telecommunications Industry Association*) has put little effort on smart grids until now. Engineering Committee TR-50 [97] addresses communication of smart devices with other devices, applications or networks; that is for various scenarios, including smart grids. It consists of a high-level interface to transfer events and information, independent of networking technologies and media and available to applications through a well-defined API. Engineering Committee TR-51 [98] provides utility companies efficient access technologies

---

[13]The Joint Technical Committee 1 is a common effort of ISO IEC to deal with all matters of information technology.

[14]ISO/IEC JWG 1, WG7. "Sensor Network and its Interface for Smart Grid System" (NP 30101). Project approved (ISO/IEC stage 10.99).

with a mesh network topography, by developing a networking stack covering layers 1 up to 4, in order to build services and applications. Currently, other Engineering Committees focusing on vehicular telematics, premise communications and telecommunication infrastructures do not explicitly take smart grids into consideration.

ATIS[15] (*Alliance for Telecommunications Industry Solutions*) examined existing standards and what is missing for smart grid applications; it has not produced documentation until now, but it actively participated in NIST initiatives [55].

SGCC (*State Grid Corporation of China*) defined its own smart grid framework and roadmap [99], based on the analysis of existing standards (domestic and international), mainly coming from IEC, and worldwide initiatives. Key issue in the SGCC vision is the new role of the grid, that will no longer be a simple carrier for transmission and distribution of electricity, but will be an integrated and intelligent platform for the Internet of things, internet network, communication network, radio and TV networks. Especially, SGCC promotes the application of Internet of things technologies in smart grids [100], by a number of national pilot projects. Although SGCC is not a standardization body, it can exert great influence on vendors and markets since China is expected to be one of the largest markets for smart grids.

7.6 Architectures for Smart Grids

Building smart grids is an evolutionary process that starts from current infrastructures and takes into account the requirements to create new applications and services. This implies breaking the whole complex system into simpler and isolated entities, and describing their internals and their interfaces, in order to have a clear understanding of the main actors playing in the system, their objectives and the relationships among them. Such description may take different forms, depending on the main perspective it is addressed to: abstract high-level conceptual models, processes and data flows, communication, information management and security, services [81].

A *conceptual model* of smart grids describes major business domains and the stakeholders involved in each of them; usually, this is a descriptive architecture useful to depict the system organization and to identify potential intra- and inter-domain interactions, applications and capabilities. Most of all, it is to understand the smart grid concept, not to define smart grid architectures. The NIST model is often taken as reference by other organizations; it is described in details in Section 7.6.1.

---

[15]ATIS is a regional standards organization that develops technical and operational standards for the telecommunication industry in North America. Web site: http://www.atis.org.

A *functional architecture* identifies a set of functions and interfaces to understand the sequencing of execution and the conditions for control or data flow. This process initially addresses the main applications (e.g., distributed energy resources, demand response, smart home automation) but it may be reiterated recursively internally to each function block. Suitable inputs for this task are the IEC TC57/SG3 mapping tool [101] and the work of the Smart Meters Coordination Group [102].

A *communication architecture* is a functional architecture focusing on connectivity. It takes into account different networking environments (home, local area, neighborhood area, wide area, wireless access) and technologies (wired, wireless, powerline, cellular), combining them to form a large set of communication scenarios.

An *information architecture* deals with the abstract representation of entities, which is essential to assure interoperability among subsystems. Several models are already available for general purposes (e.g., IEC 61970 Common Information Model) and specific application domain: smart metering (e.g., ANSI C12.22, IEC 62056), substations (IEC 61850), electrical vehicles (e.g., SAE J1772, ISO/IEC 15118), sensor networks (e.g., ZigBee Smart Energy Profile 2.0), etc.

An *information security architecture* concerns all aspects about information security; it provides guidelines for the design of systems that satisfy security requirements. Examples of initiatives in this fields are the Smart Grid Cyber Security Working Group (SGCSWG) [95] and the Smart Grid Task Force Expert Group 2 [44].

A *service-oriented architecture* is a modern paradigm that organizes system functions into standard, distributed and interoperable base services; these can be combined to build more complex services and applications. ICT technologies are essential to create service-oriented architectures; they include elements as data services, functional logic services and business logic services. Examples of interface and communication specifications are IEC 61968 and IEC 61970.

7.6.1   Conceptual Models of Smart Grids

Conceptual models of smart grids provide a high-level abstraction of these systems, in terms of actors, applications and domains. They usually are "perspective" but not "prescriptive", thus resulting in a technology-agnostic understanding without focusing on specific architectures. Conceptual models are specially useful to standardization bodies and regulators to develop policy and technological frameworks addressing the different entities encompassed in a smart grid system.

Actors make decisions and exchange information; they are mainly devices, systems and programs. Examples of actors are smart meters, power generators (solar, nuclear, hydro, etc), control systems. Actors perform *applications*, which are various kinds of tasks: e.g., home automation, energy generation, energy storage, energy management. Actors are aggregated into *domains* with similar objectives and participating in similar types of applications.

Figure 7.3 shows the NIST conceptual model. It identifies seven different domains: Customers, Markets, Service Providers, Operation, Bulk Generation, Transmission and Distribution. Bilateral relationships between actors in different domains are logical connections called *associations*; electrical associations are shown as dashed lined and communication associations are shown as solid lines.

The *Customer* domain encompasses the main stakeholders the grid is designed for, i.e., consumers. It may be split into subdomains, according to the energy profile: home, industrial, building/commercial. Interfaces of this domain mainly consists of power meters (used by distribution utilities for billing, reporting and monitoring) and energy service interfaces (entry points for applications such as remote load control, monitoring and control of distributed generators, in-home display of customer usage, reading of non-energy meters, integration with building management systems and the enterprise, auditing/logging for cyber security purposes).

Energy trading takes place in the *Markets* domain. Here, energy prices are agreed by balancing supply and demand within the power system, following similar rules to economic markets. This would hopefully result in matching consumption with production from fixed bulk and inconstant distributed generation (the latter happening in the Transmission, Distribution or Customer domains), thus making efficient use of energy resources. Controlling the electricity grid implies strict communication requirements to timely respond to dynamic conditions, thus demanding for reliable, traceable, auditable, secure and low-latency communication.

The *Service Providers* domain includes all business processes of power system producers, distributors and customers. They are traditional utility services (billing, customer accounting management), enhanced customer services (management of energy use and home energy generation), and new services to meet the requirements by the evolving smart grid (e.g., the interface allowing customers to interact with the market). Communication relies on consistent message semantics, interfaces and standards over a variety of networking technologies.

The *Operations* domain deals with smooth operation of the power system: maintenance and construction, supply chain and logistics, planning, meter reading and control, records and assets, fault analysis, reporting and statistics, load control. Today, utilities are in charge of these operations on their network segments.

The *Bulk Generation* domain supplies electrical energy from other sources (chemical/biomass combustion, nuclear fission, flowing water, wind, solar radiation, geothermal heat). Communication provides to the grid key performance and quality of service issues, such as scarcity (for inconstant sources) and generator failures.

The *Transmission* domain transfers electrical power from generation to

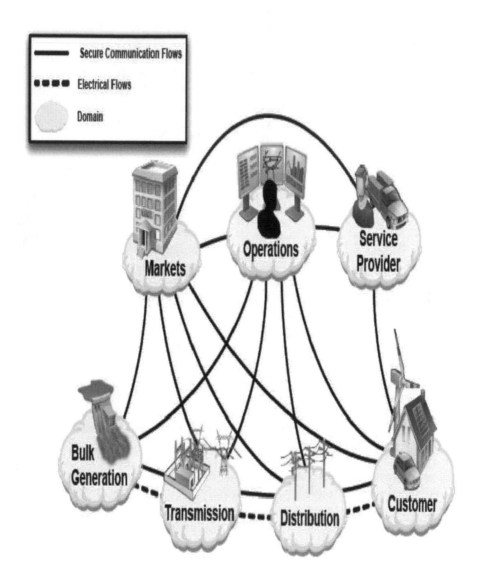

FIGURE 7.3 The NIST's conceptual model. Picture is taken from the "NIST framework and roadmap for smart grid interoperability standards" [55].

distribution points. It is responsible of grid stability, by balancing production with load, so to avoid frequency drift and voltage oscillations; most activities are carried out in substations monitored and controlled by SCADA systems, made of communication networks, monitoring and controlling devices.

Finally, the *Distribution* domain delivers electricity to end users. The grid topology may assume several structures: radial, looped, meshed. Historically, very little or even no intelligent is present in this domain; failures and break downs are usually based on human communications initiated by customer reports.

A key issue for smart grids is the distributed power generation that may occur in the Customer, Distribution, and Transmission domains; this aspect is not explicitly shown in this conceptual model.

It should be noted that organizations operating in smart grids may be involved in more domains, thus playing the role of different actors. For example, distribution system operators have components in the Distribution, Operations and Customer domains.

The NIST conceptual model has received large approval from other organizations. IEC explicitly adopted it in the SG3 roadmap [85]. IEEE adds a dimension to the NIST model, expanding each domain into three layers [89]: the power and energy layer, the communication layer and the IT/Computer layer; the first layer consists of electrical equipment and infrastructures, while the other layers make the grid "smart". Other models have been derived by making small changes. That from the Joint Working Group of CEN/CENELEC/ETSI aggregates Transmission and Distribution into a single domain, splits Customers in two distinct domains (home/building and industrial) and inserts a new domain for DER (Fig. 7.4), catching actors from the report by the Expert Group 3 of the Smart Grid Task Force [45]. Two layers are also highlighted: transaction, where interactions take place by ICT software, applications and solutions, and power, where interactions imply control and optimization of power flows.

7.6.2 Layout of the Communication System

The main foundation for deployment of smart grids is to bring intelligence into the system. IEEE meets this concepts by considering three layers orthogonal to the seven domains in the NIST's conceptual model: Power and Energy, Communication, IT/Computing (see Fig. 7.5). The Power and Energy layer includes all electrical equipment: generators, transformers, switches, feeders, transmission lines, loads, sensors and meters, etc. That builds the energy chain: production, transportation, distribution and consumption. Control and management of this infrastructure is made by intelligent processes running on suitable computing platforms at the IT/Computing layer, which includes interfaces to humans. The intermediate Communication layer allows information exchange between and within the side layers; it is a main pillar for building smart grids.

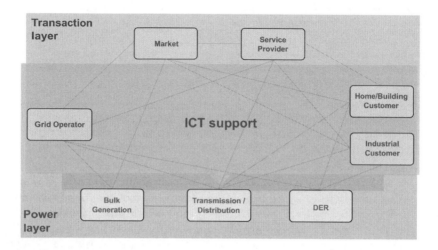

FIGURE 7.4 The conceptual model proposed by the Joint Working Group of CEN/CENELEC/ETSI; it slightly differs from the NIST's one. Picture by courtesy of M. Sánchez, in [81].

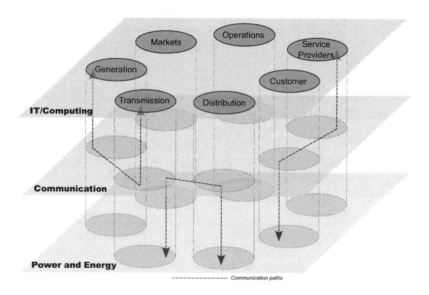

FIGURE 7.5 The IEEE three-layers architecture of the NIST's domains.

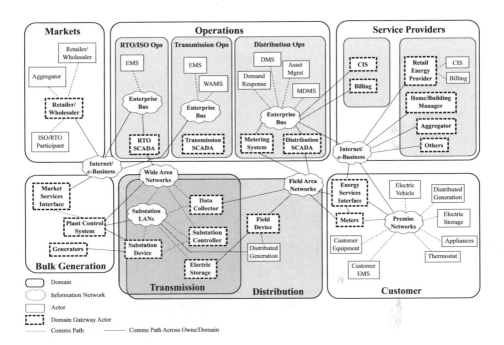

FIGURE 7.6 NIST's communication model for the 7-domains conceptual model. Picture is taken from the "NIST framework and roadmap for smart grid interoperability standards" [55].

Starting from its conceptual model, NIST developed a communication framework for the seven domains (see Fig. 7.6). Also in this case, the NIST effort was devoted to help understanding the main smart grid operational relations, and not to prescribe the communication architecture. Actors in each domain require to exchange information with other actors both in the same or with a different domain. Exchange of information takes place along communication paths. Communication within a single domain may have similar characteristics and requirements; however, due to the heterogeneity of information infrastructures and conceptual domains, different protocols may be used in the system, thus gateway actors are provided for interoperability.

Designing communication facilities in smart grids involve dealing with several aspects:

- software infrastructures to build distributed services and applications;

- syntax and semantics of information exchange;

- transport of information and networking;

- communication media and technologies.

Thorough investigations of available technologies and standards have already been carried out by several working groups [43, 81, 85]; these works provide a large census together with some discussions about possible involvement in smart grids, interoperability and missing features. The selection of the most suitable standards must happen with proper evaluation criteria; guideline for this process have already been supplied [55].

7.6.2.1   Distributed Services and Applications

Modern computer science paradigms enable to build applications by exploiting functions across the network. The *Remote Procedure Call* was initially used to let applications call remote functions through standard API, hiding all networking issues. Afterward, the interest in developing inter-platform and language-agnostic software frameworks has brought to the concept of *Service-Oriented Architecture* (SOA), where applications and new services are build by composing basic services deployed anywhere in the network. SOAs provide a more dynamic, flexible, scalable and effective way to create distributed applications with respect to RPC. SOAs require describing services with suitable languages (e.g., WSDL[16]), formatting data in a structured way (e.g., XML[17]) and transferring them with some protocol (e.g., SOAP[18]).

Frameworks to implement service-oriented architectures include REST[19]

---

[16]Web Services Description Language. URL: http://www.w3.org/TR/wsdl20/.
[17]eXtensible Markup Language. URL: http://www.w3.org/XML/.
[18]Simple Object Access Protocol. URL: http://www.w3.org/TR/soap12-part1/.
[19]REpresentational State Transfer.

[103], Web Services[20], CORBA[21], DCOM[22]. Smart grids will be populated by a large number of different actors, thus yielding to heterogeneous platforms and technologies. SOAs will likely play a strategic role to deploy uniform, extensible and interoperable services; this approach has already been used in recent projects, e.g., SmartHouse/SmartGrid [80] and Hydra [65] (see Section 7.5.2). *Model-driven* architectures may be useful as well in smart grid systems; they have been considered in eDiana and Hydra (see Section 7.5.2).

7.6.2.2 Data Models and Information Exchange

Data abstraction is needed to represent measures, events and controls of electrical systems. Many standards are already available for powerline communication (HomePlug, HomePNA, IEEE 1901, ITU G.hn), WAN operations (IEC 60870, 61850), distribution automation (IEC 61970, 61968), substation automation (IEC 60870, 61850), distributed energy resources (IEC 61850-7-420), head-end (IEC 61968-9) and cross-domain interaction (IEC 61970, 61968, 61850, ETSI TS 102690). However, other sectors currently lack any standard framework: for example, home automation and generation resources. Data and services are often represented in the Substation Configuration Language (SCL), specified by IEC 61850, which enables devices in a substations to exchange configuration files and to have complete interoperability. SCL allows the description of device capabilities, system specification, substation configuration, device configuration and system exchange. An alternative description language is the Common Information Model (CIM), adopted by IEC 61970-301, which defines an XML format and is completely developed based on UML. Both SCL and CIM address the same issues; SCL is mostly used within substation, whilst CIM is largely used in information exchanges among systems (for example, energy management system, planning, energy markets and metering). The abstract data models can be mapped to a number of protocols: MMS (Manufacturing Message Specification, ISO/IEC 9506), GOOSE (Generic Object Oriented Substation Event, IEC 61850), GSSE (Generic Substation State Event, IEC 61850) and (soon) Web Services; MMS include a complete networking protocol stack, while GSSE/GOOSE messages are directly exchanged in Ethernet frames with publish/subscribe paradigms and priority mechanisms, supporting applications which require less than 4 ms latency.

7.6.2.3 Networking

Communication networks exchange information and share resources. Coming back to the NIST communication model depicted in Fig. 7.6, the communi-

---

[20]W3C Web Services Activity. URL: http://www.w3.org/2002/ws/.

[21]Common Object Request Broker Architecture. URL: http://www.omg.org/spec/CORBA/3.1/.

[22]Microsoft Distributed Component Object Model. URL: http://msdn.microsoft.com/library/cc201989.aspx.

cation infrastructure is mainly viewed as a "network of networks". Indeed, several kinds of networks are required for smart grids: Enterprise Buses, Wide Area networks, Field Area Networks and Premises Networks. Both public and private infrastructures may be used to implement these networks; in any case, features as security and quality of service are essential design goals for targeting smart grid requirements.

Currently, management and control of energy grids rely on independent networks for different functions; the TCP/IP protocol family is usually used for enterprise data and business networks, while specific protocols are used for SCADA systems. This organization leads to various ownership and management boundaries, which would hinder the deployment of end-to-end services among stakeholders and devices in different domains.

The TCP/IP stack can play a major role in communication networks for smart grids, due to its widely adoption worldwide and to its standard, well-tested and simple communication interface. However, it may not be suitable to meet any requirement for the large set of applications in smart grids. Security concerns would arise in exploiting the public Internet infrastructure to create smart grids and should be seriously taken into account to avoid a breach in a domain or subnet will not propagate to the whole system.

7.6.2.4   Communication Media and Technologies

Smart grids include such a number of scenarios and application fields that a large set of heterogeneous transmission technologies will get involved. Wireless links and powerline communications represent a sound and economic alternative where no wired infrastructure is available: power plants and electrical substations are often located in rural or suburban areas, whereas monitoring of many household devices would require too much network equipment and too widespread wiring.

The selection of suitable standards will take into account several considerations: electromagnetic compatibility, communication paradigms, addressing schemes, quality of service, security, reliability, resilience, network extension, existing infrastructures, investments and so on. Example of technologies falling in the interest of smart grids are IEEE 802.3/Ethernet, IEEE 802.11 (WiFi), IEEE 802.16 (WiMax), IEEE 802.15.4 and ZigBee (sensor networks), IEEE 1901 (HomePlug), ITU G-series (HomePNA and HomeGrid), ITU G.992/993 (xDSL), GSM/GPRS, UMTS and most of IEC standards for electricity grids.

7.6.3   Network Architectures

A communication system for smart grids encompasses several kinds of networks, with different sizes and logical functions [87]:

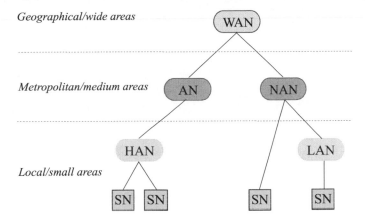

FIGURE 7.7 Relationships among local, home, neighborhood, access and wide area networks.

- **Local Area Networks** (LANs) interconnect devices and applications in small and localized areas: homes, laboratories, buildings, substations. They include sensor networks (SN) and other short-range technologies.

- **Home Area Networks** (HANs) are LANs deployed at home.

- **Access Networks** (ANs) aggregate a large number of users and provide access to the main core network.

- **Neighborhood Area Networks** (NANs) are a particular implementation of ANs that interconnect several HANs and provide connectivity to the core network.

- **Wide Area Networks** (WANs) are the main communication infrastructures over long distances and wide geographical areas.

The relationship among the different types of networks and their geographical coverage is shown in Fig. 7.7. Although it is clear there will be many local, home, access and neighborhood networks, it may be questionable whether smart grids should rely on single or multiples wide area networks; the current perspective is towards a common and unified communication infrastructure. The WAN for smart grids is likely to partially share the same infrastructure of the Internet.

7.6.3.1   Local Area Networks

LANs are usually deployed in substations, power plants and other electrical facilities. LANs provide cost-effective interconnection of devices; they include both wired (e.g., Ethernet) and wireless (e.g., WiFi) protocols. Apart the well known commercial technologies, a large set of industrial standards exist for

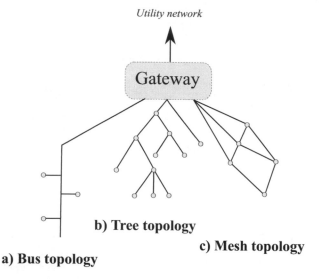

FIGURE 7.8 Different topologies for HANs.

energy and electrical applications. Most of them come from organizations as IEC, ISO, ANSI and IEEE [43, 81, 85].

7.6.3.2   Home Area Networks

HANs extend the utility network into the home. They include different devices to provide metering and control functionality; this kind of network usually exploits several cabling already available in homes: power lines, coaxial cables, twisted pairs and wireless networks. Depending on the specific medium, three different topologies may be used. In the bus topology (Fig. 7.8.a) all devices share a common transmission medium and directly communicate each other; this is typical for power lines and coaxial cables. In the tree topology (Fig. 7.8.b) all devices are rooted at a common point in a tree-like fashion; this organization is used with twisted pair cables and intermediate routers/bridges. In the mesh topology (Fig. 7.8.c), there is not a regular structure; some nodes must rely messages on behalf of other devices. The latter is used in wireless networks without any infrastructure, to extend network coverage and to provide robustness.

Figure 7.8 also shows a gateway between the HAN and the external access network. This gateway is needed to interconnect heterogeneous technologies inside the home (by bridging or routing), to protect the integrity of the utility network behind it, to handle configuration and management of home devices, to aggregate metering information collected by different sensors, to exchange information for demand/response functions, to apply utility's policies and controls on the customer appliances.

The relation between energy HANs and broadband data connections is still an open issue. There are different opinions about merging the two infrastructures, keeping them physically disjointed or allowing communication through an intermediary. Here the main concern is about security and the possible exploitation of home access to compromise the whole utility's network. The Cisco Home Energy Management solution is an example of HAN [22]; it uses wireless communication between the Home Energy Controller and domestic appliances (smart meters, sensors, thermostats) and it exploits the broadband connection to connect the Home Energy Controller with the external management system.

7.6.3.3 Access Networks

ANs provide broadband connectivity to end users and are the most widespread part of a network. End users are typically individuals or small office/home office (SOHO) customers; large companies and organizations usually have direct connection to the network core. ANs act as concentrators of traffic from low-bandwidth links. They are a critical part of a network because of the high cost of supplying each customers with a dedicated physical link compared to low revenue. Thus ANs are typically built on existing wired infrastructure or, recently, deployed as wireless solutions. Traditional wired technologies relies on PSTN networks and TV cables, by using voiceband and cable modems and xDSL protocols. Wireless access can make use of cellular networks (GSM/GPRS, UMTS), local data networks (WiFi) and metropolitan data networks (WiMax).

ANs usually spread over medium/metropolitan areas, but sometimes they cover entire nations, e.g., cellular networks.

7.6.3.4 Neighborhood Area Networks

NANs are mainly thought to implement local functions and to provide connectivity for meters and HANs in a small geographical area. For example, a NAN may aggregate data from meter readers and forward them to Head-Ends through the WAN. Network topologies are chosen according to environmental factors that affect transmission operations, resulting in different degree of reliability and performance.

When a tree topology is used (Fig. 7.9.a), NAN's nodes are organized in a tree structure; HANs attach to a single node and the NAN path may consist of one ("star" topology) or multiple hops. This structure is suitable to exploit power lines, especially between HANs and NAN's nodes; wireless links are another reasonable solution.

In a mesh topology (Fig. 7.9.b), HANs are interconnected to other HANs and/or to NAN's nodes. Multiple alternative links may be present at each HAN to access the NAN. This structure allows for extended network coverage and reliable and resilient communication paths; it is usually used in wireless environments. The availability of multiple paths requires routing strategies for messages, but this involves overhead both for the links (control traffic)

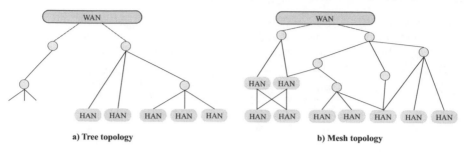

a) Tree topology   b) Mesh topology

FIGURE 7.9 Different topologies for NANs.

and nodes (routing information). Recall the main purpose of the NAN is to aggregate data and information from HANs to some management point behind the WAN, and to spread controls and configurations in the opposite direction (the NAN can be thought as a sort of tunnel); thus, mesh interconnections are likely to be mostly localized near the HAN side (left side of Fig. 7.9.b), while the internals may be organized in a hierarchical fashion (right side of Fig. 7.9.b). Packet relaying in a meshed NAN can happen by L2 or IP forwarding. Forwarding packets at the link layer makes use of simple protocols with small overhead and computation requirements, but this technique is prone to routing errors and inefficiencies. IP forwarding relies on standard IP routing and thus requires high-end nodes; this solution scales better in wide networks with hundreds of nodes. Until now there is not a clear polarization towards one of these alternatives; IETF considers both methods in its 6lowpan working group [91].

7.6.3.5   Wide Area Networks

WANs are responsible for the interconnection of remote systems over long distances. WANs represent the "core" infrastructure which provides rough transport services to large bulk of data at big aggregation points, i.e., access networks, large networks and NANs.

The Internet is today the most widespread data network; it may be a cost-effective solution to foster the deployment of country-, nation- or even continent-wide smart grids although cybersecurity must be carefully taken into account. The Internet is built upon the TCP/IP protocol stack and is composed by a number of transit networks, which provide transport services to large companies and service providers; the latter sell the service to small customers and individuals through access networks. The TCP/IP protocol stack provides an application interface that allows software designer to focus on the application protocol and information coding, without caring about the underlying network infrastructure and technologies, thus simplifying the design task.

7.7 Research Issues and Challenges for ICT

The large number of matters and issues in the smart grid ecosystem poses several challenges for the ICT world. Research efforts in the ICT community should be undertaken to investigate communication and data management infrastructures tailored to smart grid use cases: large amounts of data and information need to be collected, exchanged, processed and shared over wide geographical areas.

The Internet is a multi-protocol multi-technology general-purpose wide-area network with worldwide coverage and universal consensus; by now, its protocols and architecture are mature and well-tested, and plenty of applications and services are already available on top of it. One may wonder whether it would be suitable to accomplish many of the communication requirements in smart grid management.

Let's take into consideration the main issues depicted in Section 7.4. Dynamic load management and, mainly, distributed generation will make continuous energy supply critical. That will mostly imply reliable, redundant and resilient network operation; moreover strict upper bounds on delay, guaranteed bandwidth and quality of service will be needed by power equipment for safe, timely and synchronous functioning. Wide-area situation awareness and smart metering will collect data from virtually any remote place; this will likely result in a very large networking infrastructure with a huge number of peripheral acquisition devices, with obvious issues concerning scalability and efficiency. Information for man-machine interfaces and grid management will be available anywhere, anytime and anyway, thus requiring suitable data storage, processing, accessing and sharing over flexible and scalable infrastructures. Moreover, the expected size of the system will force the organization into smaller clusters, local domains and microgrids, thus specific protocols for data aggregation and grid cooperation will be necessary. Finally, a convergence towards common data representation and modeling, communication protocols and paradigms, and services will be the basis for a global energy market.

As a matter of fact, the core transport infrastructure and protocols of the Internet are almost the same as twenty years ago and just provide a basic delivery service, which does not meet all expectations of smart grid applications. However, the wide variety of requirements and constraints could not be accomplished through one single communications infrastructure; different architectures will likely be implemented for critical services and standard communication needs.

To understand to what extent current infrastructures may support smart grids, it would be necessary to precisely quantify communication requirements, in order to select suitable networking technologies and protocols among the many listed by standardization bodies and government agencies by now. The availability of multiple choices is positive towards an early and cost effective deployment of smart grids but, if too many options are available, interoperability issues or excessive complexity may arise due to different standards present in networking equipment and appliance interfaces and the need to bridge them. Harmonization and convergence of the efforts is therefore necessary to avoid the proliferation of different initiatives that may lead to redundant activities, poor interoperability and inhomogeneity of solutions.

Current networking infrastructures should be taken as the basis to develop and integrate new protocols and technologies, by carefully considering modern paradigms as virtualization and overlay, which can be used to build new services and features on top of the physical infrastructure. That requires a common agreement against a single reference architecture; the NIST reference model currently gets the largest interest, but it should grow in the future following the evolution of business, use cases and functionalities of smart grids. Wireless communications are another essential element that will provide ubiquitous and cost-effective network access to the huge number of sensors and devices building the grid; performance and reliability are key issues to assess for current and new technologies.

Security is expected to be one of the major challenges for ICT. It encompasses several aspects, related to safety of grid operations and to secure data exchanges. Communications in smart grids range from simple transfer of bulk information (from/to sensors, meters, actuators, ...) to critical real-time data for safety and security of the grid and its users. It will need to be guaranteed even in case of natural disasters or major network faults (black outs). ICT is requested to improve the resistance of the grid to perturbations and natural disasters, as well as its flexibility, by monitoring and protection of the power lines and environment conditions (current, voltage, frequency, wind-force, etc.). The design of the communication network should carefully consider capabilities such as sufficient energy back-up in the communication nodes, path/system redundancies, self healing capabilities, QoS and real-time constraints; these aspects should also seriously affect the choice of the various communication paths (wireless: GPRS, 3G, LTE, WiMAX, etc.; wired: optical fiber, etc.). On the other hand, sensitive information must also be protected against unauthorized access, eavesdropping, hijacking, alteration, spoofing and all other digital threats; current infrastructures should be enhanced and strengthened to allow for flexible, automatic, scalable and robust authentication for humans and machines.

ICT is also expected to heavily contribute to grid management and planning. This is not limited to data management infrastructures to measure, report and control electricity production and consumption, but also includes control systems, management and decision support tools that enable the integration of renewable energy sources (both large scale and distributed production). Other ICT challenges concern home energy controlling hubs and building energy management systems; the matter is collecting real-time or near-real time data on energy consumption/production in houses and buildings, to enable intelligent automation and control algorithms capable of learning from previous operations and situations and taking load-balance decisions in near-real time.

Finally, it is worth considering a side aspect that is (almost) always underestimated: ICT equipment also consumes energy. The implementation of smart grids will count on the presence of numerous meters, sensors, actuators: in theory, every power socket, switch, light, appliance could become an active element in a smart home. Summing up the energy footprint of all these elements could result in huge energy amounts, which may not be negligible in homes, where energy consumption is limited with respect to that one of equipment deployed in factories, substations and other large users. Until now, there has been little effort on the evaluation of the negative effect of such a global deployment (i.e. additional energy consumption) and its minimization, unless for some low power transmission technologies intended to be powered on battery (e.g. ZigBee). Thus, communication systems should implement energy-optimized techniques dynamically adapting performances to the needs, and implementing deep low power (sleep) modes, so to minimize the overall energy footprint while being able to cope with the communication requirements.

Bibliography

[1] K. A. Baumert, T. Herzog, and J. Pershing, "Navigating the numbers: Greenhouse gas data and international climate policy," World Resources Institute, Report, December 2005. [Online]. Available: http://pdf.wri.org/navigating_numbers.pdf

[2] T. Ackermann, E. Tröster, R. Short, and S. Teske, "Renewables 24/7 – Infrastructure needed to save the climate," Report, Greenpeace International, European Renewable Energy Council (EREC), November 2009. [Online]. Available: http://www.greenpeace.org/international/Global/international/planet-2/report/2010/2/renewables-24-7.pdf

[3] M. Fontela, S. Bacha, N. Hadjsaid, and C. Andrieu, "Functional specifications of electric networks with high degrees of distributed generation," CRISP project, Deliverable 1.1, June 2003. [Online]. Available: http://www.crisp.ecn.nl/deliverables/D1.1.pdf

[4] "Utilities/smart grid – industry solutions," web site, Cisco Systems. [Online]. Available: http://www.cisco.com/web/strategy/energy/external_utilities.html

[5] "Cisco Smart Grid – Solutions for the next-generation energy network," 2009. [Online]. Available: http://www.cisco.com/web/strategy/docs/energy/aag_c45_539956.pdf

[6] "Directive 96/92/EC of the european parliament and of the council," Official Journal of the European Union, January 1997.

[7] "ERGEG conclusions paper on smart grids," Position Paper, June 2010, ref: E10-EQS-38-05. [Online]. Available: http://www.energy-regulators.eu/portal/page/portal/EER_HOME/ EER_PUBLICATIONS/CEER_ERGEG_PAPERS/Electricity/2010/ E10-EQS-38-05_SmartGrids_Conclusions_10-Jun-2010_Corrige.pdf

[8] "Smart grid overview deliverable," Working document, ITU – Focus Group on Smart Grids, April 2011, Smart-O-34Rev.1.

[9] D. Wei, Y. Lu, M. Jafari, P. M. Skare, and K. Rohde, "Protecting smart grid automation systems against cyberattacks," *IEEE Transactions on smart grid*, 2011.

[10] K. Kursawe, "Some ideas on privacy preserving meter aggretation," Radboud University, Tech. Rep. ICIS-R11002, January 2011.

[11] K. Kursawe, G. Danezis, and M. Kohlweiss, "Privacy-friendly aggregation for the smart grid," in *Proceedings of the 11th Privacy Enhancing Technologies symposium*, ser. Lecture notes in Computer Sciences, S. Fisher-Hübner and N. Hopper, Eds., Waterloo, ON – Canada, July 2011, vol. 6794, pp. 175–191.

[12] "Revised text of use cases for smart grid deliverable," Working document, ITU – Focus Group on Smart Grids, April 2011, Smart-O-31Rev.4.

[13] "Interoperability knowledge base use cases," web pages, NIST – Smart Grid Interoperability Panel (SGIP), under construction. [Online]. Available: http://collaborate.nist.gov/twiki-sggrid/bin/view/ SmartGrid/IKBUseCases

[14] "Primer on demand-side management – with an emphasis on price-responsive programs," February 2005, prepared by Charles River Associates.

[15] P. Rowles, "Demand response and demand side management. whats the difference?" web page, February 2010. [Online]. Available: http://www.energyadvantage.com/blog/ 2010/02/demand-response-demand-side-management-what%E2% 80%99s-difference/

[16] R. Gustavsson and B. Ståhl, "ICT support for control and coordination," INTEGRAL project, Deliverable 3.1, September 2008. [Online]. Available: http://www.integral-eu.com/fileadmin/user_upload/ downloads/Deliverables/D3.1_final.pdf

[17] A. van der Welle, C. Kolokathis, J. Jansen, C. Madina, and A. Diaz, "Financial and socio-economic impacts of embracing the fenix concept," Fenix project, Deliverable 3.3, September 2009. [Online]. Available: http://fenix.iwes.fraunhofer.de/docs/documents/Project_Fenix_

2009-12-08_FINANCIAL_AND_SOCIO-ECONOMIC_IMPACTS_OF_
EMBRACING_THE_FENIX_CONCEPT_Jaap_Jansen_v1.pdf

[18] R. Gustavsson, "Sustainable virtual utilities based on microgrids," *Journal of Systemics, Cybernetics and Informatics*, vol. 6, no. 5, pp. 53–58, 2008.

[19] "Advanced architectures and control concepts for more microgrids," Project, 2009, MoreMicrogrids project. [Online]. Available: http://www.microgrids.eu/documents/esr.pdf

[20] "Amplex OpenCity," Product brief. [Online]. Available: http://www.amplex.dk/wp-content/uploads//2010/11/Product_brief_OpenCity_v002.pdf

[21] "OGEMA technology in brief," Open Gateway Energy Management Alliance (OGEMA). [Online]. Available: http://www.ogemalliance.org/downloads/ogema_technology-brief.pdf

[22] "Cisco home energy management," Product brief. [Online]. Available: http://www.cisco.com/web/consumer/pdf/data_sheet_c78_603194_v2.pdf

[23] "The birth of a EUropean Distributed EnErgy Partnership that will help the largescale implementation of distributed energy resources in Europe (EU-DEEP)," Research Project, 2004–2009, EC project reference: 503516. [Online]. Available: http://www.eudeep.com

[24] "Flexible Electricity Networks to Integrate the 'eXpected 'energy evolution' (fenix)," Research Project, 2005–2009, EC project reference: 518272. [Online]. Available: http://www.fenix-project.org

[25] "A european strategy for sustainable, competitive and secure energy," Green paper of the Commission of the European Communities, March 2003, COM(2006) 105 final. [Online]. Available: http://eur-lex.europa.eu/LexUriServ/LexUriServ.do?uri=COM:2006:0105:FIN:EN:PDF

[26] "Energy 2020 – A strategy for competitive, sustainable and secure energy," Communication from the Commission to the European Parliament, the Council, the European Economic and Social Committee and the Committee of the Regions, November 2010, COM(2010) 639 final. [Online]. Available: www.energy.eu/directives/com-2010-0639.pdf

[27] "Strategic Energy Technology Plan (SET Plan)," web site. [Online]. Available: http://ec.europa.eu/energy/technology/set_plan/set_plan_en.htm

[28] "Towards a european strategic energy technology plan," Communication from the Commission to the Council, the European Parliament, the European economic and social Committee and the Committee of the regions, January 2007, COM(2006) 847 final. [Online]. Available: http://eur-lex.europa.eu/LexUriServ/LexUriServ.do?uri=COM:2006:0847:FIN:EN:PDF

[29] "The European Strategic Energy Technology Plan SET-Plan – Towards a low-carbon future," Publications Office of the European Union, 2010. [Online]. Available: http://ec.europa.eu/energy/technology/set_plan/doc/setplan_brochure.pdf

[30] "The EU climate and energy package," web site, October 2010. [Online]. Available: http://ec.europa.eu/clima/policies/package/index_en.htm

[31] "Directive 2009/28/EC of the european parliament and the council," Journal of the European Union, April 2009, article 16.

[32] "Directive 2009/72/EC of the european parliament and the council," Official Journal of the European Union, July 2009, annex 1.

[33] "Directive 2009/73/EC of the european parliament and the council," Official Journal of the European Union, July 2009, annex 1.

[34] "Directive 2010/31/EC of the european parliament and the council," Journal of the European Union, June 2010, article 8.

[35] "European Technology Platform for the Electricity Networks of the Future – Smartgrids," web site. [Online]. Available: http://www.smartgrids.eu

[36] "Vision and strategy for Europe's electricity networks of the future," European Commission ref. EUR 22040, European Technology Platform Smartgrids, 2006, iSBN 92-79-01414-5. [Online]. Available: http://ec.europa.eu/research/energy/pdf/smartgrids_en.pdf

[37] "Strategic research agenda for Europe's electricity networks of the future," European Technology Platform Smartgrids, 2007. [Online]. Available: http://www.smartgrids.eu/documents/sra/sra_finalversion.pdf

[38] "Strategic deployment document for Europe's electricity networks of the future," European Technology Platform Smartgrids, April 2010. [Online]. Available: http://www.smartgrids.eu/documents/SmartGrids_SDD_FINAL_APRIL2010.pdf

[39] "Smartgrids ETP forum," web site. [Online]. Available: http://www.smartgrids.eu/?q=node/160

[40] "The european electricity grid initiative (EEGI) roadmap 2010-18 and detailed implementation plan 2010-12," May 2010, ENTSO-E, EDSO-SG. [Online]. Available: http://www.smartgrids.eu/documents/EEGI/EEGI_Implementation_plan_May%202010.pdf

[41] "Smart Grids Task force," web site. [Online]. Available: http://ec.europa.eu/energy/gas_electricity/smartgrids/taskforce_en.htm

[42] "Task Force Smart Grids - vision and work programme," March 2010. [Online]. Available: http://ec.europa.eu/energy/gas_electricity/smartgrids/doc/work_programme.pdf

[43] "Functionalities of smart grids and smart meters," Report, EU Commission Task Force for Smart Grids, Expert Group 1, December 2010. [Online]. Available: http://ec.europa.eu/energy/gas_electricity/smartgrids/doc/expert_group1.pdf

[44] "Regulatory recommendations for data safety, data handling and data protection," EU Commission Task Force for Smart Grids, Expert Group 2, February 2011. [Online]. Available: http://ec.europa.eu/energy/gas_electricity/smartgrids/doc/expert_group2.pdf

[45] "Roles and responsibilities of actors involved in the smart grids deployment," Report, EU Commission Task Force for Smart Grids, Expert Group 3, April 2011. [Online]. Available: http://ec.europa.eu/energy/gas_electricity/smartgrids/doc/expert_group3.pdf

[46] "Smart grid aspects related to gas," Report, EU Commission Task Force for Smart Grids, Expert Group 4, June 2011. [Online]. Available: http://ec.europa.eu/energy/gas_electricity/smartgrids/doc/expert_group4.pdf

[47] "Standardization mandate to CEN, CENLEC and ETSI in the field of measuring instruments for the development of an open architecture for utility meters involving communication protocols enabling interoperability," European Commission mandate, March 2009, M/441 EN.

[48] "Standardization mandate to european standardisation organisations (ESOs) to support european smart grid deployment," European Commission mandate, March 2011, M/490 EN.

[49] "Standardization mandate to CEN, CENELEC and ETSI concerning the charging of electric vehicles," European Commission mandate, June 2010, M/468 EN.

[50] D. K. Owens, B. L. Wolff, and R. McMahon, "Electric utilities: the power of opportunities," EEI 2011 Wall Street Briefing, New York, NY – USA, February 2011. [Online]. Available: http://www.eei.org/ourissues/finance/Documents/Wall_Street_Briefing_2011.pdf

[51] "Blueprint for a secure energy future," White House document, March 2011. [Online]. Available: http://www.whitehouse.gov/sites/default/files/blueprint_secure_energy_future.pdf

[52] "Federal Smart Grid Task Force," web site. [Online]. Available: http://www.oe.energy.gov/smartgrid_taskforce.htm

[53] "Office of Electricity Delivery and Energy Reliability – Smart Grids," web site. [Online]. Available: http://www.oe.energy.gov/smartgrid.htm

[54] "NIST – Smart Grid," web site. [Online]. Available: http://www.nist.gov/smartgrid/

[55] "NIST framework and roadmap for smart grid interoperability standards," NIST Special Publication 1108, Office of the National

Coordinator for Smart Grid Interoperability, January 2010, release 1.0. [Online]. Available: http://www.nist.gov/public_affairs/releases/upload/smartgrid_interoperability_final.pdf

[56] "Smart Grid Interoperability Panel (SGIP)," web site. [Online]. Available: http://www.sgipweb.org/

[57] "The Green Smart Grid Initiative," web site. [Online]. Available: http://www.greensmartgridinitiative.org/

[58] "A policy framework for the 21st century grid: Enabling our secure energy future," National Science and Technology Council, Office of Science and Technology Policy, Executive Office of the President of the United States, Washington, D.C. – USA, June 2011. [Online]. Available: http://www.whitehouse.gov/sites/default/files/microsites/ostp/nstc-smart-grid-june2011.pdf

[59] "CRISP – distributed intelligence in CRItical infrastructure for Sustainable Power," Research Project, 2002–2006, EC project reference: ENK5-CT-2002-00673. [Online]. Available: http://www.crisp.ecn.nl

[60] "ADDRESS – Active Distribution networks with full integration of Demand and Distributed energy RESourceS," Research Project, 2008–2012, EC project reference: 207643. [Online]. Available: http://www.addressfp7.org/

[61] "Smart-A – Smart Domestic Appliances in Sustainable Energy Systems," Research Project, 2007–2009, contract number: EIE/06/185/SI2.447477. [Online]. Available: http://www.smart-a.org

[62] "More microgrids," Research Project, 2006–2009. [Online]. Available: http://www.microgrids.eu

[63] "OpenNode – Open Architecture for Secondary Nodes of the Electricity SmartGrid," Research Project, 2010–2012, EC project reference: 248119. [Online]. Available: http://www.opennode.eu

[64] "OPEN METER – Open public extended network metering," Research Project, 2009–2011, EC project reference: 226369. [Online]. Available: http://www.openmeter.com

[65] "HYDRA – Networked Embedded System middleware for heterogeneous physical devices in a distributed architecture," Research Project, 2006–2010, EC project reference: 034891. [Online]. Available: http://www.hydramiddleware.eu

[66] "ADINE – Active distribution network," Research Project, 2007–2010, EC project referece: 38533. [Online]. Available: http://cordis.europa.eu/fetch?CALLER=FP6_PROJ&ACTION=D&DOC=7&CAT=PROJ&QUERY=01308ef4acce:7a9f:558cf87b&RCN=85682

[67] "MetaPV – Metamorphosis of Power Distribution: System Services from Photovoltaics," Research Project, 2009–2014, EC project reference: 239511. [Online]. Available: http://www.metapv.eu

[68] "Night Wind – Grid Architecture for Wind Power Production with Energy Storage through load shifting in Refrigerated Warehouses," Research Project, 2006–2008, EC project reference: 20045. [Online]. Available: http://cordis.europa.eu/fetch?CALLER=FP6_PROJ&ACTION= D&DOC=1&CAT=PROJ&RCN=79800

[69] "EWIS – European Wind Integration Study," Research Project, 2007–2009, EC project reference: 38509. [Online]. Available: http://www.wind-integration.eu

[70] "G4V – Grid for Vehicles," Research Project, 2010–2011, EC project reference: 241295. [Online]. Available: http://www.g4v.eu

[71] "MERGE – Mobile Energy Resources in Grid of Electricity," Research Project, 2010–2011, EC project reference: 241399. [Online]. Available: http://www.ev-merge.eu

[72] "GROW-DERS – Grid Reliability and Operability with Distributed Generation using Flexible Storage," Research Project, 2007–2010, EC project reference: 38665. [Online]. Available: http://www.growders.eu

[73] "DEHEMS – Digital Environment Home Energy Management System," Research Project, 2008–2011, EC project reference: 224609. [Online]. Available: http://www.dehems.eu

[74] "AIM – A novel architecture for modelling, virtualising and managing the energy consumption of household appliances," Research Project, 2008–2010, EC project reference: 224621. [Online]. Available: http://www.ict-aim.eu

[75] "Beywatch – Building Energy Watcher," Research Project, 2008–2011, EC project reference: 223888. [Online]. Available: http://www.beywatch.eu

[76] "IntUBE – Intelligent Use of Buildings' Energy Information," Research Project, 2008–2011, EC project reference: 224286. [Online]. Available: http://www.intube.eu

[77] "eDIANA – Embedded Systems for Energy Efficient Buildings." Research Project, 2009–2012, EC project reference: 100012. [Online]. Available: http://www.artemis-ediana.eu

[78] "BeAware – Boosting Energy Awareness," Research Project, 2008–2011, EC project reference: 224557. [Online]. Available: http://www.energyawareness.eu/beaware

[79] "INTEGRAL – Integrated ICT-platform based Distributed Control in Electricity Grids," Research Project, 2007–2010, EC project reference: 38576. [Online]. Available: http://integral-eu.com

[80] "SmartHouse/SmartGrid," Research Project, 2008–2011, EC project reference: 224628. [Online]. Available: http://www.smarthouse-smartgrid.eu

[81] CEN, CENELEC and ETSI, "Final report of the CEN/CENELEC/ETSI Joint Working Group on standards for smart grids," May 2011. [Online]. Available: ftp://ftp.cen.eu/CEN/Sectors/List/Energy/SmartGrids/SmartGridFinalReport.pdf

[82] "Machine-to-Machine communications (M2M); smart metering use cases," ETSI Technical Report, May 2010, TR 102 691 V1.1.1.

[83] "Machine to-Machine – Applicability of M2M architecture to smart grid networks – Impact of smart grids on M2M platform," Draft ETSI Technical Report, September 2010, TR 102 935 V0.0.1.

[84] "IEC – Smart grids: global standards for optimal electricity delivery," web site. [Online]. Available: http://www.iec.ch/smartgrid/

[85] "IEC smart grid standardization roadmap," June 2010, edition 1.0. [Online]. Available: http://www.iec.ch/smartgrid/downloads/sg3_roadmap.pdf

[86] "ITU – Focus group on smart grids (FG Smart)," web site. [Online]. Available: http://www.itu.int/en/ITU-T/focusgroups/smart/Pages/Default.aspx

[87] "Revised draft deliverable on smart grid architecture," Working document, ITU – Focus Group on Smart Grids, April 2011, Smart-O-33Rev.4-2.

[88] "Deliverable requirements," Working document, ITU – Focus Group on Smart Grids, April 2011, Smart-O-32Rev.3.

[89] "IEEE Smart Grid," web site. [Online]. Available: http://smartgrid.ieee.org/

[90] F. Baker and D. Meyer, "Internet protocols for the smart grid," RFC 6272, June 2011. [Online]. Available: http://tools.ietf.org/rfc/rfc6272.txt

[91] "IPv6 over Low power WPAN (6lowpan)," web site. [Online]. Available: http://datatracker.ietf.org/wg/6lowpan/charter/

[92] "Routing Over Low power and Lossy networks (roll)," web site. [Online]. Available: http://datatracker.ietf.org/wg/roll/charter/

[93] "Constrained RESTful Environments (core)," web site. [Online]. Available: http://datatracker.ietf.org/wg/core/charter/

[94] "Energy management (eman)," web site. [Online]. Available: http://datatracker.ietf.org/wg/eman/charter/

[95] "Guidelines for smart grid cyber security," The Smart Grid Interoperability Panel - Cyber Security Working Group, August 2010, nISTIR 7628, 3 volumes. [Online]. Available: http://csrc.nist.gov/publications/PubsNISTIRs.html#NIST-IR-7628

[96] "JTC 1 special working group on smart grid," web site. [Online]. Available: http://www.jtc1smartgrid.org/

[97] "TIA TR-50 smart device communications," web site. [Online]. Available: http://www.tiaonline.org/standards/committees/committee.cfm?comm=tr-50

[98] "TIA TR-51 smart utility networks," web site. [Online]. Available: http://www.tiaonline.org/standards/committees/committee.cfm?comm=tr-51

[99] "SGCC framework and roadmap for strong & smart grid standards," Report, August 2010. [Online]. Available: http://collaborate.nist.gov/twiki-sggrid/pub/SmartGrid/SGIPDocumentsAndReferencesSGAC/China_State_Grid_Framework_and_Roadmap_for_SG_Standards.pdf

[100] J. Liu, X. Li, X. Chen, Y. Zhen, and L. Zeng, "Applications of Internet of Things on smart grid in China," in *13th International Conference on Advanced Communication Technology (ICACT)*, Seoul, South Korea, Feb. 2011, pp. 13–17.

[101] "IEC smart grid standards mapping solution," web pages. [Online]. Available: http://www.iec.ch/smartgrid/mappingtool/

[102] "Final report," Smart Meter Co-ordination Group, December 2009. [Online]. Available: http://www.nbn.be/NL/SM_CG_FinalReport_2009_12_10.pdf

[103] R. Fielding and R. Taylor, "Principled design of the modern web architecture," *ACM Transactions on Internet Technology*, vol. 2, no. 2, pp. 115–150, May 2002.

Author Contact Information

Raffaele Bolla is with the Department of Naval, Electrical, Electronics and Telecommunication Engineering, DITEN-University of Genoa, Via Opera Pia 13, 16145 Genoa, Italy, Tel.: +390103532071, Fax: +390103532154, Email: raffaele.bolla@unige.it. Flavio Cucchietti is with Telecom Italia, TILab., Via Olivetti 6, 10148 Turin, Italy, Email: flavio.cucchietti@telecomitalia.it. Matteo Repetto is with CNIT – Research Unit of Genoa, Via Opera Pia 13, 16145 Genoa, Italy, Email: matteo.repetto@cnit.it.

8

Moving a Processing Element from Hot to Cool Spots: Is This an Efficient Method to Decrease Leakage Power Consumption in FPGAs?

Amor Nafkha

SCEE-SUPELEC, France

Pierre Leray

SCEE-SUPELEC, France

Yves Louet

SCEE-SUPELEC, France

Jacques Palicot

SCEE-SUPELEC, France

CONTENTS

8.1   Introduction

Even recently, Sustainable Development (SD) was only the concern of green groups. However, today, SD has become a paramount issue and an aspiration of long-term civilization development of human beings since "Resolution 42/187 of the United Nations General Assembly" was passed in December 1987. The Brundtland Commission of the United Nations defined SD as the development that "meets the needs of the present without compromising the ability of future generations to meet their own needs" [1]. From then on, several United Nations' Conferences (from Rio de Janeiro-1992 to Durban-2011) confirmed this important issue. One of the most obvious aspects and challenges of SD is the Earth climate change and the ever-growing CO_2 emission. Currently, 3% of the world-wide energy is consumed by the ICT (Information and Communications Technology) infrastructure, which causes about 2% of the world-wide CO_2 emissions and surprisingly is comparable to the world-wide CO_2 emissions due to all commercial airplanes. The ICT sector's carbon footprint is expected to quickly grow to 1.4 Giga ton CO_2 equivalents by 2020, nearly 2.7% of the overall carbon footprint from all human activities [2]. These values of carbon footprint are extremely impressive. They have been confirmed by a lot of scientific studies and reported in many relevant international conferences and workshops, such as the "Next Generation Wireless Green Networks Workshop" held in SUPELEC in November 2009 [3]. Basically, one should deal with the fundamental challenges with a twofold aim, in order to attempt to solve these problems:

- Decrease the ICT footprint itself,

- Use ICT so as to decrease the "human beings" activities footprint.

This chapter deals with electronics power consumption in nanoscale CMOS technology. Traditionally the Integrated Circuit (IC)'s power consumption was due to dynamic power consumption. Consequently, in order to decrease frequency while increasing ICs throughput, the solution was to parallelize processes. Nevertheless one of the most significant power related subjects that arose recently is static power leakage due to the leakage current. It was stated in [4] that the leakage power consumption is strongly correlated to the circuit area. Therefore a new trade-off between parallelism and power consumption should be found. This was discussed, for example, in [5]. Leakage power represents a significant share of the total power dissipation especially for 65nm technology and below. A lot of papers are dealing with leakage power consumption itself within the framework of nanoscale technology [4,6]. If, in the past, FPGAs were mainly concerned with dynamic power consumption [7], it remains no more true with new technologies (65nm and below) [8] and with the large number of transistors inside a FPGA circuit: the leakage power consumption becomes more and more important both in the used and unused part of the component. Section 2 will give more details on power consumption origins. A lot of works carried out in the recent years studied ways to reduce leakage power consumption in CMOS [9–11] and particularly for FPGA [12,13].

Real measurements in [14,15] show that dynamic power consumption may be decreased with the dynamic and partial reconfiguration (DPR), and this saved power consumption depends on the application and the design's architecture. Dynamic activation and deactivation of the FPGA's fabric clock network can be implemented to decrease dynamic power consumption. However, neglecting the reconfiguration process power consumption makes only sense if the reconfiguration time is very low compared to the application execution time's constraints. The study in [12] indicates that the polarity of the inputs in FPGA hardware structures may significantly impact leakage power consumption (average reduction of roughly 25%). In [16,17], it has been shown that the system power consumption can be decreased by exploiting the dynamic partial reconfiguration capability of FPGAs using power reduction techniques such as clock gating, hardware deactivation and removal. To the best of our knowledge, the benefit of dynamic partial reconfiguration to reduce the die temperature and then the static power consumption has not been yet evaluated.

In this work, starting with the high correlation between chip temperature and static power consumption, associated with the possibility to move a Processing Element (PE) from a hot spot to a cooler one (already published in the literature for throughput improvement [18], and for dynamic power consumption management using DPR, see previous paragraph), we study this moving possibility so as to decrease leakage power consumption. This idea was first proposed in [19]. The aim of this chapter is to conclude on the feasibility of this idea.

To summarize, we have to answer to the question:"Is there, in a FPGA, a sufficient temperature difference between hot and colder areas to decrease leakage power consumption if a PE is moved between these two zones?" If the answer is yes, then there are subsequent questions:

- Is the temperature gradient sufficiently slow to avoid having a moving process which is repeated too often?

- Is this temperature gradient's associated time greater or smaller than the reconfiguration time?

- Whatever the moving process is (DPR is good candidate, an high speed DPR technique has been proposed in [20]), it will introduce some power consumption overhead, Is the total power consumption budget positive?

The chapter is focused on temperature difference matter (the fundamental question) and is organized as follows. The next section gives a brief overview of the different power consumption sources and some definitions. Section 8.3 describes the Digital Thermal Sensor used to measure spots temperature. Section 8.4 gives both the hardware and software's set-up. In this section, the two different PEs used to increase the temperature are described. Experimental results concerning the temperature difference between hot and cool spots inside the FPGA are provided and analyzed in section 8.5. Finally, some discussion and conclusions are given in the last section.

8.2 Power Consumption Sources

The total power consumption in the FPGA can be defined as: $P_t = P_{st} + P_{sc} + P_{dy}$, where P_{st} defines the static power that a circuit consumes even when it is in the standby mode, P_{sc} denotes the short-circuit power, which is due to the short-circuit current flowing from power supply to ground when both p-network and n-network are ON, and P_{dy} defines the dynamic power consumption that occurs when the output signal of a CMOS logic cell makes a transition. Each component of the power consumption is discussed in more detail in the following subsections.

8.2.1 Static Power

The static power is the power consumed by the device due to leakage currents when there is no activity or switching in the design. Using finer semiconductor technologies has caused an increase in static power consumption in FPGAs. In fact, as transistor size decreases and lower voltages are used, a greater leakage current occurs in the transistor channel when the transistor is in the "OFF" state. The three dominant sources of leakage current in a CMOS circuit are: sub-threshold leakage, gate leakage and drain junction leakage. Those three major leakage current mechanisms are illustrated in Fig.8.1.

The value of the sub-threshold leakage current is significantly increased with technology scaling. This component increases exponentially with temperature. However, the rate of leakage reduction with temperature is diminished with technology scaling. Note that the sub-threshold leakage current is the dominant leakage mechanism. The static power of a logic circuit can be expressed as follows [21]:

$$P_{st} = \sum I_{leak} V_{dd}, \tag{8.1}$$

where I_{leak} is the current that flows between the supply rails in the absence of switching activity, V_{dd} is the supply voltage. Reducing static power dissipation is ultimately the key to maximizing stand-by time and battery life.

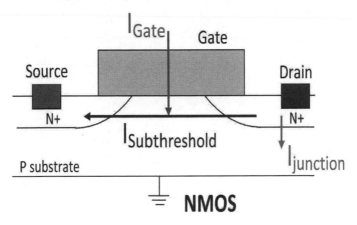

FIGURE 8.1 **Leakage currents**

8.2.2 Short-Circuit Power

During the switching operation, a conducting path exists through the pull up and pull down network of a gate and, as a consequence, a short-circuit current is occurring. In [22], the short-circuit power consumption of a simple inverter is defined as:

$$P_{sc} = \frac{\beta}{12}(V_{dd} - V_{tn} - V_{tp})^3 \frac{\tau}{T}, \tag{8.2}$$

where β denotes the gain factor of an MOS transistor, V_{tn}, V_{tp} are the threshold voltages of nMOS and pMOS transistors respectively. τ denotes the input rise and fall times, and T is the period-time of a signal $(= 1/f)$.

8.2.3 Dynamic Power

Dynamic power is the power consumed by the device during switching events in the core or in the input/output pins. The main variables affecting this power are the capacitance's charging and discharging, frequency toggling, and voltage supplying. Transistors are used for logic and programmable interconnections, so in general a large portion of the dynamic power is due to the routing fabric. The formula for dynamic power is given by:

$$P_{dy} = \frac{\alpha}{2}CV_{dd}^2 f, \tag{8.3}$$

where α denotes the percent of circuit that switches each cycle, C is the capacitance, V_{dd} and f are the voltage supply and the toggle frequency respectively. The core FPGA voltage supply and capacitance are reduced with each new process node technology. Note that dynamic power is not a function of transistor size, but rather a function of switching activity and load capacitance. Thus, it is data dependent. A very large number of papers have considered the distribution of dynamic power consumption in FPGAs (among them, [7]).

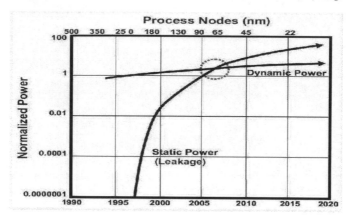

FIGURE 8.2 Static and dynamic power as a function of Node Technology [23].

8.2.4   Dynamic Versus Static Power

Until recently, dynamic power has been the dominant source of power consumption. Figure 8.2 shows the cross-over point, where static power overtakes dynamic power, to be at 65 nm. Leakage power has increased drastically in the scaled technologies because of the exponential relationship of sub-threshold current to threshold voltage. Also, since the voltage supply is not scaled down as much as the feature's size in order to have high-performance transistors, gate leakage has increased because of low gate oxide thickness. Therefore, with smaller process geometries, worsening leakage current causes static power to dominate the power consumption.

Figure 8.3 shows the total power consumption of a 15mm die fabricated, as a function of the die temperature, made of a 100 nm technology with a voltage supply of 0.7 V. Although the leakage power is about 9% of the total power at 40°C, it increases to 49% of the total power at 100°C. This clearly shows the necessity of using techniques for temperature reduction. In the rest of this chapter, we will perform a deep analysis of the temperature's behavior inside the FPGA. Of course, for that purpose, we have to measure this temperature. The next section will describe an efficient way to perform this measurement.

8.3   Digital Thermal Sensor

The self-heating of the FPGA is a very important factor that impacts the total power consumption. On-chip measurements of local temperature present an opportunity to incorporate temperature management techniques and performance optimization into the FPGA fabric. This section presents the design of

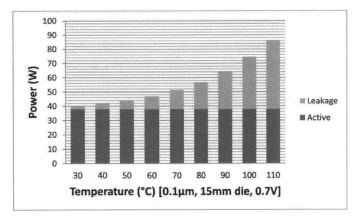

FIGURE 8.3 Power consumption of a die as a function of temperature [24] ©
2005 IEICE, Permission No.1 2RA0010.

a sensors system that continually measures the temperature values at various
locations on the FPGA. The main idea behind this work is to perform a smart
power management when an high local die temperature is detected. The sensors
system comprises of an array of digital thermal sensor implemented on
the FPGA device.

The chip's temperature can be measured using the junction forward voltage
of the clamping diodes located in the FPGA pads as suggested by [25].
This method is based on the voltage current relationship in an ideal diode.
Even though of this method has the advantage of being immune to the power
supply variations, it requires some additional external analog circuits. Recently,
Xilinx FPGA fabric series included an analog voltage and temperature
sensor connected to an A/D converter called System Monitor [26]. This sensor
is located at the center of the die, and provides an average temperature.
Recently infrared imaging used a powerful tool for thermal sensing [27]. Using
the thermal images, authors investigated the magnitude of thermal gradients
and hotspots in die. This method is able to reduce the thermal tracking errors
by 40%. However, this latter technique lack from the fact that it needs some
additional hardware equipment.

An original work in [28] proposes a simple way to measure chip's heating
using a ring oscillator associated to a frequency counter. This work has been
completed in [29,30]. A ring oscillator consists of a feedback loop that includes
an odd number of inverters needed to produce the phase shifting that maintains
the oscillation. In [31], Lopez-Buedo et al. proposed the minimization of
the operation frequency of a ring oscillator in order to minimize the problems
related to self-heating, power consumption, and counter size. Due to the lack
of knowledge of hot spot locations inside the die, all thermal sensors nodes
are spread in a uniform distribution way across the FPGA according to the
works published in [30,32].

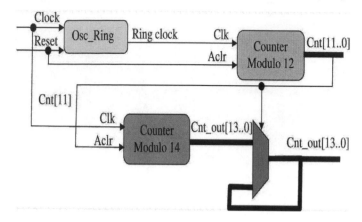

FIGURE 8.4 Temperature sensor block diagram.

Exhibiting a linear dependence of oscillation frequency on chip's temperature, at a fixed value of V_{CCINT} voltage supply, the ring oscillator circuit can be embedded inside any FPGA fabric in conjunction with a counter. It paves the way for efficient and highly accurate temperature measurement. The ring oscillator is characterised by its reduced area and power overhead which allows it to be replicated as many times as needed, so as to create a FPGA thermal map. The resulting period of the ring oscillator is defined by equation 8.4, and it is twice the sum of the delays of all inverter elements that make the loop chain.

$$T = 2Nt_p, \tag{8.4}$$

where N denotes the number of odd inverters and t_p is the propagation delay of a single inverter that is part of the loop.

Figure 8.4 shows the RTL schematic of the proposed temperature sensor based on a ring oscillator. The thermal sensor can be divided into three major modules. The first module of the sensor is made of a ring oscillator controlled by an external enabled signal, the second module is 12-bit cyclic counter and the last module is a 14-bit reference time counter which indicates the number of rising edge reference 50 MHz clock between two events.

During the process, the first 12-bit counter is clocked by the output of the ring oscillator module, and generates a boolean event to the second 14-bit reference time counter. This boolean is equal to *"true"* if the 12-bit counter value is equal to 2^{12}, otherwise the signal's value equals *"false"*. The reference 14-bit counter determines the number of rising edge reference clocks that have been counted between two *"true"* events. The reference counter value is sent back to the PC via the RS232 port (runs at $19.6kHz$). The FPGA temperature is gathered from an on-chip hardware sensor called System Monitor which is located at the center of the FPGA's core and developed around a 10 bits analog-to-digital converter.

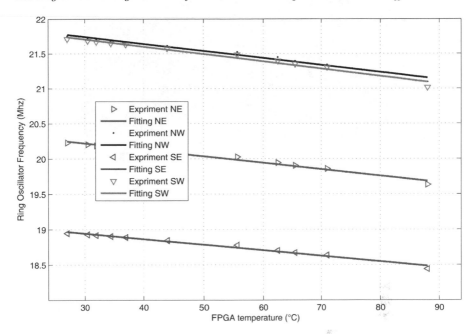

FIGURE 8.5 Measured ring oscillator's frequency in MHz(million-Hertz) versus FPGA's temperature in degree Celsius

8.3.1 Ring Oscillator vs. System Monitor

In order to analyse the relationship between the FPGA's temperature and the ring oscillator's frequency, we have used a Peltier element. To provide a large and stable FPGA's temperature range, one side of a Peltier cell is placed on the FPGA and the other side is connected to an aluminum heatsink. To keep the system flexible and responding fast when cooling down is desired, it was useful to keep the heatsink always attached. The Peltier cell is able to provide a stable temperature within our desired range (25-85°C). We can catch the desired temperature by increasing or decreasing the Peltier current/voltage supply. The ring oscillator's inverters are implemented in LUT (Look Up Table), and their RLOC (Relative location Constraints) were used to equalize the physical distances between inverters.

To have a good estimation of the ring oscillator frequency as a function of an FPGA's temperature, we have placed four sensors inside the FPGA at the following locations: Northeast (NE), Southeast (SE), Southwest (SW) and Northwest (NW) (Figure 8.6). In Figure 8.5, the y-axis represents the ring oscillators' frequencies and the x-axis represents the FPGA's temperature given by the System Monitor. We noticed that the experimental results are given when the FPGA V_{CCINT} voltage supply is equal to 0.998V. The variation of frequency between the different locations is due to the propagation delay's

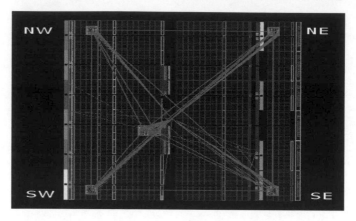

FIGURE 8.6 Ring oscillators' location inside the FPGA.

variation which exists between the ring oscillator's inverters and the various locations, and which is a Xilinx place-and-route related phase. Experimental results show that there is indeed a high correlation and proportionality between the propagation delay inside a ring oscillator and the FPGA's temperature. Linear approximation seems to be a near fit of the recorded curve. Below is the equation that expresses the mean ring oscillator's frequency value for FPGA's temperature conversion, by means of the traditional linear first-order fitting equation:

$$f = -0.0094 * T + K, \qquad (8.5)$$

Where f is the ring oscillator's frequency in MHz, T is the temperature in degree Celsius(°C), and K is a calibration constant which can be easily calculated for a given initial temperature and ring oscillator's frequency. The frequency value varies linearly with temperature. The slope of the line, equal to 9.4KHz/°C, gives us the effective change in ring oscillator frequency corresponding to an increase or decrease of temperature.

8.3.2 Ring Oscillator vs. FPGA V_{CCINT} Voltage Supply

One of the problems faced, in the experimental measurement of the ring oscillator's frequency, is a significant variation of the gates' propagation delay , which is due to the FPGA's V_{CCINT} voltage supply's change. The impact of the different voltage supply scaling techniques on the propagation delay of the CMOS gates has been widely investigated in [33–36]. It is well known that the voltage supply's scaling is one of the most used technique to reduce the power consumption. However, this approach has not been used in the present chapter. Figure 8.7 shows the mean ring oscillator's frequencies when the FPGA's V_{CCINT} supply voltage varies from 0.97V to 1.016V, for a given chip's temperature. Scaling down V_{CCINT} alleviates the ring-oscillator's propagation delay,

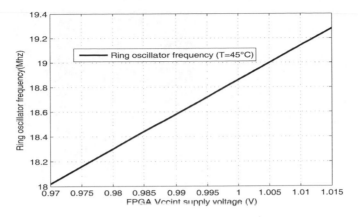

FIGURE 8.7 Change in ring oscillator frequency due to FPGA VCCINT supply voltage variation.

whose increase is an issue. As a consequence, the frequency is decreased. In order to provide various on-board V_{CCINT} voltage values, the used Xilinx ML550 Networking Interfaces platform implements a margin control circuitry providing ±2.5%, ±5%, ±7.5%, and ±10% voltage supply's adjustments. In order to overcome the problem of the supply sensitivity of our ring oscillator, we will be using a fixed value of $V_{CCINT} = 0.998$V, voltage supply, for all results given in the rest of this chapter.

8.4   Experimental Setup

We have performed experiments on the Virtex-5 FPGA to study the FPGA power consumption's behaviors. The ML550 development board, (see Figure 8.8, from Xilinx) would be well suited to power consumption's measurement for several reasons [37]. The FPGA on this board is a XC5VLX50T which is a Virtex5 LX series. This board provides us with 5 power rails (core, IOs, peripherals) and current sense resistors which could simplify the experimental measurement. The Virtex5 FPGAs generally requires at least two different voltages. The recommended Virtex5 core voltage, designated V_{CCINT}, is $1.0 ± 10\%$V. Depending on the I/O standard being implemented, the Virtex5 I/O voltage supply, designated V_{CCO}, can vary from 1.2V to 3.3V. Moreover, Xilinx defines an auxiliary voltage, V_{CCAUX}, which is recommended to operate at $2.5 ± 10\%$V to supply the FPGA's clock resources. The Virtex5 FPGA also contains a System Monitor which features on-die temperature and voltage measurement capabilities that provide valuable information for the development, evaluation and quality assurance process of power designs. . The board

FIGURE 8.8 Virtex-5 FPGA ML550 Development Board using a Peltier module and an instrumentation amplifiers [37].

hosts a two header connector which provides test points for the ML550 power regulators. Moreover, to measure different currents drained by the FPGA, the ML550 board contains a series shunt $10m\Omega \pm 1\%$ (named 3W Kelvin current sense resistors) on each voltage's regulator lines. Thus, the current will be the voltage across the shunt resistor divided by the resistance value itself. Since the sensitivity of the V_{CCINT}, V_{CCO}, and V_{CCAUX} sensors are pretty low $(0.5 - 2mV)$, voltage amplifiers are needed. To this end, we have used integrated instrumentation amplifiers AD620, from Analog Devices. The AD620 has a feature to increase gain between $1 - 10.000$ times with an external resistor. Those amplifiers permit to regulate the gain, thanks to the sole change of a single resistance, called R_G. We wanted to reach, at the output of the voltage amplifier, a voltage value ranging between $100 - 400mV$, then, Depending on the component's availability in the laboratory, gains of about 100, 275, and 2750 respectively for the V_{CCINT}, V_{CCO}, and V_{CCAUX} were established using resistances of 449, 180, and 18Ω.

Figure 8.8 shows the schematic of connections between ML550 sensors and amplifiers. The output voltage pins from the board are the differential outputs from the sensors. In order to observe the behaviour of the ring oscillator across a wide temperature range, the Peltier module is installed above the virtex5 chip. In the rest of this section, we will introduce different hardware/software modules used for different experiments.Since we would like to implement two types of PEs: a combinatorial based one and another one which is software based we define the two following PEs.

8.4.1 Processing Element Number 1: Random Number Generator

In order to increase the temperature at one of the thermal sensor location inside the FPGA region, we have designed a parametric processing unit based on the random number generator. A 512-bit random number is being generated with the help of the exclusive OR (XOR) gate. The most significant digit and the least significant digit are both processed through the XOR gate and give the first bit of the random number and then this equation runs inside the loop 512 times so as to give a 512 bits random number. So the generated number contains 512 random bits per clock cycle. To significantly increase the temperature, the used clock's frequency can be doubled by using the Digital Content Manager (DCM)'s components inside the FPGA.

This PE is a combinatorial type one whereas the next PE described in the following section is software based.

8.4.2 Processing Element Number 2: MicroBlaze

In order to increase the temperature of a predefined place in the FPGA, a MicroBlaze embedded soft-core processor has been implemented. The advantage of using this type of soft processors is that they are highly reconfigurable and can be customized according to our needs. Here, the Microblaze runs at $125MHz$, the floating point unit being enabled, and we chose a 64×64 standard matrix multiplication as an application target. MicroBlaze uses both the instruction and data sides, and its local memory buses connect it to a dual-ported primarily on-chip block random access memory, via separate interface controllers.

8.5 Experiments and Results

Power consumption and performance measurement were done for a Xilinx Virtex5 FPGA. As said in section 8.3.2 the results are based on the measurements made at a nominal VCCINT level of $0.998V$. The clock frequency was kept fixed at 100 MMz. The relationship between temperature, leakage power consumption and the on-chip temperature has been explored and the results are presented in the following subsections.

8.5.1 Temperature vs. Leakage Current Consumption

In order to measure leakage current for different die-temperatures, experimental and analytical results are carried out using Agilent 34401A multimeter and Xilinx XPower analyzer tool (XPA) [38], respectively. The XPA is part of the

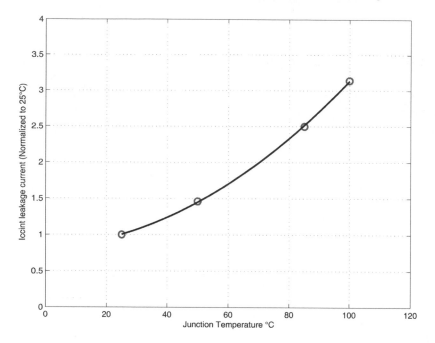

FIGURE 8.9 **Leakage current consumption versus junction temperature: curve from [38].**

Xilinx ISE design suite and provides an accurate way to analyze the power profile of post Place & Route designs. Table 8.1 shows the measured leakage current for the various die-temperature. At that level we are considering current leakage rather than power leakage (bearing in mind that flicking trough both notions is obvious) in order to be in accordance with the Xilinx general curve from [38] given in Figure 8.9.

TABLE 8.1
FPGA I_{CCINT} Leakage Current consumption using experimental and XPA tool.

T(°C)	29	36	38	40	51	53	55	59	62	68
XPA (mA)	298	325	334	343	391	401	411	433	451	488
Exp(mA)	314	336	344	350	410	420	432	468	490	540

Measuring voltages across different sense resistors enables the computation of the FPGA I_{CCINT} leakage current consumption at any time. Figure 8.10 shows the leakage current consumption measurements corresponding to different FPGA's fabic temperature. Data points were obtained via the running of each experiment until temperature stabilization. This takes several minutes.

In Figure 8.10, the continuous line presents the best-fitting quadratic func-

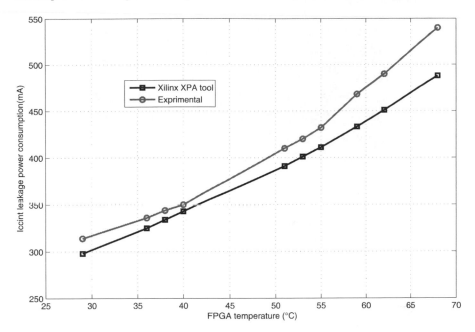

FIGURE 8.10 Leakage current's consumption versus FPGA's fabric temperature.

tion. As regard the temperature T parameter, and the data of the present experiment. It can be defined as:

$$I_{CCINT} = 0.0899T^2 - 2.889T + 321.724, \qquad (8.6)$$

where T is the junction temperature in degree Celsius (°C), and I_{CCINT} is the leakage current in milliAmper (mA). An important conclusion that can be drawn here is that for low-temperature range (30-40°C), ten-degree temperature variation can cause a small increase of the leakage current (roughly 33 mA). By contrast, the same temperature range (60-70°C) in a hot FPGA chip significantly affects the leakage current (roughly 90 mA).

8.5.2 FPGA Temperature Analysis

Figure 8.11 and figure 8.12 show the sensors and processing elements location and the experimental temperature map of the Xilinx Virtex5 FPGA. In order to perform this measurement, four temperature sensors are inserted into the four corners of the device named respectively NE, NW, SE, and SW. Data-intensive processing elements (as defined in section 8.4) have been inserted into the NW location. The core idea here is that we are able to increase locally the temperature of the NW position by running the processing elements. Moreover, the four sensors give us the temperature's evolution of the FPGA

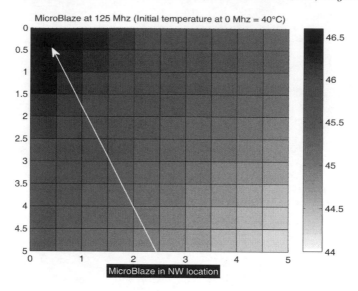

FIGURE 8.11 Thermograph obtained by an uniform interpolation of sensors responses: MicroBlaze case

at each corner when the processing elements are running.We assume that a linear temperature gradient exists inside the FPGA. Consequently, we are able to draw the termographs presented in the following figures,thanks to the four sensors. Furthermore, the temperature values are those obtained after a long time (several seconds) in the steady state behavior), this means we have not considered in this analysis the temperature peaks, which have very short duration of less than one ms. For each experiment, the temperature's steady state before the PE running is homogeneous in all the locations of the FPGA.

Figure 8.12 presents the result obtained with the PE number one (the Random Number Generator).The steady state's temperature prior to the running of the PE is equal to $32°C$.The maximum temperature's increase is $5°C$.The difference between the hottest and coolest hotspots after several seconds is roughly $2°C$.

Figure 8.11 presents the same type of results obtained with PE number 2 (MicroBlaze). The steady state's temperature prior to the running of the PE is equal to $40°C$.The maximum temperature's increase is $6.5°C$.The difference between the hottest and coolest hotspots after several seconds is roughly $2.5°C$.

Based on the FPGA's thermal imaging result, the main result we obtained here is that for any given location of the hot Processing element (MicroBlaze runs at 125 MHz/ RNG runs at 100 MHz), the temperature difference between the hot and colder FPGA regions don't exceed $3°C$ once temperature has stabilized.

To get a better information of the impact of the filling rate of the FPGA

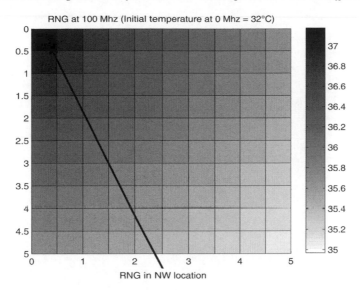

FIGURE 8.12 Thermograph obtained by an uniform interpolation of sensors responses: RNG case

on the temperature's difference between the hot and cooler FPGA regions, figures 8.13 and 8.14 show the thermal map after FPGA heating with 8 and 18 PEs number one running at 100 MHz. In the case of a filling rate of 50%, the temperature's difference increases and reaches 16°C. This latter difference is very optimistic and the mean temperature of the FPGA can not be reduced by this value. Moreover, the leakage current's consumption results given in figure 8.9 are produced for a mean FPGA's temperature. So, we can expect lower leakage current's consumption reduction ,when we reduce locally the temperature, as compared to Xilinx results.

Figure 8.15 illustrates the temperature's difference between hottest and coldest spots as a function of the FPGA's occupancy. As previously, the occupancy is performed with several PEs number one running at 100 MHz. As expected the difference increases vs the occupancy rate to a maximum when the rate is equal to 50%. Then, as this can be easily understood, the difference decreases, since the temperature of the coolest spot in this zone increases, when PEs are implemented in the free zone. From Figures 8.14 and 8.15 we conclude that the maximum temperature's difference is obtained with a filling rate of 50%. In this situation if we use this gain value of 16°C on figure 8.10 the associated power gain value obtained is lower than 100 *mw*.As already explained, this value is very optimistic because leakage power consumption gains in figure 8.10 are given for mean temperature of the FPGA.

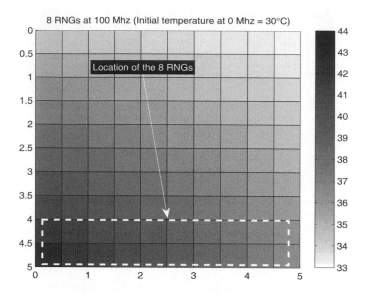

FIGURE 8.13 Thermograph obtained by an uniform interpolation of sensors responses: 8 PEs N1.

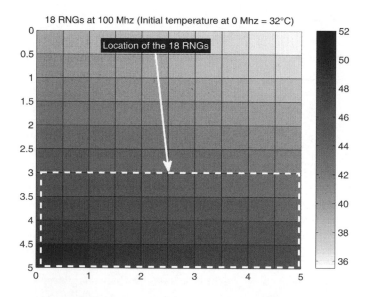

FIGURE 8.14 Thermograph obtained by an uniform interpolation of sensors responses: 18 PEs N1.

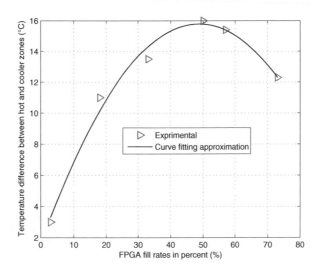

FIGURE 8.15 **Temperature's difference between hottest and coolest spots vs. FPGA occupancy.**

8.6 Conclusion

As stated above, the static power is mostly due to sub-threshold leakage current. This current, although is small for low FPGA's temperature, increases exponentially with temperature. In order to reduce power consumption by means of reducing hotspot' temperature, we have studied in deepth the temperature's difference between hot and cooler spots inside the FPGA. On the basis of the results given in the section 8.5 and the original work published by S.Liu *et al* in [39] a detailed analysis has brought us to the following conclusions:

- The temperature's differences between the hotter and colder FPGA regions do not exceed 16°C, so moving hotspots to colder regions can not save more than 100 *mW* when the die's temperature range is (35-50°C).This optimistic gain is made under the assumption than moving a PE will result in a decreasing temperature on the spot, which is not so obvious as we experimented it.

- On the basis of the original work of S.Liu *et al* in [39] dealing with partial reconfiguration to reduce Virtex4 power consumption, we are able to have a good estimation of the chip's power consumption during the partial reconfiguration's processing time (≈600 *mW*). This power consumption overhead is larger than the saved one given in the above item. Therefore, with current available technologies (mainly DPR), there is no gain in static power consumption by moving PEs from a hot to a cool spot.

The analysis performed in this chapter shows that two possibilities exits so as to decrease leakage power consumption:

- Decreasing the overall area,

- Decreasing the temperature.

The previous conclusions said that decreasing the temperature inside one FPGA is not efficient. This study provides answers to the questions raised in the title of the chapter. Taking into account these conclusions, we are currently investigating two ways to decrease leakage power consumption.

The first one meets the objective of decreasing the overall area. In fact for a given application (Radio access technology, for example), traditionally, designers used a sufficiently big FPGA to run all the PEs needed by the application. Our proposal, consists in using a small FPGA, thanks to a DPR to fill it, on the fly, with the required PEs needed by the application. To fill the required PEs at the right time will, of course, need some additional scheduling which should also be taken into account in the global power budget.

The second method we are investigating, meets the objective of decreasing the temperature. We are investigating power saving in the case of a multi-FPGA architecture. The main idea behind this study is that instead of using one high density FPGA to implement the application, it should be more efficient to use multiple low density FPGAs. In this case, we are able to maintain a high temperature's difference between adjacent FPGAs. Again we will use DPR to move PE from one FPGA to another one.

Bibliography

[1] U. Nations, "Report of the world commission on environment and development," General Assembly Resolution 42/187, Tech. Rep., Dec 1987.

[2] "The Climate Group, SMART 2020: Enabling low carbon economy in information age, june 2008," 2008, http://www.smart2020.org/assets/files/02Smart2020Report.pdf.

[3] "SUPELEC, next generation wireless green networks workshop, november 2009," 2009, http://www-prodgif.supelec.fr/d2ri/flexibleradio/Workshops/Greenworkshop.

[4] Kim, T. Austin, J. S. Hu, and M. Jane, "Leakage current: Moore's Law meets static power," 2003.

[5] N. Michael, C. Moy, A. P. Vinod, and J. Palicot, "Area-power tradeoffs for flexible filtering in green radios," *Journal of Communications and Networks*, vol. 12, no. 2, pp. 158–167, Apr 2010.

[6] J. Kao, S. Narendra, A. Chandrakasan, AnanthaCh, and I. I. S, "Sub-threshold leakage modeling and reduction techniques," in *In Proc. IC-CAD, 2002*, 2002, pp. 141–148.

[7] L. Shang, A. Kaviani, and K. Bathala, "Dynamic power consumption in the Virtex-II FPGA family," in *International Symposium on Field-Programmable Gate Arrays*, 2002, pp. 157–164.

[8] "Virtex-5 family overview: Product specification," Xilinx Inc.

[9] K. Roy, S. Mukhopadhyay, and H. Mahmoodi-Mcimand, "Leakage current mechanisms and leakage reduction techniques in deepsubmicrometer CMOS circuits," I. of Electrical and N. E.-U. Electronics Engineers, New York, Eds., vol. 91, no. 2, 2003, pp. 305–327.

[10] M. Anis, S. Areibi, M. Mahmoud, and M. Elmasry, "Dynamic and leakage power reduction in MTCMOS circuits using an automated efficient gate clustering technique," in *ACM/IEEE Design Automation*, New Orleans, 2002, pp. 480–485.

[11] J. Halter and F. N. Najm, "A gate-level leakage power reduction method for ultra-low-power cmos circuits," in *IEEE Custom Integrated Circuits Conf*, 1997, pp. 475–478.

[12] J. H. Anderson, F. N. Najm, and T. Tuan, "Active leakage power optimization for FPGAs," in *in Proc. ACM/SIGDA Int. Symp. Field Programmable Gate Arrays*, Monterey, CA, 2004, p. 3341.

[13] J. H. Anderson and F. N. Najm, "Active leakage power optimization for FPGAs," in *IEEE Trans. Computer-Aided Design of Integrated Circuits and Systems*, vol. 25, no. 3, March 2006.

[14] J. Becker, M. Hübner, and M. Ullmann, "Real-time dynamically run-time reconfiguration for power-/cost-optimized virtex fpga realizations," in *VLSI-SOC*, 2003, pp. 283–288.

[15] K. Paulsson, M. Hübner, S. Bayar, and J. Becker, "Exploitation of run-time partial reconfiguration for dynamic power management in Xilinx Spartan III-based systems," in *ReCoSoC*, 2007, pp. 1–6.

[16] C.-C. Tsang and H. K.-H. So, "Dynamic power reduction of FPGA-based reconfigurable computers using precomputation," *ACM SIGARCH Comput. Archit. News*, vol. 38, pp. 87–92, January 2011.

[17] M. G. Lorenz, L. Mengibar, M. Garca-Valderas, and L. Entrena, "Power consumption reduction through dynamic reconfiguration." in *FPL*, ser. Lecture Notes in Computer Science, J. Becker, M. Platzner, and S. Vernalde, Eds., vol. 3203. Springer, 2004, pp. 751–760.

[18] A. Gupte and P. Jones, "Hotspot mitigation using dynamic partial reconfiguration for improved performance," in *Proceedings of the 2009 International Conference on Reconfigurable Computing and FPGAs*, ser. RECONFIG '09.   Washington, DC, USA: IEEE Computer Society, 2009, pp. 89–94.

[19] J. Palicot, X. Zhang, P. Leray, and C. Moy, "Cognitive radio and green communications: power consumption consideration," in *Proc. ISRSSP 2010 - 2nd International Symposium on Radio Systems and Space Plasma*, Sofia, Bulgarie, Aug 2010.

[20] J. Delorme, A. Nafkha, P. Leray, and C. Moy, "New OPBHWICAP interface for realtime partial reconfiguration of FPGA." in *Proc. International conference on ReConFigurable computing and FPGAs*, 2009, pp. 386–391.

[21] J. M. Rabaey, A. Chandrakasan, and B. Nikolic, *Digital integrated circuits A design perspective.*   Prentice Hall, 2004.

[22] H. Veendrick, "Short-circuit dissipation of static CMOS circuitry and its impact on the design of buffer circuits," *IEEE Journal of Solid-State Circuits*, vol. SC-19, no. 4, pp. 468–473, 1984.

[23] "Power consumption in 65 nm FPGAs," Xilinx, Inc.

[24] F. Fallah and M. Pedram, "Standby and active leakage current control and minimization in CMOS VLSI circuits," *IEICE Trans. Electronics, Special Section on Low-Power LSI and Low-Power IP*, vol. E88-C, no. 4, pp. 509–519, Apr. 2005.

[25] R. F. Wolffenbuttel, *Silicon Sensors and Circuits: On-chip compatibility*, Springer, Ed.   London: Chapman & Hall, 1996.

[26] "Xilinx UG192 Virtex-5 FPGA system monitor user guide," Xilinx Inc.

[27] A. N. Nowroz and S. Reda, "Thermal and power characterization of field-programmable gate arrays," in *Proceedings of the 19th ACM/SIGDA international symposium on Field programmable gate arrays*, ser. FPGA '11.   New York, NY, USA: ACM, 2011, pp. 111–114.

[28] N. P. G. Quenot and B. Zavidovique, "A temperature and voltage measurement cell for VLSI circuits," E. A. Conf., Ed.   Piscataway, N.J.: IEEE Press, pp. 334–338.

[29] K. M. Zick and J. P. Hayes, "On-line sensing for healthier FPGA systems," in *Proceedings of the ACM/SIGDA 18th International Symposium on Field Programmable Gate Arrays, FPGA 2010, Monterey, California, USA, February 21-23, 2010*, P. Y. K. Cheung and J. Wawrzynek, Eds., 2010, pp. 239–248.

[30] R.-O. E. Boemo, E. Boemo, and S. Lpez-buedo, "Thermal monitoring on FPGAs using," in *Proc. FPL 1997 Workshop, Lecture Notes in Computer Science 1304*.   Springer-Verlag, 1997, pp. 69–78.

[31] S. Lopez-Buedo, J. Garrido, and E. I. Boemo, "Thermal testing on recon-figurable computers." *IEEE Design & Test of Computers*, vol. 17, no. 1, pp. 84–91, 2000.

[32] S. Lopez-Buedo and E. Boemo, "Making visible the thermal behaviour of embedded microprocessors on FPGAs: a progress report," in *Proceedings of the 2004 ACM/SIGDA 12th international symposium on Field programmable gate arrays*, ser. FPGA '04. New York, NY, USA: ACM, 2004, pp. 79–86.

[33] F. Li, Y. Lin, L. He, and J. Cong, "Low-power FPGA using pre-defined dual-Vdd/dual-Vt fabrics," in *Proceedings of the 2004 ACM/SIGDA 12th international symposium on Field programmable gate arrays*, ser. FPGA '04. New York, NY, USA: ACM, 2004, pp. 42–50.

[34] D. J. Frank, P. Solomon, S. Reynolds, and J. Shin, "Supply and threshold voltage optimization for low power design," in *Proceedings of the 1997 international symposium on Low power electronics and design*, ser. ISLPED '97. New York, NY, USA: ACM, 1997, pp. 317–322.

[35] T. Wu, K. Mayaram, and U. Moon, "An on-chip calibration technique for reducing supply voltage sensitivity in ring oscillators," *IEEE Journal of Solid-State Circuits*, vol. 42, no. 4, pp. 775–783, 2007.

[36] "Minimizing the supply sensitivity of CMOS ring oscillator by jointly biasing the supply and control voltage," *IEEE Custom Integrated Circuits Conference*, pp. 531–534, Sept. 2008.

[37] "Virtex-5 FPGA ML550 networking interfaces platform user guide," Xilinx Inc.

[38] "Virtex-5 FPGA system power design considerations: White paper," Xilinx Inc.

[39] S. Liu, R. N. Pittman, and A. Forin, "Energy reduction with run-time partial reconfiguration (abstract only)," in *Proceedings of the 18th annual ACM/SIGDA international symposium on Field programmable gate arrays*, ser. FPGA '10. New York, NY, USA: ACM, 2010, pp. 292–292. [Online]. Available: http://doi.acm.org/10.1145/1723112.1723189

Author Contact Information

Amor Nafkha, Pierre Leray, Yves Louet, and Jacques Palicot, SCEE-SUPELEC, France, Email: anafkha@rennes.supelec.fr, pierre.leray@supelec.fr, yves.louet@supelec.fr, jacques.palicot@supelec.fr.

9

Cloud Computing - a Greener Future for IT

Jinzy Zhu

Huawei Technologies Co. Ltd., China

Xing Fang

Huawei Technologies Co. Ltd., China

CONTENTS

9.1   Introduction

One trend of communication industry is the widely adoption of IT technology. For example, the communication network infrastructure is evolving into IP network. More and more telecom devices are built upon commercial IT devices. So, when we are discussing green communication, we must talk about green IT.

Green IT has become a common choice for both non-IT and IT enterprises, especially against the backdrop of energy shortage, worsening climatic environment and long-lasting economic recession. IT has become the most important part of the informatized society, the energy consumption and environmental pressure are increasing, and the financial crisis is exerting huge influences, so the green IT, as a marked trend in the development of the current IT industry, has been attached great importance to by the industry.

The investment on green IT in the future has a huge potential. For example, the energy consumption in China is growing by 10% annually. The research reported that 50% of the government energy cost in China (80 billion RMB) comes from IT products. In the last climate change conference in Copenhagen, China has promised to reduce carbon emission intensity per unit of GDP by 40% to 45% and this intensity is closely related to energy consumption. This means that China will limit the energy consumption in various areas strictly, including the IT area. Against this backdrop, green IT products whose standard core values are energy-saving, eco-friendly and high-efficient have become a core requirement in the next revolution in the IT industry. China, the world's largest IT producer, has an ever-increasing demand on green products.

Cloud Computing, then, becomes a crucial element in green technology in its rise to international prominence in the data centers - the factories for IT.

Some recent work has done some research on the green characteristic of cloud computing. [1] analyzed the overall energy consumption in cloud data center. [2] and [3] analyzed the energy aware management approach for a large scale IaaS cloud infrastructure to reduce overall power consumption. The paper [4] compared the energy consumption of cloud and the traditional usage model in which workload is executed on PC. It also breakdown the energy consumption on transport, processing and storage part. In this paper, we will illustrate the green technologies in a cloud data center and show how energy will be saved by applying them.

9.2   Why IT Needs to be Green

Datacenter spaces can consume up to 100 to 200 times as much electricity as standard office spaces. The energy, space and cost consumed by traditional data centers have been growing rapidly in recent years. Due to the explosion of data and the growing need for communication, data centers are continuously in need of upgrades and expansions.

According to a study performed by IDC [5], as shown in Figure 9.1, worldwide IT spending on new server procurement keeps flat. In the meantime, the cost on server power and cooling are growing fast. Power and cooling costs in data centers have risen sharply over the past several years as much as 4 times [6] in the span of a decade.

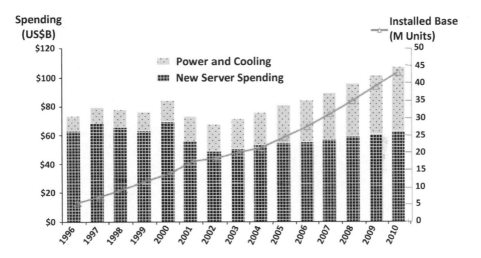

FIGURE 9.1 Worldwide IT spending trend

Global carbon emissions from IT have been estimated to be about 2% of world totals [7]. It is similar to the airline industry. McKinsey forecasts that the IT sector's carbon footprint will triple during the period from 2002 to 2020. For office buildings, IT typically accounts for more than 20% of the energy used, and in some offices up to 70%. According to US EPA's (Environmental Protection Agency) report, data centers will account for nearly 3% of total US electricity consumption by 2011.

From economic perspective, a typical enterprise data center whose size is 1000 square meters will cost $100,000 on power consumption each year. IT consumption on power could not be ignored anymore. Reducing energy consumption have been one of the top priority for enterprises to develop in a sustainable manner. Because the IT demands and costs are growing rapidly, energy conservation and environment protection are becoming ever more crucial in the difficult economic situation of global integration.

If we can improve IT efficiency by 20%, we can save 36 billion kWh or 22 million tons of CO2. It is equal to remove 3 million cars and light trucks from the road. Save 20 billion kWh per year by 2015 worth $2 billion. It is equal to annual electricity use in 1.8 million American homes.

If we want to make IT greener, where to start? We must first analyze where the energy goes. Some rough study shows between 30% and 60% of the electricity consumed in server rooms is wasted. By using current and emerging technologies, it possible to reduce power consumption in data centers by 50-80%.

A study made by US energy departments shows the detailed energy use within data center [8]. It is shown in Figure 9.2. It is found that 45% of energy is consumed by cooling devices, 10% is consumed by Power conversion and distribution units, only 45% is really spent on actual IT devices. On IT devices, only 30% of the power is spent on processor, other 70% is spent on peripheral components, including power supply, fans, memory, disk. Even the power used by processor is not fully utilized. The average processor utilization is always below 20%. It means in most of the time, the server is idle. By considering all these numbers, out of 100 unit consumed power, only 100*45%*30%*20%=2.7 unit is spent on the actual workload. There is huge room to improve about IT energy efficiency.

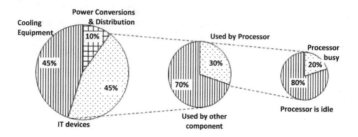

FIGURE 9.2 Datacenter energy use break down

Suppose the actual workload is not changed, according to the figure, we can have following reasoning:

- 1% improvement in energy efficiency of data center (cooling and power) will result in 0.5% save to overall energy consumption;

- 1% improvement in energy efficiency of server will result in 3.2% save to overall energy consumption;

- 1% improvement in average server utilization will result in 4.7% save to overall energy consumption.

9.3 Cloud Computing, the Future of IT

Cloud Computing is an inevitable trend in IT industry. It will help IT to become green by nature.

9.3.1 What is Cloud Computing?

Cloud computing is the delivery of computing as a service rather than a product, whereby shared resources, software, and information are provided to computers and other devices as a utility (like the electricity grid) over a network (typically the Internet) [9] as illustrated in Figure 9.3.

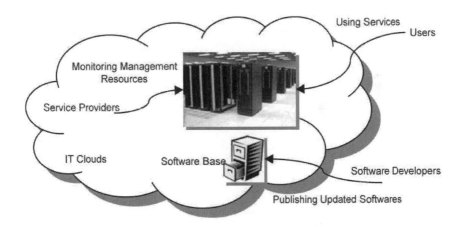

FIGURE 9.3 Cloud computing illustration diagram

There are 3 main roles: cloud computing users, service providers and software developers. Service provider is the party to provide computing service and maintain the entire infrastructure. It will central manage a large number of computing resources. User is the subscriber of the computing service. User does not need own or build IT resources. He just uses the provided service through the network. Software developers are indispensable, because cloud is an open platform whose existing features can be enhanced as the time going.

Looking at the evolution of computing mode shown in Figure 9.4, we can conclude it as centralized-distributed-centralized. In the early stages, few enterprises can afford to have computing capacity due to the limited technology development and high costs, so the centralized computing mode prevailed. Later, it became distributed mode as computers are getting smaller and costs are reduced. Currently it is becoming centralized again.

Cloud computing is growing fast. It is said that year 2008 is "the start of the cloud computing application era". From this year onwards, almost all ma-

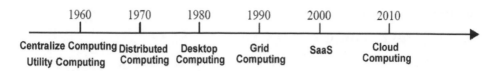

FIGURE 9.4 The Evolution of Computing Modes

jor IT companies have begun to discuss it, including hardware manufactures (such as IBM, HP, Intel, Cisco, and Sun), software vendors (such as Microsoft, Oracle, and VMware), Internet service providers (such as Google, Amazon, and Salesforce) and telecommunication operators (such as China Mobile, and AT&T). Besides these IT giants, some small-scale IT enterprises have also focused their efforts on cloud computing, including RightScale, Skytap, Enomaly, GoGrid, Rackspace and so on. These companies cover the entire IT industry chain, and make up a complete cloud computing ecosystem. Various kinds of new products and services which have names related to "cloud" spring up, such as cloud computing, cloud storage, cloud software, cloud security, cloud backup and cloud disaster recovery. These new services have brought innovations in usage, user experiences and business modes.

In 2010 and 2011, the well-known market research company Gartner identified 10 strategic technologies. Cloud computing is in the top 1 position in both year [10]. Gartner predicted that the global market value of cloud computing services would increase by a growth rate of 16.6% from $58.9 billion in 2009 to $68.3 billion in 2010 and increase to $148.8 billion by 2014 [11].

The popularity of cloud computing is driven by several key facts:

1. The green requirement of data center. As mentioned earlier, data center operators are meeting several key challenges:

   - With the growing of business requirement, the data center are becoming bigger and bigger. The management complexity is increasing fast, which leads to high management cost.

   - Energy consumption is becoming a big part of operation cost.

   - The IT resource usage is hard to optimize, which leads to the waste of computing power, energy and hardware investment.

2. Increase in network nandwidth. Bandwidth is a critical factor in the promotion and growth of cloud computing. Users need to access data conveniently, now that computing and storage is performed on the other side of the network. In recent years, with the development of the Internet, major network operators are keeping investing to improve the infrastructure of the Internet. On the one hand, the bandwidth of core networks expands rapidly; on the other hand, the network access for family and enterprise

users have also witnessed substantial changes. For family users, the Internet speed grows from about 50 Kbit/s with dial-up, to 512 K/1 M/2 Mbit/s with ADSL and finally to 10 M/20 Mbit/s with latest fiber cables. The growth in bandwidth has changed the mode for network usage and the type of network application.

3. The fast growing of mobile internet users. In recent years, the highlight in the communication industry is the development of 3G. Major telecommunication operators saw great potential in this. Besides the traditional mobile device manufacturers, consumer electrics and home appliances manufacturers have also entered this market. The telecommunication technology, computer technology and home appliance technology are integrated and traditional equipment can be connected with each other. Mobile phones can be used to have conversations and connected to the Internet just like a PC. For other devices, such as MP3, media player, pad computer, digital cameras, game consoles and other intelligent home devices, all support 3G or Wi-Fi and can be connected to become a new network terminal. All these require the computing power in the backend, which is a huge increase in computing demand.

4. The related technologies become mature. There are different types of cloud service. For each cloud service, it may require different technologies. The Figure 9.5 shows the major technologies used to implement different cloud services.

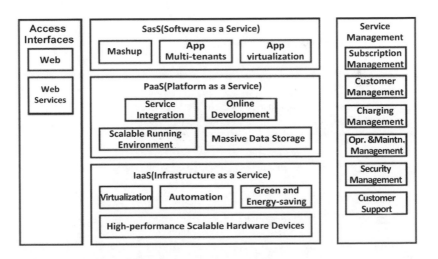

FIGURE 9.5 Cloud computing technology map

9.3.2   Cloud Computing Classification

Cloud has different types. From deployment perspective (Figure 9.6), there are public cloud, private cloud, and hybrid cloud. From technology perspective (Figure 9.7), there are infrastructure-as-a-service, platform-as-a-service, and software-as-a-service (SaaS).

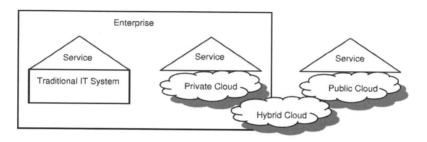

FIGURE 9.6 Cloud deployment type

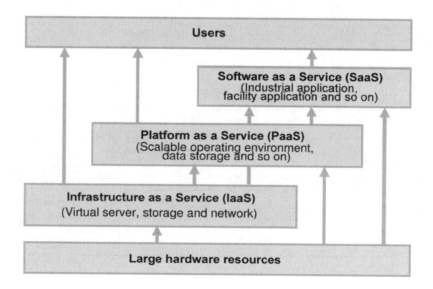

FIGURE 9.7 Cloud technology type

For public cloud, its service is provided to all the users across the internet. The cloud supplier can offer installation, management, deployment and maintenance for various IT resources ranged from application program and software operating environment to physical infrastructure, etc. End users accomplish their own goals through shared IT resources and only need to pay for the used resources (pay-as-you-go), obtaining their required IT resource services through such economical way.

In the public cloud, the end users do not know about the other users who share resources with or how specific underlying resource is implemented, and is even unable to control the physical infrastructure. Therefore, the cloud service provider should ensure the nonfunctional requirements (e.g. security and reliability) of the resources provided.

Private cloud, on the contrary, only provides service to internal users, for example, within an enterprise firewall. It is owned by a commercial business or other social organization. Compared with the traditional data center, the cloud data center can reduce the complexity of the IT architecture. Different from public cloud, the user and service provider of private cloud may come from same company or even same department. The risk on security is much lower. User does not need worry about data leaking.

Hybrid cloud combines the model of public cloud and private cloud. Some enterprise may adopt the both model at the same time to gain the benefits of them. In hybrid cloud, user could store sensitive data within the private cloud to ensure security and performance because the data is processed and stored locally. The other non-confidential data could be stored and processed in the public cloud to further reduce the CAPAX and OPEX. Data and workload could also dynamically migrate between the two cloud according to local resource availability and business requirement.

IaaS (Infrastructure as a Service) offers basic IT capability like computing, storage and network. The two key technologies to provide IaaS service is virtualization and automation. Virtualization enabled resource sharing on a single hardware. Automation enables self-service to virtualized resources. It allows user to create, change, remove virtualized resource dynamically. One example of IaaS is Amazon Webservice [12].

PaaS (Platform as a service) provides the application development, execution and storage environment. It is the middle tier between the infrastructure and the application. The provided service is not the raw resource, like CPU and memory, but some logical capability such as databases, file systems, and messaging queue, etc. With PaaS, application programming and management become much easier, since application developer and operator do not need care about how to handle application deployment in a distributed environment. The PaaS will native support application deployment, scalability and optimization. Distributed data store and data processing cluster are common technologies in this layer. Google AppEngine [13] and Microsft Azure [14] is good example of PaaS.

SaaS (Software as a Service) is the most common cloud computing service. It provides software over the Internet through the standard Web browser. The service provider is responsible for the maintenance and management of the software and hardware facilities. Google Calendar and Gmail are SaaS example for consumer user. Salesforce.com is example for enterprise user.

9.3.3 Cloud Computing Characteristics

Cloud computing has 9 key characteristics.

1. Resource abstraction and network access. Cloud service is abstracted to the end user. The user only knows the interface of service, but has no idea about the cloud backend. Through this abstraction, the cloud service is standardized. Anyone could access to the cloud service through network, with location independence.

2. Large scale resource pool and resource sharing. Many cloud data centers are equipped with a rather large scale and supported by thousands of servers. The "clouds" of Amazon, Google, Facebook, Microsoft and Yahoo all have more than 100,000 servers. Under such scale, huge computing power could be provided. By resource sharing, different workloads could use as many resources as they require. In the end, the cost of single workload could be deduced to minimal. This is so-called economics of scale.

3. Self-service by end user. The user with only basic IT skill could browse the service catalog and subscribe various cloud services. One can use cloud service without support from professional people.

4. Dynamical scalability. Different workloads are running on same cloud. They do not only share resource, but also share it dynamically. It means when a workload need more resource, the cloud could fit the requirement just in time, without human intervention. In case of a change on business development, cloud computing can make rapid response through flexible mechanism and automation to adapt to the change on business.

5. Resource metering and pay as you go. Just as water and electricity charges, various IT resource services (e.g. storage, CPU, bandwidth, software license) in the cloud can be monitored, controlled and metered. The meter result could be used to charge user.

6. Standardization of interface. Cloud service could not only be provided to end user, but also be provided to integrate with other service. The interface standard will make the integration easier and stable.

7. Automation: The service management in cloud is performed automatically in a centralized mode without manual intervention. It is a key capability since the cloud scale is huge. On the other hand, automation increases the efficiency of management. User request could be met quickly by automated response.

8. Online service upgrade: Cloud service is always on. It is not like traditional IT systems which only cover a small number of users. Since cloud service is accessed by worldwide users, it must have the capability to upgrade itself without interrupting end user visit.

9. Service incubation: User requirement is changing with time going, so cloud need be able to incubate new service. Otherwise the cloud platform will end its life in a few years. A good cloud platform could form a big cloud ecosystem which enables the self-evolution.

9.3.4 Why does Cloud Computing Make IT Green?

Cloud computing is a quite eco-friendly business mode. For public cloud, the reasons that cloud computing can be so green include: data is stored in the cloud side; users can get what they need from the data pool; the usage and allocation of data resources are unlimited and reasonable; there is no need to invest hardware devices and pay fee for managing them; of course, the labor resources can been saved due to the proper usage of resources. As for private cloud, virtualization technologies have greatly improved the eco-friendly application of data centers and will been applied widely in the future because they can deliver greater processing capacity with fewer hardware and less power. Many devices with low utilization rates can be integrated into a single server through the proper analysis and centralized management, which in turn will reduce the energy consumption and costs.

The cloud computing model with the features of resource abstraction and flexible, rapid service is an important driver for 'green' applications in the data center. The resource abstraction technology, which delivers greater processing capabilities with less hardware and power consumption, is certain to be widely used in the future. Through business model analysis and centralized infrastructure management, a lot of machines with low utilization rate is optimized into multiple business services, reducing both the costs and the power consumption of these equipments. The technologies of resource abstraction, automation and service management help companies promote the base virtualization of giant servers, memory and networks into services, thus enabling the flexible utilization of all the resources from an upper-level through one interface. In this way, companies can successfully improve the resource utilization rate, ensure resilient management of data centers and set up green data centers with higher efficiency and less energy consumption.

So it is clear that the features of cloud computing can help to promote energy conservation and sustainable development of the society.

According to traditional approaches, infrastructural equipments such as a single firewall, switch, data storage and backup systems are required for utilizing IT resources within each company. However, with the resource sharing approach of cloud computing, companies no longer need such variety of equipments. Instead, only a limited number of people are necessary for constructing, maintaining and developing the IT infrastructure, which significantly reduces costs.

Cloud computing helps companies decrease power consumption as well as e-waste and carbon emissions. Now, IT companies, as among the largest energy consumers, have to acknowledge the ever-lasting importance of green development in aspects of servers, switches and large data centers. What essential is to take advantage of new technologies to promote green development.

In a changing world where application requirements become increasingly complex, cloud computing will help promote the vision of a greener IT by continuously facilitating business model changes and innovations.

9.4 Cloud Foundation: Green Data Center

Green data center is an inevitable trend along the path of development and covers its overall system problems, including IT devices (such as buildings, computer rooms, air-conditioning and servers), application systems and data management efficiency. It is the foundation of cloud computing. In cloud era, data center size and complexity is unprecedented. It is becoming cost center for a cloud service provider. Several industry-leading cloud service providers have made a lot of progress on making data center become green.

9.4.1 Green Measurement, PUE

To build a green data center, we must have a measurement first. PUE is the most common metric to measure the energy efficiency of a data center. PUE is Power Usage Effectiveness. It is defined as the ratio of the total power to run the data center facility to the total power drawn by all IT equipment: PUE= Total Facility Power/IT Equipment Power. Datacenter energy distribution is illustrated in Figure **??**. The value is higher, means more energy is consumed by non-IT facility. Since the purpose of a data center is to provide IT service, we expect all power could be used by IT equipment. The PUE can range from 1.0 to infinity. Ideally, a PUE value approaching 1.0 would indicate 100% efficiency. In practice, some power is consumed by lighting, cooling and power conversion. An average data center has a PUE of 1.8, according to a survey of more than 500 data centers conducted by The Uptime Institute [15].

Some industry leading cloud service provider has achieved great deduction on PUE. Facebook reported one of its data center has achieved an initial PUE ratio of 1.07 [16]. This value is achieved by some innovations in server design, power supply, server chassis, server rack, battery cabinets, and data center electrical and mechanical construction. To help industry partners also become green. Facebook announced the formation of the Open Compute Project [17], an industry-wide initiative to share specifications and best practices for creating the most energy efficient and economical data centers.

There are also other measurements of data center efficiency. The DCPE (Datacenter Performance Efficiency) is the natural evolution from PUE: DCPE= Useful Work/ Total Facility Power. For PUE, we only know the

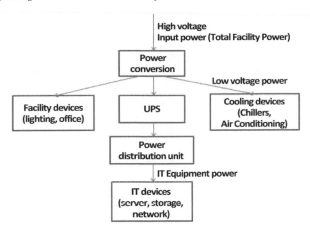

FIGURE 9.8 Datacenter energy distribution

efficiency of energy spent on IT devices, but we do not know whether or not the IT device itself is efficient. Obviously, DCPE is much closer to describe the green degree than PUE. However, the DCPE is much more difficult to determine.

9.4.2 Data Center Site Selection

Select a right place to construct data center is the first step towards a green data center. From green perspective, site selection needs to consider several criteria:

1. The price of electricity The electricity price varies a lot in different locations. Following is a simple comparison:

Location	Price per KWH
Idaho, US	3.6
California, US	10.0
Hawaii, US	18.0
China	16.0
Ukraine	3.05
Russia	9.49

The price varies widely from country to country, and may vary significantly from locality to locality within a particular country. There are many reasons that account for these differences in price. The price of power generation depends largely on the type and market price of the fuel used, government subsidies, government and industry regulation, and even local weather patterns. The locations having easy power supply always has a low price. Some locations, like Hawaii, must rely on fuel shipping from outside. Then the price is must higher.

2. The replacement of electricity to supply power. Traditional data center fully rely on electricity to supply power. As new energy is put into practice, data center has new choice. Dat acenter could utilize wind energy or solar energy directly to generate power, which is required by IT devices. Obviously, these clean energies will generate zero carbon emission and reduce the impact to local environment to minimal.

3. The natural condition, like climate. Data center needs to maintain a constant temperature for the internal IT devices. It is always between 18 °C-27 °C. Since IT devices will always generating heat, if we can find a place whose full year temperature is always under 18 °C, it is possible to eliminate the need for cooling. Since cloud computing has centralized computing requirement to few data centers, it is possible for cloud service provider to select the best location for their data center in the world range. The Facebook data center, which achieved the lowest PUE, is located at Oregon in the northwest of US. The electricity price is only 5.45 per KWH(kilowatts per hour). The average temperature in full year is only 13 °C. By utilizing naturally cool outside air to maintain the inside temps, very little energy is wasted to cool the data center.

9.4.3   Datacenter Construction

In data center construction stage, green technology could also be used. There is also rating system for this. The Leadership in Energy and Environmental Design (LEED) Green Building Rating System(TM) is a third-party certification program and the widely accepted benchmark for the design, construction and operation of high performance green buildings. LEED promotes a whole-building approach to sustainability by recognizing performance in five key areas of human and environmental health: sustainable site development, water savings, energy efficiency, materials selection and indoor environmental quality. The six major environmental categories of review include: Sustainable Sites, Water Efficiency, Energy and Atmosphere, Materials and Resources, Indoor Environmental Quality and Innovation and Design. Certified, Silver, Gold, and Platinum levels of LEED green building certification are awarded based on the total number of points earned within each LEED category. LEED can be applied to all building types including new construction, commercial interiors, core & shell developments, existing buildings, homes, neighborhood developments, schools and retail facilities. Incentives for LEED are available at the state and local level and LEED has also been adopted nationwide by federal agencies, state and local governments, and interested private companies.

9.4.4 Power Conversion and UPS

Before power is distributed to IT devices, electricity needs to go through power conversion and UPS devices. Power conversion device will convert power voltage from high voltage to the level required by IT devices. This conversion will typically consume 2% of total power supply. Uninterruptible Power Supply (UPS) is to provide battery based power for a limited time in the event of loss of the primary electricity source. UPS always consume 5%-14% of overall power supply. Battery-based UPS has a characteristics that the higher the load, the higher the efficiency. It is shown in Figure 9.9. It is better to keep the UPS load factor as high as possible. When a full data center equipment load is served through a UPS system, even a small improvement in the efficiency of the system can yield a large annual cost savings. For example, a data center with IT equipment requires 20,000 MWh(million-watts per hour) of energy annually for the IT equipment. If the UPS system supplying that power has its efficiency improved from 90% to 95%, the annual energy bill will be reduced by 1.17 MWh, or about \$117,000 at \$0.1/kWh(kilo-watts per hour), plus significant additional cooling system energy savings from the reduced cooling load.

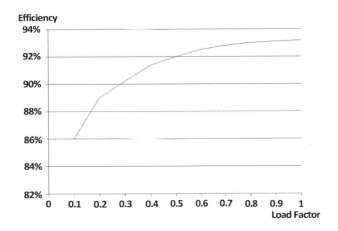

FIGURE 9.9 UPS efficiency

Several techniques could be used to reduce the power consumption in this part:

- Use higher voltage electrical distribution system to reduce energy loss. So the difference between the source and target voltage level is reduces, resulting in efficiency improvement. However, this require IT devices to be adjusted to adapt to the higher input voltage.

- Eliminate the need for a central UPS. For example, add a small UPS battery to each IT devices so that each device could supply itself during power outage. By doing this, each UPS only need to maintain the power supply for each device. The efficiency could be improved greatly. When using a central UPS, when most of the devices are powered of, the UPS load would be quite low, resulting low efficiency. When considering redundancy, more UPS devices are required, which drag down the efficiency more.

- Use new UPS technologies, such as flywheel systems. Such systems eliminate replacement and disposal concerns associated with conventional lead-acid battery based UPS systems. It also reduce the costs of special ventilation systems and conditioning systems which is used to maintain temperature to ensure battery life.

9.4.5   Cooling

Cooling energy consumption is always the largest part beside IT device consumption. Assuming the total power of IT devices is 100 KW, this energy will finally be converted to heat. Plus the heat transmitted from external environment (if the temperature in the data center is cooler than external), the cooling system needs to have the capability of exchanging 100+ KW heat.

Besides exchanging heat, air management is also important. A bad air management will result in unbalanced heat distribution, resulting in server over-heat in some areas.

There are 3 key approaches to improve cooling efficiency:

1. Better air management. Traditional air management is depending on experience (eg. design of hot aisle and cold aisle). That will lead to wrong result in practice. Recent technology has been able to simulate air flow and find out heat map during design phase. So that heat point could be identified before building the infrastructure. It is shown in Figure 9.10 In production phase, temperature sensor could be deployed to data center to monitor the temperature in real time. So operator could adjust data center quickly and reduce unnecessary investment in cooling devices.

2. Another technology is modular data center. The idea is to ship data center as pre-designed, pre-configured unit. The cooling devices are already in place before shipping. User doesn't need manage air anymore because it is already well managed as designed.

3. Use liquid cooling. Traditional cooling process: IT devices generate heat air - heat air is cooled down by CRAC (Computer Room Air Conditioning) within server room - CRAC sent the heated cooling medium (air or liquid) to chiller, which is outside the data center - chiller cools down the cooling medium and send back the medium to CRAC. In the entire process, air is a key cooling medium. However, the efficiency for air to act as cooling medium is quite low (the coefficient of thermal conductivity of water is 30 times to air. The thermal capacity of water is 4000 times to air). So the idea of liquid cooling is to use liquid to directly cool IT devices. This

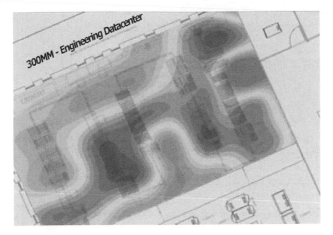

FIGURE 9.10 Cooling simulation and monitor

removes the heat generated from the back of IT systems before the heat enters the server room. IBM has introduced liquid cooling to its iDataplex and Power server product. This design reduces server heat output in data center up to 60 percent. Server uses 40% less power while providing five times as much computing power [18]. The total power of a single rack is only about 3 KW because of inefficient air cooling. With liquid cooling, the power of a rack could reach as high as 30 KW, which is 10 times than before.

4. Use of free cooling. As mentioned earlier, Facebook uses free air to cool its data center by carefully design the ventilation path between data center and external. Google's approach is also innovative. It moves its data center to sea. The cooling system for data center is utilizing sea water (as opposed to a chiller plant system) which will be pumped from around 7 meters below sea level. Once the naturally chilled sea water enters the data center, it will travel through specially designed channels throughout the infrastructure made out of granite.

9.5   Cloud Building Block: Green Hardware

Computer servers are at the core of data center operations, and therefore contribute greatly to this overall consumption. Deploying more energy efficient servers can be a key strategy to reduce total data center energy consumption. In a cloud data center, the server number may reach 10,000 or even more. A slight improvement in the server efficiency will greatly reduce overall energy consumption. The green IT trend has led the hardware vendor to put more resource on the development of energy saving technology.

9.5.1   Energy Saving Chip

In recent years, both client device and server vendor are developing energy saving technology. For client device vendor, their major interest is to increase the device working hour when using battery. For server vendor, their target is as same as data center operator, trying to provide more computing capability with less power. Whatever, they achieved a common result, which generates more efficient hardware.

In server area, there are several major vendors: Intel (x86), IBM (PowerPC), Sun (SPARC). Their products are following the same trend:

- Multi-core: Traditionally, processor vendor are pursuing higher frequency to gain better performance. But today multi-core has become a standard configuration for processor, even for mobile processor. There are two major reasons. The first reason is higher frequency cannot improve performance much because of the bottleneck between CPU and memory. The second reason is the higher frequency may create huge heat (the power will grow exponentially when the frequency is growing) within the processor which will damage the circuit. With multi-core and SMP, a single server could have hundreds of cores. These cores share same server periphery devices, like fans, memory, disk, motherboard, etc. A single server could support workload which would be supported by several servers before. This means energy saving.

- Low energy consumption: There are two major energy saving technologies within CPU: dynamic voltage scaling and dynamic frequency scaling. These will reduce CPU voltage and frequency when the workload is low. As the result, the energy is saved. Since the performance will impacted in low frequency and voltage, such features are always working with control software. Server administrator could set energy saving policy to get a best combination of performance and energy consumption. For example, the policy could be, reduce CPU frequency to 80% when the CPU utilization is under 5% for 30 minutes. IBM power server could even hibernate or shut down the entire CPU when the server is running.

Compared to server CPU, client device CPU is more caring about CPU energy consumption. It is because of huge sales volume of mobile devices on which standby time is a key sale point. ARM CPU is a fast growing as a mobile CPU. Its power consumption is only 1/10 of Intel CPU. Several hardware vendors, including HP, has announced server product which is based on ARM CPU.

Besides CPU, other server components also consume power, especially on high end server which has lots of components. IBM EnergyScale technology could dynamically shut down unused components to save energy. These components include: memory, PCI device, etc.

9.5.2   Energy Saving Hardware

Energy Star is an international standard for energy efficient consumer products. It has release a specification for standalone servers with one to four processor sockets. The Energy Star program aids consumers by recognizing high-efficiency servers. Servers that meet Energy Star efficiency requirements will, on average, be 30% more efficient than standard servers. The new ENERGY STAR server consumed 54% less power than older model servers [19].

Servers can earn the ENERGY STAR by offering the following features:

- Efficient power supplies that limit power conversion losses and generate less waste heat, which reduces the need for excess air conditioning in the data center facility where they are housed;

- Improved power quality which provides building-wide energy efficiency benefits;

- Capabilities to measure real time power use, processor utilization, and air temperature, which improves manageability and lowers total cost of ownership;

- Advanced power management features and efficient components to save energy across various operating states, including idle

- A Power and Performance Data Sheet for purchasers that standardizes key information on energy performance, features, and other capabilities.

Historically, a typical server's power supply converted AC power to DC power at efficiencies of around 60% to 70%. Today, through the use of higher-quality components and advanced engineering, it is possible to find power supplies with efficiencies up to 95%. Using higher efficiency power supplies will directly lower a data center's power bills and indirectly reduce cooling system cost and rack overheating issues.

Besides optimizing power supply and chip, server could fine-tune its design in air path and fan position to further improve efficiency. Other IT devices, like storage and network, are following same trend as server. One technology in cloud era is cloud storage. It eliminates traditional storage design concept. The entire storage is built upon commercial server cluster. The storage function is relying on software. This unifies the server and storage infrastructure and makes it possible to deploy same energy saving design to all IT devices.

9.6   Cloud Core: Intelligent Management Software

Efficient data center and hardware is not the end of efficient IT. According to
energy consumption break down, better utilize server resource is the best way
to save energy. Cloud is not only putting server together (consolidation), but
also employing a lot of latest software technologies to improve average server
utilization. These technologies are the core of cloud computing.

9.6.1   Virtualization

Server virtualization is a technology to create virtual machine (VM) that
acts like a real computer with an operating system. Software executed on
these virtual machines is separated from the underlying hardware resources.
For example, a computer that is running Redhat Linux may host a virtual
machine that looks like a computer Microsoft Windows operating system.

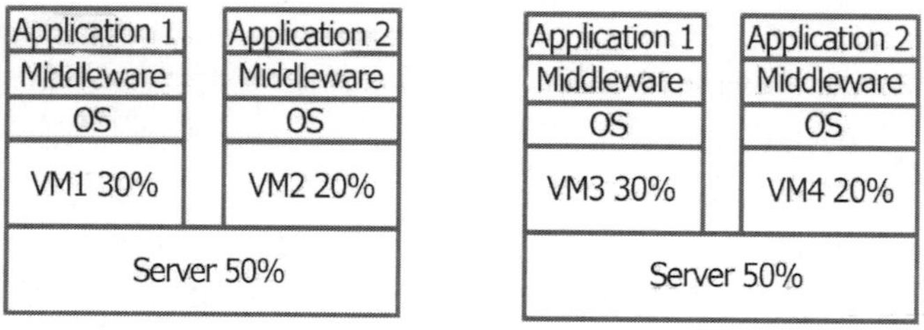

FIGURE 9.11 **Server virtualization technology**

Virtualization is widely used in cloud era for several reasons:

- Each server has many cores due to multi-core technology. A typical 4-
  socket, six-core server will have 24 cores in total. However, many applications do not require so many cores at the most of the time. With virtualization, different applications could be consolidated to a single server,
  each running in a separate virtual machine. This brings great deduction
  in server investment. It is also quite flexible because user does not need
  dedicated hardware for each application or workload.

- Virtualization provides abstraction and standardization over hardware.
  When a user subscribe server from an IaaS cloud service provider, he
  cares more about the server capacity. He does not care much about the
  underlying hardware. He also needs a standard server configuration. By
  using virtualization, a standard 2CPU, 4G memory, 150G disk virtual machine could be created on any available server in resource pool. When user
  finishes using the virtual machine, he could also save the virtual machine
  to storage and restore it in other cloud service provider.

- Virtualization enables resource sharing. A single server could support many virtual servers. These virtual servers share CPU, memory, disk and IO device of the same physical server. With virtualization, the server resource utilization could be improved from 20% to 60%. It may result in 66% overall energy reduction because of deduction of server number and cooling requirement.

Besides server virtualization, there is also storage virtualization, network virtualization and middleware virtualization. Storage virtualization and network virtualization will achieve similar result as server virtualization.

Middleware virtualization (Figure 9.12) will further improve overall efficiency. Middleware virtualization is a general term to describe PaaS technologies, include high volume distributed data storage, parallel computing framework, etc. Such technologies turn a big server cluster into a single logical server. It is like a distributed OS to application. Different applications could share the same hardware infrastructure and middleware environment. The middleware will provide isolation and scheduling to different applications, so that each application is not aware of existence of other application.

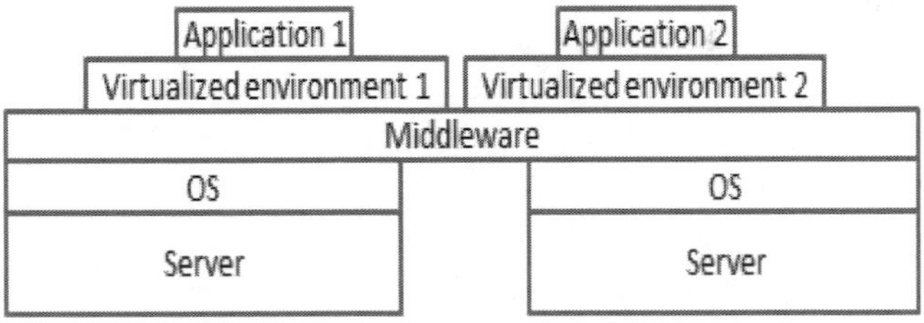

Application 1		Application 2	
Virtualized environment 1		Virtualized environment 2	
Middleware			
OS		OS	
Server		Server	

FIGURE 9.12 Middleware virtualization technology

In server virtualization scenario, each virtual machine will consume some resource like memory and CPU. Even though the overall resource is shared, each virtual machine is occupying a small portion of resource at any given of time. These resources are dedicated to a single virtual machine. With middleware virtualization, the resource sharing is reaching a new level because the VM boundary is break down. There is no application dedicated resources. Server resource could be used by each application in a more efficient manner. For example, an application will consume 4G memory in average. To handle the peak resource usage, a virtual machine containing this application needs to have at least 6G memory. A physical server with 96G memory could support 16 such applications when using server virtualization. If we use middleware virtualization, there is no need to allocate dedicated memory to each application. It can support 20 applications at the same time, leaving 16G memory as single buffer to handle any peak workload. By doing this, the server efficiency is improved by 25%.

9.6.2   Automation

Infrastructure administration is one of the major challenges in a large scale virtualized environment. Simply building a virtualized environment without the proper approach to administration can increase complexity and thus generate added costscosts high enough to cancel out the cost savings derived from virtualization in the first place. Automation is the key to managing these problems. It is critical that a cloud be equipped with tools that facilitate, simplify and enable management of the physical environment that provides the virtual resources.

Automation turns sequential manual steps into automated execution workflows. It is like a pipeline which finishes every complicated task in pre-defined sequence. Each data center requires common maintenance task, including service monitoring, data backup, problem resolving, configuration management, process management, etc. With automation, the human effort could be greatly reduced. In a traditional data center, a single administrator could only manage 100 servers. In a cloud data center with automation, over 1000 servers could be managed by a single person. A typical large scale data center with tens of thousands servers will be equipped with 40-50 staffs.

Automation is a key to become green. For a large data center, when the workload is low or online user is few, most of the servers could be in powered off state to save energy. When a new user comes, an automation process could be triggered to boot a new server and do some preparation to make sure the new workload could be handled well.

Automation is also helpful to control energy saving policy. Each application workload has repeatable pattern. The pattern may appear daily, weekly, monthly or yearly. For example, every Monday will emerge a peak workload in the whole week. If we can identify all such patterns, we can set policy to automation system to control the server readiness in different time slot. Then the servers could be automated started or stopped according to the policy, saving a great deal of energy.

9.6.3   Autonomic Management

Autonomic management is the final step towards intelligent management. It brings intelligence to automation. Automation only repeats action as pre-defined policy or sequence. Autonomic, on the other hand, could be aware of business change and adapt to it dynamically. It could not only understand what is going on, but also predict what future will be based on the trend analysis. It could also adapt to unpredictable changes. An autonomic system makes decisions on its own, using high-level policies; it will constantly check and optimize its status and automatically adapt itself to changing conditions.

Administrator could define the policies and service levels in an autonomic environment. This creates an application-aligned infrastructure that can be scaled up or down based on the needs of each application. When an application's SLA is broken, the autonomic control engine will put more resources to the application to recover the SLA.

IBM started to study autonomic computing from 2001. The goal is to develop computer systems capable of self-management, to overcome the rapidly growing complexity of computing systems management, and to reduce the barrier that complexity poses to further growth.

Energy-aware management is the current trend of cloud computing management. The goal is the enable the management stack be aware of energy consumption during the autonomic management. A typical scenario is, when the data center utilization is not full, the workload could be consolidated to few servers. Thus other idle servers could be shut down to save energy. For IaaS, virtual machine live migration could be used to consolidate workload without interrupting it. In fact, the management system could keep monitoring overall system utilization and balancing the workload. Then it could keep powering-off or powering-on server as time going. One challenge in this is the performance downgrade during the migration or after the workload is consolidated.

9.7   Outlook to a Green Cloud Future

To have a clear understanding of what green cloud will be, we can have an assumption.

Assume there are 10000 data center today. These data centers are owned by different companies, enterprise and service providers. The typical size of the data center is 500 servers each. The power of each server is 200W (2CPU, 4 cores). The average server utilization is 20%. The average PUE is 2.0. Then the overall power consumption is 200*500*2.0*10000=2000 MWH.

With cloud, the data center number is reduced. Assume each data center contains 50000 servers. The power of each server is 500W (4CPU, 32 cores). The server utilization could be maintained at 60%. Then we need 500*4*20%*10000/32/60%/50000=5 such data centers. If we can improve PUE to 1.2, then the total energy consumption is 50000*500*1.2*5=150MWH.

It means we can reduce the total IT energy consumption by 90% if all the users adopt cloud computing model. In other words, if we don't spend more energy, we can increase the processing power by 10 more times.

Bibliography

[1] A. Berl, E. Gelenbe, M. di Girolamo, and et al., "Energy-Efficient Cloud Computing," *The Computer Journal*, vol. 53, 2010.

[2] R. Buyya, A. Beloglazov, and J. Abawajy, "Energy-Efficient Management of Data Center Resources for Cloud Computing: A Vision, Architectural Elements, and Open Challenges," in *Proc. the 2010 International Conference on Parallel and Distributed Processing Techniques and Applications*, 2010.

[3] L. Liu, H. Wang, X. Liu, and et al., "GreenCloud: A New Architecture for Green Data Center," in *Proc. the 6th international conference industry session on Autonomic computing and communications industry session*, 2009.

[4] J. Baliga, R. W. A. Ayre, K. Hinton, and R. S. Tucker, "Green Cloud Computing: Balancing Energy in Processing, Storage, and Transport," *Proceedings of the IEEE*, vol. 99, 2011.

[5] "IDC website: http://www.idc.com/."

[6] D. Filani, J. He, S. Gao, and et al., "Dynamic data center power management: Trends, issues, and solutions," *Intel Technology Journal*, p. 59, 2009.

[7] "Green IT: A New Industry Shock Wave," *Gartner Symposium/ITxpo*, 2007.

[8] P. Scheihing, "Creating energy-efficient data centers," in *Proc. Data Center Facilities and Engineering Conference*, 2007.

[9] "Cloud computing definition from Wiki: http://en.wikipedia.org/w/index.php?title=Cloud_computing."

[10] "Gartner Study: http://www.gartner.com/it/page.jsp?id=1454221."

[11] "Gartner Study: http://www.gartner.com/it/page.jsp?id=1389313."

[12] "Amazon Webservice: http://aws.amazon.com."

[13] "Google AppEnginee: http://appengine.google.com."

[14] "Microsoft Azure: http://www.microsoft.com/windowsazure/."

[15] "Uptime Institute: http://uptimeinstitute.com/."

[16] "Facebook Open Compute: http://www.facebook.com/notes/facebook-engineering/building-efficient-data-centers-with-the-open-compute-project/10150144039563920."

[17] "Open Compute Project: http://opencompute.org/."

[18] "IBM EnergyScale: http://www-03.ibm.com/systems/x/solutions/infrastructure/energy/save.html."

[19] "ENERGY   SAVINGS   FROM   ENERGY   STAR-QUALIFIED SERVERS," u.S. EPA ENERGY STAR. [Online]. Available: http:// www.energystar.gov/ia/products/downloads/ES_server_case_study.pdf

Author Contact Information

Jinzy Zhu and Xing Fang are with Huawei Technologies Co. Ltd., China, Email: JINZY.ZHU@huawei.com and shaun.fangxing@huawei.com.

10

Energy-Efficient Management of Campus PCs

Luca Chiaraviglio

Politecnico di Torino, Torino, Italy

Marco Mellia

Politecnico di Torino, Torino, Italy

CONTENTS

10.1   Introduction

Energy consumption has become a key challenge in the last few years. According to several studies [1, 2], the Information and Communication Technology (ICT) sector alone is responsible of a percentage which varies widely from 2% to 10% of the worldwide energy consumption, and several projects, both from industries and universities, are trying to reduce the power consumption of electronic devices.

Considering ICT, the energy consumption can be coarsely split into three main power-hungry classes: telecommunication systems, data-centers, and users' terminals. For each class, energy saving approaches have been studied and proposed. For example, considering telecommunication systems, in [3–7] the authors investigate the energy saving obtained by finding the minimal subset of network devices that guarantees the telecommunication system to match the current traffic demand. Both WiFi, backbone or UMTS networks are considered, showing that large power saving can be obtained during off-peak periods of time, i.e., when traffic is low. Considering data-centers, energy efficiency has become one of the primarily performance metric, with players like Google, IBM, Amazon, eBay, etc. that are investing a lot of resources to improve the energy efficiency of their server farms. To mention some recent research papers, technological solutions have been studied as well (see [8] for example), and more innovative software design is being investigated (see [9] for example), in which the performance of an algorithm/application is compared in terms of energy consumption instead of CPU time.

While reducing the power required to run a network or a server farm is a must, today the largest majority of power consumption in the ICT is due to the billions of terminals in both households and companies. Furthermore, with the current proliferation of networked devices that are continuously power on, it is becoming urgent to think about a simple and effective way to reduce their power consumption, not only by reducing the energy consumed while they are active, but by turning them off when left unused.

Considering the personal computers, manufacturers have moved their attention to offer energy efficient devices, e.g., CPUs, memory, video, or hard-disks, proposing "green component" as a competitive gain. From a system point of view, even commercial solutions like [10] are becoming to be adopted, and solutions that rely on the idea of protocol proxying [11] and virtualization techniques (including cloud computing) to concentrate the number of PCs (or functionalities) on to a small set of devices are being investigated. However, the power consumption of a PC, even if used as a "dumb" terminal, is far from being negligible, and today a simple desktop PC requires about 100W to be simply up, despite its much more energy efficient design. However, people generally leave their PC on, even if not used. To this extent, in [12], authors show results of computer usage in 20 households, and observe that the economic advantages offered by powering off PCs at home are too little to compensate frustrations like long boot up times.

In this chapter we experimentally investigate the users' habits in our Campus. We answer to questions like "What is the energy cost that today a medium University has to face due to the number of PCs that are present?" "Is that a negligible cost, or is it possible to reduce it, by implementing some smart energy efficient solutions?" Focusing on PC usage, we find out that most people prefer to leave their PCs always on, causing an energy waste that overall can correspond to more than 250,000 Euros per year. This is mainly due to two dominant factors: i) the little sensibility people have versus the cost of keeping a PC on, and ii) the cost both in terms of time and technical skill to properly and quickly power down and up a PC. These somehow surprising facts suggested us to design a solution that controls the power state of PCs in the Campus, explicitly targeting the ease of use. The result is PoliSave, a centralized web-based architecture which allows users to schedule power state of their PCs; the server remotely triggers power-up and power-down events by piloting a custom software which has to be installed in each PC. The client software handles all the tasks of correct PC configuration, enabling Wake-On-Lan (WoL) on network cards and hibernation feature on the OS, according to which the current PC state is saved on the hard disk for quick recovery at bootstrap. While the proposed scheme follows a traditional approach, its implementation faced several issues that we describe and discuss in the next sections. Differently from our previous work [13], in this chapter we present a comprehensive description of PoliSave features, and a more complete set of results.

All the functionalities of PoliSave have been implemented, and a deployment trial has been studied. Results show that the possible saving is huge, with negligible impacts on users' habit.

The chapter is organized as follow: Sec. 10.2 reports the motivations of our work, while Sec. 10.3 details the methodology and results to assess the impact of electronic devices power consumption in our Campus. Then, Sec. 10.4 reports the description of PoliSave architecture. Results are instead presented in Sec. 10.5. Sec. 27.6 describes the related works. Finally, conclusions are drawn in Sec. 10.7.

10.2 Motivations

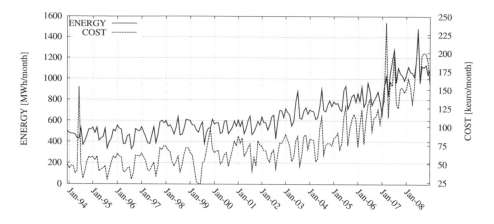

FIGURE 10.1 Energy and electricity bill at Politecnico di Torino.

10.2.1   Reducing PCs Power Consumption: Why Now?

Despite in the past years the Computer Industries have spent lot of time improving energy efficiency of PCs, today a typical desktop PC consumes about 65 to 250 watts when active. To confirm this figure, several measurements are available from the Internet, made both by enthusiastic, independent firms and manufacturers. For example, [14] reports the time variability of the power consumed by two desktop PCs, equipped by a dual core CPU from Intel and AMD respectively, which were running standard benchmarking software. At first, the power optimization a modern PC can achieve by finely controlling the power state of internal components is excellent, so that when no computation power is required, the energy consumption is halved; nonetheless, the idle-state power consumption is still in the order of 100W.

Considering that today most of office-related jobs require a PC with which the worker interacts, it is easy to foresee that the power consumption of PCs in any Company, University, Public Institution, etc. can be an important component of the total energy consumption. But the actual quantification of this figure is hard to obtain, since up to recent years, power cost (and optimization) has always been considered to be marginal compared to the total cost. However, the recent increase of energy price and the increased sensibility toward the global warming problem have made this an important issue.

10.2.2   Impact of PCs Power in Campus Environments

To answer the question "How much does the energy consumed by PCs cost in a Campus?", we started collecting data from Politecnico di Torino. Our institution is the second largest technical University in Italy, with about 1800

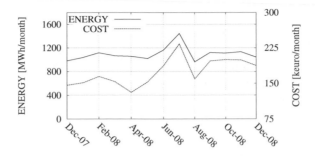

FIGURE 10.2 Energy and electricity bill at Politecnico di Torino (2008 detail).

staff members, and 28.000 students. The main Campus in Torino is divided into 14 departments, the Electrical Engineering (EE) and Computer Science (CS) being the two largest ones; more than 9000 total PCs are registered in our DNS database (DNS-db). Thanks to available historical data, Fig. 10.1 details the total energy consumption and the electricity bill for our Campus since December 1993. It shows the energy consumption and its actual cost on a monthly basis, reporting the scale on the left and right axis, respectively. Not surprisingly, the energy consumption has almost doubled, passing from 500 MWh/month to more than 1 TWh/month, with a percentage growth of more than 116%. The electricity price has experienced an even larger increment, with an increase of 218% since 1993, so that now the monthly bill is always more than 150k Euros per month. This is due to the recent increase of the energy cost, which went from 0.009 to the 0.017 Euro per kWh. It is easy to assume that these phenomena are due to proliferation of electronic systems in our Campus, which started to happen since 1996. Fig. 10.2 details the 2008 statistics, which show the typical season trend, with the highest consumptions in July, where the costs of cooling and air conditioning become predominant. Considering the electricity costs, the peak is also located in July, while the overall increasing trend is due to the oil speculation experienced in 2008.

These numbers suggest that saving on energy consumption has also beneficial impacts on cost reduction. But "how much of the energy consumption is due to PCs, and how much is *wasted* when PCs are left powered on but idle?" This second question motivated the development of a methodology to monitor PC usage in our Campus.

10.3   Monitoring PC Usage

To assess the impact of networked electronic devices, we design and implement a monitoring tool to check the power state of all the terminals in the central area of our Campus.

Different solutions can be taken into account to monitor the number of networked devices actually powered on. For example, most of modern OSes keeps information about the PC activity in system logs. This information may be reliable, yet difficult to collect in a heterogeneous enterprise environment where not all the PCs (and devices in general) are under the control of a central administrator. Moreover, the OS does not update the system logs during a period of inactivity, e.g., during the night; unfortunately, this information is crucial to decide if the system is powered off or on.

Another possibility is to remotely *scan the network* interface of devices, e.g., by "pinging" each of them. The implementation may be very simple, yet it is inadequate, since by default most of modern OSes (e.g., Windows XP) does not reply to the ICMP `echo request` messages, so the number of active devices ends up to be largely underestimated. Using an "ARP scan" would be much more reliable, since all networked devices have to support the ARP protocol. The main disadvantage is that this system works only if the network is a single Ethernet LAN, since ARP requests are not routed by IP routers: this is a limitation in our scenario, in which hundreds of IP subnets are interconnected at the IP layer.

Finally, another technique to detect active hosts is performing a "port-scan" using tools like `nmap` [15]. Intuitively, some of the most commonly used TCP ports are checked to see if any reply is received as a proof of host alive state. The main disadvantage is that this method can be quite intrusive, so a careful evaluation of ports and sub-networks to scan must be considered before running the scan.

10.3.1   Methodology

For our purposes, we used `nmap` version 4.85beta9 to find the number of devices actually powered on. The port-scanning routine is scheduled every $\Delta T = 15m$ using a `cron` entry, to allow `nmap` to complete a scan without time overlap. The port-scanning is performed targeting subnets of each Department/Center of our Campus, including all public and private IP addresses. This allows us to monitor and to track each department device activity.

Table 10.1 details the minimum subset of ports which have been selected to reduce the intrusiveness of the scanning while having the highest accuracy. For UNIX systems, it is usually easy to detect them by scanning the SSH port; for Windows machines, we selected the Remote Desktop Protocol port, Active Directory, NetBios and file sharing services. Finally, the classical port 80 is

TABLE 10.1 Ports Scanned by `nmap`

Number	Use
22	SSH
80	HTTP/Skype
135	Active Directory
139	NetBios
445	File Sharing
3389	Remote Desktop Protocol

checked since it is typically used for remote management services of devices like routers, Access Points, VoIP phones, etc. On normal hosts, port 80 is also used by some P2P applications like Skype. This subset of ports proved to be very reliable; results are practically equal to the ones obtained by performing a port scan covering the entire range of ports, which however requires more than 3 hours to be completed and causes several alarms on firewalled hosts, being it very intrusive. Note that the number of TCP packets generated during the scan of a single TCP port is reduced to 2 or 3 packets – corresponding to three possible cases: i) SYN+RST for a closed port, ii) SYN+SYN/ACK+RST for an opened port, iii) SYN+SYN+SYN for a silent PC. The total probing overhead amounts to less than $3 \times 64 \times 6 \times 9000/(15 \times 60) = 11.52$kBps to complete the scan assuming 64B long Ethernet messages, 6 selected ports, 9000 IP addresses scanned in a $\Delta T = 15 \ m$ time interval.

10.3.2 Results

10.3.2.1 Active Devices vs. Energy Consumption

The final monitoring of our Campus network has been up and running since mid of June 2009. From the same period, we collected also data regarding the power consumption of all our Campus, and of specific departments as well.[1] Therefore, the comparison between the Campus power consumption and the number of active devices is possible. Fig. 10.3 reports the average power required by our Campus, P_{TOT}, and the total number of networked devices powered on, N_{ALL}. Results refer to Friday the 26th of June. Both curves follow a typical day-night trend. In particular, the power consumption during the night is 38% of the peak hour demand, i.e., 1MW versus 2.5MW of power. Considering N_{ALL}, during the day more than 3500 devices are powered on. Astonishingly, during the night between Thursday and Friday more than 2000 terminals are left on, and on Friday night, no less than 1840 hosts are left up and running for the whole week-end. Update data are available from [13] which tracks the power state of networked devices in real time.

---

[1]Power consumption data are collected every 15 minutes using probes connected to the main power cabinets. We thank the Electrical Engineering Department for sharing these data.

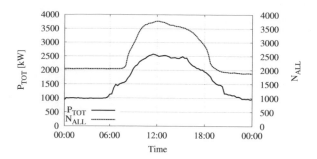

FIGURE 10.3 Total power consumption vs devices powered on during a day ([16] © 2010 University of Split, Faculty of Electrical Engineering, Mechanical Engineering and Naval Architecture)

10.3.2.2   Device Classification

A natural question arises: "How many of the active devices are actually PC used as terminals and can be potentially turned off during inactivity intervals?" To answer this question, Fig. 10.4 reports the breakdown of the active devices, detailing different OS architectures. In particular, for each IP address we perform a double check using both the information registered in the Campus DNS-db and the OS fingerprint feature of `nmap`. IP addresses are then grouped considering the different OS categories. From the bottom, the plot reports: network devices (e.g., switches and routers), networked printers, VoIP phones and other small network boxes (e.g., Access Points, small routers, etc.). All these devices are always powered on, with only printers that are seldom powered off at night. Considering Unix like OSes, we define two classes: hosts running Linux and other Unix hosts (mainly BSD/SUN systems). Also in this case, most of Unix hosts are left up and running, possibly due to their "server" capabilities, even if a large fraction of the 350 hosts running Linux could be actually used as simple terminals. Finally, the largest fraction of devices is due to personal computers running Windows like OS, representing about 30% of active hosts during the night and more than 40% during the day. Moreover, we notice that this estimation is a lower-bound, since it is very likely that unclassified machines (labeled UNCL) belong to this category as we verified by manually checking a random samples of them.[2] Windows machines are characterized by a more pronounced variability, yet during off-peak periods about 50% of them are left on. We argue that a large fraction of Windows hosts that are powered off are actually laptops, while the majority of PCs that are left on are regular (and more power hungry) desktop PCs. These data confirm that of the 2000 hosts left up and running during night and week-ends, up to 75% could be effectively turned off to save energy.

Fig. 10.5 details the daily variation of active devices considering only the

---

[2]OS fingerprint is unreliable in case a firewall is present.

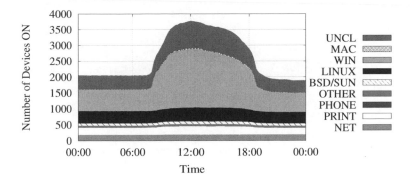

FIGURE 10.4 **Variation of devices powered on [16] © 2010 University of Split, Faculty of Electrical Engineering, Mechanical Engineering and Naval Architecture.**

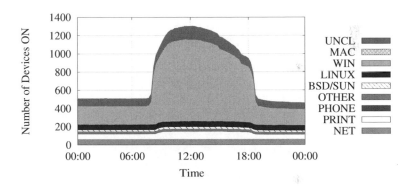

FIGURE 10.5 **Variation of devices powered on for the administration building.**

administrative and technical offices. This area comprises most of Campus servers (email, web, etc.), as well as student offices and laboratories. Interestingly, the number of active devices follows a stronger day-night trend, and more than 66% of hosts are turned off during night-time. This reflects the fact that hundreds of PCs located in the student laboratories are manually powered down during the laboratory closing time. Despite this fact, still more than 500 devices are left on, causing a clear waste of power. Notice also that in this environment, Windows architectures are predominant, representing more than 80% of hosts during the day.

Fig. 10.6 details the device variation in the CS/EE building. In this case, the day-night fluctuation is much less evident, so that 80% of devices are always on. Interestingly, the unclassified devices amount to about 30% of the total, due to the extensive use of firewalls which limit nmap OS fingerprint

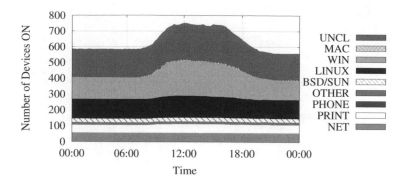

FIGURE 10.6 **Variation of devices powered on for the CS/EE building.**

effectiveness. Finally, nearly the same number of Windows and Linux machines are present.

10.3.2.3   Energy Consumption of Networked Devices

To assess the impact on the energy consumption of the networked devices, we estimate the power required by each category of devices according to figures that are publicly found. Table 10.2 reports the average power consumption estimation we adopted. In particular, we assume that desktop computers consume about 150W to account for the power required by the monitor too. We assume that BSD/SUN systems are used as servers, for which the power footprint is higher than desktop PCs. Finally, the power consumption of network devices like routers is assumed to be 100W, which is possibly a low figure; however routers typically have several interfaces/IP addresses, so that the same device is counted several times during a scan process. We therefore model the power consumption of a single interface/IP address, so that the power consumption of the router is proportional to the number of interfaces it has [4]. Fig. 10.7 reports the power consumption projection, considering the same dataset of Fig. 10.3. During the day the total energy required is more than 500 kWh, representing around 26% of the total power required by our Campus. During the night, about 300kWh are still consumed, corresponding to 35-40% of the total power consumption (see Fig.10.3). Moreover, the power consumption of computers (both Windows and Linux) is predominant, suggesting that further improvements have to be considered to reduce their power consumption. Notice also that the power figures represent average values, which give only a coarse prediction of the actual power consumption due to hosts in our Campus. As already shown in [14], the power consumption varies over the time according to the actual load of the system. This possibly justifies the higher total power consumption seen during the day, besides other sinks of power, like illumination, air conditioning, power hungry laboratories, etc.

TABLE 10.2 Power Consumption of Devices ([16], © 2010 University of Split, Faculty of Electrical Engineering, Mechanical Engineering and Naval Architecture)

Type	Power [W]
Win	150
Linux	150
UNCL	150
Network	100
Printer	50
BSD/SUN	200
Other	50
Mac	100
Phone	10

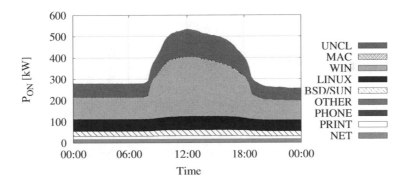

FIGURE 10.7 Estimation of the total power consumed by devices [16] © 2010 University of Split, Faculty of Electrical Engineering, Mechanical Engineering and Naval Architecture.

10.3.3 Characterization of Users' Habit

In the previous section we detailed the aggregate number of devices and their total power consumption. In this section, we instead provide a more detailed characterization of each device power state, i.e., the uptime of a device. We assume that time is divided in slot of duration $\Delta T = 15m$, corresponding to the measurement instants, i.e. $i\Delta T$ is the current slot, $i \in \{1, 2, \ldots, 96\}$ to consider single day. For each device x, we define the function

$$I_x(i) = \begin{cases} 0 & \text{if } x \text{ is OFF} \\ 1 & \text{if } x \text{ is ON} \end{cases} \tag{10.1}$$

Then, we compute the total amount of times x is on in a given day $\mathcal{D}$:

$$ON_x(\mathcal{D}) = 15m \sum_{i \in \mathcal{D}} I_x(i) \tag{10.2}$$

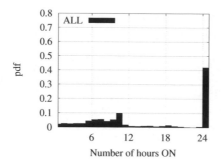

FIGURE 10.8 Pdf of the host daily average uptime.

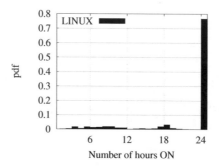

FIGURE 10.9 Pdf of the Linux machines.

Finally, we average over the set $\mathcal{K}$ of days in the dataset, obtaining the average time the device x is active per day, i.e., the host daily average uptime:

$$ON_x = \frac{\sum_{\mathcal{D}\in\mathcal{K}} ON_x(\mathcal{D})}{|\mathcal{K}|} \qquad (10.3)$$

Fig. 10.8 reports the probability density function (pdf) of ON_x considering all devices in the dataset, evaluated considering $\mathcal{K} = \{Mon, Tue, Wed, Thu, Fri\}$ during a normal working week. Interestingly, more than 40% of devices are always on, with other 40% that are alive for less that 12 hours, and the remaining 20% of devices that have a day lifetime between 13 and 20 hours. In particular, Fig. 10.9 details the host daily uptime for Linux devices: in this case, the percentage of always on devices grows to 76%, as intrinsically a Unix machine tends to be left on due its server nature. Considering instead only the Windows machines, Fig. 10.10 shows that 18% of them is powered on for 12 hours, which corresponds to the subset of the PCs of the student laboratories, while 27% of terminals are always left on. Recall that, although the percentage of Windows PCs always on is lower than the Linux machines, Windows devices represent about 70% of the total.

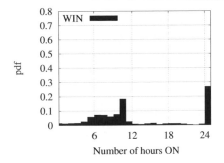

FIGURE 10.10 Pdf of the Windows machines.

10.3.4 Discussion

These results confirm the intuition that people usually leave their PCs up and running for most of the time, causing a considerable energy waste. We conducted a survey among users to investigate what are the reasons that refrain users to turn off their PC. First, the economic incentive is little or totally absent in the context of a Campus, since energy costs are not split among users. Second, the frustration of long power down and bootstrap times of today PCs typically discourages the adoption of energy wise policies by users. Third, the loss of state a reboot causes has also been found to be annoying (if not upsetting), since users are used to leave the office with applications and documents still opened on their desktops. Fourth, administration tasks (e.g., software updates or deployment of new software) can be scheduled at night. And fifth, some users want to access the applications and data on their office PCs even when they are at home. While technical solutions to the previous issues are already available (e.g., the "hibernate" feature that allows to freeze and recover the state of the PC storing its state on a file), people are not aware of them. In addition, the OS configuration has been identified as a complex task that typical user prefers to avoid. Note that some of these motivations were already pointed out [12], in which authors consider instead a set of home users.

10.4 Reducing PC Energy Consumption

The design of PoliSave has to face a complex and very heterogeneous scenario:

User heterogeneity Different *classes of users* exist in Campus environment, and roughly divisible into three main categories: researchers (including professors and PhD students), administrative staff, and bachelor/master students. Researchers usually have administration privileges on the Operating

System (OS), and they often tend to remotely access their office PC. Moreover, their PC is usually a single-user machine, registered in the Campus DNS-db with updated information about the user. Administrative people instead have restricted privileges on their PCs, and typically do not need to access their PCs from home. Administration PCs are registered within the DNS-db with the name of an inspector, who has also administration privileges over a set of machines. Finally, students have access to laboratory PCs with restricted privileges. Each lab is managed by an administrator who manually turns off the PCs at the end of each working day.

Therefore, the implementation of the software needs to take into account the people heterogeneity in our Campus, allowing the maximum freedom of PCs operation to professors/researchers/PhD students while at the same time offering a high level of automation for administration and students' PCs, since these terminals are managed by few administrators.

Operating System heterogeneity As shown by the previous sections, *different OSes* are typically present in a Campus. Nevertheless, the largest part of PCs are Windows and Linux machines, so we decided to primarily focus on them. Among the Windows machines, different versions exist, ranging from Windows 2000 to the most recent Windows 7. Similarly, different Linux distributions are used. The variegate popularity of OSes inevitably impacts on the design of PoliSave, since we need to invoke primitives at the system level to power on and off the PC. For example, considering the powering off technique, nearly all OSes offer primitives to put the system in a quiet state, but performing remotely these actions requires specific permission and tools. For example, Windows machines can be remotely shut down by their administrator using the Active Directory services if properly configured. Linux offers control to remote users through the SSH service, but root access must be granted to manage power state, and different distributions adopt different mechanisms to support advanced features like hibernation. Therefore, it is clear that a common solution among different OSes requires a program that acts as *middle-ware* between the user policies and the system primitives.

Considering instead the requirements, several constraints have to be considered:

Remote Control The first constraint is that the power management actions need to be performed remotely. For example, a professor would like to power on his office PC even from home, and lab administrators have to power on at the same time an entire set of PCs to perform software update operations. Similarly, scheduled maintenance operations of the power cabinets force to power off large groups of PCs; default scheduling policies can be enforced too, thus allowing users to forget about energy issues. Finally, offering a unified interface for the power control preferences avoids the user to understand and to familiarize with each OS interface.

Simple GUI Our application has then to offer a simple GUI, since the complexity of the control panel offered by OSes was identified as one of the major problem. Technical details must be as much as possible hidden, so that the user has not to be aware of even the IP address of his PC. The software needs to be transparent to other applications already installed and running on the machine, and its installation must follow the typical "double-click and forget" paradigm.

Flexible Management Yet, the system has to be fully controlled by users with administrative privileges, while all the other categories, such as students, have to be aware of the adopted policies. Moreover, lab administrators need to simultaneously perform actions over a set of PCs.

Custom Deployment Another requirement is the possibility to manually install the software, or to deploy it using remote installation. The first case is suited to users with administrative privileges, while the second case is useful for administrators of labs and administration PCs. Therefore, it is important to have custom installers, which allows to assign default policies, e.g., turn off by default the PCs at 22:00.

Security Due to the fact that actions are remotely performed, the system has to meet a reasonable level of security, e.g., to guarantee secure authentication and encrypted communications.

User and PC Information Update Finally, PoliSave has to handle the association between a PC and his user/owner, e.g., a professor or an inspector. In a dynamic scenario like a Campus, tracking this association is not easy, while regulations require it. For example, IP addresses can be allocated in a dynamic fashion, users can manually change them, or mobile terminals can change IP address while moving along the Campus. Similarly, when a PC changes his owner, then also the PoliSave settings for that particular machine have to be changed, e.g., granting the new user to change the scheduled policies. Another issue arises when the OS is reinstalled or upgraded: also in this case the user has to be informed how to reinstall/update PoliSave.

10.4.1 PoliSave Architecture

Given the previous presented list of requirements and constraints, here we detail the final *PoliSave architecture*. It is based on three main components: a Server, a Client, and a Communication Protocol, following a traditional and well understood architecture. Additionally, the live monitoring tool described in Sec. 10.3 has been implemented, and it is now integrated among the services of our Campus.

10.4.1.1 Client Architecture

The client manages the actual powering off mechanism. Two different client architectures have been developed, for Windows and Linux systems respectively.

Considering the Windows version, the client has been implemented as a multi-threaded background service. One thread manages the server communication, while a second thread is used to display pop-up messages to com-

```
bool ClientSetup(bool silent,
                 INFO info) {
    RetrieveInfo(&info);
    SendInfo("START",info,&s_ans,&ip,&mac);
    if((s_ans=="NO") && (!silent)) {
            ShowHelp(info);
            return false;
    }
    ADAPTER adapt=FindAdapt(ip,mac);
    if(info.vectAdapt[adapt].wol==false)
        SetWol(adapt);
    if(info.Hib==false)
        SetHib(info);
    InstallPolisave(info);
    RetrieveInfo(&info);
    SendInfo("END",info,&s_ans);
    if((s_ans=="NO") && (!silent)) {
            ShowHelp(info);
            return false;
    }
    ShowWebPage(info);
    return true;
}
```

FIGURE 10.11 The installation steps of the PoliSave client [16], © 2010 University of Split, Faculty of Electrical Engineering, Mechanical Engineering and Naval Architecture.

municate with the user. Pop-ups are used to warn the user when an action is going to be performed, allowing him to override a power-off action by canceling it. To perform the PC shutdown, the client exploits the standard Windows API, which requires the program to be executed as administrator: we force the execution of the client with high privileges during the installation phase.

The Linux version of PoliSave is composed by a background daemon, named `polisaved`. `polisaved` communicates with the server through socket function calls and displays the informations to the user by means of pop-up windows. The pop-up is activated using `dbus` and the `x11` system, eventually opening multiple pop-ups in case several users are remotely connected. Actions are instead performed by invoking the primitives of the Hardware Abstraction Layer `HAL`, which allows to list hardware properties, and control power state of the PC.

During setup, the client software performs some preliminary actions to optimally configure user's PC. In particular, the installer gathers system information, including WoL state, Hibernation capabilities, IP and MAC addresses of all network interfaces, OS type and version, and DNS name. This information is sent to the server, using either HTTP or HTTPS protocol. Then, once the server answer is received, the client processes it: if the server does not grant the client, the setup is aborted and a web page with detailed information is presented to the user. In case the server grants the installation, the IP and MAC address of the selected network interface (the one actually used to contact the server) are given back. The installer then enables the WoL for this interface and activates the OS hibernation feature; finally, the PoliSave service (or daemon) are installed. Result of each operation is sent back to the server, which then finally records the successful client installation, and present the user its intranet private area to show the installation result. For example, in case the WoL or the Hibernation have not been activated, detailed instructions are displayed to help the user. The client can be also installed in silent mode: in this case an eventual failure is reported directly to the PC administrator and no information is displayed to the user. Fig. 10.11 reports the main steps of the installer procedure.

At the moment of writing this chapter, the client supports Windows OS (Windows 7, Vista and XP, 32 or 64 bits, and earlier versions like Windows 2000 and Windows 98). Considering Linux, a complete version for the Ubuntu distribution has been tested.

10.4.1.2 Server Architecture

The server is the core of PoliSave. It performs client remote power control and it manages the database of clients, which includes the scheduled events of users. Both power-off and power-on operations are managed by the server to which users can access using a web interface. Users are free to specify a timetable that stores scheduled *actions* like *stand-by, hibernation, power-off, power-on* of a PC. The server then automatically performs the operations,

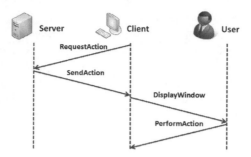

FIGURE 10.12 The powering off protocol of PoliSave.

TABLE 10.3 Structure of the messages ([16], © 2010 University of Split, Faculty of Electrical Engineering, Mechanical Engineering and Naval Architecture)

RequestAction	IP list, MAC list, Counter, Host, OS info
SendAction	Action, Time, *URL*, *Message*

periodically querying the database to look for actions to be performed. In addition, the user can perform real time actions, e.g. to immediately turn on (or off) the PC, triggering a "manual action".

The web interface allows users to interact with the server, from which users can change/add/remove entries from the PC power scheduling. Web pages are integrated in the Intranet personal pages, so that the user can simply login using his credentials. The web interface shows the list of PCs the user has administrative or user credentials, showing the information contained in the Campus DNS-db, i.e., IP and MAC addresses, administrator's and PC name. The web site is publicly available to remotely control PCs even from outside the University Intranet.

To update the Campus PC DNS-db, during installation the server verifies the client IP and MAC addresses in the DNS-db. If an entry is returned and the data are consistent, the installation proceeds. If instead there is some mismatch, an email is sent to warn both PC user and administrator, and installation of the client program is aborted. Once the client software has been successfully installed, the DNS-db is verified and eventually updated every time a client contact the server, thus allowing to track eventual modification to IP or MAC addresses of the client.

The server is implemented using simple Visual Basic scripts, with a SQL Server back-end to store the power scheduling and additional information. As in any server based architecture, the server represents a single point of failure, and standard solutions are available to eventually handle server failures.

10.4.1.3 Communication Protocols

The *power-on* mechanism relies on the WoL protocol standard. According to the standard, the PC to be woken is shut down (Sleeping, Hibernating or Soft Off, i.e. ACPI state G1 or G2), with power reserved for the network card. The network card keeps listening for a specific packet containing a message with 6 copies of its MAC address, called the "Magic Packet," broadcasted on the LAN. The magic packet has to be routed over the network of our Campus which is a L3 network. However, for security reason, routers typically do not forward broadcast messages to the destination subnet, so that proper configuration is required on all L3 routers, since the broadcast packet must be forwarded to the destination subnet too.

The *power-off* technique is instead implemented by a proprietary protocol, as Fig. 10.12 shows. More in depth, the client performs a *periodic polling* with the server, by sending a *request* message, whose fields are detailed in Table 10.3. The information sent to the server includes the list of IP and MAC addresses for all the network interfaces of the host, the PC name as it appears in the DNS and the type of OS in use, including the system version. These fields are used to keep the Campus DNS-db updated. Finally, a counter, increased at any polling and reset after any power on operation, is reported too. This counter allows to track the PC power state, and eventually detects missing poll operations.

When the server receives a *RequestAction* message, it sends back a response containing a *SendAction* among {*Hibernation, Stand-by, Power-Off, Wait, Message*}. Besides the power control actions, the *Wait* action forces the client to simply return idle and wait for the next polling event. The *Message* action allows the server to send a string *Message* that is displayed to the user screen. This feature can be used for example to warn users to update the client. The optional *URL* field contains the new server URL that has to be used for future polling (useful to relocate the server). In all cases, the response message includes a *Time* value which details when the client has to perform the next poll. This parameter trades between timely management and server load. In particular, low *Time* values lead to quick response times, but tend to increase the load on the server.

When an action has to be performed, the client displays a warning message to the user via a pop-up window. The user can cancel or grant the action, up to a maximum timeout of one minute. If the event is granted (or the timeout has expired), then the action is performed by invoking the system APIs and finally the PC is shut down.

All protocol messages are encapsulated using HTTP protocol, and `OpenSSL` libraries are included to guarantee privacy and authenticate the server. In this case the HTTPS protocol is used, and the public key of the server is distributed with the setup package, together with the server's certificate.

Thanks to the fact that the communication is started by the client, the powering off mechanism works even if the client is behind a NAT. Another advantage is that the client OS does not need to keep any port open, avoiding both security issues, and firewall configurations. The power-on procedure still relies on WoL messages sent by the server, so that the eventual NAT box has to appropriately route them. Finally, a possible drawback of the communication protocol is that a manually triggered power-off event will be actually delayed until the next client polling phase starts.

10.4.2   Discussion and Implementation Issues

We faced several design and implementation issues during the development of PoliSave, and we tried to efficiently solve them during the implementation phases. First of all, we had to choose among a local application scheduling its own actions, or allow a remote server to have full control on the client. In the first case, the work would be simpler because all modern OSes allows the user to schedule powering off actions after periods of inactivity. However, the power on operation can not be performed by the local computer, thus limiting the transparency of PoliSave since the user has then to manually bootstrap the PC. Moreover, local solutions do not scale well in laboratories, where hundreds of PCs needs to be managed. Finally, the user interface that each OS offers to modify the power properties is not standard and is often difficult to use. As previously noted, users are restrained by system complexity. Motivated by these issues we therefore implemented PoliSave as a client-server application.

The second issue regards the possible actions to allow, and the policies to activate them. We decided to restrict the range of possible actions to two categories (power on and power off), and to offer only a scheduled activation policy. Indeed, modern OSes offer sophisticated techniques to track the user activity and then to automatically decide *when* to enter in power saving mode, e.g., hibernating the PC. Such techniques, though may appear quite prominent, have some drawbacks. Detecting the inactive phase is a hard task, solved typically by identifying absence of user interaction with the keyboard or mouse, i.e., using a simple "inactive time" threshold. However, setting this threshold is not straightforward: a short interval frustrates the users due to unwanted powering off events, while large intervals lead to possible waste of power. Moreover, some user may leave their PC powered on while doing some computation. Others may like their PC to raise an alarm in case some emails arrives, even if they are not currently using the PC, etc. We therefore excluded such techniques and designed a system in which the users have explicitly full control on the scheduled actions, eventually including some default actions.

Given the choice of a client-server architecture, we had to choose which communication protocol to support communication between the client and the server. In particular, the choice is between a novel solution, or one of the already available protocols for remote PC administration. For example, the Active Directory service is available to manage windows OS. This latter choice would limit the support to only Windows PCs correctly managed in the Active Directory service. Moreover, in our scenario the presence of Network

Address Translators (NAT) poses additional issues, since this would prevent the server to initiate a communication with a client. Therefore, to overcome these issues, we decided to build a proprietary solution for the powering off mechanism, in which the client starts the communication with the server.

Considering the PC powering on action, the WoL mechanism has to be activated in the BIOS and/or in the OS settings of the PC. By analyzing the default settings over a significant set of desktop PCs, we have found that in most cases the WoL is already enabled in the BIOS but not in the OS, which then causes the WoL to fail. We have therefore decided to automatically activate the WoL feature at the OS level during the installation of PoliSave, to limit the number of manual operations required by users. However, for a small number of PCs the WoL has to be activated also through the BIOS, and in this case a manual configuration is imposed. Notice also that the primitive offered by OSes to activate WoL are not standard, and even change with different versions of the same OS (e.g., Windows XP or Vista or Seven show different behaviors). Finally, in some cases, the PC configuration has to be manually checked to find the correct configuration of the hardware to support WoL features. This denotes that the WoL feature is not completely supported and correctly implemented in today systems, so that hardware manufacturers and OS producers should improve its support.

Another configuration problem arises when supporting the Hibernation feature. Also in this case, the OS has to be properly configured, and hardware must support it as well. In particular, a temporary file to contain the data has to be created, and the OS must be properly configured. Also in this case, OS primitives change among different OS versions and distributions. During the setup phase, we try to automatically enable hibernation, and report its availability to the server.

Another key problem is how to associate the user to the PC. To this purpose we exploit the information contained in our DNS-db, which stores this information for historical reasons (but is rarely updated). In particular, we use the IP address and MAC address to identify the PC and the Intranet user-ID to identify the user. However, different network cards can be present, and the identification of the one actually connected to the Campus LAN is not trivial. To solve this problem, the client always send all available IP and MAC addresses back to the server, which then can select the correct one (e.g., by verifying the actual IP address of the sender). To prevent malicious installation of the software, during the installation of PoliSave client, the user must login into the University Intranet personal home page to finalize the setup. Note also that the PoliSave web interface is integrated with the already available Intranet pages, avoiding the creation of a new login on a different server, explicitly addressing the ease of use. Finally, as detailed in the previous section, all communication is encrypted to prevent possible attacks.

At last, during installation we impose some default scheduling of the powering on and switching off actions. In particular, the powering off is scheduled for every week day at 22:00, and by default the system shuts down completely.

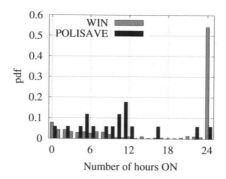

FIGURE 10.13 Host daily average uptime for PCs in the trial and for other PCs [16], © 2010 University of Split, Faculty of Electrical Engineering, Mechanical Engineering and Naval Architecture.

This point is crucial since otherwise most of the users would not modify the default setting, i.e., never activating the power off schedule. The user is then clearly free to modify the default scheduling later.

10.5   Results

10.5.1   Testbed Description

To test the effectiveness of PoliSave, we installed the system on a trial of 70 users of the Electrical Engineering Department (EEDept). The aim was i) to test the software implementation on a real environment, ii) to assess the possible power-saving and iii) to collect feedback from actual users. In particular, we have computed both $ON_x(\mathcal{K})$ for $x \in \{PoliSave\}$ and $ON_x(\mathcal{K})$ for $x \notin \{PoliSave\} \cap \{EEDept\}$; $\mathcal{K} = \{Mon, Tue, Wed, Thu, Fri\}$.

10.5.2   Minimizing the Number of PCs On

Fig. 10.13 shows the comparison among the two pdfs. Normally, about 53% of PCs without PoliSave are always on, while with PoliSave this percentage falls to less than 6%, with most of PCs that are alive for less than 12h. The average daily uptime of PCs managed by PoliSave is $9.7h$, while the average daily uptime for other PCs is $15.9h$. This corresponds to an average savings of more than $6h$ per working day, or an annual saving of about 219kW/year (about 100,000 Euros per year using current electricity costs). The savings achieved including weekends, for which PoliSave PCs are powered off for the whole day with probability 0.93, amount to more than 250,000 Euros per year.

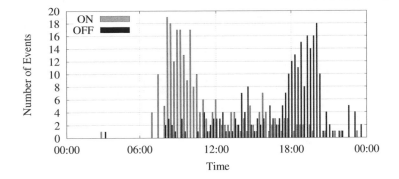

FIGURE 10.14 **Number of on and off events recorded every 15minutes [16], ©**
2010 University of Split, Faculty of Electrical Engineering, Mechanical Engi-
neering and Naval Architecture.

10.5.3 Automatic vs. Manual Operations

To give some more details on the users'habit, Fig. 10.14 reports the number
of ON and OFF events recorded during one week, computed as

$$E_{OFF}(i) = \sum_{x \in \mathcal{X}} |(I_x(i) == 1) \,\&\&\, (I_x(i-1) == 0)|$$

$$E_{ON}(i) = \sum_{x \in \mathcal{X}} |(I_x(i) == 0) \,\&\&\, (I_x(i-1) == 1)|$$

with $\mathcal{X} = \{PoliSave\}$. Most of PCs running PoliSave are powered on in
the morning and turned off during the late evening, with the variability of
the measurements that suggests each user has customized the action from the
web-interface. Interestingly, some rare events are recorded at night, suggesting
that the option of manually turning on/off the PC via the server web interface
is seldom used.

10.5.4 Different Users' Profiles

For each PC running PoliSave we compute the probability to be alive during
a time slot i as

$$Prob\{ON_x(i)\} = \frac{\sum_{k \in \mathcal{K}} I_x(i + k24)}{|\mathcal{K}|} \quad (10.4)$$

Examples of $Prob\{ON_x(i)\}$ are plotted in Fig. 10.15, which shows three
different profiles. In particular PC2 exploits automatic power on feature of
PoliSave during all working morning at 8.30am. Manual power off is instead
used. The probability of finding PC2 on during the day is then $5/7 = 0.71$. On
the contrary, the profile of PC1 reveals that PoliSave automatically turns off
PC1 at 19:30, so that the probability to be on at night is equal to 0. Manual

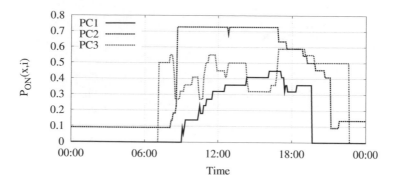

FIGURE 10.15 **Probability to be on for three PC using PoliSave [16], © 2010 University of Split, Faculty of Electrical Engineering, Mechanical Engineering and Naval Architecture.**

power on is adopted, with the user being in the office (its PC being alive) smaller than 0.5. PC3 leverages instead both the power on and off features, and a very aggressive policy is adopted by the user to turn off the PC when possibly idle.

10.5.5   Users' Characterization

Considering the category of people that volunteered to take part in the PoliSave trial, Fig. 10.16 details their distribution among PhD students, Researchers, Professors and Administrative members. Not surprisingly, PhD enthusiastically joined the trial, while administrative members represent a very small percentage, suggesting that further incentive needs to be considered to extend and to encourage the adoption of PoliSave. Finally, considering the users' feedback, no major complaint was returned, but generally a very positive experience was returned.

10.5.6   Discussion

These very favorable results actually encourage our effort, and, at the time of writing, PoliSave is being extended to the whole set of Campus PCs, and other Italian Universities are studying how to deploy it.

  Despite the very good results of the initial PoliSave trial, some open issues need to be considered. First of all, incentive has proved to be a key challenge, because users are typically not conscious about energy waste. The fact that energy costs are not divided among users definitively limits the awareness of people. To this extent, we are working on the customization of the monitor capability of PoliSave, so that individual users can track their power consumption. The idea is to precisely account the energy waste, introducing a

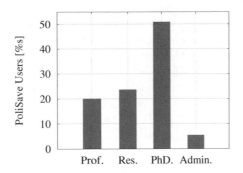

FIGURE 10.16 Distribution of PoliSave users among classes.

competition inside the Campus, e.g. awarding the Department or group which proves to be the more energy wise.

10.6 Related Work

Several projects are currently trying to reduce the energy consumption of networked devices. In the last years, researchers have extensively studied solutions to reduce the power consumption of data center, [17–19]. Our solution is instead intended to mitigate the power consumption of clients when not in use.

In [20] the authors collected data on the after-hours power state of networked devices in office buildings, showing that most of devices are left powered on during night. However, the proposed measurement technique is manual, thus limiting the number of measurements over time and the applicability in large buildings. In this chapter instead we have applied a fully automatic technique that scales well also for large networks and tracks the number of devices powered on in real time.

A complementary approach to put the device in power save mode relies on the idea of proxying. In [5] the authors propose the proxying technique for network elements. In [21, 22] the authors extend their technique to end-user PCs, analyzing which protocols and applications require proxying. Moreover, connectivity issues are considered in [11]. All of these works show that the possible power saving derived from proxying can be huge. However, this technique requires the modification of the hardware on PCs, which can be a hard task in large Campuses. Our solution instead is completely software-based to explicitly address the ease of management.

Commercial solutions like [10] are proposing green solutions to reduce the power consumption waste for enterprise networks. Cisco Systems and IBM have recently announced a solution to remotely monitor and control the network devices such as APs, routers, phones and even terminals, exploiting the idea of a central administration [23]. Unfortunately, at the time of writing, this solution supports only Cisco network devices, and server operating system, while ignoring Desktop PCs.

Researchers from both Universities [24] and industries [25] are developing new programming techniques to reduce the computational load and consequently the energy consumption of the machine. Their idea is to adapt expensive functions with approximate versions, while guaranteeing an admissible Quality of Service. This technique could be integrated with our solution to further increase the energy savings.

Considering academia, several projects [26, 27] are trying to face the problem of energy consumption as a whole, by putting together experts from networking and energy areas to study energy-saving multi-disciplinary approaches. Our work can be integrated in these contexts.

Current OSes offer inbuilt power management functions, whereby a system with no activity is gradually moved to sleep state (S1 S4) as defined in ACPI specifications. As an example, by default a Windows 7 based system goes to sleep due to inactivity after 30 minutes. Our solution can work also in such context, since the server can be eventually informed from the client that to system is going to sleep, or alternatively detect the sleep state if polling messages are not received.

Finally, carbon offsets approaches are being proposed [28]. Carbon offset relies on the idea of trading to compensate the emissions, e.g. by buying electricity from renewable sources or by abatement of pollutants. The possible savings obtained through energy-efficiency of PCs devices can lead the universities to be the main players in the carbon-offset economy, since they could acquire a zero-carbon condition.

10.7   Conclusion

In this chapter we have presented PoliSave, a software that reduces the power consumption in computer networks. We have first quantified the energy waste that PCs generate since they are not turned off during nights and weekend in our Campus network. To this extent, we have implemented a monitor methodology that has shown that more than 50% of PCs are always on, i.e., more than 2000 PCs are wasting about 350kWh each night and weekends. We have then described and discussed the architecture of PoliSave, a web-based service that allows users to schedule power on and off of their PCs, avoiding the frustration of long boot-up and power-down times. Extensive measures on a trial involving more than 70 users have proved the effectiveness of our solution, with an enthusiastic feedback. PoliSave is being extended to all PC in our Campus, with the goal of saving about 250,000 Euros from the University energy bill.

Acknowledgment

The research leading to these results has received funding from the European Union Seventh Framework Programme (FP7/2007-2013) under grant agreement n. 257740 (Network of Excellence TREND.

Bibliography

[1] G. Plan, "An inefficient truth," *Global Action Plan, London, http://www. globalactionplan. org. uk/upload/resource/Full-report. pdf,* 2009.

[2] *SMART 2020: Enabling the low carbon economy in the information age,* 2008, the Climate Group, Global eSustainability Initiative.

[3] A. Jardosh, K. Papagiannaki, E. Belding, K. Almeroth, G. Iannaccone, and B. Vinnakota, "Green wlans: on-demand wlan infrastructures," *Mobile Networks and Applications*, vol. 14, no. 6, pp. 798–814, 2009.

[4] J. Chabarek, J. Sommers, P. Barford, C. Estan, D. Tsiang, and S. Wright, "Power awareness in network design and routing," in *Proc. IEEE INFO-COM 2008, the 27th IEEE Conference on Computer Communications*, 2008, pp. 457–465.

[5] S. Nedevschi, L. Popa, G. Iannaccone, S. Ratnasamy, and D. Wetherall, "Reducing network energy consumption via sleeping and rate-adaptation," in *Proceedings of the 5th USENIX Symposium on Networked Systems Design and Implementation.* USENIX Association, 2008, pp. 323–336.

[6] L. Chiaraviglio, M. Mellia, and F. Neri, "Reducing power consumption in backbone networks," in *Proc. IEEE International Conference on Communications.* IEEE, 2009, pp. 1–6.

[7] L. Chiaraviglio, D. Ciullo, M. Meo, and M. Marsan, "Energy-efficient management of umts access networks," in *Teletraffic Congress, 2009. ITC 21 2009. 21st International.* IEEE, 2009, pp. 1–8.

[8] X. Fan, W. Weber, and L. Barroso, "Power provisioning for a warehouse-sized computer," in *Proceedings of the 34th annual international symposium on Computer architecture.* ACM, 2007, pp. 13–23.

[9] A. Kansal and F. Zhao, "Fine-grained energy profiling for power-aware application design," *ACM SIGMETRICS Performance Evaluation Review*, vol. 36, no. 2, pp. 26–31, 2008.

[10] "Verdiem surveyor," in *http://www.verdiem. com.*

[11] M. Jimeno, K. Christensen, and B. Nordman, "A network connection proxy to enable hosts to sleep and save energy," in *Performance, Computing and Communications Conference, 2008. IPCCC 2008. IEEE International.* IEEE, 2008, pp. 101–110.

[12] M. Chetty, A. Brush, B. Meyers, and P. Johns, "It's not easy being green: understanding home computer power management," in *Proceedings of the 27th international conference on Human factors in computing systems.* ACM, 2009, pp. 1033–1042.

[13] PoliSave web site. [Online]. Available: http://www.polisave.polito.it/hosts_new.shtml

[14] "The truth about pc power consumption." [Online]. Available: http://www.tomshardware.com/reviews/truth-pc-power-consumption,1707.html

[15] nmap - Network exploration tool and security/port scanner. [Online]. Available: http://nmap.org

[16] L. Chiaraviglio and M. Mellia, "Polisave: Efficient power management of campus pcs," in *Software, Telecommunications and Computer Networks (SoftCOM), 2010 International Conference on.* IEEE, 2010, pp. 82–87.

[17] C. Bash and G. Forman, "Cool job allocation: Measuring the power savings of placing jobs at cooling-efficient locations in the data center," in *Proc. the USENIX Annual Technical Conference.* USENIX Association, 2007, p. 29.

[18] J. Stoess, C. Lang, and F. Bellosa, "Energy management for hypervisor-based virtual machines," in *2007 USENIX Annual Technical Conference on Proceedings of the USENIX Annual Technical Conference.* USENIX Association, 2007, p. 1.

[19] K. Goda and M. Kitsuregawa, "Power-aware remote replication for enterprise-level disaster recovery systems," in *USENIX 2008 Annual Technical Conference on Annual Technical Conference.* USENIX Association, 2008, pp. 255–260.

[20] J. Roberson, C. Webber, M. McWhinney, R. Brown, M. Pinckard, and J. Busch, "After-hours power status of office equipment and inventory of miscellaneous plug-load equipment."

[21] S. Nedevschi, J. Chandrashekar, J. Liu, B. Nordman, S. Ratnasamy, and N. Taft, "Skilled in the art of being idle: reducing energy waste in networked systems," in *Proceedings of the 6th USENIX symposium on Networked systems design and implementation.* USENIX Association, 2009, pp. 381–394.

[22] Y. Agarwal, S. Hodges, R. Chandra, J. Scott, P. Bahl, and R. Gupta, "Somniloquy: augmenting network interfaces to reduce pc energy usage," in *Proceedings of the 6th USENIX symposium on Networked systems design and implementation*. USENIX Association, 2009, pp. 365–380.

[23] "Cisco energy wise," in http://www.cisco.com/en/US/products/ps10195/index.html.

[24] W. Baek and T. Chilimbi, "Green: A system for supporting energy-conscious programming using principled approximation," Technical Report TR-2009-089, Microsoft Research, Tech. Rep., 2009.

[25] H. Hoffmann, S. Misailovic, S. Sidiroglou, A. Agarwal, and M. Rinard, "Using code perforation to improve performance, reduce energy consumption, and respond to failures," *Massachusetts Institute of Technology, Tech. Rep. MIT-CSAIL-TR-2209-042*, 2009.

[26] GreenLight Project. [Online]. Available: http://greenlight.calit2.net/projects.php

[27] "Sustainable it at stanford." [Online]. Available: http://sustainablestanford.stanford.edu/

[28] "The carbon centric computing initiative," in http://www.ccci.uow.edu.au/.

Author Contact Information

Luca Chiaraviglio and Marco Mellia are with Electronics Department, Politecnico di Torino, Corso Duca degli Abruzzi 24, Torino, Italy, Email: luca.chiaraviglio@polito.it, mellia@tlc.polito.it.

Part II

Focus on Wireless Communications

11

C-RAN: A Green RAN Framework

Kuilin Chen

China Mobile Research Institute, China

Chunfeng Cui

China Mobile Research Institute, China

Yuhong Huang

China Mobile Research Institute, China

Bill Huang

China Mobile Research Institute, China

CONTENTS

11.1   Introduction

Today's mobile operators are facing a strong competition environment. The cost to build, operate and upgrade the Radio Access Network (RAN) is becoming more and more expensive while the revenue is not growing at the same rate. With the rapid adaptation of wide screen smart phones in the market, the mobile Internet traffic is surging, while the ARPU (Average Revenue Per User) is flat or even decreasing slowly, which impacts the ability to build out the networks and offer services in a timely fashion. To maintain profitability and growth, mobile operators must find solutions to reduce cost as well as to provide better services to the customers.

On the other hand, the proliferation of mobile broadband Internet also presents a unique opportunity for developing an evolving network architecture that will enable new applications and services, and become more energy efficient.

China Mobile Research Institute (CMRI) has started preliminary research on these topics back to as early as 2008. In April 2010, after two-year research efforts a new RAN architecture by the name of C-RAN which features centralized processing, collaborative radio, real-time cloud computing and clean system is proposed by CMRI. In the following sections we will introduce the motivation, the concept, advantages, challenges, and current research and deployment status of C-RAN in details.

11.2 Challenges of Today's RAN

11.2.1 Explosive Network Data Traffic

Data rate of mobile broadband network grows significantly with the introduction of air-interface standards such as 3G(the third generation) and B3G(beyond the third generation) and wide screen smart phones; this in turn speeds up end user's mobile data explosion. In the past 3 years after 2007, a number of operators already witnessed 20 times mobile data traffic growth after introducing iPhone in their networks [1]. Moreover, one major mobile operator predicted that in the next decade from 2011 to 2020, the mobile data traffic will double every year and eventually increase to about 1000 times compared to today's [2]. This new trend will become a serious challenge to future RAN.

To meet the requirement of 1000 times data traffic increase in the next decade, there are a few ways. First, more spectrum resources can be allocated to carry more traffic. However, spectrum is a kind of "natural resource" that is very limited. Secondly, improvement of spectrum efficiency can be expected through more advanced transmission (e.g. modulation) schemes. But, there is theoretic limit and it becomes harder and harder as the relevant technologies are already approaching the limit. Finally, we can increase the number of cells in the system, i.e., splitting the cells into smaller ones and increasing the reuse factor of system. Giving that the first two facts cannot be changed at will, the last one becomes a must for the operator to achieve the 1000 times data traffic growth in the next decade.

11.2.2 Large Number of BS and High Power Consumption

As telecommunication operators constantly introduce new air interface and increase the number of base stations to offer broadband wireless services, the power consumption gets a dramatic rise. As illustrated in Figure 11.1, in the past 5 years, China Mobile has almost doubled its number of BS, in order to provide better network coverage and capacity. As a result, the total power consumption has also doubled [3]. The higher power consumption is translated directly to the higher OPEX (Operating Expenditure) and a significant environmental impact, both of which are now increasingly unacceptable.

Figure 11.2 expresses the key components of the power consumption of China Mobile [3], which shows the majority of power consumption is from BS in the radio access network (RAN). Inside the BS, only half of the power is used by the RAN equipment; while the other half is consumed by air conditioners and the other facilitate equipment.

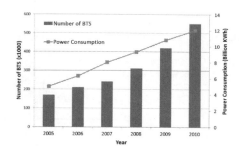

FIGURE 11.1 Rapid growth of BS numbers and power consumption of China Mobile

11.2.3   Rapid Increasing CAPEX and OPEX

The TCO (Total Cost of Ownership) including the CAPEX (Capital Expenditure) and the OPEX comes from the network construction and operation. The CAPEX is mainly associated with network infrastructure establishment, while the OPEX is mainly associated with network operation and management.

In general, up to 80% CAPEX of a mobile operator is spent on the RAN. This means that most of the CAPEX is related to building up cell sites for the RAN. The historical CAPEX expenditure of 2007-2012 are shown in Figure 11.3. Because 3G/B3G radio signals (transmitted at 2GHz) have higher path loss and penetration loss than 2G radio signals (transmitted at 900MHz), multiple cell sites are needed for the similar coverage level of 2G mobile cellular network. Thus, dramatic increase was found in the CAPEX when building up a 3G network. The CAPEX is mainly spent at the stage of cell site constructions and consists of purchase and construction expenditures. Purchase expenditures include the purchases of BS and supplementary equipment, such as power and air conditioning equipment etc. Construction expenditures include network planning, site acquisition, civil works and so on. As shown in Figure 11.3, it is noticeable that the cost of major wireless equipment makes up only 35% of CAPEX, while the cost of the site acquisition, civil works, and equipment installation is more than 50% of the total cost. Essentially, this means that more than half of CAPEX is not spent on productive wireless functionality. Therefore, finding ways to reduce the cost of the supplementary equipment and the expenditure on site installation and deployment is important to lower the CAPEX of mobile operators [3].

The OPEX in network operation and maintenance stage plays a significant role in the TCO. Operational expenditure includes the expense of site rental, transmission network rental, operation/maintenance and bills from the power supplier. Given a 7-year depreciation period of BS equipment, as shown in Figure 11.3, an analysis of the TCO shows that the OPEX accounts for over

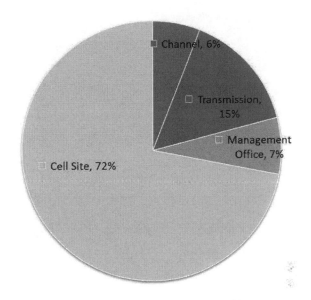

FIGURE 11.2 Power consumption of typical BS

60% of the TCO, while the CAPEX only accounts for about 40% of the TCO. The OPEX is a key factor that must be considered by mobile operators in building the future RAN.

11.2.4   Dynamic Mobile Network Load

One characteristic of the mobile network is that subscribers are frequently moving from one place to another. From measurement data based on real operation network, we have noticed that the movement of subscribers shows a very strong time-geometry pattern. Around the beginning of working time, a large number of subscribers move from residential areas to central office areas for work; when the work hour ends, subscribers move back to their homes. Consequently, the network load moves in the mobile network with a similar patternso called "Tidal Effect". As shown in Figure 11.4, during working hours, the core office area's Base Stations are the busiest; in the non-work hours, the residential or entertainment area's Base Stations are the busiest [3].

Today, each Base Station's processing capability today can only be used by the active users within its cell range, causing idle BS in some areas/times and oversubscribed BS in other areas. When the subscribers are moving to the other areas, the Base Station may just stay in idle with a large part of

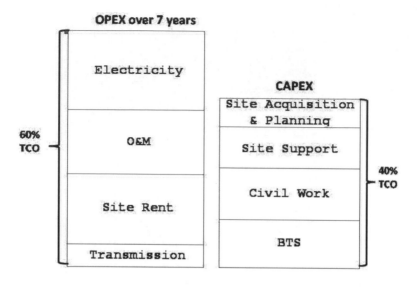

FIGURE 11.3 CAPEX and OPEX analysis of cell sites

its processing power wasted. Because generally mobile operators must provide 7 × 24 coverage, these idle Base Stations consume almost the same level of energy as they do in busy hours. Even worse, the Base Stations are often dimensioned to be able to handle a maximum number of active subscribers in busy hours, thus they are designed to have much more capacity than the average needed, which means that most of the processing capacity is wasted in non-busy time. Therefore, sharing the processing capacity and thus the power between the different cell areas is a way to utilize these BS more efficiently.

11.2.5   Growing Internet Service Pressure on the Core Network

The exponential growth of mobile broadband data puts heavy pressure on operators' existing packet core elements such as SGSNs (Serving GPRS Support Node) and GGSNs (Gateway GPRS Support Node), increasing mobile Internet delivery cost and challenging the flat-rate data service models. The majority of this traffic is either Internet bound or sourced from the Internet. Catering to this exponential growth in mobile Internet traffic by using traditional 3G deployment models, the older 3G platform is resulting in huge CAPEX and OPEX cost while adding little benefit to the ARPU. Additional issues are the continuous CAPEX spending on older SGSNs & GGSNs, the

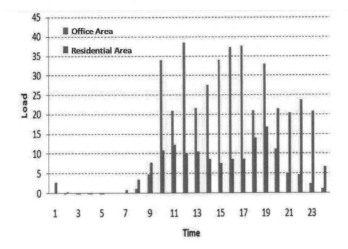

FIGURE 11.4 **Mobile network traffic load in daytime**

higher Internet distribution cost, the congestion on backhaul and the congestion on limited shared capacity of base stations. Therefore, offloading the Internet traffic, as close to the base stations as possible, can be an effective way to reduce the mobile Internet delivery cost.

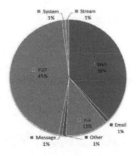

FIGURE 11.5 **Wireless data traffic on a commercial 3G**

Meanwhile it is interesting to understand how people are using today's mobile Internet. A recent research paper [4] published by one major TEM (Telecom Expense Management) may give us a glimpse of the most popular mobile applications. It is surprising to see that people are gradually using mobile Internet just like they use the fixed broadband network. Content services which include content delivered through web and P2P are actually dominating the network traffic. Figure 11.5 is an example of wireless data traffic on a commercial 3G mobile network. Considering this usage pattern, do we have a better choice than just blindly spending billions of dollars to upgrade the back-haul and core network?

11.3   Architecture of C-RAN

11.3.1   Overall Concept

We believe Centralized processing, Cooperative radio, Cloud, and Clean (Green) infrastructure Radio Access Network (C-RAN) is the answer to solve the fundamental challenges mentioned above. It's a natural evolution of the distributed BTS, which is composed of the baseband unit (BBU) and remote radio head (RRH). According to the different function splitting between BBU and RRH, there are two kinds of C-RAN solutions: one is called "full centralization", where baseband (i.e. layer 1), the layer 2 and layer 3 BTS(base station transceiver system) functions are located in the BBU; the other is called "partial centralization", where the RRH integrates not only the radio function but also the baseband function, while all the other higher layer functions are still located in BBU. For the solution 2, although the BBU doesn't include the baseband function, it is still called BBU for simplicity. The different function partition method is described in Figure 11.6.

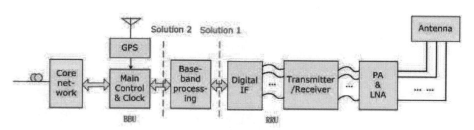

FIGURE 11.6 Different separation method of BTS functions

11.3.1.1   Different Technical Approaches

Based on these two different function splitting methods, there are two C-RAN architectures. Both of them are composed of three main parts: first, the distributed radio units which can be referred to as RRHs plus antennas which are located at the remote site; second, the high bandwidth low-latency optical transport network which connects the RRHs and BBU pool; and third, the BBU that is composed of high-performance programmable processors and real-time virtualization technology.

The "fully centralized" C-RAN architecture, as shown in Figure 11.7, has

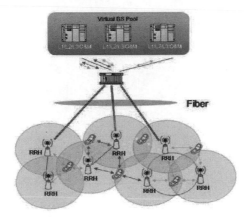

FIGURE 11.7 C-RAN architecture 1 (fully centralized solution)

the advantages of easy upgrading and network capacity expansion; it also has better capability for supporting multi-standard operation and maximum resource sharing, and it's more convenient towards support of multi-cell collaborative signal processing. Its major disadvantage is the high bandwidth requirement between the BBU and RRH to carry the baseband I/Q signal. In the extreme case, a TD-LTE 8 antenna with 20MHz bandwidth will need a 10Gpbs transmission rate.

The "partial centralized" C-RAN architecture, as shown in Figure 11.8, has the advantage of requiring much lower transmission bandwidth between BBU and RRH, by separating the baseband processing from BBU and integrating it into RRH. Compared with the "full centralized" one, the BBU-RRH connection only need to carry demodulated data, which is only $1/20 \sim 1/50$ of the original baseband I/Q sample data. However, it also has its own shortcomings. Because the baseband processing is integrated into RRH, it has less flexibility in upgrading, and less convenience for multi-cell collaborative signal processing.

With either one of these C-RAN architectures, mobile operators can quickly deploy and make upgrades to their network. The operator only needs to install new RRHs and connect them to the BBU pool to expand the network coverage or split the cell to improve capacity. If the network load grows, the operator only needs to upgrade the BBU pool's HW (hardware) to accommodate the increased processing capacity. Moreover, the "fully centralized solution," in combination with open platform and general purpose processors, will provide an easy way to develop and deploy software-defined radio (SDR) which enables upgrading of air interface standards by software only, and makes it easier to upgrade RAN and support multi-standard operation.

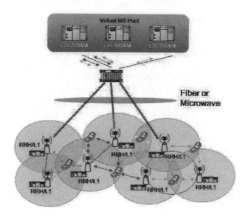

FIGURE 11.8 C-RAN architecture 2 (partial centralized solution

11.3.2   Relationship with Other Standards

Although C-RAN is an implementation and deployment solution for RAN, it may still have impact on the existing standards. Here is some potential impact that C-RAN may have.

First, C-RAN architecture relies on the fiber optical link between the BBU and RRU. This link has been standardized be various industry organizations like CPRI, OBSAI, CCSA and ETSI. There are already some standards specifying this interface to achieve inter-operation between the BBU and RRU from different telecommunication vendors.

However, most of the standards of the BBU-RRU interface mainly focus on point-to-point functionality. This is sufficient for local or short distance separation of BBU and RRU in the case of distributed BS system. However, for C-RAN system which may cover a vast area with RRUs deployed many kilometers away from the centralized BBU pool, the existing BBU-RRU standards need to posses sufficient functionalities to provide the features necessary for C-RAN system; to name a few, the failure protection mechanism of BBU-RRU links in various topologies; the ability to transport multiport air interface standards in the same logic link; the O&M ability to monitor and manage large scale fiber optical links without the help of active transport devices but only the BBU and RRU. Not to mention that there are a few incompatible standards in the industry that makes the situation even complicated.

C-RAN's vision is to have multi-mode, low cost, and smooth operation of the whole RAN system. Its nature requires a BBU-RRU standard that can support not only point-to-point BBU-RRU topology but also ring, ring with chain and others. It also requires reliable failure protection mechanism that prevents service interruption from single fiber failure. It also needs to have strong O&M capacity to monitor and manage the long distance fiber optical links as well as the remote RRU sites. Finally, a unified standard that can support all the air interfaces on C-RAN is highly wanted.

Some standard organizations like ETSI is already working on a standard called ORI (Open Radio Equipment Interface) which has similar object as specified above. C-RAN clearly would have impact on the ORI feature definition and inter-operation requirement.

Another major standard C-RAN may have impact on is 3GPP (3rd Generation Partnership Project). At the first thought, 3GPP only define the logic functions of RAN elements and the interface among these elements. It doesn't specify how these functions should be implemented. On the contrary, C-RAN is an implementation and deployment solutions, which is design to support all 3GPP air interfaces like UMTS (Universal Mobile Telecommunications System) and LTE (Long Term Evolution). It seems that it should not have any impact on the standard itself. However, the implementation always has impact on the standard as it actually defines what is possible and impossible in system. The centralized processing feature of C-RAN architecture allows much larger bandwidth and much lower latency among different BS in the same BBU pool. This would give more advantages to the collaboration technologies like CoMP (Coordinated multipoint transmission and reception), eICIC (Enhanced inter-cell interference coordination) and the others as defined in 3GPP LTE. It is foreseeable that if C-RAN is taken as one of the possible RAN implementation and deployment solutions, the corresponding co-ordination capacities will enable some new, advanced schemes to be defined in 3GPP.

11.4 Advantages of C-RAN

11.4.1 Energy Efficient Perspective

C-RAN is an eco-friendly infrastructure. Firstly, with centralized processing of the C-RAN architecture, the number of BS sites can be reduced several folds. Thus the air conditioning and other site support equipment's power consumption can be largely reduced. Secondly, the distance from the RRHs to the UEs (user equipment) can be decreased since the cooperative radio technology can reduce the interference among RRHs and allow a higher density of RRHs. Smaller cells with lower transmission power can be deployed while the network coverage quality is not affected. The energy used for signal transmission will be reduced, which is especially helpful for the reduction of power consumption in the RAN and extend the UE battery stand-by time. Lastly, because the BBU pool is a shared resource among a large number of virtual BS, it means a much higher utilization rate of processing resources and lower power consumption can be achieved. When a virtual BS is idle at night and most of the processing power is not needed, they can be selectively turned off (or be taken to a lower power state) without affecting the 7×24 service commitment.

11.4.2   RAN Cost Structure Optimization Perspective

Because the BBUs and site support equipment are aggregated in a few big rooms, it is much easier for centralized management and operation, saving a lot of the O&M cost associated with the large number of BS sites in a traditional RAN network. Secondly, although the number of RRHs may not be reduced in a C-RAN architecture, its functionality is simpler, the size and power consumption are both reduced, and they can sit on poles with minimum site support and management. The RRH only requires the installation of the auxiliary antenna feeder systems, enabling operators to speed up the network construction to gain a first-mover advantage. Thus, operators can get large cost saving on site rental and O&M.

11.4.3   Enhanced Network Performance Perspective

In C-RAN, virtual BS's can work together in a large physical BBU pool and they can easily share the signaling, traffic data and channel state information (CSI) of active UE's in the system. It is much easier to implement joint processing & scheduling to mitigate inter-cell interference (ICI) and improve spectral efficiency. For example, cooperative multi-point processing technology (CoMP in LTE-Advanced), can easily be implemented under the C-RAN infrastructure.

11.4.4   Service on the Edge

Through enabling the smart breakout technology in C-RAN, the growing Internet traffic from smart phones and other portable devices, can be offloaded from the core network of operators. The benefits are as follows: reduced backhaul traffic and cost; reduced core network traffic and gateway upgrade cost; reduced latency to the users; differentiating service delivery quality for various applications. The service overlapping the core network also supplies a better experience to users.

11.5   Challenges of C-RAN

Although C-RAN has many benefits and advantages in deployment and operation as stated above, it also has many technical challenges that need to be solved before it could be largely applied in commercial mobile network. These challenges are discussed in the following sections.

11.5.1 Efficient Transmission of Radio over Optical Transport Networks

In C-RAN architecture 1, the optical fiber between the BBU pool and the RRHs has to carry a large amount of baseband sampling data in real time. Due to the wideband requirement of LTE/LTE-A system and multi-antenna technology, the bandwidth of optical transport link to transmit multiple RRHs' baseband sampling data is at 10 gigabit level with strict requirements of transportation latency and latency jitter.

Air interface is upgrading rapidly, new technologies like multiple antenna technology (2 ∼ 8 antenna in every sector) and wide bandwidth (10 MHz ∼ 20 MHz every carrier) have been widely adopted in LTE/LTE-A, thus the bandwidth of CPRI (Common Public Radio Interface)/Ir/OBRI (Open BBU-RRH Interface) link is much higher than the 2G and 3G era. In general, the system bandwidth, the MIMO antenna configuration and the RRH concatenation levels are the main factors which have an impact on the OBRI bandwidth requirement. For example, the bandwidth for 200 kHz GSM systems with 2Tx/2Rx antennas and 4 times sampling rate is up to 25.6Mbps. The bandwidth for 1.6MHz TD-SCDMA systems with 8Tx/8Rx antennas and 4 times sampling rate is up to 330Mbps. The transmission on bandwidth of this level on fiber link is matured and economic. However, with the introducing of multi-hop RRH and high orders MIMO supporting 8Tx/8Rx antenna configuration, the baseband signal bandwidth between BBU-RRH would rise to dozens of Gbps. Therefore, exploring different transport schemes for the BBU-RRH baseband signal is very important for C-RAN.

There are also strict requirements in terms of latency, jitter and measurement. In CPRI/Ir/OBRI transmission latency, due to the strict requirements of LTE/LTE-A physical layer, delay processing also affects the baseband wireless signal transmission delay jitter and requirements indirectly. Not including the transmission medium between the round-trip time (i.e., regardless of delays caused by the cable length), for the user plane data (IQ data) on the CPRI/Ir/OBRI links, the overall link round-trip delay should not exceed 5s. The OBRI interface requires periodic measurement of each link or multi-hop cable length. In terms of calibration, the accuracy of round trip latency of each link or hop should satisfy 16.276ns(nanoseconds) [5].

At the same time, under the traditional RAN architecture, the transmission network which consists of SDH (Synchronous Digital Hierarchy)/PTN(Packet Transport Network) also provides the unified optical fiber network management ability for the access ring. This includes unified management of the access ring's fiber optic link of the entire network, supervisory control of the access ring's optical fiber breakdown, etc., BBU-RRH wireless signal transport directly on the access ring, whose CPRI/Ir/OBRI interface should also provides similar management ability and fit into the unified optical fiber network management.

There are some technology candidates to address this challenge today. One of them is I/Q data compressing. Some companies claim to be able to compress the LTE I/Q data to 1/2 or even 1/3 which would largely reduce the bandwidth requirement [6]. However, the performance loss is not clear and not verified in OTA (Over-the-Air) test yet. Another solution is to use PON (Passive Optical Network) technology to transport multiple wavelengths in one fiber to reduce the fiber resource required for BBU-RRU links in C-RAN architecture. It sounds promising but also faces challenges of complex wavelength management and limited range due to the loss of insertion of ODM (Optical Dispersion Measurement) in the optical fiber path. The management of a passive optical system is also a challenge. Yet another candidate is to use WDM/OTN (Optical Transport Network) at each node of the RRU. This solution has the maximum capacity and strongest management ability as it uses active transport devices to support the BBU-RRU link. The major drawbacks are that WDM/OTN is still too expensive and it is not suitable for outdoor deployment.

11.5.2   Cooperative Transmission and Reception on C-RAN

Joint processing is the key to achieve higher system spectrum efficiency. To mitigate interference of the cellular system, multi-point processing algorithms that can make use of special channel information and harness the cooperation among multiple antennas at different physical sites should be developed. Joint scheduling of radio resources is also necessary to reduce interference and increase capacity.

To support the above Cooperative Multi-Point Joint processing algorithms, both end-user data and UL(uplink)/DL(downlink) channel information needs to be shared among the virtual BSs. The interface between the virtual BSs to carry these information should support high bandwidth and low latency to ensure real time cooperative processing. The information exchanged in this interface includes one or more of the following types: end-user data package, UE channel feedback information, and the virtual BS's scheduling information. Therefore, the design of this interface must meet the real-time joint processing requirement with low backhaul transportation delay and overhead.

China Mobile has evaluated the performance of the joint processing scheme with intra-cell collaboration, and the performance with inter-cell collaboration in C-RAN architecture in a TDD system. We assume that full DL channel state information (CSI) can be obtained ideally at the eNB (eNodeB) side. The numerical results of downlink throughput and spectrum efficiency with different schemes in both 2 antenna and 8 antenna configuration are shown in Tables 11.1, 11.2, 11.3, and 11.4. More detailed simulation parameters can be found in [7–10].

From the simulation results we can see, compared to the non-cooperative transmission mechanism (MU-BF in LTE-A), the spectrum efficiency of intra-cell collaboration and inter-cell collaboration under C-RAN architecture could achieve 13% and 20% gain, respectively, while the cell edge user's spectrum

TABLE 11.1 Average cell spectrum efficiency (bps/Hz): 3GPP Case 1 (TDD)

Antenna Configurations	SU-MIMO	MU-MIMO	Intra-site CoMP	C-RAN CoMP
2Tx/2Rx	1.9	2.47	2.81	3.01
8Tx/2Rx	2.54	5.46	6.15	6.58

TABLE 11.2 Cell-edge spectrum efficiency (bps/Hz): 3GPP Case 1 (TDD)

Antenna Configurations	SU-MIMO	MU-MIMO	Intra-site CoMP	C-RAN CoMP
2Tx/2Rx	0.056	0.047	0.078	0.101
8Tx/2Rx	0.098	0.183	0.227	0.266

TABLE 11.3 Average cell spectrum efficiency (bps/Hz): ITU UMi (TDD)

Antenna Configurations	SU-MIMO	MU-MIMO	Intra-site CoMP	C-RAN CoMP
2Tx/2Rx	1.44	1.81	1.93	1.97
8Tx/2Rx	1.97	3.78	4.54	5.35

TABLE 11.4 Cell-edge spectrum efficiency (bps/Hz): ITU UMi (TDD)

Antenna Configurations	SU-MIMO	MU-MIMO	Intra-site CoMP	C-RAN CoMP
2Tx/2Rx	0.041	0.039	0.07	0.075
8Tx/2Rx	0.052	0.092	0.161	0.202

efficiency, from the above two mechanisms can get 75% and 119% gain respectively.

Cooperative transmission/reception (CT/CR) has great potentials in reducing interference and improving spectrum efficiency of system. However, this technology has many problems that need to be further studied before it can be applied to the practical networks. There are many challenges like: advanced joint processing schemes, DL channel state information (CSI) feedback mechanism, user pairing and joint scheduling algorithms for multi-cells and coordinated radio resource allocation for multi-cells.

11.5.3 Real-time Cloud and BS Virtualization

Driven by Moore's law in semiconductor industry, Digital Signal Processor (DSP) and General Purpose Processors (GPP) have made a lot of progresses in the architecture, performance and power consumption in recent years. This provides more choices for SDR base stations. Multi-core technology is widely used in DSP and $3 \sim 6$ cores processors have been commercially available.

At the same time, DSP floating-point processing capacity is also improving at a fast pace. The emergence of the DSP system based on SoC (System-on-Chip) architecture combining the traditional DSP core and communication accelerator together has improved the BBU processing density and improved the power efficiency. Moreover, real-time OS running on DSP pave the path to virtualization of DSP processing resources. On the other hand, DSP from different manufacturers and even a same manufacturer cannot guarantee backwards compatibility. The real-time operating systems are different from each other, and there is no de fact standard yet. Generally the BBUs based on DSP platform are proprietary platforms, and it is still difficult to achieve smooth upgrading and resource virtualization.

Once the large scale BBU pool with high-speed, low-latency interconnection, plus the common platform of DSP/GPP and open SDR solution could be realized, it would set the base for a virtual BS.

Virtualization is a term that refers to the abstraction of computer resources. It hides the physical characteristics of a computing platform from users, instead showing another abstract computing platform. If such a concept can be utilized in a base station system, the operator can dynamically allocate processing resources within a centralized baseband pool to different virtualized base stations and different air interface standards. This allows the operator to efficiently support a variety of air interfaces, and adapt to the "Tidal Effect" in different areas and fluctuating demands. At the same time, the common hardware platform will provide cost effectiveness to manage, maintain, expand and upgrade the base station. Therefore, we believe real time virtualized baseband pools will be key part of the next generation mobile network. Within a given centralized baseband pool, all the physical layer processing resources will be managed and allocated by a real time virtualized operating system. Therefore, for instance a base station can be easily built up through the flexible resource combination. The real time virtualized OS would adjust, allocate and re-allocate resources based on each virtualized base station requirements, in order to meet its demands.

After the baseband processing units have been put in a centralized pool, it is essential to design virtualization technologies to distribute/group the processing units into virtual BS entities. The major challenges of virtualization are: real-time processing algorithm implementation, virtualization of the baseband processing pool, and dynamic processing capacity allocation to deal with the dynamic cell load in system.

11.5.4 Performance and Power Challenge of SDR on General Purpose Processor

Software-defined radio (SDR) provides the maximum flexibility in multistandard support as well as maximum savings in resource sharing and virtualization, since it allows the use of the same hardware to support different air interfaces, and it facilitate the convergence of processing resources and virtualization of resources.

SDR on general purpose processor further increase the flexibility as general purpose processor can be used for both L1, L2, L3 functions [11] as well as more complicated high level functions like Packet Forwarding, Deep Packet Inspection, Content Distribution, etc. This will allow the convergence of not only the RAN, but also the transport network, the core network and service network.

However, the flexibility comes with a price. The major disadvantages of SDR and SDR on general purpose processor is its performance power ratio. It is well understood that the if the HW is specified for certain functions, it will be most power efficient as no other overhead in HW/SW(software) is involved at the cost of non-flexibility. When we add more flexibility, like programmability, there will be sparer HW and the associated SW overheads to complete the same job. Thus there will be more power needed to complete the same function. When SDR is implemented on general purpose processor, it will pay higher price in power consumption for all the flexibility it enjoys.

Initial PoC (Prototype of Concept) of SDR on general purpose processor shows that for the same wireless digital processing tasks like LTE physical layer, SDR system would need about 10 times higher power consumption compared with ASIC solutions [12]. For example, the SDR system based on IT platform may need $100 \sim 150W$ to process one carrier of 20MHz LTE signal with 2x2 function, while the existing commercial solutions only need $10W \sim 15W$ for the same functionality in BBU. Although virtualization technology and smart power management can help to alleviate the problem, it is still a big gap that the SDR solutions need to overcome to be commercially attractive.

11.6   China Mobile's View on RAN Evolution Path

The major focus of this section is on point-to-point MIMO communications. The focus on single user communication systems may be surprising considering that power control is one of the primary problems of interest. However, there are two important reasons for this choice. First of all, the single-user case has most of the main effects we want to emphasize and allows us to describe the proposed approach in a clear and concise manner. Secondly, once the single-user case problem is solved, reasonably more complex multi-user channels can be easily solved (the multiple access channel is one of them). The multi-user scenario becomes tractable provided some conditions are met. One of these conditions being the quasi-concavity of the proposed energy-efficiency metric as we will see in the multi-user section.

The energy-efficiency metric is defined at first in a very general way. Then, we specialize, depending on the specific situation under consideration and study it as a function of various model parameters such as the transmit power.

11.6.1   Trials on GSM/TD-SCDMA Network

China Mobile conducted the first C-RAN trial with partners in 2010 [12]. It is a C-RAN centralized deployment field trial within the commercial TD-SCDMA system in Zhuhai city, Guangdong province. After that, there has been multiple GSM field trials conducted in multiple cities throughout China, include Changsha, Baoding, Jilin, Dongguan, Zhaotong, etc. Rest of the section discusses the pros and cons of the C-RAN centralized deployment solution's pros and cons in different scenarios. For the ease of discussion, two typical cases, TD-SCDMA trial in Zhuhai city and GSM trial in Changsha city are shown here.

11.6.1.1   Overall Situation

The first trial in Zhuhai city only took 3 months to complete. The commercial trial has 18 TD-SCDMA macro sites covering about 30 km2 area. This trial has verified some centralized deployment technologies feasibility. The construction and operation of a commercial mobile network clearly highlighted the C-RAN's advantage over tradition RAN in cost, flexibility and energy savings. At the same time, it also exposed challenges on fiber resource, as well as transmission construction.

After that, there have been several trials on centralized deployment solutions of GSM system. The network layout is mainly consisted of replacing and upgrading the existing cell sites. There are total 15 cell sites covering 15 km2 in the trial, where only 2 of them are new cell sites. Compared with TD-SCDMA network, GSM solutions have unique features; for example, it could support daisy-chain of 18 RRHs with only 1 pair of fiber. This could significantly reduce the number of fiber resources needed in C-RAN centralized deployment with dark fiber solution.

The following sections will describe the network status before and after C-RAN deployment, the key technology introduced, the field test results and the challenges observed.

11.6.1.2   Field Trial Area

The trial area in Zhuhai city is mainly consisted of a national high-tech development zone, a residential community, and a few college campuses. The data traffic in this area is growing rapidly, as the customers there are well-educated and early adapters of new services. Part of the trial areas has demonstrated the "Tidal Effect" of traffic loading, with predictable traffic loading pattern associated with time, location or event. For example, the national high-tech development zone has most people during the working hours. The same group of customers usually returns to the nearby community after work. Students in colleges tend to stay away from using wireless devices during school hours, while they tend to make a lot of calls in night.

Traditionally, network planning must support the peak traffic load at each individual site, which is usually 10 times higher than the down time. This results in a very low average utilization rate of the BTS (Base Transceiver Station) devices. It also introduces difficulties in network planning, construction and optimization. It is suitable to adopt the baseband pool with dynamic carrier allocation. In the trial field, there are 9 cell sites co-located with existing GSM site, while another 9 cell sites are new. All these 9 cell sites have to be connected with new fiber channels and they are spread in 30 km2, which is a challenge for fiber construction.

The trial area in Changsha city is consisted of a few campuses near Yuelu Mountains. The traffic load and traffic density are quite high there. In addition, there are a lot of dormitories and local residential apartments. The propagation environment is very complex and the coverage KPI (Key Performance Indicator) still has room to be improved. This makes it suitable to verify C-RAN's capacity in urban city environment. Finally, since most of the trial sites are reusing or upgrading the existing ones, there are plenty of fiber resources.

11.6.1.3   Overall Solution

The solution starts with planning of system capacity in centralized deployment. In the Zhuhai trial, each TD-SCDMA site's configuration is 4/4/4, which means that there are 3 sectors in each cell site, and every sector has 4 carriers. Overall, the 18 trial sites need 216 carriers. When considering the BBU pool capacity, the total BBU pool can be planned to support the maximum co-current traffic for the same area.

There are two kinds of TD-SCDMA carriers, R4 carrier is mainly used for voice traffic, and HSDPA (High-Speed Downlink Packet Access) carrier is mainly used for data traffic. Based on China Mobile's planning requirements, every site's traffic load should not exceed 75

Similarly, the trial in Changsha also has used the co-current capacity to decide the total capacity of the BBU pool.

The second part of the solution involves dynamic carrier allocation. In TD-SCDMA system, each RRH/sector can support maximum 6 R4 and HSDPA carriers. In the idle situation, each RRH/sector has only one R4 carrier and one HSDPA carrier. There are different carrier allocation decision criteria whether more R4 and HSDPA carriers should be added. Whenever the existing R4 carrier's loading rate is above a threshold, there should be more R4 carriers allocated in this site. For HSDPA carrier, the similar rule applies. When there is not enough load in multiple R4 or HSDPA carrier, it is also possible to reduce the number of R4 and HSDPA carriers in one sector.

For GSM system, the similar rule also applies but the criteria is the utilization rate of each GSM carrier.

The third portion of the solution involves RRH daisy chain and fiber failure protection technologies. These technologies are derived from the distributed BBU-RRH deployment method which usually uses point-to-point dark fiber connections. When BBU-RRHs are separated by significant distance, it is important to consider the saving of fiber resource and protection against unpredictable fiber failure caused by external factors. In TD-SCDMA, each fiber link can handle up to 6.144Gpbs transmission, enough to support 15 TD-SCDMA carriers. Thus, one pair of fiber is able to support one site with 3 sectors and maximum carrier of 15. In the Zhuhai trial, each access ring has 9 sites and uses 9 pair of fibers to support the 9 sites connected to the ring.

On the other hand, GSM has far less baseband requirement due to its narrow band nature; therefore it can support more capacity in daisy-chain configuration. There are commercial products that can support 18 to 21 RRH daisy chained on one pair of dark fiber. We can calculate the fiber resource required per access ring as following: usually, each access ring has $8 \sim 12$ physical sites and each site has 3 sectors, and has 900M and 1800M dual bands. This means, each access ring may has up to $16 \sim 24$ logical sites, which is 48 to 72 sectors/RRH. To connect all the RRH in daisy chain, we would need $4 \sim 5$ pair of fibers in the ring.

Lastly, the field trial has also verified the key technology for outdoor deployment, like power supply for remote sites. In the Zhuhai trial, there is no BTS equipment room in the 9 new sites. Thus the traditional DC power supply is not available. External power booth is used instead. The existing outdoor power solution met the need of network deployment: with sufficient operation temperature range, -40+70, C-level anti-flash capacity and theft-proof solution to ensure the safety of device without on-site attendance. GSM and TD-SCDMA remote site both can apply this outdoor power solution.

11.6.2   Progress in C-RAN PoC system

China Mobile, in collaboration with IBM, ZTE, Huawei, Intel, Datang Mobile, France Telecom Beijing Research Center, Beijing University of Posts and Telecommunications, China Science Institute, jointly developed the C-RAN prototype supporting multiple air interfaces, entirely using platform based on the general purpose processor [12]. The prototypes supporting GSM and TD-SCDMA have successfully completed interoperability with the commercial end user devices, while the TD-LTE version has gone through testing with UE simulator. The prototypes have proved the feasibility of implement GSM/TD-SCDMA/TD-LTE physical layer signal processing on the general purpose processor based platform, and a step closer to achieve greater software implementation and upgrade flexibility.

The following sections will describe hardware and software architecture of the prototype. As shown below, the PCI Express interface is connected to the CPRI/Ir interface converter, which then carries GSM/TD-SCDMA/TD-

LTE signals to the commercial RRHs. IQ samples of all three standards are processed by the commercial server in real time.

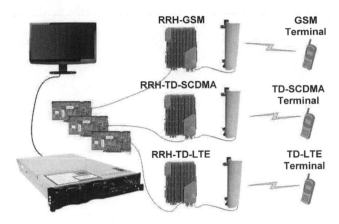

FIGURE 11.9 IT server platform topology

The proof-of-concept of C-RAN focuses on baseband processing feasibility on IT server, as illustrated in Figure 11.9, therefore, the software development does not cover any core network functions. The baseband processing software is developed on Linux, and has implemented Layer 1, 2 and 3 on GSM and TD-SCDMA, and Layer 1 on TD-LTE, with plan to add MAC scheduling in the near future. As a result, the system currently only supports single UE. In the future, the TD-LTE system will support MAC, L2, L3, LTE-A features like CoMP, and completes interoperability with the commercial devices.

Signal processing carries stringent real time requirements which pose challenges to the IT servers. GSM protocol requires each frame being processed within 40ms (millisecond); TD-SCDMA frame is within 5ms while TD-LTE protocol requires every frame has to be completely processed within 1ms.

Typical IT operating system is not designed to meet the real time requirements of telecom grade, therefore subframe scheduling delay and resource management typically are not guaranteed to complete within fewer than 1ms. In addition, IT platform generally lacks the stringent timing required by base station. Lastly, traditional signal processing algorithm is typically designed to be implemented on ASIC (Application-specific integrated circuit), FPGA (Field-programmable gate array), and DSP (Digital Signal Processing). Therefore, many believe that IT server is not capable of handling complex signal processing such those of LTE.

However, the C-RAN trial so far has proved that IT server can meet the aforementioned challenges with technology innovations. First step is to expand the real time capability on IT server to meet the subframe processing timing and accuracy demand. In addition, by adding hard real time synchronization on the CPRI/Ir interface card, we can separate the RRH's "hard real time"

CPRI/Ir functions from the IT signal processing tasks which only require soft real time.

Finally, significant efforts have been spent to optimize LTE algorithm on the general purpose processor, fully utilizing every available instruction set and memory to the maximum advantages, therefore significantly increasing the CPU processing efficiency. We were able to implement 3GPP release 8 TD-LTE physical layer entirely on software running on the general purpose processor and meeting all the timing and delay benchmarks. The TD-LTE implementation parameters are: 20MHz bandwidth, 2x2 MIMO downlink, 1x2 SIMO uplink, 64QAM/15QAM/QPSK modulation, Turbo decoder with adaptive early termination. Under peak throughput, every subframe was being processed under 1ms TTI (Transmission Time Interval), meeting the most stringent HARQ (Hybrid automatic repeat request) processing latency requirements in TD-LTE. As expected, GSM and TD-SCDMA processing met the timing requirements raised in Section 11.5.4.

SDR provides the maximum flexibility in multi-standard support as well as maximum savings in resource sharing and virtualization, since it allows the use of the same hardware to support different air interfaces and it facilitate the convergence of processing resources and virtualization of resources. SDR on the general purpose processor further increase the flexibility as the general purpose processor can be used for both L1, L2, L3 functions as well, such as more complicated high level functions like Packet Forwarding, Deep Packet Inspection, Content Distribution, etc. This would allow the convergence of not only the RAN, but also the transport network, the core network and the service network devices. However, the flexibility comes with a price. The major disadvantages of SDR and SDR on the general purpose processor is its performance power ratio. It is well understood that the if the HW is specified for certain functions, it will be most power effective since no other overhead in HW/SW is involved at the cost of non-flexibility. When we add more flexibility, like programmability, there will be sparer HW and the associated SW overhead to complete the same job. Thus there will be more power needed to complete the same function. When SDR is implemented on the general purpose processor, it will pay higher price in power consumption for all the flexibility it enjoys.

Initial PoC of SDR on the general purpose processor shows that for the same wireless digital processing tasks like LTE physical layer, SDR system would need about 10 times higher power consumption compared with ASIC solutions. For example, the SDR system based on IT platform may need 100 ~ 150W to process one carrier of 20MHz LTE signal with 2×2 function, while the existing commercial solutions only need 10W ~ 15W for the same functionality in BBU. Although virtualization technology and smart power management can help to alleviate the problem, it is still a big gap that SDR solutions need to overcome to be commercially attractive.

11.6.3 Potential Evolution Path

Considering the technical challenges as well as the limitation in current optical network resources, it is clear that C-RAN cannot be widely applied in a short time frame. Instead, a stepped plan should be used to gradually construct the centralized network: first, centralized deployment can be applied to some green field or replacement of old network in a small scale. Dark fiber can be used as the BBU-RRH transmission solution. One access ring that connects 8 ∼ 12 macro cell sites can be centralized together, with a maximum ring range of 40km. In the future, a larger number of macro BS in various deployment scenarios can be further tested.

The novel C-RAN architecture is a revolution of the traditional RAN deployment. It is impossible to replace today's RAN overnight. Moreover, the technical challenges of C-RAN should be carefully addressed and tested in labs and field environments to ensure its reliability. This naturally leads to a step-by-step evolution path of C-RAN to gradually replace the traditional RAN. The following is our vision on how the evolution would take place:

11.6.3.1 C-RAN Centralized Base Station Deployment

In the first step, Base Stations can be implemented by separating the RRH and BBU, and the baseband processing resources between multiple BBUs in a centralized Base Station can be scheduled at carrier level. The RRHs are small and light weight for easier deployment. They receive/transmit digital radio signals from/to the BBU via fiber links such as CPRI/Ir/OBRI. The BBU is the core of the radio signal processing. The RRHs can be deployed in remote sites far from the physical location of the BBU (e.g. 1 ∼ 10km). The optical fiber transmission network between the RRH and BBU will have the corresponding loop protection and management functions. The fiber link between the RRHs and BBUs can be standardized like CPRI/Ir/OBRI so that the RRHs and BBUs from multiple venders can be connected together.

The centralized BS has a high bandwidth, low latency switch matrix and the corresponding protocol to support the inter-connection of carrier processing units among the multiple BBUs in order to constitute a large-scale baseband pool. The signals from the distributed RRH can be switched to any BBU inside the centralized baseband pool. Thus, the centralized baseband pool can realize carrier load balance to avoid some BBUs overloaded while some BBUs idle, and realize fault-tolerance to avoid that the fault of a single BBU affect the overall functions and coverage of the whole wireless network. The above technologies can improve the usage efficiency of devices, reduce power consumption and improve system reliability.

11.6.3.2 Multi-Standard SDR and Joint Signal Processing

In the second step, on the basis of centralized BS deployment, the BBU' baseband processing functions can be fully implemented by SDR based on a unified, open platform. By moving the baseband processing to SDR, it is much easier to support multiple standards, upgrade the SW/HW, introduce new standard and increase system capacity.

Meanwhile, with multiple RRHs attached to the centralized BBU pool, it is easier to implement coordinated beamforming (CBF) and cooperative multipoint processing in this platform. Multiple BBUs can coordinate with each other to share the scheduling information, channel status and user data efficiently to improve the system capacity as well as reduce interferences in the system.

11.6.3.3   Virtual BS on Real-time Cloud Infrastructure

Once the centralized baseband pool consisted of a lot of standardized BBUs is built on a unified, open platform and the baseband processing is implemented by SDR, the virtual BS based on real-time cloud infrastructure is the next step of C-RAN evolution.

The centralized baseband pool consisted of large-scale BBUs by a high bandwidth, low latency network, combined with some system software, can constitute a large "real time baseband cloud", just like the cloud computing environment in IT industry. The difference is that the baseband processing tasks are real-time computing tasks in a real time baseband pool. Through the cooperation of BBU in the baseband pool and RRH to send and receive wireless signals, it can be expected that multi-standard wireless network functions in the same platform. In the system software instructions of the baseband pool based on real-time cloud architecture, the CPRI/Ir/OBRI optical fiber transmission network and the optical Internet architecture in large-scale centralized baseband pool can send the baseband signal signals generated by RRH to the virtual base station running on the designated BBU. Then the virtual base station uses the calculation resources of the designated BBU to finish the real-time processing of the wireless baseband signals. Moreover, in a C-RAN system which has several baseband pools, the CPRI/Ir/OBRI optical fiber transmission network should have the ability to forward the baseband signals from RRHs to the other baseband pools in order to improve system reliability and realize load balance across the different baseband pools.

11.7   Conclusions

With the arrival of the mobile Internet era, today's RAN architecture is facing more and more challenges that the mobile operators need to solve: mobile data flow increase drastically caused by the proliferation of smart terminals, very hard to improve spectrum efficiency, lack of flexibility to multi-standard, dynamic network load because of "Tidal Effect", and expensive to provide ever increasing Internet service to the end users. Mobile operators must consider the evolution of the RAN to high efficient and lost cost architecture.

C-RAN is a promising solution to the challenges mentioned above. By using new technologies, we can change the network construction and deployment approaches, fundamentally change the cost structure of mobile operators, and provide more flexible and efficient services to end users. With the distributed RRH and centralized BBU architecture, advanced multipoint transmission/reception technology, SDR with multi-standard support, virtualization technology on general purpose processor, more efficient way of dealing with the "Tidal Effect" and service on the edge of the RAN, C-RAN will be able to provide today's mobile operator with a competitive infrastructure to keep profitable growth in the dynamic market environment.

Acknowledgments

We would like to thank Ling Shao, Yonghua Lin from IBM China Labs, Caroline Chang, John Magon, Shawn LI from Intel Corporation, Haiyun Luo, Qixing Wang, Peng Jiang, Dong Wang, Ran Duan, Lifeng He, Jinri Huang, Dechao Zhang, Shiguang Wang, Hong Liu from China Mobile and all the other colleagues and partners for their commitment to C-RAN project, suggestions and comments on the contents and other contributions to this book.

Bibliography

[1] "The mobile Internet report," December 2009. [Online]. Available: http://www.morganstanley.com/institutional/techresearch/

[2] H. Taoka, *Views on 5G*. DoCoMo, WWRF21, Dusseldorf, Germany, Oct. 2011.

[3] X. Wang, "C-RAN:the road towards green RAN," *China Commun. Journal*, Jun 2010.

[4] G. Szabo, D. Orincsay, B. P. Gero, S. Gyori, and T. Borsos, "Traffic analysis of mobile broadband networks," in *Proc. the Third Annual International Wireless Internet Conf.*, Oct. 2007.

[5] "Common public radio interface (CPRI) specification v4.1," Feb 2009.

[6] *Light Radio Portfolio: Technical Overview*. Alcatel Lucent, Feb. 2011.

[7] *SRS feedback mechanism based CoMP schemes in TD-LTE-Advanced*. 3GPP, R1-093273, Aug 2009.

[8] Q. Spencer, A. Swindlehurst, and M. Haardt, "Zero-forcing methods for downlink spatial multiplexing in multiuser MIMO channels," IEEE Trans. Sig. Proc., pp. 461–471, Feb. 2004.

[9] L. Choi and R. Murch, "A transmit preprocessing technique for multiuser MIMO systems using a decomposition approach," *IEEE Trans. Wireless Commun.*, Jan 2004.

[10] J. Zhang, R. Chen, J. Andrews, and R. Heath, "Coordinated multi-cell MIMO systems with cellular block diagonalization," in *Proc. 41st Asilomar Conference on Signals, Systems and Computers*, Nov. 2007, pp. 1669–1673.

[11] R. Gadiyar and J. Mangan, "Using Intel architecture for implementing SDR in wireless basesations," in *Proc. SDR 2009, SDRForum*, 2009.

[12] *C-RAN White Paper.* China Mobile Research Institute, Oct. 2011.

Author Contact Information

Kuilin Chen, Chunfeng Cui, Yuhong Huang, and Bill Huang are with China Mobile Research Institute, China Mobile Communications Corporation, China, Email: chenkuilin@chinamobile.com, cuichunfeng@chinamobile.com, huangyuhong@chinamobile.com, bill.huang@chinamobile.com.

12

Green Ad Hoc and Sensor Networks

Ting Zhu

State University of New York, Binghamton, USA

Abedelaziz Mohaisen

University of Minnesota, USA

Ping Yi

Shanghai Jiaotong University, China

Jianqing Ma

Fudan University, China

CONTENTS

12.1   Introduction

With the increasing demand of ubiquitous sensing and cyber-physical interaction, ad hoc and sensor networks have emerged as one of the key technologies for many promising applications. Compared with competing high-end technologies (e.g., WiFi and cellular networks), ad hoc and sensor networks are low-cost, low-profile, and easy to deploy. These design characteristics, however, imply that resources available to individual nodes are severely limited. Although it is highly possible that the constraints on computation and storage will disappear along with the fast development of fabrication techniques, energy will continue to be the victim of Moore's law, i.e., more transistors indicate more power consumption. According to R.A. Powers, battery capacity only doubles in 35 years. On the other hand, there is a growing need for the sustainable and green deployment of ad hoc and sensor systems to reduce operational costs and ensure service continuity. To bridge such an expanding gap between limited energy supply and increasing energy demands in many long-term applications, developers have proposed three options: energy harvesting, energy conservation, and energy synchronization. In energy harvesting, researchers have designed various types of energy harvesting technologies to collect ambient energy from the environments. These include solar [1–7], wind [8], kinetic [9], piezoelectric strain [10], and vibrational [11] energy. In energy conservation, numerous solutions have been proposed for energy efficiency at various levels of the system architecture, ranging from energy-efficient hardware [12,13], low power listening link layer [14–16], topology management [17], node placement [18,19], sensor clustering [20], network routing [21, 22], flooding [23–25], sensing coverage [26–29], data dissemination [30–32], data aggregation [33, 34], in-network caching and storage [35], up to application-level energy-aware designs [36, 37]. In energy synchronization, researchers take a holistic and systematic approach to synchronize sensor network activities with dynamic energy supply from the environments. Feedback-based energy synchronization techniques [38, 39] and energy synchronized communication protocols [40] have been proposed. In this chapter, we will discuss these design options and study the design principles for green ad hoc and sensor networks. This study deals with both hardware platforms and algorithms for achieving green and energy efficient operations.

12.2 Energy Harvesting Techniques

Energy harvesting is the conversion of ambient energy into usable electrical energy. Several technologies have been developed for extracting energy from the environment, including solar [7], wind [8], kinetic, and vibrational [11] energy. With these energy-harvesting technologies, researchers have designed various types of platforms to collect ambient energy from human activity or environments [1, 2, 7, 8]. Notable ones designed specially for sensor networks are Heliomote [2, 41], Prometheus [7], Trio [1], AmbiMax [8] and PUMA [42]. According to the type of energy storage used, these platforms can be separated into three categories: (i) rechargeable battery-based platforms [41], (ii) designs combining ultra-capacitors and rechargeable batteries [1,7], and (iii) capacitor-only designs.

- In rechargeable battery designs, such as Heliomote [2, 41], the energy harvesting panel is directly connected to its battery. As the primary energy storage device, the rechargeable battery is charged and discharged frequently, leading to low system lifetimes due to the physical limitation on the number of recharge cycles.

- In designs that combine ultra-capacitors and rechargeable batteries such as Prometheus [7], the solar energy is first stored in the primary energy buffer, which is one or multiple ultra-capacitors. The rechargeable batteries are then used as the secondary energy buffer. This design inherits both the advantages and limitations of batteries and capacitors. Rechargeable batteries have limited recharge cycles due to cyclic memory and crystalline formation (e.g., a Li-ion battery has 500 cycles, NiMH 300 cycles). Moreover, sophisticated recharging circuits and electro-chemical conversion could reduce energy efficiency. It is difficult to predict remaining energy because of the inclusion of batteries, and the lifetime of energy storage subsystem is decided by the shelf time of batteries (in the order of a few years). Compared with rechargeable batteries, capacitors possess a set of advantages: they (i) have more than 1 million recharge cycles; (ii) have predicable remaining energy independent of discharge modes; (iii) are robust to temperature changes, shock, and vibration; and (iv) have high charging and discharging efficiency. Within this category, several similar systems have been built with a few enhanced features. For examples, AmbiMax [8] harvests energy from multiple ambient power sources (e.g., solar and wind generators), and PUMA [42] uses a power routing switch to route multiple power sources to multiple subsystems. The higher utilization of ambient power is achieved through a combination of maximum power point tracking (MPPT) and power defragmentation. These works have intentionally avoided capacitor-only design, citing the leakage issue [7]. To alleviate leakage, small capacitors are normally used, which makes secondary energy storage (batteries) necessary. However, the development of battery capacities is very slow and still leakage-prone. In addition, charging efficiency for batteries is comparatively low [43]. For example, according to the Natural Resources Defense Council [43], common battery chargers provide an efficiency between 6% and 40%.

- In capacitor-only designs, such as TwinStar [38], use ultra-capacitors as the only energy storage to overcome the intrinsic limitations of batteries (e.g., energy uncertainty, limited recharge cycles, low conversion efficiency, and environment unfriendliness). The architecture of TwinStar is shown in Figure 12.1. It consists of (i) the solar panels and peripheral circuit for energy harvesting; (ii) the power measurement switch; (iii) the ultra-capacitor-based energy storage; and (iv) the smart power supply circuit with a DC/DC converter for powering the working node attached to the TwinStar board. The corresponding printed circuit board is shown in Figure 12.2. The TwinStar node has a battery-free hardware structure using a combination of an ultra-capacitor and solar panels. It has several unique features:

  1. Stability: It is the first design that considers the zero-energy boot up

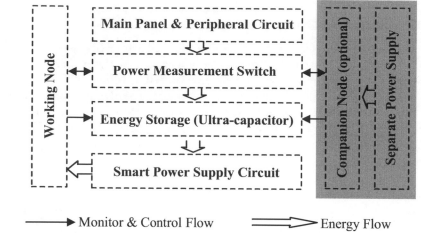

FIGURE 12.1 TwinStar Hardware Architecture

challenge and uses a dual-panel structure comprising a boot panel and a harvesting panel. The boot panel ensures stable power harvesting under extremely low ambient energy situations. The harvesting panel is used as the main charging device for the ultra-capacitor. DC/DC converter provides stable output to power the working node.

2. Visibility: As an experimental platform, the TwinStar node provides monitoring/debug interfaces for attaching an additional companion node (as shown in Figure 12.2), which can help the user understand the behaviors of the running working node and facilitate system debugging and performance evaluation. The separate power supply for the companion node ensures accurate characterizing of the energy profile of the working node.

3. Efficiency: The design of the TwinStar node is concise, effective, and efficient. Without any sophisticated ICs or circuits, the design achieves necessary functions, energy efficiency, and system stability with a short BOM (Bill of Materials).

12.3   Energy Conservation Techniques

Energy conservation is an intensively studied area. Many solutions have been proposed at different layers, including high-efficiency hardware design [44], link layer design [45], topology management [46], node placement [19], network routing [21], sensing coverage [47], data aggregation [33], data placement [48, 49], operating system [50], and application-level energy-aware designs [51]. In

FIGURE 12.2 TwinStar Platform

the rest of this section, we will discuss (i) Operating System Design for Energy Management, (ii) Energy-Efficient Communication, and (iii) Energy-Efficient Sensing.

Operating System Design for Energy Management: Operating system designs for wireless sensor networks have drawn a significant amount of attention in recent years. Many excellent operating systems have been developed, including such notable ones as TinyOS/T2 [52,53], LiteOS [54], Contiki [55], SOS [56], Mantis [57], t-Kernel [58], and Nano-RK [59]. Popular TinyOS supports a single-thread mode with two-level scheduling. SOS [56] is also a single-thread, even-driven architecture with dynamically loaded modules and a common kernel. Differently, LiteOS [54], Contiki [55], and Mantis [57] provide multi-threading support with different features in the file system, dynamic memory and debugging supports. In the area of mobile computing, the closest related projects are Odyssey [60] and ECOSystem [61] for laptops and PDAs. A few recent works provide resource-aware adaptations such as Pixie [62] and Eon [63] specifically for data-flow-oriented applications. DEOS [64] introduces a dynamic energy-oriented scheduling method for sustainable sensor networks. In DEOS, energy is treated as a first-class schedulable resource. Tasks are dynamically scheduled based on their energy consumption and the system's real-time available energy.

Energy-Efficient Communication: Due to the growing gap between limited energy and increasing energy demand in many long-term applications, there has been a surge of research interest in low-duty-cycle networking. In [65], the authors present a Streamlined Forwarding (SF) strategy to reduce delay in end-to-end communication. In [66], the authors provide optimal forwarding solutions for low-duty-cycle networks with tree and ring topologies. To reduce energy waste due to low link quality, the authors introduce the expected transmission count metric (ETX) to find highly efficient paths on multi-hop wireless networks in [67]. In [68], the authors propose a cost-based routing using a minimum expected transmission metric. In [69], the authors propose a scalable, opportunistic, and energy efcient routing protocol (E^2R) that uses wireless broadcast nature to opportunistically deliver both control messages and data packets in a multi-hop wireless network. In addition, many MAC protocols, such as B-MAC [14], S-MAC [16], FPA [70] and SCP-MAC [71], effectively reduce energy consumption by reducing duty-cycle through the Low-Power-Listening (LPL) [14] technique and time synchronization [72].

Energy-Efficient Sensing: Sensing is an indispensable research component for sensor networks. In [73], the authors support full surveillance coverage based on an off-duty eligibility rule. In [74], surveillance coverage is achieved through probing. In [75], the authors identify a critical bound for k-coverage in a network, assuming a node is randomly turned on with a certain probability. In [76], the authors investigate the k-barrier coverage problem, identifying the critical condition for weak k-barrier coverage. In [77, 78], the authors provide a theoretical analysis and simulation of the delay (or stealth distance) before a target is detected. Several algorithms are designed based on the concept of set cover. In [79], the authors propose two heuristic algorithms to identify a maximum number of set covers to monitor a set of static targets at known locations. In [80], the authors propose three approximation algorithms for a relaxed version of the previously defined SET K-COVER problem [81].

12.4   Energy Synchronization Techniques

As discussed in previous sections, hundreds of related works have ingrained a belief among many sensor network researchers that *maximum* energy harvesting and *minimum* energy consumption are always beneficial. For the sake of simplicity, both of these two important research directions have usually been regarded as orthogonal, and hence studied separately. However, energy management should synchronize the supply with demand. It is not always beneficial to conserve energy when a network can harvest excessive energy from the environment, because energy storage devices (e.g., batteries or capacitors) are always limited in capacity and usually energy-leakage-prone. Energy synchronization is very crucial for supporting many long-term unattended sensor network applications such as military surveillance, habitat monitoring, infrastructure protection, and scientific exploration. Without long-term sustainability, these applications cannot be deployed practically at a reasonable maintenance cost. Energy synchronization is also very challenging because that (i) both sensor network activities (workloads) and energy harvested are driven dynamically and unpredictably by the environment, and (ii) energy synchronization is a system-wide concern that can not be addressed with an isolated design. In this section, we introduce a feedback-based energy synchronization control technique and an energy synchronized communication protocol.

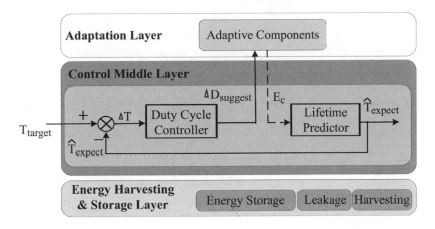

FIGURE 12.3 Overview of Control Layer Design

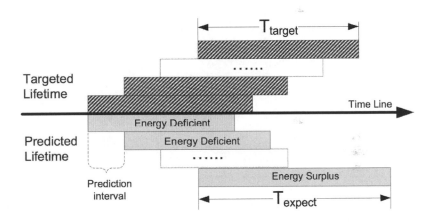

FIGURE 12.4 Sustainability

12.4.1 A Feedback-Based Energy Synchronization Control

The feedback-based energy synchronization control technique has been proposed in [38]. The design objective of the control technique is to *consume as much energy as possible while maintaining sustainability*. Sustainability T_{target} is defined as the duration a node must survive without ambient energy. T_{target} is a configurable parameter, set by users according to the environment in which sensors are deployed. For example, in energy-rich environments such as deserts, users can set a smaller T_{target} to consume energy aggressively.

On the other hand, in energy-poor and unpredictable environments, a larger T_{target} is desired to ensure the aliveness of sensor nodes. Once T_{target} is chosen, it is used as the set point in the feedback control based design as shown in Figure 12.3.

As shown in Figure 12.4, to maintain sustainability, the lifetime predictor periodically predicts the expected lifetime $\widehat{T}_{expect}$ and compares $\widehat{T}_{expect}$ with the target lifetime T_{target}. In the case of energy surplus (i.e., $\widehat{T}_{expect} > T_{target}$), a node can increase the activity accordingly for application performance improvement. In the case of energy deficit (i.e., $\widehat{T}_{expect} < T_{target}$), a node is expected to run out of energy after $\widehat{T}_{expect}$ seconds, if it maintains current level of activity. As shown in Figure 12.3, the change of activity is achieved by using the difference between $\widehat{T}_{expect}$ and T_{target} as the input to the duty cycle controller, which suggests a certain percentage of duty cycle change to the adaptation layer. Unlike conventional control designs, the duty cycle is *suggested* by the control layer to the adaptation layer. The adaptation layer adjusts the duty cycle based on the application requirement. This allows more flexible design of the energy adaptation layer. The rationale behind this is that the short-term duty cycle available might not always be synchronized with the activity demanded by the applications. For example, a node should not stop sending critical control messages disruptively simply because the control layer suggests to reduce duty cycle due to the brief drop in energy supply. Short-term mismatching is tolerable because the energy storage unit (e.g., ultra-capacitor) can serve as a buffer.

12.4.2   An Energy Synchronized Communication Protocol

Communication is a major source of energy consumption for small sensor nodes. For example, the RF component(Chipcon CC2420 [82]) in a Telos mote draws 18.8mA when receiving or idle listening, while its CPU (MSP430) draws only 1.8mA when active, which is about 75 times less. In this part of our work, we shall investigate how to build energy-synchronized communication that efficiently uses energy allocated by operating systems.

It is known that in low-duty-cycle sensor networks, end-to-end communications cannot afford to maintain an always-awake communication backbone. This is because that RF idle listening [14, 71] is a major source of energy consumption. The most effective way to reduce RF energy consumption is duty-cycling (i.e., turn on/off radio according to a certain schedule). Many energy management protocols have been proposed [28, 29, 83, 84] for scheduling communication to enable a duty cycle of 1% or less. Essentially, during the operation of sensor applications, sensor nodes activate very briefly and stay in a dormant state for a very long period of time. Consequently, to forward a packet, a sender may experience *sleep latency* — the time spent waiting for the receiver to wake up. In addition to sleep latency, the communication links between low-power sensor devices have been proven highly unreliable [68]. To provide reliable communication, link-layer retransmission is normally used, which introduces additional delay in waiting for the forwarding nodes to wake up again.

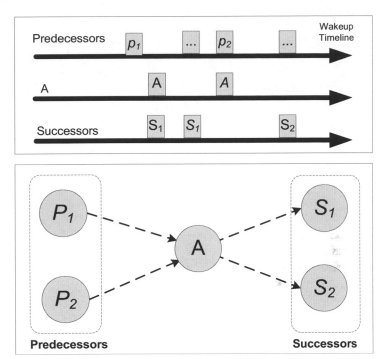

FIGURE 12.5 Cross-Traffic Delay Modeling

In [40], an Energy Synchronized Communication (ESC) protocol is designed as a generic middleware service for supporting existing network protocols. The forwarding delay is modeled at low-duty-cycle sensor nodes. The preliminary idea is shown in Figure 12.5. Let *predecessors* of Node A be the set of nodes that use Node A as forwarder (e.g., P_1, P_2 in the example) and *successors* of Node A be the set of nodes that Node A forwards to (e.g., S_1, S_2 in the example). Note that a node can be both a predecessor and successor of Node A, if they exchange data bi-directionally. The delay experienced by Node A is modeled as the sum of the delay of both incoming and outgoing packets. To illustrate the main idea without loss of generality, let's consider the delay in one link between P_i and A. Suppose the bi-directional link quality q denotes the success ratio of a round-trip transmission (DATA and ACK) between node A and node P_i. The probability that the packet is reached node A at its ith attempt can be expressed as $P(i) = (1 - q)^{i-1}q$. For a packet ready time t at node P_i, the expected transmission delay to reach node A is the sum of the product of probability that the packet reaches node A at its ith attempt and corresponding sleep latency $L_t^P(i)$ (where $i = 1, 2, \dots$). $L_t^P(i)$ can be calculated according to the wake-up schedule of Node A. Consequently, the link delay between P_i and A can be formulated as: $D(t) = \sum_{i=1}^{\infty} P(i)L_t^P(i)$. Similarly, we can model the delay between A and one of the successor nodes S_i. To obtain the average delay, a weighted average can be used based on traffic rates at individual links.

The research reveals that cross-traffic delay through a duty-cycled node is determined only by the number of active instances at intervals, partitioned by active instances of predecessor and successor nodes. Evaluation result demonstrates that ESC can effectively reduce delay and increase delivery ratios, while synchronizing radio activity with available energy.

Bibliography

[1] P. Dutta, J. Hui, J. Jeong, S. Kim, C. Sharp, J. Taneja, G. Tolle, K. Whitehouse, and D. Culler, "Trio: Enabling Sustainable and Scalable Outdoor Wireless Sensor Network Deployments," *Information Processing in Sensor Networks, 2006. IPSN 2006. The Fifth International Conference on*, pp. 407–415, 19-21 April 2006.

[2] K. Lin, J. Yu, J. Hsu, S. Zahedi, D. Lee, J. Friedman, A. Kansal, V. Raghunathan, and M. Srivastava, "Heliomote: Enabling Long-Lived Sensor Networks Through Solar Energy Harvesting," in *SenSys '05: Proceedings of the 3rd international conference on Embedded networked sensor systems.* New York, NY, USA: ACM, 2005, pp. 309–309.

[3] V. Raghunathan, A. Kansal, J. Hsu, J. Friedman and M. Srivastava, "Design Considerations for Solar Energy Harvesting Wireless Embedded Systems," in *IPSN*, 2005.

[4] T. Zhu, Y. Gu, T. He, and Z. Zhang, "Achieving Long-Term Operation with a Capacitor-Driven Energy Storage and Sharing Network," in *ACM Transactions on Sensor Networks (TOSN)*, Vol. 8, No. 3, Aug. 2012.

[5] S. Wright, D. Scott, J. Haddow, and M. Rosen, "The Upper Limit to Solar Energy Conversion," in *IECEC*, 2000.

[6] T. Zhu, Z. Zhong, T. He, and Z. Zhang, "Energy-Synchronized Computing for Sustainable Sensor Networks," in *Elsevier Ad Hoc Networks Journal*, 2010.

[7] X. Jiang, J. Polastre and D. Culler, "Perpetual Environmentally Powered Sensor Networks," in *IPSN*, 2005.

[8] C. Park and P. Chou, "AmbiMax: Autonomous Energy Harvesting Platform for Multi-Supply Wireless Sensor Nodes," *Sensor and Ad Hoc Communications and Networks, 2006. SECON '06. 2006 3rd Annual IEEE Communications Society on*, vol. 1, pp. 168–177, 28-28 Sept. 2006.

[9] J. Paradiso and M. Feldmeier, "A Compact, Wireless, Self-powered Push-button Controller," in *Ubicomp'01*, 2001.

[10] N. Shenck and J. Paradiso, "Energy Scavenging with Shoe-Mounted Piezoelectrics," *Micro, IEEE*, vol. 21, no. 3, pp. 30–42, May/Jun 2001.

[11] S. Meninger, J. Mur-Miranda, R. Amirtharajah, A. Chandrakasan, and J. Lang, "Vibration-to-Electric Energy Conversion," in *ISLPED*, 1999.

[12] *Mica2 data sheet*, CrossBow, available at http://www.xbow.com.

[13] P. Dutta, M. Grimmer, A. Arora, S. Biby, and D. Culler, "Design of a Wireless Sensor Network Platform for Detecting Rare, Random, and Ephemeral Events," in *IPSN'05*, 2005.

[14] J. Polastre and D. Culler, "Versatile Low Power Media Access for Wireless Sensor Networks," in *Second ACM Conference on Embedded Networked Sensor Systems (SenSys 2004)*, November 2004.

[15] T. van Dam and K. Langendoen, "An Adaptive Energy-Eaaaaaafficient MAC Protocol for Wireless Sensor Networks," in *SenSys 2003*, November 2003.

[16] W. Ye, J. Heidemann, and D. Estrin, "An Energy-Efficient MAC Protocol for Wireless Sensor Networks," in *INFOCOM*, 2002.

[17] B. Chen, K. Jamieson, H. Balakrishnan, and R. Morris, "Span: An Energy-Efficient Coordination Algorithm for Topology Maintenance in Ad Hoc Wireless Networks," in *6th ACM MOBICOM Conference*, 2001.

[18] A. Bogdanov, E. Maneva, and S. Riesenfeld, "Power-Aware Base Station Positioning for Sensor Networks," in *IEEE Infocom*, march 2004.

[19] D. Ganesan, R. Cristescu, and B. Berefull-Lozano, "Power-Efficient Sensor Placement and Transmission Structure for Data Gathering under Distortion Constraints," in *IPSN'04*, 2004.

[20] W. Heinzelman, A. Chandrakasan, and H. Balakrishnan, "Energy-Efficient Communication Protocol for Wireless Microsensor Networks," in *HICSS*, 2000.

[21] K. Seada, M. Zuniga, A. Helmy, and B. Krishnamachari, "Energy Efficient Forwarding Strategies for Geographic Routing," in *Second ACM Conference on Embedded Networked Sensor Systems (SenSys 2004)*, November 2004.

[22] Y. Xu, J. Heidemann, and D. Estrin, "Geography-Informed Energy Conservation for Ad Hoc Routing," in *MobiCom*, 2001.

[23] S. Guo, S. M. Kim, T. Zhu, Y. Gu, and T. He, "Correlated Flooding in Low-Duty-Cycle Wireless Sensor Networks," in *ICNP*, 2011.

[24] T. Zhu, Z. Zhong, T. He, and Z. Zhang, "Achieving Efficient Flooding by Utilizing Link Correlation in Wireless Networks," in *to appear in IEEE/ACM Transactions on Networking (TON)*, 2012.

[25] T. Zhu, Z. Zhong, T. He and Z.-L. Zhang, "Exploring link correlation for efficient flooding in wireless sensor networks," in *Proceedings of the 7th USENIX conference on Networked Systems Design and Implementation (NSDI '10), Berkeley, CA*, 2010.

[26] M. Cardei, M. Thai, and W. Wu, "Energy-Efficient Target Coverage in Wireless Sensor Networks," in *IEEE Infocom*, march 2005.

[27] M. L. Sichitiu, "Cross-Layer Scheduling for Power Efficiency in Wireless Sensor Networks," in *IEEE Infocom*, march 2004.

[28] X. Wang, G. Xing, Y. Zhang, C. Lu, R. Pless, and C. Gill, "Integrated Coverage and Connectivity Configuration in Wireless Sensor Networks," in *SenSys'03*, 2003.

[29] T. Yan, T. He, and J. Stankovic, "Differentiated Surveillance Service for Sensor Networks," in *First ACM Conference on Embedded Networked Sensor Systems (SenSys 2003)*, November 2003.

[30] M. Agarwal, J. H. Cho, L. Gao, and J. Wu, "Energy-efficient Broadcast in Wireless Ad hoc Networks with Hitch-hiking," in *IEEE Infocom*, march 2004.

[31] W. Choi and S. Das, "A Novel Framework for Energy-Conserving Data Gathering in Wireless Sensor Networks," in *IEEE Infocom*, march 2005.

[32] Y. Yu, B. Krishnamachari, and V. K. Prasanna, "Energy-Latency Trade-offs for Data Gathering in Wireless Sensor Networks," in *IEEE INFOCOM*, 2004.

[33] S. Madden, M. Franklin, J. Hellerstein, and W. Hong, "TAG: A Tiny Aggregation Service for Ad-Hoc Sensor Networks," in *Operating Systems Design and Implementation*, December 2002.

[34] N. Shrivastava, C. Buragohain, S. Suri, and D. Agrawal, "Medians and Beyond: New Aggregation Techniques for Sensor Networks," in *Second ACM Conference on Embedded Networked Sensor Systems (SenSys 2004)*, November 2004.

[35] S. Bhattacharya, H. Kim, S. Prabh, and T. Abdelzaher, "Energy-Conserving Data Placement and Asynchronous Multicast in Wireless Sensor Networks," in *The First International Conference on Mobile Systems, Applications, and Services (MobiSys)*, May 2003.

[36] R. Szewczyk, A. Mainwaring, J. Anderson, and D. Culler, "An Analysis of a Large Scale Habit Monitoring Application," in *Second ACM Conference on Embedded Networked Sensor Systems (SenSys 2004)*, November 2004.

[37] N. Xu, S. Rangwala, K. K. Chintalapudi, D. Ganesan, A. Broad, R. Govindan, and D. Estrin, "A Wireless Sensor Network for Structural Monitoring," in *Second ACM Conference on Embedded Networked Sensor Systems (SenSys 2004)*, November 2004.

[38] T. Zhu, Z. Zhong, Y. Gu, T. He, and Z. Zhang, "Leakage-Aware Energy Synchronization for Wireless Sensor Networks," in *MobiSys*, 2009.

[39] ———, "Feedback Control-based Energy Management for Ubiquitous Sensor Networks," in *IEICE Transactions on Communications*, Vol. E93-B, No.11, pages 2846-2854, Nov. 2010.

[40] Y. Gu, T. Zhu, and T. He, "ESC: Energy Synchronized Communication in Sustainable Sensor Networks," in *Proc. of the 17th International Conference on Network Protocols (ICNP '09)*, October 2009.

[41] A. Kansal, J. Hsu, S. Zahedi, and M. B. Srivastava, "Power Management in Energy Harvesting Sensor Networks," *TECS*, vol. 6, no. 4, 2007.

[42] C. Park and P. Chou, "Power Utility Maximization for Multiple-Supply Systems by a Load-Matching Switch," *Low Power Electronics and Design, 2004. ISLPED '04. Proceedings of the 2004 International Symposium on*, pp. 168–173, 2004.

[43] Suzanne Foster, "The Energy Efficiency of Common Household Battery Charging Systems: Results and Implications," in *Natural Resources Defense Council*, http://www.efficientproducts.org/reports/bchargers/NRDC-Ecos_Battery_Charger_Efficiency.pdf.

[44] D. Lymberopoulos and A. Savvides, "XYZ: a motion-enabled, power aware sensor node platform for distributed sensor network applications," in *IPSN '05*, 2005.

[45] G. Zhou, T. He, and J. A. Stankovic, "Impact of Radio Irregularity on Wireless Sensor Networks," in *The Second International Conference on Mobile Systems, Applications, and Services (MobiSys)*, June 2004.

[46] S. Lin, J. Zhang, G. Zhou, L. Gu, T. He, and J. A. Stankovic, "ATPC: Adaptive Transmission Power Control for Wireless Sensor Networks," in *Fourth ACM Conference on Embedded Networked Sensor Systems (SenSys 2006)*, November 2006.

[47] S. Kang, J. Lee, H. Jang, H. Lee, Y. Lee, S. Park, T. Park, and J. Song, "Seemon: scalable and energy-efficient context monitoring framework for sensor-rich mobile environments," in *MobiSys '08*, 2008.

[48] S. Bhattacharya, H. Kim, S. Prabh, and T. Abdelzaher, "Energy-conserving data placement and asynchronous multicast in wireless sensor networks," in *MobiSys '03*, 2003.

[49] C. M. Sadler and M. Martonosi, "Dali: a communication-centric data abstraction layer for energy-constrained devices in mobile sensor networks," in *MobiSys '07*, 2007.

[50] C.-C. Han, R. Kumar, R. Shea, E. Kohler, and M. Srivastava, "A dynamic operating system for sensor nodes," in *MobiSys '05*, 2005.

[51] T. He, S. Krishnamurthy, J. A. Stankovic, T. Abdelzaher, L. Luo, R. Stoleru, T. Yan, L. Gu, J. Hui, and B. Krogh, "Energy-efficient surveillance system using wireless sensor networks," in *MobiSys '04*, 2004.

[52] J. Hill, R. Szewczyk, A. Woo, S. Hollar, D. E. Culler, and K. S. J. Pister, "System Architecture Directions for Networked Sensors," in *Proc. of Architectural Support for Programming Languages and Operating Systems (ASPLOS)*, 2000, pp. 93–104.

[53] P. Levis, D. Gay, V. Handziski, J. Hauer, M. T. Be. Greenstein, J. Huio, K. Klues, C. Sharp, R. Szewczyk, J. Polastre, P. Buonadonnao, L. Nachman, G. Tolleo, D. Cullero, and A. Wolisz, "T2: A Second Generation OS For Embedded Sensor Networks," in *Technical Report TKN-05-007*, 2005.

[54] Q. Cao, T. Abdelzaher, J. Stankovic, and T. He, "LiteOS, A Unix-like Operating System and Programming Platform for Wireless Sensor Networks," in *International Conference on Information Processing in Sensor Networks(IPSN'08)*, 2008.

[55] A. Dunkels, B. Gronvall, and T. Voigt, "Contiki - A Lightweight and Flexible Operating System for Tiny Networked Sensors," in *LCN '04: Proceedings of the 29th Annual IEEE International Conference on Local Computer Networks*. Washington, DC, USA: IEEE Computer Society, 2004, pp. 455–462.

[56] C.-C. Han, R. Kumar, R. Shea, E. Kohler, and M. Srivastava, "A Dynamic Operating System for Sensor Nodes," in *MobiSys '05: Proceedings of the 3rd international conference on Mobile systems, applications, and services*. New York, NY, USA: ACM, 2005, pp. 163–176.

[57] S. Bhatti, J. Carlson, H. Dai, and et al., "MANTIS OS: an Embedded Multithreaded Operating System for Wireless Micro Sensor Platforms," *Mob. Netw. Appl.*, vol. 10, no. 4, pp. 563–579, 2005.

[58] L. Gu and J. A. Stankovic, "t-kernel: Providing Reliable OS Support to Wireless Sensor Networks," in *SenSys '06: Proceedings of the 4th international conference on Embedded networked sensor systems*. New York, NY, USA: ACM, 2006, pp. 1–14.

[59] A. Eswaran, A. Rowe, and R. Rajkumar, "Nano-RK: An Energy-Aware Resource-Centric RTOS for Sensor Networks," in *RTSS '05: Proceedings of the 26th IEEE International Real-Time Systems Symposium*. Washington, DC, USA: IEEE Computer Society, 2005, pp. 256–265.

[60] J. Flinn and M. Satyanarayanan, "Managing Battery Lifetime with Energy-Aware Adaptation," *ACM Trans. Comput. Syst.*, vol. 22, no. 2, pp. 137–179, 2004.

[61] H. Zeng, C. Ellis, and A. Lebeck, "Experiences in Managing Energy with ECOSystem," *Pervasive Computing, IEEE*, vol. 4, no. 1, pp. 62–68, Jan.-March 2005.

[62] K. Lorincz, B. rong Chen, J. Waterman, G. Werner-Allen, and M. Welsh, "Pixie: An Operating System for Resource-Aware Programming of Embedded Sensors," in *In Proceedings of the Fifth workshop on Embedded Networked Sensors (HotEmNets 2008)*, 2008.

[63] J. Sorber, A. Kostadinov, M. Garber, M. Brennan, M. D. Corner, and E. D. Berger, "Eon: a Language and Runtime System for Perpetual Systems," in *SenSys '07: Proceedings of the 5th international conference on Embedded networked sensor systems.* New York, NY, USA: ACM, 2007, pp. 161–174.

[64] T. Zhu, A. Mohaisen, Y. Ping, and D. Towsley, "DEOS: Dynamic Energy-Oriented Scheduling for Sustainable Wireless Sensor Networks," in *INFOCOM*, 2012.

[65] Q. Cao, T. Abdelzaher, T. He, and J. Stankovic, "Towards Optimal Sleep Scheduling in Sensor Networks for Rare Event Detection ," in *The Fourth International Conference on Information Processing in Sensor Networks*, 2005.

[66] G. Lu, N. Sadagopan, B. Krishnamachari, and A. Goel, "Delay Efficient Sleep Scheduling in WIreless Sensor Networks," in *INFOCOM'05*, 2005.

[67] D. S. J. D. Couto, D. Aguayo, J. Bicket, and R. Morris, "A High Throughput Path Metric for Multi-Hop Wireless Routing," in *MOBICOM'03*, 2003.

[68] A. Woo, T. Tong, and D. Culler, "Taming the Underlying Challenges of Reliable Multihop Routing in Sensor Networks," in *First ACM Conference on Embedded Networked Sensor Systems (SenSys 2003)*, November 2003.

[69] T. Zhu, and D. Towsley, "E2R: Energy efficient routing for multi-hop green wireless networks," in *Workshop on Green Communications and Networking*, 2011.

[70] Y. Li, W. Ye, and J. Heidemann, "Energy and Latency Control in Low Duty Cycle MAC Protocols," in *WCNC'05*, 2005.

[71] W. Ye, F. Silva, and J. Heidemann, "Ultra-low Duty Cycle MAC with Scheduled Channel Polling," in *SenSys '06: Proceedings of the 4th international conference on Embedded networked sensor systems.* New York, NY, USA: ACM, 2006, pp. 321–334.

[72] J. Elson and D. Estrin, "Time Synchronization for Wireless Sensor Networks," in *Proc. of Workshop on Parallel and Distributed Computing Issues in Wireless Networks and Mobile Computing*, April 2001.

[73] D. Tian and N. Georganas, "A Node Scheduling Scheme for Energy Conservation in Large Wireless Sensor Networks," *Wireless Communications and Mobile Computing Journal*, May 2003.

[74] F. Ye, G. Zhong, S. Lu, and L. Zhang, "PEAS: A Robust Energy Conserving Protocol for Long-lived Sensor Networks," in *Proc. of International Conference on Distributed Computing Systems (ICDCS)*, May 2003.

[75] S. Kumar, T. H. Lai, and J. Balogh., "On k-Coverage in a Mostly Sleeping Sensor Network," in *Mobicom*, 2004.

[76] S. Kumar, T. H. Lai, and A. Arora, "Barrier Coverage With Wireless Sensors," in *MobiCom 2005*, 2005.

[77] C. Gui and P. Mohapatra, "Power Conservation and Quality of Surveillance in Target Tracking Sensor Networks," in *MobiCom'04*, 2004.

[78] S. Ren, Q. Li, H. Wang, X. Chen, and X. Zhang, "Analyzing Object Tracking Quality under Probabilistic Coverage in Sensor Networks," *ACM Mobile Computing and Communications Review*, vol. 9, no. 1, January 2005.

[79] M. Cardei, M. T. Thai, Y. Li, and W. Wu, "Energy-Efficient Target Coverage in Wireless Sensor Networks," in *IEEE INFOCOM*, 2005.

[80] Z. Abrams, A. Goel, and S. Plotkin, "Set K-Cover Algorithms for Energy Efficient Monitoring in Wireless Sensor Networks," in *IEEE IPSN*, 2004.

[81] S. Slijepcevic and M. Potkonjak, "Power Efficient Organization of Wireless Sensor Networks," in *IEEE ICC*, 2001.

[82] "2.4 GHz IEEE 802.15.4 / ZigBee-Ready RF Transceiver (Rev. B) ," 2007, texas Intruments. [Online]. Available: http://focus.ti.com/docs/prod/folders/print/cc2420.html

[83] Y. Gu, J. Hwang, T. He, and D. H. Du, "uSense: A Unified Asymmetric Sensing Coverage Architecture for Wireless Sensor Networks," in *International Conference on Distributed Computing Systems (ICDCS)*, June 2007.

[84] J. Liu, F. Zhao, P. Cheung, and L. Guibas, "Apply Geometric Duality to Energy-efficient Non-local Phenomenon Awareness using Sensor Networks," *IEEE Wireless Communications*, vol. 11, no. 6, 2004.

Author contact information

Ting Zhu is with State University of New York, Binghamton, USA, Email: tzhu@binghamton.edu. Abedelaziz Mohaisen is with Computer Science and Engineering Department, University of Minnesota - Twin Cities, Minneapolis, Minnesota 55455, USA, Email: mohaisen@cs.umn.edu. Ping Yi is with School of Information Security Engineering, Shanghai Jiao Tong University, China. Jianqing Ma, Network and Information Engineering Center, Fudan University, China.

13

Green Wireless Communications under Quality of Service Constraints

Deli Qiao

University of Nebraska-Lincoln, USA

Mustafa Cenk Gursoy

Syracuse University, USA

Senem Velipasalar

Syracuse University, USA

CONTENTS

13.1   Introduction

Two key characteristics of wireless communications that greatly impact system design and performance are 1) the randomly-varying channel conditions and 2) limited energy resources. In wireless systems, the power of the received signal fluctuates randomly over time and distance due to multipath fading, mobility, and changing environment. These random changes in the received signal strength lead to variations in the instantaneous data rate that can be supported by the channel. In addition, mobile wireless systems can only be equipped with limited energy resources, and hence energy efficient operation is a crucial requirement in most mobile applications. Also, with emerging applications in wireless ad-hoc and wireless sensor networks, energy-efficient operation is becoming a vital issue, especially when the replenishment of energy resources is not easy. Finally, there has been a new surge of interest in green communications in order to reduce the carbon footprint of communication systems and lower the energy costs. In summary, energy efficiency is a critical concern in wireless systems.

In many wireless communication systems, in addition to energy-efficient operation, satisfying certain quality-of-service (QoS) requirements is of paramount importance in providing acceptable performance and quality. For instance, in voice over IP (VoIP), interactive-video (e.g., video-conferencing), and streaming-video applications in wireless systems, latency is a key QoS metric. In such cases, information has to be communicated within the delay constraints. On the other hand, wireless channels, as described above, are characterized by random changes in the channel, and such volatile conditions generally present significant challenges in providing QoS guarantees in wireless communications.

Therefore, the central design challenge in next-generation wireless systems, which are envisioned to provide communications anytime, anywhere in a reliable and robust fashion, is how to provide the best possible performance levels while making efficient use of energy resources and satisfying certain QoS constraints (e.g., latency, packet loss, buffer violation probability). Information theory provides the ultimate performance limits and identifies the most efficient use of resources. Due to this fact, wireless fading channels have been extensively studied from an information-theoretic point of view, considering different assumptions on the availability of the channel side information (CSI) at the receiver and transmitter. While providing powerful results, information-theoretic studies generally do not address delay and QoS constraints. On the other hand, satisfying QoS requirements is crucial for the successful deployment and operation of most communication networks. As a result, in the networking literature, how to handle and satisfy QoS constraints has been one of the key considerations for many years. In addressing this issue, the theory of effective bandwidth of a time-varying source has been developed to identify the minimum transmission rate needed to satisfy the statistical QoS requirements. In this development, the key tools are provided by the large deviations and queueing theories.

In this chapter, in order to address the problem of energy-efficient operation under QoS constraints in the form of limitations on the queue length or queueing delay, we present a unified analysis by employing tools from both information theory and queueing theory as studied in [1], [2], and [3]. The throughput metric we employ is the *effective capacity*, which provides the maximum constant arrival rate that a wireless channel can sustain while satisfying statistical QoS constraints in the form of limitations on buffer overflow probability. We investigate the energy efficiency by identifying the minimum bit energy requirements and the wideband slope, which can be regarded as the information-theoretic performance measures of energy efficiency. We focus on point-to-point links and characterize the spectral efficiency vs. bit energy tradeoff for different transmission techniques. We conduct the analysis in low-power and wideband regimes. Unlike some other works that consider more practical settings [4], the material covered in this chapter lays the theoretical basis for the design of energy efficient wireless systems.

13.2 Preliminaries - Effective Capacity and Energy Per Information Bit

This section introduces the preliminary concepts by describing the effective capacity and the bit energy metric.

13.2.1 Effective Capacity

In wireless communications, the instantaneous channel capacity varies randomly depending on the channel conditions. Hence, in addition to the source, the transmission rates for reliable communication are also time-varying. The time-varying channel capacity can be incorporated into the theory of effective bandwidth by regarding the channel service process as a time-varying source with negative rate and using the source multiplexing rule ([5, Example 9.2.2]). Wu and Negi in [6] defined the effective capacity as a dual concept to effective bandwidth. Effective capacity is defined as the maximum constant arrival rate[1] that a given service process can support in order to guarantee a statistical QoS requirement specified by the QoS exponent θ. If we define Q as the stationary queue length, then θ is the decay rate of the tail distribution of the queue length Q:

$$\lim_{q \to \infty} \frac{\log \Pr(Q \ge q)}{q} = -\theta. \tag{13.1}$$

Therefore, for large $q_{\max}$, we have the following approximation for the buffer violation probability: $\Pr(Q \ge q_{\max}) \approx e^{-\theta q_{\max}}$. Hence, while larger θ corresponds to more strict QoS constraints, smaller θ implies looser QoS guarantees. Similarly, if D denotes the steady-state delay experienced in the buffer, then $\Pr(D \ge d_{\max}) \approx e^{-\theta \delta d_{\max}}$ for large $d_{\max}$, where δ is determined by the arrival and service processes [7].

Let $\{s[i], i = 1, 2, \ldots\}$ denote the discrete-time stationary and ergodic stochastic service process and $S[t] \triangleq \sum_{i=1}^{t} s[i]$ be the time-accumulated process. Assume that the Gärtner-Ellis limit of $S[t]$, expressed as

$$\Lambda_C(\theta) = \lim_{t \to \infty} \frac{1}{t} \log_e \mathbb{E}\{e^{\theta S[t]}\}. \tag{13.2}$$

exists. The effective capacity is given by $C_E(\mathrm{SNR}, \theta) = -\frac{\Lambda_C(-\theta)}{\theta} = -\lim_{t \to \infty} \frac{1}{\theta t} \log_e \mathbb{E}\{e^{-\theta S[t]}\}$.

If the fading process is constant during the frame duration T and changes independently from frame to frame, then the effective capacity simplifies to

$$C_E(\mathrm{SNR}, \theta) = -\frac{1}{\theta T} \log \mathbb{E}\{e^{-\theta T R[i]}\} \quad \text{bits/s.} \tag{13.3}$$

where $R[i]$ is the instantaneous service rate in the ith frame duration $[iT; (i+1)T]$, which will be discussed in detail later for specific transmission strategies. (13.3) is obtained using the fact that instantaneous rates $\{R[i]\}$ vary independently.

13.2.2   Spectral Efficiency vs. Bit Energy

If we denote the effective capacity normalized by bandwidth or equivalently the spectral efficiency in bits per second per Hertz by

$$\mathsf{C}_E(\mathrm{SNR}, \theta) = \frac{C_E(\mathrm{SNR}, \theta)}{B} = -\frac{1}{\theta T B} \log \mathbb{E}\{e^{-\theta T R[i]}\}, \tag{13.4}$$

---

[1] For time-varying arrival rates, effective capacity specifies the effective bandwidth of the arrival process that can be supported by the channel.

then it can be easily seen that the minimum energy per bit $\frac{E_b}{N_0}_{\min}$ under QoS constraints can be obtained from [8]

$$\frac{E_b}{N_{0\,\min}} = \lim_{\text{SNR}\to 0} \frac{\text{SNR}}{C_E(\text{SNR})} = \frac{1}{\dot{C}_E(0)}. \tag{13.5}$$

At $\frac{E_b}{N_0}_{\min}$, the slope $\mathcal{S}_0$ of the spectral efficiency versus E_b/N_0 (in dB) curve is defined as [8]

$$\mathcal{S}_0 = \lim_{\frac{E_b}{N_0}\downarrow\frac{E_b}{N_0}_{\min}} \frac{C_E(\frac{E_b}{N_0})}{10\log_{10}\frac{E_b}{N_0} - 10\log_{10}\frac{E_b}{N_0}_{\min}} 10\log_{10} 2. \tag{13.6}$$

Considering the expression for normalized effective capacity, the wideband slope can be found from[2]

$$\mathcal{S}_0 = -\frac{2(\dot{C}_E(0))^2}{\ddot{C}_E(0)} \log 2 \tag{13.7}$$

where $\dot{C}_E(0)$ and $\ddot{C}_E(0)$ are the first and second derivatives, respectively, of the function $C_E(\text{SNR})$ in bits/s/Hz at zero SNR [8].

13.3   Variable-Rate/Variable-Power and Variable-Rate/Fixed-Power Transmissions

We first study the variable-rate/variable-power and variable-rate/fixed-power transmission schemes with different assumptions on the availability of channel side information (CSI) at the transmitter and receiver. We obtain the minimum bit energy and wideband slope expressions, and in the variable-power case, we analyze the impact of power control policies on energy efficiency.

13.3.1   Channel Model

We consider a point-to-point communication system in which there is one source and one destination. The general system model is depicted in Fig. 13.1, and is similar to the one studied in [9]. In this model, it is assumed that the source generates data sequences which are divided into frames of duration T. These frames are initially stored in the buffer before being transmitted over the wireless channel. The discrete-time channel input-output relation in the

---

[2]We note that the expressions in (13.5) and (13.7) differ from those in [8] by a constant factor due to the fact that we assume that the units of C_E is bits/s/Hz rather than nats/s/Hz.

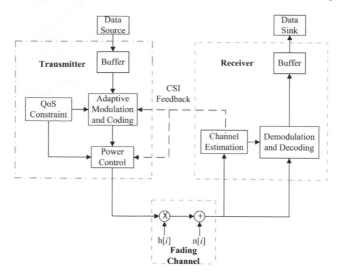

FIGURE 13.1 The system model

i^{th} symbol duration is given by

$$y[i] = h[i]x[i] + n[i] \quad i = 1, 2, \ldots. \tag{13.8}$$

where $x[i]$ and $y[i]$ denote the complex-valued channel input and output, respectively. The channel input is subject to an average power constraint $\mathbb{E}\{|x[i]|^2\} \leq \bar{P}$ for all i, and we assume that the bandwidth available in the system is B. In eq. (13.8), $n[i]$ is a zero-mean, circularly symmetric, complex Gaussian random variable with variance $\mathbb{E}\{|n[i]|^2\} = N_0$. The additive Gaussian noise samples $\{n[i]\}$ are assumed to form an independent and identically distributed (i.i.d.) sequence. Finally, $h[i]$ denotes the channel fading coefficient, and $\{h[i]\}$ is a stationary and ergodic discrete-time process. We assume that perfect channel state information (CSI) is available at the receiver while the transmitter has either *no* or *perfect* CSI. The availability of CSI at the transmitter is facilitated through CSI feedback from the receiver. Note that if the transmitter knows the channel fading coefficients, it employs power and rate adaptation. Otherwise, the signals are sent with constant power.

Note that in the above system model, the average transmitted signal-to-noise ratio is $\text{SNR} = \bar{P}/(N_0 B)$. We denote the magnitude-square of the fading coefficient by $z[i] = |h[i]|^2$, and its distribution function by $p_z(z)$. When there is only receiver CSI, instantaneous transmitted power is $P[i] = \bar{P}$ and instantaneous received SNR is expressed as $\gamma[i] = \bar{P} z[i]/(N_0 B)$. Moreover, the maximum instantaneous service rate $R[i]$ is

$$R[i] = B \log_2 \left(1 + \text{SNR} z[i]\right) \quad \text{bits/s.} \tag{13.9}$$

We note that although the transmitter does not know $z[i]$, recently developed rateless codes such as LT [10] and Raptor [11] codes enable the transmitter to adapt its rate to the channel realization and achieve $R[i]$ without requiring CSI at the transmitter side [12], [13]. For systems that do not employ such codes, service rates are smaller than that in (13.9), and the results here serve as upper bounds on the performance. Indeed, a scenario in which the transmitter sends the information at a fixed rate is analyzed in Section 13.4.

When the transmitter also has CSI, the instantaneous service rate is

$$R[i] = B \log_2 \left(1 + \mu_{\text{opt}}(\theta, z[i]) z[i] \right) \quad \text{bits/s} \tag{13.10}$$

where $\mu_{\text{opt}}(\theta, z)$ is the power-adaptation policy that maximizes the effective capacity, which has been discussed in Section 13.2.1. The optimal power policy is determined in [9]:

$$\mu_{\text{opt}}(\theta, z) = \begin{cases} \dfrac{1}{\alpha^{\frac{1}{\beta+1}} z^{\frac{\beta}{\beta+1}}} - \dfrac{1}{z} & z \geq \alpha \\ 0 & z < \alpha \end{cases} \tag{13.11}$$

where θ is the QoS exponent defined in (13.1), $\beta = \frac{\theta TB}{\log_e 2}$ is the normalized QoS exponent and α is the channel threshold chosen to satisfy the average power constraint:

$$\text{SNR} = \mathbb{E}\{\mu_{\text{opt}}(\theta, z)\} = \mathbb{E}\left\{ \left[\frac{1}{\alpha^{\frac{1}{\beta+1}} z^{\frac{\beta}{\beta+1}}} - \frac{1}{z} \right] \tau(\alpha) \right\} \tag{13.12}$$

where $\tau(\alpha) = 1\{z \geq \alpha\} = \begin{cases} 1 & \text{if } z \geq \alpha \\ 0 & \text{if } z < \alpha \end{cases}$ is the indicator function. Note that $\mu_{\text{opt}}(\theta, z)$ depends on the average power constraint only through the threshold α. Moreover, the power allocation strategy $\mu_{\text{opt}}(\theta, z)$, while varying with the instantaneous values of the fading coefficients, depends only statistically on the queueing constraints through the QoS exponent θ, and hence is not a function of the instantaneous queue lengths.

We finally note that since the maximum service rates are equal to the instantaneous channel capacity values, we assume through information-theoretic arguments that when the transmitter transmits at or below rate $R[i]$ given in (13.9) and (13.10), the information is reliably received at the receiver and no retransmissions are required.

13.3.2 Energy Efficiency in the Low-Power Regime

As discussed in the previous section, the minimum bit energy is achieved as $\text{SNR} = \frac{\bar{P}}{N_0 B} \to 0$, and hence energy efficiency improves if one operates in the low-power regime in which $\bar{P}$ is small, or the high-bandwidth regime in which B is large. From the Shannon capacity perspective, similar performances are achieved in these two regimes, which therefore can be seen as equivalent. However, as we shall see in this chapter, considering the effective capacity leads to different results at low power and high bandwidth levels. In this section, we consider the low-power regime for fixed bandwidth, B, and study the spectral efficiency vs. bit energy tradeoff by finding the minimum bit energy and the wideband slope.

13.3.2.1   CSI at the Receiver Only

We initially consider the case in which only the receiver knows the channel conditions. Substituting (13.9) into (13.4), we obtain the spectral efficiency given θ as a function of SNR as

$$C_E(\text{SNR}) = -\frac{1}{\theta TB} \log_e \mathbb{E}\{e^{-\theta TB \log_2(1+\text{SNR}z)}\} = -\frac{1}{\theta TB} \log_e \mathbb{E}\{(1+\text{SNR}z)^{-\beta}\} \tag{13.13}$$

where again $\beta = \frac{\theta TB}{\log_e 2}$. Note that since the analysis is performed for fixed θ, we henceforth express the effective capacity only as a function of SNR to simplify the expressions. The following result provides the minimum bit energy and the wideband slope.

Theorem 13.1. *When only the receiver has perfect CSI, the minimum bit energy and wideband slope are*

$$\frac{E_b}{N_0}_{min} = \frac{\log_e 2}{\mathbb{E}\{z\}} \text{ and } S_0 = \frac{2}{(\beta+1)\frac{\mathbb{E}\{z^2\}}{(\mathbb{E}\{z\})^2} - \beta}, \text{ respectively.} \tag{13.14}$$

From the result above, we immediately see that $\frac{E_b}{N_0}_{min}$ does not depend on θ and the minimum *received* bit energy is $\frac{E_b^r}{N_0}_{min} = \frac{E_b}{N_0}_{min} \mathbb{E}\{z\} = \log_e 2 = -1.59$ dB. Note that if the Shannon capacity is used in the analysis, i.e., if $\theta = 0$ and hence $\beta = 0$, $\frac{E_b^r}{N_0}_{min} = -1.59$ dB and $S_0 = 2/(\mathbb{E}\{z^2\}/\mathbb{E}^2\{z\})$. Therefore, we conclude from Theorem 13.1 that as the average power $\bar{P}$ decreases, energy efficiency approaches the performance achieved by a system that does not have QoS limitations. However, we note that wideband slope is smaller if $\theta > 0$. Hence, the presence of QoS constraints decreases the spectral efficiency or equivalently increases the energy requirements for fixed spectral efficiency values at low but nonzero SNR levels.

13.3.2.2   CSI at Both the Transmitter and Receiver

We now consider the case in which both the transmitter and receiver have perfect CSI. Substituting (13.10) into (13.3), we have

$$C_E(\text{SNR}) = -\frac{1}{\theta TB} \log_e \mathbb{E}\left\{e^{-\theta TB \log_2\left(1+\mu_{\text{opt}}(\theta,z)z\right)}\right\}$$

$$= -\frac{1}{\theta TB} \log_e \left(F(\alpha) + \mathbb{E}\left\{\left(\frac{z}{\alpha}\right)^{-\frac{\beta}{\beta+1}} \tau(\alpha)\right\}\right) \tag{13.15}$$

where $F(\alpha) = \mathbb{E}\{1\{z < \alpha\}\}$. We note that the normalized effective capacity expression in (13.15) is obtained assuming that the optimal power control policy $\mu_{\text{opt}}(\theta, z)$ given in (24.22) is employed in the system. Maximizing the effective capacity, this optimal power allocation policy minimizes the bit energy requirements. Following an approach similar to that in [14], we obtain the following result for this case.

Theorem 13.2. *When both the transmitter and receiver have perfect CSI, the minimum bit energy with optimal power control and rate adaptation becomes*

$$\frac{E_b}{N_0}_{min} = \frac{\log_e 2}{z_{\max}} \qquad (13.16)$$

where $z_{\max}$ is the essential supremum of the random variable z, i.e., $z \leq z_{\max}$ with probability 1.

Note that for distributions with unbounded support, we have $z_{\max} = \infty$ and hence $\frac{E_b}{N_0}_{min} = 0 = -\infty$ dB. In this case, it is easy to see that the wideband slope is $S_0 = 0$.

Example 13.1. *Specifically, for the Rayleigh fading channel, as in [15], it can be shown that $\lim_{SNR \to 0} \frac{C_E(SNR)}{SNR \log_e(\frac{1}{SNR}) \log_e 2} = 1$. Then, spectral efficiency can be written as $C_E(SNR) \approx SNR \log_e(\frac{1}{SNR}) \log_e 2$, so $\frac{E_b}{N_0}_{min} = \lim_{SNR \to 0} \frac{SNR}{C_E(SNR)} = \lim_{SNR \to 0} \frac{1}{\log_e(\frac{1}{SNR}) \log_e 2} = 0$ which also verifies the above result.*

Remark: We note that as in the case in which there is CSI at the receiver, the minimum bit energy achieved under QoS constraints is the same as that achieved by the Shannon capacity [14]. Hence, the energy efficiency again approaches the performance of an unconstrained system as power diminishes. Searching for an intuitive explanation of this observation, we note that arrival rates that can be supported vanishes with decreasing power levels. As a result, the impact of buffer occupancy constraints on the performance lessens. Note that in contrast, increasing the bandwidth increases the arrival rates supported by the system. Therefore, limitations on the buffer occupancy will have significant impact upon the energy efficiency in the wideband regime as discussed next.

13.3.3 Energy Efficiency in the Wideband Regime

In this part, we study the performance at high bandwidths while the average power $\bar{P}$ is kept fixed. We investigate the impact of θ on $\frac{E_b}{N_0}_{min}$ and the wideband slope S_0 in this wideband regime. Note that as the bandwidth increases, the average signal-to-noise ratio $\text{SNR} = \bar{P}/(N_0 B)$ and the spectral efficiency decreases. Note further that the analysis also applies if the wideband channel is broken into subchannels, each with a bandwidth that is equal to the coherence bandwidth, and the coherence bandwidth grows with increasing bandwidth due to multipath sparsity while the number of subchannels remains bounded. If both the coherence bandwidth and the number of sub-channels grow without bound with increasing bandwidth, then the minimum bit energy and wideband slope values can be obtained from the results in Section 13.3.2 by letting B and hence $\beta = \frac{\theta T B}{\log_e 2}$ go to infinity when $\theta > 0$. A more detailed analysis of decomposition of the wideband channel into narrowband subchannels and the impact of multipath sparsity is provided in Sections 13.4 and 13.5.

13.3.3.1 CSI at the Receiver Only

We define $\zeta = \frac{1}{B}$ and express the spectral efficiency (13.13) as a function of ζ:

$$C_E(\zeta) = -\frac{\zeta}{\theta T} \log_e \mathbb{E}\{e^{-\frac{\theta T}{\zeta} \log_2(1 + \frac{P\zeta}{N_0} z)}\}. \tag{13.17}$$

The bit energy is again defined as

$$\frac{E_b}{N_0} = \frac{\text{SNR}}{C_E(\text{SNR})} = \frac{\frac{\bar{P}\zeta}{N_0}}{C_E(\zeta)} = \frac{\frac{\bar{P}}{N_0}}{C_E(\zeta)/\zeta}. \tag{13.18}$$

It can be readily verified that $C_E(\zeta)/\zeta$ monotonically increases as $\zeta \to 0$ (or equivalently as $B \to \infty$). Therefore

$$\frac{E_b}{N_0 \text{ min}} = \lim_{\zeta \to 0} \frac{\bar{P}\zeta/N_0}{C_E(\zeta)} = \frac{\bar{P}/N_0}{\dot{C}_E(0)} \tag{13.19}$$

where $\dot{C}_E(0)$ is the first derivative of the spectral efficiency with respect to ζ at $\zeta = 0$. The wideband slope S_0 can be obtained from the formula (13.7) by using the first and second derivatives of the spectral efficiency $C_E(\zeta)$ with respect to ζ.

Theorem 13.3. *When only the receiver has CSI, the minimum bit energy and wideband slope, respectively, in the wideband regime are given by*

$$\frac{E_b}{N_0 \text{ min}} = -\frac{\frac{\theta T \bar{P}}{N_0}}{\log_e \mathbb{E}\{e^{-\frac{\theta T \bar{P}}{N_0 \log_e 2} z}\}}, \quad \text{and} \tag{13.20}$$

$$S_0 = 2 \left(\frac{N_0 \log_e 2}{\theta T \bar{P}}\right)^2 \frac{\mathbb{E}\{e^{-\frac{\theta T \bar{P}}{N_0 \log_e 2} z}\} \left(\log_e \mathbb{E}\{e^{-\frac{\theta T \bar{P}}{N_0 \log_e 2} z}\}\right)^2}{\mathbb{E}\{e^{-\frac{\theta T \bar{P}}{N_0 \log_e 2} z} z^2\}}. \tag{13.21}$$

It is interesting to note that in contrast to the low-power regime results, we now have

$$\frac{E_b}{N_0 \text{ min}} = \frac{-\frac{\theta T \bar{P}}{N_0}}{\log_e \mathbb{E}\{e^{-\frac{\theta T \bar{P}}{N_0 \log_e 2} z}\}} \geq \frac{-\frac{\theta T \bar{P}}{N_0}}{\mathbb{E}\{\log_e e^{-\frac{\theta T \bar{P}}{N_0 \log_e 2} z}\}} = \frac{\log_e 2}{\mathbb{E}\{z\}}$$

where Jensen's inequality is used. Therefore, we will be operating above -1.59

dB unless there are no QoS constraints and hence $\theta = 0$. For the Rayleigh fading channel, from (13.20) and (13.21) we obtain

$$\frac{E_b}{N_0}_{\min} = \frac{\frac{\theta T \bar{P}}{N_0}}{\log_e(1 + \frac{\theta T \bar{P}}{N_0 \log_e 2})} \quad \text{and}$$

$$S_0 = \left(\frac{N_0 \log 2}{\theta T \bar{P}} \log(1 + \frac{\theta T \bar{P}}{N_0 \log 2}) + \log(1 + \frac{\theta T \bar{P}}{N_0 \log 2}) \right)^2. \quad (13.22)$$

It can be easily seen that in the Rayleigh fading channel, the minimum bit energy increases monotonically with increasing θ.

13.3.3.2   CSI at Both the Transmitter and Receiver

To analyze $\frac{E_b}{N_0}_{\min}$ in this case, we initially obtain the following result and identify the limiting value of the threshold α as the bandwidth increases to infinity.

Theorem 13.4. *In wideband regime, the threshold α in the optimal power adaptation scheme (24.22) satisfies*

$$\lim_{\zeta \to 0} \alpha(\zeta) = \alpha^* \quad (13.23)$$

where α^* is the solution to

$$\mathbb{E}\left\{ \left[\log\left(\frac{z}{\alpha^*} \right) \frac{1}{z} \right] \tau(\alpha^*) \right\} = \frac{\theta T \bar{P}}{N_0 \log 2}. \quad (13.24)$$

Moreover, for $\theta > 0$, $\alpha^*_{opt} < \infty$.

Remark: As noted before, wideband and low-power regimes are equivalent when $\theta = 0$. Hence, we can easily see in the wideband regime that the threshold α approaches the maximum fading value $z_{\max}$ as $\zeta \to 0$ when $\theta = 0$. Hence, for fading distributions with unbounded support, $\alpha \to \infty$ with vanishing ζ. The threshold being very large means that the transmitter waits sufficiently long until the fading assumes very large values and becomes favorable. That is how arbitrarily small bit energy values can be attained. However, in the presence of QoS constraints, arbitrarily long waiting times will not be permitted. As a result, α approaches a finite value (i.e., $\alpha^*_{opt} < \infty$) as $\zeta \to 0$ when $\theta > 0$. Moreover, from (13.24), we can immediately note that as θ increases, α^*_{opt} has to decrease.

The spectral efficiency with optimal power adaptation is now given by

$$C_E(\zeta) = -\frac{\zeta}{\theta T} \log \left(F(\alpha) + \mathbb{E}\left\{ \left(\frac{z}{\alpha} \right)^{-\frac{\theta T}{\theta T + \zeta \log 2}} \tau(\alpha) \right\} \right) \quad (13.25)$$

where again $F(\alpha) = \mathbb{E}\{1\{z < \alpha\}\}$ and $\tau(\alpha) = 1\{\tau \geq \alpha\}$.

Theorem 13.5. *When both the receiver and transmitter have CSI, the minimum bit energy and wideband slope in the wideband regime are given by*

$$\frac{E_b}{N_0}_{\min} = -\frac{\frac{\theta T \bar{P}}{N_0}}{\log \xi} \quad and \quad \mathcal{S}_0 = \frac{\xi (\log \xi)^2 \log 2}{\theta T (\frac{\bar{P} \alpha_{opt}^*}{N_0} + \dot{\alpha}(0) \mathbb{E}\{\frac{1}{z} \tau(\alpha_{opt}^*)\})} \quad (13.26)$$

where $\xi = F(\alpha^*) + \mathbb{E}\{\frac{\alpha^*}{z} \tau(\alpha^*)\}$, and $\dot{\alpha}(0)$ is the derivative of α with respect to ζ, evaluated at $\zeta = 0$.

It is interesting to note that the minimum bit energy is strictly greater than zero for $\theta > 0$. Hence, the wideband regime is in stark contrast to low-power regime in which the minimum bit energy is zero for fading distributions with unbounded support. Generally speaking, due to power and rate adaptation, $\frac{E_b}{N_0}_{\min}$ in this case is smaller compared to that in the case in which only the receiver has CSI. We note that the presence of CSI at the transmitter is especially beneficial for very small and also large values of θ. Whereas the bit energy in the CSIR case approaches -1.59 dB with vanishing θ, it decreases to $-\infty$ dB when the transmitter knows the channel as well. We note from (13.20) and (13.26) that the minimum bit energy expressions have a common expression in the numerator while the expressions in the denominator are proportional to the asymptotic value of C_E.

13.4   Fixed-Rate/Fixed-Power Transmissions

In this section, we assume that the transmitter does not have channel knowledge and it sends the information at a fixed rate and fixed power. For this fixed rate scenario, we adopt a two state (ON-OFF) transmission model, where information is transmitted reliably at a fixed rate in the ON state while no transmission occurs in the OFF state. We investigate the wideband regime in sparse multipath fading, in which the number of subchannels remains bounded as bandwidth increases, and also in rich multipath fading, in which the number of non-interacting subchannels increases without bound with increasing bandwidth. The minimum bit energy and wideband slope expressions are found for the wideband regime with multipath sparsity. The expressions for bit energy required at zero spectral efficiency and wideband slope are quantified for the low-power regime, which is also equivalent to the wideband regime with rich multipath fading. It is shown for a certain class of fading distributions that the bit energy required at zero spectral efficiency is indeed the minimum bit energy for reliable communications.

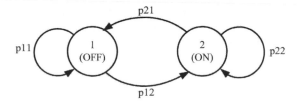

FIGURE 13.2 ON-OFF state transition model

13.4.1 Channel Model

We consider the scenario in which the receiver has perfect channel side information and hence perfectly knows the instantaneous values of $\{h[i]\}$ while the transmitter has no such knowledge. In this case, the instantaneous channel capacity with channel gain $z[i]=|h[i]|^2$ is

$$C[i] = B \log_2(1 + \text{SNR}z[i]) \text{ bits/s} \tag{13.27}$$

where $\text{SNR} = \bar{P}/(N_0B)$ is the average transmitted signal-to-noise ratio. Since the transmitter is unaware of the channel conditions, information is transmitted at a fixed rate of r bits/s. When $r < C$, the channel is considered to be in the ON state and reliable communication is achieved at this rate. From information-theoretic arguments, this is possible if strong codes with large blocklengths is employed in the system. Since there are TB symbols in each block, we assume TB is large enough to establish reliable communication. If, on the other hand, $r \geq C$, outage occurs. In this case, channel is in the OFF state and reliable communication at the rate r bits/s cannot be attained. Hence, the effective data rate is zero and information has to be resent. We assume that a simple ARQ mechanism is incorporated in the communication protocol to acknowledge the reception of data and to ensure that the erroneous data is retransmitted [16].

Fig. 13.2 depicts the two-state transmission model together with the transition probabilities. We assume that the channel fading coefficients stay constant over the frame duration T. Hence, state transitions occur every T seconds. Now, the probability of staying in the ON state, p_{22}, is defined as follows[3]:

$$p_{22} = Pr\{r < C[i+TB] \,|\, r < C[i]\}$$
$$= Pr\{z[i+TB] > \alpha \,|\, z[i] > \alpha\} \tag{13.28}$$

where

$$\alpha = \frac{2^{\frac{r}{B}} - 1}{\text{SNR}}. \tag{13.29}$$

Note that p_{22} depends on the joint distribution of $(z[i+TB], z[i])$. For the

---

[3]The formulation in (13.28) assumes as before that the symbol rate is B symbols/s, hence we have TB symbols in a duration of T seconds.

Rayleigh fading channel, the joint density function of the fading amplitudes can be obtained in closed form [17]. Here, aiming at simplifying the analysis and providing results for arbitrary fading distributions, we assume that fading realizations are independent for each frame[4]. Hence, we again consider a block-fading channel model. Note that in block-fading channels, the duration T over which the fading coefficients stay constant can be varied to model fast or slow fading scenarios.

Under the block fading assumption, we now have $p_{22} = Pr\{z[i + TB] > \alpha\} = Pr\{z > \alpha\}$. Similarly, the other transition probabilities become

$$p_{11} = p_{21} = Pr\{z \leq \alpha\} = \int_0^\alpha p_z(z)dz \quad \text{and}$$

$$p_{22} = p_{12} = Pr\{z > \alpha\} = \int_\alpha^\infty p_z(z)dz \qquad (13.30)$$

where p_z is the probability density function of z. Throughout, we assume that both $p_z(z)$ and the cumulative distribution function $P\{z \leq \alpha\}$ are differentiable. We finally note that rT bits are successfully transmitted and received in the ON state, while the effective transmission rate in the OFF state is zero.

Noting that $p_{11} + p_{22} = 1$ in our model, we express the effective capacity normalized by the frame duration T and bandwidth B, or equivalently spectral efficiency in bits/s/Hz, for a given statistical QoS constraint θ, as

$$\mathsf{R}_E(\mathrm{SNR}, \theta)$$

$$= \frac{1}{TB} \max_{r \geq 0} \left\{ -\frac{\Lambda(-\theta)}{\theta} \right\} \qquad (13.31)$$

$$= \max_{r \geq 0} \left\{ -\frac{1}{\theta TB} \log \left(p_{11} + p_{22} e^{-\theta Tr} \right) \right\} \qquad (13.32)$$

$$= \max_{r \geq 0} \left\{ -\frac{1}{\theta TB} \log \left(1 - P\{z > \alpha\}(1 - e^{-\theta Tr}) \right) \right\} \qquad (13.33)$$

$$= -\frac{1}{\theta TB} \log \left(1 - P\{z > \alpha_{\mathrm{opt}}\} \left(1 - e^{-\theta Tr_{\mathrm{opt}}} \right) \right) \text{ bits/s/Hz} \qquad (13.34)$$

where r_{opt} is the maximum fixed transmission rate that solves (13.33) and $\alpha_{\mathrm{opt}} = (2^{\frac{r_{\mathrm{opt}}}{B}} - 1)/\mathrm{SNR}$. Note that both α_{opt} and r_{opt} are functions of SNR and θ.

The normalized effective capacity, R_E, provides the maximum throughput under statistical QoS constraints in the fixed-rate transmission model. It can easily be shown that

$$\lim_{\theta \to 0} \mathsf{R}_E(\mathrm{SNR}, \theta) = \max_{r \geq 0} \frac{r}{B} P\{z > \alpha\}. \qquad (13.35)$$

---

[4]This assumption also enables us to compare the results of this section with those of Section 13.3 in which variable-rate/variable-power and variable-rate/fixed-power transmission schemes are studied for block fading channels.

Hence, as the QoS requirements relax, the maximum constant arrival rate approaches the average transmission rate. On the other hand, for $\theta > 0$, $\mathsf{R}_E < \frac{1}{B} \max_{r \geq 0} r P\{z > \alpha\}$ in order to avoid violations of QoS constraints.

13.4.2   Energy Efficiency in the Wideband Regime

In this part, we consider the wideband regime in which the bandwidth is large. We assume that the average power $\bar{P}$ is kept constant. Note that as the bandwidth B increases, $\mathsf{SNR} = \frac{\bar{P}}{N_0 B}$ approaches zero and we operate in the low-SNR regime.

Following the approach typically employed in information-theoretic analyses, we assume that the wideband channel is decomposed into N parallel subchannels. We further assume that each subchannel has a bandwidth that is equal to the coherence bandwidth, B_c. Therefore, each subchannel experiences independent flat-fading, and we have $B = NB_c$. Similar to (13.8), the input-output relation in the kth subchannel can be written as

$$y_k[i] = h_k[i] x_k[i] + n_k[i] \quad i = 1, 2, \dots \text{ and } k = 1, 2, \dots, N. \tag{13.36}$$

The fading coefficients $\{h_k\}_{k=1}^n$ in different subchannels are assumed to be independent. The signal-to-noise ratio in the kth subchannel is $\mathsf{SNR}_k = \frac{\bar{P}_k}{N_0 B_c}$ where $\bar{P}_k$ denotes the power allocated to the kth subchannel and we have $\sum_{k=1}^N \bar{P}_k = \bar{P}$. Over each subchannel, the same transmission strategy as described in Section 13.4.1 is employed. Therefore, the transmitter, unaware of the fading coefficients of the subchannels, sends the data over each subchannel at the fixed rate of r. If $r < B_c \log(1 + \mathsf{SNR}_k z_k[i])$ where $z_k = |h_k|^2$, then transmission over the kth subchannel is successful. Otherwise, retransmission is required. Hence, we have an ON-OFF state model for each subchannel. On the other hand, for the transmission over N subchannels, we have a state-transition model with $N+1$ states because we have overall the following $N+1$ possible total transmission rates: $\{0, rT, 2rT, \dots, NrT\}$. For instance, if all N subchannels are in the OFF state simultaneously, the total rate is zero. If j out of N subchannels are in the ON state, then the rate is jrT.

Now, assume that the states are enumerated in the increasing order of the total transmission rates supported by them. Hence, in state $j \in \{1, \dots, N+1\}$, the transmission rate is $(j - 1)rT$. The transition probability from state $i \in \{1, \dots, N + 1\}$ to state $j \in \{1, \dots, N + 1\}$ is given by

$$p_{ij} = p_j = P\{(j - 1) \text{ subchannels out of } N \text{ subchannels are in the ON state}\}$$

$$= \sum_{\mathcal{I}_{j-1} \subset \{1,\dots,N\}} \left(\prod_{k \in \mathcal{I}_{j-1}} P\{z_k > \alpha_k\} \prod_{k \in \mathcal{I}_{j-1}^c} (1 - P\{z_k > \alpha_k\}) \right) \tag{13.37}$$

where $\mathcal{I}_{j-1}$ denotes a subset of the index set $\{1, \dots, N\}$ with $j - 1$ elements.

The summation in (13.37) is over all such subsets. Moreover, in (13.37), $\mathcal{I}_{j-1}^c$ denotes the complement of the set $\mathcal{I}_{j-1}$, and $\alpha_k = \frac{2^{\frac{r}{B_c}}-1}{\text{SNR}_k}$. Note in the above formulation that the transition probabilities, p_{ij}, do not depend on the initial state i due to the block-fading assumption. If, in addition to being independent, the fading coefficients and hence $\{z_k\}_{k=1}^N$ in different subchannels are identically distributed, then p_{ij} in (13.37) simplifies and becomes a binomial probability:

$$p_{ij} = p_j = \binom{N}{j-1} \left(P\{z > \alpha\}\right)^{j-1} \left(1 - P\{z > \alpha\}\right)^{N-j+1}. \qquad (13.38)$$

Note that if the fading coefficients are i.i.d., the total power should be uniformly distributed over the subchannels. Hence, in this case, we have $\bar{P}_k = \frac{\bar{P}}{N}$ and therefore $\text{SNR}_k = \frac{\bar{P}_k}{N_0 B_c} = \frac{\bar{P}/N}{N_0 B/N} = \frac{\bar{P}}{N_0 B} = \text{SNR}$ which is equal to the original SNR definition used in (13.27). Now, we have the same $\alpha = \frac{2^{\frac{r}{B_c}}-1}{\text{SNR}}$ for each subchannel.

The effective capacity of this wideband channel is given in the following theorem.

Theorem 13.6. *For the wideband channel with N parallel noninteracting subchannels each with bandwidth B_c and independent flat fading, the normalized effective capacity in bits/s/Hz is*

$$\mathsf{R}_E(\text{SNR}, \theta)$$

$$= \max_{\substack{r \geq 0 \\ \bar{P}_k \geq 0 \ s.t. \ \sum \bar{P}_k \leq \bar{P}}} \left\{ -\frac{1}{\theta T B} \log \left(\sum_{j=1}^{N+1} p_j \, e^{-\theta(j-1)rT} \right) \right\} \qquad (13.39)$$

where p_j is given in (13.37). If $\{z_k\}_{k=1}^N$ are identically distributed, then the normalized effective capacity expression simplifies to

$$\mathsf{R}_E(\text{SNR}, \theta)$$

$$= \max_{r \geq 0} \left\{ -\frac{1}{\theta T B_c} \log \left(1 - P\{z > \alpha\}(1 - e^{-\theta T r}) \right) \right\}. \qquad (13.40)$$

where $\alpha = \frac{2^{\frac{r}{B_c}}-1}{\text{SNR}}$ and $\text{SNR} = \frac{\bar{P}}{N_0 B}$.

Theorem 13.6 shows that the effective capacity of a wideband channel with N subchannels each with i.i.d. flat fading has an expression similar to (13.33), which provides the effective capacity of a single channel experiencing flat fading. The only difference between (13.33) and (13.40) is that B is replaced in (13.40) by B_c, which is the bandwidth of each subchannel.

In this part, we consider the wideband regime in which the overall bandwidth of the system, B, is large. In particular, we analyze the performance in the scenario of sparse multipath fading. Motivated by the recent measurement studies in the ultrawideband regime, the authors in [18] and [19] considered sparse multipath fading channels and analyzed the performance under channel uncertainty, employing the Shannon capacity formulation as the performance metric. In particular, [18] and [19] noted that the number of independent resolvable paths in sparse multipath channels increase at most sublinearly with the bandwidth, which in turn causes the coherence bandwidth B_c to increase with increasing bandwidth. To characterize the performance of sparse fading channels in the wideband regime, we assume in this section that $B_c \to \infty$ as $B \to \infty$. We further assume that the the number of subchannels N remains bounded, hence the degrees of freedom are limited. For instance, this case arises if the number of resolvable paths are bounded even at infinite bandwidth. Such a scenario is considered in [20] where the capacity and mutual information are characterized under channel uncertainty in the wideband regime with bounded number of paths.

The case of rich multipath fading in which the B_c remains fixed and N grows without bound and the case in which both B_c and N increase to infinity are treated in Section 13.4.3 because each subchannel in these scenarios operates in the low-power regime as N increases.

We first introduce the notation $\zeta = \frac{1}{B_c}$. Note that as $B_c \to \infty$, we have $\zeta \to 0$. Moreover, with this notation, the normalized effective capacity in (13.40) given for i.i.d. fading can, after maximization, be expressed as

$$R_E(\text{SNR}) = -\frac{\zeta}{\theta T} \log\left(1 - P\{z > \alpha_{\text{opt}}\}\left(1 - e^{-\theta T r_{\text{opt}}}\right)\right). \tag{13.41}$$

Note that α_{opt} and r_{opt} also in general depend on B_c, and hence ζ. The following result provides the expressions for the minimum bit energy, which is achieved at zero spectral efficiency (i.e., as $B \to \infty$ and $B_c \to \infty$), and the wideband slope, and characterizes the spectral efficiency-bit energy tradeoff in the wideband regime when multipath fading is sparse, the number of subchannels is bounded, and the fading coefficients are i.i.d. in different subchannels.

Theorem 13.7. *In sparse multipath fading wideband channels with bounded number of subchannels each with i.i.d. fading coefficients, the minimum bit energy and wideband slope are given by*

$$\frac{E_b}{N_0}_{\min} = \frac{-\delta \log 2}{\log \xi} \quad \text{and} \tag{13.42}$$

$$S_0 = \frac{2\xi \log^2 \xi}{(\delta \alpha_{\text{opt}}^*)^2 P\{z > \alpha_{\text{opt}}^*\} e^{-\delta \alpha_{\text{opt}}^*}}, \tag{13.43}$$

respectively, where $\delta = \frac{\theta T \bar{P}}{N N_0 \log 2}$ and $\xi = 1 - P\{z > \alpha_{\text{opt}}^*\}(1 - e^{-\delta \alpha_{\text{opt}}^*})$. α_{opt}^* is defined as $\alpha_{\text{opt}}^* = \lim_{\zeta \to 0} \alpha_{\text{opt}}$ and α_{opt}^* satisfies

$$\delta \alpha_{\text{opt}}^* = \log\left(1 + \delta \frac{P\{z > \alpha_{\text{opt}}^*\}}{p_z(\alpha_{\text{opt}}^*)}\right). \tag{13.44}$$

Remark: Theorem 13.7, through the minimum bit energy and wideband slope expressions, quantifies the bit energy requirements in the wideband regime when the system is operating subject to statistical QoS constraints specified by θ. Note that both $\frac{E_b}{N_0}_{\min}$ and $\mathcal{S}_0$ depend on the QoS exponent θ through δ. More specifically, $\frac{E_b}{N_0}_{\min}$ and the bit energy requirements at nonzero spectral efficiency values generally increase with increasing θ. Moreover, when compared with the results in Section 13.4.3, it will be seen that sparse multipath fading and having a bounded number of subchannels incur energy penalty in the presence of QoS constraints while performances do not depend on the multipath sparsity when there are no such constraints and thus $\theta = 0$.

13.4.3   Energy Efficiency In The Low-Power Regime

In this section, we investigate the spectral efficiency - bit energy tradeoff in a single flat-fading channel as the average power $\bar{P}$ diminishes. We assume that the bandwidth allocated to the channel is fixed. Note that $\text{SNR} = \bar{P}/(N_0 B)$ vanishes with decreasing $\bar{P}$, and we again operate in the low-SNR regime similarly as in Section 13.4.2. Note further from (13.34) that the effective capacity of a flat-fading channel is given by

$$\text{R}_E(\text{SNR}) = -\frac{1}{\theta T B} \log\left(1 - P\{z > \alpha_{\text{opt}}\}\left(1 - e^{-\theta T r_{\text{opt}}}\right)\right). \tag{13.45}$$

On the other hand, we remark that the results derived here also apply to the wideband regime under the assumption that the number of non-interacting subchannels increases without bound with increasing bandwidth. Note that in that case, each subchannel operates in the low-power regime.

The following result provides expressions for the bit energy at zero spectral efficiency and the wideband slope.

Theorem 13.8. *In the low-power regime, the bit energy at zero spectral efficiency and wideband slope are given by*

$$\left.\frac{E_b}{N_0}\right|_{\text{R}_E=0} = \frac{\log 2}{\alpha_{opt}^* P\{z > \alpha_{opt}^*\}} \quad and \tag{13.46}$$

$$\mathcal{S}_0 = \frac{2P\{z > \alpha_{opt}^*\}}{1 + \beta(1 - P\{z > \alpha_{opt}^*\})}, \tag{13.47}$$

respectively, where $\beta = \frac{\theta T B}{\log 2}$ is the normalized QoS constraint. In the above formulation, α_{opt}^* is defined as $\alpha_{opt}^* = \lim_{\text{SNR}\to 0} \alpha_{opt}$, and α_{opt}^* satisfies

$$\alpha_{opt}^* p_z(\alpha_{opt}^*) = P\{z > \alpha_{opt}^*\}. \tag{13.48}$$

Corollary 13.1. *The same bit energy and wideband slope expressions as in (13.46) and (13.47) are achieved in the wideband regime as $B \to \infty$ if the fading coefficients in different subchannels are i.i.d. and also if the number of subchannels N increases linearly with increasing bandwidth (as in rich multipath fading channels), keeping the coherence bandwidth fixed.*

Under the assumptions stated in Corollary 13.1, the effective capacity is given by (13.40). Moreover, as $B \to \infty$, we have B_c fixed while $N \to \infty$. Hence, $\text{SNR} = \frac{\bar{P}/N}{N_0 B_c} \to 0$. This setting is exactly the same as the low-power regime considered in Theorem 13.8. Therefore, the results of Theorem 13.8 apply immediately.

Next, we show that equation (13.48), which needs to be satisfied by α_{opt}^*, has a unique solution for a certain class of fading distributions.

Theorem 13.9. *Assume that the probability density function of z, denoted by $p_z(\cdot)$, is differentiable, and that both $p_z(\cdot)$ and its derivative $\dot{p}_z(\cdot)$ do not contain impulses or higher-order singularities at the origin and are finite. Assume further that the support of $p_z(\cdot)$ is $[0, \infty)$. Under these assumptions, if $2p_z(x) + x\dot{p}_z(x) = 0$ has a solution at a single point $x_0 > 0$ among all $x \in (0, \infty)$, then the equation $\alpha_{\text{opt}}^* p_z(\alpha_{\text{opt}}^*) = P\{z > \alpha_{\text{opt}}^*\}$ has a unique solution.*

Remark: The conditions of Theorem 13.9 are satisfied by a general class of distributions, including the Gamma distribution,

$$p_z(z) = \frac{z^{\alpha-1} e^{-\frac{z}{\beta}}}{\beta^\alpha \Gamma(\alpha)},$$

where $z, \alpha, \beta > 0$, and Log-normal distribution,

$$p_z(z) = \frac{1}{\sigma z \sqrt{2\pi}} e^{-\frac{(\log z - m)^2}{2\sigma^2}},$$

where $z > 0$, $-\infty < m < \infty$, and $\sigma > 0$. Note that in Nakagami-m and Rayleigh fading channels, the distribution of $z = |h|^2$ can be seen as special cases of the Gamma distribution.

Remark: Theorem 13.8 shows that the $\left.\frac{E_b}{N_0}\right|_{R_E=0}$ for any $\theta \geq 0$ depends only on α_{opt}^*. From Theorem 13.9, we know under certain conditions that α_{opt}^* is unique and hence is the same for all $\theta \geq 0$. We immediately conclude from these results that $\left.\frac{E_b}{N_0}\right|_{R_E=0}$ also has the same value for all $\theta \geq 0$ and therefore does not depend on θ for the class of distributions and channels given in the previous Remark.

Moreover, using the results of Theorem 13.9 above and Theorem 13.7 in Section 13.4.2, we can further show that $\left.\frac{E_b}{N_0}\right|_{R_E=0}$ is the minimum bit energy. Note that this implies that the same minimum bit energy can be attained regardless of how strict the QoS constraint is. On the other hand, we note that the wideband slope $\mathcal{S}_0$ in general varies with θ.

Corollary 13.2. *In the low-power regime, when $\theta = 0$, the minimum bit energy is achieved as $\bar{P} \to 0$, i.e., $\left.\frac{E_b}{N_0}\right|_{R_E=0} = \frac{E_b}{N_0}_{\min}$. Moreover, if the probability density function of z satisfies the conditions stated in Theorem 13.9, then the* *minimum bit energy is achieved as* $\bar{P} \to 0$, *i.e.* $\left.\frac{E_b}{N_0}\right|_{R_E=0} = \frac{E_b}{N_0 \min}$, *for all* $\theta \geq 0$.

Remark: From the result of Corollary 13.1, we note that the analytical results in this section apply to wideband channels with rich multipath fading. Through numerical results provided in [2], we can make several insightful observations. We find that in the absence of QoS constraints, multipath sparsity or richness has no effect, which confirms the claim in the proof of Corollary 13.2 that low-power and wideband regimes are equivalent when $\theta = 0$. However, we see a stark contrast when $\theta > 0$. We observe that multipath sparsity and the bound on the number of subchannels in the wideband regime increases the bit energy requirements significantly, especially when θ is large.

In Section 13.4.2, the number of subchannels is assumed to be bounded. In this section, we have considered the rich multipath fading channels in which the number of subchannels increases linearly with bandwidth. A scenario in between these two cases is one in which the number of subchannels N increases only sublinearly with increasing bandwidth. As N increases, less power is allocated to each subchannel and the subchannels operate in the low-power regime. At the same time, since N increases sublinearly with B, the coherence bandwidth $B_c = B/N$ also increases. Therefore, the minimum bit energy and wideband slope expressions for this scenario can be obtained by letting B in the results of Theorem 13.8 go to infinity. Note that under the conditions of Theorem 13.9, α^*_{opt} is unique and does not depend on the bandwidth.

Corollary 13.3. *In the wideband regime, if the number of subchannels N increases sublinearly with B and if fading coefficients in different subchannels are i.i.d. and the probability density function p_z satisfies the conditions in Theorem 13.9, then the minimum bit energy and wideband slope are given by*

$$\frac{E_b}{N_0 \min} = \frac{\log 2}{\alpha^*_{\mathrm{opt}} P\{z > \alpha^*_{\mathrm{opt}}\}} \quad and \tag{13.49}$$

$$\mathcal{S}_0 = \begin{cases} 2P\{z > \alpha^*_{\mathrm{opt}}\} & \theta = 0 \\ 0 & \theta > 0 \end{cases}. \tag{13.50}$$

From this result, we see that although the same minimum bit energy is attained for all $\theta \geq 0$, approaching this minimum energy level is extremely slow and demanding when $\theta > 0$ due to zero wideband slope.

████████

13.5　Transmissions over Imperfectly-Known Wireless Channels

In this section, we consider the scenario in which neither the transmitter nor the receiver has CSI prior to transmission and the channel coefficients are estimated at the receiver via minimum mean-square-error (MMSE) estimation with the aid of training symbols. For this scenario, we identify the optimal fraction of power allocated to training. We show that the bit energy increases without bound in the low-power regime as the average power vanishes. A similar conclusion is reached in the wideband regime if the number of non-interacting subchannels grows without bound with increasing bandwidth. On the other hand, it is proven that if the number of resolvable independent paths and hence the number of noninteracting subchannels remain bounded as the available bandwidth increases, the bit energy diminishes to its minimum value in the wideband regime.

13.5.1 Training and Data Transmission Phases

As mentioned above, we assume that the transmitter and the receiver initially have no channel side information. While the transmitter remains unaware of the actual realizations of the fading coefficients throughout the transmission, the receiver attempts to learn them through training.

13.5.1.1 Training Phase

The system operates in two phases: training phase and data transmission phase. In the training phase, known pilot symbols are transmitted to enable the receiver to estimate the channel conditions, albeit imperfectly. We assume that minimum mean-square-error (MMSE) estimation is employed at the receiver to estimate the channel coefficient $h[i]$. Since the MMSE estimate depends only on the training energy and not on the training duration [21] and the fading coefficients are assumed to stay constant during the frame duration of T seconds, it is easily seen that transmission of a single pilot at every T seconds is optimal. Note that in every frame duration of T seconds, we have TB symbols and the overall available energy is $\bar{P}T$. We now assume that each frame consists of a pilot symbol and $TB - 1$ data symbols. The energies of the pilot and data symbols are

$$\mathcal{E}_p = \rho \bar{P}T, \quad \text{and} \quad \mathcal{E}_s = \frac{(1 - \rho)\bar{P}T}{TB - 1}, \tag{13.51}$$

respectively, where ρ is the fraction of total energy allocated to training. Note that the data symbol energy $\mathcal{E}_s$ is obtained by uniformly allocating the remaining energy among the data symbols.

In the training phase, the transmitter sends the pilot symbol $x_p = \sqrt{\mathcal{E}_p} = \sqrt{\rho \bar{P}T}$ and the receiver obtains[5]

$$y[1] = h\sqrt{\mathcal{E}_p} + n[1]. \tag{13.52}$$

---

[5]Since the analysis in this section focuses on a single frame in which the fading stays constant, we drop the time index in $h[i]$ and express the fading coefficient as h. In (13.52), $y[1]$ and $n[1]$ denote the received symbol and noise sample, respectively, in the training phase. Note that the first symbol duration in each frame is allocated to the training phase in which a single pilot symbol is sent.

Based on the received signal in this phase, the receiver obtains the MMSE estimate $h_{est} = \mathbb{E}\{h|y[1]\}$ which as can easily be seen is a circularly symmetric, complex, Gaussian random variable with mean zero and variance $\frac{\gamma^2 \mathcal{E}_p}{\gamma \mathcal{E}_p + N_0}$, i.e., $h_{est} \sim \mathcal{CN}\left(0, \frac{\gamma^2 \mathcal{E}_p}{\gamma \mathcal{E}_p + N_0}\right)$ [22]. Now, the channel fading coefficient h can be expressed as $h = h_{est} + h_{err}$, where h_{err} is the estimate error and $h_{err} \sim \mathcal{CN}(0, \frac{\gamma N_0}{\gamma \mathcal{E}_p + N_0})$.

13.5.1.2   Data Transmission Phase and Capacity Lower Bound

Data transmission follows the training phase. Since the receiver is now equipped with the channel estimate, the channel input-output relation in one frame in the data transmission phase can be expressed as

$$y[i] = h_{est}x[i] + h_{err}x[i] + n[i] \quad i = 2, 3, \ldots, TB. \tag{13.53}$$

Since finding the capacity of the channel in (13.53) is a difficult task [22], a capacity lower bound is generally obtained by treating $h_{err}x[i] + n[i]$ as Gaussian distributed noise with variance $\mathbb{E}\{|h_{err}x[i] + n[i]|^2\} = \sigma_{h_{err}}^2 \mathcal{E}_s + N_0$, where $\sigma_{h_{err}}^2 = \mathbb{E}\{|h_{err}|^2\} = \frac{\gamma N_0}{\gamma \mathcal{E}_p + N_0}$ is the variance of the estimate error. Under these assumptions, a lower bound on the instantaneous capacity is given by [21], [22]

$$C_L = \frac{TB - 1}{T} \log_2 \left(1 + \frac{\mathcal{E}_s}{\sigma_{h_{err}}^2 \mathcal{E}_s + N_0} |h_{est}|^2 \right)$$

$$= \frac{TB - 1}{T} \log_2 \left(1 + \text{SNR}_{\text{eff}}|w|^2\right) \text{ bits/s} \tag{13.54}$$

where the effective SNR is

$$\text{SNR}_{\text{eff}} = \frac{\mathcal{E}_s \sigma_{h_{est}}^2}{\sigma_{h_{err}}^2 \mathcal{E}_s + N_0}, \tag{13.55}$$

and $\sigma_{h_{est}}^2 = \mathbb{E}\{|h_{est}|^2\} = \frac{\gamma^2 \mathcal{E}_p}{\gamma \mathcal{E}_p + N_0}$ is the variance of the estimate h_{est}. Note that the expression in (13.54) is obtained by defining $h_{est} = \sigma_{h_{est}} w$, where w is a standard complex Gaussian random variable with zero mean and unit variance, i.e., $w \sim \mathcal{CN}(0, 1)$. Henceforth, we base our analysis on C_L to understand the impact of the imperfect channel estimate.

13.5.1.3   Fixed-Rate Transmission and ON-OFF Model

Since the transmitter is unaware of the channel conditions, it is assumed that information is transmitted at a fixed rate of r bits/s. When $r < C_L$, the channel is considered to be in the ON state and reliable communication is achieved at this rate. Note that under the block-fading assumption, the channel stays in the ON state for T seconds and the number of bits transmitted in this duration is rT. If, on the other hand, $r \geq C_L$, we assume that outage occurs. In this case, channel is in the OFF state during the frame duration and no reliable communication at the rate of r bits/s can be attained. Hence, the effective data rate is zero and information has to be resent. The probability of the channel being in the OFF state is

$$p_{\text{off}} = \Pr\{r \geq C_L\} = 1 - e^{-\alpha} \tag{13.56}$$

where

$$\alpha = \frac{2^{\frac{rT}{TB-1}} - 1}{\text{SNR}_{\text{eff}}}. \tag{13.57}$$

The rightmost expression in (13.56) follows from the fact that $|w|^2$ is an exponential random variable with mean 1. Since $|w|^2$ is the normalized estimated channel strength, the channel is in the OFF state if this channel strength is less than the threshold α. Similarly, the probability of being in the ON state is

$$p_{\text{on}} = \Pr\{r < C_L\} = \Pr\{|w|^2 > \alpha\} = e^{-\alpha}. \tag{13.58}$$

We finally remark that since the fading coefficients (and consequently h_{est}, w, and C_L) change independently from one frame to another under the block-fading assumption in any given frame is either in the ON or OFF state irrespective of its previous state.

As shown in Section 13.4.1, the effective capacity normalized by the frame duration T and bandwidth B, or equivalently spectral efficiency in bits/s/Hz, for a given QoS delay constraint as a function of θ is given by

$$
\begin{aligned}
R_E(\text{SNR}, \theta) &= \max_{\substack{r \geq 0 \\ 0 \leq \rho \leq 1}} -\frac{1}{\theta TB} \log\left(p_{\text{off}} + p_{\text{on}} e^{-\theta Tr}\right) \\
&= \max_{\substack{r \geq 0 \\ 0 \leq \rho \leq 1}} -\frac{1}{\theta TB} \log\left(1 - e^{-\alpha}(1 - e^{-\theta Tr})\right) \\
&= -\frac{1}{\theta TB} \log\left(1 - e^{-\alpha_{\text{opt}}}(1 - e^{-\theta Tr_{\text{opt}}})\right)
\end{aligned} \tag{13.59}
$$

where r_{opt} and α_{opt} are the optimal values of r and α, respectively, and p_{on} and p_{off} are the probabilities of the channel being in the ON and OFF states, respectively. Note that the optimal values r_{opt} and α_{opt} are functions of SNR in general. Note further that R_E is obtained by optimizing both the fixed transmission rate r and the fraction of power allocated to training, ρ. The dependence of the normalized effective capacity on ρ is through the threshold

α which depends on SNR_{eff}. Also, it can easily be seen that

$$R_E(\text{SNR}, 0) = \lim_{\theta \to 0} R_E(\text{SNR}, \theta)$$

$$= \max_{\substack{r \geq 0 \\ 0 \leq \rho \leq 1}} \frac{r}{B} \Pr\left\{ |w|^2 > \frac{2^{\frac{rT}{TB-1}} - 1}{\text{SNR}_{\text{eff}}} \right\}$$

$$= \max_{\substack{r \geq 0 \\ 0 \leq \rho \leq 1}} \frac{r}{B} exp\left\{ -\frac{2^{\frac{rT}{TB-1}} - 1}{\text{SNR}_{\text{eff}}} \right\}. \tag{13.60}$$

Hence, as the QoS requirements relax, the maximum constant arrival rate approaches the average transmission rate. On the other hand, for $\theta > 0$, $R_E < \frac{1}{B} \max_{\substack{r \geq 0 \\ 0 \leq \rho \leq 1}} re^{-\alpha}$ in order to avoid violations of buffer constraints.

13.5.1.4 Optimal Training Power

Before performing the energy efficiency analysis, we first obtain the following result on the optimal value of ρ (the fraction of the total energy allocated to training) in the presence of QoS constraints.

Theorem 13.10. *At a given SNR level, the optimal fraction of power ρ_{opt} that maximizes expression (13.59) does not depend on the QoS exponent θ and the transmission rate r, and is given by*

$$\rho_{opt} = \sqrt{\eta(\eta + 1)} - \eta, \tag{13.61}$$

where

$$\eta = \frac{\gamma T B SNR + TB - 1}{\gamma TB(TB - 2)SNR} \quad and \quad SNR = \frac{\bar{P}}{N_0 B}. \tag{13.62}$$

13.5.2 Energy Efficiency in the Low-Power Regime

In this section, we analyze the spectral efficiency-bit energy tradeoff in the low power regime, in which the average power of the system, $\bar{P}$, is small.

With the optimal value of ρ given in Theorem 13.10, we can now express the normalized effective capacity as

$$R_E(\text{SNR}, \theta) = \max_{r \geq 0} -\frac{1}{\theta TB} \log\left(1 - exp\left\{ -\frac{2^{\frac{rT}{TB-1}} - 1}{\text{SNR}_{\text{eff,opt}}} \right\} (1 - e^{-\theta Tr}) \right) \tag{13.63}$$

$$= -\frac{1}{\theta TB} \log\left(1 - exp\left\{ -\frac{2^{\frac{r_{opt}T}{TB-1}} - 1}{\text{SNR}_{\text{eff,opt}}} \right\} (1 - e^{-\theta Tr_{opt}}) \right) \tag{13.64}$$

where r_{opt} is the optimal r in (13.63), and

$$\text{SNR}_{\text{eff,opt}} = \frac{\phi(\text{SNR})\text{SNR}^2}{\psi(\text{SNR})\text{SNR} + TB - 1}, \tag{13.65}$$

and

$$\phi(\text{SNR}) = \rho_{\text{opt}}(1 - \rho_{\text{opt}})\gamma^2 T^2 B^2,$$
$$\psi(\text{SNR}) = (1 + (TB - 2)\rho_{\text{opt}})\gamma TB. \tag{13.66}$$

With these notions, we obtain the following result that shows us that operation at very low power levels is extremely energy inefficient and should be avoided.

Theorem 13.11. *In the presence of channel uncertainty, the bit energy for all $\theta \geq 0$ increases without bound as the average power $\bar{P}$ and hence SNR vanishes, i.e.,*

$$\left.\frac{E_b}{N_0}\right|_{R_E=0} = \lim_{SNR \to 0} \frac{E_b}{N_0} = \lim_{SNR \to 0} \frac{SNR}{R_E(SNR)} = \frac{1}{\dot{R}_E(0)} = \infty. \tag{13.67}$$

Remark: Theorem 13.11 shows that $\left.\frac{E_b}{N_0}\right|_{R_E=0} = \infty$ for any $\theta \geq 0$. Note that this result should be interpreted with caution. Energy efficiency still improves if one operates at low power levels. However, if the power is reduced below a certain threshold, bit energy requirements start increasing and the required bit energy level grows without bound as power vanishes. One reason for this behavior is that although channel estimation at very low power levels does not provide reliable estimates, the receiver regards this estimate as perfect. Hence, in the low power regime, we have both diminishing power and deteriorating channel estimate, which affect the performance adversely. The result of Theorem 13.11 also indicates that the minimum bit energy is achieved at a non-zero power level, and we observe through numerical analysis that both the minimum required bit energy and the other bit energy values required at a given level of spectral efficiency increase as the QoS constraints become more stringent.

13.5.3 Energy Efficiency in the Wideband Regime

In this section, we consider the wideband regime in which the bandwidth is large. We assume that the average power $\bar{P}$ is kept constant.

13.5.3.1 Decomposing the Wideband Channel into Narrowband Subchannels

Similarly as in Section 13.4.2, we decompose the wideband channel into N parallel subchannels, and assume that each subchannel has a bandwidth equal to the coherence bandwidth, B_c. Then, we can assume independent flat-fading in each subchannel. Note that we have $B = NB_c$. Similar to (13.8), the input-output relation in the kth subchannel can be written as

$$y_k[i] = h_k[i]x_k[i] + n_k[i] \quad i = 1, 2, \ldots \quad \text{and} \quad k = 1, 2, \ldots, N. \tag{13.68}$$

The fading coefficients $\{h_k\}_{k=1}^N$ in different subchannels are assumed to be independent zero-mean Gaussian distributed with variances $\mathbb{E}\{|h_k|^2\} = \gamma_k$.

The signal-to-noise ratio in the kth subchannel is $\text{SNR}_k = \frac{\bar{P}_k}{N_0 B_c}$ where $\bar{P}_k$ denotes the power allocated to the kth subchannel and satisfy $\sum_{k=1}^{N} \bar{P}_k = \bar{P}$ [6]. Over each subchannel, the transmission strategy described in Section 13.5.1.3 is employed. Therefore, the transmitter, unaware of the fading coefficients of the subchannels, sends the data over each subchannel at the fixed rate r. Now, we can find that $C_{L,k}$ for each subchannel is given by $\frac{TB_c - 1}{T} \log_2 \left(1 + \text{SNR}_{\text{eff},k} |w|^2 \right)$ bits/s, in which

$$\text{SNR}_{\text{eff},k} = \frac{\mathcal{E}_{s,k} \sigma^2_{h_{k,est}}}{\sigma^2_{h_{k,err}} \mathcal{E}_{s,k} + N_0} \tag{13.69}$$

where $\mathcal{E}_{s,k} = \frac{(1-\rho_k)T\bar{P}_k}{TB_c - 1}$, $\mathcal{E}_{p,k} = \rho_k T \bar{P}_k$, $\sigma^2_{h_{k,err}} = \frac{\gamma_k N_0}{\gamma_k \mathcal{E}_{p,k} + N_0}$ and $\sigma^2_{h_{k,est}} = \frac{\gamma_k^2 \mathcal{E}_{p,k}}{\gamma_k \mathcal{E}_{p,k} + N_0}$. As described above, if $r < C_{L,k}$, then transmission over the kth subchannel is successful. Otherwise, retransmission is required. Hence, we have ON and OFF states for each subchannel. As also discussed in Section 13.4.2, for the transmission over N subchannels we have a state model with $N + 1$ states, because we have overall the following $N+1$ possible total transmission rates: $\{0, rT, 2rT, \ldots, NrT\}$.

We again assume that the states are enumerated in increasing order of the total transmission rates supported by them. Hence, in state $j \in \{1, \ldots, N+1\}$, the transmission rate is $(j-1)rT$. The probability of being in state

$$j \in \{1, \ldots, N+1\}$$

is given by

$$q_j = \Pr\{(j-1) \text{ subchannels out of } N \text{ subchannels are in the ON state}\}$$

$$= \sum_{\mathcal{I}_{j-1} \subset \{1,\ldots,N\}} \left(\prod_{k \in \mathcal{I}_{j-1}} \Pr\{|w|^2 > \alpha_k\} \prod_{k \in \mathcal{I}_{j-1}^c} (1 - \Pr\{|w|^2 > \alpha_k\}) \right) \tag{13.70}$$

$$= \sum_{\mathcal{I}_{j-1} \subset \{1,\ldots,N\}} \left(\prod_{k \in \mathcal{I}_{j-1}} e^{-\alpha_k} \prod_{k \in \mathcal{I}_{j-1}^c} (1 - e^{-\alpha_k}) \right) \tag{13.71}$$

where $\mathcal{I}_{j-1}$ denotes a subset of the index set $\{1, \ldots, N\}$ with $j-1$ elements. The summation in (13.71) is over all such subsets. Also, in (13.71), $\mathcal{I}_{j-1}^c$ denotes the complement of the set $\mathcal{I}_{j-1}$, and $\alpha_k = 2^{\frac{rT}{TB_c - 1}} - 1/\text{SNR}_{\text{eff},k}$. Note that in the above formulation, similarly as in Section 13.5.1.3, the probability of currently being in state j, i.e., q_j does not depend on the state in the previous

---

[6]While not equipped with the knowledge of the instantaneous values of the fading coefficients, the transmitter is assumed to know the statistics of the fading coefficients, and possibly allocate different power levels to different subchannels with this knowledge.

frame, again due to the block-fading assumption. Moreover, the product form inside the summation in (13.70) is due to the noninteracting subchannels. If fading in different subchannels is correlated, q_j can be written as

$$
q_j = \sum_{\mathcal{I}_{j-1} \subset \{1,\dots,N\}} \left(\Pr \left\{ \left(\bigcap_{k \in \mathcal{I}_{j-1}} \{|w_k|^2 > \alpha_k\} \right) \cap \left(\bigcap_{k \in \mathcal{I}_{j-1}^c} \{|w_k|^2 \le \alpha_k\} \right) \right\} \right)
$$
$$(13.72)$$

which, in general, depends on the joint distribution of $\{|w_1|^2, \dots, |w_N|^2\}$.

If, in addition to being independent, the fading coefficients h_k in different subchannels are identically distributed (i.e., the variances $\{\gamma_k\}_{k=1}^N$ are the same) and provided that the total power is uniformly distributed over the subchannels and ρ_k (the fraction of power allocated to training) is the same in each subchannel, then q_j in (13.71) simplifies to a binomial probability:

$$
q_j = \binom{N}{j-1} \left(\Pr\{|w|^2 > \alpha\} \right)^{j-1} \left(1 - \Pr\{|w|^2 > \alpha\} \right)^{N-j+1}
$$
$$
= \binom{N}{j-1} \left(e^{-\alpha} \right)^{j-1} \left(1 - e^{-\alpha} \right)^{N-j+1}. \tag{13.73}
$$

Note that with equal power allocation, we have $\bar{P}_k = \frac{\bar{P}}{N}$ and therefore $\mathrm{SNR}_k = \frac{\bar{P}_k}{N_0 B_c} = \frac{\bar{P}/N}{N_0 B/N} = \frac{\bar{P}}{N_0 B} = \mathrm{SNR}$, which is equal to the original SNR used in (13.62). Since $\{\mathrm{SNR}_{\mathrm{eff},k}\}_{k=1}^N$ are also equal due to having equal ρ_k's, we have the same $\alpha = \frac{2^{\frac{rT}{TB_c-1}} - 1}{\mathrm{SNR}_{\mathrm{eff}}}$ for each subchannel.

The effective capacity of the wideband channel with N subchannels is given by the following result.

Corollary 13.4. *For the wideband channel with N parallel noninteracting subchannels, each with bandwidth B_c and independent flat fading, the normalized effective capacity in bits/s/Hz is given by*

$$
R_E(SNR, \theta) = \max_{\substack{r \ge 0 \\ \bar{P}_k \ge 0 \ s.t. \ \sum \bar{P}_k \le \bar{P} \\ 0 \le \rho_k \le 1 \ \forall k}} \left\{ -\frac{1}{\theta TB} \log \left(\sum_{j=1}^{N+1} q_j \, e^{-\theta(j-1)rT} \right) \right\} \tag{13.74}
$$

where q_j is given in (13.71). If $\{h_k\}_{k=1}^N$ are identically distributed Gaussian random variables with zero mean and variance γ and the data and training energies are uniformly allocated over the subchannels, then the normalized effective capacity expression simplifies to

$$
R_E(SNR, \theta) = \max_{\substack{r \ge 0 \\ 0 \le \rho \le 1}} \left\{ -\frac{1}{\theta TB_c} \log \left(1 - e^{-\alpha}(1 - e^{-\theta Tr}) \right) \right\}. \tag{13.75}
$$

where $\alpha = \frac{2^{\frac{rT}{TB_c-1}} - 1}{\mathrm{SNR}_{\mathrm{eff}}}$ and $SNR_{\mathrm{eff}} = \frac{\rho(1-\rho)\gamma^2 T^2 B_c^2 SNR^2}{\rho\gamma TB_c(TB_c-2)SNR + \gamma TB_c SNR + TB_c - 1}$, in which $SNR = \frac{\bar{P}}{N_0 B} = \frac{\bar{P}}{N N_0 B_c}$.

Remark: Although we concentrate on noninteracting subchannels, the effective capacity result in (13.74) is general and holds for the case in which the fading in different subchannels are correlated and q_j is given as in (13.72).

Remark: Corollary 13.4 shows that if the fading coefficients in different subchannels are i.i.d. and the data and training energies are uniformly allocated over the subchannels, then the effective capacity of a wideband channel has an expression similar to that in (13.59), which provides the effective capacity of a single channel experiencing flat fading. The only difference between (13.59) and (13.75) is that B is replaced in (13.75) by B_c, which is the bandwidth of each subchannel.

13.5.3.2   Rich and Sparse Multipath Fading Scenarios

After the characterization in Corollary 13.4, we henceforth limit our analysis to the case in which the effective capacity is given by (13.75) because optimization over the power allocation schemes and obtaining closed-form expressions are in general difficult tasks in the wideband regime in which the number of subchannels is potentially high. Under these assumptions, we investigate two scenarios:

1. *Rich multipath fading*: In this case, we assume that the number of independent resolvable paths increases linearly with the bandwidth. This in turn implies that as the bandwidth B increases, the number of noninteracting subchannels N increases while B_c stays fixed.

2. *Sparse multipath fading*: In this case, we assume that the number of independent resolvable paths increases *at most sublinearly* with the bandwidth. This assumption implies the coherence bandwidth $B_c = \frac{B}{N}$ increases with increasing bandwidth B [18], [19]. We can identify two subcases:

    (a) If the number of resolvable paths remains bounded in the wideband regime (as considered for instance in [20]), then N remains bounded while B_c increases linearly with B.

    (b) If the number of resolvable paths increases but only sublinearly with B, then both N and B_c grow without bound with B.

We first consider scenario (1) where rich multipath fading is assumed. In this case, as B increases, the signal-to-noise ratio $\mathrm{SNR} = \frac{\bar{P}}{N_0 B} = \frac{\bar{P}}{N N_0 B_c}$ approaches zero while B_c stays fixed. From these facts and the similarity of the formulations in (13.59) and (13.75), we immediately conclude that the wideband regime analysis of the rich multipath case is the same as the low-power regime analysis conducted in Section 13.5.2. Therefore, as $B \to \infty$ in the rich multipath fading scenario, we have $\frac{E_b}{N_0}\big|_{R_E=0} = \lim_{\mathrm{SNR}\to 0} \frac{E_b}{N_0} = \infty$ for all $\theta \geq 0$. Therefore, the minimum bit energy is attained at a high but finite bandwidth that can be identified through numerical analysis. If the bandwidth is further increased, a penalty in energy efficiency starts to be experienced due to increased uncertainty. Note that we have high diversity in rich multipath fading as the number of noninteracting subchannels increases linearly with bandwidth. On the other hand, since independent fading coefficients are only imperfectly known and moreover the receiver's ability to estimate the subchannels diminishes with decreasing SNR, we have high uncertainty as well. Hence, uncertainty becomes the more dominant factor and extreme energy-inefficiency is experienced in the limit as $B \to \infty$.

Next, we analyze the performance in the scenario of sparse multipath fading. Note that we consider channel uncertainty and queueing constraints jointly and use the effective capacity to identify the performance. We first consider scenario (2a) where the the number of subchannels N remains bounded and the degrees of freedom are limited. The following result provides the expressions for the bit energy at zero spectral efficiency and the wideband slope, and characterize the spectral efficiency-bit energy tradeoff in the wideband regime when N is fixed and B_c grows linearly with B. It is shown that the bit energy required at zero spectral efficiency is indeed the minimum bit energy.

Theorem 13.12. *For sparse multipath fading channel with bounded number of independent resolvable paths, the minimum bit energy and wideband slope in the wideband regime are given by*

$$\frac{E_b}{N_0}_{min} = \frac{E_b}{N_0}\bigg|_{R_E=0} = \lim_{SNR \to 0} \frac{E_b}{N_0} = \frac{-\delta}{\log \xi} \quad and \quad (13.76)$$

$$\mathcal{S}_0 = \frac{\xi \log^2 \xi \log 2}{\theta T \alpha^*_{opt}(1-\xi)\left(\frac{1}{T}\left(\sqrt{1+\frac{\gamma \bar{P} T}{N N_0}}-1\right)+\frac{\varphi \alpha^*_{opt}}{2}\right)}, \quad (13.77)$$

respectively, where $\delta = \frac{\theta T \bar{P}}{N N_0}$, $\xi = 1 - e^{-\alpha^*_{opt}}(1 - e^{-\frac{\theta T \varphi \alpha^*_{opt}}{\log 2}})$, *and* $\varphi = \frac{\gamma \bar{P}}{N N_0}\left(\sqrt{1+\frac{N N_0}{\gamma \bar{P} T}}-\sqrt{\frac{N N_0}{\gamma \bar{P} T}}\right)^2$. α^*_{opt} *is defined as* $\alpha^*_{opt} = \lim_{\zeta \to 0} \alpha_{opt}$ *and* α^*_{opt} *satisfies*

$$\alpha^*_{opt} = \frac{\log 2}{\theta T \varphi} \log\left(1 + \frac{\theta T \varphi}{\log 2}\right). \quad (13.78)$$

Above, we have defined $\zeta = \frac{1}{B_c}$.

Remark: We note that the minimum bit energy in the sparse multipath case with bounded degrees of freedom is achieved as $B \to \infty$ and hence as SNR $\to 0$. This is in stark contrast to the results in the low-power regime and rich multipath cases in which the bit energy requirements grow without bound as SNR vanishes. This is due to the fact that in sparse fading with bounded number of independent resolvable paths, uncertainty does not grow without bound because the number of subchannels N is kept fixed as $B \to \infty$.

Remark: Theorem 13.12 quantifies the bit energy requirements in the wideband regime (through the minimum bit energy and wideband slope expressions) when the system is operating subject to both statistical QoS constraints specified by θ and channel uncertainty. Note that both $\frac{E_b}{N_0}_{min}$ and $\mathcal{S}_0$ depend

on θ through δ and ξ. More specifically, $\frac{E_b}{N_0}_{\min}$ and the bit energy requirements at nonzero spectral efficiency values generally increase with increasing
θ. Moreover, when compared with the results in Section 13.5.2, it can be seen
that sparse multipath fading and having a bounded number of subchannels
incur energy penalty whether there are QoS constraints or not ($\theta = 0$), which
is in stark contrast with the results presented in Section 13.4 where there is
perfect CSI at the receiver.

We finally consider the sparse multipath fading scenario (2b) in which
the number of subchannels N increases but only sublinearly with increasing
bandwidth. Note that in this case, the bit energy required as $B \to \infty$ can be
obtained by letting N in the result of Theorem 13.12, where N is assumed to
be fixed, go to infinity.

Corollary 13.5. *In the wideband regime, if the number of subchannels N
increases sublinearly with B, then the bit energy required in the limit as $B \to
\infty$ is*

$$\left.\frac{E_b}{N_0}\right|_{\mathsf{R}_E=0} = \infty. \tag{13.79}$$

Remark: As N increases, each subchannel is allocated less power and operate in the low-power regime. Therefore, it is not surprising that we obtain
the same bit energy result as in the low-power regime. Additionally, since the
number of subchannels N increases without bound, uncertainty in the wideband channel increases as well. Hence, similarly as in rich multipath fading,
extreme energy-inefficiency is experienced as $B \to \infty$.

13.6   Conclusion

In this chapter, we analyzed the energy efficiency in single-input single-output
fading channels under QoS constraints by considering the effective capacity as
a measure of the maximum throughput under certain statistical queueing constraints, and analyzing the bit energy levels. Our analysis provided a characterization of the energy-bandwidth-delay tradeoff. In particular, we investigated
the spectral efficiency vs. bit energy tradeoff in the low-power and wideband
regimes under QoS constraints. We elaborated the analysis under different scenarios: perfect CSI available at both the transmitter and the receiver, perfect
CSI at the receiver side only, and imperfect CSI at the receiver only. We obtained expressions for the minimum bit energy and wideband slope. Through
this analysis, we have quantified the increased energy requirements in the
presence of delay-QoS constraints. We have also identified the impact of different transmission strategies, namely variable-rate/variable-power, variable-
rate/fixed-power, and fixed-rate/fixed-power transmissions, on the energy efficiency.

Bibliography

[1] M. Gursoy, D. Qiao, and S. Velipasalar, "Analysis of energy efficiency in fading channels under qos constraints," *Wireless Communications, IEEE Transactions on*, vol. 8, no. 8, pp. 4252 –4263, August 2009.

[2] D. Qiao, M. Gursoy, and S. Velipasalar, "The impact of qos constraints on the energy efficiency of fixed-rate wireless transmissions," *Wireless Communications, IEEE Transactions on*, vol. 8, no. 12, pp. 5957 –5969, Dec. 2009.

[3] ——, "Energy efficiency in the low-snr regime under queueing constraints and channel uncertainty," *Communications, IEEE Transactions on*, vol. 59, no. 7, pp. 2006 –2017, july 2011.

[4] Y. Chen, S. Zhang, S. Xu, and G. Li, "Fundamental trade-offs on green wireless networks," *Communications Magazine, IEEE*, vol. 49, no. 6, pp. 30 –37, June 2011.

[5] C.-S. Chang, *Performance Guarantees in Communication Networks*. London, UK: Springer-Verlag, 2000.

[6] D. Wu and R. Negi, "Effective capacity: a wireless link model for support of quality of service," *Wireless Communications, IEEE Transactions on*, vol. 2, no. 4, pp. 630 – 643, July 2003.

[7] J. Tang and X. Zhang, "Cross-layer modeling for quality of service guarantees over wireless links," *Wireless Communications, IEEE Transactions on*, vol. 6, no. 12, pp. 4504 –4512, December 2007.

[8] S. Verdu, "Spectral efficiency in the wideband regime," *Information Theory, IEEE Transactions on*, vol. 48, no. 6, pp. 1319 –1343, June 2002.

[9] J. Tang and X. Zhang, "Quality-of-service driven power and rate adaptation over wireless links," *Wireless Communications, IEEE Transactions on*, vol. 6, no. 8, pp. 3058 –3068, August 2007.

[10] *LT Codes*, ser. FOCS '02. Washington, DC, USA: IEEE Computer Society, 2002. [Online]. Available: http://dl.acm.org/citation.cfm?id=645413.652135

[11] A. Shokrollahi, "Raptor codes," *Information Theory, IEEE Transactions on*, vol. 52, no. 6, pp. 2551 –2567, June 2006.

[12] J. Castura and Y. Mao, "Rateless coding and relay networks," *Signal Processing Magazine, IEEE*, vol. 24, no. 5, pp. 27 –35, Sept. 2007.

[13] ——, "Rateless coding over fading channels," *Communications Letters, IEEE*, vol. 10, no. 1, pp. 46 – 48, Jan 2006.

[14] S. Shamai and S. Verdu, "The impact of frequency-flat fading on the spectral efficiency of cdma," *Information Theory, IEEE Transactions on*, vol. 47, no. 4, pp. 1302 –1327, May 2001.

[15] S. Borade and L. Zheng, "Wideband fading channels with feedback," *Information Theory, IEEE Transactions on*, vol. 56, no. 12, pp. 6058 –6065, Dec. 2010.

[16] L. Liu, P. Parag, J. Tang, W.-Y. Chen, and J.-F. Chamberland, "Resource allocation and quality of service evaluation for wireless communication systems using fluid models," *Information Theory, IEEE Transactions on*, vol. 53, no. 5, pp. 1767 –1777, May 2007.

[17] P. Sadeghi and P. Rapajic, "Capacity analysis for finite-state markov mapping of flat-fading channels," *Communications, IEEE Transactions on*, vol. 53, no. 5, pp. 833 – 840, May 2005.

[18] D. Porrat, D. Tse, and S. Nacu, "Channel uncertainty in ultra-wideband communication systems," *Information Theory, IEEE Transactions on*, vol. 53, no. 1, pp. 194 –208, Jan. 2007.

[19] V. Raghavan, G. Hariharan, and A. Sayeed, "Capacity of sparse multi-path channels in the ultra-wideband regime," *Selected Topics in Signal Processing, IEEE Journal of*, vol. 1, no. 3, pp. 357 –371, Oct. 2007.

[20] I. Telatar and D. Tse, "Capacity and mutual information of wideband multipath fading channels," *Information Theory, IEEE Transactions on*, vol. 46, no. 4, pp. 1384 –1400, July 2000.

[21] B. Hassibi and B. Hochwald, "How much training is needed in multiple-antenna wireless links?" *Information Theory, IEEE Transactions on*, vol. 49, no. 4, pp. 951 – 963, April 2003.

[22] M. Gursoy, "On the capacity and energy efficiency of training-based transmissions over fading channels," *Information Theory, IEEE Transactions on*, vol. 55, no. 10, pp. 4543 –4567, Oct. 2009.

Author Contact Information

Deli Qiao is with Department of Electrical Engineering, University of Nebraska-Lincoln, NE 68588, USA, Email: dqiao726@huskers.unl.edu. Mustafa Cenk Gursoy and Senem Velipasalar are with Department of Electrical Engineering and Computer Science, Syracuse University, Syracuse, NY 13244, USA, Email: mcgursoy@syr.edu, svelipas@syr.edu.

14

On the Energy Efficiency-Spectral Efficiency Trade-off in Cellular Systems

Fabien Héliot

Centre for Communication Systems Research, University of Surrey, UK

Efstathios Katranaras

Centre for Communication Systems Research, University of Surrey, UK

Oluwakayode Onireti

Centre for Communication Systems Research, University of Surrey, UK

Muhammad Ali Imran

Centre for Communication Systems Research, University of Surrey, UK

CONTENTS

14.1   Introduction

Energy efficiency (EE) is becoming a central research focus in communication in the current context of growing energy demand and increasing energy price [1, 2]. In the past, this area has already been thoroughly investigated but only through the prism of power-limited applications such as battery-driven systems [3], e.g. mobile terminals, underwater acoustic telemetry [4], or wireless ad-hoc and sensor networks [5,6]. Nowadays, this area is being revisited for unlimited power applications such as cellular networks [7,8]. This shift of focus in the research agenda from power-limited to power-unlimited applications is mainly driven by two factors: environmental, i.e. reducing the carbon footprint of communication systems, and; commercial, i.e. reducing the ever-growing operational cost of network operators. And yet, the spectral efficiency (SE) remains a key metric for assessing the performance of communication systems. The SE, as a metric, indicates how efficiently a limited frequency spectrum is used but fails to provide any insight on how efficiently the energy is consumed. In a context of energy saving, the latter will become as important as the former and, therefore, it has to be included in the performance evaluation framework.

Maximizing the EE, or equivalently minimizing the consumed energy, while maximizing the SE are conflicting objectives which implies the existence of a trade-off. The concept of EE-SE trade-off, has first been introduced for power-limited system and accurately defined for the low-power (LP)/low-SE regime in [9]. With the recent emergence of the EE as a key system design criterion alongside the established SE criterion, the EE-SE trade-off will soon become the metric of choice for efficiently designing future communication systems. As we previously mentioned, research on EE is currently shifting from power-limited to power-unlimited applications and, as a result, the concept of EE-SE trade-off must be generalized for power-unlimited systems and accurately defined for a wider range of SE regime, as initiated by the works in [10, 11].

The EE of a communication system is obviously closely related to its power consumption. In most of the past theoretical studies [4, 9, 12, 13], the EE-SE trade-off has been defined by considering that the total consumed power of the system is solely the transmit power, which is a fair assumption for power-limited applications such as sensor networks but is clearly not realistic for power-unlimited applications such as cellular systems. For instance, in cellular systems, the main power-hungry component is the base station (BS) and its total consumed power accounts for various power elements such as cooling, processing and amplifying powers. Consequently, in order to get a full picture of the total consumed power in a cellular system and evaluate fairly its EE, a realistic power consumption model (PCM) must be defined for each node, such as the PCMs proposed in [14–16] for different types of BSs as well as for backhaul links [15, 17]. In addition, these realistic PCMs can be simplified and incorporated in the EE-SE trade-off formulation for turning it into a simple and reliable performance evaluation metric.

In this chapter, we first introduce the EE-SE trade-off concept in Section 14.2 and explain the relevance and importance of this tool as a key performance evaluation and design criterion for future communication systems. Then, we will proceed with a survey of its usage from its genesis to its current development, mainly for evaluating power-limited application performance, and also provide an overview of its future usage, for instance, for fully assessing cooperation, coordination and cognition in future cellular systems. Next, we will discuss about the latest development in PCM for cellular systems and provide some insights on how to incorporate a realistic EE framework into the performance evaluation framework. In Section 14.3, we focus on the formulation of the EE-SE trade-off at the link level, and start by recalling the explicit expressions of this trade-off for the additive white Gaussian noise (AWGN) and deterministic channels. We then present novel and accurate explicit formulations of the EE-SE trade-off both for the single-input single-output (SISO) and multiple-input multiple-output (MIMO) ergodic Rayleigh fading channels and explain how they have been used in [10] for analytically evaluating the potential of MIMO in terms of EE. Next, we study the EE-SE trade-off at the cell level in Section 14.4 and provide explicit expressions of this trade-off for the SISO multi-user orthogonal and residual interference channels. In a context of multi-user communication, we design optimal resource allocation schemes based on EE by using our explicit expressions as objective functions in a multi-constraint optimization problem for both the multi-user orthogonal and residual interference channels. We compare these EE-based resource allocation schemes against sum-rate-based and fairness-based schemes. In Section 14.5, we explore the EE-SE trade-off of cellular systems employing BS cooperation to overcome inter-cell interference (ICI), i.e. coordinated multi-point (CoMP) systems. We discuss the PCM of CoMP and its implications on the overall consumed power of the system, and show how it has been incorporated in [11] for formulating the EE-SE trade-off of such systems. We use this expression for comparing the uplink performance of the idealistic global BS cooperation against the traditional non-cooperative system. We also study the more practical scenario of clustered cooperation based on the work of [18] and review the transmit EE (TxEE) and SE gains that can be achieved when using an efficient power control scheme. We finally conclude our analysis of the EE-SE trade-off in Section 14.6 by summarizing and discussing the valuable insights that have been obtained throughout this chapter.

14.2   Background Literature

14.2.1   Spectral Efficiency

The SE is the traditional metric for measuring the efficiency of a communication system. It is defined as the ratio of the information rate that can be transmitted over a channel to the channel bandwidth. It measures how efficiently the limited frequency resource (spectrum) is utilized, and it is usually expressed in terms of: bit/s/Hz, bit/Symbol, bit/channel use. Given a data rate R (bit/s) over a channel with bandwidth B (Hz), the achievable SE S is simply given by

$$S = \frac{R}{B} \qquad \text{(bit/s/Hz)}. \qquad (14.1)$$

14.2.2   Energy Efficiency

The SE as a metric does not provide any insight on how efficiently the energy is utilized in a system. Hence, EE metrics such as the bit-per-Joule (bit/J) capacity [4] have been defined to provide this insight. According to [19], the bit-per-Joule capacity of an energy limited wireless network is the maximum amount of bits that can be delivered by the network per joule of consumed energy in the network, i.e., the ratio of the capacity of the system to the total consumed power P_Σ, such that

$$\mathcal{C}_J = \frac{R}{P_\Sigma} \qquad \text{(bit/Joule)}. \qquad (14.2)$$

In addition, EE can be measured in terms of the rate-per-energy [20–22] or the capacity per unit cost [23]. When the cost represents the average power, the capacity per unit cost can be viewed as a special case of bit-per-joule capacity. Furthermore, EE can be inferred from the energy consumption index E_b, which is either expressed in terms of Joule/bit or energy-per-bit and is equivalent to $1/C_J$. In [24], Gallager gave a capacity definition for reliable communication under energy constraint as the maximum number of bits per unit energy that can be transmitted so that the probability of error goes to zero with energy.

14.2.3 Energy Efficiency-Spectral Efficiency Trade-Off

The EE-SE trade-off concept can simply be described as how to express the EE in terms of SE for a given available bandwidth. According to the famous Shannon's capacity theorem [25], the maximum achievable SE or equivalently the channel capacity per unit bandwidth C (bit/s/Hz) is a function of the signal-to-noise ratio (SNR), γ, such that

$$C = f(\gamma), \tag{14.3}$$

where $\gamma = P/N$ is the ratio between the transmit power P and the noise power N, and $N = N_0 B$ with N_0 (Joule) being the noise spectral density. In the general case, $f(\gamma)$ can be described as an increasing function of γ mapping SNR values in $[0, +\infty)$ to capacity per unit bandwidth values in $[0, +\infty)$. As long as $f(\gamma)$ is a bijective function, $f(\gamma)$ would be invertible such that

$$\gamma = f^{-1}(C), \tag{14.4}$$

where $f^{-1} : C \in [0, +\infty) \mapsto \gamma \in [0, +\infty)$ is the inverse function of f. For instance, over the AWGN channel $f(\gamma)$ and $f^{-1}(C)$ are simply given in [25] and [4] as

$$f(\gamma) = \log_2(1 + \gamma) \text{ and } f^{-1}(C) = 2^{C-1}, \tag{14.5}$$

respectively. As it has been explained in [4], the transmit power P can be expressed as RE_b and hence the SNR, γ, can be re-expressed as a function of both the achievable SE, S, and EE, C_J, such that

$$\gamma = \frac{P}{N_0 B} = \frac{R\,E_b}{B\,N_0} = \frac{S}{N_0 C_J}. \tag{14.6}$$

Inserting (14.6) into (14.4), the EE-SE trade-off expression in the general case can simply be formulated as

$$C_J = \frac{B}{N} \frac{S}{f^{-1}(C)}. \tag{14.7}$$

The last equation describes the EE-SE trade-off for the case of $P_\Sigma = P$, i.e. the idealistic PCM, however, for more generic PCM such that $P_\Sigma = g(P)$, the EE-SE trade-off can be reformulated as follows

$$C_J = \frac{B}{N} \frac{S}{g(P)/N}. \tag{14.8}$$

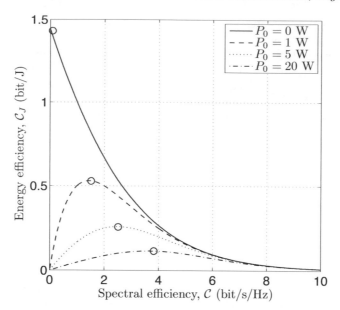

FIGURE 14.1　EE-SE trade-off over the AWGN channel for different values of the overhead power.

In order to provide some insights on the EE-SE trade-off, we plot in Figure 14.1 the EE-SE trade-off expression in (14.7) and (14.8) for $g(P) = P + P_0$ by considering $f^{-1}(\mathcal{C})$ as given in (14.5), $S = \mathcal{C}$ and $B = N = 1$. The results indicate that maximizing the EE while maximizing the SE are conflicting objectives and, hence, a EE-SE trade-off exists between these two metrics [4]. Indeed, in the case of $P_0 = 0$ W, the maximum EE is achieved when $\mathcal{C} = 0$ bit/s/Hz and, conversely, the maximum SE on the graph, i.e. $\mathcal{C} = 10$ bit/s/Hz is achieved for very low EE. Consequently, the most energy efficient policy over the AWGN channel is to not transmit anything at all when $P_0 = 0$ W. However, when the total consumed power P_Σ is not only restricted to the transmit power P and an overhead power P_0 is consumed, the existence of an optimal SE-EE trade-off operation point becomes apparent, which is circled in Figure 14.1 for different values of P_0. The existence of such a point illustrates the growing importance of the EE-SE trade-off as a system design criterion.

14.2.4　Explicit Formulation vs. Approximation

In general, the problem of defining a closed-form expression (CFE) for the EE-SE trade-off is equivalent to obtaining an explicit expression for the inverse function of the channel capacity per unit bandwidth, as it is explained in the previous section. This has so far been proved feasible only for the AWGN channel and the deterministic channel with colored noise in [4] and [9], respectively, and it explains why approximation has been widely used for formulating this trade-off in more complex communication scenarios.

14.2.4.1 Low-Power Approximation Approach

Based on the fact that the EE of a communication system depends mainly on its SE in the LP/low-SE regime, Verdú et al. introduced the concept of LP approximation for the EE-SE trade-off in [9]. This method is in effect quite generic and, thus, it can be used to approximate the EE-SE trade-off of any communication channel or system for which an explicit expression of its capacity per unit bandwidth as a function of γ, i.e. $f(\gamma)$, exists and is twice differentiable. Because of its simplicity, this approach has gained popularity and it has been extended over the years for formulating the EE-SE trade-off of several communication scenarios and schemes such as point-to-point [12], multi-user [26, 27], interference channel [26], single relay [20–22, 28–30], relay networks [13, 31, 32] and BS cooperative networks [33, 34]. At low γ, the achievable SE is described by the minimum energy-per-bit required for reliable communication $E_b/N_{0\min}$ and the slope S_0 at $E_b/N_{0\min}$. The LP analysis is valid for $\mathcal{C} \ll 1$, i.e. the wideband regime, whenever a very large bandwidth is used for the transmission of a given data rate or a very small data rate is transmitted through a given bandwidth. The SE $f(\gamma)$ being in general a monotonically increasing concave function, the minimum energy per bit $\frac{E_b}{N_0 \min}$ required for reliable communication is given as

$$
\begin{aligned}
\frac{E_b}{N_0 \min} &= \lim_{\gamma \to 0} \frac{f^{-1}(\mathcal{C})}{f(\gamma)} = \lim_{\gamma \to 0} \frac{\gamma}{f(\gamma)} \\
&= \frac{\log_e 2}{\dot{f}(0)},
\end{aligned}
\tag{14.9}
$$

where $\dot{f}(0)$ is the first order derivative of $f(\gamma)$ at $\gamma = 0$. The EE-SE trade-off based on the LP approximation can be expressed as follows [9]

$$
10 \log_{10} \frac{E_b}{N_0}(\mathcal{C}) \approx 10 \log_{10} \frac{E_b}{N_0 \min} + \frac{\mathcal{C}}{S_0} 10 \log_{10} 2,
\tag{14.10}
$$

where $S_0 = \frac{2[\dot{f}(0)]^2}{-\ddot{f}(0)}$ is the slope of SE in bit/s/Hz/(3 dB) at the point $\frac{E_b}{N_0 \min}$ and $\ddot{f}(0)$ is the second order derivative of $f(\gamma)$. Equivalently (14.10) can be expressed as

$$
f^{-1}(\mathcal{C}) \approx S \frac{E_b}{N_0 \min} \, 2^{\frac{\mathcal{C}}{S_0}},
\tag{14.11}
$$

based on the formulation of (14.7). Figure 14.2 compares the exact and nearly-exact EE-SE trade-off over the AWGN and SISO Rayleigh fading channels, respectively, with their LP approximations, i.e. LP approx.. Results show the fair accuracy of the LP approximation method in the LP/low-SE regime and its versatility since it can be used for different scenarios by using the same

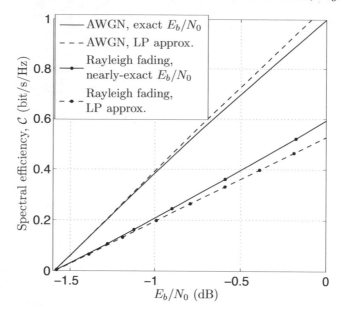

FIGURE 14.2 EE-SE curves of AWGN channel and Rayleigh channel with their respective LP approximations.

formulation as in (14.10). However, the main shortcoming of this approach is its rather limited range of SE values for which it is accurate. Indeed, it is by design limited to the low-SE regime and, thus, it cannot be used for assessing the EE of future communication systems such as long term evolution (LTE) which are meant to operate in the mid-high SE region.

14.2.4.2   EE-SE Closed-Form Approximation

Until recently, the two main approaches for obtaining explicit expression of the EE-SE trade-off have been either to use the explicit expression of $f(\gamma)$ for finding an explicit solution to $f^{-1}(\mathcal{C})$ as for instance in [4] or to use the explicit expression of $f(\gamma)$ for approximating $f^{-1}(\mathcal{C})$ as it is explained in Section 14.2.4.1. Another approach would be to use an accurate closed-form approximation (CFA) of $f(\gamma)$, i.e. $f(\gamma) \approx \widetilde{f}(\gamma)$ for finding an explicit solution to $f^{-1}(\mathcal{C})$, as it was recently proposed in [35] and [10] for the SISO and MIMO Rayleigh fading channels, respectively, as well as in [36] and [11] for the uplink of a symmetrical cellular system with BS cooperation when assuming the Wyner or uniformly distributed UT model, respectively. More details about this new approach for tightly approximating the EE-SE trade-off will be given in Sections 14.3 and 14.5.

14.2.5 Idealistic vs. Realistic Power Consumption Model

14.2.5.1 Limitations on the Idealistic Power Model

In most of the theoretical works related to the EE-SE trade-off [4, 9, 12, 13], the total consumed power of any transmitting node has always been idealized such that it is equal to its transmit power, i.e $P_\Sigma = P$, however, this is not the case in a real system. Although the idealistic PCM provides a good framework for the analysis of the EE-SE trade-off, it does not capture the realistic total amount of power that is consumed in the network and, hence, may lead to misguiding conclusions for designing energy efficient systems. For instance, schemes such as BS cooperation, user terminal (UT) cooperation, multiple antenna techniques can improve at the same time the spectral and energy efficiencies when assuming an idealistic PCM. Yet, this is not necessarily true when considering a more realistic PCM.

14.2.5.2 Realistic Power Model Perspective

The initial framework for EE-SE trade-off analysis was based on the idealistic approach which considers only the transmit power as consumed power. Clearly, this approach does not give a true measure of the EE. A good measure of the EE must incorporate the total consumed power, hence, a new approach for analyzing the EE-SE trade-off based on a more realistic PCM is needed. Initial work in this direction was based on the assumption that the total consumed power is the sum of a fixed power usually termed the circuit power, which is independent of the transmit power, and the power consumption of all the power amplifiers (PAs) [6, 37]. The circuit power usually accounts for the processing powers at both the transmitter and receiver ends, i.e. the power consumed by components such as the: digital to analog converter, low noise amplifier, mixer, active filters at the transmit and receiver sides, intermediate frequency amplifier, analog to digital converter and frequency synthesizer. This approach has the clear advantage of being computationally tractable since all elements that make up the circuit power are assumed to be independent of the transmit power. Whereas, its limitation is that the circuit power is rate dependent as far as the UT is concerned [38, 39] and, moreover, it usually does not include losses due to cooling, direct current (DC)-DC regulation, main supply and feeder components, which account for a fair proportion of the BS power consumption. Recently in [14], the authors presented a linear PCM for BS, which takes into account the signal processing power, amplifier inefficiency and other power losses. An extension of this model for BS cooperation, which incorporates the backhauling power as well, was presented in [15] where the backhauling power was assumed to be dependent of the backhaul requirement of each BS. A refined backhaul power model can also be found in [17], whereas, the power consumption of extra BS components such as the alternating current(AC) - DC and DC-DC converters have recently been considered in [16]. The latter work shows that the relationship between the transmit power and the BS power consumption is nearly linear.

14.2.5.3   Power Model Mathematical Framework

The EE of a communication system is obviously closely related to its PCM. From a top-level perspective, a PCM describes how much total consumed power P_Σ is needed by a node or a system for transmitting information given a transmit power P. As already mentioned in Section 14.2.3, this relation can simply be mathematically described as

$$P_\Sigma = g(P), \qquad (14.12)$$

where $g(P)$ is the function that relates P to P_Σ. The function $g(P)$ can be more or less complex according to different types of applications: For simplicity reasons, most of the theoretical studies about EE considered only the transmit power as consumed power in the system such that $g(P) = P$. Whereas in reality, the input power of any equipment in a communication system, e.g. BS or UT, is composed of various fixed as well as variable power components and, moreover, amplifier efficiency is generally not perfect. Therefore, more realistic PCMs have recently been proposed in [8, 16] and [39, 40] for the BS and UT nodes, respectively, which take into account the PA efficiency, signal processing overhead, cooling and power supply (PS) losses as well as current conversion for the former and PA as well as circuit power for the latter. Note also that in reality, the different equipments are power-limited in terms of transmit power such that $0 \leq P \leq P_{\max}$, where $P_{\max}$ is the maximum transmit power.

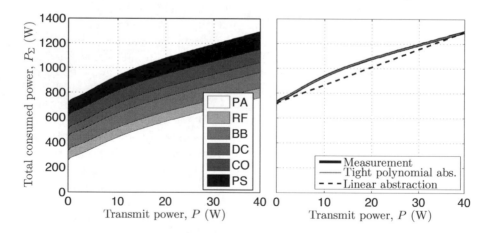

FIGURE 14.3 Total consumed power dependency on transmit power for a 2 Tx macro BS with 3 sectors and $B = 10$ MHz based on measurement (left) and two types of abstraction for this dependency relation (right). Legend: PA=Power Amplifier, RF=small signal radio frequency transceiver, BB=Baseband processor, DC: DC-DC converters, CO: Cooling, PS: AC/DC Power Supply.

In order to put things into perspective, we depict in Figure 14.3, the relation between the transmit power P and the total consumed power P_Σ for a LTE macro BS with 2 transmit antennas and $P_{max} = 80$ W, i.e. 40 W per transmit antenna, according to the measurement-based PCM of [16]. On the left side of the figure, the variation of the various power components of the model as a function P are detailed. It shows that a fair amount of power is still consumed even when no power is used for transmitting and that the PA clearly becomes the main power-hungry component when the transmit power increases. On the right side of the figure, we depict again the overall relation between the total consumed power and transmit power based on measurement and on two types of approximation for this measurement: a tight polynomial approximation where $g(P) \approx -3.52E^{-7}P^5 + 7.23E^{-5}P^4 - 4.60E^{-3}P^3 + 3.50E^{-2}P^2 + 11.28P + 711.13$ and a linear approximation where $g(P) \approx 7.25P + 712$, which have a relative approximation error of 0.12% and 4.88% in comparison with the measured P_Σ, respectively.

In most cases, using a complex definition of $g(P)$ will not be practical for formulating in a simple way the EE-SE trade-off in (14.8). The result in Figure 14.3 indicates that using a linear abstraction for approximating real measurements can be a practical solution without sacrificing too much accuracy. One can also think about quadratic or even high-degree polynomial abstraction depending on the level of desired accuracy. For instance in [16], a linear abstraction has been utilized and $g(P)$ has been defined for different

TABLE 14.1 Different types of PCM abstraction and their relevant parameter values

PCM types	Function definitions $g(P) =$	Node types	Parameter values			
			$P_{max}(W)$	Δ_P	$P_0(W)$	$P_1(W)$
Idealistic	P	—	∞	—	—	—
Linear	$\Delta_P P + P_0$	maBS [16]	$2*40$	$14.5/2$	712	—
		miBS [16]	$2*6.31$	$6.35/2$	106	—
		piBS [16]	$2*0.25$	$8.4/2$	14.9	—
		feBS [16]	$2*0.1$	$15/2$	10.1	—
		UT [40]	$0.32/\Delta_P$	$[1,\infty]$	0.1	—
Linear [39]	$\Delta_P P + P_0(R)$	UT	N/A	N/A	N/A	—
DbL	$\Delta_P P + tP_0$	maBS [8]	$20t$	4.7	130	—
		RRH [8]	$20t$	2.8	84	—
		miBS [8]	$6.3t$	2.6	56	—
		piBS [8]	$0.13t$	4.0	6.8	—
		feBS [8]	$0.05t$	8.0	7.8	—
DbL [10]	$\Delta_P P + tP_0 + P_1$	maBS	80	7.25	244	225

types of BSs as $g(P) \approx \Delta_P P + P_0$, where Δ_P accounts for the amplifier inefficiency and P_0 is the overhead power. In addition, this abstraction has been extended into a double linear (DbL) abstraction, i.e linear both in terms of P and t, for various types of BSs in [8], such that $g(P) = \Delta_P P + t P_0$. However, as it is explained in [16], it is expected that some of the power components like DC-DC / AC-DC converters and cooling unit do not grow linearly with t and, hence, the previous approximation gives an upper bound on the macro BS power consumption with t transmit antennas. A more realistic DbL PCM should take into account that only one part of the overhead power grows linearly with t and one part remains fixed such that $g(P) = \Delta_P P + t P_0 + P_1$. For the reader convenience, we summarize in Table 14.1 the different types of PCM discussed in this section along with the numerical values of their parameters. In Table 14.1, maBS, miBS, piBS and feBS stand for macro, micro, pico and femto BS, whereas RRH stands for remote radio head.

14.3   EE-SE Trade-Off on a Link

14.3.1   AWGN Channel

As it is already mentioned in Section 14.2.3, $\mathcal{C} = f(\gamma) = \log_2(1 + \gamma)$ over the AWGN channel and, hence, $f^{-1}(\mathcal{C})$ can be explicitly formulated as [4,9]

$$f^{-1}(\mathcal{C}) = 2^{\mathcal{C}} - 1. \tag{14.13}$$

The CFE of the EE-SE trade-off over the AWGN channel can then be simply obtained by inserting (14.13) into (14.7).

14.3.2   Deterministic Channel with Colored Gaussian Noise

According to the work of Verdú in [9], in the case that the channel matrix $\mathbf{H} \in \mathbb{C}^{r \times t}$ is known both at the transmitter as well as receiver sides and is constant over time, and the noise covariance is given by $E\{\mathbf{nn}^\dagger\} = N_0 \mathbf{\Sigma}$, then $f^{-1}(\mathcal{C})$ can be explicitly formulated as

$$f^{-1}(\mathcal{C}) = \frac{1}{r} \min_{i \in \{1, \ldots, \widehat{r}\}} i 2^{r\mathcal{C}/i} \prod_{j=1}^{i} \rho_j^{1/i} - \sum_{j=1}^{i} \rho_j \tag{14.14}$$

over the deterministic channel with colored Gaussian noise. Note that r and t denote the number of transmit and receive antennas, respectively, $\rho_1, \rho_2, \ldots, \rho_{\widehat{r}}$ are the ordered version of the reciprocals of the non-zero eigenvalues of the matrix $\mathbf{H}^\dagger \mathbf{\Sigma}^{-1} \mathbf{H}$, $\widehat{r}$ is the number of non-zero eigenvalues and $\{.\}^\dagger$ is the complex conjugate operator.

14.3.3 Ergodic Rayleigh Fading Channel

14.3.3.1 Channel Capacity: CFE vs. CFA

The generic expression of the ergodic channel capacity per unit bandwidth for the Rayleigh fading channel is usually given by [41]

$$C = f(\gamma) \triangleq \mathbf{E_H} \left\{ \log_2 \left| \mathbf{I}_r + \frac{\gamma}{t} \mathbf{HH}^\dagger \right| \right\}, \tag{14.15}$$

where $\mathbf{H} \in \mathbb{C}^{r \times t}$, $\mathbf{I}_r$ is a $r \times r$ identity matrix, $|.|$ is the determinant operator, and $\mathbf{E_H}$ is the expectation over $\mathbf{H}$. Using (14.15) as a starting point, two main paths have been followed in the literature for explicitly formulating either CFEs or approximations of the ergodic channel capacity per unit bandwidth for the Rayleigh fading channel.

In [41], the expression of C has been simplified into an analytical formula by computing the expectation of the ordered eigenvalues of the Wishart matrix $\mathcal{W} \triangleq \mathbf{HH}^\dagger$ or $\mathbf{H}^\dagger \mathbf{H}$ if $r < t$ or $r \geq t$, respectively. This work has attracted a lot of interest in this area of research and as a result proper CFEs of the MIMO channel capacity have been obtained in various independent works [42–45]. Although these expressions are perfectly accurate, their formulations are not as simple as in the AWGN or deterministic channel cases. For instance in [42], $f(\gamma)$ is given by

$$f(\gamma) = \sum_{k=0}^{m-1} \frac{k!}{(k+d)!} \sum_{l_1=0}^{k} \sum_{l_2=0}^{k} (-1)^{l_1+l_2} A_{l_1}(k,d) A_{l_2}(k,d) \widehat{C}_{l_1+l_2+d}\left(\frac{t}{\gamma}\right), \tag{14.16}$$

where $d \triangleq n-m$, $n \triangleq \max(t,r)$, $m \triangleq \min(t,r)$ and $A_l(k,d) \triangleq \frac{(k+d)!}{(k-l)!(d+l)!l!}$. In addition,

$$\widehat{C}_i(x) \triangleq \frac{1}{\ln(2)} \sum_{j=0}^{i} \frac{i!}{(i-j)!} \left[(-x)^{i-j} e^x E_1(x) + \sum_{k=1}^{i-j} (k-1)! (-x)^{i-j-k} \right], \tag{14.17}$$

where $E_1(x) = \int_x^\infty \frac{e^{-t}}{t} dt$ is the exponential integral function. Consequently, finding an explicit expression for $f^{-1}(C)$ based on (14.16) will prove extremely challenging especially since even the simplest case of $f(\gamma) = e^{t/\gamma} E_1(t/\gamma)$, i.e. $f(\gamma)$ in (14.16) for $t = r = 1$, does not have to the best of our knowledge an explicit formulation for its inverse function.

Meanwhile, Biglieri and Taricco in [46] have proposed a CFA of (14.15) based on asymptotical analysis and random matrix theory. Their CFA is obviously less accurate than the CFE in (14.16) but its formulation is far more simplified [46]

$$C \approx \widetilde{f}(\gamma) = -\frac{t}{\ln(2)} \left[-(1+\beta) \ln(\sqrt{\gamma}) + q_0 r_0 + \ln(r_0) + \beta \ln\left(\frac{q_0}{\beta}\right) \right], \tag{14.18}$$

where $q_0 \triangleq \frac{\gamma(\beta-1) - 1 + \sqrt{(\gamma(\beta-1)-1)^2 + 4\gamma\beta}}{2\sqrt{\gamma}}$, $r_0 \triangleq \frac{\gamma(1-\beta) - 1 + \sqrt{(\gamma(\beta-1)-1)^2 + 4\gamma\beta}}{2\sqrt{\gamma}}$ and

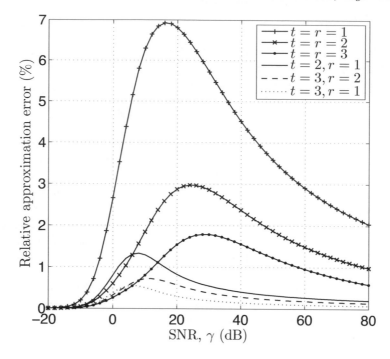

FIGURE 14.4 Relative approximation error in percentage between the CFE in (14.16) and CFA in (14.18) of the ergodic channel capacity per unit bandwidth as a function of the SNR for various antenna configurations.

$\beta \triangleq r/t$. The fact that (14.18) is more simplified than (14.16) is not an end in itself. Above all, the main advantage of $\widetilde{f}(\gamma)$ over $f(\gamma)$ is the fact that the inverse function of $\widetilde{f}(\gamma)$, i.e. $\widetilde{f}^{-1}(\mathcal{C})$, can be expressed into a closed-form, as it has been recently demonstrated in [10] and is reported in the next subsection.

In order to illustrate the relative accuracy of $\widetilde{f}(\gamma)$ in (14.18) against $f(\gamma)$ in (14.16), we depict in Figure 14.4 their relative approximation errors in percentage, i.e. $100|f(\gamma) - \widetilde{f}(\gamma)|/|f(\gamma)|$, as a function of the SNR γ (dB) for symmetric, i.e. $t = r$, and asymmetric, i.e. $t \neq r$, MIMO configurations. We have only considered here the case of $t > r$ for the asymmetric configuration, however, note that similar results are obtained for the case of $t < r$. In the asymmetric scenario, the accuracy of $\widetilde{f}(\gamma)$ is already acceptable for $t = 2, r = 1$ such that $f(\gamma)$ and $\widetilde{f}(\gamma)$ differ on average by less than 0.5% for γ between - 20 to 80 dB. Moreover, it can be remarked that the accuracy increases as the antenna configuration becomes more asymmetric since the curve of $t = 3, r = 1$ is lower than $t = 3, r = 2$. Whereas, in the symmetric scenario, the results indicate that $\widetilde{f}(\gamma)$ becomes more accurate as the number of antennas increases. For instance, $f(\gamma)$ and $\widetilde{f}(\gamma)$ differ on average by less than 2% when $t = r = 2$ and then by less than 1% when $t = r = 3$.

TABLE 14.2 Parameters η_0 and η_1 values as a function of $\overline{\beta}$

$\overline{\beta} \in [2,+\infty)$		$\overline{\beta} \in (1,2)$						
$\overline{\beta}$	any	10/9	9/8	8/7	7/6	6/5	5/4	9/7
η_0	1	0.377	0.373	0.366	0.365	0.369	0.384	0.508
η_1	$\eta(\overline{\beta})$	3.914	3.835	3.705	3.515	3.266	2.968	3.059
	$\overline{\beta} \in (1,2)$							
$\overline{\beta}$	4/3	7/5	10/7	3/2	8/5	5/3	7/4	9/5
η_0	0.4285	0.528	0.608	0.1315	0.1621	0.1808	0.2028	0.2153
η_1	$\varphi = \eta(\overline{\beta})$ $+ \log_2(\eta_0)$	2.682	2.751	φ	φ	φ	φ	φ

14.3.3.2 CFA of the MIMO EE-SE Trade-Off

Based on equation (14.18), it has been recently proved in [10] that the EE-SE trade-off for the ergodic MIMO Rayleigh fading channel can be explicitly formulated by means of an accurate CFA as

$$f^{-1}(\mathcal{C}) \approx \widetilde{f}^{-1}(\mathcal{C}) = \frac{\left\{-1 + \left(1 + [W_0\left(g_t(\mathcal{C})\right)]^{-1}\right)\left(1 + [W_0\left(g_r(\mathcal{C})\right)]^{-1}\right)\right\}}{2(1+\beta)},$$
(14.19)

where $W_0(x)$ denotes the real branch of the Lambert function [47], whereas the functions $g_t(\mathcal{C})$ and $g_r(\mathcal{C})$ are defined as

$$g_t(\mathcal{C}) \triangleq -2^{-\left(\frac{C+h(\mathcal{C})}{2t}+1\right)}e^{-\frac{1}{2}} \text{ and } g_r(\mathcal{C}) \triangleq -2^{-\left(\frac{C-h(\mathcal{C})}{2r}+1\right)}e^{-\frac{1}{2}},$$
(14.20)

respectively, and the function $h(\mathcal{C})$ in (14.20) is expressed as

$$h(\mathcal{C}) \triangleq \zeta m \log_2\left(1 - \eta_0\left[1 - \cosh\left(\frac{C\ln(2)}{m[\eta(\overline{\beta}) + \log_2(\eta_0)]}\right)^{\eta_1}\right]\right),$$
(14.21)

with $\zeta \triangleq \text{sgn}(\ln(\beta))$ and $\text{sgn}(x) \triangleq -1, 0$ or 1 if $x < 0$, $x = 0$ or $x > 0$ such that $h(\mathcal{C}) = 0$ when $\beta = 1$. In addition,

$$\eta(\overline{\beta}) = \frac{1}{\ln(2)}\left[-1 + 2\overline{\beta}\ln\left(\frac{\overline{\beta}}{\overline{\beta}-1}\right)\right]$$
(14.22)

in (14.21), where $\overline{\beta} \triangleq n/m$, $\overline{\beta} \in [1,+\infty)$, and the values of the parameters η_0 and η_1 are given in Table 14.2.

Furthermore, $\beta = 1$ in the case of $t = r$ and, hence, $\zeta \triangleq \text{sgn}(\ln(\beta)) = 0$, which in turn implies that $g_t(\mathcal{C}) = g_r(\mathcal{C})$ such that equation (14.19) simplifies as [10]

$$\widetilde{f}^{-1}(\mathcal{C}) = \frac{1}{4}\left\{-1 + \left(1 + \left[W_0\left(-2^{-\left(\frac{C}{2t}+1\right)}e^{-\frac{1}{2}}\right)\right]^{-1}\right)^2\right\}$$
(14.23)

in the symmetric antenna configuration.

14.3.3.3   CFA of the SISO EE-SE Trade-Off

Figure 14.4 clearly indicates that the relative approximation error between $\widetilde{f}(\gamma)$ in (14.18) and $f(\gamma)$ in (14.16) for $t = r = 1$ is rather large for γ in between 5 to 40 dB and, consequently, $\widetilde{f}^{-1}(\mathcal{C})$ in (14.23) is not a good approximation of $f^{-1}(\mathcal{C})$. In the SISO case, $f(\gamma)$ simplifies to

$$f(\gamma) = e^{\gamma^{-1}} \mathrm{E}_1\left(\gamma^{-1}\right) / \ln(2), \qquad (14.24)$$

whereas $\widetilde{f}(\gamma)$ in (14.18) simplifies to

$$\widetilde{f}(\gamma) = \frac{2}{\ln(2)}\left[-[1/2 + \ln(2)] + \frac{1}{1 + \sqrt{1+4\gamma}} + \ln(1 + \sqrt{1+4\gamma})\right]. \quad (14.25)$$

Using (14.25) as a starting point, we have obtained in [35] a very tight CFA of $f(\gamma)$ in (14.24) such that its inverse can be simply and accurately formulated as

$$f^{-1}(C) \approx \left[\alpha(b,\phi)^b - 1 - b\phi\left[W_0\left(-b\phi e^{-b\phi} 2^{-\frac{bC}{n}}\right)\right]^{-1}\right]^{\frac{1}{b}} - \alpha(b,\phi), \quad (14.26)$$

where

$$\alpha(b,\phi) = e^{\frac{1}{1-b}\ln(1-b\phi)}, \qquad (14.27)$$

$\phi = 0.57721...$ denotes the Euler-Mascheroni constant [48] and $b = 0.71435$.

14.3.3.4   Accuracy of the CFAs: Numerical Results

In order to illustrate the accuracy of the CFA of the EE-SE trade-off for the ergodic Rayleigh fading channel in (14.19), (14.23) and (14.26), we compare them in Figure 14.5 with the approximation method of [9] and the nearly-exact $\mathcal{C}_J$ as a function of $\mathcal{C}$ that has been obtained via (14.16). Indeed, equation (14.16) returns the values of SE $\mathcal{C}$ for a given SNR γ; then, since f is a bijective function, one can easily obtain the SNR $\gamma = f^{-1}(\mathcal{C})$ for a given SE $\mathcal{C}$ by using (14.16) in conjunction with a simple line search algorithm where the target $\mathcal{C}$ is set to differ by less than 10^{-8} from the actual $\mathcal{C}$. Using this approach, we have obtained $f^{-1}(\mathcal{C})$ for $\mathcal{C} = 10^{-2}$ to 40 bit/s/Hz with an incremental step of 0.5 bit/s/Hz; then, by inserting $f^{-1}(\mathcal{C})$, $S = \mathcal{C}$ and $B = N = 1$ in (14.7), the nearly-exact $\mathcal{C}_J$ has been plotted as a function of $\mathcal{C}$. Regarding the LP approximation method of [9], note that the values of $\frac{E_b}{N_0}{}_{\min}$ and S_0 are given in equations (213) and (215) of [9], respectively, such that $\frac{E_b}{N_0}{}_{\min} = \ln(2)/r$ and $S_0 = \frac{2tr}{t+r}$ when equal power allocation is assumed and the MIMO Rayleigh fading channel is unknown at the transmitter. The results in Figure 14.5 clearly demonstrate the tight fitness between the nearly-exact $\mathcal{C}_J$ curves and $\mathcal{C}_J$ obtained via the CFAs, hence, they graphically confirm

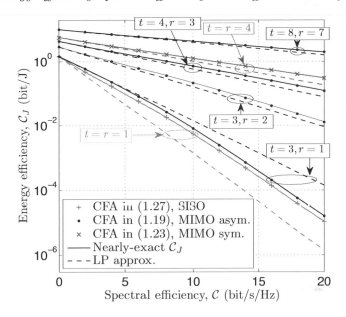

FIGURE 14.5 Comparison of the EE-SE trade-off CFAs in (14.19), (14.23) and (14.26) with the LP approx. method and the nearly-exact $\mathcal{C}_J$ for various antenna configurations.

the great accuracy of the latter. They also confirm the poor accuracy of the LP approximation method of [9] for $\mathcal{C} \geq 1$ and $\mathcal{C} \geq 4$ bit/s/Hz in the SISO and MIMO cases, respectively. Note that accuracy results for extra antenna configurations can be found in [10].

14.3.4 MIMO vs. SISO: An Energy Efficiency Analysis

EE can be interpreted as a ratio between the transmission rate and consumed power. Consequently, the EE gain between two systems can either result from an increase of SE (one of the systems providing a better rate than the other, for a given transmit power) or a decrease in consumed power (one of the systems consuming less power than the other, for a given transmission rate). The former definition of the EE gain is actually equivalent to the definition of the SE gain and, thus, it can be seen as an indirect EE gain, since the concept of EE is implicitly linked with power consumption and cost reductions. The latter definition of the EE gain, or in short the direct EE gain, is obviously more suitable for EE-based analysis. Here, we summarize some results of [10] in which this metric has been utilized to analyze MIMO effectiveness for reducing power consumption over the Rayleigh fading channel.

The EE gain of MIMO over SISO can simply be defined as

$$G_{\text{EE}} \triangleq \frac{\mathcal{C}_{J,\text{MIMO}}}{\mathcal{C}_{J,\text{SISO}}}, \tag{14.28}$$

14.3.4.1   MIMO vs. SISO EE Gain: Idealistic PCM

Assuming an idealistic PCM where $P_\Sigma = P$ and, hence, using the definition of the EE-SE trade-off $\mathcal{C}_J$ in (14.7) for simplifying (14.28), the EE gain due to a reduction in consumed power, i.e. G_{PR} where PR stands for power reduction, can be expressed as

$$G_{\text{PR}} = f_{\text{SISO}}^{-1}(\mathcal{C}) / f_{\text{MIMO}}^{-1}(\mathcal{C}) \tag{14.29}$$

for a fixed rate, where $f_{\text{SISO}}^{-1}(\mathcal{C})$ and $f_{\text{MIMO}}^{-1}(\mathcal{C})$ are approximated in equations (14.26) and (14.19)/(14.23), respectively. In order to get some insight about this EE gain in the low and high-SE regimes, limits of G_{PR} have been derived at low and high SEs, i.e. G_{PR}^0 and G_{PR}^∞, respectively, in [10] such that

$$\begin{aligned} G_{\text{PR}}^0 &= r \\ G_{\text{PR}}^\infty &= (\bar{\beta} - 1)^{(1-\bar{\beta})} \bar{\beta}^{\left(\bar{\beta} - \frac{1-\varsigma}{2}\right)} e^{(\phi - 1)} 2^{C\left(1 - \frac{1}{m}\right)} \end{aligned} \tag{14.30}$$

Moreover, notice that $G_{\text{PR}}^\infty = e^{(\phi-1)} 2^{C\left(1 - \frac{1}{m}\right)}$ in the symmetric antenna configuration.

Similarly, using the definition of the EE-SE trade-off $\mathcal{C}_J$ in (14.7) for simplifying (14.28), the EE gain due to an increase of SE, i.e. G_{SE}, can be expressed as

$$G_{\text{SE}} = f_{\text{MIMO}}(\gamma) / f_{\text{SISO}}(\gamma) \tag{14.31}$$

for a fixed transmit power, where $f_{\text{MIMO}}(\gamma)$ and $f_{\text{SISO}}(\gamma)$ are given in (14.16) and (14.24), respectively. The limits of this gain at low and high SEs can be given by

$$\begin{aligned} G_{\text{SE}}^0 &= r \\ G_{\text{SE}}^\infty &\propto m = \min\{t, r\} \end{aligned} \tag{14.32}$$

Comparing equations (14.30) with (14.32) indicates that, at low SE, reducing the transmit power while keeping the same rate is equivalent to increasing the rate while keeping the same transmit power. This is consistent with the fact that in this SE region the rate scales linearly with the power and the number of receive antennas. However, in the high-SE regime, the rate scales in a logarithm manner with the power and, hence, a larger EE gain can be achieved by reducing power instead of increasing SE. For instance, G_{PR}^∞ increases with the SE (exponentially) as well as the number of antennas, whereas G_{SE}^∞ increases only with the number of antennas. In order to cross-validate these analytical insights with numerical results, we plot in Figure 14.6, G_{PR} and G_{SE} as a function of the SE and number of antenna elements when $n_{\text{ant}} = t = r$. The results show that $G_{\text{PR}} \geq G_{\text{SE}}$ and confirm that G_{PR} grows both with the

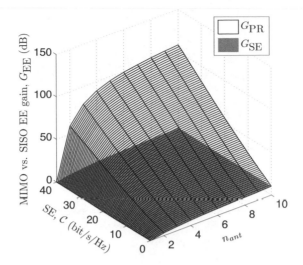

FIGURE 14.6 Idealistic EE gain due to a reduction in consumed power vs. EE gain due to a SE improvement as a function of the SE and number of antenna elements when $n_{\text{ant}} = t = r$.

SE and number of antenna elements n_{ant}. Overall, these results indicate that MIMO has a huge potential for EE improvement when an idealistic PCM is considered.

14.3.4.2 MIMO vs. SISO EE Gain: Realistic PCM

Without loss of generality, we consider in this subsection one of the realistic PCMs of Table 14.1, i.e.,

$$g(P) = t(\Delta_P P/t + P_0) + P_1, \tag{14.33}$$

for analyzing the EE gain of MIMO vs. SISO.

Under the assumption that both SISO and MIMO systems are affected by the same level of noise, G_{PR} in (14.29) can be simplified as

$$G_{\text{PR}} = P_{\text{SISO}}/P_{\text{MIMO}}, \tag{14.34}$$

where P_{SISO} and P_{MIMO} are the respective SISO and MIMO transmit powers. Using this definition for G_{PR} and inserting (14.33) in (14.7), the EE gain due to a reduction in consumed power can then be re-expressed as [10]

$$\widehat{G}_{\text{PR}} = \frac{\Delta_P P_{\text{SISO}} + P_0 + P_1}{\Delta_P (P_{\text{SISO}}/G_{\text{PR}}) + tP_0 + P_1} = \frac{\psi + 1 + P_1/P_0}{\psi/G_{\text{PR}} + t + P_1/P_0}, \tag{14.35}$$

where ψ is the power ratio given by $\psi \triangleq \frac{\Delta_P P_{\text{SISO}}}{P_0}$. Inserting (14.30) into (14.35), we can easily obtain the realistic EE gain of MIMO against SISO system in the low and high-SE regimes, i.e., $\widehat{G}_{\mathrm{PR}}^0$ and $\widehat{G}_{\mathrm{PR}}^\infty$, respectively. It can be remarked in (14.30) that $G_{\mathrm{PR}}^\infty \gg 1$ as long as $m > 1$ and, hence, $\widehat{G}_{\mathrm{PR}}^\infty$ can be formulated as [10]

$$
\begin{aligned}
\widehat{G}_{\mathrm{PR}}^\infty = (\psi + 1 + P_1/P_0) \Big([1 - \mathrm{sgn}(m-1)]\psi(n-1)^{n-1}n^{-(n-\frac{1-\varsigma}{2})} \\
\times e^{(1-\phi)} + t + P_1/P_0 \Big)^{-1}
\end{aligned} \tag{14.36}
$$

Similarly, the EE gain due to an increase of SE can be re-expressed as

$$
\widehat{G}_{\mathrm{SE}} = G_{\mathrm{SE}} \frac{\Delta_P P_{\mathrm{SISO}} + P_0 + P_1}{\Delta_P P_{\mathrm{SISO}} + tP_0 + P_1} = G_{\mathrm{SE}} \frac{\psi + 1 + P_1/P_0}{\psi + t + P_1/P_0}. \tag{14.37}
$$

Inserting (14.32) into (14.37), we can easily obtain the realistic EE gain due to SE improvement of MIMO against SISO system in the low and high-SE regimes, i.e., $\widehat{G}_{\mathrm{SE}}^0$ and $\widehat{G}_{\mathrm{SE}}^\infty$, respectively.

Comparing the second equation of (14.30) with (14.36), an interesting paradox can be observed in the high-SE regime for the EE gain based on power reduction. In the idealistic PCM, G_{PR} increases both with the SE and number of antennas; whereas, equation (14.36) shows that $\widehat{G}_{\mathrm{PR}}$ decreases with the number of transmit antennas when considering a realistic PCM and $m > 1$. Hence, it implies that, as long as $r > 1$, $t = 2$ is the most energy efficient number of transmit antennas in the high-SE regime for a realistic PCM. Moreover, equation (14.36) reveals that if $\psi < 1$ and $m > 1$, then MIMO cannot be more EE than SISO. Concerning the EE gain due to SE improvement, it also decreases as the number of transmit antennas increases according to (14.37). These analytical results also indicate that contrarily to the idealistic PCM case where $G_{\mathrm{PR}} \geq G_{\mathrm{SE}}$, it is likely that improving the SE would be more EE than reducing the transmit power for certain SE and antenna configurations depending on the values of the parameters Δ_P, P_0 and P_1. Given a set of PCM parameters, it would be energy efficient to deploy a MIMO system instead of a SISO if at least $\widehat{G}_{\mathrm{PR}} \geq 1$ and even more if $\widehat{G}_{\mathrm{PR}} \geq \widehat{G}_{\mathrm{SE}}$.

According to equation (14.36), ψ can be a simple indicator for assessing whether or not to use MIMO for improving the EE. Setting $\widehat{G}_{\mathrm{PR}} \geq 1$ in (14.35), we can express ψ as function of the SE and the number of antennas via the EE gain $\widehat{G}_{\mathrm{PR}}$ such that

$$
\psi \geq \frac{t-1}{1 - 1/\widehat{G}_{\mathrm{PR}}}. \tag{14.38}
$$

This inequality indicates the value of ψ that is required for MIMO to be more energy efficient than SISO via power reduction. Similarly, we can define the value of ψ for ensuring that the EE gain via power reduction is greater than the EE gain due to an increase of SE by setting $\widehat{G}_{\mathrm{PR}} \geq \widehat{G}_{\mathrm{SE}}$ such that

$$
\psi \geq \frac{(t + P_1/P_0)(\widehat{G}_{\mathrm{SE}} - 1)}{1 - \widehat{G}_{\mathrm{SE}}/\widehat{G}_{\mathrm{PR}}}. \tag{14.39}
$$

Using the values of the double linear PCM parameters given in Table 14.1, i.e. $P_{\text{SISO}} = P_{\max} = 80$ W, $\Delta_P = 7.25$, $P_0 = 244$ W and $P_1 = 225$ or 0 W, we plot in the left and right sides of Figure 14.7, $1/\psi$ as the function of the SE and the number of antenna elements, respectively, when $n_{\text{ant}} = t = r$. The idea behind this graph is to show for a given set of PCM parameters whether or not to use MIMO for improving the EE. The different regions depicted on this graph are as follows: the white region represents the area where EE gain can only be obtained via SE improvement; the yellow region represents the area where EE gain can be obtained via power reduction, but more EE gain can be obtained via SE improvement. This region extends to the pale green area when $P_1 = 225$ W; the dark green region represents the area where EE gain is mainly obtained via power reduction and it expands to the pale green area when $P_1 = 0$ W. From an EE perspective, this dark green region is obviously the most desirable area for MIMO to operate in. Moreover, the brown, green, and blue curves, which have been plotted by using equations (14.38) and (14.39) represent the limits between the regions. Finally, the red curve represents the actual value of $1/\psi$ according to the value of the PCM parameters given above. Considering this particular set of parameters, the results indicate that using a 2x2 MIMO system instead of a SISO system can help to reduce power usage, especially if the value of P_1 is close to 0 W and for SE above 5 bit/s/Hz, since in that case the red curve will lie in the dark green area. The results also clearly indicate that using a MIMO system with more than three antenna elements is unlikely to be energy efficient via power saving.

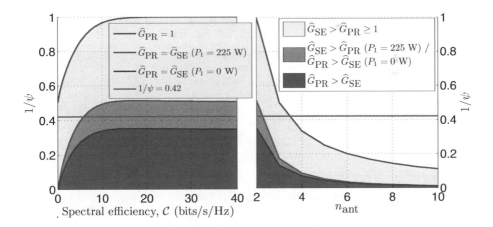

FIGURE 14.7 **MIMO EE indicator as the function of the SE and the number of antenna elements when $n_{\text{ant}} = t = r$, $P_{\text{SISO}} = 80$ W, $\Delta_P = 7.25$, $P_0 = 244$ W and $P_1 = 225$ or 0 W.**

14.4   EE-SE Trade-Off in a Cell

As it has been demonstrated at the link level, formulating the EE-SE trade-off into a CFE or CFA is a first step towards understanding how to reduce the consumed power while keeping an acceptable SE or quality of service (QoS). Moreover, the concept of trade-off itself is implicitly linked to optimization as indicated in Figure 14.1. At the cell level, where different users compete with each other for a limited set of resources, optimization based methods, such as link adaptation [49] and resource allocation [50] algorithms, have been extensively developed for taking a full advantage of the channel conditions and distributing the resources in an effective manner. An optimization method is in general only optimal for a given problem with a certain set of criteria; changing the criteria is likely to change the problem as well as the method. Until recently, the most popular criterion for designing efficient resource allocation algorithms has been the rate or SE. In order to make the resource allocation process 'fairer' and allow for QoS, fairness has also been used as a criterion but often in conjunction with SE. Power has also been considered in resource allocation but mainly as a constraint instead of a criterion. With the emergence of the EE as a key system design criterion, resource allocation based on EE is becoming very popular, especially in the uplink of a single-cell system for increasing the battery autonomy of UTs [39, 40, 51]. Moreover, we are currently witnessing a shift of research focus from battery-limited to unlimited power applications and resource allocation is not immune to this trend. For instance, the work in [52] has recently introduced a framework for optimizing the EE in the downlink of a single-cell system and the work in [53] proposed an algorithm for optimizing the EE-SE trade-off in the scalar broadcast channel. In the following, explicit expressions of the EE-SE trade-off are given for the SISO multi-user orthogonal and residual interference channels. These expressions are used as objective functions in an optimization problem, and the results of this EE-based optimization are compared against SE-based optimization.

14.4.1   SISO Orthogonal Multi-User Channel

Considering that K parallel subchannels are used for transmission and each of them has a different channel gain, i.e. equivalent to an orthogonal frequency division multiplexing (OFDM) transmission over a frequency-selective channel. Moreover, assuming block fading and that perfect channel state information (CSI) is available at both transmitter and receiver, the k-th user maximum achievable SE can be expressed as [39, 40]

$$
\mathcal{C}_k = \log_2 \left(1 + \frac{g_k p_k}{N\Gamma} \right), \tag{14.40}
$$

where p_k is the k-th user transmit power, $g_k = |h_k|^2$ is the k-th user channel gain, h_k represents the k-th user fading channel coefficient and Γ denotes the SNR gap between the channel capacity and the performance of a practical coding and modulation scheme as in [39, 40]. Conversely, from (14.40), p_k can be expressed as

$$p_k = \left(2^{\mathcal{C}_k} - 1\right) g_k^{-1} N\Gamma. \tag{14.41}$$

Knowing that the transmit power P is such that $P = \sum_{k=1}^{K} p_k$, it implies that

$$f^{-1}(\mathcal{C}) = \Gamma \sum_{k=1}^{K} \left(2^{\mathcal{C}_k} - 1\right) g_k^{-1}, \tag{14.42}$$

where $\mathcal{C} = [\mathcal{C}_1, \dots, \mathcal{C}_K]$. Inserting (14.42) into (14.7) yields the CFE of the EE-SE trade-off over the SISO orthogonal multi-user channel. Note that this formulation is the same for both uplink and downlink.

14.4.2 SISO Multi-User Channel with Residual Interference

We first consider the derivation of the EE-SE trade-off for the downlink of a single-cell single-antenna multi-user system, or equivalently the scalar broadcast (BC-S) model. Assuming that dirty paper coding [54] is employed at the BS and that the users are ordered as in [55], i.e. the user with the strongest channel is denoted as user 1 and it does not see the interference from other users, the maximum achievable SE of the k-th user can be expressed as [55]

$$\mathcal{C}_k = \log_2 \left(1 + \frac{g_k p_k}{N + g_k \sum_{j=1}^{k-1} p_j}\right). \tag{14.43}$$

Conversely, from (14.43), the transmit power of the k-th user can be given by

$$p_k = \left(2^{\mathcal{C}_k} - 1\right) \left(\sum_{j=1}^{k-1} p_j + g_k^{-1} N\right). \tag{14.44}$$

Moreover, assuming that $g_1 \geq g_2 \geq \dots \geq g_K > 0$, or conversely that $g_K^{-1} \geq g_{K-1}^{-1} \geq \dots \geq g_1^{-1} > 0$ and defining $\alpha_k = g_{K+1-k}^{-1} - g_{K-k}^{-1} \geq 0$ for $k \in \{1, \dots, K-1\}$ and $\alpha_K = g_1^{-1} > 0$, we obtain [53]

$$f^{-1}(\mathcal{C}) = -g_K^{-1} + \sum_{k=1}^{K} \alpha_k \prod_{j=1}^{k} 2^{\mathcal{C}_{K+1-j}}. \tag{14.45}$$

Inserting (14.45) into (14.7) yields the CFE of EE-SE trade-off over the BC-S channel.

Similarly, in the uplink, assuming that successive interference cancelation

(SIC) is employed at the BS and that the users are ordered as in [55], the maximum achievable SE of the k-th user can be expressed as [55]

$$C_k = \log_2 \left(1 + \frac{g_k p_k}{N + \sum_{j=k+1}^{K} g_j p_j} \right). \tag{14.46}$$

Conversely from (14.46), the transmit power of the k-th user can be given after simplifications by

$$p_k = N g_k^{-1} \left(2^{C_k} - 1 \right) \prod_{j=k+1}^{K} 2^{C_j}. \tag{14.47}$$

Then, it can be easily proved that $f^{-1}(\mathcal{C})$ is also expressed as in (14.45) for the uplink case, when considering the same ordering of the user gains as in the downlink case.

14.4.3   EE-Based Resource Allocation

14.4.3.1   Single User Optimal EE

In order to illustrate the optimization process in terms of EE, we derive the optimal EE point, C_J^*, when considering a single user, $S = C_1$ and the linear PCM of Table 14.1, i.e. $P_\Sigma = \Delta_P p_1 + P_0$ with $p_1 = P$. According to (14.42) and (14.7), the EE-SE trade-off is given by

$$C_J = \frac{B}{N} \frac{C_1}{\Delta_P f^{-1}(C_1) + P_0/N} \tag{14.48}$$

in this case. Consequently, the optimal EE point, C_J^*, is attained when $\frac{\partial f^{-1}(C_1)}{\partial C_1}\big|_{C_1=C_1^*} = \frac{\Delta_P f^{-1}(C_1^*)+P_0/N}{\Delta_P C_1^*}$ is fulfilled and the optimal corresponding SE, C_1^*, and transmit power, p_1^*, are given by

$$C_1^* = \frac{1}{\ln(2)} \left[W_0 \left(e^{-1} \left[P_0 g_1 (N\Gamma\Delta_P)^{-1} - 1 \right] \right) + 1 \right] \text{ and}$$
$$p_1^* = N\Gamma g_1^{-1} \left[\frac{P_0 g_1 (N\Gamma\Delta_P)^{-1} - 1}{W_0 \left(e^{-1} \left[P_0 g_1 (N\Gamma\Delta_P)^{-1} - 1 \right] \right)} - 1 \right], \tag{14.49}$$

respectively. Considering $\Delta_P = 4.7$, $P_0 = 130$ W and $P_{\max} = 20$ W, i.e. DbL maBS PCM of Table 14.1 for $t = 1$, as well as $N\Gamma = B = 1$ and $g_1 = 0.1$, we plot the EE as a function of the SE and transmit power, in the left and right sides of Figure 14.8, respectively. We depict both on these graphs the optimal unconstraint points, i.e. C_1^* and p_1^* and the optimal constraint points according to the PCM of Table 14.1, i.e. $p_1 = P_{\max} = 20$ W and $C_{1\max} = 1.585$ bit/s/Hz which is the value of C_1 when $p_1 = P_{\max}$ in (14.40). Note that in the multi-user case, it has recently been shown in [56] that the k-th user EE-optimal channel

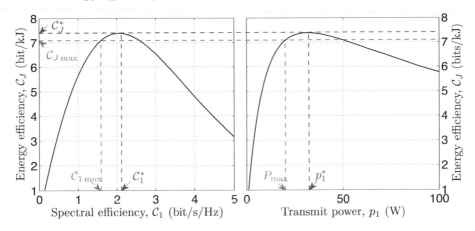

FIGURE 14.8 Optimal EE for the single user SISO channel. The optimal EE point is obtained at C_1^* and p_1^* if $C_1^* < C_{1\,\mathrm{max}}$ or at P_{max} if $C_1^* \geq C_{1\,\mathrm{max}}$.

capacity per unit bandwidth is given by

$$C_k^* = \frac{1}{\ln(2)}\left(W_0\left[\frac{\left(P_0(N\Gamma\Delta_P)^{-1} - a\right)g_k e^{\frac{b_k}{K^\star}-1}}{K^\star}\right] + 1 - \frac{b_k}{K^\star}\right), \quad (14.50)$$

where $a = \sum_{k=1}^{K^\star} g_k^{-1}$, $b_k = \sum_{j=1}^{K^\star} \ln(g_k^{-1} g_j)$ and $K^\star \geq 1$ is the EE-optimal number of active users in the system.

14.4.3.2 EE vs. SE Resource Allocation

In order to study the trade-off between energy, rate and fairness, the EE-based resource allocation strategy is compared here with the SE and fairness based strategies in terms of five different metrics: the transmit power P, the cell total consumed power P_Σ, the cell total sum-rate Σ_R, the cell total bit-per-Joule Σ_{C_J}, and the Jain's fairness index $\mathcal{J}$ given by [57]

$$\mathcal{J}(\mathcal{C}) = \frac{\left(\sum_{k=1}^K C_k\right)^2}{K \sum_{k=1}^K C_k^2}, \quad (14.51)$$

such that $\mathcal{J}(\mathcal{C}) \in [0, 1]$. We consider the following EE-based resource allocation strategy as in [53]

$$\begin{aligned}
\max_{\mathcal{C}} \ & \Sigma_{C_J}(\mathcal{C}) = \frac{B}{N}\frac{\sum_{k=1}^K C_k}{\Delta_P f^{-1}(\mathcal{C}) + P_0/N} \\
\text{s.t.} \ & C_k \geq 0, \forall k \in \{1, \ldots, K-1\} \text{ and } C_K > 0 \\
& P = \sum_{k=1}^K p_k \leq P_{\mathrm{max}} \\
& \mathcal{J}(\mathcal{C}) \geq \mathcal{J}_{\mathrm{min}}
\end{aligned} \quad (14.52)$$

which aims at maximizing the EE while keeping $P \in [0, P_{\max}]$ and ensuring a minimum of fairness. We denote this EE-based resource allocation strategy as $\text{RA}_{\Sigma C_J}$ if $\mathcal{J}_{\min} = 0$, i.e. no fairness constraint, as $\text{RA}_{\Sigma C_J, \mathcal{J}}$ if $\mathcal{J}_{\min} = 1$, i.e. full fairness constraint, and as $\text{RA}_{\Sigma C_J, \mathcal{J} \geq \mathcal{J}_{\min}}$ otherwise. Note that this optimization problem has been proved to be convex for both the orthogonal and residual interference channel scenarios in [39,40] and [53], respectively, i.e. if $f^{-1}(\mathcal{C})$ is either as in (14.42) or as in (14.45). This EE resource allocation strategy is compared here against the sum-rate and min-max fairness based resource allocation methods subject to a total power constraint, which are denoted as RA_{Σ_R} and $\text{RA}_{\mathcal{J}}$, as well as defined as

$$\max_{\mathbf{p}} \ \Sigma_R = B \sum_{k=1}^{K} \mathcal{C}_k(\mathbf{p})$$

$$\text{s.t.} \quad p_k \geq 0, \forall k \in \{1, \ldots, K\}, \text{ and } \sum_{k=1}^{K} p_k \leq P_{\max} \tag{14.53}$$

and

$$\max_{\mathbf{p}} \ \min_{\{k\}} \{\mathcal{C}_k(\mathbf{p})\}$$

$$\text{s.t.} \quad p_k \geq 0, \forall k \in \{1, \ldots, K\}, \text{ and } \sum_{k=1}^{K} p_k \leq P_{\max} \tag{14.54}$$

respectively, where $\mathbf{p} = [p_1, \ldots, p_K]$.

Considering the downlink of a single-cell single-antenna multi-user system, the resource allocation strategies $\text{RA}_{\Sigma C_J}$, $\text{RA}_{\Sigma C_J, \mathcal{J} \geq 0.5}$, $\text{RA}_{\Sigma C_J, \mathcal{J}}$, RA_{Σ_R} and $\text{RA}_{\mathcal{J}}$ are evaluated in Figures 14.9 and 14.10 in terms of various metrics and for $K = 10$ users uniformly distributed within the cell. Moreover, we consider that the k-th user channel gain is given by

$$g_k = 10^{(G_{\text{TxRx}} - PL(d_k))/10}, \tag{14.55}$$

where G_{TxRx} is the antenna gain of the BS-UT transmission. In addition, $PL(d_k) = P_{\text{LOS}}(d_k) PL_{\text{LOS}}(d_k) + (1 - P_{\text{LOS}}) PL_{\text{NLOS}}(d_k)$ is the path-loss as a function of the distance d_k between the BS and the k-th user, P_{LOS} is the line-of-sight (LOS) probability, and $PL_{\text{LOS}}(d_k)$ and $PL_{\text{NLOS}}(d_k)$ are the LOS and non-LOS (NLOS) path-loss functions, respectively. According to [58], we set $G_{TxRx} = 14$ dBi, $\text{PL}_{\text{LOS}}(d) = 24.8 + 20 \log_{10}(f_c) + 24.2 \log 10(d)$ (d in meter), $\text{PL}_{\text{NLOS}}(d) = -3.3 + 20 \log_{10}(f_c) + 42.8 \log 10(d)$, $P_{\text{LOS}} = \max\{1, e^{(-(d-10)/200)}\}$, $f_c = 2.1$ GHz, $B = 10$ MHz and $N_0 = -165.2$ dBm/Hz in our simulation. Furthermore, a capacity approaching coding and modulation scheme such that $\Gamma \simeq 1$ is assumed as in [39]. Concerning the PCM, the DbL maBS PCM of Table 14.1 with $t = 1$, $\Delta_P = 4.7$ and $P_0 = 130 + KP_{\text{UT}}$ W is used, where P_{UT} is the consumed power by each UT for reception and processing and it is set to 100 mW [40].

The results in Figure 14.9 focus on the orthogonal channel scenario. They show that $RA_{\Sigma C_J}$ provides a direct EE gain in comparison with RA_{Σ_R} and RA_J, since this strategy increases the EE by reducing the total transmit and cell total consumed powers. However, this comes at a cost of a lower sum-rate and level of fairness than RA_{Σ_R} and RA_J, respectively. The same conclusion can be made from Figure 14.10, which depicts the residual interference channel scenario, where $RA_{\Sigma C_J}$ can reduce the total transmit power by about 90% in comparison with RA_{Σ_R} and RA_J. Moreover, the results in Figure 14.10 also indicate that increasing the EE while keeping an acceptable level of fairness is possible, since $RA_{\Sigma C_J, J \geq 0.5}$ can increase the fairness from 0.1 to 0.5 while using less than a third of the total transmit power and keeping a near-optimal EE. High EE can still be achieved for small cell sizes, i.e. $r \leq 300$ m, even when the level of fairness is further increased from 0.5 to 1. Comparing the results in Figures 14.9 and 14.10, we observe that both the sum-rate and EE are higher in the interference rather than in the orthogonal case, but at the expense of fairness. It can also be remarked that the relative difference in terms of bit-per-Joule capacity between RA_{Σ_R} and $RA_{\Sigma C_J}$ is smaller and RA_{Σ_R} is fairer than $RA_{\Sigma C_J}$ in the orthogonal than in the BC-S channel scenario.

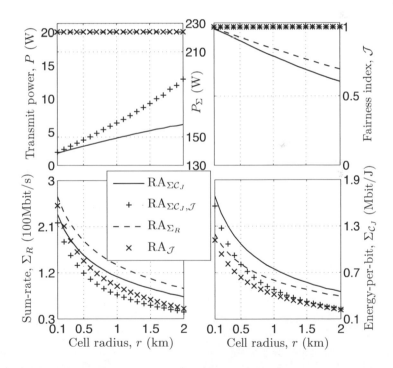

FIGURE 14.9 Performance comparison of various resource allocation strategies in terms of different metrics over the SISO orthogonal channel with $K = 10$ users.

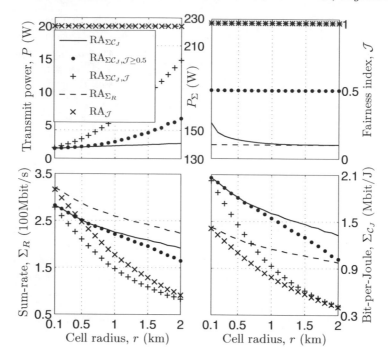

FIGURE 14.10 Performance comparison of various resource allocation strategies in terms of different metrics over the BC-S channel with $K = 10$ users.

14.5   EE-SE Trade-Off in Cellular Systems

The EE-SE trade-off in cellular systems is investigated in this section. More specifically, we focus our analysis on cellular networks employing BS cooperation. BS cooperation is a very promising multi-cell concept for delivering high data rate to users by exploiting or mitigating ICI and by ensuring a homogenous rate distribution among these users, which is essential for meeting the demand of future cellular systems. In the following, we first introduce the CoMP concept and the relevant existing contributions aiming at evaluating the EE-SE potential of futuristic cellular networks. We then discuss various works on PCM and their implications for the PCM of CoMP systems before proposing EE-SE trade-off expressions for CoMP with realistic PCMs. We analyze the EE-SE performance of CoMP by means of the LP approximation [9], our CFA [11] as well as numerical approaches. Focusing on the uplink scenario, we compare the EE-SE performances of the idealistic global BS cooperation, which is the information theoretic SE optimal approach for cooperative cellular systems, with the traditional approach of no cooperation and single user decoding at each BS. In order to make our analysis more tractable, we consider the circular Wyner model, which is a simplified yet mathematically more tractable architecture for modeling a cellular system. Moreover, we investigate the more practical case of clustered cooperation. We review the transmit EE (TxEE) and SE gains that can be achieved in the uplink when using an efficient power control scheme that properly manages inter-cluster interference (ICLI). We then identify the key design parameters and evaluate their effects on system performance by interpreting our results in a practical system scenario.

14.5.1   The Key Role of CoMP in Cellular Systems

From relaying schemes to mixed cell size overlay-underlay techniques, numerous innovative deployment strategies based on MIMO and *network densification* concepts are currently being investigated to address the SE-EE challenges in cellular systems. On the one hand, providing antenna diversity and decreasing the effective distance that radio signals have to travel between transceivers eventually promise both a substantial increase in SE and decrease in transmit power [59]. On the other hand, the densification process may lead to significant losses in throughput and degradations in fairness due to an increased amount of ICI. In this regard, CoMP can be seen as a highly promising technology for mitigating and even exploiting ICI through signal coordination at the BSs. BS cooperation by its nature has the potential to provide higher and more homogeneous data rate distribution for users [60, 61] and is currently being considered for wide-spread implementation by 3rd generation partnership project (3GPP) in LTE-Advanced networks [62]. In particular, the joint signal processing CoMP scheme, i.e. distributed MIMO (DMIMO) scheme, where transmit or receive information data are exchanged between BSs, can exploit ICI by transforming it into useful information and at the same time virtually densify the network by taking advantage of its inherent MIMO diversity gain [63]. However, additional energy burden is introduced by CoMP schemes due to the 1) extra signal processing at the BSs and 2) extra backhauling in order to obtain high speed, low-latency, low-error connectivity between cooperating BSs [15]. Consequently, investigating the EE-SE trade-off for auspicious technologies like CoMP is essential for better understanding the relationship between the throughput gain and the energy consumption of futuristic cellular architectures.

14.5.1.1   Relevant Background on CoMP

Looking beyond the original aim of CoMP techniques, i.e. improving SE, some recent studies have explored the viability of CoMP in terms of EE. As an initial step, the authors in [64] introduced a comprehensive cost evaluation model for CoMP systems, i.e. including both manufacturing and operational costs. However, this model only considers the expenditure from the BSs part since BSs have been identified as the major contributor to the total power consumption, e.g. see [65,66]. This comprehensive cost per bit analysis of [64], in which the BSs' operating power has been modeled for a specific LTE-based network as a function of each BS transmitted power and the cooperation cluster size, showed that additional backhauling requirements could limit the potential of CoMP in terms of EE. However, it has recently been argued in [67] that recent standards such as evolved high-speed packet access (HSPA+) and LTE require anyway high capacity backhaul links and since the cost of backhauling increases less than linearly with its capacity, these additional backhauling requirements should not be an issue for deploying relatively small-scale cooperative systems. In the same direction, the work of [15] studied the throughput-energy tradeoff of CoMP technology by taking into account the various components of a LTE-based network's radio access part along with the additional backhaul and signal processing power requirements for defining an extended mathematical PCM as a function of the inter-site distance (ISD). This analysis indicated that CoMP schemes with an appropriate cooperation size may only have a moderate positive effect on the bit-per-joule efficiency.

However, the aforementioned studies did not examine an important EE-related aspect. Indeed, BS cooperation can be used to minimize a part of the overall system energy consumption by ensuring that most of the energy spent by BSs and UTs is used to transport data. Therefore, a more inclusive PCM is needed for providing safe conclusions on CoMP technology's viability in terms of EE. Such a holistic PCM should also include the energy consumption savings due to reduction in transmit power needs into the energy performance framework.

14.5.2   Power Model Implications

As discussed above, the CoMP strategy may eventually result in higher signal-to-interference-plus-noise ratio (SINR) at each BS/UT which directly translates into an increase in SE for the network. Therefore, potential savings on the overall network energy resources may also be available when an operator is interested in satisfying certain rate constraints. However, additional energy is required to maintain a successful BS cooperation scheme. A general model for the power consumption in CoMP enabled systems can be given by

$$P_{\text{CoMP}} = P_{\text{SCP}} + \Delta P_{\text{Bh}} + \Delta P_{\text{SP}} - \Delta P_{\text{Tx}}, \qquad (14.56)$$

where P_{SCP}, ΔP_{Bh}, ΔP_{SP} and ΔP_{Tx} stand for the consumed power of the conventional single cell processing, the extra power requirements for back-hauling, the additional processing power needs at the BS and the savings on power requirements related to BSs/UTs transmit power, respectively.

Based on (14.56), a generic PCM for the uplink of cellular systems is described in the following and, without loss of generality, can also be employed for the downlink. Considering a cellular system with K active UTs per-cell where each of them transmits with a power of P, the total uplink transmit power is equal to KP. On top of the total UT transmit power, the UT circuit power and the BS processing and backhaul powers (when backhauling is utilized) must be taken into account in a realistic PCM for the uplink of cellular system. Adapting the PCMs of [15, 16, 40] to the cellular uplink scenario, we can express the realistic total consumed power per cell as

$$P_\Sigma = g(P) = K\left(\Delta_P P + P_c\right) + bP_{\text{SP}} + cP_{\text{Bh}}, \qquad (14.57)$$

where $0 \leq P \leq P_{\max}$, P_c and Δ_P are the circuit power and amplifier efficiency of each UT, respectively, while P_{SP} and P_{Bh} denote the BS signal processing and backhauling induced powers, respectively. In addition, the parameter $b = (1 + c_c)(1 + c_{dc})(1 + c_{ms})$ accounts for the cooling, DC-DC and main supply losses [14], i.e. c_c, c_{dc} and c_{ms} respectively, and c is the node degree, i.e. the ratio of the number of outgoing backhaul links to the number of BSs [15]. Coming back to the formulation of the EE-SE trade-off, its expression can be reformulated as

$$\mathcal{C}_J = \frac{BS}{NK\Delta_P f^{-1}(\mathcal{C}) + KP_c + bP_{\text{SP}} + cP_{\text{Bh}}} \qquad (14.58)$$

in the cellular uplink, by inserting (14.57) into (14.8). Note that a similar expression can be obtained for the cellular downlink by considering the power consumption related to the BS transmit power instead of the UTs' transmit power.

An in-depth understanding of the role of the different power elements in (14.57) is essential for improving the EE of cellular systems. In the following sections we review in more details the BS backhauling and signal processing induced powers.

14.5.2.1 Backhaul

In CoMP system, the increase in required backhauling power is related to the increase in required information exchange between BSs. CSI, scheduling and signaling information or even user data may have to be exchanged, depending on the type of cooperation, i.e. from limited coordination to full cooperation. In comparison with the conventional backhauling implementations currently standardized, e.g. the X2 interface for LTE [68], CoMP will require more links' capacity and less latency. According to the detailed analysis in [67], various powerful backhaul technologies are already available nowadays. For example, fiber-based Ethernet, passive optical network (PON), microwave and very-high-speed digital Subscriber Line 2 (VDSL2), are already or can be made suitable to support CoMP in terms of capacity and latency needs. Of course, a main controlling parameter for planning the backhaul requirements is the cooperation cluster size Q: how many BSs can communicate with each other to form a cooperative cluster given a certain type of cooperation. For instance in [67], it has been shown that the backhaul load requirement of even only three cooperating BSs can vary between few Mbps to few Gbps depending on the type of cooperation.

From an EE point of view, backhaul links are expected to play a significant role on the total network power consumption. Hence, it is imperative to examine the efficiency of the various existing backhauling options. Here, we focus our discussion on the two main candidates for backhauling; microwave and fiber links can be utilized to reliably support backhaul capacities far beyond few Mbps per link. The main shortcomings of a microwave backhaul include its high maintenance cost, licensed spectrum fee and LOS restrictions, i.e. limited both in terms of reach (few kilometers depending on weather conditions) and in terms of data rates (up to 1.25 Gbps). Therefore, a single point-to-point microwave link may not be sufficient for transporting reliably vast loads of data between BSs. Another restricting factor is the low end-to-end latency requirements which can be hardly fulfilled unless relaxation methods are applied [67]. However, the microwave technology allows for cheap deployment costs (no fiber digging) and relatively good power consumption figures (40-50W per 1.25 Gbps load). In [15], the backhauling power consumption was modeled as a set of $R_{\text{mw-link}} = 100$ Mbps microwave links where each link consumes $P_{\text{mw-link}} = 50$ W

$$P_{\text{Bh}} = \frac{R_{\text{Bh}}}{R_{\text{mw-link}}} \cdot P_{\text{mw-link}}, \qquad (14.59)$$

where R_{Bh} represents the average backhaul load requirement for each BS. Of course, the backhaul PCM in (14.59) can be generalized for any type of backhaul links.

On the other hand, even though optical fiber solutions have high installation cost, there is no LOS requirement and the maintenance cost is minimal when compared to the microwave backhaul case. As a result, power requirements for optical fiber solution is of great research interest [69, 70], while optical technology in general has recently attracted the attention of many operators for its high capacity and low power consumption [71–73]. In particular, PON technology is an emerging technology that can be used to establish optical routes between cooperating BSs; it promises very high data rates at

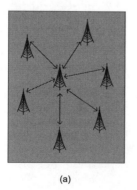
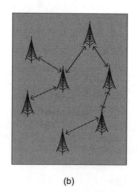
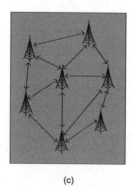

(a)   (b)   (c)

FIGURE 14.11 Backhaul Topologies: (a) Star, (b) Tree, (c) Mesh.

TABLE 14.3 Backhaul main categories Pros & Cons

Backhaul type	Advantages	Disadvantages
Microwave	- Low deployment cost - Low power consumption	- Licensed spectrum fee - High maintenance cost - LOS & distance restrictions
Fiber optic	- No LOS requirement - Low maintenance cost	- High installation cost (proportional to distance) - Unsuitable for mesh topology

higher power efficiency compared to the correspondent backhaul topologies implemented through microwave links. A comparison overview of the strengths and weaknesses of each backhauling technology is provided in Table 14.3. An overall cost evaluation based on the advantages and drawbacks of each backhauling architecture should be undertaken for deciding which one is the most desirable given a specific cooperation implementation. Although this type of analysis is out of the scope of this chapter, it is an interesting topic for future research.

Another important factor controlling the backhaul capacity requirements is the backhaul topology. The three main topology categories, namely tree, star and mesh, are illustrated in Figure 14.11. Mesh topology enables more BS connections than the two other topologies and, therefore, provides more cooperation capabilities between BSs, i.e. higher cluster sizes, adaptive clustering, and also increases the likelihood of finding alternative low latency routes. However, if the energy cost of a single link is as high as in PON, this topology would certainly lead to a very high increase in power consumption. Thus, PON technology can mainly be deployed by using a tree or star topology. In order to illustrate the power consumption of various backhaul topologies for different backhaul density (BD), we summarize in Table 14.4 the main results of [73]. Table 14.4 lists the values of consumed power per served user for various backhaul types and BD factors of 1.0, 1.25 and 1.5. The BD factor indicates the node degree, i.e. the average number of outgoing links at each BS. The higher the value of this factor is, the higher is the number of links connecting a certain BS to its neighbors. For instance, a BD factor of 1.0 in the mesh topology corresponds to an average node degree of 2.54 at each BS.

14.5.2.2 Signal Processing

BS cooperation also implies an increase in signal processing complexity and, therefore, power consumption. This is due to 1) the increased amount of channel estimations regarding UT receive signals that have to be processed by cooperating BSs; 2) the increased uplink and downlink MIMO processing. In [15,67], it has been reported that around $a_{\text{csi}} = 10\%$ of the total processing

386   *Green Communications: Theoretical Fundamentals, Algorithms ...*

TABLE 14.4 Backhaul topologies and their respective BD power consumptions

Backhaul type & Topology	Power consum. (Watt/User) BD = 1.0	Power consum. (Watt/User) BD = 1.25	Power consum. (Watt/User) BD = 1.5
PON - Star	10	18	19
PON - Tree	18	22	22
Microwave - Star	25	42	50
Microwave - Tree	45	62	62
Microwave - Mesh	95	175	180

power consumption in a cooperative system is due to extra channel estimation needs while $a_{\mathrm{mimo}} = 1 - 10\%$ is due to extra MIMO processing needs, based on results from an LTE-Advanced test bed simulator implemented by EASY-C project. It is also discussed in [15,67] that the signal processing load will increase with the increase of the cooperative cluster size Q. It is expected that the former effect will scale linearly with Q while the latter will scale quadratically with Q when assuming minimum mean-square error (MMSE) filter operation. Accordingly, the CoMP signal processing power consumption is given in [15] by

$$P_{\mathrm{SP}} = r.p_{\mathrm{SP}} \left[(1 - a_{\mathrm{csi}} - a_{\mathrm{mimo}}) + a_{\mathrm{csi}}Q + a_{\mathrm{mimo}}Q^2 \right], \qquad (14.60)$$

where p_{SP} denotes a signal processing power base value and r stands for the number of antennas per BS.

14.5.3   Global Cooperation: CFA of EE-SE Trade-Off

We first study the case of *global cooperation* where communication is possible among all the M BSs such that the cluster size $Q = M$. We consider the uplink of a cellular network where L UTs and all the M BSs are in different locations and can communicate with each other. Assuming that each BS is associated with K UTs, such that $L = KM$, where the jth BS is equipped with r_j antennas and the lth UT with t_l antennas, then the signal received at the jth BS is given by

$$y_j = \sum_{l=1}^{L} \alpha_{jl} \mathbf{H}_{jl} \mathbf{x}_l + \mathbf{n}_j, \qquad (14.61)$$

where $\mathbf{x}_l \in \mathbb{C}^{t_l}$ is the transmitted vector signal by the lth user and $\mathbf{H}_{jl} \in \mathbb{C}^{r_j * t_l}$ is the channel matrix between the lth UT and the jth BS. The gain elements in $\mathbf{H}_{jl}$ are independent and identically distributed random variables with zero mean and unit variance. Note that in (14.61), α_{jl} is the average channel gain between the lth user and the jth BS, $\mathbf{n}_j$ is the AWGN at the jth BS with zero mean and σ^2 variance. In addition, the signal transmitted by the lth user must satisfy the following power constraint : $tr(\mathbb{E}(\mathbf{x}_l \mathbf{x}_l^h)) \leq P_l$. The parameter

$\gamma_l = P_l/\sigma^2$ represents the transmit power of the lth user normalized by the noise at the BS. When the BSs cooperate to receive data from UTs, the overall system model can be represented by

$$\mathbf{y} = \tilde{\mathbf{H}}\mathbf{x} + \mathbf{n}, \tag{14.62}$$

where $\mathbf{y} = [\mathbf{y}_1^T \cdots \mathbf{y}_M^T]^T$ is the joint received signal vector, $\mathbf{x} = [\mathbf{x}_1^T \cdots \mathbf{x}_L^T]^T$ is the transmitted signal vector and $\mathbf{n} = [\mathbf{n}_1^T \cdots \mathbf{n}_M^T]$ is the joint received noise vector. The channel matrix can be expressed as

$$\tilde{\mathbf{H}} = \mathbf{\Omega}_V \odot \mathbf{H}_V, \tag{14.63}$$

$$\mathbf{H}_V = \begin{bmatrix} \mathbf{H}_{11} & \cdots & \mathbf{H}_{1L} \\ \vdots & \ddots & \vdots \\ \mathbf{H}_{M1} & \cdots & \mathbf{H}_{ML} \end{bmatrix}, \mathbf{\Omega}_V = \begin{bmatrix} \alpha_{11} & \cdots & \alpha_{1L} \\ \vdots & \ddots & \vdots \\ \alpha_{M1} & \cdots & \alpha_{ML} \end{bmatrix}, \tag{14.64}$$

where $\mathbf{\Omega}_V$ is a $Mr \times Lt$ deterministic matrix while $\mathbf{H}_V$ is a $Mr \times Lt$ matrix with independent and identically distributed random variables with zero mean and unit variance. As a result of the collocation of the multiple antennas at the UT and the BS, $\mathbf{\Omega}_V = \mathbf{\Omega} \otimes \mathbf{J}$, where $\mathbf{J}$ is a $r \times t$ matrix with all its elements equal to one and $\mathbf{\Omega}$ is a $M \times L$ deterministic matrix.

14.5.3.1 EE-SE Analysis of the Symmetrical Cellular Model

For simplicity reasons, we assume equal transmit power and an equal number of antennas for all UTs such that $\gamma_l = \gamma$ and $t_l = t$, $\forall l \in \{1, ..., L\}$ as well as an equal number of antennas at all BSs such that $r_j = r$, $\forall j \in \{1, ..., M\}$. We consider here the generic symmetrical cellular model introduced in [74], in which the sum of squared elements of the columns and rows of matrix $\mathbf{\Omega}_V$ can be given by

$$\Upsilon_j = \sum_{l=1}^{L} \alpha_{jl}^2 = \Upsilon, \quad \forall\, j \in \{1, ..., M\}, \tag{14.65}$$

$$\Theta_l = \sum_{j=1}^{M} \alpha_{jl}^2 = \Theta, \quad \forall\, l \in \{1, ..., L\}, \tag{14.66}$$

such that $L\Theta = M\Upsilon$. Examples of cellular models in which this assumption holds include the Wyner circular model and the Wyner two dimensional hexagonal array [75]. Notice that for the Wyner circular model $\Upsilon = 1 + 2\alpha^2$, where α is the attenuation scaling factor of the adjacent (next neighboring) cells, while for the Wyner two dimensional hexagonal array (Planar model) $\Upsilon = 1 + 6\alpha^2$.

It it has been recently demonstrated in [36] that the EE-SE trade-off for the uplink of the symmetrical MIMO CoMP cellular system can be obtained

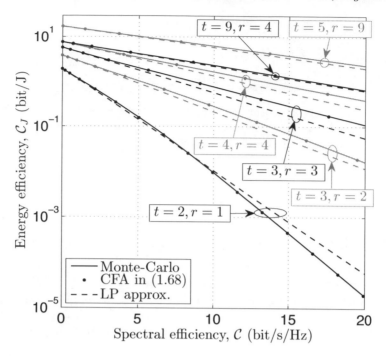

FIGURE 14.12 Comparison of the CFA, Monte-Carlo simulation and LP approximation based on the idealistic PCM.

as

$$f^{-1}(\mathcal{C}) \approx \widetilde{f}^{-1}(\mathcal{C}) = \frac{\left\{ -1 + \left(1 + [W_0\left(g_t(\mathcal{C})\right)]^{-1}\right)\left(1 + [W_0\left(g_r(\mathcal{C})\right)]^{-1}\right)\right\}}{2\Upsilon(1+\beta)},$$

(14.67)

where $\beta = \frac{r}{Kt}$ and $m = \min(Kt, r)$ instead of $\beta = \frac{r}{t}$ and $m = \min(t, r)$ in Section 14.3.3, $g_t(\mathcal{C}) \triangleq -2^{-\left(\frac{C+h((\mathcal{C}))}{2Kt}+1\right)}e^{-\frac{1}{2}}$, and $g_r(\mathcal{C})$ as well as $h(\mathcal{C})$ are expressed in equations (14.20) and (14.21), respectively. In the low-SE/power regime, i.e. when assuming that $\mathcal{C} \sim 0$, equation (14.67) can be simplified such that the minimum energy-per-bit is given by

$$\frac{E_b}{N_0}_{min} \approx \frac{\ln(2)}{2\Upsilon(1+\beta)Kt}.$$

(14.68)

In the symmetrical Wyner model with no BS cooperation and intra cell time-division multiple access (TDMA), the average per-cell sum-rate is given by (see (56) and (57) in [76])

$$
\begin{aligned}
R_{Psud} = \ & Bt \log\left(1 + \beta K\gamma\kappa_1\right) + Zt \log\left(\frac{1 + \beta\alpha^2 K\gamma\kappa_1}{1 + \beta\alpha^2 K\gamma\kappa_2}\right) \\
& + r \log\left(\frac{\kappa_2}{\kappa_1}\right) + r(\kappa_1 - \kappa_2)\log(e),
\end{aligned}
\tag{14.69}
$$

when the single user decoding approach is applied and where κ_1 and κ_2 satisfy

$$
\begin{aligned}
\kappa_1 + \frac{K\gamma\kappa_1}{1 + \beta KP\kappa_1} + Z\frac{K\gamma\alpha^2\kappa_1}{1 + \beta KP\alpha^2\kappa_1} &= 1, \\
\kappa_2 + Z\frac{K\gamma\alpha^2\kappa_2}{1 + \beta K\gamma\alpha^2\kappa_2} &= 1.
\end{aligned}
\tag{14.70}
$$

Given that Z is the number of interfering cells, which is two for the Wyner circular model and six for the planar model, the EE-SE trade-off expression for intra cell TDMA with no cooperation based on the per-cell sum-rate (equal rate in all the cells) and the per-cell transmit power is thus given by

$$
C_{j_{nc}} = \frac{R_{Psud}}{KP}.
\tag{14.71}
$$

Whereas, the realistic EE for the non-cooperative case is expressed as $C_{j_{ef-nc}} = \frac{R_{Psud}}{P_\Sigma}$, where P_Σ is given in equation (14.57).

In Figure 14.12, the trade-off between EE and SE in the circular Wyner model is depicted by inserting (14.67) into (14.7) for various antenna combinations, $\alpha = 0.4$ $B/N = 1$ and the idealistic PCM. Results demonstrate that our CFA closely matches Monte-Carlo simulation results, whereas the LP approximation approach of [9] is mainly accurate in the low-SE regime. Increasing the number of antennas at the UT or BS nodes results in an increase of both the EE and SE of the system since the slope of the trade-off curve becomes less steep in this case.

Figure 14.13 compares full cooperation, i.e. joint user decoding (JUD), with the non-cooperative scheme, i.e. single user decoding (SUD), in terms of EE based on the idealistic PCM for $P = 27$ dBm, $N = 1$, $B = 5$ MHz and α between 0 and 1. This figure shows that increasing α leads to an increase in EE for the full BS cooperation scheme as a result of the increase in diversity gain. On the other hand, increasing α leads to a reduction in EE for the non-cooperative scheme due to the increase in the interference.

The EE performance based on the realistic PCM, which incorporates the signal processing and backhaul powers is illustrated in Figure 14.14 for $N = 1$, $B = 5$ MHz and the PCM parameters for the uplink of CoMP in [15], i.e. $c_c = 0.12$, $c_{dc} = 0.08$, $c_{ms} = 0.09$, $p_{SP} = 58$ W, $C_{Bh} = 100$ Mbit/s and $P_{mw-link} = 50$ W. Increasing the number of cooperating BSs results in a loss in EE as no gain in per-cell sum-rate is achieved by increasing M beyond three. When M increases then the backhaul power increases at the same time, thus,

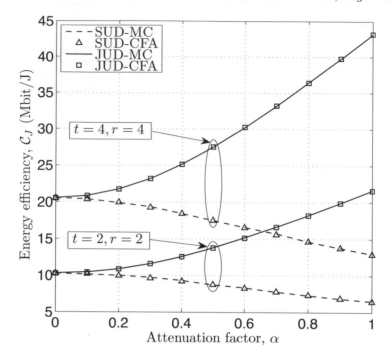

FIGURE 14.13 Comparison of the EE performance of the non-cooperative BS scheme vs. the M =-BS cooperative scheme based on the idealistic PCM, as a function of the attenuation factor α for $P = 27$ dBm.

leading to a loss in EE. Consequently, for very large M, the SUD scheme can outperform the JUD scheme over a significant range of the attenuation scaling factor.

14.5.4   Clustered CoMP: The Transmit EE-SE Relationship

Although CoMP is not an entirely new area of academic research, its realistic performance in terms of both SE and EE has yet to be assessed. The CoMP research agenda is gradually shifting from theoretical analysis towards more practical implementation studies where only a limited number of BSs cooperate, i.e. *clustered cooperation*, such that the increased backhaul infrastructure and the additional BSs' processing requirements are made affordable for real-world deployment. Numerous works have recently studied clustered cooperation, which is often referred to as locally performed *network MIMO*, however, these works mostly focus on SE aspects [77–80]. An initial attempt to investigate the effect of clustering on the overall power consumption of cellular system has been performed in [15]. This work indicated that the processing and backhaul power contributions become dominant especially for small site distances, where CoMP strategy is expected to deliver higher SE gains, when

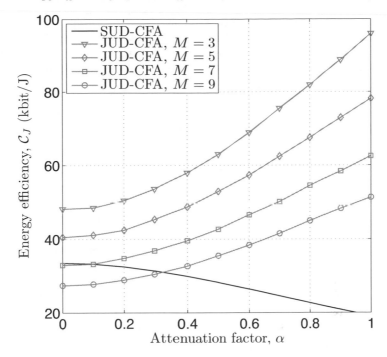

FIGURE 14.14 Comparison of the EE-SE performance of the non-cooperative BS scheme vs. various M-BS cooperative schemes based on a realistic PCM, as a function of the attenuation factor α for $r = 2, t = 2$, and $P = 27$ dBm.

the cooperation cluster size increases. It was concluded that cooperation between co-located BSs may be more appropriate in some cases to avoid the extra backhaul burden, whereas, only a few BSs should cooperate in real systems such that the effective SE gains and extra power consumption are kept in reasonable levels.

In this section, we focus on identifying how much energy gain can be provided by cooperation and efficient clustering in the uplink of cellular systems. Note that our analysis can easily be reproduced for the downlink case. Therefore, we evaluate the efficiency of various UT power management (PM) strategies under clustered CoMP. We put aside the consideration of energy dissipation at BSs and focus on the idealistic ΔP_{Tx} conservation in (14.56). As far as UT is concerned, one cannot ignore the fact that emerging data demanding applications deplete the batteries of mobile UT faster and faster; moreover, UT operational power is a function of its data transmission rate [38]. Furthermore, the total operational power consumption of UTs is becoming a significant proportion of the overall information and communications technology (ICT) industry power consumption, when considering the already huge (in the order of billions [81]) and massively growing number of UTs around the world. Therefore, the efficient use of UTs' energy to transmit information bits is becoming a crucial system design criterion. In this direction, it was observed in an initial study [18] that an efficient power allocation for the uplink of BS cooperation can indeed increase both the SE and EE. In the following, we concentrate our effort on evaluating and improving the per cluster average TxEE defined as

$$\text{TxEE}_Q \triangleq \frac{R_Q}{\sum_q \sum_k P_{q,k}}, \ (\text{bit/Joule}) \tag{14.72}$$

where R_Q is the cluster sum-rate (sum of the rates of all UTs in the cluster) and $P_{q,k}$ denotes the transmit signal power of UT k in cell q of any cooperating cluster. Note that the usage of the bit-per-joule metric is more appropriate for evaluating the EE in capacity limited situations, where the capacity of the network is an important design criterion, which is the case for future networks due to the rise of multi-media applications.

In order to model the network, we also consider as in Section 14.5.3 a linear system of M cells divided into M_Q clusters of cells but each with Q cells with $Q \ll M$, as it is depicted in Figure 14.15. The BSs are uniformly distributed (i.e. the ISD between neighboring BSs is the same for any two BSs) across a linear grid, each one at the center of each linear segment forming a cell. K UTs are distributed across each linear cell. The cooperation among the BSs is now limited only to those in cells that belong to the same cluster and hence a *Joint Processor* (JP) in each cluster of cells jointly decodes all the received signals from UTs of that cluster.

Following an information theoretic analysis and considering the strong law of large numbers and multipath fading with independent, uniformly distributed random phase on the specular path between UTs and BSs as in [82], the approximated ergodic achievable cell sum-rate can be given by

$$R_Q = B \log_2 \prod_{q=1}^{Q} \left[1 + \frac{\sum_{\dot{q}=1}^{Q} \sum_{k=1}^{K} P_{\dot{q},k} \left(\varsigma_{m,\dot{q},k}^{m,q} \right)^2}{\sigma^2 + \sum_{\dot{m}} \sum_{\dot{q}=1}^{Q} \sum_{k=1}^{K} P_{\dot{q},k} \left(\varsigma_{\dot{m},\dot{q},k}^{m,q} \right)^2} \right], \ (\text{bit/s})$$

$$\tag{14.73}$$

where $\varsigma_{\dot{m},\dot{q},k}^{m,q} = \sqrt{L_0} \left(1 + d_{\dot{m},\dot{q},k}^{m,q} \right)^{-\eta/2}$ denotes the squared distance dependent

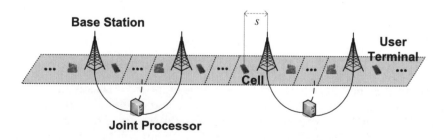

FIGURE 14.15 Linear clustered cellular system model

path loss coefficient with L_0 specifying the power received at a unit reference distance for a unit transmit power, η denoting the path loss exponent and $d_{\dot{m},\dot{q},k}^{m,q}$ defining the distance between user k in cell $\dot{q}$ of cluster $\dot{m}$ from the reference point, essentially co-located with the BS, in cell q of cluster m. Furthermore, σ^2 stands for the noise power at the receiver end. Note that $\varsigma_{\dot{m},\dot{q},k}^{m,q}$ is a more detailed formulation of the Wyner parameter α used in Section 14.5.3.

Clustering can be implemented in numerous ways; for instance, frequency, time and space division schemes can be utilized for creating and isolating clusters from each other as well as mitigating ICLI [83]. However, these schemes limit the available resources in the system and inevitably will lead to a reduction of the cooperation gain without providing any reduction in the total consumed energy for transmitting. As an alternative solution, an *ICLI allowance scheme* can be considered, where UTs and BSs in all clusters can exploit at any time the full amount of resources allocated to the system. In this case, the cells of every cluster experience ICLI since there is no complete isolation among clusters and signals transmitted from any UT may cause interference to BSs in other cooperating clusters of the system. To tackle the ICLI, power control on each UT's transmission can be performed. Since we study the ergodic capacity of the system, we assume that all UT signals during a long enough period of time experience all possible fading states and, hence, the parameter that defines the strength of a signal over that period of time is, effectively, the UT position in the cellular system. For that reason, we consider a variant UT power allocation according to the position of each UT on its respective cell and cluster. Since, cluster symmetry is assumed in our cellular system model and UT distribution is the same for each cells (and subsequently at each cluster), the power allocation outcome will be the same for any cluster.

Following the *sum rate optimization* approach in [82, 83], we adopt the linear cell-based power allocation for UTs such that $P_{q,k}$ will be a function of s, i.e. the distance between each UT and its respective BS. Hence, values of $P_{q,k}$ are limited to the following set

$$
\left\{
\begin{array}{cc}
\alpha_3 P_{\max} & \frac{1}{2} \cdot \mathrm{ISD}\,(1 - \alpha_1) \le s \le \frac{1}{2} \cdot \mathrm{ISD} \\
P(s) & \frac{1}{2}\alpha_2 \cdot \mathrm{ISD} \le s \le \frac{1}{2} \cdot \mathrm{ISD}\,(1 - \alpha_1) \\
P_{\max} & 0 \le s \le \frac{1}{2}\alpha_2 \cdot \mathrm{ISD}
\end{array}
\right\} ,
\tag{14.74}
$$

where the power function $P(s)$ is defined as

$$
P(s) \triangleq P_{\max}\left(\alpha_3 + (1 - \alpha_3) \frac{\left| s - \frac{1}{2}\mathrm{ISD}\,(1 - \alpha_1) \right|}{\frac{1}{2}\mathrm{ISD}\,(1 - \alpha_1 - \alpha_2)} \right)
\tag{14.75}
$$

with $P_{\max}$ and $P_{\min}$ denoting the maximum and minimum power constraints, respectively, ensuring that all UTs are able to perform their basic and/or emergency communication needs. The parameters $\alpha_{1,2,3}$ $((\alpha_1 + \alpha_2) \le 1)$ are defined to specify the power allocation of each user according to its location, as follows

$$0 \leq \alpha_1 \leq 1 \text{ - } \textit{edge-UTs} \text{ with } P_{\min}, \qquad (14.76\text{a})$$

$$0 \leq \alpha_2 \leq 1 \text{ - } \textit{center-UTs} \text{ with } P_{\max}, \qquad (14.76\text{b})$$

$$0 \leq \alpha_3 \leq 1 \text{ - defines } P_{\min} = \alpha_3 P_{\max}. \qquad (14.76\text{c})$$

The terms *edge-* and *center-* refer to the respective UT location, either in the cell or cluster.

In the following, we aim at interpreting the information theoretic results in a practical system scenario and evaluating the spectral and energy performances in a real-world network. In this regard, propagation parameters suggested by 3GPP in [62] are utilized. The value of the power loss L_0 at the unit reference distance is set according to the "Urban Macro - LOS" empirical scenario. Table 14.5 summarizes the various parameter values that have been utilized for obtaining our simulation results.

The analytical expressions for the average network SE and EE were validated through numerical Monte-Carlo simulations by using Matlab tools for generating 100 random channel matrix instances when assuming that only cells from adjacent clusters can interfere with each other. Three system density scenarios were defined and examined: 1) "Dense" (ISD= 100 m, $\eta = 2$); 2) "Average" (ISD= 600 m, $\eta = 3$); and 3) "Sparse" (ISD= 2 Km, $\eta = 3.5$). After an exhaustive search, the most efficient PM strategy of *opportunistic transmission* [18], both in terms of throughput and energy, was adopted. In this strategy, few "best" channel UTs in each cell (i.e. the UTs close to their respective BS in our case) are allowed to transmit with high power while the rest use lower power or can even be switched to a 'silent' mode during that communication slot. Figure 14.16 illustrates the TxEE-SE relationship for different PM strategies, where $P_{\min} = \alpha_3 P_{\max}$. Results for the three system density scenarios and for various cluster sizes (i.e. $Q = 1 - 6$) are obtained and also compared with the conventional case of *no Interference Management*, where all UTs are transmitting with 200 mW. It is observed that: 1) When ev-

TABLE 14.5 System Model Parameters

Parameters	Symbols	Values & Ranges
Frequency Carrier	f_c	2.1 GHz
Channel Bandwidth	B	10 MHz
Thermal Noise Density at BS	N_0	-169 dBm/Hz
UTs per cell	K	20
Inter-Site Distance	ISD	100 m to 5 Km
Reference Distance	d_0	1 m
Power Loss at Reference Distance	L_0	-34.5 dB
Path Loss Exponent	η	2 | 3 | 3.5
UT Max Transmit Power	$P_{\max}$	23 dBm

ery cell follows the same PM scheme, the relationship between SE and TxEE for any Q remains linear (as expected considering (14.72)).

The various PM schemes alter the slope of this linear relationship (i.e. steeper slope for lower total power and vice versa) as well as the TxEE-SE performance areas for each density scenario; 2) Higher level of cooperation, i.e larger cluster size, leads in general to better performance. However, it can be remarked that a higher Q value implies a larger performance gain in the average density scenario while there is no gain in the sparse scenario; 3) Average density scenario achieves higher performance and thus seems to be the most viable scenario for implementation of CoMP schemes; 4) It is preferable for UTs with the best channel conditions to transmit with less than the maximum available power (i.e. $P_{max} \leq 200$ mW in our case) as long as this does not affect the system throughput performance while the rest of the UTs should remain "silent". For example, Figure 14.16 indicates that when UTs transmit with a maximum power of $P_{max} = 150$ mW (instead of 200 mW) and a minimum power of $P_{min} = 0$, then high TxEE can be achieved without compromising the system's SE. In contrast, when the UTs' minimum power is set to $P_{min} = 50$ mW, then the system performance is significantly degraded.

In order to quantify the combined gains due to cooperation and UT power control, we plot in Figure 14.17 the SE and TxEE as a function of various cluster sizes for the average density scenario. While the improvement in SE due

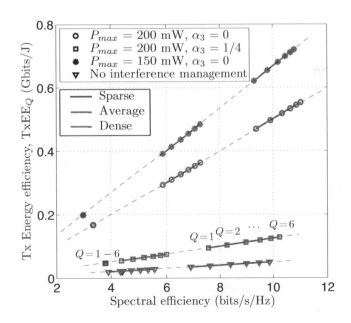

FIGURE 14.16 TxEE vs. SE for various UT power strategies, density systems, cluster sizes and $K = 20$.

to the combined effect of the two strategies is quite decent, a very large gain in TxEE, which is mainly due to power control, can be observed. A performance comparison between no cooperation (cluster size of $Q = 1$), cooperation with no interference management and cooperation with PM for a cluster size of $Q = 3$ is given as an example. In this case, the SE is increased by 0.6 dB due to cooperation and an extra 1 dB gain is achieved by managing the ICLI. Moreover, cooperation also improves the TxEE by 0.6 dB, whereas, UTs' power management provides a very large TxEE gain of 9.2 dB.

14.6   Conclusion and Outlook

In this chapter, we have introduced the EE-SE trade-off concept and explained why the usage of this metric is getting momentum in the current energy-aware context. We have surveyed the different theoretical approaches that can be used for defining this metric in any communication scenario. We have also emphasized on the importance of the PCM for getting meaningful results in EE related studies, and have reviewed the latest developments in PCMs for cellular systems.

In general, the problem of defining a closed-form expression for the EE-

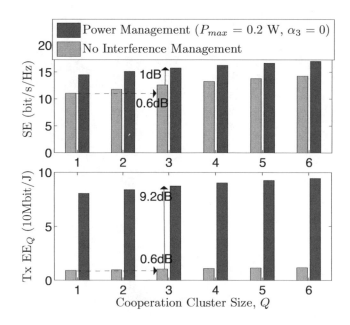

FIGURE 14.17 TxEE and SE gains due to the combined effects of cooperation and power management for $K = 20$, ISD $= 600$m and $\eta = 3$.

SE trade-off is equivalent to obtaining an explicit expression for the inverse function of the channel capacity per unit bandwidth. In this regard, we have reported two examples of explicit formulation for EE-SE trade-off over the AWGN and deterministic channels and have presented our accurate and generic CFAs for the EE-SE trade-off over the SISO and MIMO Rayleigh fading channels. Using these expressions, the EE potential of MIMO over SISO has been analytically and numerically assessed by means of an EE gain metric for both idealistic and realistic PCMs. In the high-SE regime, the theoretical EE gain increases with the SE as well as the number of antennas, whereas the practical EE gain decreases with the number of transmit antennas. Analytical insights have been confirmed by numerical results that have indicated the large discrepancy between the theoretical and practical MIMO/SISO EE gains; in theory, MIMO has a great potential for EE improvement over the Rayleigh fading channel; in contrast, when a realistic PCM is considered, a MIMO system with two transmit antennas is not necessarily more EE than a SISO system and utilizing more than three transmit antennas is likely to be energy inefficient, which is consistent with the findings in [6] for sensor networks.

In the context of multi-user communication, we have designed EE-optimal resource allocation schemes by using our explicit expressions for the SISO multi-user orthogonal and residual interference channels as objective functions in a multi-constraint optimization problem. We have then compared our EE-based resource allocation methods against the traditional methods based on sum-rate and fairness. Results indicate that our methods provide a large EE improvement via a significant reduction of the total consumed power in comparison with the conventional methods. Consequently, our methods always outperform the two other methods in terms of EE as well as transmit and cell total consumed powers. Moreover, our results also show that near-optimal EE and average fairness can be achieved at the same time.

The EE-SE trade-off of cellular systems employing BS cooperation has also been studied and we have discussed how to define a realistic PCM for CoMP systems. We have utilized this realistic PCM for formulating the EE-SE trade-off in the uplink of CoMP systems. The EE of CoMP has been compared against the traditional non-cooperative system for both the idealistic and realistic PCMs and results have pointed out that increasing the number of antennas at the UT or BS nodes results in an increase in both the EE and SE when considering the idealistic PCM. Moreover, BS cooperation with joint signal decoding approach always outperforms the non-cooperative approach. However, increasing the number of cooperating BS can lead to a reduction of EE when considering a realistic PCM. It also turns out that using more than 3 BSs for cooperation is unlikely to be beneficial in terms of EE. Finally, we have studied the more practical scenario of clustered cooperation and have reviewed the transmit EE and SE gains that can be achieved when using an efficient UT PM strategy. Our results show that the relationship between SE and transmit EE remains linear for any cluster size when all the cells rely on the same PM strategy. Larger cluster size leads to higher performance, especially in the average density scenario, while there is almost no gain in the sparse scenario. Moreover, CoMP achieves higher performance in average density scenario and, thus, we recommend the usage of CoMP in conjunction with an efficient UT power control for getting improved TxEE performance in such cellular deployments.

The performance evaluation framework for EE-based studies has been set up in this chapter and initial valuable insights have been obtained from the analysis of specific system models. The holistic EE-SE evaluation renders a very promising field for future research and further investigations are expected to shed more light into the actual effect of systems' key design parameters.

Acknowledgment

The research leading to these results has received funding from the European Commission's Seventh Framework Programme FP7/2007-2013 under grant agreement n°247733-project EARTH.

Bibliography

[1] A. Fehske, G. P. Fettweis, J. Malmodin, and G. Biczok, "The Global Carbon Footprint of Mobile Communiations: The Ecological And Economic Perspective," vol. 49, no. 8, pp. 55–62, Aug. 2011.

[2] I. Humar, X. Ge, L. Xiang, M. Jo, M. Chen, and J. Zhang, "Rethinking Energy Efficiency Models of Cellular Networks with Embodied Energy," *IEEE Network*, vol. 25, no. 2, pp. 40–49, Mar. 2011.

[3] K. Lahiri, A. Raghunathan, S. Dey, and D. Panigrahi, "Battery-driven System Design: A New Frontier in Low Power Design," in *Proc. Intl. Conf. on VLSI Design*, Bangalore, India, Jan. 2002, pp. 261–267.

[4] H. M. Kwon and T. G. Birdsall, "Channel Capacity in Bits per Joule," *IEEE J. Ocean. Eng.*, vol. OE-11, no. 1, pp. 97–99, Jan. 1986.

[5] C. Bae and W. E. Stark, "End-to-End Energy–Bandwidth Tradeoff in Multihop Wireless Networks," *IEEE Trans. Inf. Theory*, vol. 55, no. 9, pp. 4051–4066, Sep. 2009.

[6] S. Cui, A. J. Goldsmith, and A. Bahai, "Energy-Efficiency of MIMO and Cooperative MIMO Techniques in Sensor Networks," *IEEE J. Sel. Areas Commun.*, vol. 22, no. 6, pp. 1089–1098, Aug. 2004.

[7] L. M. Correia, D. Zeller, O. Blume, D. Ferling, Y. Jading, I. Godór, G. Auer, and L. V. D. Perre, "Challenges and Enabling Technologies for Energy Aware Mobile Radio Networks," *IEEE Commun. Mag.*, vol. 48, no. 11, pp. 66–72, Nov. 2010.

[8] G. Auer, V. Giannini, I. Godór, P. Skillermark, M. Olsson, M. A. Imran, D. Sabella, M. J. Gonzalez, C. Desset, and O. Blume, "How Much Energy is Needed to Run a Wireless Network ?" *IEEE Wireless Commun. Mag.*, vol. 18, no. 5, pp. 40–49, Oct. 2011.

[9] S. Verdú, "Spectral Efficiency in the Wideband Regime," *IEEE Trans. Inf. Theory*, vol. 48, no. 6, pp. 1319–1343, Jun. 2002.

[10] F. Héliot, M. A. Imran, and R. Tafazolli, "On the Energy Efficiency-Spectral Efficiency Trade-Off over the MIMO Rayleigh Fading Channel," *IEEE Trans. Commun.*, vol. 60, no. 5, pp. 1345–1356, May 2012.

[11] O. Onireti, F. Héliot, and M. A. Imran, "On the Energy Efficiency-Spectral Efficiency Trade-Off in the Uplink of CoMP System," *IEEE Trans. Wireless Commun.*, vol. 11, no. 2, pp. 556–561, Feb. 2012.

[12] A. Lozano, A. M. Tulino, and S. Verdu, "Multiple-antenna Capacity in the Low-power Regime," *IEEE Trans. Inf. Theory*, vol. 49, no. 10, pp. 2527–2544, Oct. 2003.

[13] O. Oyman and A. J. Paulraj, "Spectral Efficiency of Relay Networks in the Power Limited Regime," in *Proc. 42th Annual Allerton Conf. on Communication, Control and Computing*, Allerton, USA, Sep. 2004.

[14] O. Arnold, F. Richter, G. P. Fettweis, and O. Blume, "Power Consumption Modeling of Different Base Station Types in Heterogeneous Cellular Networks," in *Proc. ICT Future Network & Mobile Summit*, Florence, Italy, Jun. 2010.

[15] A. Fehske, P. Marsch, and G. P. Fettweis, "Bit per Joule Efficiency of Co-operating Base Stations in Cellular Networks," in *Proc. IEEE Globecom Workshops (GC Wkshps)*, Miami, USA, Dec. 2010.

[16] G. Auer et al., "D2.3: Energy Efficiency Analysis of the Reference Systems, Areas of Improvements and Target Breakdown," INFSO-ICT-247733 EARTH (Energy Aware Radio and NeTwork TecHnologies), Tech. Rep., Nov. 2010.

[17] S. Tombaz, P. Monti, K. Wang, A. Vastberg, M. Forzati, and J. Zander, "Impact of backhauling power consumption on the deployment of heterogeneous mobile networks," in *Proc. IEEE Globecom*, Houston, USA, Dec. 2011.

[18] E. Katranaras, M. A. Imran, and R. Hoshyar, "Energy Aware Transmission in Cellular Uplink with Clustered Base Station Cooperation," in *Proc. IEEE VTC-Spring*, Budapest, Hungary, May 2011.

[19] V. Rodoplu and T. H. Meng, "Bits-per-Joule Capacity of Energy-limited Wireless Networks," *IEEE Trans. Wireless Commun.*, vol. 6, no. 3, pp. 857–865, Mar. 2007.

[20] J. Gómez-Vilardebó and A. I. Pérez-Neira, "Duplexing and Synchronism for Energy Efficient Communication over a Relay Channel," in *Proc. IEEE ISIT 2008*.

[21] ——, "The Energy Efficiency of the Ergodic Fading Relay Channel," in *Proc. EUSIPCO*, Glasgow, UK, Aug. 2009.

[22] J. Gómez-Vilardebó, A. I. Pérez-Neira, and M. Nájar, "Energy Efficient Communications over the AWGN Relay Channel," *IEEE Trans. Wireless Commun.*, vol. 9, no. 1, pp. 32–37, Jan. 2010.

[23] S. Verdu, "On Channel Capacity per Unit Cost," *IEEE Trans. Inf. Theory*, vol. 36, no. 5, pp. 1019–1030, Sep. 1990.

[24] R. G. Gallager, "Energy Limited Channels: Coding, Multiaccess and Spread Spectrum," Tech. Report LIDS-P-1714, Tech. Rep., Nov. 1987.

[25] C. Shannon, "A Mathematical Theory of Communication," *Bell Syst. Tech. J.*, vol. 27, pp. 379–423, 623–656, July-Oct. 1948.

[26] G. Caire, G. Taricco, and E. Biglieri, "Suboptimality of TDMA in the Low-power Regime," *IEEE Trans. Inf. Theory*, vol. 50, no. 4, pp. 608–620, Apr. 2004.

[27] A. Lapidoth, I. E. Telatar, and R. Urbanke, "On Wide-band Broadcast Channels," *IEEE Trans. Inf. Theory*, vol. 49, no. 12, pp. 3250–3258, Dec. 2003.

[28] X. Cai, Y. Yao, and G. B. Giannakis, "Achievable Rates in Low-power Relay Links over Fading Channels," *IEEE Trans. Commun.*, vol. 53, no. 1, pp. 184–194, Jan. 2005.

[29] Y. Yao, X. Cai, and G. B. Giannakis, "On Energy Efficiency and Optimum Resource Allocation of Relay Transmissions in the Low-power Regime," *IEEE Trans. Wireless Commun.*, vol. 4, no. 6, pp. 2917–2927, Nov. 2005.

[30] J. Gómez-Vilardebó and A. I. Pérez-Neira, "Duplexing and Synchronism for Energy Efficient Source Cooperation Communications," in *Proc. IEEE SPAWC*, Recife, Brazil, Jul. 2008, pp. 451–455.

[31] O. Oyman and S. Sandhu, "Non-Ergodic Power-Bandwidth Tradeoff in Linear Multi-hop Networks," in *Proc. IEEE ISIT*, Seattle, USA, Jul. 2006, pp. 1514–1518.

[32] ——, "A Shannon-Theoretic Perspective on Fading Multihop Networks," in *Proc. IEEE CISS*, Princeton, USA, Mar. 2006, pp. 525–530.

[33] O. Somekh, B. M. Zaidel, and S. Shamai, "Sum Rate Characterization of Joint Multiple Cell-site Processing," *IEEE Trans. Inf. Theory*, vol. 53, no. 12, pp. 4473–4497, Dec. 2007.

[34] O. Simeone, O. Somekh, Y. Bar-Ness, and U. Spagnolini, "Throughput of Low-lower Cellular Systems with Collaborative Base Stations and Relaying," *IEEE Trans. Inf. Theory*, vol. 54, no. 1, pp. 459–467, Jan. 2008.

[35] F. Héliot, M. A. Imran, and R. Tafazolli, "A Very Tight Approximation of the SISO Energy Efficiency-Spectral Efficiency Trade-Off," *IEEE Commun. Lett.*, vol. 16, no. 6, 2012.

[36] O. Onireti, F. Héliot, and M. A. Imran, "Closed-form Approximation for the Trade-off between Energy Efficiency and Spectral Efficiency in the Uplink of Cellular Network," in *Proc. European Wireless Conference*, Vienna, Austria, Apr. 2011.

[37] T. Shu, M. Krunz, and S. Vrudhula, "Joint Optimization of Transmit Power-Time and Bit Energy Efficiency in CDMA Wireless Sensor Networks," *IEEE Trans. Wireless Commun.*, vol. 5, no. 11, pp. 3109–3118, Nov. 2006.

[38] Y. Hou and D. I. Laurenson, "Energy Efficiency of High QoS Heterogeneous Wireless Communication Network," in *proc. IEEE VTC-Fall*, Ottawa, Canada, Sep. 2010.

[39] C. Isheden and G. P. Fettweis, "Energy-Efficient Multi-Carrier Link Adaptation with Sum Rate-Dependent Circuit Power," in *Proc. IEEE Globecom*, Miami, USA, Dec. 2010.

[40] G. Miao, N. Himayat, and G. Y. Li, "Energy-Efficient Link Adaptation in Frequency-Selective Channels," *IEEE Trans. Commun.*, vol. 58, no. 2, pp. 545–554, Feb. 2010.

[41] I. E. Telatar, "Capacity of Multi-antenna Gaussian Channels," *Europ. Trans. Telecommun. and Related Technol.*, vol. 10, no. 6, pp. 585–596, Nov. 1999.

[42] M. Dohler, "Virtual Antenna Arrays," Ph.D. dissertation, King's College London, University of London, Nov. 2003.

[43] H. Shin and J. Lee, "Closed-form Formulas for Ergodic Capacity of MIMO Rayleigh Fading Channels," in *Proc. IEEE ICC'03*, Anchorage, USA, May 2003, pp. 2996–3000.

[44] M. Kang and M. Alouini, "On the Capacity of MIMO Rician Channels," in *Proc. 40th Annual Allerton Conf. on Communcation, Control and Computing*, Allerton, USA, Oct. 2002, pp. 936–945.

[45] M. Dohler and H. Aghvami, "On the Approximation of MIMO Capacity," *IEEE Trans. Wireless Commun.*, vol. 4, no. 1, pp. 30–34, Jan. 2005.

[46] E. Biglieri and G. Taricco, *Transmission and Reception with Multiple Antennas: Theoretical Foundations*. Now Publishers Inc., 2004.

[47] R. M. Corless, G. H. Gonnet, D. E. G. Hare, D. J. Jeffrey, and D. E. Knuth, "On the LambertW Function," *Adv. Comput. Math.*, vol. 5, pp. 329–359, 1996.

[48] X. Gourdon and P. Sebah, "The Euler constant: γ." [Online]. Available: http://numbers.computation.free.fr/Constants/Gamma/gamma.html

[49] S. Sampei, S. Komaki, and N. Morinaga, "Adaptive Modulation/TDMA Scheme for Personal Multimedia Communication Systems," in *Proc. IEEE Globecom*, San Francisco, USA, Nov. 1994, pp. 989–993.

[50] R. G. Gallager, *Information Theory and Reliable Communication.* John Wiley & Sons Inc., 1969.

[51] F. Meshkati, H. V. Poor, S. C. Schwartz, and N. B. Mandayam, "An Energy-Efficient Approach to Power Control and Receiver Design in Wireless Networks," *IEEE Trans. Commun.*, vol. 5, no. 1, pp. 3306–3315, Nov. 2006.

[52] Z. Chong and E. Jorswieck, "Analytical Foundation for Energy Efficiency Optimisation in Cellular Networks with Elastic Traffic," in *Proc. MOBI-LIGHT 2011*, Bilao, Spain, May 2011.

[53] F. Héliot, M. A. Imran, and R. Tafazolli, "Energy-Efficiency based Resource Allocation for the Scalar Broadcast Channel," in *Proc. IEEE WCNC*, Paris, France, Apr. 2012.

[54] M. H. M. Costa, "Writing on Dirty Paper," *IEEE Trans. Inf. Theory*, vol. 29, no. 3, pp. 439–441, May 1983.

[55] P. Viswanath and D. N. C. Tse, "Sum Capacity of the Vector Gaussian Broadcast Channel and Uplink-Downlink Duality," *IEEE Trans. Inf. Theory*, vol. 49, no. 8, pp. 1912–1921, Aug. 2003.

[56] F. Héliot, M. A. Imran, and R. Tafazolli, "Energy-Efficiency based Resource Allocation for the Orthogonal Multi-User Channel," in *Proc. IEEE VTC-Fall 2012 (to appear)*, Québec city, Canada, Sep. 2012.

[57] R. Jain, D. M. Chiuand, and W. Hawe, "A Quantitative Measure of Fairness and Discrimination for Resource Allocation in Shared Systems," DEC Research Report, Tech. Rep. TR-301, 1984.

[58] A. Ambrosy et al., "D2.2: Definition and Parameterization of Reference Systems and Scenarios," INFSO-ICT-247733 EARTH (Energy Aware Radio and NeTwork TecHnologies), Tech. Rep., Jun. 2010.

[59] Z. Roth et al., "Vision and Architecture Supporting Wireless GBit/sec/km2 Capacity Density Deployments," in *Proc. ICT Future Network & Mobile Summit*, Florence, Italy, Jun. 2010.

[60] M. K. Karakayali, G. J. Foschini, and R. A. Valenzuela, "Network Ccoordination for Spectrally Efficient Communications in Cellular Systems," *IEEE Wireless Commun. Mag.*, vol. 13, no. 4, pp. 56–61, Aug. 2006.

[61] O. Somekh, O. Simeone, Y. Bar-Ness, A. Haimovich, U. Spagnolini, and S. Shamai, *Distributed Antenna Systems: Open Architecture for Future Wireless Communications.* Auerbach Publications, CRC Press, 2007, ch. An Information Theoretic View of Distributed Antenna Processing in Cellular Systems.

[62] 3GPP TSG-RAN, "Further Advancements for E-UTRA, Physical Layer Aspects (Release 9)," 3GPP Std. TR 36.814 v1.5.1 (2009-12), Tech. Rep., 2009.

[63] H. Dai and H. Poor, "Asymptotic Spectral Efficiency of Multicell MIMO Systems with Frequency-flat Fading," *IEEE Trans. Signal Process.*, vol. 51, no. 11, pp. 2976–2988, Nov. 2003.

[64] P. Marsch, A. Fehske, and G. P. Fettweis, "Increasing Mobile Rates While Minimizing Cost per Bit; Cooperation vs. Denser Deployment," in *Proc. ISWCS'10*, York, UK, Sep. 2010, pp. 636–640.

[65] G. P. Fettweis and E. Zimmermann, "ICT Energy Consumption - Trends and Challenges," in *Proc. 11th International Symposium on WPMC*, Lapland, Finland, Sep. 2008.

[66] G. Auer et al., "Enablers for Energy Efficient Wireless Networks," in *Proc. IEEE VTC-Fall*, Ottawa, Canada, Sep. 2010.

[67] P. Marsch and G. P. Fettweis, *Coordinated Multi-Point in Mobile Communications - From Theory to Practice.* Cambridge University Press, 2011.

[68] *General Packet Radio System (GPRS) Tunnelling Protocol User Plane (GTPv1-U)*, 3GPP TS 29.281 v9.3.0 Std., Jun. 2010.

[69] O. Tipmongkolsilp, S. Zaghloul, and A. Jukan, "The Evolution of Cellular Backhaul Technologies: Current Issues and Future Trends," *IEEE Commun. Surveys Tuts.*, vol. 13, no. 1, pp. 97–113, Feb. 2011.

[70] S. R. Abdollahi, H. S. Al-Raweshidy, R. Nilavalan, and A. Darzi, "Future Broadband Access Network Challenges," in *Proc. 7th IFIP International Conference on WOCN*, Colombo, Sri Lanka, Sep. 2010.

[71] B. G. Bathula and J. M. H. Elmirghani, "Green Networks: Energy Efficient Design for Optical Networks," in *Proc. 6th IFIP International Conference on WOCN*, Cairo, Egypt, Apr. 2009.

[72] K. Sato, "Optical Technologies that Enable Green Networks," in *Proc. 12th ICTON*, Munich, Germany, Jun. 2010.

[73] L. Scalia, T. Biermann, C. Changsoon, K. Kozu, and W. Kellerer, "Power-efficient Mobile Backhaul Design for CoMP Support in Future Wireless Access Systems," in *Proc. IEEE INFOCOM Workshops*, Shanghai, China, Apr. 2011.

[74] D. Aktas, M. N. Bacha, J. S. Evans, and S. V. Hanly, "Scaling Results on the Sum Capacity of Cellular Networks with MIMO Links," *IEEE Trans. Inf. Theory*, vol. 52, no. 7, pp. 3264–3274, Jul. 2006.

[75] A. D. Wyner, "Shannon-theoretic Approach to a Gaussian Cellular Multiple-Access Channel," *IEEE Trans. Inf. Theory*, vol. 40, no. 6, pp. 1713–1727, Nov. 1994.

[76] A. Lozano and A. M. Tulino, "Capacity of Multiple-Transmit Multiple-Receive Antenna Architectures," *IEEE Trans. Inf. Theory*, vol. 48, no. 12, pp. 3117–3128, Dec. 2002.

[77] A. Papadogiannis, D. Gesbert, and E. Hardouin, "A Dynamic Clustering Approach in Wireless Networks with Multi-Cell Cooperative Processing," in *Proc. IEEE ICC'08*, Beijing, China, May 2008, pp. 4033–4037.

[78] S. Venkatesan, "Coordinating Base Stations for Greater Uplink Spectral Efficiency in a Cellular Network," in *Proc. IEEE PIMRC*, Athens, Greece, Sep. 2007.

[79] M. Sawahashi, Y. Kishiyama, A. Morimoto, D. Nishikawa, and M. Tanno, "Coordinated Multipoint Transmission/Reception Techniques for LTE-advanced [Coordinated and Distributed MIMO]," *IEEE Wireless Commun. Mag.*, vol. 17, no. 3, pp. 26–34, Jun. 2010.

[80] R. Irmer, H. Droste, P. Marsch, M. Grieger, G. P. Fettweis, S. Brueck, H.-P. Mayer, L. Thiele, and V.Jungnickel, "Coordinated Multipoint: Concepts, Performance, and Field Trial Results," *IEEE Commun. Mag.*, vol. 49, no. 2, pp. 102–111, Feb. 2011.

[81] P. Somavat, S. Jadhav, and V. Namboodiri, "Accounting for the Energy Consumption of Personal Computing Including Portable Devices," in *IFIP e-Energy '10*, Passau, Germany, Apr. 2010.

[82] E. Katranaras, D. Kaltakis, M. A. Imran, and R. Hoshyar, "Interference Allowance in Clustered Joint Processing and Power Allocation," in *Proc. IWCMC 2010*, Caen, France, Jun. 2010.

[83] E. Katranaras, M. A. Imran, and R. Hoshyar, "Sum Rate of Linear Cellular Systems with Clustered Joint Processing," in *Proc. IEEE VTC-Spring*, Barcelona, Spain, Apr. 2009.

Author Contact Information

Fabien Héliot, Efstathios Katranaras, Oluwakayode Onireti, and Muhammad Ali Imran are with Centre for Communication Systems Research, University of Surrey, UK, Email: F.Heliot@surrey.ac.uk, Efstathios.Katranaras@surrey.ac.uk, O.Onireti@surrey.ac.uk, M.Imran@surrey.ac.uk.

15

Energy Savings for Mobile Communication Networks through Dynamic Spectrum and Traffic Load Management

Adnan Aijaz

King's College London

Oliver Holland

King's College London

Paul Pangalos

King's College London

Hamid Aghvami

King's College London

Hanna Bogucka

Poznan University of Technology

CONTENTS

15.1   Background

The cellular industry has seen tremendous growth over the past few years. Information and communication technology (ICT) usage has increased at an exponential rate worldwide with approximately 5.6 billion users in 2010, and the data volumes growing by an order of magnitude every 5 years. Of this, mobile communications use has rocketed from 12% of all worldwide inhabitants subscribing to a service in the year 2000, to 78% in 2010 [1].To meet these growing traffic demands, operators have to deploy more base stations per area for higher capacity. As per one survey, every year 120,000 new base stations are deployed worldwide [2].

This increase in the number of base stations results in increased energy consumption and carbon footprint, particularly in remote areas where base stations rely on inefficient and polluting diesel generators for power. Recent studies show that ICT systems cause 2% global CO_2 emissions, which exceeds emissions produced by entire aviation industry [3]. Moreover, ICT is responsible for 10% of world energy consumption and it is expected that ICT energy consumption will increase by 15-20% per year [4]. For the cellular operator, 57% of energy is used in the access network as the base station, as shown in Figure 15.1, turns out to be the most energy intensive component of a mobile network. A typical 3G base station consumes between 800W(Watts) and 1500W for a transmission power of 40-120W. A 3G network with 12,000 base stations will consume over 50GWh per annum [5]. This huge energy consumption not only affects the environment, but also raises operational expenditure (OPEX) for operators.

With the ongoing increases in fuel prices worldwide, the awareness of harmful effects of CO_2 emissions on environment and the depletion of non-renewable energy sources, there is increasing trend towards more energy-efficient or green communications. The basic idea behind green communications, in a cellular sense, is to mitigate the inefficiencies in the cellular network operation, particularly in the access network. For the cellular operator, reducing energy consumption is not just a matter of corporate responsibility but is also very much an economically important issue.

Existing efforts towards green communications are broadly focusing on the following areas: Power efficient hardware design, energy efficient network

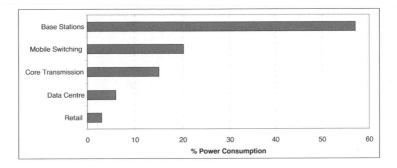

FIGURE 15.1 Power consumption of a typical cellular network (adapted from [5])

operation, and integration of alternate energy sources such as solar and wind power. This chapter focuses on energy efficient network operation through dynamic spectrum and traffic load management techniques. Section 15.2 presents different concepts for power saving in cellular networks through dynamic spectrum and traffic load management. A detailed performance assessment of these concepts is then carried out in Section 15.3. Finally, Section 15.4 concludes the chapter.

15.2 Dynamic Spectrum and Traffic Load Management

Radio spectrum has been managed in a rigid manner in the past, with the primary objective being to minimize interference among systems and users to maintain the viability of spectrum. Such management is inefficient, since inactivity of a system at any particular time will leave the associated spectrum unused as well. In recent years, spectrum management has entered new paradigms where more freedom has been created for systems to be able to dynamically assess which spectrum bands to use and use them. Technological advancements like software defined radios and adaptive radios are facilitating this freedom from a technical point of view. Moreover, it is widely expected that telecommunication operators worldwide will have access to increasing range of spectrum bands in future. Already, many operators are providing services in multiple bands, such as GSM 900MHz, GSM 1800MHz and UMTS 2GHz. This will lead the operators to dynamically distribute their traffic loads among a wider range of spectrum bands.

This chapter uses such advancements to save energy for cellular communication systems. In this section, the proposed dynamic spectrum and traffic load management concepts for reducing energy consumption in cellular access networks are discussed. The concepts may also provide alternate benefits, such as improved capacity.

15.2.1   Power Saving by Dynamically Powering Down Radio Network Equipment

The first concept, depicted in Figure 15.2 is reallocating traffic loads to particularly active bands from other bands to allow radio network equipment in those other bands to be switched off or put into standby mode, at times of low load [6], [7], [8]. The reduction in traffic load in some parts of a cellular network at some times might occur due to a number of effects, including typical day-night behavior of users, daily swarming of users from residential to corporate areas and back [9] and movement of users to/from some areas at weekend and vacation times, for example. This concept proposed by us is extremely promising as it implies guaranteed from-the-mains power saving.

There are two possibilities considered regarding the dynamic powering down of radio network equipment: (i) turning off base stations entirely in one network or spectrum band at that time/location, through traffic being sufficiently carried by single network or spectrum band and (ii) removing sectorization for a network, e.g., using spare capacity of one network/frequency to cover the required drop in load of another network/frequency in order to enable that other network/frequency to operate in omni-directional mode instead of tri-sectorized. These two techniques can be employed together in sectorized networks [6], [7], [8].

15.2.1.1   Cellular - Cellular Offloading

For the purpose of this chapter, the concept of dynamically powering down radio network equipment is investigated for the case where an operator has 3 available bands. The motivation for selecting this 3-bands scenario is the fact that most of the operators worldwide are already providing 2G/3G services in 900MHz, 1800MHz and 2GHz bands. Firstly, the scenario of removing sectorization is considered as follows. Given the threshold T_{switch} as the number of users per cell at which the omni-directional mode is switched to tri-sectorized mode. Taking L_i, L_j and L_k to be the cell loads for 3 networks respectively, we have the following cases:

(a) $L_i \geq T_{switch}$, $L_j \geq T_{switch}$, $L_k \geq T_{switch}$ $i \neq j \neq k,$
(b) $L_i \geq T_{switch}$, $L_j \geq T_{switch}$, $L_k < T_{switch},$
(c) $L_i \geq T_{switch}$, $L_j < T_{switch}$, $L_k < T_{switch},$
(d) $L_i < T_{switch}$, $L_j < T_{switch}$, $L_k < T_{switch}.$

In cases (a) and (d) removing sectorization is not possible since all the three networks must remain operating in tri-sectorized or would already be omni-directional mode respectively. However in cases (b) and (c), it is possible to

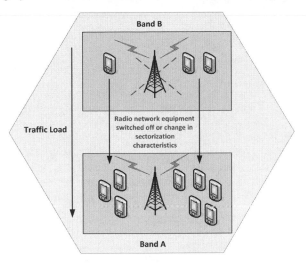

FIGURE 15.2 Reallocating traffic load between bands to enable radio network equipment to be switched off

remove sectorization for one or two of the networks and hence to reduce the power consumption.

Considering the other scenario of power saving by turning off one or more networks entirely through moving users to other networks, in case of 3 networks, it is possible to switch off either one or two networks when:

(e) $L_i + L_j < T_{switch}$,

(f) $L_i + L_k < T_{switch}$,

(g) $L_j + L_k < T_{switch}$,

(h) $L_i + L_j + L_k < T_{switch}$.

15.2.1.2 Cellular - Wi-Fi Offloading

The concept of dynamically powering down radio network equipment can be extended to opportunistically shift users from cellular to Wi-Fi networks. The area of cellular and Wi-Fi integration has been extensively investigated in recent years. There are two architectures for coupling Wi-Fi and cellular (3G) networks: loose coupling and tight coupling. In a loose coupling architecture, the networks are independent requiring no major co-operation between them. Service continuity is provided by roaming between the two networks. On the other hand, in a tightly coupled system the networks share a common core and the majority of network functions such as vertical handover, resource management, billing and security are controlled and managed centrally.

We consider a tightly coupled cellular/Wi-Fi system and for simplicity assume a single 3G cell with multiple Wi-Fi access points (or a cluster of

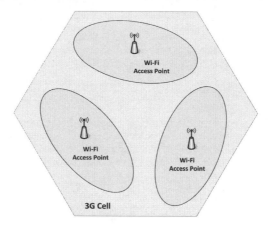

FIGURE 15.3 Single 3G cell having multiple Wi-Fi access points within its coverage area

access points) deployed therein. Typically, a Wi-Fi access point has much smaller coverage area than a 3G base station.

Let the maximum number of users supported by each Wi-Fi access point to be N. If a total of M access points exist in a 3G cell then the number of users collectively supported would be $N \times M$. This important parameter is named N_{tot}. For the cellular network, an important parameter is T_{switch}, which is the threshold (in terms of load on the cell) at which the omni-directional mode is switched to tri-sectorized mode. The conditions for removing sectorization and turning off the base station are shown in 15.1 and 15.2 respectively,

$$L_{3G} - T_{switch} + L_{Wi-Fi} \leq N_{tot}, \tag{15.1}$$

$$L_{3G} + L_{Wi-Fi} \leq N_{tot}, \tag{15.2}$$

where L_{3G} and L_{Wi-Fi} are the loads (in terms of number of users) on 3G and Wi-Fi networks respectively.

15.2.1.3   Cellular - White-Fi Offloading

The cellular and Wi-Fi offloading scenario can be further extended to White-Fi offloading. White-Fi, as currently being standardized by IEEE 802.11af, is the proposal for using conventional Wi-Fi in the TV white spaces (TVWS) using geolocation database technology. TVWS spectrum comprises large portions of UHF/VHF spectrum that are already unused or becoming unused on a geographical basis due to the completion of the digital TV switch-over program. White-Fi has many benefits such as increased range due to better propagation characteristics and potentially the availability of much larger bandwidths in

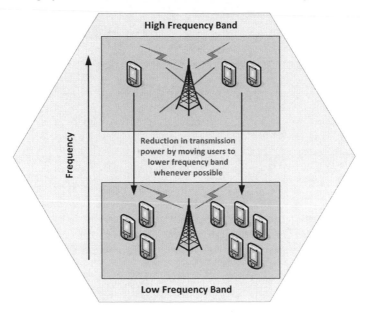

FIGURE 15.4 Reallocating user/links to improve propagation

many areas. However, channel aggregation will be necessary to provide data rates comparable to Wi-Fi operating in 2.4 and 5 GHz bands.

15.2.2   Power Saving by Propagation Improvement

The second concept is the opportunistic usage of more appropriate spectrum given local path loss characteristics and base station density to improve propagation or reduce inter-cell interference in frequency reuse scenarios. This concept is illustrated in a simplified representation in Figure 15.4.

Considering a GSM/DCS system where both 900MHz and 1800MHz bands are used to provide services over a coverage area. For a wide range of path loss models, the 900MHz band experiences at least 4 times (6 dB) less path loss as compared with the 1800MHz band. This simple model (ignoring aspects such as antenna gains and heights), implies that at least 75% less transmission power can be used in 900MHz band compared to 1800MHz band. Thus the opportunistic usage of 900MHz band under periods of low utilization, particularly in early morning hours for example, might allow the switching off of half of the base stations for 1800MHz network [6].

It should be noted that this concept of power saving by propagation improvement can be employed together with the concept of reallocation to power down equipment, yielding further improvement in power efficiency.

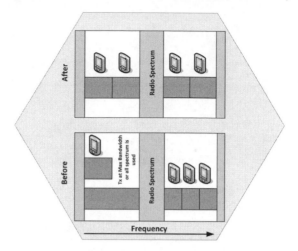

FIGURE 15.5 Reallocating user to increase or better balance channel bandwidths

15.2.3   Power Saving by Channel Bandwidth Increase or Better Balancing

The third concept is dynamic sharing of spectrum to facilitate channel bandwidths being increased or better balanced. Again, considering a multi-band case, if the spectrum that can be allocated in transmissions to users among the band is unequal due to bands experiencing different loadings, better power efficiency can be achieved by moving users among the bands to better balance the traffic loading to bandwidth ratio among the bands. This is simply a consequence of the power saving proportion against bandwidth increase factor being an increasing function of decreasing gradient [6], [7], [8].

This concept might alternatively increase average channel bandwidths in transmissions to users. Considering the case in Figure 15.5, where the bandwidth allocated to one user is already at the maximum bandwidth for radio access technology being used, and there is still spare capacity in the band. The opportunistic reallocation of one user between the bands could increase the bandwidth by 50% for three out of four users, while the bandwidth would remain the same for the other user. Such a bandwidth increase could give a significant decrease in necessary transmission power to three of the four users.

15.3   Performance Assessment

Performance evaluation of the above mentioned concepts has been carried out in Matlab, using a custom-written code. This section details those performance assessments and their results.

15.3.1   Power Saving by Dynamically Powering Down Radio Network Equipment

Here we assess power savings achievable under both of the concepts for dynamically powering down cells. We consider a simple power consumption model for the base station, where it is assumed that each active radio chain consumes the same unit power. This model is not far from reality, particularly in terms of macro-cells where the from-the-mains power consumption only varies marginally against transmission power and traffic loading.

15.3.1.1   Cellular - Cellular Offloading

It is assumed that the average traffic load over a 24 hour period, for the cellular network can be seen as varying according to scaled and shifted sine or cosine cycle [10].

$$L(t) = \frac{BusyLoad + QuietLoad}{2} + \frac{BusyLoad - QuietLoad}{2} \cos(2\pi(t - \phi)/24), \tag{15.3}$$

where the *BusyLoad* is the busy hour load and *QuietLoad* is the quiet hour load, in terms of number of users in the cell, ϕ, is the busy hour and t is the time of day, both on a 24-hour clock. The *QuietLoad* is taken to be 25% of *BusyLoad* in order to match the statistics, as shown in Figure 15.3.1.1, for a 3G network in London, UK, obtained by interaction with Vodafone representatives within the Mobile VCE Green Radio research program. It has been observed through a wide range of simulations that results for the real Vodafone traffic representation and since cycle alternative closely match.

In addition to this, the statistical number of active users in the cell is assumed to be Poisson distributed, the mean of which can be taken from the average load at that time of the day, $L(t)$. The probability that k number of active users is present at any time of the day t is expressed as:

$$P(k, t) = \frac{L(t)^k e^{-L(t)}}{k!}. \tag{15.4}$$

The numerical assessment uses the value of $L(t)$ at each hourly time unit to parameterize Equation 15.4. In the outer loops it cycles through a 24 hour period in steps of t of one hour while in inner loops, it cycles through each possible value of k (representing the number of users in the cell) for each participating frequency band and ascertains the power consumption that would be

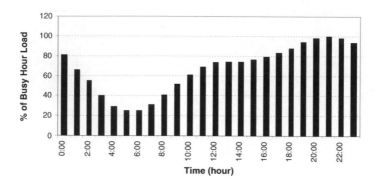

FIGURE 15.6 Traffic load as a fraction of busy hour load over a 24 hour period (Vodafone)

required by applying the selected power saving solution (sectorization switching or powering down). The actual power consumption is then given as this power consumption, multiplied by the probability of its occurrence, which is the product of probabilities of chosen values of k occurring for the participating networks/frequencies, i.e., $P_1(k1, t) \cdot P_2(k2, t) \cdot P_3(k3, t)$. The result is then summed with equivalent results for all possible chosen values of k to obtain the overall power consumption at time t. The same operation is performed over all hourly time units in the 24-hour period and average power consumption is calculated among all time units.

Exactly the same process is then performed to find the average power consumption of a conventional system without applying any power saving solution. The power saving percentage is then calculated as

$$100 \times \left(1 - \frac{\text{Power after applying power saving solution}}{\text{Power before applying power saving solution}} \right) \qquad (15.5)$$

The configuration parameters are shown in Table 15.1 where it is assumed that three spectrum bands are available among the GSM, GPRS and UMTS systems. The capacity calculations are made at 1% blocking probability (grade of service). With the given parameters as shown in Table 15.1, value of T_{switch} as calculated, is 25 users.

The power savings achievable under different scenarios are shown in Figures 15.7 - 15.10. The busy hour load for all three networks/bands is taken to be equal. For each value of busy load, the graphs show percentage of from-the-mains power savings averaged for a 24-hour period. From these results, it is clear that significant power savings can be achieved by applying sectorization switching and powering down solutions together. Maximum savings of up to

TABLE 15.1 Configuration Parameters

Parameter	Value
System configuration	Reflecting GSM/GPRS/UMTS system
No of channels available to the system (GSM/GPRS)	252 [11]
Tswitch	25
No of carriers per cell (UMTS)	1
Bandwidth per carrier (UMTS)	5 MHz
Busy hour load per band (in terms of average number of users per cell)	varied

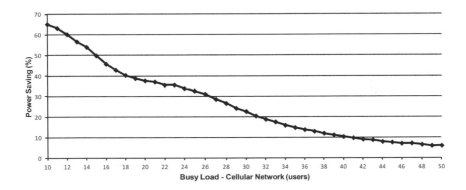

FIGURE 15.7 Power saving against busy hour load (sectorization switching and powering down together)

65% can be achieved when the networks are lightly loaded. It should be noted that throughout this chapter, the achievable savings are represented in terms of percentage. The from-the-mains power saving percentage can be translated into actual power saving given the actual power consumption of the base station. However, the base station power consumption depends on its design and varies among different manufacturers, so is not considered here.

The power savings by applying the sectorization switching solution only are shown in Figure 15.8. Maximum savings of up to 11% are possible in this case. With 3 networks/spectrum bands, it is possible to switch two networks / spectrum bands to omni-directional mode. The savings in this case, as shown in Figure 15.9, are not significant as it requires all networks to be lightly loaded at the same time i.e., $(L_i - T_{switch} + L_j - T_{switch} + L_k) \leq T_{switch}$, which statistically under our Poisson representation rarely happens.

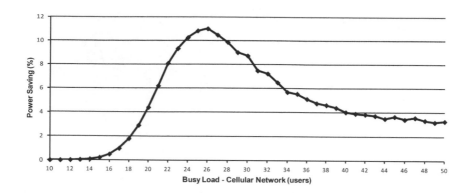

FIGURE 15.8 Power saving against busy hour load (sectorization switching only)

The power savings achievable under powering down solution are shown in Figure 15.10. This shows that significant savings are possible under low load conditions.

Sectorization switching is applicable when the load is above T_{switch}. This is the reason why this solution yields little savings under low load conditions. One might argue that the savings should be zero for all the values of load less than or equal to T_{switch}. This is not the case, as the active number of users is Poisson distributed with the mean taken from $BusyLoad$. For example, if the $BusyLoad$ is 20 users, the active number of users would be Poisson distributed around an average of 20 at the busy hour load, leading to an effective range of users from, for example, 10 to 30 (the closer to the mean, the higher the probability). Thus we see savings from the sectorization switching solution even if the load is below T_{switch}. The location of the peak power saving against $BusyLoad$ is close to the value of T_{switch}.

In order to carry out a comparison with two networks / spectrum bands, surface plots have been generated where busy hour loads of two bands are varied while the busy hour load for a third band is taken to be the same as first band. As it can be seen from Figure 15.11, significant power savings of up to 65% can be achieved compared to 50% in case of 2 bands when both sectorization switching and powering down solutions are applied. The average/overall savings are higher as well. These extra savings are due third band which can be opportunistically used to offload users.

A comparison of multiple bands scenario is shown in Figure 15.12. Busy hour load of all the bands is taken to be same and the network powering down solution is applied. Also it is assumed that all the bands have same power consumption per active radio chain. It is clear that the 3 bands scenario gives

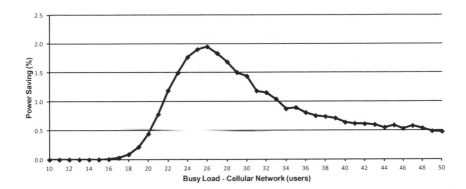

FIGURE 15.9 Power saving against busy hour load (two networks switched to omni)

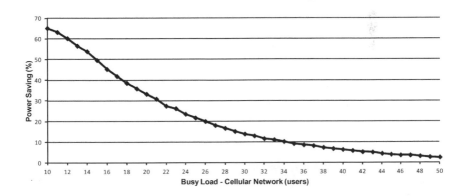

FIGURE 15.10 Power saving against busy hour load (powering down only)

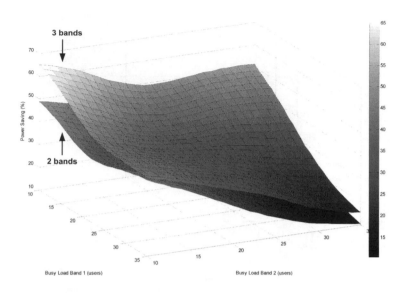

FIGURE 15.11 Comparison of power savings for 3 networks/bands with 2 networks/bands)

significant performance improvement over the 2 bands however the 4 and 5 bands scenarios show limited further improvement.

15.3.1.2 Cellular - Wi-Fi Offloading

We assume the user association rate, $\lambda(t)$, for the Wi-Fi network, over a 24 hour period to be varying approximately according to a sine wave [12] as shown in Figure 15.13, which resembles closely with the observations made in [13]. For the cellular network, the traffic load $L(t)$ over a 24 hour period is assumed to be varying according to the Vodafone distribution as shown in Figure 15.3.1.1. The traffic load is scaled by the BusyLoad (busy hour load) at any time of the day; for example if the BusyLoad is 10 users then the number of users at 10:00 is 60% of BusyLoad i.e., 6 users.

In addition to this, the statistical number of active users in the 3G cell are assumed to be Poisson distributed, the mean of which can be taken from the average load, $L(t)$ at that time of the day, as given by Equation 15.3. The probability that k number of active users is present at any time of the day t is given by Equation 15.4 as before.

Similarly, we can assume the number of active users in a Wi-Fi cell to be Poisson distributed, the mean of which at any time of day can be taken from the user association rate, $\lambda(t)$, at that time of the day and calculate the active number of users i.e., using $\lambda(t)$ in Equation 15.4 instead of $L(t)$.

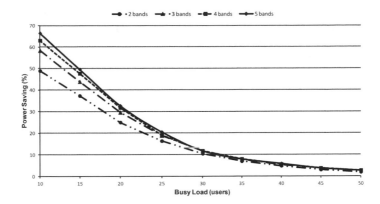

FIGURE 15.12 Power savings achievable for multiple bands (powering down solution only)

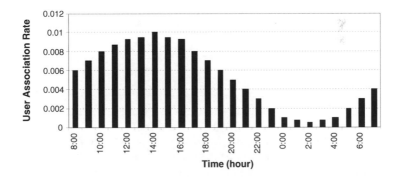

FIGURE 15.13 User association rate (Wi-Fi) over a 24 hour period

The numerical assessment follows a similar approach as before and cycles in outer loops through a 24 hour period in steps of t of one hour, using the value of $L(t)$ or $\lambda(t)$ to parameterize Equation 15.4. In inner loops it cycles through each possible value of k (number of active users) for 3G and Wi-Fi networks and ascertains the power consumption that would be required given the proposed offloading approach. The actual power consumption is then given as the product of this power consumption and the probabilities of the chosen values of k occurring for 3G and Wi-Fi, i.e., $P(k, t)_{3G} \cdot P(k, t)_{Wi-Fi}$. This result is then summed with the equivalent values for all possible chosen values of k to obtain overall power consumption at time t. The operation is repeated over all hourly time units in the 24 hour period, and the average power consumption is calculated. Exactly the same process is performed to find the average power consumption of a conventional system without applying any power saving solution. The percentage of power saving is calculated by Equation 15.5 as before.

It should be noted that power consumed by Wi-Fi access points is infinitesimal in comparison to cellular provision, so power saving calculations are omitted. Further, from power savings perspective, the combined effect of shifting users to different access points for load balancing purposes is the same as shifting user to a single access point supporting the maximum number of users, N_{tot}.

TABLE 15.2 Simulation configuration parameters for Cellular and Wi-Fi Case

Network	Parameter	Value
Cellular	System configuration	Reflecting GPRS/EDGE system
	No of channels available to the system	252 [11]
	No of users supported per cell in omni-directional mode	25 [11]
	No of users supported per cell in tri-sectorized mode	38 [11]
	Busy hour load (in terms of users per cell)	varied
Wi-Fi	N_{tot}	40, 36

A series of simulations is performed with different cellular systems and under different traffic types. For the first case, power savings solutions are applied are for a system with configuration parameters as shown in Table 15.2. Results obtained for different solutions are shown in Figures 15.14 - 15.16.

The graphs shown in Figures 15.14 - 15.16 give the percentage of from-the-mains power savings for the cellular network. It can be seen that very significant savings can be achieved by applying sectorization switching and powering down solutions together. The power savings by turning the base station off at low loads are most notable (up to 85%). At high loads the primary

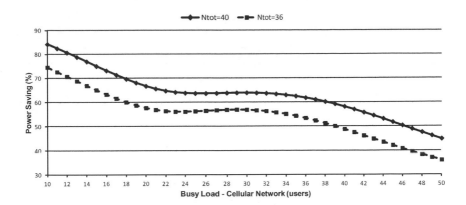

FIGURE 15.14 Power saving against busy hour load for Poisson model (sectorization switching and powering down together)

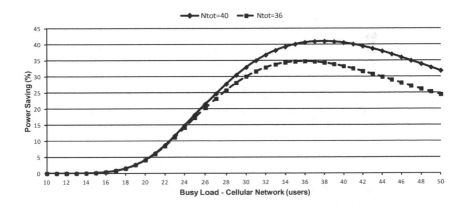

FIGURE 15.15 Power saving against busy hour load for Poisson model (sectorization switching only)

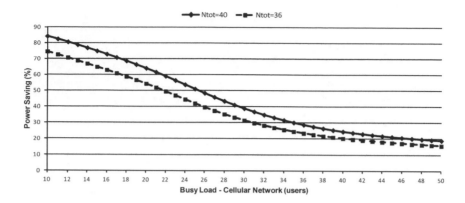

FIGURE 15.16 Power saving against busy hour load for Poisson model (powering down only)

contribution is from sectorization switching solution which gives savings up to 40%.

The power savings are primarily dependent on the capacity of Wi-Fi network (no. of access points available and the maximum no. of users supported per access point). As it can be seen, savings are reduced if the capacity decreases ($N_{tot} = 36$ instead of 40). Another important factor is the user association rate on the Wi-Fi network. If the busy hour of user association rate is changed to somewhere near the busy hour of traffic on cellular network, reduced savings are achieved as shown in Figure 15.17.

Performance assessment has also been carried out using ON/OFF Traffic Models for the cellular network. The simulation approach assumes a separate ON/OFF traffic flow to each user parameterized as a combination of FTP, HTTP and video streaming traffic. The number of users receiving traffic flows varies according to the Vodafone distribution as shown in Figure 15.3.1.1. The chosen ON/OFF model parameterizations are widely used in literature, taken from [14]. For the Wi-Fi network, the same Poisson model is used as before. A separate simulation is performed for each hour in the 24 hour period for the given number of users being present (obtained from Figure 15.3.1.1) where at each second in the simulation duration, the simulation tallies the numbers of users present according to the ON/OFF model being applied to each user. At the same time, the value of $\lambda(t)$ is used to parameterize Equation 15.4 and cycles in inner loops through each possible value of k (active number of users on Wi-Fi) network and performs the power saving solution according to the number of users present according to ON/OFF model, and ascertains the power consumption in before and after power saving solution cases. This power consumption is then multiplied by $P(k, t)_{Wi-Fi}$ to get the actual power consumption. These results are then averaged over results achieved at each

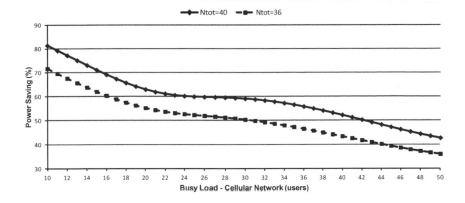

FIGURE 15.17 **Effect of changing the user association rate (sectorization switching and powering down together)**

second in the simulation duration, where all such simulations are typically performed over tens of millions of seconds. Finally, the results are averaged over all simulations performed at each hour in the 24 hour period.

The simulations are performed for a mix of traffic with percentages of FTP, HTTP and video streaming carefully selected as 15%, 45% and 40% respectively in order to reflect a practical scenario. The performance of power saving against busy hour load for mix of traffic is shown in Figures 15.18, 15.20, and 15.20. Results again show a significant power saving potential of up to 90% under low load conditions. In high loads savings of 10% are achieved. It should be noted that per user load for video streaming traffic is highest, followed by FTP and HTTP traffics respectively.

15.3.1.3 Cellular - White-Fi Offloading

The cellular and Wi-Fi scenario, as investigated in previous section is further extended to cellular and White-Fi with the primary objective of comparing energy savings with conventional Wi-Fi operating in 2.4GHz and 5GHz. We use the coverage results obtained from a recent study [15] that explores the feasibility of providing broadband wireless access in TVWS for different residential environments such as dense urban, urban and rural.

The coverage results (over a square kilometer) for different deployment densities in different environments are shown in Table 15.4. Other simulation parameters are the same as given in Table 15.3, whereby we simulate a mix of traffic with the respective percentages of FTP, HTTP and video streaming traffic taken as 15%, 45% and 40%. Results in Figures 15.21 and 15.22 show the average power savings achievable by employing both sectorization switching and powering down solutions in tandem.

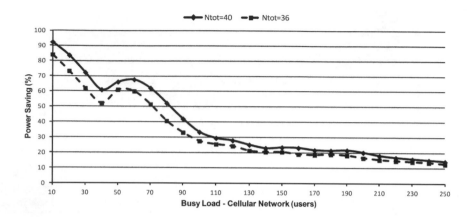

FIGURE 15.18 Power saving against busy hour load for mix of traffic(sectorization switching and powering down together)

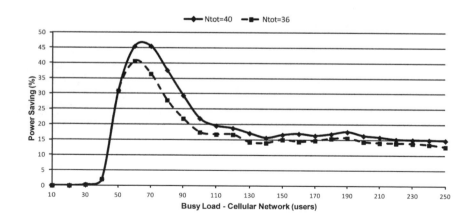

FIGURE 15.19 Power saving against busy hour load for mix of traffic(sectorization switching only)

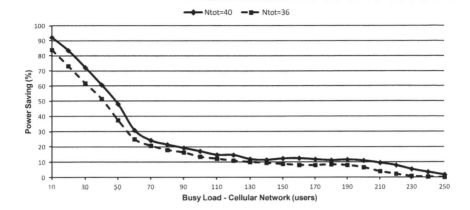

FIGURE 15.20 **Power saving against busy hour load for mix of traffic(powering down only)**

Highest savings are achievable for Wi-Fi operating in TVWS as this provides almost blanket coverage due to better propagation characteristics. Comparatively lower savings are achievable for Wi-Fi operating in 2.4GHz and 5GHz bands because of the reduced savings primarily due to interference effects. Operation in the 5GHz band solves the interference problem to some extent but implies higher attenuation losses.

15.3.2   Power Saving by Propagation Improvement

Next we move into performance evaluation of the propagation improvement concept. We assume the traffic load over a 24 hour period to be varying according to the cosine distribution (as given by Equation 15.3) and the active number of users to be Poisson distributed (as given by Equation 15.4). The COST Walfisch-Ikegami non-line-of-sight (COST W-I NLOS) path loss model is selected for the system configuration given in Table 15.1. Other simulation parameters are shown Table 15.5. The propagation distance is assumed to be 1Km; however the results are valid for a range of propagation distances.

The performance comparisons through opportunistic use of better propagation bands for 2 and 3 band cases are shown in Figure 15.23. For the case of 2 bands (900MHz and 1800MHz), the results show significant power saving potential of up to 80%. For the case of 3 bands, where the upper band is of higher frequency (poorer propagation), savings of up to 90% are achievable.

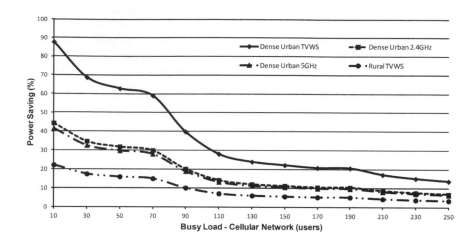

FIGURE 15.21 Power saving against busy hour load for different environments (20% deployment density)

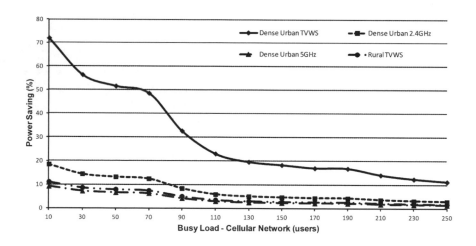

FIGURE 15.22 Power saving against busy hour load for different environments (5% deployment density)

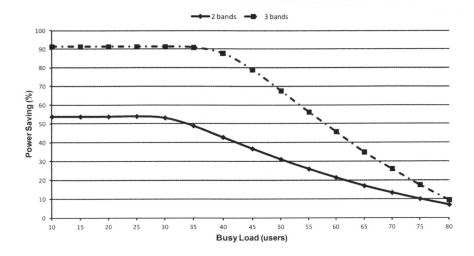

FIGURE 15.23 Power saving against busy hour load through opportunistic use of better propagation bands

15.3.3 Power Saving by Channel Bandwidth Increase or Better Balancing

The power saving potential through channel bandwidth increase can be assessed by a simple analysis, using the famous Shannon capacity equation:

$$C = B log_2(1 + S/N), \tag{15.6}$$

where increasing channel bandwidth by a factor A, with the same required capacity, increases both B and N by the same factor i.e., after an increase of bandwidth by the factor A,

$$S = (2^{C/(B \cdot A)} - 1)A. \tag{15.7}$$

The power saving (proportion) achieved by increasing channel bandwidth by the factor A is therefore

$$Saving = 1 - \frac{(2^{C/(B \cdot A)} - 1)A}{2^{C/B} - 1}. \tag{15.8}$$

The savings, as a function of bandwidth increase factor are plotted against both SNR and required channel capacity in Figures 15.24 and 15.25 respectively.

An example calculation is given as follows. Considering the average traffic load for the cellular network, over a 24 hour period to be varying according to the cosine distribution and taking the quiet hour load to be 25% of busy hour load yields an average traffic load throughout the day of 0.625 of the busy

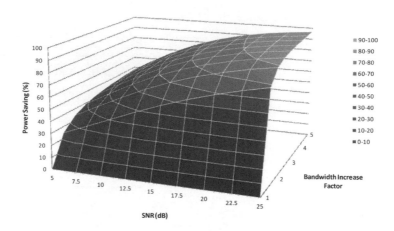

FIGURE 15.24 Transmission power savings achievable through increasing channel bandwidth (as a function of SNR)

hour load. This leads to an average bandwidth increase by a factor of 1.6. If the required SNR and bandwidth for a conventional system(i.e., without any bandwidth increase) are assumed to be 10dB and 5MHz respectively and using the aforementioned bandwidth increase factor of 1.6, gives transmission power savings of 44.3% [6].

On the other hand, if we consider the case of balancing bandwidths assuming that the leftmost user in Figure 15.5 (in the 'before' case) is already able to maximize its bandwidth to cover the entire available spectrum in the leftmost band, the 'after' case in Figure 15.5 leads to a reduction of 50% in bandwidth for that user, and an increase of 50% for the other users. For the 10^7 b/s (Shannon) capacity case, this gives a necessary transmission power increase of 50% for the leftmost user in Figure 5, and a decrease of 36% for the other users, which leads to an overall transmission power saving of 25% among all users. Consideration of $3 \cdot 10^7$ b/s capacity case, for example, gives a necessary transmission power increase of 450% for the leftmost user and a transmission power decrease of 82% for the other users. This evaluates to an overall power saving of 76%. Hence as the required capacity increases, the achievable savings also increase.

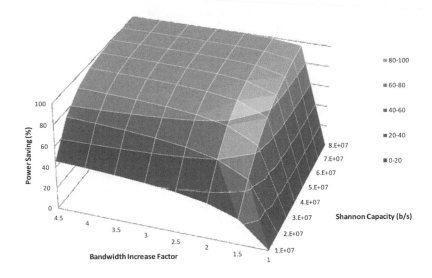

FIGURE 15.25 Transmission power savings achievable through increasing channel bandwidth (as a function of shannon capacity)

15.4 Conclusion

Energy consumption in mobile communications is an important issue, which like in all areas of technology must be reduced to aid the environment. Mobile network operators are increasingly looking to enhance energy efficiency of their networks through enabling power saving modes, and are also more frequently offering connectivity over heterogeneous networks, blending licensed (3G) and unlicensed (Wi-Fi) spectrum for economically optimal capacity. The trend for small cell networks increases the challenges of energy management.

These observations have been combined to discuss different power saving options for cellular operators. The concept of opportunistic reallocation of loads among a range of available frequency bands to dynamically power down cellular radio network equipment is initially discussed. Results have shown considerable power saving potential of up to 50% or more in base station power. This concept is extended further extended to investigate the dynamic movement of traffic loads from cellular to Wi-Fi networks to power down the cellular radio equipment. Results have shown that significant savings of up to 65% or more can be achieved and point to a viable solution for reducing energy consumption in cellular networks through coverage and service provisioning supplemented by Wi-Fi.

Significant transmission power savings of 50% or more through opportunistically reallocating users between bands to improve propagation and through moving users between bands to increase or better balance transmission bandwidths have also been shown.

Acknowledgement

The work in this Chapter has been partially supported by the Green Radio Core Research Program of the Virtual Centre of Excellence in Mobile & Personal Communications, Mobile VCE, www.mobilevce.com, the ICT-ACROPOLIS Network of Excellence, FP7 project number 257626, www.ict-acropolis.eu, and by COST Actions IC0902 and IC0905 "TERRA".

Bibliography

[1] I. T. U. (ITU). (2010) The world telecommunication/ict indicators database. [Online]. Available: http://www.itu.int/ITU-D/ict/statistics/

[2] H. Sistek. (2010) Green-tech basestations cut diesel usage by 80 percent. [Online]. Available: http://news.cnet.com/8301-11128_3-9912124-54.html

[3] Incisive Financial Publishing Limited. (2010) Business green blog. [Online]. Available: http://blog.businessgreen.com

[4] G. Fettweis and E. Zimmerman, "Ict energy consumption; trends and challenges," in *Proc. IEEE Symposium on Wireless Multimedia Personal Communications (WPMC'08)*, Lapland, Finland, 2008, pp. 1–3.

[5] Mobile VCE. (2010) Uk mobile virtual centre of excellence - green radio. [Online]. Available: http://mobilevce.co.uk/green-radio

[6] O. Holland, V. Friderikos, and A. H. Aghvami, "Green spectrum management for mobile operators," in *Proc. IEEE (Globecom'10)*, Miami, FL, USA, Dec. 2010, pp. 1458–1463.

[7] O. Holland and et al., "Opportunistic load management for mobile communications energy efficiency," in *Proc. IEEE Symposium on Personal, Indoor and Mobile Radio Communications (PIMRC'11)*, Toronto, Canada, Sept. 2011.

[8] O. Holland, T. Dodgson, A. H. Aghvami, and H. Bogucka, "Intra operator dyanmic spectrum management for energy efficiency," *IEEE Communications Magazine*, in press.

[9] M. A. Marsan and et al., "Optimal energy savings in cellular access networks," in *Proc. IEEE First International Workshop on Green Communications (GreenComm'09)*, Dresden, Germany, Jun. 2009, pp. 1–5.

[10] S. Thajchayapong and J. Peha, "Mobility patterns in microcellular wireless networks," *IEEE Trans. Mobile Computing*, vol. 5, no. 1, pp. 52–63, Jan. 2006.

[11] J. Eberspcher, H.-J. Vgel, C. Bettstetter, and C. Hartmann, GSM Architecture, Protocols, and Services. Wiley, 2009.

[12] M. A. Marsan and et al., "A simple analytical model for energy efficient activation of access points in dense wlans," in *Proc. IEEE First International Conference on Energy Efficient Computing and Networking (e-Energy'10)*, NY, USA, Jun. 2010.

[13] A. Mahanti, C. Williamson, and M. Arlitt, "Remote analysis of a distributed WLAN using passive wireless side measurement," *ACM Journal of Performance Evaluation*, vol. 64, no. 9-12, Oct. 2007.

[14] 3rd Generation Partnership Project (3GPP), "Technical specification group radio access network; dual-cell HSDPA operation," TR 25825, May 2008.

[15] S. Kawade and M. Nekovee, "Broadband wireless delivery using an inside-out TV white space network architecture," in *Proc. IEEE (Globecom'11)*, Houston, TA, USA, Dec. 2011.

Author Contact Information

Adnan Aijaz, Oliver Holland, Paul Pangalos, and Hamid Aghvami are with King's College London, UK, Email: adnan.aijaz@kcl.ac.uk, oliver.holland@kcl.ac.uk, paul.pangalos@kcl.ac.uk, hamid.aghvami@kcl.ac.uk. Hanna Bogucka is with Poznan University of Technology, Poznan, Poland, Email: hbogucka@ET.PUT.Poznan.PL.

TABLE 15.3 Simulation configuration parameters for ON/OFF Traffic Models

Network	Parameter	Value
Cellular	System configuration	Reflecting HSDPA system
	Bandwidth per carrier	5 MHz
	Per user rate in ON durations	64 Kbps
	FTP traffic; OFF duration	Exponentially distributed, mean 180s [14]
	FTP traffic; ON duration	Pareto distributed file size of mean 2MB [14], $\alpha=1.5$ (unless otherwise stated) and k calculated from the mean and α, with ON duration calculated from each sampled file size assuming a fixed data rate per user
	HTTP traffic; reading time (OFF duration)	Exponentially distributed, mean 30s [14]
	HTTP traffic; parsing time (OFF duration)	Exponentially distributed, mean 0.13s [14]
	HTTP traffic; main object size (contributes to ON duration)	Truncated Lognormally distributed, $\sigma=1.37$, $\mu=8.35$, min=100B, max=2MB [14]
	HTTP traffic; embedded object size (contributes to ON duration)	Truncated Lognormally distributed, $\sigma=2.36$, $\mu=6.17$, min=50B, max=2MB [14]
	HTTP traffic; number of embedded objects per page (contributes to ON duration)	Truncated Pareto distributed, $\alpha=1.1$, $k=2$, max=55 (k subtracted from each sampled value) [14]
	Video Streaming traffic; inter-arrival time between beginning of each frame	Deterministic (based on 10 fps), 100ms [14]
	Video Streaming traffic; number of packets/slices in a frame	Deterministic, 8 [14]
	Video Streaming traffic; packet/slice size (contributes to ON duration)	Truncated Pareto distributed, Min = 50B, Max = 125B, $\alpha = 1.2$, k = 20B [14]
	Video Streaming traffic; inter-arrival times between packets/slices in a frame (contributes to OFF duration)	Truncated Pareto distributed, Min = 6ms, Max = 12.5ms, $\alpha = 1.2$, k = 2.5 m [14]
Wi-Fi	User association rate	As shown in figure 13
	N_{tot}	40

TABLE 15.4 Network coverage with increasing access point penetration level

Access Point Penetration	Dense Urban TVWS	Dense Urban 2.4GHz	Dense Urban 5GHz	Rural TVWS
5%	78%	20%	10%	12%
10%	95%	30%	20%	19%
20%	95%	48%	45%	24%

TABLE 15.5 Simulation Configuration Parameters for Propagation Improvement Concept

Parameter	Value
System configuration	Reflecting GSM/GPRS/UMTS system
Frequency Bands	900MHz, 1800MHz, 2GHz
Height of Transmitter	30m
Height of Receiver	1.5m
Propagation distance (Tx-Rx separation)	1Km
Mean building height	10m
Mean street width	20m
Mean building separation	5m
Angle between street orientation and propagation path	0°

16

Toward Energy-Efficient Operation of Base Stations in Cellular Wireless Networks

Kyuho Son

University of Southern California, Los Angeles, USA

Hongseok Kim

Sogang University, Seoul, Korea

Yung Yi

Korea Advanced Institute of Science and Technology, Daejeon, Korea

Bhaskar Krishnamachari

University of Southern California, Los Angeles, USA

CONTENTS

Energy use of base stations (BSs) in cellular networks has lately become a vital design consideration, due to increased awareness of environmental and economic issues for wireless network operators. This chapter introduces a range of techniques to reduce energy consumption in BSs, referred to as green cellular networks. In particular, the chapter focuses on presenting details on recent developments from analytical algorithms to practical applications regarding: (i) energy-aware heterogeneous deployment and (ii) joint BS on/off and user association problem. Further, the chapter presents several open problems and other directions including new paradigm and architecture towards green cellular networks.

16.1 Introduction

For decades, abundant research in wireless networks has contributed to the improvement of energy-efficiency for battery-operated devices [1], e.g., prolonging the lifetime of sensor nodes and mobile terminals. However, more recently, potential harmful effects to the environment caused by CO_2 emissions and the depletion of non-renewable resources bring renewed focus on the need to develop more energy-efficient underlying network infrastructures that account for heavy energy usage [2].

It has been estimated that 3% of the world's annual electrical energy consumption and 2% of CO_2 emissions are caused by the information and communication technology (ICT) industry [3]. According to this estimate, about a tenth of this can be attributed to cellular mobile communication systems. As of 2008, the energy consumption corresponded to 60 billion kWh of electricity usage annually, about 40 million metric tons of CO_2 emissions each year. To put it into perspective this is equivalent to annual greenhouse gas emissions from about 8 million cars. Another consistent estimate states that there were over 600,000 base stations (BSs) in China deployed by three major operators which consumed about 20 billion kWh in 2007 [2].

From the perspective of cellular network operators, reducing energy consumption is not only a matter of *social environmental responsibility* towards being green and sustainable but also tightly related to their *business survivability* in coming years. They are spending huge operational expenditures (OPEX) to pay electricity bills. Moreover, it is expected to grow due to explosive growth in data demand and the possible increase of energy price in the near future. It is estimated that the energy consumption rises at 15-20% per year, doubling every five years in the field of ICT. This will result in a collective cellular network OPEX of \$22 billion in 2013, according to a study from ABI Research [4]. Thus, reining back the spiraling OPEX is crucial to the continuing success of operators.

16.2   Overview of Green Cellular Network Design

Pushed by such needs of energy reduction, the operators have been seeking ways to improve energy-efficiency in all components of cellular networks including mobile terminals [5–7], BSs, and mobile backhaul networks. In particular, it has been reported that BSs are the key source of energy usage, contributing to 60-80% of the total power consumption[1] in cellular networks [8]. In this chapter, we therefore focus on energy-efficient operation of cellular BSs, which we refer to as *green cellular base stations operation*.

The goal of green cellular networks can be achieved in various ways:

- *Component level*: novel designs and hardware implementations, e.g., energy-efficient power amplifiers [9] and fanless cooler, or even cooling based on natural resources [10].

- *Link level*: energy-aware transmission and resource management schemes, e.g., power control [11] and user association [12].

- *Network level*: topological approaches from deployment to operation, e.g., smart deployment at the stage of network planning by using micro BSs [13–16] or relays [17, 18], and traffic-aware dynamic BS on/off [2, 8, 12, 19–21].

In this chapter, we focus on *network/link level solutions* with emphasis

---

[1]Throughout this chapter, we interchangeably use the terms 'energy' and 'power' (really meaning average power) unless clear distinction is needed.

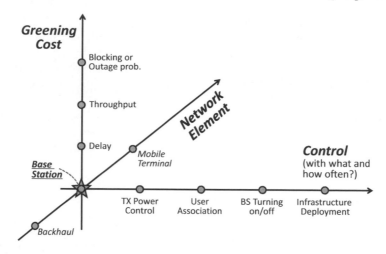

FIGURE 16.1 Problem space of green cellular networks.

on BS operations and deployment topology designs, whose problem space can be seen from a variety of angles. As depicted in Figure 16.1, we explore the entire problem space based on the following two axis: *(i)* what we can control under what time scale, and *(ii)* what performance we need to sacrifice in saving energy. We used the word "sacrifice" because the key of green BS design lies in how to tune the tradeoff between energy and efficiency, where both energy and efficiency are measured by various metrics. A research paper in the community typically has corresponded to one or multiple points in this space. We will provide a taxonomy of the state-of-the-art research in the context of this problem space later in Table 16.1. We now elaborate on each axis here.

- *(i) Control and time-scales.* Saving energy in the cellular networks is differentiated by the control mechanisms we handle as well as their operating time scales. For example, deployment of energy-aware base stations may take the order of months or even years time scale, whereas the time-scale of transmission power control of mobile terminals or BSs is on the order of milliseconds.

- *(ii) Efficiency metric and greening cost.* Major system performance metrics of applications in cellular networks, which are also the ones to be sacrificed by power reduction, include throughput, delay, and blocking or outage probability. Throughput is an appropriate metric for best-effort data traffic, whereas delay and blocking/outage probability are mainly for real-time traffic.

Table 16.1 shows the survey of the recent papers which has studied greening effect and algorithms in wireless cellular networks, which are classified

TABLE 16.1 Classification of green cellular network problems.

Work	Techniques						Metric		Keyword
	Deployment	BS On/Off	Association	Power control	Relay	CoMP	System Performance $(T, D, O$ or $ASE)$	Energy Consumption (E, APC)	
[22–26]	x	-	-	-	-	-	T	E	Macro-micro deployment
[14, 27–29]	x	-	-	-	-	-	ASE	APC	
[13, 30]	x	x	x	-	-	-	ASE or T	E	CAPEX-OPEX tradeoff
[2, 16, 19–21]	-	x	-	-	-	-	T	E	Dynamic BS operation
[31, 32]	-	x	-	x	-	-	O	E	Cell-zooming
[12, 22]	-	x	x	-	-	-	D or T	E	Energy-delay tradeoff
[11]	-	-	-	x	-	-	T	E	Spatio-temporal power sharing
	-	-	-	x	-	-	T	E	
[17, 18]	-	-	-	-	x	-	T	E	Relaying
[33, 34]	-	-	-	-	-	x	T or O	E	Diversity

according to which metrics for system performance and energy consumption are taken with which focus on greening techniques.

Most of the papers use a model with two objectives chosen from the perspectives of system performance and energy consumption. A classical method to deal with such multi-objective cases are: maximizing [system performance] - $\eta \cdot$ [energy consumption] (similarly minimizing [energy consumption] -$1/\eta \cdot$[system performance]), or maximizing [system performance]/[energy consumption], where η is the weight that controls the tradeoff between two perspectives.

The *system performance metrics* they studied are denoted, in Table 16.1, by T (Throughput, bit/sec), D (Delay, sec), O (Outage probability, %), and ASE (Area Spectral Efficiency, bit/sec/Hz/m^2). The *energy consumption metrics* consist of APC (Area Power Consumption, W/m^2), and E (Energy, J). Dividing one metric by the other, the *unified metrics* such as EE (Energy Efficiency, bit/J) and AEE (Area Energy Efficiency, bit/J/m^2) are also often used in literature to capture the tradeoff between system performance and energy consumption.

Under the techniques categorized in Figure 16.1 and Table 16.1, we summarize several control mechanisms that are mainly discussed in this chapter.

(a) Deployment. Future cellular networks are expected to have a mixture of large and small cells. In such environments, the general problem of deployment is to determine *where* (site locations), *how many* and *which type* (macro/micro/pico/femto) of BSs need to be deployed in an energy-efficient manner.

(b) BS on/off. BSs are typically deployed and operated on the basis of peak traffic volume and also stayed turned-on irrespective of traffic load. Recent temporal traffic traces [2, 8] point out that BSs are largely under-utilized in late night or early morning over the daily time-scale and also under-utilized in weekends over the weekly time-scale. Dynamic BS operation which turns on and off depending on the imposed traffic may lead to huge energy saving.

(c) User association. Turning BSs on/off is naturally coupled with user association because a user may need to change its associated BS. A proper association mechanism, which determines the most energy-efficient BS among a set of active[2] BSs, is necessary to fully exploit the amount of energy savings from traffic-aware BS operations.

(d) Power control. To mitigate interference, the current and future cellular systems are expected to employ dynamic interference management (IM) algorithms that perform dynamic BS transmission power control depending on the scheduled users over the multiple neighboring cells. It is also possible to achieve huge greening gain by reducing the transmitted power of BSs intelligently while maintaining the system performance.

In the following sections, we present details on recent developments from analytical algorithms to practical applications regarding: (i) energy-aware heterogeneous deployment in section 16.3 and (ii) joint BS On/Off and user association problem in section 16.4. Finally, in section 16.5, we present several open problems and other directions including a new paradigm and architecture towards green cellular networks.

16.3   Energy-Aware Heterogeneous Deployment

Generally speaking, the problem of energy-aware deployment is to determine *where* (site locations), *how many* and *which type* (macro/micro/pico/femto) of BSs need to be deployed in an energy-efficient manner. There have been many efforts in literature [13, 14, 22–30] that deal with the general problem. In particular, this section introduces two case studies[3] in an energy-aware heterogeneous deployment.

The first case study, that is presented in section 16.3.1, considers the use of micro BSs on top of the pre-existing deployment of macro BSs to upgrade the network capacity in a cost-effective way. A theoretical implication and a practical solution for the following questions are provided: what is the minimum number of additional micro BSs to meet the quality of service requirement

---

[2]Within the scope of this chapter, *active* means that a BS is *ON* state.
[3]The key results of the two case studies we present in sections 16.3.1 and 16.3.2 were drawn from [13] and [14, 28], respectively.

and where should they be deployed. The second case study in section 16.3.2 focuses on the predefined topology of regular grid, where only inter-site distance is a variable, and considers two deployment strategies: a conventional homogeneous and heterogeneous deployments. Though the model is simple, it allows us to systematically investigate the energy efficiency of the deployment strategies and to answer under what conditions, which deployment strategy is better.

16.3.1   Micro Base Station Deployment Strategy

16.3.1.1   System Model

Network model. A wireless cellular network where the sets of macro and micro BSs, denoted by $\mathcal{B}_M$ and $\mathcal{B}_m$, respectively, is considered. Throughout the section, subscript M is used for macro BSs, and m is for micro BSs. Let us denote by $b \in \mathcal{B} = \mathcal{B}_M \cup \mathcal{B}_m$ the index of BSs. Our main focus is on downlink communication that is a primary usage mode for the mobile Internet, i.e., from BSs to mobile terminals (MTs). However, we would like to emphasize that some aspects of our work can be applied to the uplink as well.

Link model. The received signal strength from BS b to MT at location x can be expressed as $E_b(x) = p_b \cdot g_b(x)$, where p_b denotes the transmission power of BS b, $g_b(x)$ denotes the channel gain from BS b to location x, including path loss attenuation, shadowing and other factors if any. Note, however, that fast fading is not considered here because the time scale for measuring $g_b(x)$ is assumed to be much larger. Accordingly, the signal to interference plus noise ratio (SINR) at location x can be written as:

$$\Gamma(x, \mathcal{B}) = \frac{E_{b(x,\mathcal{B})}(x)}{\sum\limits_{b \in \mathcal{B},\ b \neq b(x,\mathcal{B})} E_b(x) + \sigma^2}, \tag{16.1}$$

where σ^2 is noise power and $b(x, \mathcal{B})$ denotes the index of the BS at location x that provides the highest signal strength, i.e., $b(x, \mathcal{B}) = \arg\max_{b \in \mathcal{B}} E_b(x)$. Following Shannon's formula, spectral efficiency at location x is given by $C(x, \mathcal{B}) = \log_2(1 + \Gamma(x, \mathcal{B}))$.

Area spectral efficiency. The notion of the area spectral efficiency (ASE), firstly introduced in [35], is considered as a performance metric. It is defined as the summation of the spectral efficiency over the reference area $\mathcal{A}$:

$$S(\mathcal{B}) \doteq \frac{\sum_{x \in \mathcal{X}} C(x, \mathcal{B}) \cdot Pr(x)}{|\mathcal{A}|}, \quad [\text{bps/Hz/m}^2] \tag{16.2}$$

where $Pr(x)$ is the probability of the MT being at a specific location x; $\mathcal{X}$ is the set of locations included in the area $\mathcal{A}$ satisfying $Pr(x) > 0$ for all $x \in \mathcal{X} \subset \mathcal{A}$. For simplicity, in this section, the user distribution is assumed to be homogeneous[4] such that the discrete set $\mathcal{X}$ is a rectangular lattice with a small grid size and the probability of each location is the same.

16.3.1.2　Problem Formulation

Consider an area of interest $\mathcal{A}$ served by a wireless network operator whose access network consists of only macro BSs $\mathcal{B}_M$. Suppose that the maximum required ASE $S_{th}^{t^*}$ at the peak time $t^* = \arg\max_t S_{th}^t$ during a day $t \in [t_0, t_0 + D)$ almost approaches to the one that can be provided by turning on all the macro BSs $\mathcal{B}_M$, i.e., $S(\mathcal{B}_M) \simeq S_{th}^{t^*}$. Thus, the operator wants to upgrade its access network by micro BSs which are considered as the cost-effective way of incrementally increasing capacity inside the initial macro cell deployment.

Minimal deployment problem. The objective is to find a minimal deployment of micro BSs (i.e., minimizing the additional power consumption) while providing $\zeta \geq 1$ times higher ASE than before the upgrade. This can be mathematically formulated as the following optimization problem:

$$\textbf{(MDP1)}\quad \min_{\mathcal{B}_m}\quad P_m \cdot |\mathcal{B}_m| \tag{16.3}$$

$$\text{s.t.}\quad S(\mathcal{B}_M \cup \mathcal{B}_m) \geq \zeta \cdot S(\mathcal{B}_M) = \zeta \cdot S_{th}^{t^*}, \tag{16.4}$$

where P_m is the total operational power consumption of micro BSs. Note that **(MDP1)** can be also interpreted as CAPEX minimization. It is basically a combinatorial problem, and that makes it difficult to find an optimal solution, especially, when the number of candidate locations is large.

16.3.1.3　Key Observations and Algorithm

Several interesting observations[5] can be made from various real topologies which help us to gain insight and develop an efficient algorithm. Given the area that is covered by the existing set of macro BSs, let us focus on the deployment of *one new micro BS*. The contour plot in Figure 16.2(a) shows how much ASE a micro BS can improve according to the location of deployment.

Observation 16.1. *As long as a new micro BS is placed not too close to the one of existing BSs that would interfere with each other, ASE can be expected to increase before the upgrade. Especially, the ASE increment becomes large as the distance from macro BSs increase.*

The wireless network operators are supposed to deploy a micro BS at the location where ASE can be improved. Therefore, only such locations are considered as candidate positions for the micro BS deployment:

$$\forall k \in \mathcal{K}, \quad S(\mathcal{B} \cup \{k\}) > S(\mathcal{B}), \tag{16.5}$$

---

[4]Please refer to [13] for the further results of the heterogeneous user distribution.

[5]The results provided here is from the topology of Korea [36], similar trends could be observed in the other topologies [2, 37] as well. Please refer to [13] for more details.

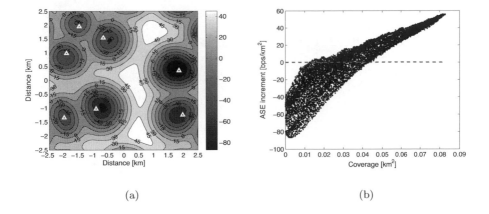

(a)   (b)

FIGURE 16.2 Several interesting observations on the ASE increment by one new micro BS: (a) contour plot of ASE increment and (b) scatter plot between coverage and ASE increment [13], ©2011 IEEE.

Let us denote by $\mathcal{A}_{i<j}$ the set of locations that have better SINR from BS i than j, and further denote by $\mathcal{A}_{i=j}$ the set of boundaries having the same SINR from both BSs i and j. Then, the set of locations covered by BS k (or simply, coverage) can be written as:

$$\mathcal{A}_k(\mathcal{B}) \doteq \{x|x \in \mathcal{A} \text{ s.t. } b(x, \mathcal{B}) = k\} = \bigcap_{b \in \mathcal{B},\, b \neq k} \mathcal{A}_{k>b}. \tag{16.6}$$

Figure 16.2(b) examines how much area the micro BS can cover according to the location of deployment and investigates the correlation with ASE increment. As can be seen, ASE increment has a distinct tendency to increase with the coverage of micro BS. Interestingly, it becomes sharper (i.e., smaller variance) as the coverage increases. In such locations giving high increment, the most likely candidates for the deployment, coverage and ASE increment almost surely has a near-monotonic relationship.[6]

Observation 16.2. *The larger area can be covered by the new micro BS, the higher ASE increment is likely to be expected.*

Motivated by this observation, the following monotone relationship is assumed to be hold.

$$|\mathcal{A}_k(\mathcal{B} \cup \{k\})| \geq |\mathcal{A}_{k'}(\mathcal{B}' \cup \{k'\})|$$
$$\Rightarrow S(\mathcal{B} \cup \{k\}) - S(\mathcal{B}) \geq S(\mathcal{B}' \cup \{k'\}) - S(\mathcal{B}'), \tag{16.7}$$

where k (or k') is the index of the micro BS.

---

[6]Results from monotone test, which randomly picks two points having positive ASE increments in Figure 16.2(b) and checks whether the slope between these points are positive or not, (90.4~97.0% depending on the topologies) also support the following observation.

These two observations are intuitively understandable. Consider the area covered by the micro BS far from existing macro BSs. Since the signals from the macro BSs are weak, the micro BS will provide the highest SINR to a large extent area. In addition to this large coverage, the area originally had low spectral efficiency, resulting in the high increment of ASE.

Constant-factor approximation greedy deployment algorithm. Prior to introducing a natural greedy algorithm for **(MDP1)**, a real-valued set function $F : \mathcal{B}_m \to \mathbb{R}$ is defined as follows:

$$F(\mathcal{B}_m) \doteq S(\mathcal{B}_M \cup \mathcal{B}_m) - S(\mathcal{B}_M), \tag{16.8}$$

which returns the ASE increment by additionally installing the micro BSs $\mathcal{B}_m$.

Greedy deployment algorithm for (MDP1)

1: Initialize $\mathcal{B}_m^{\text{greedy}} = \emptyset$
2: **do while** $S(\mathcal{B}_M \cup \mathcal{B}_m^{\text{greedy}}) < \zeta \cdot S_{th}^{t^*}$
3:     $k^* = \arg \max_{k \in \mathcal{K}} F(\mathcal{B}_m^{\text{greedy}} \cup \{k\}) - F(\mathcal{B}_m^{\text{greedy}})$,
4:     $\mathcal{B}_m \leftarrow \mathcal{B}_m \cup \{k\}$
5: **end do**

The greedy algorithm starts with the empty set $\mathcal{B}_m^{\text{greedy}} = \emptyset$, and iteratively adds the micro BS location having the highest increment among the set of candidate locations $\mathcal{K}$ until ASE reaches a target value, i.e., satisfying the constraint (16.4).

Theorem 16.1. *The ASE increment achieved by an optimal placement with the same number of micro BSs as the greedy algorithm cannot be more than a factor of $e/(e-1)$ from the ASE increment achieved by the greedy algorithm.*

$$\max_{|\mathcal{B}_m| = |\mathcal{B}_m^{greedy}|} F(\mathcal{B}_m) \leq \frac{e}{e-1} F(\mathcal{B}_m^{greedy}), \tag{16.9}$$

where the constant e is base of the natural logarithm.

This theorem can be proved following a similar procedure as in [38] taking into account the fact that the ASE increment function F is submodular based on assumptions (16.5) and (16.7). For proof, please refer to [13]. In order to further reduce the computational complexity of the greedy algorithm, it is possible to check the locations near cell boundaries instead of all candidate locations.

16.3.1.4   Numerical Results

For simulations, the deployment of macro BSs, shown in Figure 16.3(a), is considered. There are 10 macro BSs in $8 \times 8\text{km}^2$. In order to avoid edge effects, one the area of $5 \times 5\text{km}^2$ in the center is observed. To model the power consumptions, an affine power model [27] developed based on data sheets of

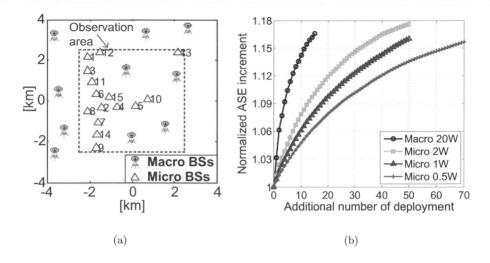

(a)   (b)

FIGURE 16.3 Deployment results: (a) snapshot after the deployment of 15 micro BSs and (b) normalized ASE increment according to the deployment of different types of BSs. (from [13], ©2011 IEEE)

several GSM (Global System for Mobile Communications) and UMTS (Universal Mobile Telecommunications System) BSs with focus on component-level is adopted. Denote p_M^{tx} and p_m^{tx} are the average transmit (radiated) power for macro and micro BSs, respectively. Then the total operational power consumptions for macro and micro BSs, P_M and P_m, are given by

$$P_M = a_M \cdot p_M^{tx} + b_M,$$
$$P_m = a_m \cdot p_m^{tx} + b_m,$$

(16.10)

where the relationship between the total operational power consumption and the transmit power is modeled in a linear fashion. The coefficient a accounts for the power consumption that scales with the average radiated power due to amplifier and feeder losses as well as cooling of sites. The term b denotes power offsets which are consumed independently of the average transmit power. These offsets are, amongst others, due to signal processing, battery backup, as well as site cooling. In Table 16.3, typical transmission and total operational powers for a macro BS with three sectors and two antennas per sector and a micro BS with a single sector with one omni-directional antenna are summarized.

On top of the deployment of macro BSs, Figure 16.3(a) illustrates the snapshots after 15 micro BSs additionally deployed by the proposed greedy deployment algorithm. It should be noted that the number associated with each of new micro BS are the order of the greedy deployment. As expected, the micro BSs tend to be placed in the boundaries of the cell because this makes the micro BSs cover the larger area, resulting in the more ASE increment.

Figure 16.3(b) shows the performance improvement according to the ad-

TABLE 16.2 Additional total power consumption required for the target ASE increment. (from [13], ©2011 IEEE)

BS type		Macro	Micro	Micro	Micro
Tx power		20W	2W	1W	0.5W
Target ASE	10%	4325W	645W	836W	1050W
increment	15%	9515W	1476W	1672W	2240W

ditional deployment of BSs having different transmission powers. Four types of BSs are considered: the macro BS with the transmit power of 20W and the micro BSs with transmit power of 0.5W, 1W and 2W, respectively. As can be clearly seen, there are diminishing returns on the normalized ASE increment. This is not only because the coverage of newly deployed BS will shrink but the amount of interference in the network increase as the number of BSs increases.

To meet the target ASE increment of 10%, while only five additional macro BSs are needed, 15, 22 or 30 micro BSs (three to six times more than macro BSs) are needed depending on their transmission powers. Nevertheless, the transmission power consumptions (p_M^{tx} and p_m^{tx}) of additional micro BSs are much less than that of additional macro BSs. For example, while 100W is consumed by the macro BSs, only 30W, 22W, or 15W is consumed by the micro BSs. Reflecting the total power consumptions (P_M and P_m), the advantage of micro BSs becomes more clear. Table 16.2 shows the required the additional total power consumptions for different target ASE increment. Compared to the case of macro BSs, deploying micro BSs can reduce more than 3kW and 6kW for the target ASE increments of 10% and 15%, respectively. This corresponds to about 70% energy savings.

16.3.2   Deployment Structure: Homogeneous vs. Heterogeneous

In terms of energy consumption, it is often believed that network topologies with the high density deployment of small, low power BSs yield strong improvements compared to the low density deployment of few high power BSs [39]. However, when the static offset powers for site cooling, power supply, battery backup and so on are taken into account, it is no longer true, i.e., making cell sizes too small may lead to energy-inefficient solutions in some cases. This section summarizes the results given in [14, 28] where it is considered a conventional *homogeneous* deployment (macro BSs only) and an *heterogeneous* deployment (macro + micro BSs). The performance of both deployments are compared under conditions where the inter-site distance D varies in a predefined regular grid topology.

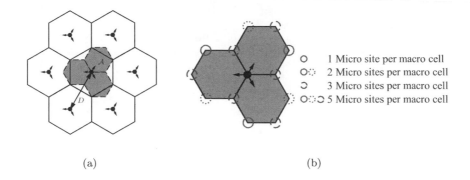

○	1 Micro site per macro cell
○∶∷	2 Micro sites per macro cell
⊃	3 Micro sites per macro cell
○∷⊃	5 Micro sites per macro cell

(a)   (b)

FIGURE 16.4 Regular grid deployment model: (a) regular grid of macro BSs and corresponding cell geometry with inter-site distance D and cell area $\mathcal{A}$, and (b) location of micro BSs within the macro deployment [28] © 2010 IEEE.

16.3.2.1 System Model

Deployment types. The homogeneous macro network is simply modeled as a cloverleaf network layout consisting of three-sectorized macro BSs as shown in Figure 16.4(a). The layout of regular grid deployment can be characterized by an arbitrary inter-site distance D varying in a certain range. Each macro BS serves an area denoted by $\mathcal{A}$ (corresponding to the grey-shaded region). The area $\mathcal{A}$ is referred to as *cell*, whereas the geographic location of a BS is denoted as cell site or simply *site*.

In the heterogeneous network, as depicted in Figure 16.4(b), a certain number of micro BSs are additionally placed in a specific location at the top of the above homogeneous macro network. For instance, if a micro BS is placed only at each corner marked in Figure 16.4(b) as a solid circle, then this can be considered as the particular case of single micro BS per macro cell.

α-percentile area spectral efficiency. Earlier definition of ASE in (16.2) only considers the expectation of the achievable rates but is not concerned with the distribution of rates around in the system. In order to incorporate a fairness aspect into the notion of ASE, *α-percentile area spectral efficiency* is defined as the α-quantile of the overall spectral efficiency in the reference cell divided by the cell size as follows:

$$\mathcal{S}_\alpha = \frac{Q_\alpha[S]}{|\mathcal{A}|}. \tag{16.11}$$

By scaling area spectral efficiency with the subcarrier bandwidth B_{sc}, the notion of *α-percentile area throughput* per subcarrier is sometimes used as a more practical relevant measure, i.e., $\mathcal{T}_\alpha = \mathcal{S}_\alpha \cdot B_{sc}$.

Area power consumption. It may be unsuitable to observe only power consumption for comparing the networks with different site densities. This is because they may have different coverages. In order to assess the power consumption of the network relative to its size, the notion of *area power consumption* (APC) is introduced as the total power consumption in a reference cell divided by the corresponding reference area. With an average of N micro BSs in a reference cell of size $\mathcal{A}$, APC can be written by

$$\mathcal{P} = \frac{P}{|\mathcal{A}|} = \frac{P_M + N \cdot P_m}{|\mathcal{A}|}, \quad [\text{Watt/m}^2]. \tag{16.12}$$

Now let us investigate the effect of inter-site distance D on APC, under a simplified propagation model [40] without considering shadowing and fading. The received signal strength decreases exponentially as the propagation distance d increases, i.e., $p^{rx}/p^{tx} = K \cdot d^{-\beta}$, where β is path loss exponent, and K is a unitless constant which depends on the antenna characteristics. In order to guarantee the minimum signal level p^{min} at the distance R from the cell site, the required transmit power p^{tx} can be given by

$$p^{tx} \propto p^{min} \cdot R^{\beta}. \tag{16.13}$$

By substituting $D = \sqrt{3}R$ into (16.13), it can be obtained that the transmit power required for a certain coverage increases to D^{λ}. For the case of path loss exponent $\lambda = 2$, APC is not affected by the site distance D because both numerator and denominator in (16.13) increase with D^{λ}. In general, for $\lambda > 2$, the following asymptotic results can be obtained [14]:

$$\lim_{D \to \infty} \mathcal{P}(D) = \infty \quad \text{as well as} \quad \lim_{D \to 0} \mathcal{P}(D) = \infty, \tag{16.14}$$

where the latter holds due to the nonzero constant terms b_M and b_m in (16.10). Thus, there exists an optimal inter-site distance D^*, which minimizes the area power consumption of the network.

16.3.2.2 Problem Formulation

In order to find the minimal area power consumption required for a certain target area throughput, the following optimization problem is considered in [27,28]. The problem is to determine an optimal site distance D^* that minimize the area power consumption while achieving a given target 10-percentile area throughput $\mathcal{T}_{10}^{target}$ at least.

$$\textbf{(MDP2)} \quad \min_{D \in [\underline{D}, \infty)} \quad \mathcal{P}(D)$$
$$\text{s.t.} \quad \mathcal{T}_{10}(D) \geq \mathcal{T}_{10}^{target}, \tag{16.15}$$

where $\underline{D}$ is the minimum inter-site distance.

Remark 16.3.1. *It is worthwhile mentioning that making the network denser (i.e., reducing D) helps to satisfy the target area throughput constraint because*

$\mathcal{T}_{10}(D)$ *is strictly monotonically decreasing function*[7]. *However, as can be seen in* (16.14), *the area power consumption* $\mathcal{P}(D)$ *goes to infinity as D goes to zero. More specifically, it can be shown later in [14] that* $\mathcal{P}(D)$ *is a convex function. Thus, reducing D beyond a certain point will definitely increase the area power consumption.*

16.3.2.3   Optimal   Area   Power   Consumption   vs.   Target   10-Percentile Area Throughput

Based on Remark 16.3.1, the following simple approach is able to solve **(MDP2)**. For different numbers of micro BSs $k = 0, 1, 2, \ldots$ per reference area $\mathcal{A}$ ($k=0$ means the case of macro BSs only),

---

Algorithm for (MDP2)

---

1. Determine the respective maximum inter-site distance $\widehat{D}_k$ that achieves the target 10-percentile area throughput $\mathcal{T}_{10}^{target}$. Note that since the curves are strictly monotonically decreasing and thus all $D_k \geq \widehat{D}_k$ are feasible.

2. Determine the optimal distance D_k^* that minimize the area power consumption $\mathcal{P}(D)$.

3. If $\widehat{D}_k$ is larger than D_k^*, then D_k^* is an optimal solution for problem (16.15). Otherwise, the optimal solution happens at the boundary, i.e., $\widehat{D}_k$.

Figure 16.5 shows the minimum area power consumption for different deployment strategies by varying the target 10-percentile area throughput. For each deployment strategy, the optimal power consumption remains constant until a certain target (i.e., the constraint in (16.15) is not active) and then increase almost linearly as the target area throughput increases. Note that for very low area throughput targets ($< 40\text{kbit/s/km}^2$), the pure macro BSs is most efficient. However, as the target area throughput increases, the benefit of micro BSs becomes more clear, i.e., the more micro BSs installed the more energy-efficiency the network becomes.

16.4   Base Station On/Off and User Association

Recall that so far we have considered energy-aware deployment, which is an offline problem on the time-scale of months or even years. Once the BS de-

---

[7]If D goes below a certain distance, say $\underline{D}$, then the area throughput $\mathcal{T}_\alpha(D)$ may decrease due to severe interference. However, such a region (i.e., $D \in (0, \underline{D})$ is not our of interest because it makes both the area throughput and power consumption worse.

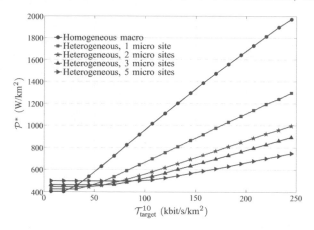

FIGURE 16.5 Minimum area power consumption with respect to a target 10-percentile area throughput for different deployment strategies [28] ©2010 IEEE.

ployment is done, the next problem is how to efficiently operate the BSs for energy conservation during the off-peak period. The solutions for the operation problem should be online distributed algorithms in order to be implemented in practical systems. There have been many efforts in literature [2,12,16,19–22,31,32] that deal with the operation problem. In particular, this section introduces our theoretical framework[8] that encompasses dynamic BS operation and the related problem of user association together. For the total cost minimization problem that allows for a flexible tradeoff between flow-level performance and energy consumption, an efficient yet practical solution will be provided: an optimal energy-efficient user association policy and simple greedy-on and greedy-off BS switching on/off algorithms.

16.4.1   A Quantitative Case Study

We shall begin with a motivational example [2] to provide an insight how much energy savings that could be obtained by turning off redundant BSs in cellular networks.

In order to quantitatively estimate potential energy savings, *two sets of real data* are combined together: the first is an anonymized temporal traffic trace of a period of one week from a cellular operator in a metropolitan area as shown in Figure 16.6; and the second is BS locations (total of 139 BSs in 128 locations in the area of 3.5×3.5km) from a central part of the city of Manchester, United Kingdom, as illustrated in Figure 16.7(a), obtained from a UK government-sponsored website [37]. As can be seen in Figure 16.7(b), which assumes an ideal circular range of 700 meters for each BS, there could be sig-

---

[8]The key results of this section were drawn from [12].

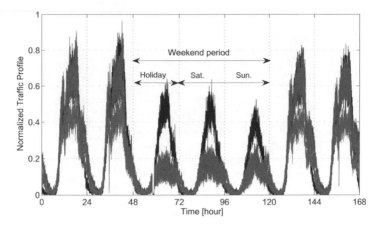

FIGURE 16.6 **Temporal traffic trace of a period of one week from a cellular operator in a metropolitan area (different color curves correspond to the traffic profile for different BSs). Note that this real dataset here is available online: [41] for the normalized cellular traffic trace [2] ©2011 IEEE.**

nificant redundant overlap in cellular coverage. Putting together the temporal and spatial data-set, it has been shown that individual operators are able to save between about 8 to 22 percent of energy in such an urban deployment. Sharing BS resources together, a total reduction of about 29 percent can be achieved for the energy expended. This corresponds to about 78000-530000 kWh of total electricity savings for the region over a year (assuming the single BS power is between 800W and 1500W). According to [42], the cost in $ per KWh for electricity for industry is about 0.068 and for homes is about 0.116, so these translate to between about $5000 to more than $61000 of annual savings for the electricity bill for operating BSs. According to [43], these also translate to between 53.7 metric tons to 365 metric tons of CO_2 emissions. This is a substantial reduction in green-house gas emission as well as in the cost of operation.

16.4.2   System Model and Formulation

Consider a wireless cellular network in an area $\mathcal{R} \subset \mathbb{R}^2$ served by a set of BSs $\mathcal{B}$. Let $x \in \mathcal{R}$ denote a location and we use $i \in \mathcal{B}$ to index a typical i-th BS. Our focus is downlink communication, from BSs to MTs. File transfer requests are assumed to arrive following an inhomogeneous Poisson point process with arrival rate per unit area $\lambda(x)$ and file sizes which are independently distributed with mean $1/\mu(x)$ at location $x \in \mathcal{R}$, so the traffic load density is defined as $\gamma(x) = \frac{\lambda(x)}{\mu(x)} < \infty$. This captures spatial traffic variability. For example, a hot spot can be characterized by a high arrival rate and/or possibly large file sizes.

When the set of BSs $\mathcal{B}_{\mathrm{on}}$ is turned on, the transmission rate of a user

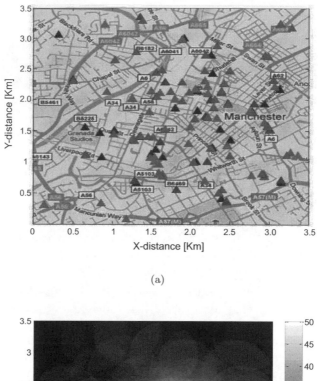

(a)

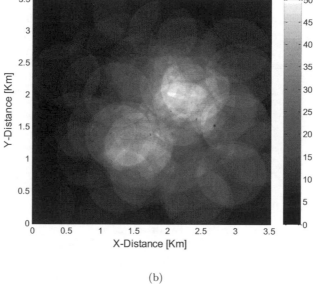

(b)

FIGURE 16.7 Real BS layout: (a) BS location data from a part of Manchester, UK (some BSs are collocated, in which case just one triangle is shown on the map), and (b) redundancy in the cellular coverage of Manchester area at 700m BS coverage. Note that this real dataset here is available online: [37] for the location of BSs [2] ©2011 IEEE.

located at x and served by BS $i \in \mathcal{B}_{on}$ is denoted by $c_i(x, \mathcal{B}_{on})$. For analytical tractability, it is assumed that $c_i(x, \mathcal{B}_{on})$ does not change over time, i.e., without considering fast fading or dynamic inter-cell interferences. Instead, $c_i(x, \mathcal{B}_{on})$ can be considered as an *time-averaged* transmission rate. This assumption is reasonable in the sense that the time scale of user association is much larger than the time scale of fast fading or dynamic inter-cell interferences. Hence, the inter-cell interference is considered static Gaussian-like noise, which is feasible under interference randomization or fractional frequency reuse [44–46]. It should be noted, however, that $c_i(x, \mathcal{B}_{on})$ is *location-dependent* but not necessarily determined by the distance from the BS i. Hence, $c_i(x, \mathcal{B}_{on})$ can capture shadowing effect as well.

The *system-load density* $\varrho_i(x, \mathcal{B}_{on})$ is then defined as $\varrho_i(x, \mathcal{B}_{on}) = \frac{\gamma(x)}{c_i(x, \mathcal{B}_{on})}$, which denotes the fraction of time required to deliver traffic load $\gamma(x)$ from BS $i \in \mathcal{B}_{on}$ to location x. A routing function $p_i(x)$ is further introduced to specify the *probability* that a flow at location x is associated with BS i. Intuitively, $p_i(x)$ can be interpreted as the time fraction that a flow arrived at location x is routed to BS i. We will see that, however, the optimal $p_i(x)$ will be either 1 or 0, i.e., deterministic routing (or user association) is the solution of our optimization problem, which is defined in (16.22). Then, the utilization of BS i, $0 \le \rho_i \le 1 - \epsilon$ is defined as

$$\rho_i = \int_{\mathcal{R}} \varrho_i(x, \mathcal{B}_{on}) p_i(x) dx, \forall i \in \mathcal{B}, \qquad (16.16)$$

and where $\epsilon > 0$ is an arbitrarily small constant. We further denote the vectors containing the utilizations of all BSs by $\rho = (\rho_1, \cdots, \rho_{|\mathcal{B}|})$.

Two performance metrics are considered: the cost of flow-level performance such as file transfer delay and the cost of energy. The problem is to find an optimal set of active BSs ($\mathcal{B}_{on}$) and BS loads ρ (i.e., user association) that minimize the total system cost function, given by

$$\min_{\mathcal{B}_{on}, \rho} \left\{ d_\alpha(\rho, \mathcal{B}_{on}) + \eta e(\rho, \mathcal{B}_{on}) \mid \rho \in \mathcal{F}(\mathcal{B}_{on}), \mathcal{B}_{on} \subseteq \mathcal{B} \right\}, \qquad (16.17)$$

where $\mathcal{F}(\mathcal{B}_{on})$ is a feasible set of the load factor ρ given the set of active BSs $\mathcal{B}_{on} \subseteq \mathcal{B}$ and $\eta \ge 0$ is the parameter that balances the *tradeoff between the flow-level performance $d_\alpha(\rho, \mathcal{B}_{on})$ and the energy consumption $e(\rho, \mathcal{B}_{on})$*, both of which will be explained shortly. In [12] we have shown that the feasible set $\mathcal{F}(\mathcal{B}_{on})$ is a convex set of ρ. The implication of η is as follows; when η is zero, the focus is only on the flow-level performance, however, as η grows, more emphasis is given to energy conservation.

(i) The cost function of flow-level performance: A generalized α-delay

performance function in [47] is adopted.

$$
d_\alpha(\rho, \mathcal{B}_{\text{on}}) =
\begin{cases}
\displaystyle\sum_{i \in \mathcal{B}_{\text{on}}} \frac{(1 - \rho_i)^{1-\alpha} - 1}{\alpha - 1}, & \alpha \neq 1 \\[2ex]
\displaystyle\sum_{i \in \mathcal{B}_{\text{on}}} \log\left(\frac{1}{1 - \rho_i}\right), & \alpha = 1
\end{cases}
\tag{16.18}
$$

where $\alpha \geq 0$ is a parameter specifying the desired *degree of load balancing*.

(ii) The cost function of energy: We use a general model for BSs consisting of two types of power consumptions: fixed power consumption and adaptive power consumption.

$$
e(\rho, \mathcal{B}_{\text{on}}) = \sum_{i \in \mathcal{B}_{\text{on}}} \left[(1 - q_i)\rho_i P_i + q_i P_i \right],
\tag{16.19}
$$

where $q_i \in [0, 1]$ is the portion of the fixed power consumption of BS i, and P_i is the maximum operational power of BS i when it is fully utilized, i.e., $\rho_i = 1$, which includes power consumptions of the transmit antennas, power amplifiers, cooling equipment, signal processor, battery backup, power supply, etc. Note that the first and second terms in (16.19) are the fixed and adaptive (i.e., proportional to the utilization) power consumptions, respectively.

When $q_i = 0$, BSs are assumed to consist of only energy-proportional devices. Such BSs would ideally consume no power when idle, and gradually consume more power as the activity level increases. This type of BSs will be referred to as *energy-proportional* BS. However, energy-proportional BSs are still far from reality because several devices in the BSs dissipate standby power while inactive. As an example, a class-A amplifier [48], which is a typical power amplifier for macro BSs and one of the most power consuming devices in BSs, has the maximum theoretical efficiency of 50%. This type of BSs, which consume the fixed power irrespective of its activity unless they are totally turned off, i.e., $q_i > 0$, will be referred to as *non-energy-proportional* BS. Note that when $q_i = 1$, this model can also capture a constant energy consumption model, which is widely used in many works in the literature [2, 8, 49–51].

Solving the general problem given in (16.17) is very challenging due to highly complex coupling of BS operation and user association. For analytical tractability, we shall make an assumption on time-scale separation that flow arrival and departure process and the corresponding user association process are much faster than the period on which the set of active BSs are determined.

Under this assumption, our general problem given in (16.17) can be decomposed into two subproblems, in which BS operation problem is solved at a slower time scale than user association problem. For any given set of active BSs $\mathcal{B}_{\text{on}}$, the problem in (16.17) reduces to the following load balancing problem by ignoring the constant fixed power consumption term $\sum_{i \in \mathcal{B}_{\text{on}}} q_i P_i$. This problem can be also interpreted as user association problem because it finds the optimal BS utilization vector ρ by determining which BS should be associated to each MT.

User association problem [P-UA]:

$$\min_{\rho \in \mathcal{F}(\mathcal{B}_{\mathrm{on}})} d_\alpha(\rho, \mathcal{B}_{\mathrm{on}}) + \eta \sum_{i \in \mathcal{B}_{\mathrm{on}}} (1 - q_i)\rho_i P_i. \tag{16.20}$$

Then, the problem to solve is the following BS operation that finds the optimal set of active BSs $\mathcal{B}_{\mathrm{on}}$ on the longer time-scale.

BS operation problem [P-BO]:

$$\min_{\mathcal{B}_{\mathrm{on}} \subseteq \mathcal{B}} G(\mathcal{B}_{\mathrm{on}}) + \eta \sum_{i \in \mathcal{B}_{\mathrm{on}}} q_i P_i, \tag{16.21}$$

where the function $G(\mathcal{B}_{\mathrm{on}})$ is defined as the obtaining optimal value from the underlying user association in (16.20), i.e., $G(\mathcal{B}_{\mathrm{on}}) = \min_{\rho \in \mathcal{F}(\mathcal{B}_{\mathrm{on}})} d_\alpha(\rho, \mathcal{B}_{\mathrm{on}}) + \eta \sum_{i \in \mathcal{B}_{\mathrm{on}}} (1 - q_i)\rho_i P_i$.

Remark 16.4.1. *Note that* [P-UA] *and* [P-BO] *may have conflicting interest.* [P-UA] *tries to distribute traffic loads to improve the flow-level performance* $d_\alpha(\rho, \mathcal{B}_{on})$. *On the other hand, to minimize* $\sum_{i \in \mathcal{B}_{on}} q_i P_i$, [P-BO] *tries to concentrate traffic loads to a subset of BSs* $\mathcal{B}_{on}$ *and turn off the other BSs.*

Each of the these two problems will be discussed in the consequent Sections 16.4.3 and 16.4.4.

16.4.3   Energy-Efficient User Association

Given the set of active BSs $\mathcal{B}_{\mathrm{on}}$, we focus on solving [P-UA] in (16.20), i.e., associating users with BSs in an energy-efficient manner, considering load-balancing. Let us denote the optimal BS load vector by $\rho^* = (\rho_1^*, \cdots, \rho_{|\mathcal{B}|}^*)$, i.e., solution to the problem [P-UA], and further denote the optimal user association at location x by $i^*(x)$. We now present the optimality condition of the problem that describes an optimal user association policy.

Theorem 16.2. *[12] If the problem* [P-UA] *is feasible, then the optimal user association made by the MT located at x to join BS $i^*(x)$ is given by*

$$i^*(x) = \operatorname*{argmax}_{j \in \mathcal{B}_{on}} \frac{c_j(x, \mathcal{B}_{on})}{(1 - \rho_j^*)^{-\alpha} + \eta(1 - q_j)P_j}, \quad \forall x \in \mathcal{R}. \tag{16.22}$$

An online distributed algorithm that achieves the global optimum of [P-UA] in an iterative manner involves the following two parts.

Mobile terminal: At the start of the k-th iteration period, MTs receive BS loads $\rho^{(k)}$, e.g., through broadcast control messages from BSs.[9] Then, a new

---

[9] IEEE 802.16m facilitates this type of message structure [52, 53].

flow request from a MT located at x simply selects the BS $i(x)$ using the deterministic rule given by

$$i^{(k)}(x) = \operatorname*{argmax}_{j \in \mathcal{B}_{\mathrm{on}}} \frac{c_j(x, \mathcal{B}_{\mathrm{on}})}{\left(1 - \rho_j^{(k)}\right)^{-\alpha} + \eta(1 - q_j)P_j}, \ \forall x \in \mathcal{R} \qquad (16.23)$$

Base station: During the k-th period, BSs *measure* their average utilizations after some period of time, i.e., when the system exhibits stationary performance. Then, BSs broadcast the average utilization vector $\rho^{(k+1)}$ for the next iteration.

This simple iteration provably converges to the global optimal point with a simple modification of the proof in [47].

16.4.4 Traffic-Driven Base Station On/Off

BSs typically consume large amounts of energy in power amplifier circuit, air conditioning unit, etc, irrespective of offered loads. As a simple intuition, the ratio of the overhead power to the total power $(= q_i P_i / [q_i P_i + (1 - q_i)\rho_i P_i])$ is close to 100% for small loads ρ_i. Thus, it would be definitely beneficial to turn off BSs with low activity, in conjunction with energy-efficient user association. In this section, we show algorithms that reduce the energy consumption by solving the BS operation problem **[P-BO]** in (16.21) determining the set of BSs that can be switched off.

Recall that the function $G(\mathcal{B}_{\mathrm{on}}) = \min_{\rho \in \mathcal{F}(\mathcal{B}_{\mathrm{on}})} d_\alpha(\rho, \mathcal{B}_{\mathrm{on}}) + \eta \sum_{i \in \mathcal{B}_{\mathrm{on}}} (1 - q_i)\rho_i P_i$ is obtained by the optimal user association policy in (16.22). The objective function in (16.20) is convex in ρ given $\mathcal{B}_{\mathrm{on}}$, but becomes a nonconvex and also discontinuous function when $\mathcal{B}_{\mathrm{on}}$ is considered as a variable. Thus, this BS operation problem is a challenging combinatorial problem with $O(2^{|\mathcal{B}|})$ possible cases, which makes it very difficult to find an optimal solution through exhaustive search, especially, when the number of BSs is large. Thus, we show greedy-style heuristic algorithms, each of which has slightly different design rationale.

16.4.4.1 Greedy Turning on Algorithm

We first describe a greedy turning on algorithm, called GON, that iteratively finds BSs that have some benefit of delay reduction per their power usages.

Greedy on algorithm (GON)

1: Initialize $\mathcal{B}_{\mathrm{on}} = \mathcal{B}_{\mathrm{init}}$
2: **while** $\mathcal{B}_{\mathrm{on}} \neq \mathcal{B}$
3:    Calculate $M_{\mathrm{GON}}(i) = \frac{G(\mathcal{B}_{\mathrm{on}}) - G(\mathcal{B}_{\mathrm{on}} \cup i)}{q_i P_i}, \forall i \in \mathcal{B} \setminus \mathcal{B}_{\mathrm{on}}$
4:    Find the BS $i^* = \operatorname{argmax}_{i \in \mathcal{B} \setminus \mathcal{B}_{\mathrm{on}}} M_{\mathrm{GON}}(i)$,
5:    **if** $M_{\mathrm{GON}}(i^*) > \eta$, **then** $\mathcal{B}_{\mathrm{on}} \leftarrow \mathcal{B}_{\mathrm{on}} \cup \{i^*\}$,
6:    **else**, stop the algorithm.
7: **end while**

We introduce a metric $M_{\text{GON}}(i)$ for BS i that represents the *turn-on benefit per fixed power consumption* for BS i. The GON starts with a initial set of BSs $\mathcal{B}_{\text{init}}$ and iteratively finds the best BS as a candidate among the set of inactive BSs $\mathcal{B} \setminus \mathcal{B}_{\text{on}}$ that has the highest M_{GON} (step 4). Then, the algorithm finally adds the selected BS to the list of BSs to turn on, only if its metric is greater than η (step 5), or stops otherwise. Note that the criterion $M_{\text{GON}}(i^*) > \eta$ is directly obtained from the condition that additionally turning on BS i^* is beneficial (i.e., minimizing the total system cost), given by:

$$G(\mathcal{B}_{\text{on}}) + \eta \sum_{i \in \mathcal{B}_{\text{on}}} q_i P_i > G(\mathcal{B}_{\text{on}} \cup \{i^*\}) + \eta \bigg(\sum_{i \in \mathcal{B}_{\text{on}}} q_i P_i + q_{i^*} P_{i^*} \bigg).$$

16.4.4.2 Design Rationale

Consider the following problem that is closely related to (16.21):

$$\min_{\mathcal{B}_{\text{on}} \subseteq \mathcal{B}} \ G(\mathcal{B}_{\text{on}}) \ \text{subject to} \ \sum_{i \in \mathcal{B}_{\text{on}}} q_i P_i \le C, \qquad (16.24)$$

where we essentially move the power consumption cost in the objective function into the constraint of power consumption with some nonnegative budget C. For a given η, we can find $C = C(\eta)^{10}$, such that the same optimal solutions are achieved for (16.21) and (16.24), in which η is interpreted as a Lagrange multiplier of the dual formulation of (16.24).

We transform (16.24) into:

$$\max_{\mathcal{A} \subseteq \mathcal{B} \setminus \mathcal{B}_{\text{init}}} \ H(\mathcal{A}) \ \text{subject to} \ c(\mathcal{A}) = \sum_{i \in \mathcal{A}} c(i) \le \tilde{C}, \qquad (16.25)$$

where $\mathcal{A} = \mathcal{B}_{\text{on}} \setminus \mathcal{B}_{\text{init}}$, $H(\mathcal{A}) = G(\mathcal{B}_{\text{init}}) - G(\mathcal{B}_{\text{init}} \cup \mathcal{A}) = G(\mathcal{B}_{\text{init}}) - G(\mathcal{B}_{\text{on}})$, $c(i) = q_i P_i$ and $\tilde{C} = C - \sum_{i \in \mathcal{B}_{\text{init}}} c(i)$. If it can be shown that H is a non-decreasing submodular set function, then a variant greedy algorithm of GON, where the only difference lies in the stopping condition (step 5), can be shown to achieve a constant factor $(1 - 1/e)$ approximation[11] of the optimal value of the problem (16.25).

Submodularity, informally, is an intuitive notion of *diminishing returns*, which states that adding an element to a small set helps more than adding that same element to a larger set. Formally, it is defined as follows.

Definition 16.1. *A real-valued set function H, defined on subsets of a finite*

---

[10]$C(\eta)$ is a non-increasing function of η.

[11]The submodular maximization problem (SMP) in (16.25) is in general a NP-hard problem. It has been proved that the greedy algorithm of SMP can achieve a constant factor $(1 - 1/e)$ approximation and its ratio is an optimal in the sense that no other polynomial algorithms with better constant approximation ratio exist. We refer the readers to [38,54,55] for the details.

set $\mathcal{S}$ is called submodular *if for all* $\mathcal{A}_1 \subseteq \mathcal{A}_2 \subseteq \mathcal{S}$ *and for all* $s \in \mathcal{S} \setminus \mathcal{A}_2$, *if it satisfies that*

$$H(\mathcal{A}_1 \cup s) - H(\mathcal{A}_1) \geq H(\mathcal{A}_2 \cup s) - H(\mathcal{A}_2). \qquad (16.26)$$

16.4.4.3   Greedy Turning off Algorithm

One can think of another greedy algorithm, called GOFF (Greedy Off), which can be interpreted as the opposite of GON. The GOFF, unlike GON, starts from the entire BSs $\mathcal{B}$ and finds a solution by iteratively removing the BS with the lowest *turn-off detriment per fixed power consumption*. Note that GOFF does not have the issue of choosing $\mathcal{B}_{\mathrm{init}}$.

Greedy off algorithm (GOFF)

1: Initialize $\mathcal{B}_{\mathrm{on}} = \mathcal{B}$
2: **while** $\mathcal{B}_{\mathrm{on}} \neq \emptyset$
3:    Calculate $M_{\mathrm{GOFF}}(i) = \frac{G(\mathcal{B}_{\mathrm{on}} \setminus \{i\}) - G(\mathcal{B}_{\mathrm{on}})}{q_i P_i}, \forall i \in \mathcal{B}_{\mathrm{on}}$
4:    $i^* = \arg\min_{i \in \mathcal{B}_{\mathrm{on}}} M_{\mathrm{GOFF}}(i)$
5:    **if** $M_{\mathrm{GOFF}}(i^*) < \eta$, **then** $\mathcal{B}_{\mathrm{on}} \leftarrow \mathcal{B}_{\mathrm{on}} - \{i^*\}$,
6:    **else**, stop the algorithm.
7: **end while**

16.4.5   Discussion: GON and GOFF

16.4.5.1   Interpretation

We now discuss the implication of the metric $M_{\mathrm{GOFF}}(i)$ used in GOFF (or $M_{\mathrm{GON}}(i)$ used in GON). Note that this metric can be interpreted as the *network-wide impact* per unit power cost. GOFF tends to choose and remove the BS that will bring the small impact on the network when turned off, whereas GON tends to choose and add the BS that will bring the large impact on the network when turned on. From the BS i's perspective, the following internal and external factors, coupled in a complex manner each other, affect the metric $M_{\mathrm{GOFF}}(i)$[12] and the choice of the final set B_{on}.

- *Internal factors of BS i:* *Traffic loads* imposed on BS i is one of the dominant factors. Turning-off BS i with high utilization will cause high impact on neighboring BSs because the large amount of traffic loads needs to be transferred (or handed over) to its neighboring BSs with potentially low signal strengths.

- *External factors around BS i:* When turning-off the BS i, its network-wide impact also depends on the neighboring environment, e.g, the *number* of,

---

[12]The case of $M_{\mathrm{GON}}(i)$ can be understood similarly.

the *distance* to, and the *utilization* of the neighboring BSs. As the number is small, the distance is far, and/or the utilization is high, we can expect high network-wide impact.

It should be noted that GON and GOFF require information about spatial system-load density ϱ in order to compute their metrics. This is because G inside the metrics depends on ρ, where ρ is defined as the integral (or summation) of ϱ over the space in (16.16). This information can be obtained by exchanging signaling messages or use the predetermined traffic profile over a period (e.g., one day) as in [8]. In the next subsection, we introduce other purely heuristic algorithms that are more *operator-friendly* in the sense that no signaling or measurement overhead is necessary, yet with possibly slight performance degradation compared to GON and GOFF.

16.4.5.2   Other Heuristic Algorithms

For other heuristic algorithms, one can exploit the distances between BSs, such that BSs distant from (resp. close to) each other are turned on (resp. off). Motivated by this fact, another algorithm is proposed, i.e., distance-based greedy heuristics based on GON and GOFF, called GON-DIST and GOFF-DIST, by simply modifying the metrics in the step 3 of GON and GOFF as follows:

$$M_{\text{GON-DIST}}(i) = \left[\prod_{j \in \mathcal{B}_{\text{on}}} d(i,j) \right]^{1/|\mathcal{B}_{\text{on}}|}, \ \forall i \in \mathcal{B} \setminus \mathcal{B}_{\text{on}}, \qquad (16.27)$$

$$M_{\text{GOFF-DIST}}(i) = \left[\prod_{j \in \mathcal{B}_{\text{on}}, j \neq i} d(i,j) \right]^{1/|\mathcal{B}_{\text{on}}|}, \ \forall i \in \mathcal{B}_{\text{on}}, \qquad (16.28)$$

where the geometric mean of the distances to the other BSs are used for the distance metric.

Another greedy algorithm is called GOFF-UTIL[13], that chooses the most underutilized BS by modifying the metric in the step 3 of GOFF as follows:

$$M_{\text{GOFF-UTIL}}(i) = \rho_i, \ \forall i \in \mathcal{B}_{\text{on}}, \qquad (16.29)$$

16.4.6   Numerical Results

The energy-efficient user association and BS operation algorithms are verified through extensive simulations under various practical configurations. A network topology composed of five macro BSs and five micro BSs in 2×2 km^2 as shown in Figure 16.9 is considered for our simulations. A real 3G BS deployment topology consisting of heterogeneous environments (urban, suburban and rural areas) is also considered in subsections 16.4.6.2 in order to

---

[13]Note that it does not make sense to have GON-UTIL policy since BSs that are turned off cannot have utilizations by the definition.

TABLE 16.3 Typical transmit power and total power consumption for macro BSs (3 sectors/2 antennas) and micro BSs (1 sector/1 antenna).

Macro BS	p_M^{tx}	10W	20W	40W
$(a_M, b_M) = (22.6, 412.4W)$	P_M	638W	865W	1317W
Micro BS	p_m^{tx}	0.5W	1W	2W
$(a_m, b_m) = (5.5, 32.0W)$	P_m	35W	38W	43W

provide more realistic simulation results. Among several typical levels of maximum transmission powers for BSs given in [27], the intermediate values are used for the simulations, i.e., 43dBm and 30dBm for macro and micro BSs, respectively. Based on the linear relationship between transmission and operational power consumptions in Table 16.3, the maximum operational powers for BSs could be calculated, i.e., 865W and 38W for macro and micro BSs, respectively.

For the traffic model, we assume that file transfer request follows a Poisson point process with an arrival rate $\lambda(x)$. Each request has exactly one file that is log normally distributed with mean $1/\mu(x) = 100$ kbyte. In modeling propagation environment, the modified COST 231 path loss model with macro BS height $h = 32$m and micro BS height $h = 12.5$m is used. Other parameters for the simulations follow the suggestions in the IEEE 802.16m evaluation methodology document [56]. We consider the average delay experienced by a typical flow as our system performance metric, i.e., setting the degree of load balancing parameter as $\alpha = 2$ for the cost function of level performance. For the cost function of energy, the portion of fixed power consumption q_i ranges between 0 and 1 to include several types of BSs from energy-proportional BSs to non-energy-proportional BSs.

16.4.6.1   Energy-Delay Tradeoff for Energy-Proportional BSs

Energy-proportional BSs ($q_i = 0$) is considered to investigate the performance obtained purely by the proposed energy-efficient user association algorithm. Figure 16.8 shows the energy-delay tradeoff curves by varying the energy-delay tradeoff parameter from $\eta = 10^{-5}$ to 10^0 for the different values of arrival rate $\lambda(x)$. As can be expected, energy savings at the cost of delay increase when η grows. The percentage of maximum energy saving (moving from $\eta = 10^{-5}$ to $\eta = 10^0$) is about 50%. This result is obtained under homogeneous traffic distribution, i.e., $\lambda(x) = \lambda$ for all $x \in \mathcal{R}$. Note that similar trends can be also observed in inhomogeneous traffic distribution, please refer to [12] for more details.

In order to examine the details of where these energy savings come from, Figs. 16.9 (a) and (b) illustrate the snapshots of cell coverage for two extreme cases: low $\eta = 10^{-5}$ and high $\eta = 10^0$. By comparing these two figures, we can clearly see that the micro BS, which is more energy-efficient than the macro BSs, will have large coverage for the case of high η (i.e., giving more emphasis

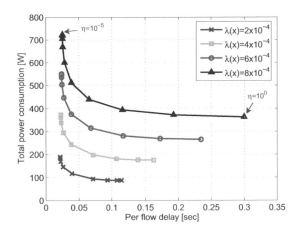

FIGURE 16.8 Energy-delay tradeoff for the case of energy-proportional BSs ($q_i = 0$) by varying the energy-delay tradeoff parameter $\eta = 10^{-5} \sim 10^0$. As η increases, energy saving can be obtained at the cost of delay increase. (from [12], ©2011 IEEE)

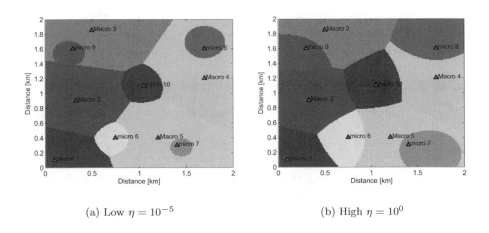

(a) Low $\eta = 10^{-5}$ (b) High $\eta = 10^0$

FIGURE 16.9 Snapshots of cell coverage by the energy-efficient user association algorithm. As η increases, the energy-efficient micro BS (indexed by 6 to 10) will have larger coverage [12] ©2011 IEEE.

on conserving energy). In other words, more MTs are likely to be associated with and served by the energy-efficient micro BSs that are indexed by 6 to 10 in the figures. However, as the traffic loads are concentrated in the micro BSs, the large utilizations at the micro BSs will results in the increase of per-flow delay.

16.4.6.2   Energy-Delay Tradeoff for Non-Energy-Proportional BSs

As a numerical example, non-energy-proportional BSs ($q_i = 0.5$) is considered and the performance is obtained by both the proposed energy-efficient user association and BS operation algorithms. Consider GON, GOFF, GON-DIST, GOFF-DIST and GOFF-UTIL as the BS operation algorithm, and compare their performance with the optimal solution obtained by exhaustive search. Figure 16.10 depicts the map of BS layout [36] that is chosen for more realistic simulations. It is a part of real 3G network operated by one of the major mobile network operators in Korea (the source has to be anonymized to preserve confidentiality). There are totally 30 BSs within 20×10 km^2 rectangular area. Particularly, this partial map is chosen to include a scenario that three environments (urban, suburban and rural) coexit together.

Figure 16.11(a) shows the energy-delay tradeoff curves of different algorithms by varying the energy-delay tradeoff parameter $\eta = 10^{-5} \sim 10^0$ in urban area with the homogeneous traffic distribution of $\lambda(x) = 10^{-4}$ for all $x \in \mathcal{R}$. This offered load corresponds to about 10% of BSs utilizations when all BSs are turned on. Recall the real traffic measurement report [2] showing that the time fraction when the traffic is below 10% of peak during the day is about 30% in weekdays and about 45% in weekends. As can be seen from Figure 16.11(a), energy is saved at the cost of per-flow delay increase. The greedy algorithms perform close to the optimal solution when η is small. For example, from $\eta = 10^{-5}$, up to $\eta = 10^{-2}$ for GON and GOFF, $\eta = 10^{-3}$ for GOFF-UTIL, and $\eta = 10^{-4}$ for GON-DIST and GOFF-DIST, respectively, the solution is very close to the optimal (i.e., Exhaustive $\geq$ GON = GOFF $\geq$ GOFF-UTIL $\geq$ GON-DIST = GOFF-DIST). However, there is a performance gap when η becomes large.

Figure 16.11(b) shows the energy-delay tradeoff curves under the inhomogeneous traffic distribution. As an example of inhomogeneous traffic loads, a linearly increasing load along the diagonal direction from right bottom to left top is considered. They are normalized over the space so as to have the same amount of total traffic as the homogeneous traffic loads have. Similar tradeoff curve can be observed in inhomogeneous traffic distribution as well. The greedy algorithms GON, GOFF and GOFF-UTIL still perform close to the optimal solution up to $\eta = 10^{-3}$, however, GON-DIST and GOFF-DIST start to deviate much from the optimal solution after $\eta = 10^{-4.5}$.

There is a reason why such GON-DIST and GOFF-DIST based on the distance do not work well under the inhomogeneous traffic distribution: turning on (resp. off) the BSs distant from (resp. close to) each other is no longer

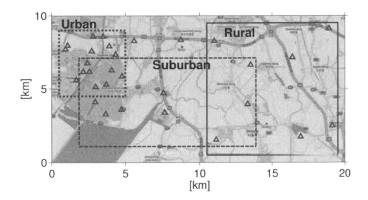

FIGURE 16.10 Real 3G BS deployment map (30 BSs in 20 x 10 km^2). (from [12], ©2011 IEEE)

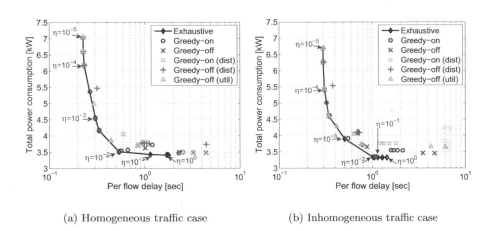

(a) Homogeneous traffic case

(b) Inhomogeneous traffic case

FIGURE 16.11 Energy-delay tradeoff of different algorithms for the case of non-energy-proportional BSs ($q_i = 0.5$) by varying the energy-delay tradeoff parameter from $\eta = 10^{-5}$ to 10^0. While greedy algorithms perform close to the optimal solution when η is small, there is a gap when η is large. Especially, GON-DIST and GOFF-DIST have large performance gaps under the inhomogeneous traffic distribution [12] ©2011 IEEE.

TABLE 16.4 Effect of BS density on energy savings.

BS Density		Urban	Suburban	Rural
Maximum	Greedy-on	66.2%	32.4%	0.1%
Energy Savings	Greedy-off	72.8%	32.5%	4.2%

reasonable because the BSs distant from (resp. close to) each other but located in the area of low (resp. high) traffic loads may not be beneficial to turn on (resp. off). It is noteworthy that GOFF-UTIL has the almost comparable performance to that of GON and GOFF under both homogeneous and inhomogeneous traffic distribution. This is a desirable observation for wireless network operators who do want to implement a simple yet efficient BS operation algorithm.

16.4.6.3   Effect of BS Density on Energy Savings

We also examine how much energy savings can be achieved according to the density of BS deployment. To this end, we vary the BS density by adopting the BS topology from three different environments in Figure 16.10: urban (15 BSs in 4.5×4.5 km^2), suburban (15 BSs in 12×6 km^2) and rural (8 BSs in 9×9 km^2) areas. Table 16.4 shows the effect of BS deployment density on the maximum energy savings. As expected, much energy saving in the urban and suburban environments, but almost no or low energy saving in the rural environment. This is because the degradation of signal strength is significant in the rural environment when traffic loads are transferred from the switched-off BS to neighboring BSs.

16.5   Other Issues and Discussions

16.5.1   Open Problems and Other Directions Towards Green Cellular Networks

Even though recent papers have started to investigate ways to increase the energy efficiency of cellular networks, we are still at an early stage of this research. Therefore, we would like to encourage the community to give it greater attention by pointing out several open problems and other directions including new paradigm and architecture towards green cellular networks.

Multi-operator cooperation. There is also room for greater energy saving if different operators can pool their BSs together and accept each other's traffic as roaming traffic when the BSs are switched off [2,57]. This kind of operator cooperation to ensure pooled coverage is particularly helpful in metropolitan areas where each operator sets up a dense deployment. Such a scenario has also previously been examined by Marsan *et al.* [57], who consider the simple setting of a single cell with two operators and show that considerable energy savings can be obtained in such a setting. In realizing cooperation, physical sharing may not be a substantial concern. In many cases, in urban areas, operator's BSs tend to be closely situated, or even co-located. The main challenge is the complexity in network operation, with respect to issues such as cross-operator user authentication and billing, introduced by a novel fine-granularity roaming. This yields a rich set of problems to be investigated from *a game theoretic and economic perspective*. Under what conditions would self-interested operators agree to cooperate with others? What kind of profit-sharing agreements will provide an adequate incentive for all participants? It also needs to be examined whether such operator agreements can be potentially abused to create oligopolies that hurt customers and how these can be ameliorated in the interest of green operation that offer other benefits to society.

Component-level deactivating or deceleration. As discussed in section 16.4, shutting down some underutilized BSs has substantial potential to obtain energy savings. However, despite very low traffic, there is a certain group of BSs that should not be turned off not to create coverage holes.[14] Further, switching BSs off may bring degradation in user experience, for example, more uplink power consumption for file uploading. Due to these technical challenges of entirely turning off BSs, component-level techniques [58, 59] has been recently proposed, which are more conservative yet can reduce conserve energy consumption effectively. In [58], the authors adjusted the number of radio units and transmit antennas in BSs by deactivating them during low load period. In [59], the authors considered cooperation of a component-level deceleration technique into BS operation. This technique, called dynamic voltage frequency scaling (or simply speed-scaling) [60], allows a central processing unit (CPU) to adapt its speed for energy conservation based on incoming processing demand.[15] They also investigated its impact on the design of network protocols in cellular networks.

Coordinated multi-point transmission and reception (CoMP). The primary constraint of green cellular operation is to preserve coverage and service quality although certain BSs are turned off during low activity period. One can potentially increase transmission power when some BSs are turned off to

---

[14]According to the classification in [22], there are two different classes of BSs: class-A (non-redundant) and class-B (redundant) BSs. The former are the BSs that have been deployed to provide the coverage to the users, usually at an early stage of the deployment. The latter BSs that have been additionally deployed to mainly increase the capacity of the network at a later stage. While class-B BSs may be turned off, class-A BSs cannot because their contribution to the coverage is particularly high.

[15]Note that examples of in-BS processing are increasingly abundant from flow classification, signal processing (OFDM modulation, coding, etc.) to even security and multimedia conversion. It is also worthwhile mentioning that DVFS lowers heat dissipation as well. As a consequence, it can reduce the power consumption in cooling equipment (not to mention in CPU) contributing to a considerable amount of total energy consumption, where this exerting influence is often linear.

increase the coverage area of the remaining BSs. Another novel approach that can play a role in maintaining coverage is the use of coordinated multi-point transmission and reception (CoMP) being developed in the context of LTE-Advanced systems [61]. The basic idea in CoMP, a macro-diversity scheme based on network MIMO (multiple-input and multiple-output), is to improve the signal quality by coordinating the transmission and reception at neighboring cells. CoMP can improve the coverage of high data rates and/or the cell-edge throughput, which may allow the systems to turn off more BSs [34]. However, overheads such as message exchanges between cooperating BSs via backhaul and complex signal processing at the BSs are the major challenges to overcome. Taking into account additional energy consumption for such overheads, it is essential to investigate the net-energy-efficiency of CoMP [33].

Multi-hop relaying. The future cellular networks such as LTE-Advanced systems have very challenging requirements because there is a growing demand for coverage and capacity. Instead of deploying additional BSs to meet such requirements, it is sometimes beneficial to install relays. The rationale of relaying is that the path loss between BSs and MTs can be reduced significantly by breaking a long weak single-hop link into shorter strong multi-hop links. They have low power consumption due to their small size and have a wireless backhaul, which enables deployment flexibility and is much cheaper than deploying new BSs with a fixed wired backhaul. Furthermore, multi-hop relaying will clearly be useful to ensure that the dynamic shutting down of BSs, while saving energy, does not leave coverage holes. All these reasons allow relays to become one of candidate technologies for green cellular networks. Examples of recent studies on relaying in cellular networks include: an investigation into how relays can increase the spatial reuse and therefore provide the required data rates while reducing energy consumption [62], a joint optimization of relay placement and sleep/active problem [17], and a relay caching mechanism [63] to improve the energy efficiency, especially for multimedia applications.

Transmission power control. The current and future cellular networks are expected to become more heterogeneous with a mixture of macro and small cells, where interference is a major obstacle that can impair the potential gain of small cells and its pattern is highly diverse [64]. As the number of small cells increases, the portion of users at cell edges also grows, resulting in the increase of the number of users suffering from low throughput due to severe interference. To mitigate interference, starting from a static interference management (IM) having specific reuse patterns, e.g., a traditional frequency reuse, fractional frequency reuse (FFR) [65], or its variation [45, 66], the research community has paid extensive research attention to dynamic IM algorithms that performs dynamic BS power control depending on the scheduled users over the multiple neighboring cells [67–73]. Their proposals are more performance-oriented in the sense that with a given power budget the objective is to maximize the efficiency (typically measured by throughput or utility). However, in fact with significantly less amount of powers, it may be possible to achieve a near-optimal performance mainly because the environment is typically interference-limited and/or operated under the high SINR regime, where, for example, the achievable rate for users under such regimes slowly decreases as SINR decreases. Recently, in [11], Kwak *et al.* studied the impact of sharing the power budget, and concluded that sharing the power budget more spatially and temporally enables us to save significant amount of energy.

Radio over fiber. Radio over fiber (RoF) refers to a technology where *analog* radio signal is directly transmitted over an optical fiber link for wireless access deployment [74, 75]. To elaborate, a base station server is connected to antennas at remote sites via optical fiber, and each remote site serves as a (small) cell. RoF is well aligned with the upcoming wireless access network in several aspects such as network capacity, energy efficiency and economic perspective. To increase the network capacity, the cell size needs to be small, and then the base station can be made simple by moving the main functionality, i.e., the scheduler, into a base station server. Then, the base station server controls a number of (small) base stations. Hence, the base station at the remote site can be made of only antennas and power amplifiers without complex functionalities. When a base station server controls a number of remote sites, it further contributes to making multi-cell cooperation easier, for example, in realizing inter-cell coordinated scheduling and interference management. From greening perspective, RoF is energy efficient because removing the scheduling and digital processing part of the base station can significantly reduce the energy consumption at each cell site by reducing the circuit power. From economic perspective, RoF lowers CAPEX of base station deployment and OPEX of base station operation; simple infrastructural architecture reduces CAPEx and low electricity bill thanks to energy efficiency decreases OPEX. The way of realizing RoF depends on how to transport radio signal over optical fiber; RoF is classified either intermediate frequency (IF) over fiber or radio frequency (RF) over fiber depending on whether up/down conversion at the remote site is required (IF over fiber) or not (RF over fiber).

16.5.2   Ongoing Projects and Consortiums

This section summarizes major international projects and consortia pertaining to green networks conducted by research institutes, universities and industries in recent years.

- *GreenTouch* [76], led by Bell Labs, is a worldwide research consortium comprised of 15 founder members from leading research institutions and university (e.g., INRIA, CEA-LETI, Stanford, MIT, etc.) and world's largest equipment manufacturers and network operators (e.g., Samsung, Freescale, AT&T, China Mobile, Telefonica, etc.). Their goal is to increase energy efficiency in networks by a factor of 1,000 within five years by designing fundamentally new network architectures and creating enabling technologies on which they are based.

- *EARTH* (Energy-Aware Radio and neTwork tecHnology) [77, 78] is a major European research project starting in 2010 with 15 partners from 10 countries, focusing on the energy efficiency in the next generation wireless access networks. The goal is to achieve at least 50% reduction within the next two-and-a-half years to provide directly applicable solutions for environmentally sustainable and cost efficient broadband wireless services.

- *GreenRadio* [79] is one of the three working areas under Mobile VCE (Virtual Centre of Excellence). The goal of GreenRadio is to secure 100x reduction in energy requirements for high data rate services by investigating both architectural and individual technological approaches.

- *Cool Silicon* [80] is a research cluster supported by German government since September 2008. It currently focuses on on three lead projects, CoolReader, CoolSensornet and CoolComputing, all aiming at massively increasing the energy efficiency in computing, cellular communications, and sensor networks.

Acknowledgments

We would like to acknowledge collaborations on green cellular network papers with Eunsung Oh at Korea Institute of Energy Technology Evaluation and Planning, Xin Liu at the University of California at Davis and Zhisheng Niu at Tsinghua University. We thank Mahasweta Sarkar and Santosh Nagaraj at San Diego State University for their helpful discussions. We also would like to thank Fred Richter, Albrecht J. Fehske, Patrick Marsch, and Gerhard P. Fettweis at the Dresden University of Technology, Germany for their generosity in allowing us to use their energy-aware deployment results in Section 16.3.2.

This work was supported in part by NSF NetSE grant 1017881, CNS-1116874, the Sogang University Research Grant of 2011, and Basic Science Research Program through the National Research Foundation of Korea (NRF) funded by the Ministry of Education, Science and Technology (2011-0015042).

Bibliography

[1] C. E. Jones, K. M. Sivalingam, P. Agrawal, and J. C. Chen, "A survey of energy efficient network protocols for wireless networks," *Wireless*

Networks, vol. 7, no. 4, pp. 343–358, Aug. 2001.

[2] E. Oh, B. Krishnamachari, X. Liu, and Z. Niu, "Towards dynamic energy-efficient operation of cellular network infrastructure," *IEEE Comm. Mag.*, pp. 56–61, Jun. 2011.

[3] G. Fettweis and E. Zimmermann, "ICT energy consumption–trends and challenges," in *Proc. IEEE WPMC*, Lapland, Finland, Sep. 2008.

[4] "Mobile network energy OPEX to rise dramatically to $22 billion in 2013," Article appearing in Cellular News, Jul. 3, 2008.

[5] H. Kim, C.-B. Chae, G. de Veciana, and R. W. Heath Jr., "A cross-layer approach to energy efficiency for adaptive mimo systems exploiting spare capacity," *IEEE Trans. Wireless Commun.*, vol. 8, no. 8, pp. 2745–53, Aug. 2009.

[6] H. Kim and G. de Veciana, "Leveraging dynamic spare capacity in wireless systems to conserve mobile terminals energy," *IEEE/ACM Trans. Networking.*, vol. 18, no. 3, pp. 802–815, Jun. 2010.

[7] G.-W. Miao, N. Himayat, G. Y. Li, and A. Swami, "Cross-layer optimization for energy-efficient wireless communications: A survey," *Wiley Journal WCMC*, vol. 9, no. 4, pp. 529–542, Apr. 2009.

[8] M. A. Marsan, L. Chiaraviglio, D. Ciullo, and M. Meo, "Optimal energy savings in cellular access networks," in *Proc. GreenComm*, Dresden, Germany, Jun. 2009.

[9] "Optimizing performance and efficiency of PAs in wireless base stations," White Paper, Texas Instruments, Feb. 2009.

[10] "Sustainable energy use in mobile communications," White Paper, Ericsson, Aug. 2007.

[11] J. Kwak, K. Son, Y. Yi, and S. Chong, "Impact of spatio-temporal power sharing policies on cellular network greening," in *Proc. WiOpt*, Princeton, NJ, May 2011, pp. 167–174.

[12] K. Son, H. Kim, Y. Yi, and B. Krishnamachari, "Base station operation and user association mechanisms for energy-delay tradeoffs in green cellular networks," *IEEE Jour. Select. Areas in Comm.*, pp. 1525–1536, Sep. 2011.

[13] K. Son, E. Oh, and B. Krishnamachari, "Energy-aware hierarchical cell configuration: from deployment to operation," in *Proc. IEEE INFOCOM 2011 Workshop on Green Communications and Networking*, Shanghai, China, Apr. 2011, pp. 289–294.

[14] F. Richter, A. J. Fehske, and G. P. Fettweis, "Energy efficiency aspects of base station deployment strategies for cellular networks," in *Proc. IEEE VTC*, Anchorage, AK, Sep. 2009.

[15] P. Rost and G. Fettweis, "Green communications in cellular networks with fixed relay nodes," *Cooperative Cellular Wireless Networks*, Sep. 2010.

[16] S. Bhaumik, G. Narlikar, S. Chattopadhyay, and S. Kanugovi, "Breathe to stay cool: adjusting cell sizes to reduce energy consumption," in *Proc. the first ACM SIGCOMM workshop on Green networking*, New Delhi, India, Aug. 2010, pp. 41–46.

[17] S. Zhou, A. J. Goldsmith, and Z. Niu, "On optimal relay placement and sleep control to improve energy efficiency in cellular networks," in *Proc. IEEE ICC*, Kyoto, Japan, Jun. 2011.

[18] W. Yang, L. Li, Y. Wang, and W. Sun, "Energy-efficient transmission schemes in cooperative cellular systems," in *Proc. GreenComm*, Miami, FL, Dec. 2010.

[19] S. Ramanath, V. Kavitha, and E. Altman, "Open loop optimal control of base station activation for green networks." in *Proc. WiOpt*, May 2011, pp. 161–166.

[20] C. Peng, S.-B. Lee, S. Lu, H. Luo, and H. Li, "Traffic-driven power saving in operational 3g cellular networks," in *ACM MobiCom*, Las Vegas, NV, Sep. 2011, pp. 121–132.

[21] E. Oh and B. Krishnamachari, "Energy savings through dynamic base station switching in cellular wireless access networks," in *Proc. IEEE Globecom*, Miami, FL, Dec. 2010.

[22] K. Dufkova, M. Popovic, R. Khalili, and et al., "Energy Consumption Comparison Between Macro-Micro and Public Femto Deployment in a Plausible LTE Network," in *e-Energy 2011*, New York, NY, USA, Jun. 2011, pp. 1–10.

[23] W. Wang and G. Shen, "Energy efficiency of heterogeneous cellular network," in *Proc. IEEE VTC Fall*, Ottawa, Canada, Sep. 2010, pp. 1–5.

[24] B. Badic, T. O'Farrell, P. Loskot, and J. He, "Energy efficient radio access architectures for green radio: large versus small cell size deployment," in *Proc. IEEE VTC Fall*, Anchorage, AK, 2009, pp. 1–5.

[25] F. Richter and G. Fettweis, "Cellular mobile network densification utilizing micro base stations." in *Proc. IEEE ICC*, Cape Town, South Africa, 2010, pp. 1–6.

[26] F. Richter, M. Gruber, O. Blume, and V. S. T. U. Dresden, "Micro base stations in load constrained cellular mobile radio networks," in *Proc. IEEE PIMRC Workshop: W-Green*, Instanbul, Turkey, Sep. 2010, pp. 357–362.

[27] A. J. Fehske, F. Richter, and G. P. Fettweis, "Energy efficiency improvements through micro sites in cellular mobile radio networks," in *Proc. GreenComm*, Honolulu, HI, Dec. 2009.

[28] F. Richter, A. J. Fehske, P. Marsch, and G. P. Fettweis, "Traffic demand and energy efficiency in heterogeneous cellular mobile radio networks," in *Proc. IEEE VTC Spring*, Taipei, Taiwan, May 2010, pp. 1–6.

[29] T. T. Tesfay, R. Khalili, J.-Y. L. Boudec, and et al., "Energy saving and capacity gain of micro sites in regular lte networks: Downlink traffic layer analysis," in *Proc. ACM MSWiM Poster*, Instanbul, Turkey, Sep. 2010, pp. 357–362.

[30] S. Boiardi, C. Antonio, and S. Brunilde, "Radio planning of energy-aware cellular networks," *Technical Report*, 2010.

[31] Z. Niu, Y. Wu, J. Gong, and Z. Yang, "Cell zooming for cost-efficient green cellular networks," *IEEE Comm. Mag.*, vol. 48, pp. 74–79, Nov. 2010.

[32] X. Weng, D. Cao, and Z. Niu, "Energy-efficient cellular network planning under insufficient cell zooming," in *Proc. IEEE VTC Spring*, Budapest, Hungary, May 2011, pp. 1–5.

[33] A. Fehske, P. Marsch, and G. Fettweis, "Bit per joule efficiency of cooperating base stations in cellular networks," in *Proc. GreenComm*, Miami, FL, Dec. 2010, pp. 1406–1411.

[34] D. Cao, S. Zhou, C. Zhang, and Z. Niu, "Energy saving performance comparison of coordinated multi-point transmission and wireless relaying," in *Proc. IEEE Globecom*, Miami, FL, Dec. 2010, pp. 1–5.

[35] M. S. Alouini and A. J. Goldsmith, "Area spectral efficiency of cellular mobile radio systems," *IEEE Trans. Veh. Technol.*, vol. 48, no. 4, pp. 1047–1066, Jul. 1999.

[36] K. Son, S. Lee, Y. Yi, and S. Chong, "REFIM: A practical interference management in heterogeneous wireless access networks," *IEEE Jour. Select. Areas in Comm.*, vol. 29, no. 6, pp. 1260–1272, Jun. 2011.

[37] "Sitefinder: Mobile phone base station database," ofcom. [Online]. Available: http://www.sitefinder.ofcom.org.uk/

[38] G. Nemhauser, L. Wolsey, and M. Fisher, "An analysis of the approximations for maximizing submodular set functions-I," *Mathematical Programming*, vol. 14, no. 1, pp. 265–294, Dec. 1978.

[39] C. Bianco, F. Cucchietti, and G. Griffa, "Energy consumption trends in the next generation access network - a telco perspective," in *Proc. INTELEC*, Rome, Italy, Sep. 2007.

[40] A. Goldsmith, *Wireless communications.* Cambridge Univ Press, 2005.

[41] "Normalized cellular traffic trace recorded during one week," aNRG. [Online]. Available: http://anrg.usc.edu/www/index.php/Downloads/

[42] "Energy information administration (update: June, 2010)." [Online]. Available: http://www.epa.gov/cleanenergy/energy-resources/calculator.html

[43] "Greenhouse gas equivalencies calculator." [Online]. Available: http://www.eia.gov/emeu/international/elecprii.html

[44] A. Sang, M. Madihian, X. Wang, and R. D. Gitlin, "Coordinated load balancing, handoff/cell-site selection, and scheduling in multi-cell packet data systems," in *Proc. ACM Mobicom*, Philadelphia, PA, Sept. 2004, pp. 302–314.

[45] K. Son, S. Chong, and G. de Veciana, "Dynamic association for load balancing and interference avoidance in multi-cell networks," *IEEE Trans. Wireless Commun.*, vol. 8, no. 7, pp. 3566–357, Jul. 2009.

[46] P. Hande, S. Patil, and H. Myung, "Distributed load-balancing in a multi-carrier system," in *Proc. IEEE Wireless Comm. and Net. Conf.*, 2009.

[47] H. Kim, G. de Veciana, X. Yang, and M. Venkatachalam, "Distributed α-optimal user association and cell load balancing in wireless networks," *IEEE/ACM Trans. Networking.*, vol. 10, no. 1, pp. 177–190, Feb. 2012.

[48] A. Grebennikov, *RF and Microwave Power Amplifier Design*. New York: McGraw-Hill, 2004.

[49] L. Chiaraviglio, D. Ciullo, M. Meo, M. A. Marsan, and I. Torino, "Energy-aware UMTS access networks," in *Proc. WPMC Symposium*, Lapland, Finland, Sep. 2008.

[50] S. Zhou, J. Gong, Z. Yang, Z. Niu, and P. Yang, "Green mobile access network with dynamic base station energy saving," in *Proc. ACM MobiCom (Poster)*, Beijing, China, Sep. 2009, pp. 1–3.

[51] E. Oh and B. Krishnamachari, "Energy savings through dynamic base station switching in cellular wireless access networks," in *Proc. IEEE Globecom*, Miami, FL, Dec. 2010.

[52] "IEEE P802.16m-2007 draft standards for local and metropolitan area networks part 16: Air interface for fixed broadcast wireless access systems," *IEEE Standard 802.16m*, 2007.

[53] H. Kim, X. Yang, M. Venkatachalam, Y.-S. Chen, K. Chou, I.-K. Fu, and P. Cheng, "Handover and load balancing rules for 16m," IEEE C802.16m-09/0136r1, Jan. 2009.

[54] M. Sviridenko, "A note on maximizing a submodular set function subject to knapsack constraint," *Operations Research Letters*, vol. 32, pp. 41–43, 2004.

[55] D. Golovin and A. Krause, "Adaptive submodularity: A new approach to active learning and stochastic optimization," in *Proc. the 23rd Annual Conference on Learning Theory*, Haifa, Israel, Jun. 2010, pp. 333–345.

[56] *IEEE 802.16m-08/004r5: IEEE 802.16m Evaluation Methodology Document (EMD)*, IEEE Std. 802.16m, 2009.

[57] M. A. Marsan and M. Meo, "Energy efficient management of two cellular access networks," in *Proc. ACM GreenMetrics*, Seattle, WA, Jun 2009.

[58] M. Hedayati, M. Amirijoo, P. Frenger, and J. Moe, "Reducing energy consumption through adaptation of number of active radio units," in *Proc. IEEE VTC Spring*, Budapest, Hungary, May 2011, pp. 1–6.

[59] K. Son and B. Krishnamachari, "Speedbalance: Speed-scaling-aware optimal load balancing for green cellular networks," in *Proc. IEEE INFO-COM*, Orlando, FL, Mar 2012.

[60] J. R. Lorch and A. J. Smith, "Improving dynamic voltage scaling algorithms with PACE," in *Proc. ACM SIGMETRICS 2001*, Annapolis, MD, June 2001, pp. 50–61.

[61] S. Parkvall and D. Astely, "The evolution of LTE towards IMT-advanced," *Journal of Communications*, vol. 4, no. 3, pp. 146–154, Apr. 2009.

[62] P. Rost and G. Fettweis, "Green communications in cellular networks with fixed relay nodes," Technical Report, Sep. 2010. [Online]. Available: http://www.vodafonechair.com/staff/rost/Rost.Fettweis.2010-09. Book.pdf

[63] X. Wang, Y. Bao, X. Liu, and Z. Niu, "On the design of relay caching in cellular networks for energy efficiency," in *IEEE INFOCOM 2011 Workshop on Green Communications and Networking*, Apr. 2011, pp. 259–264.

[64] "3G home nodeb study item technical report," *3rd Generation Partnership Project (3GPP), TR25.820, v8.2.0*, Aug. 2008.

[65] R. Giuliano, C. Monti, and P. Loreti, "WiMAX fractional frequency reuse for rural environments," *IEEE Commun. Mag.*, vol. 15, pp. 60–65, Jun. 2008.

[66] *Soft frequency reuse scheme for UTRAN LTE*, 3GPP Std. R1-050 507, May 2005.

[67] S. Das, H. Viswanathan, and G. Rittenhouse, "Dynamic load balancing through coordinated scheduling in packet data systems," in *Proc. IEEE INFOCOM*, San Francisco, CA, Apr. 2003, pp. 786–796.

[68] K. Son, Y. Yi, and S. Chong, "Utility optimal multi-pattern reuse in multi-cell networks," *IEEE Trans. Wireless Commun.*, vol. 10, no. 1, pp. 142–153, Jan. 2011.

[69] A. Gjendemsj, D. Gesbert, G. E. Øien, and S. G. Kiani, "Binary power control for sum rate maximization over multiple interfering links," *IEEE Trans. Wireless Commun.*, vol. 7, no. 8, pp. 3164–3173, Aug. 2008.

[70] L. Venturino, N. Prasad, and X. Wang, "Coordinated scheduling and power allocation in downlink multicell OFDMA networks," *IEEE Trans. Veh. Technol.*, vol. 58, no. 6, pp. 2835–2848, Jul. 2009.

[71] A. Stolyar and H. Viswanathan, "Self-organizing dynamic fractional frequency reuse for best-effort traffic through distributed inter-cell coordination," in *Proc. IEEE INFOCOM*, Rio de Janeiro, Brazil, Apr. 2009, pp. 1–9.

[72] Home NodeB Output Power, 3GPP TSG Working Group 4 meeting TSG-RAN WG1 Contribution R4-070 969, 2007.

[73] V. Chandrasekhar and J. Andrews, "Uplink capacity and interference avoidance for two-tier femtocell networks," *IEEE Trans. Wireless Commun.*, vol. 8, no. 7, pp. 3498–3509, Jul. 2009.

[74] A. Ng'oma, *Radio over fiber technology for broadband wireless communication systems.* Ph.D. Thesis, Eindhoven University of Technology, Eindhoven, 2005.

[75] D. Opatic, "Radio over fiber technology for wireless access." Ericsson, 2009.

[76] "GreenTouch^tm." [Online]. Available: http://www.greentouch.org/.

[77] "EARTH." [Online]. Available: https://www.ict-earth.eu/.

[78] R. Tafazolli, "EARTH - energy aware radio and network technologies," in *Proc. Next Generation Wireless Green Networks Workshop*, Pairs, France, Nov. 2009.

[79] "Mobile VCE." [Online]. Available: http://www.mobilevce.com/

[80] "Cool Silicon." [Online]. Available: http://www.cool-silicon.de/.

Author Contact Information

Kyuho Son is with Department of Electrical Engineering, University of Southern California, Los Angeles, CA 90089, USA, Email: kyuhoson@usc.edu. Hongseok Kim is with Department of Electronic Engineering, Sogang University, Seoul 121-742, Korea, Email: hongseok@sogang.ac.kr. Yung Yi is with Department of Electrical Engineering, Korea Advanced Institute of Science and Technology (KAIST), Daejeon 305-701, Korea, Email: yiyung@kaist.edu. Bhaskar Krishnamachari is with Department of Electrical Engineering, University of Southern California, Los Angeles, CA 90089, USA, Email: bkrishna@usc.edu.

17

Green Wireless Access Networks

István Gódor & Pål Frenger
Ericsson Research, Hungary and Sweden

Oliver Blume
Alcatel-Lucent Bell Labs, Germany

Hauke Holtkamp
DOCOMO Euro-Labs, Germany

Muhammad Imran
University of Surrey, UK

Attila Vidács & Péter Fazekas
Budapest University of Technology and Economics, Hungary

Dario Sabella
Telecom Italia, Italy

Emilio Calvanese Strinati & Rohit Gupta
CEA-LETI, France

Pekka Pirinen
University of Oulu, Finland

Albrecht Fehske
Technische Universität Dresden, Germany

CONTENTS

17.1 Introduction

17.1.1 Motivation

Scientific findings have indicated that the CO_2 emission of the ICT industry has contributed to 2% to the emission budget of the world energy consumption [1–3]. Nevertheless, indirect impact of Information and Communications Technologies (ICT) on global CO_2 footprint could be far more considerable when ICT applications replace tasks from other industrial segments like transportation. In addition, the sharp rising cost of the energy resources in the past few years has motivated major operators like Vodafone and Orange to strive for reduction in energy consumption by 20% to 50% [4]. In the development of mobile systems like Global System for Mobile Communications (GSM) and Universal Mobile Telecommunications System (UMTS), the main target was to increase performance. Growing energy consumption did not draw much attention and therefore was not addressed. However, for future mobile systems, e.g., Long Term Evolution (LTE) and LTE-Advanced, the energy consumption will be a major issue due to, e.g., increasing energy costs and denser networks.

All segments of the ICT community including industry and academia recognized this responsibility and new projects come to life worldwide and in the EU, as well. Several EU projects target to improve the energy efficiency of existing systems and identify the design principles for the future in all aspects of green networking. The Towards Real Energy-efficient Network Design (TREND) project aims to establish the integration of the EU research community in green networking with a long term perspective to consolidate the European leadership in the field and laying down the bases for a new holistic approach to energy-efficient networking [5]. The low Energy COnsumption NETworks (ECONET) project aims to exploit dynamic adaptive technologies for wired network devices that allow saving energy when devices are not (fully) used [6]. The Cognitive radio and Cooperation strategies for POWER saving in multi-standard wireless devices (C2POWER) project aims to develop and demonstrate energy saving technologies for multi-standard wireless mobile devices, exploiting the combination of cognitive radio and cooperative strategies while still enabling the required performance in terms of data rate and Quality of Service (QoS) to support active applications [7]. The Energy Aware Radio and neTwork tecHnologies (EARTH) project aims to address the global environmental challenge by investigating and proposing effective mechanisms to drastically reduce energy wastage and improve energy efficiency of mobile broadband communication systems, without compromising user perceived quality of service and system capacity [1].

17.1.2 Energy Efficiency and Network Technologies

This chapter discusses promising techniques and solutions to reduce the system level energy consumption of broadband mobile access networks, based on the work of the EARTH project focusing on 3rd Generation Partnership Project (3GPP) networks. These techniques include deployment strategies, network management concepts, radio resource management techniques and some proposals for future architectures that are designed to be inherently energy efficient.

The increasing usage of mobile broadband communications requires rollout of new Radio Access Technologies (RATs) like LTE and LTE-Advanced, thus the density of base stations has to be increased in order to provide high level of data transmission capability and coverage that subscribers request from their service providers. The fact that base stations are the most energy-intensive component of mobile networks calls for reconsidered and more energy efficient deployment strategies. Such strategies are discussed in Section 17.2 focusing on cell deployment strategies, Section 17.3 focusing on relaying techniques and Section 17.4 focusing on multi-RAT aspects.

As a consequence of the increased number of network elements, the human driven (half-manual) network management solutions are gradually replaced by the techniques providing Self-Organizing Networks (SON). These techniques also enable adaptation to requirements of performance and quality. The fact that current techniques are focusing on the traditional three quality criteria (i.e., better throughput, capacity or coverage) calls for novel network management concepts with primary focus on energy efficiency. Such concepts are discussed in Section 17.5 focusing on Base Station (BS) coordination & cooperation and in Section 17.6 focusing on the adaptive reconfiguration of the network according to daily traffic variation.

A key functionality to provide QoS and to maximize the system capacity is the Radio Resource Management (RRM), particularly the packet scheduling operation of mobile networks. However, when maximum capacity is not required, scheduling schemes operate with low efficiency. This fact calls for new techniques minimizing the power consumption by managing time, frequency and transmit power during the scheduling operation. Such techniques are discussed in Section 17.7.

There are complementary solutions in the literature investigating, e.g., the energy saving technologies for Wireless Local Area Network (WLAN) based access networks, the hardware solutions supporting the network level management techniques or possible cooperative network sharing schemes between operators. Such complementary solutions are not in the main scope of the chapter, however, to give some insight to the reader, some interesting aspects of these solutions are shortly summarized in Section 17.8.

Beyond the above highlighted networking ideas, a new research direction is to extend, renew or replace the existing mobile network architecture towards an energy efficient network operation. E.g., multi-hop extension of the networks and network coding can be utilized to save energy. Moreover, networks can be operated more intelligently and efficiently by departing from the cellular paradigm. Such future architectures are discussed in Section 17.9.

17.1.3   Energy Efficiency Evaluation Framework

In order to evaluate and quantify the performance and energy consumption of mobile systems, an evaluation framework including reference system settings, power models for different network elements and traffic models applicable for the near future is needed. The most comprehensive solution that has been published so far is the Energy Efficiency Evaluation Framework (E^3F) defined by the EARTH project [8]. Its reference system settings are proposed to be according to the most relevant 3GPP and International Telecommunication Union (ITU) recommendations (e.g., [9–13]) summarized in [14].

The approach of the E^{3}F is to assess the total energy consumption of a large scale (i.e., countrywide) network by a break down into a set of typical deployment types. E^{3}F allows to study the effect of changes in hardware, deployment or management in separate scenarios, e.g., for dense urban areas or rural areas and for different times of the day. Each scenario is characterised by the type and deployment density of BSs, by the density of users and by their activity pattern. The overall consumption is computed from the relative distribution of the scenarios in Europe. The most important elements of the E^{3}F are summarized in the following.

The E^{3}F incorporates State-of-the-Art (SOTA) power models for base stations which map a set of parameters to supply power consumption of the entire device (see TABLE 17.1 for illustration and [8,15] for details). In the available linear approximation of the power model, three values describe the power consumption of the different nodes: the sleep mode power, the minimum and the maximum power consumption during normal operation. This model is given for 10 MHz bandwidth, two transmit antenna configurations and 3 sectors (only Macro) scenario based on the 2010 SOTA estimation. Note that beyond the generic model, special models are used in relay related studies (see Section 17.3 for details).

TABLE 17.1 Linear power models

Model	Node type	Sleep power [W]	Minimum power [W]	Maximum power [W]
Generic	Macro	450	780	1344
model	Micro	78	112	145
Relay	Relay (SOTA)	–	84	116
study	Relay (advanced)	–	14	116
	Macro (advanced)	–	156	1292

Furthermore, the E^{3}F defines realistic deployment & traffic models with long-term and short-term traffic models [8,15]. The daily traffic variation is illustrated in FIGURE 17.1. The traffic profile allows to map instantaneous energy consumption to average daily consumption.

In addition to the variation in the time of day, networks are inhomogeneous in geography which is represented by the deployment models. The most important specifics of deployment areas are summarized in TABLE 17.2. Note that deployment structure of Nordic countries and Russia are substantially less densely populated than the European average, so TABLE 17.2 excludes them. Also note that "special cases" like central districts of a metropolis (where the population density may exceed 20,000 citizen/km^2) and sparsely populated areas (few citizen/km^2) are omitted from the investigated model because the former only contribute a minor fraction of the network area and the latter will probably not be covered by high data rate services.

The availability of such traffic models allows the assessment of energy saving measures in different temporal or geographical settings.

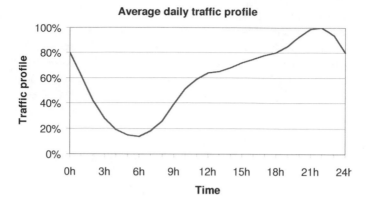

FIGURE 17.1 Daily variation of the traffic compared to peak values (European scenario)

TABLE 17.2 Deployment model for Europe (excluding Nordic countries and Russia)

Deployment	Population density [citizen/km^2]	Covered area
Dense urban	3000	1%
Urban	1000	2%
Suburban	500	4%
Rural	100	36%

Finally, the E^{3}F defines precise metrics upon which energy efficiency in cellular networks of various types should be measured and compared. In particular, it proposes "power per area unit" and "energy per bit" as energy consumption metrics (i.e., how to provide the same service with less energy), and follows the 3GPP recommendations on the quality performance metrics (e.g., [9, 10]).

This framework has been applied to calculate the power consumption of SOTA LTE networks and provides a basis for calibration and comparison of energy efficiency research, see [8] and references therein. Due to the comprehensive nature of the E^{3}F, the evaluations of energy saving potentials presented in the chapter are based on this framework.

17.2   Cell Deployment Strategies

In traditional network planning the Inter Site Distance (ISD) between BSs is adjusted to provide the maximum requested system performance. A lower ISD implies more sites, thus increases network capacity, but also increases Capital Expenditures (CAPEX) costs for site acquisition and equipment. The issue of lowering network energy consumption adds another dimension to the optimization problem. How does the ISD impact the energy cost? Can the deployment of additional cells (e.g. an underlay of micro cells) save power?

Broadband transmission requires a strong signal even at the cell edge. However, due to the high path loss coefficient ($\alpha > 3$) of Radio Frequency (RF) propagation the necessary transmission power at the antenna grows faster with the ISD than the covered area, i.e., larger cells use excessive transmission power per area. This is often interpreted in a way that small cells lower the system power consumption. However, each BS consumes power for basic operation and for signaling purposes. The selection of the most energy efficient ISD therefore is non-trivial and depends on the type of BS (e.g., power class, antenna height), on the traffic demand, and on the detailed behavior of power consumption over traffic load.

First, the impact of different power models will be discussed. Using the detailed EARTH power model [8], derived from actual hardware components, the optimum ISD of ideal hexagon deployments is computed. After that, more advanced deployments with cells of different sizes are considered.

17.2.1   Macro Cell Deployment

Due to the high path loss over the air smaller transmission ranges offer high savings in transmit power. On the other hand, this obviously requires more base stations. Each additional cell adds components to the network that consume power, mainly due to the large offset power macro sites already consume when they are empty. The saving potential thus needs to be analyzed with respect to the full network power consumption. This requires a load depend power model mapping the RF transmission power to the total power taken from the supply grid.

TABLE 17.3 summarizes four different linear models, representing different behaviors of the output power P (expressed in W) depending on load L (expressed in percentage [0%, 100%]). The models described a SOTA hardware with approx. 50% offset power for the empty BS, a constant power model that is completely independent of load and an offset-free power model (note that the sum of the two latter models yields the simplified version of a SOTA model (see TABLE 17.1 for the reference SOTA numbers). Finally, a hypothetical model mimics the case of a limited offset with high load dependency.

In a system level simulation the cell capacity, the cell edge rate and the power consumption can be computed. A typical 3GPP scenario is applied with a hexagon arrangement of 3-sectorized macro cells with 2x2 MIMO antennas in the 2.6 GHz band. The ISD is varied, subject to a minimum capacity of 20 Mbps/km^2 system throughput and of 2 Mbps cell edge bit-rate. For the sake of simplicity it is assumed that an LTE cell is either transmitting with 40 W per sector or not transmitting at all. The probability of transmitting is calculated from the traffic load per cell (see FIGURE 17.2), neglecting the power needed to transmitted reference symbols and the broadcast channel.

TABLE 17.3 Simplified linear power models

Powermodel			Offset		Load dependency
SOTA	P_1	=	700 W	+	$L \cdot 600$ W
Fixed power	P_2	=	700 W		
Zero offset	P_3	=			$L \cdot 600$ W
Strong slope	P_4	=	355 W	+	$L \cdot 2145$ W

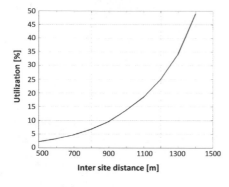

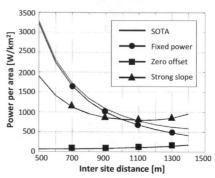

(a) **Average resource utilization over ISDs**

(b) **Average power per area over ISD**

FIGURE 17.2 Load and area power consumption as a function of ISD

Using the SOTA power model the power consumption is a decreasing function of the ISD (see FIGURE 17.2). The contribution of the load dependent part is rather low, and the power consumption is dominated by the fixed part of the power model. Thus, there are two complimentary ways of reducing the total system power: decreasing the offset power of the base stations or decreasing the number of base stations in the system. The latter way matches exactly the traditional deployment paradigm which considers the optimum deployment as the one with the minimum number of nodes satisfying the coverage/capacity conditions. On the contrary, for the power model with stronger load dependency an optimum ISD exists. This effect also occurs when the transmit power of the BS is scaled with the ISD (see the following section). Therefore, the choice of the BS power model is critical for the validity of the results. In the remainder of this chapter the EARTH power model, which corresponds closely to the SOTA model in TABLE 17.3, will be used as reference model and compared to a power model with improved hardware designed for higher energy efficiency.

17.2.2   Heterogeneous Cell Deployment

At smaller ISD the required cell edge rate can be achieved already at lower transmit power than 40 W, which also reduces the interference from neighbor cells. Therefore, we now study the optimum BS ISD when the transmit power of the macro base stations is adapted to provide coverage in 95% of the area. In the range of inter site distances of 500 m to 1500 m this yields macro cell transmit powers of about 0.7 W to 87.3 W [16].

At the vertices of the hexagonal macro cells the signal quality from the macro base stations is comparably low and, thus, the data rates in these areas are low. Remote users require more transmission resources for a given data rate than users close to the antenna. Therefore, potentially much power can be saved by offloading cell edge users into micro cells dedicatedly serving the cell edge (see FIGURE 17.3). The micro cells are considered as an additional under-layer, i.e. they add capacity but the macro cells have to provide the 95% coverage target. Micro base stations are designed for comparably short distances, hence their transmit power is fixed to 1 W. Also the amount of power consumed by signal processing and cooling is lower, thus, they consume much less power than macro cells.

This section investigates how the network power consumption depends on the macro cell ISD and on the micro deployment strategy. The different deployment strategies are compared by means of area power consumption, area throughput, and coverage. The throughput is calculated for indoor users with full buffer, considering the interference from all base stations. The system throughput is the sum of the throughputs in all non-empty cells of the network.

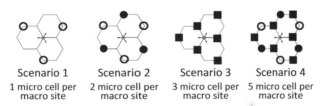

| Scenario 1 | Scenario 2 | Scenario 3 | Scenario 4 |
| 1 micro cell per macro site | 2 micro cell per macro site | 3 micro cell per macro site | 5 micro cell per macro site |

FIGURE 17.3 Placement of micro cells in a hexagonal macro cell network

Each deployment strategy has an optimal ISD yielding its minimum power consumption. For homogeneous macro deployment the optimum is between 500 m and 1500 m, which are the ISDs for urban and dense urban environments. The addition of micro cells shifts the optimum ISD to larger inter site distances, but always increases the total power consumption, i.e., the lower transmission power of micro base stations is overcompensated by the additional offset power of the added cells. For a network consisting of micro cells only, the optimum ISD is around 100 m for the full load case.

The addition of small BS is thus not saving power per se. Rather, the higher area power consumption of deployments with micro cells has to be related to its superior throughput. The best deployment can be taken from FIGURE17.4 where the gain in area power consumption is plotted over the target area throughput. Note that every point on each curve belongs to an optimal inter site distance of the corresponding network topology. It can be seen that the energy saving effect of the small cell deployment strongly depends on the traffic demand: the higher the throughput requirements, the more micro cells should be used for yielding an optimal area power consumption. For up to 70 Mbps/km^2 the conventional macro cell scenario is best. For traffic density of up to 100 Mbps/km^2 one micro cell per macro site is the optimum deployment and can yield up to 3% of saving compared to the homogeneous macro scenario optimized for the corresponding area throughput target. At very high densities above 150 Mbps/km^2, which can be expected due to quickly rising data rates in the near future, deployments with five or more micro cells are becoming interesting with a saving potential of more than 10%. Another promising case for micro cells occurs in situations with inhomogeneous traffic density. In this case micro cells would not only be placed at the cell edge but also at hotspots inside the macro cell.

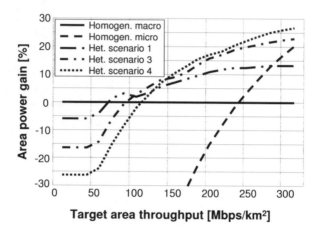

FIGURE 17.4 Area power consumption gain over area throughput target

17.3   Relaying Techniques

Relaying is being studied as part of the LTE-Advanced study item as a technology that offers the possibility to extend coverage and increase capacity, allowing more flexible and cost-effective deployment options [9]. Relaying is also a well-known technique for many radio technologies (WiFi, WiMAX, etc.), and in general the concept is to split the transmission into smaller hops, where the link between base stations (or a transmitter) and relays may have better channel quality compared to the direct communication with the user terminals (or a receiver). Relays can be used to improve coverage and/or capacity (see FIGURE 17.5 showing two possible example of relays' deployment in a cellular system); by applying relaying technology, some key properties of mobile systems, such as Spectral Efficiency (SE) and Energy Efficiency (EE), can be improved. The SE is improved because of the increase of the achievable rate, while the EE improvement stems from the fact that the distance between the transmitter and receiver is decreased compared with the direct transmission case, hence, the transmit power is expected to be reduced. This fact, jointly with a relay power consumption model that is more efficient compared to a base station model, leads to potential benefits in terms of energy efficiency as well. The power model used for relays is given in TABLE 17.1.

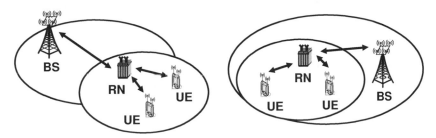

FIGURE 17.5 **Relay deployment approaches for coverage extension and capacity increase**

17.3.1 Relaying Schemes

The relay model was first introduced by van der Meulen [17] and further developed by Cover and El Gamal [18]. While the optimal relaying strategy in wireless networks has not yet been fully understood, several relaying schemes have been developed, such as Amplify-and-Forward (AF) and Decoded-and-Forward (DF) [19, 20] and Compress-and-Forward (CF) [21, 22]. In AF, the relay simply amplifies what it has received and forwards it; while in DF, the relay tries to decode the received message and forwards the re-encoded message. In CF, the relay quantizes its observation, compresses the quantization index and forwards the compressed information. Unlike DF, which benefits from transmit diversity, CF provides its observation to the destination and thus enjoys receive diversity. Spectral and Energy efficiency perspective of more advanced relaying strategies have recently been discussed in detail [23].

17.3.2 Two Hop vs. Multicast Cooperative Relaying

Many possible types of relay nodes are derived in 3GPP. In-band relays share the same frequency band for BS, Relay Node (RN) and the RN-User Equipment (UE) links. Two cases are distinguished in 3GPP:

- "Type 1" relays in 3GPP notation [9] are RN operating on Layer 3 (L3), i.e. protocol layer up to L3 for user data packets is available at the RN. Such a L3 relay has all functions that a BS has, and it can receive and forward Internet Protocol (IP) packets (incl. Packet Data Convergence Protocol (PDCP) Subscriber Data Units (SDUs));

- "Type 2" relays may be both L3 or Layer 2 (L2) relay nodes, depending on the particular solution/implementation.

Coherently with 3GPP terminology, Type 1 relays are considered visible to the mobile devices while Type 2 relays are transparent relay nodes.

The *two hop* relaying scheme (Type 1 relay) and the *multicast cooperative* scheme (a possible implementation of a Type 2 relay) have also been compared for their energy consumption. While the former scheme is relative to a time division transmission of data packets, the later envisages a first phase where the BS transmits to the UE in the direct link and the RN "eavesdrops" in the backhauling link, and a second phase where the RN starts helping the BS transmission to the terminal using the access link. In this sense the cooperation is realized by means of exploitation of spatial diversity at the receiver side.

First performance evaluations were performed for full load conditions and showing results considering only the radiated power. These results highlighted the potential of relay nodes as a tool to improve the energy efficiency in a network, due to a more efficient usage of the RF power. The results of a deeper analysis presented below takes into account the power models of the whole transmission node based on E^3F framework proposed in EARTH [8].

TABLE 17.4 Energy efficiency comparison of different deployment strategies for relay nodes

Deployment scenario 1 Reference: Macro (SOTA) w/o relays	Two hop relaying	Multicast coopera-tive relaying
Macro (SOTA) + RN SOTA	-5.1%	-9.3%
Macro (SOTA) + RN advanced	+6.3%	-0.5%
Deployment scenario 2 Reference: Macro (advanced) w/o relays	Two hop relaying	Multicast coopera-tive relaying
Macro (advanced) + RN advanced	+13.2%	+0.3%

TABLE 17.4 show relative gains in case of uniform user distribution compared to a reference deployment of macro nodes without relays (positive values mean savings, negative values mean the need of more energy). We can conclude that the multicast scheme is less efficient than the 2-hop scheme, because in the multicast case the BS is transmitting all of the time and for a certain period additional power is consumed by the relay.

Nevertheless, the power needs of relays should be well-scaling with the traffic load to provide any gain [24]. If one has such an "advanced" relay, the 2-hop scheme can provide some saving (+6.3%). Once the BSs have similar well-scaling power needs, then the whole system can be even better with the 2-hop scheme providing 13.2% gain compared to reference deployment without relays. These values correspond to a gain of up to 68% in the area covered by relays.

The above analysis considered uniform user distribution and further studies are needed to analyze non-uniform traffic distribution, which is more favorable for relays. In such case, relays can capture hot spots, provide attional capacity and are utilized more often. Further details can be found in [24].

17.3.3   Hybrid Schemes for Two Hop Relaying

Focusing on the two-hop relaying schemes, several relaying techniques are possible: like AF, DF and CF and their hybrid schemes in the sense that the most suitable strategy for a given scenario is adapted. The energy saving potential of the relays, in comparison to direct (single hop) transmission is highly dependent on the original separation between the source and the destination node. For short to medium separation of source and destination nodes (in the order of 100 meters) the direct transmission consumes less energy for every transmitted bit (in terms of Joules/Bit metric) in comparison to any relaying strategy. For moderate to long distances of the direct hop (source to destination hop), the relaying strategies show better performance considering the transmit energy only. This is depicted in the performance comparison given in TABLE 17.5.

TABLE 17.5 Energy per bit consumption index of different relaying schemes

Relay schemes	Source-destination separation (m)						
	100	200	300	500	700	900	1000
Direct Tx	0.037	0.061	0.096	0.171	0.284	0.416	0.524
DF, optimum α	0.039	0.056	0.079	0.134	0.182	0.224	0.248
CF/DF $\alpha = 0.5$	0.037	0.055	0.065	0.086	0.114	0.142	0.164
AF/DF $\alpha = 0.5$	0.039	0.057	0.082	0.141	0.202	0.263	0.303
DF $\alpha = 0.5$	0.038	0.058	0.086	0.149	0.223	0.301	0.358

DF is a simple relaying strategy and can be used by splitting the available time for communication of each data unit over two phases (source to relay transmission phase, retransmission of decoded message by the relay to the destination). The ratio of the length of these phases can be expressed as a factor (α). If both channels do not have same quality, we can optimize the duration of the two phases (i.e., the parameter α) to maximize spectral efficiency or minimize the energy consumption (or in other words to minimize joule per bit metric). The flexibility to control α can provide further reduction in energy consumption index, as depicted by the results in TABLE 17.5. Decode and forward scheme can be combined with two other strategies like AF and CF. The motivation is to use the most suitable scheme for any given scenario when the scenario changes as channels' gains vary. The hybrid between DF and CF schemes shows lowest energy consumption index in terms of joules per bit. The performance comparison of different relaying schemes is given in TABLE 17.5. Further details can be found in [23, 25].

17.4 Multi-RAT Deployment

When several types of access networks are available in an environment, further aspects can be considered beyond what was presented in the previous sections. In this section, the most important aspects are collected focusing on how to deploy LTE roll-out over legacy networks like GSM and UMTS. After that a topology planning method is presented targeting the less energy consuming network layout serving the coverage and QoS requirements. The presented method is able to take into consideration the advantages of co-siting in a multi-RAT environment. Finally, it is shown which frequency band to allocate for given services, where to deploy LTE and what to do if the legacy networks are anchored to lower frequency bands (like 900 MHz allocated for GSM services today).

17.4.1 Aspects of LTE Roll-Out over Legacy Networks

Since new LTE base stations will be gradually deployed in an environment where legacy mobile communication systems (GSM and UMTS) will still be operational for a long time in the future, multi-RAT deployment evolution should be taken into account within the framework of energy efficiency. To do so, it is necessary to focus on a capacity distribution among RATs that takes into account the annual increase of the traffic demand, the different capabilities of UEs available at given time, the energy efficiency aspects and the possible reuse of legacy sites.

In order to obtain a complete picture of the evolving system, it is important to know the energy consumption parameters of all different nodes deployed, or to-be-deployed in the analyzed network. To perform an energy efficiency analysis it is necessary to have a power model of the new and legacy RATs in relation with the data throughput. Such a model has been proposed in [8].

Another important aspect of an LTE roll-out analysis is the co-location of different RATs within the same site. By sharing a location, the fixed components of the power consumption can be shared among the different RATs, while the energy efficiency of other components like the cooling system can benefit by the sheer economy of scale and technological advancement of new and bigger equipment.

The above considerations could help operators to perform an analysis for energy aware network evolutions.

17.4.2 Topology Planning In Multi-RAT Environment

Traditional cellular network planning algorithms usually do not take energy efficiency into account, rather the goal is to minimize the number of required base stations, while satisfying coverage and user perceived quality requirements. Basically, this approach leads to an energy efficient solution, too. Since large part of base station power consumption is always present, regardless of transmission power, targeting minimum number of stations lowers this fixed consumption. However, topology planning should consider advanced energy efficiency features and metrics in order to provide network layouts optimized for energy consumption as well.

A possible planning solution is given in [26] and can be outlined as follows. Suppose that the network is capable of performing advanced network management and reconfiguration actions, in order to adapt the network layout to changing traffic conditions (and save energy via these mechanisms). The considered actions are complete switching off of sectors (cells) or base stations (having three cells), changing of transmit power level or changing the main lobe direction of sector antennae. The deployment method optimizes the network knowing that these reconfiguration mechanisms will work during the operation of the network.

The planning considers spatial and temporal changes in data traffic demands as well. This is supposed to be given by traffic demand maps that indicate the amount of traffic spread over the considered area, characterized by the positions where traffic might originate from (demand positions) and the average traffic amount. Temporal changes are taken into account by having several maps representing different times of the day. The elementary planning is based on the K-means clustering algorithm. This clustering first tries to put base stations close to the highest demands, then using again that clustering turns main lobe of antennae to have higher traffic demands near the middle of the main lobe. At last, an optimization for setting the power level of each sector is executed. This topology planning is carried out for the joint traffic map, containing the daily maximum values from the temporal maps at each part of the area. This results in a network that is capable of serving traffic needs.

To evaluate the potential of network reconfiguration mechanism, for each temporal traffic map the same planning is executed, but with restrictions on the base station positions (allowed only where there are base stations after the basic planning steps). Main lobe and power level settings then follow again. Hence the resulting network topology shows configuration optimal for serving the particular temporal map, and the rest of the devices is assumed to be turned off to reduce the total consumption. Detailed studies in [26] show that the amount of power saving depends very much on the actual traffic maps, but the bigger the difference between temporal maps the more energy reduction can be achieved by the optimized planning and reconfiguration. As an example, TABLE 17.6 shows power savings when the planning based adaptation mechanism is active, compared to the same planning method without adaptation. In the first case higher temporal variation was assumed by placing 100, 50 and 25 demand positions, while the second case assumes less significant temporal variations with 60, 40 and 30 positions. It is apparent that higher variability of the traffic allows for more energy saving with the adaptation.

TABLE 17.6 Energy saving of adaptation mechanism in case of temporal traffic variation

Traffic per demand position	2 Mbps	4 Mbps	6 Mbps	8 Mbps
Power saving 1st case	21%	33%	36%	40%
Power saving 2nd case	8%	18%	22%	23%

Another typical deployment question for operators is where to put novel radio networks, when there is an already existing deployment. An intuitive answer is to apply co-siting, that is, to put the new technology into existing base station sites. This solution has additional energy saving gain, due to consumption of common elements. However, it is shown in [27] that there are scenarios where co-siting is not the optimal solution from energy efficiency perspective.

17.4.3 Comparison of LTE and Legacy Networks

When combining legacy RAT (Wideband Code Division Multiple Access (WCDMA)/High Speed Packet Access (HSPA)) with LTE, the first question is which frequency band to allocate to which RAT. Results show that if LTE is deployed in the lower frequency band (with smaller path loss exponent) and legacy RAT is deployed in the higher frequency band (with greater path loss exponent), then it is always more energy efficient than putting LTE in the higher band and have legacy in the lower band.

However, in may cases it is not an option for operators to allocate the lower band for LTE. Nevertheless, the proper combination of legacy RAT and LTE can still provide some energy saving.

FIGURE 17.6 shows that in case of Voice over IP (VoIP) (or voice) traffic, WCDMA in the lower band can be even 50% more efficient solution compared to LTE for conversational services in the higher band. Note that the main reason behind this is that LTE uses relatively high bit rate modulation and thus leaves sub-bands empty when there is no data to send. On the contrary, WCDMA uses the whole available bandwidth for spreading and thus the spreading gain can be interpreted as an additional transmit power gain, extending the maximal cell size.

FIGURE 17.6 also shows that Mobile Broadband (MBB) HSPA access (i.e., requiring at least 2 Mbps user throughput) can only be more energy-efficient than LTE when the active user population is very rare (i.e., there are only a few active users per km^2). The potential gain to keep HSPA is inversely proportional to the density of active users (equivalent to the traffic demand of the given area).

As a conclusion, it is apparent that whenever carrier frequencies in the lower bands become available (e.g., GSM is gradually phased out), then these frequencies should be reallocated to LTE. However, if legacy RATs should be kept or coverage/conversational services are enough to be provided, then LTE should be primarily deployed in urban areas, where capacity is needed.

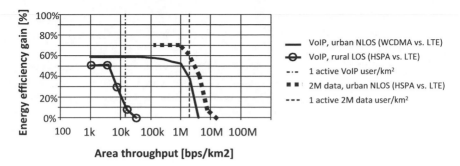

FIGURE 17.6 Achievable gains in multi-RAT deployments

This is the current practice for spectrum availability reasons as well; also this deployment strategy is the proper solution when energy efficiency is concerned.

17.5 Base Station Coordination and Cooperation

After the layout of the network is selected according to the roll-out plans based on the predicted evolution of traffic, the operators main concern is to utilize the already deployed resources as efficiently as it is possible.

First of all, the interference between neighboring BSs needs to be coordinated. Fractional Frequency Reuse (FFR) is identified as a promising technique to coordinate the interference by allocating a well-selected portion of available bandwidth to cell-edge users in harmony between the neighboring BSs [28, 29]. The efficiency of this technique is analyzed in Section 17.5.1.

Once the systems gets fully utilized due to interference limitations, it is beneficial for BSs to actively cooperate to avoid interference. In such cases, Coordinated Multi-Point (CoMP) schemes provide more throughput both at cell edge and in general. However, there is a compromise between the increased throughput and the extra energy needed for backhaul connections and signal processing required by the cooperation. These aspects are discussed in Section 17.5.2.

17.5.1 Fractional Frequency Reuse

FFR has been proposed as competitive strategy to limit the interference in Orthogonal Frequency Division Multiple Access (OFDMA) networks compared to single frequency reuse deployments [28–31]. The basic idea of FFR is to divide the whole bandwidth into two types of sub-bands. One sub-band is dedicated to the users that are close to their base station (inner zone of each cell). This sub-band is exploited with single frequency reuse planning. The other sub-bands are occupied with a larger frequency reuse factor of (2, 3 or more) and are attributed to the users that are located in the cell edge area (outer zone of each cell).

The original idea was to handle BSs equally without considering the difference in the traffic load between BSs. This case can be handled easily with a static solution. However, if the load is unbalanced between the neighboring BSs then the overloaded cell starts to increase the transmit power in order to compensate the lack of spectrum resource. This will generate more interference and practically waste the energy.

Thereby we propose to use a dynamic solution following the number of users located in each cell and allocate the power for each sub-band based on the Relative Narrowband Transmission Power (RNTP) indicators available in LTE networks [32]. The basic idea of this adaptation is to give more bandwidth to the users that are located in the edge of overloaded cells. In order to be efficient, the adaptation has to be performed as fast as possible. Thereby this method can be considered as a type of "coordinated scheduling".

FIGURE 17.7 illustrates a simplified downlink scenario with two neighboring BSs denoted by BS_1 and BS_2. The bandwidth is divided into three slices: N_0 Physical Resource Blocks (PRBs) are utilized with a frequency reuse factor of 1 for both sites and N_1 (respectively N_2) PRBs are utilized by BS_1 (respectively BS_2) exclusively for users that are located in the cell edge.

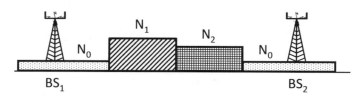

FIGURE 17.7 Adaptive FFR scenario with two base stations

In this setup, a dynamic optimization scheme determines the size of each sub-band (N_0, N_1 and N_2) as follows. First BS_1 allocates its power and frequency resources (e.g., based on the method proposed in [33]) and then BS_2 estimates interference generated by BS_1 based on the RNTP indicator and allocates the resources accordingly. After a few iteration the system will be stabilized. FFR based coordination is more effective in loaded cells, where this technique provides up to 8% gain in energy per bit in average BSs, meanwhile providing up to 15% gain in the most heavily loaded cells compared to the static case (when N_0, N_1 and N_2 are fixed).

The proposed approach can be used for each pair of neighboring base stations or pair of sectors with sectorized antennas. When neighboring sectors experience the same load there is no interest in FFR adaptation. However, when a sector is experiencing a high load compared to it nearest one, FFR adaptation is triggered using LTE signaling and the resource usage is optimized based on LTE RNTP indicators. Using this adaptation the loaded sector will "borrow" a certain amount of frequency resources from its neighbor which results in a lower energy consumption compared to the static case.

17.5.2 Cooperating Base Stations

CoMP schemes are able to significantly increase the achievable total cell throughput as well as cell edge throughput. However, these schemes introduce a certain amount of additional overhead and power consumption into the system due to additional:

- Pilots to estimate channels to multiple UEs at each BS;

- Feedback to establish channel state information at BSs, which is required for cooperative transmission of user signals in Downlink (DL);

- Signal processing due to more elaborate Uplink (UL) channel estimation as well as UL and DL Multiple Input Multiple Output (MIMO) processing;

- Backhauling capacity to exchange information between cooperating BSs.

In the following, the impact of the cooperation on the energy per bit performance is analyzed as the function of distance and number of cooperating BSs.

Power and System Models for CoMP

In order to include energy needs of CoMP related Baseband (BB) processing and backhauling in the BS, we propose the following additions to models provided in [34]. The digital BB functionalities mainly affected by CoMP are UL channel estimation and DL MIMO precoder computation. The amount of energy that processing of these tasks consume is technology and implementation dependent. As a rough estimate, UL channel estimation is assumed to be 10% of total digital BB processing and its energy needs are assumed to scale linearly with the cooperation cluster size N_c. Energy needs of MIMO processing are assumed to scale quadratically with N_c resulting in strong impact on total energy. Assessing the impact of processing in more detail, we compare two cases where MIMO processing accounts for either 1% or 10% of total BB processing. Energy needs of backhauling is assumed to scale linearly with N_c.

We consider cooperation in clusters of 2 up to 7 neighboring cells positioned in a regular grid of 57 cells with standard path loss and fading models according to 3GPP [9, 10]. The system is fed by full buffer users providing full load in the investigated BSs and the transmit power is adapted to the inter site distance with a 95% coverage requirement. The additional pilots in UL and the channel information in DL are assumed to scale linearly with N_c. Detailed description of models and assumptions are provided in [35, 36].

Efficiency of CoMP schemes

FIGURE 17.8 shows the comparison of energy needs of BS cooperation[1] to the non-cooperative reference ($N_c = 1$) as a function of network density, i.e., ISD. The results are grouped according to how much energy the MIMO processing needs (1% or 10% of BB processing, depicted by solid and dashed lines, respectively).

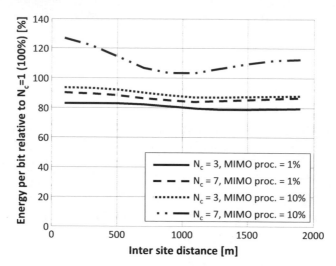

FIGURE 17.8 Energy per bit need of different cooperation sizes relative to non-cooperative communication

If MIMO processing accounts for only a small portion of BB processing (in the order of 1%), then 10% to 20% of the energy per bit can be saved by cooperation; and the most efficient cooperation is among three co-located BSs.

If MIMO processing accounts for more than 10% of BB processing, then the cooperation sizes of four and above are less efficient than the non-cooperative case independently of the network density. That is, the quadratic energy scaling of MIMO processing eventually outgrows the throughput gain provided by the cooperation.

Note that the lower the load is the less gain can be achieved by cooperation. In such case, it is more beneficial to employ schemes that orthogonalize transmission among neighboring cells such as FFR presented above.

---

[1]For the sake of clarity, only the most relevant cooperation sizes are shown in FIGURE 17.8.

17.6   Adaptive Network Reconfiguration

Analysis of traffic in current networks and the predictions of its growth trend show that BSs will keep using only a small fraction of their capacity [8]. On one hand, the strong requirements on low latency and low blocking probability demand that even during busy hour resources are on average not fully utilized. Besides, the analysis of daily variation of the data traffic in today's networks [8, 37] shows that there are long periods during a day in which the average load of the network can be 5 or 10 times smaller than the peak values in the busy/peak hours (see FIGURE 17.1). A good deal of the daily energy consumption is thus spent for providing the full system capacity, even when the actual traffic demand is much lower. Smart management tries to reduce such overprovisioning in order to save energy.

A promising way to reduce the energy consumption of the mobile networks is to dynamically reduce the number of active network elements to follow adaptively the daily variation of the traffic. Adaptation methods benefit from empty resources by putting the transceiver hardware to a lower energy mode of operation. It is key to manage resource utilization in a way that maximizes the opportunities for such hardware gains. This idea is discussed in Section 17.6.1.

According to today's best practice, only those cells are the target of energy reduction which were added as extra capacity on top of existing cells (e.g., by using extra carriers in HSPA or extra frequencies in GSM). Such extra cells could be switched on/off on a daily basis following the typical traffic variation (e.g., extra capacity needed to serve office areas during office hours is most probably not needed for the night). Nevertheless, even this technique is not yet a worldwide daily practice. This topic is discussed in Section 17.6.2.

17.6.1   Bandwidth Adaptation

The Bandwidth Adaptation (BW-Adapt) approach is adjusting the cell bandwidth to the required traffic load. During medium or low traffic the bandwidth is downscaled, and so lower numbers of PRBs are allocated. This relaxes the maximum transmit power level and thus the RF input level of the Power Amplifier (PA). The transceiver can stepwise lower the supply voltage of the PA to yield power savings. For a lower bandwidth also less reference signals/pilots have to be sent. In LTE the transmission of Orthogonal Frequency Division Multiplexing (OFDM) reference symbols makes up for around 10% of the total resource elements, i.e., pilot overhead is significant at low user load. However, in currently deployed LTE releases a change of the system bandwidth requires a reconfiguration of LTE BS system parameters and has to be signaled to the mobile terminals. Latest standardization contributions for LTE release 11 are proposing advanced carrier structures (carrier aggregation and extension carrier) that may ease implementing bandwidth adaptation.

Meanwhile, the Capacity Adaptation (CAP-Adapt) approach is an alternative to BW-Adapt, where the cell bandwidth and the number of pilots remain unchanged. The adaptation to lower load is solely performed by scheduling, i.e. only a limited number of LTE subcarriers are scheduled at any given time. This approach is transparent to the mobile terminals but still allows for lowering the PA supply voltage. Furthermore, frequency diversity is fully maintained. This can be exploited by frequency selective scheduling, which is especially attractive for low to medium resource utilization, when there are large opportunities for exploiting channel diversity.

These adaptation approaches play on the frequency dimension of the OFDM resource scheduling. On the other hand, the time dimension can be used to aggregate unused resources in a way, that creates subframes without any data transmission. Transceiver hardware that can be deactivated between two LTE reference symbols can save energy during micro-sleeps of up to 214μsec (the duration of three OFDM symbols). Even though this is not strictly a network reconfiguration method it compares well to the two adaptation methods.

All three approaches have benefits and limitations, simulation analysis is used to check which approach has the higher saving potential [38]. The study was using the EARTH traffic model with 60% peak hour resource utilization in dense urban areas and 30% in rural areas. It compared a state-of-the-art BSs to adaptive BSs leveraging the described PA features and a load adaptive BB processing. The reduced power consumption is further exploited by passive cooling.

The overall saving potential of the approaches turns out to be quite similar. During the night at lowest traffic load bandwidth adaptation achieves the highest energy savings. The CAP-Adapt approach yields slightly lower energy savings along a day due to the higher number of pilot symbols, but is most flexible and has the benefit that there is no impact on standardization. Micro-Discontinuous Transmission (DTX) is more beneficial at busy hour scenarios, because at low traffic the PA operates inefficiently when sending just the pilots between micro-sleeps. With the proper scheduler strategy and adaptive transceiver hardware energy savings of around 50% can be achieved over a full day by using micro-DTX in dense urban or BW-Adapt in rural scenarios (see TABLE 17.7).

TABLE 17.7 Daily saving potential of network adaptation strategies

Deployment area	BW-Adapt	CAP-Adapt	Micro-DTX
Dense urban	44.6%	42.6%	46.7%
Rural	57.4%	51.4%	56.0%

Further gains are possible when bandwidth adaptation is combined with interference mitigation strategies. Neighboring cells can coordinate which part of the frequency band is used, so that the Signal to Interference plus Noise

Ratio (SINR) value improves for all mobile terminals and thus data can be sent with less resource utilization. It will be further studied in EARTH, how much additional power saving can be achieved by combining the above solutions with frequency reuse schemes.

17.6.2 Multi-RAT Management

Most operators already operate two parallel networks, i.e., the GSM network mainly used for conversational services and 3rd Generation (3G) network (including both WCDMA and HSPA) mainly used for data services. A new LTE network is a third option to be deployed in order to serve the rapidly increasing bandwidth demand of the mobile data services. Such multi-RAT environment allows to extend the analysis of a) adapting the number of active BSs, b) choosing small cell or large macro cell to serve the users, and with a third possibility of c) deciding which network is needed to provide the required quality.

Adapting the Number of Base Stations

Thinking further the static analysis done in Section 17.4.3, we can calculate how much energy could be saved if the number of active base stations/cells was decreased adaptively to the actual traffic demand. The subject of this investigation is an ideal system, which can dynamically determine where and what base stations are required to provide the required minimum service quality and coverage.

In the era of LTE, it can be a minimum service requirement towards mobile networks to provide bandwidth for HDTV service for mobile PC users (e.g., 2 Mbps is required for 720p HD). This continuous traffic demand accumulates to 900 MByte/hour, which accounts as an average monthly data usage today. Let us assume homogeneously distributed, identically-behaving users in a "dense" urban area[2]. The daily variation of traffic profiles is according to FIGURE 17.1. This traffic demand approximately corresponds to 120 Mbps/km^2 in busy hour, and the particular value is scaled linearly with the population density. The results of the energy consumption optimization compared to the static dimensioning to the peak hours are shown in TABLE 17.8. In the table frequency band "high" denotes that LTE works at 2100 MHz and legacy network works at 900 MHz, while "low" denotes that LTE works at 800 MHz and legacy network works at 1800 MHz.

It can be concluded that it is worth to operate LTE in low frequency bands where more than 40% of the energy can be saved. In case of this traffic level, there is relatively more space to decrease the energy consumption in non-urban areas, but the power consumption is already very small compared to urban areas. The country-wide average of the absolute power needed to

---

[2]Note that the core of the cities and hot spot areas with higher population density are omitted from this analysis.

TABLE 17.8 Area power consumption of the system

Type of area	Frequency band of LTE	Power w/o optimization [W/km^2]	Optimized power [W/km^2]	Saved power
Urban	High	1670	1360	18.6%
	Low	1030	755	26.7%
Non-urban	High	38	26.4	30.5%
	Low	32.7	20.1	38.5%
Countrywide	High	151	119	21.2%
	Low	103	71.4	30.7%

serve the users increases with the square root of the traffic, and it is between 70 and 120 W/km^2 in the presented scenarios. The country-wide average energy saving potential is about 30% in addition to the selection of appropriate frequency band.

If it is not possible to switch off complete base stations (e.g., keep only one out of four or seven base stations), we still have a possibility to achieve gains by dynamic sectorization of base stations. From the three sectors used in normal circumstances, we can switch off some and reconfigure the antennas to provide a 2-sector setup or an omni cell solution (with one active sector). These solutions can still provide considerable energy saving in those base stations where one or two cell/sector can serve the user traffic demand [39].

Adaptive HetNets

The traffic and MBB penetration in cellular mobile radio networks greatly increases year-by-year, so the Heterogeneous Networks (HetNets) are in the focus of network deployments and modernization especially in densely populated urban environments. In such networks there is a *coverage layer* serving as the main connection towards the users, dimensioned basically to serve conversational services and low data traffic. The *capacity layer* of HetNets is dimensioned to serve high data traffic demands including needs for large user throughput. As a consequence, if the coverage layer is not limited to some very specific hot spots, but forms a solid layer covering the given urban area, then one can expect that the capacity layer is inevitably underutilized and it is worth to apply power saving techniques to reduce the energy used by the capacity layer of the HetNet.

In the following analysis, the coverage layer is comprised of macro BSs with three sectors and based on HSPA system operating on 900 MHz and dimensioned to provide 1 Mbps cell edge throughput corresponding to 357 m ISD. The capacity layer is comprised of small BSs with one omni cell and based on LTE system operating on 2100 MHz and dimensioned to provide 10 Mbps cell edge throughput corresponding to 112 m ISD. Thus, the area covered by a single macro BS is ten times larger than that of the omni BS.

In order to evaluate how much the capacity layer of the HetNet is utilized, the users' distribution over the deployment area needs to be modeled, assuming that the MBB penetration is 50%. Two cases are considered, namely, the homogeneous and the inhomogeneous user distributions. The uniform spatial user distribution in the area is modeled as a homogeneous Poisson point process. The spatial traffic distribution in the inhomogeneous case is modeled according to a modified Matern process [40]. In the Matern process cluster centers are determined first according to a homogeneous Poisson process. Then, the number of points (i.e., users) around each cluster center is again Poisson distributed, and these points are uniformly distributed within a predefined radius (r) around the center. It can be shown that the probability that a cell does contain active users in the inhomogeneous case $(P_I \approx 1 - \exp(1 - e^{-\rho R^2 \pi}))$, where R is the cell radius and ρ is the user density) is always below than in the homogeneous case $(P_H = 1 - \exp(-\rho R^2 \pi))$ and a lower limit for P_I compared to P_H is $P_I \approx 0.63 P_H$. Given that how many cells of the capacity layer are empty or non-empty at a given time, it can be calculated that how much of the sites can be put into power saving mode if there is no user to be served. When considering the energy saving potential the daily average and peak utilizations are good measures. These measures as network averages are given in TABLE 17.9 for the case of 80 m^2 cluster/hotspot size (e.g., the size of a café).

TABLE 17.9 Utilization of small cells and energy saving potential by switching off empty small cells

User distribution	Utilization		Energy saving	
	daily average	peak hour	daily average	peak hour
Homogeneous	45%	67%	30%	17%
Inhomogeneous	40%	54%	33%	24%

It can be seen that more than half of the cells are empty on average and even in peak hours a great portion of the cells are not utilized, and there is a significant difference between homogeneous and inhomogeneous user distributions.

In order to calculate the potential energy savings, the linear approximation of the power model developed by the EARTH project [8] is used. In this model, the power consumption of the macro BSs varies between 780 W and 1344 W, while the power consumption of the small BSs varies between 112 W and 145 W in the function of the load. Based on the above utilization figures and the power model, the energy saving potential by switching off empty small cells of the capacity layer is shown in TABLE 17.9.

The above results illustrate that if the coverage layer is not limited to some very specific hot spots, but forms a solid layer covering an urban area, then considerable energy can be potentially saved not just in daily average, but even in peak hours.

17.7   Radio Resource Management

Traditionally, RRM is concerned with maximizing outcome at a fixed power cost using a set of levers that include transmit power, radio channels, transmission delay, modulation and coding. In contrast, 'green' RRM provides fixed outcome at a minimal power cost leveraging the same set of parameters.

Most existing RRM strategies and algorithms are designed around the fullload or peak traffic assumption. They apply when large number of terminals are competing for high data rates. In fact, some techniques specifically address situations in which the capacity requirement is higher than the available capacity (overload). While these techniques can still be applied in low load they carry little benefit in this setting and do not address power consumption. In contrast, an RRM strategy that is sub-optimal or spectrally inefficient at high loads, may be able to fulfill capacity targets at much higher EE.

In this section, we address four aspects of green RRM: Theoretical considerations, high load scheduling, low load scheduling and hardware optimal scheduling.

17.7.1   Theoretical Considerations

Abstractly, downlink radio resource scheduling has a total power constraint. This power budget can be utilized to optimize any of the following objectives: *Maximize sum rate, Maximize fairness, Minimize Energy per Bit.*

Since there is an inherent trade-off between some of these conflicting objectives, there is a need to achieve energy efficient performance subject to some constraints on other desirable performance metrics. Fairness is one of such metrics. If our target is only fairness maximization or sum rate maximization, then the system has a tendency to utilize all available power budget. If the target is to minimize energy per bit, the power budget is partially used and the amount of power used depends on the cell size and fairness constraint. Fairness of sum rate maximization and energy bit minimization is in general low unless we have specific constraints to guarantee minimum fairness targets. Improvement in fairness and minimization of energy per bit has a cost in terms of reduced sum-rate performance; however, this cost is reasonably moderate (see TABLE 17.10). Focusing on the radio resource scheduling in a single cell setting and using simulations to solve the optimization problem for different channel scenarios, we compare the performance of different techniques in TABLE 17.10 (P_{in} is total consumed power, $\sum R$ is sum rate, $\sum E_b$ denotes energy per bit, $\mathcal{J}$ denotes fairness index). A reasonable compromise performance for sum rate, energy per bit and fairness can be obtained with a target fairness index of 0.5 while minimizing the energy per bit.

TABLE 17.10 Performance of different radio resource management approaches

	P_{in} (kW)	$r = 2\ km$ $\sum R$ (Mbits/s)	$\sum E_b$ (J/Mbits)	$\mathcal{J}$
max $\sum R$	0.740	1908	3.87	0.1
max $\mathcal{J}$	0.740	1142	6.47	1
min $\sum E_b$	0.478	1624	2.94	0.1
min $\sum E_b$ ($\mathcal{J} = 1.0$)	0.602	982	6.11	1
min $\sum E_b$ ($\mathcal{J} \geq 0.5$)	0.501	1574	3.17	0.5

17.7.2 High Load Scheduling

When approaching RRM from the power consumption perspective, it is useful to revisit existing technologies. In fact, while some existing RRM strategies may be preferred from the capacity or fairness point of view, it is necessary to be aware of the power consumption behaviors of existing technologies. Hence, we study the performance of Max C/I, proportional fair (PF) and Round Robin scheduling algorithms with regard to their power consumption under high system loads. The performances of these packet schedulers have been analyzed in terms of EE, and can be considered as a basis for the assessment of further innovative energy aware algorithms. System level simulations (based on full load assumption) showed a fundamental trade-off between EE and fairness.

The Max C/I scheduler is often used as a comparison benchmark in performance evaluation studies (see [41], for instance) since it gives an upper bound of the throughput that can be obtained.The PF algorithm was originally proposed in the context of CDMA HDR systems [42], so as to provide the users of a cell with a fair long-term rate. Finally, Round Robin is the simplest scheduler that can be conceived and it is often used as a performance reference for comparative analysis. First versions of this scheduler have been proposed in the context of scheduling in wire-line routers, since it provides short-term weighted fairness at $O(1)$ computational complexity [43].

In particular, it has been found that under full loads, EE of the system is limited by the SE of the packet scheduler, and relative gains (with respect to the baseline) do not depend on the particular linear power model considered. On the other side, when maximizing the capacity at full load, fairness among served users can suffer, and could give lower system performance in terms of QoS. We see from the simulation results of FIGURE 17.9(a) and FIGURE 17.9(b) that the EE performance of Max C/I scheduler is the best, however it does not perform very well from the fairness point of view [44]. Hence, in this case PF scheduler is a good compromise from both EE and fairness point of view.

17.7.3   Low Load Scheduling

When relaxing the high capacity requirement, new scheduling strategies become available. Low load implies unused spectrum and/or time slots which can be exploited opportunistically. Delay-tolerant data can then be scheduled to radio channels which are power efficient, e.g., by the well-known Earliest Deadline First (EDF) algorithm applied to OFDMA [45]. The main idea of EDF scheduling is to prioritize packet transmissions based on their deadline. The longer the packet deadline, the lower is the priority for packet transmission. EDF scheduling rule allocates packets according to their remaining Time to Lives (TTLs), thus granting priority to traffic flows with stringent QoS time constraints. Hence, the goal here is to reduce the overall downlink energy consumption while adapting the target of SE to the actual load of the system in order to meet the QoS requirements.

Thinking further on this idea, an efficient multi-user scheduling algorithm can be designed, which can be applied to heterogeneous traffic scenarios. In [46], the Green Scheduling algorithm is proposed which splits the resource allocation process into four steps. In the first step, it is identified which entities (packets) are rushing and which are not rushing. In step two, the resources are assigned only to entities that have high probability of missing their QoS requirements regardless of their momentary link quality and their potential to save energy. Then, if any chunks are still unscheduled, in a third step resources are allocated to users (non rushing) with highest momentary link quality, regardless of their QoS constraints. Finally, in the fourth step energy efficient link adaptation is performed to save downlink energy. We trade throughput (lowering the transmission SE and allocating a larger number of chunks to UEs) with downlink power by limiting the power budget on each chunk. In this way, downlink transmission power is minimized over a time window, which provides significant additional flexibility to the scheduling algorithm. In addition to throughput, both latency and SE enter in the trade-off. We show in FIGURE 17.9(c) that the transmit power consumption reduction from such an approach is around 8-18% [46], or 4-10% based on the EARTH system power model [8].

17.7.4   Hardware Optimal Scheduling

Ultimately, power is not consumed by a link or transmission, but by base station hardware. Understanding the underlying hardware leads to choosing the right parameters for energy efficient RRM. For example, the power consumption of a base station monotonically depends on the total power transmitted [34]. This initially leads to the conclusion that the transmission power should be minimized in order to reduce supply power consumption – resulting in transmit power control strategies. However, even at minimum transmit power operation, a base station has static circuit power consumption which is present at any transmit power level. Lowering the circuit power is only pos-

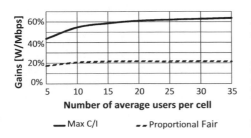

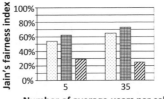

(a) **Energy efficiency gains over Round Robin** [44]

(b) **Fairness of scheduling techniques** [44]

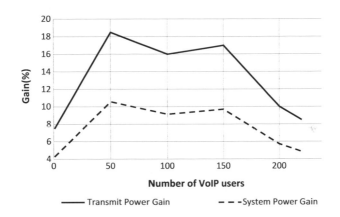

(c) **Transmit- and system power gain of Green scheduler over EDF**

FIGURE 17.9 Comparison of scheduling techniques

sible via disabling hardware components in so-called 'sleep modes' which are limited in time and in the number of components affected. During sleep mode, transmission is impossible creating an important trade-off between power control on the one hand and maximizing sleep mode duration on the other hand. An energy efficient RRM strategy has thus to decide whether the overall power consumption is lower when minimizing transmission power during transmission, maximizing transmission power followed by some sleep duration or a combination thereof.

This problem is convex within a cell and can be solved analytically [47]. We provide only the general findings here: The effectiveness of power control depends on the absolute values of the load-dependent power component. Only if transmit power is large compared to circuit power, does power control have significant benefits. For example, this is the case in high-traffic situations in LTE when the power spectral density is high. In situations of low load, it is much more beneficial to keep transmission power (and thus data rates) high to allow for long sleep durations.

While in theory this technique is capable of saving more than 30% in operating power it is not yet readily applicable in SOTA systems. Today's LTE standards do not foresee downlink power control due to unsolved technical challenges in power amplifier linearity and resulting adjacent carrier leakage. Sleep modes are available for installation as long as they do not violate control signaling requirements, which in turn inhibits their applicability to around 74% of an LTE frame [48]. Future systems are expected to loosen these constraints, especially in light of the current energy efficiency emphasis in standardization activities.

Just as traditional RRM is capable of boosting cell capacities and fairness, so is green RRM capable of increasing energy efficiency by aligning power optimized hardware, link and scheduling techniques for an overall energy efficient system.

17.8 Complementary Research Activities for Access Networks

Beyond the above described network technologies, there are complementary solutions to save energy in access networks like cooperative network sharing between operators, energy saving technologies for WLAN based access networks and the hardware solutions supporting the network level management techniques. To give some insight to the reader, these solutions are described shortly in the following subsections.

17.8.1 Operator Cooperation

One of the most straightforward solutions to save energy is based on the fact that in most countries there are several operators providing services to their subscribers with overlapping service areas. Such networks can be cooperatively managed and only some of the networks (preferably only one) should be kept active during low traffic hours. An appropriate national roaming agreement can support that the users can access the active networks seamlessly. Such cooperation can provide 20-30% energy saving in case of two cooperating networks or even more in case of several operators, especially in low populated areas, where the deployment area has less effect on the achievable indoor service quality. From management point of view, this technique can be considered as a special case of those network reconfiguration techniques, where only those cells are the target of energy reduction which were added as extra capacity on top of existing cells (here the extra capacity is represented by the overlapping cells of different operators). More information about operator cooperation schemes can be found, e.g., in [49,50] and the references therein.

17.8.2   Management of WLANs

Mobile BSs are the most significant energy consumers in the wide area wireless access networks. However, the overall number of individual Access Points (APs) in small and medium size WLANs increases exponentially every year [51]. Although average BS energy consumption is much higher in comparison to those of APs, vast number of WLAN network devices installed continuously worldwide contribute to the energy consumption in wireless access networks [51].

Corporate WLANs of today are dense, they comprise hundreds to thousands of APs connected to a wired corporate Local Area Network (LAN) (typically Ethernet). In such dense WLANs, APs can be grouped into clusters [52], where APs belonging to the same cluster provide overlapping coverage. Vendors offer centralized solutions for the management of such dense WLANs, concentrating in a central controller the management capabilities for the whole set of APs. The information about the global state of the WLAN is available at the controller, which can then implement appropriate algorithms to decide which APs are actually needed to achieve the desired QoS, and switch off the other APs [53].

In [52,54], the authors propose the adoption of resource on-demand strategies to power on or off WLAN APs dynamically, based on the volume and location of user demand, thus avoiding to always power on all the resources that are necessary to serve users during peak traffic periods. However, to ensure complete coverage, resources should be powered off in only those areas served by multiple layers of APs so that a single layer of complete coverage can be maintained at all times [52]. When an AP is switched off, user terminals need to be transferred to another AP via an association procedure similarly to mobile networks. Recently IEEE 802.11v provides procedures for seamless handover via explicit re-association commands, i.e., APs can explicitly ask users to re-associate to another AP [53].

In [52], the authors propose and implement a strategy called SEAR (Survey, Evaluate, Adapt, and Repeat), which is a practical policy-driven resource-on-demand strategy for high-density WLANs. Their experiments show that SEAR can reduce power consumption with roughly 50%. Other experiments presented in [53] show that even more than 80% energy saving is possible during low traffic periods in dense WLANs. For a comprehensive list of related research results and proposed solutions, see [51] and references therein.

17.8.3   Hardware Techniques

Beyond the above detailed green network technologies, there are extensive research activities on development of more energy efficient hardware components. According to [8], the most power consuming part of macro BSs is the transceiver chain. The application of a signal load adaptive transceiver technology in macro BSs achieves significant energy savings for medium and low loads according to the optimization of operating points and the deactivation of certain components [48, 55, 56].

The power consumption of small cell BSs is more equally distributed among the different components. BB signal processing algorithms can be redesigned to save energy especially in low loads. The efficiency of RF transceiver can be increased with Signal-to-Noise and Distortion ratio (SiNAD) adaptation and time/frequency duty-cycling. In case of medium and low loads, similar techniques can improve the efficiency of the power amplifier like in case of macro BSs, meanwhile in case of high loads, tunable matching network provides an energy efficient solution [57]. The antenna interface can be enhanced with a duplexing antenna allowing lower RF front-end insertion loss [58].

The energy efficiency of printed antenna arrays used in BSs can be increased with 15% by using low loss foam substrates compared to the conventional substrates used today.

The radio interface technologies can exploit the interaction between higher protocol layers and the hardware to improve the energy efficiency. The adaptability to system dynamics solution allows selecting the most power-efficient modulation as a function of the fast changing channel quality.

By controlling the radiation pattern and so the spatial filter, beamforming active antennas can focus the energy in the direction of interest, i.e., reconfigurable antenna systems can take care of moving "hotspots" in cells [59].

MIMO solutions was originally designed to exploit the channel properties to improve the throughput. It has been found that the number of receiver and transmitter antennas to be used in the system should be dynamically set according to the traffic demand to be served [60].

These hardware techniques can provide savings in the order 30% or even beyond depending of the use case. More information about hardware techniques can be found, e.g., in [61] and the references therein.

17.9   Future Architectures

Despite the success of the existing cellular architecture, there remains great potentials for improvement in several directions resulting in energy saving. In order to add more "flexibility and intelligence" to the systems, we need to adapt, re-plan or even rethink the existing systems including a) how to decouple data and system information, b) how to extend existing systems with multi-hops and c) how to utilize the benefits of network coding.

17.9.1 De-coupling of System Information and User Data

In order to design an energy efficient system there are two challenges that needs to be addressed: i) be energy efficient when transmitting data, and ii) be energy efficient when not transmitting data. When data is being transmitted, high data rates (that enables us to stop transmitting and receiving quickly) and beam-forming (that enables us to focus the transmitted energy to the intended user) are key improvement areas for enhancing energy efficiency. When data is not being transmitted, the key thing is to enable the network to operate in a low power mode with DTX and Discontinuous Reception (DRX) as much as possible.

Current cellular systems (GSM, WCDMA/HSPA, and LTE) was never designed with the second challenge in mind, and consequently the amount of energy a cellular network consumes depends very little on the amount of data traffic in the system. Instead energy consumption in current cellular systems is by far dominated by mandatory transmissions of system overhead when no data is transmitted. If we are to effectively reduce energy consumption in cellular networks then we must take a closer look on what type of system overhead we have today and what can be done to reduce it. The reason why we need to broadcast system overhead at all in cellular systems is that we need to provide in-active mobile stations with information on how they can access the system. An in-active mobile station needs at least the following support from the network:

- *Access information:* A mobile station needs to know how to contact the network whenever it wants to initiate a service. This procedure is commonly known as random access, and the mobile stations need to know how the random access channel is configured (e.g. which pre-ambles, time slots etc. that are used).

- *Mobile station measurements:* In-active mobiles need to perform mobility measurements and possibly also measurements related to positioning. Based on such measurements the mobile stations perform location update signaling with the network which is needed for paging to work.

- *System presence:* In principle the mobile station could test if a system is available by transmitting a probe and checking for a network response. However, it is typically not considered acceptable with mobile stations transmitting in any frequency band before they know that they are allowed to do so (the mobile station may e.g. be in a country where these frequencies are used for another service). Therefore, in practice, in-active mobile stations need to first detect that a system is present before they can transmit anything.

System overhead in current cellular systems is transmitted from each individual cell in the entire network. This is a legacy from the first and second generation voice centric systems (Nordic Mobile Telephone (NMT), GSM),

but for third and forth generation data centric systems (HSPA, LTE) this is
starting to become a problem. Distribution of system information can be much
more efficiently solved if we view it as a broadcast problem. In OFDM based
systems, such as LTE, the natural solution to provide broadcast services is to
make use of single frequency network transmission techniques.

Furthermore, the concept of having fixed cells becomes more and more
difficult to motivate with Beam-forming, MIMO, multi-carrier, CoMP, multi-
RAT, and reconfigurable antenna systems. We argue that this calls for a re-
interpretation of the cell concept. It is more fruitful to view the cell as something that is specific for a single mobile station rather than something that is
common for all mobile stations in a given area. By also letting go of the idea
that the coverage of the system information broadcast channel and the data
channels need to be identical we can enable a much more efficient system operation. This is illustrated in FIGURE 17.10, where all mobile station receives
system information from a Broadcast Channel (BCH). The mobile station MS_1
is configured to received data from a MIMO capable cell provided by RBS_1;
MS_2 communicates with a CoMP cell corresponding of signals from the base
stations RBS_1, RBS_2, and RBS_3; the mobile station MS_3 is communicating
with a cell with an omni-directional antenna pattern provided by RBS_4; the
base stations RBS_5, RBS_6, RBS_7 are not used for transmission of system information and are idle in this example. The decision to set up a CoMP cell
for MS_2 and a MIMO cell for MS_1 can then take into account the amount of
traffic that MS_2 and MS_1 want to communicate.

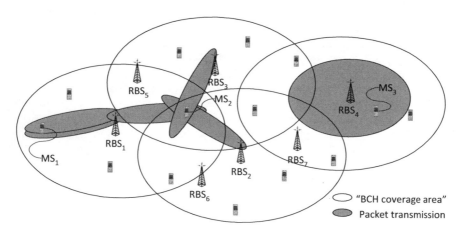

**FIGURE 17.10 System operation enabling separate coverage of system broad-
cast channels (BCH) and packed data transmissions**

By logically de-coupling system information from data transmissions we
can enable base station sleep mode in an efficient way since not all nodes
need to participate in the transmission of system information. We can also
optimize the cells that we provide to the active users based on their current

needs. This enables a system where the first challenge is effectively addressed by flexible cells providing high packet data rates to active users only and the second challenge is addressed by designing system overhead transmissions with maximum DTX in both time and space.

17.9.2 Coverage Extension by Multihop Relaying

During the past decades enormous research activities targeted ad-hoc networking, yet the widespread application of such networking techniques are not foreseen in the near future. However, utilizing various solutions from ad-hoc networks might be useful in future network architectures, targeting reduced energy usage.

A particular example is the architecture where mobile devices are capable of relaying others' traffic towards/from the base station (cf. relays being part of the main infrastructure presented in Section 17.3). The motivation of this is that by means of such multihop relaying, service can be provided to terminals dwelling in uncovered areas, or parts of the infrastructure might be turned off temporally (to save energy) and coverage could be maintained. There are lots of protocol, routing and control questions to be solved for such operation to be implemented, however, here only the basic capabilities (in terms of coverage extension) of multihop relaying is treated. Moreover, there are severe security threats to be solved and economic incentives to be applied, before such an operation can be utilized.

For coverage extension it is assumed that there is a limited area (e.g. a settlement) covered by a number of base stations. However, there might be a few customers that are (temporally) outside the coverage and new infrastructure is not worth deploying for their service. A similar scenario is when coverage is not contiguous, yet there might be customers in the area with no service (e.g. between settlements). In this case however, due to the generally random placement and existence of terminals, the coverage extension caused by relaying is also of probabilistic nature. It is worth noting that due to the problem of sharing radio capacity between relayed and "normal" traffic, this extension is feasible only in case of low traffic hours.

TABLE 17.11 reveals the increase in the average covered area, for a single base station coverage scenario, as function of the number of terminals in the original coverage cell as well as allowed hop numbers. The results are obtained by static snapshot simulations, where terminal-terminal and base station-terminal maximum link distances were calculated for suburban propagation and 1 Mbps achievable throughput as the edge of coverage. The result shows the average coverage increase for each parameter setting. As it is apparent, the number of allowed hops, as well as the number of terminals in the cell dramatically increase the coverage area, 50% extension is achievable by allowing 5 hop forwarding.

TABLE 17.11 Coverage extension of single cell system

Number of hops	Number of UEs per cell				
	120	220	320	420	520
1 hop	3.5%	6.0%	7.0%	9.0%	10.5%
2 hops	5.0%	9.5%	14.0%	17.5%	20.5%
3 hops	5.5%	12.5%	19.0%	25.0%	30.3%
4 hops	5.5%	13.3%	23.0%	32.5%	40.7%
5 hops	5.0%	14.2%	26.5%	38.0%	50.0%

17.9.3 Network Coding

Traditionally, network coding has been used to maximize information flow in the network. However, there are new perspectives to cut down the energy consumption of wireless networks with the help of network coding both on physical layer and datalink layer. Physical layer Network Coding (PNC) and MIMO are potential enablers of energy efficient network design and could be applied together to enhance the performance of two-way relay systems [62] (cf. relays analyzed in Section 17.3). On the datalink layer, network coding provides an alternative to "routing" and reliable distributed packet data delivery in decentralized wireless architectures of the future. E.g., Random Linear Network Coding (RLNC) forms a practical network coding subcategory that allows distributed implementation.

The reliable delivery is challenge for "sensitive" wireless networks, where RLNC can be used as an alternative to the Automatic Repeat reQuest (ARQ) schemes also from energy efficiency point of view. Let us assume a simple topology comprising of a single source that multicasts a finite amount of data packets to a set of destinations and compare RLNC and ARQ schemes in erasure channel impaired by Additive White Gaussian Noise (AWGN) [63].

We have found that in the absence of Forward Error Correction (FEC), RLNC approach outperforms retransmissions up to 40% when the links are highly unreliable, i.e., in low power per bit regime. When the average link quality gets better the retransmissions become less frequent, ARQ turns to be slightly more energy efficient than RLNC.

When FEC is incorporated the comparison between RLNC and retransmission schemes can be evaluated as a function of the number of FEC symbols. In that case the largest energy saving from network coding is achieved when the number of FEC symbols is relatively small resulting in gains exceeding 30%. Increasing the number of FEC symbols gradually reduces the performance gap between these techniques to the crossover point of 28 FEC symbols, after which the retransmission policy becomes better by a narrow margin.

17.10   Summary and Outlook

We have studied the network power consumption in wireless communications with a load dependent power model comprising not only transmit power but total grid power. The analysis identified large potentials for savings, especially in situations of low load. We have discussed improvements in deployment, resource management and network management. Each single techniques can provide savings of up to 30%, depending of the use case. Further improvements (not considered in this chapter) are possible for the underlaying hardware components. However, these savings cannot simply be added. It has to be analyzed whether two or more techniques can cooperate or even provide synergy. On the other hand, different techniques can be applied in different areas of the network (dense urban vs. rural areas) or at different times of the day (e.g. by network reconfiguration between busy hour and night times).

The following conclusions and trends have already been identified [64]:

- Heterogeneous networks, multi-RAT deployments and relays can be beneficial in high load but are costly in low load. For example, small cells do not save power per se but can provide high capacity at reasonable power. Therefore additional equipment should only be deployed if justified by the traffic demand.

- Traffic patterns with strong busy hours call for management methods that turn off hotspot cells, relays or one out of multiple RATs when there is no need for the extra capacity. Different standby schemes can be efficiently employed, like changing sectorized cells to omni cells or switching off complete sites.

- A trade-off occurs between spectral efficiency and energy efficiency. An effective scheduling solution controls the transmit power per resource element, balancing transmission duration, transmit power and channel diversity. Base station cooperation can avoid interference to reduce the energy consumption in cell-edge communication.

- Energy aware RRM can provide additional savings when they are combined with energy saving modes of the hardware. If the hardware provides sleep modes it is beneficial to bundle the scheduled traffic load in order to maximize the times where the hardware can go to sleep. If the hardware can adapt to different load levels scheduling knowledge can be exploited to control the degree of resource utilization and to select an adaptation mode.

- Once energy efficiency becomes a major criterion, more radical approaches are possible for rethinking architectures beyond today's LTE system. Terminals could support multi-hop transmission, packets can be handled as "bits" on finite field operations (aka network coding) or even the very paradigm of cellular networks can be challenged. However, these approaches are not yet mature and require further research before applicable in commercial systems.

Research work for an integrated solution with the optimum combination of saving techniques is ongoing [1]. The total saving potential is not yet evaluated, however, overall savings of 50% seem achievable within an LTE-based network.

―――

Acknowledgments

The authors gratefully acknowledge the invaluable insights and visions received from partners of the EARTH consortium, especially the contributions of Yinan Qi, Fabien Héliot, Mohamed Kamoun, Giorgio Calochira, Sergio Barberis, Fred Richter, Luca Scalia and László Hévizi.

The authors would like to acknowledge the contributions of Prof. Gerhard Fettweis from the Technical University of Dresden towards results presented in Section 17.5 and Prof. Harald Haas from the University of Edinburgh towards results presented in Section 17.7.

The work of the project EARTH that has been referenced in this chapter has received founding from the European Community's 7th Framework Programme FP7/2007–2013 under grant agreement n° 247733 — project EARTH.

―――

Bibliography

[1] EARTH, "Project summary leaflet," EARTH, Tech. Rep., 2010. [Online]. Available: https://www.ict-earth.eu

[2] EU, "Eu commissioner calls on ICT industry to reduce its carbon footprint by 20% as early as 2015," EU Press Release, MEMO/09/140, Tech. Rep., 2009.

[3] Ericsson, "Long term evolution (LTE): an introduction," White Paper, Tech. Rep., 2007.

[4] Vodafone, "Vodafone corporate responsibility report," Report, Tech. Rep., 2008.

[5] TREND, "Project description," TREND, Tech. Rep., 2010. [Online]. Available: http://www.fp7-trend.eu

[6] ECONET, "Project description," ECONET, Tech. Rep., 2010. [Online]. Available: https://www.econet-project.eu

[7] C2POWER, "Project description," C2POWER, Tech. Rep., 2010. [Online]. Available: http://www.ict-c2power.eu

[8] G. Auer, V. Giannini, I. Gódor, P. Skillermark, and et al., "How much energy is needed to run a wireless network?" *IEEE Wireless Communications Magazine Special Issue on Technologies for Green Radio Communication Networks*, October 2011.

[9] *3GPP TR 36.814 v1.5.1 (2009-12), Further Advancements for E-UTRA, Physical Layer Aspects*, 3GPP Std.

[10] *Physical layer aspects for evolved Universal Terrestrial Radio Access (UTRA)*, 3GPP Std. TR 25.814, 2006.

[11] *Technical Specifications and Technical Reports for a UTRAN-based 3GPP system*, 3GPP Std. TS 21.101, 2009.

[12] *Guidelines for evaluation of radio transmission technologies for IMT-2000*, ITU Std. M.1225, 1997.

[13] *Guidelines for evaluation of radio interface technologies for IMT-Advanced*, ITU Std. M.2135, 2008.

[14] EARTH, "D2.2, definition and parameterization of reference systems and scenarios," EARTH Project Deliverable, Tech. Rep., dec 2010. [Online]. Available: https://www.ict-earth.eu/publications/deliverables/deliverables.html

[15] ——, "D2.3, energy efficiency analysis of the reference systems, areas of improvements and target breakdown," EARTH Project Deliverable, Tech. Rep., dec 2010. [Online]. Available: https://www.ict-earth.eu/publications/deliverables/deliverables.html

[16] O. Arnold, F. Richter, G. Fettweis, and O. Blume, "Power consumption modeling of different base station types in heterogeneous cellular networks," in *Proceedings of Future Network & Mobile Summit 2010*, Florence, Italy, 06 2010.

[17] E. van der Meulen, "Three-terminal communication channel," *Adv. Appl. Prob.*, vol. 3, pp. 120–154, 1971.

[18] T. Cover and A. Gamal, "Capacity theorems for the relay channel," *IEEE Trans. Inform. Theory*, vol. 25, pp. 572–584, Sep 1979.

[19] M. Janani, A. Hedayat, T. Hunter, and A. Nosratinia, "Coded cooperation in wireless communications: space-time transmission and iterative decoding," *IEEE Trans. Signal Processing*, vol. 52, pp. 362–371, Feb 2004.

[20] T. Hunter and A. Nosratinia, "Diversity through coded cooperation," *IEEE Trans. Wireless Commun.*, vol. 5, pp. 283–289, Feb 2006.

[21] M. Katz and S. Shamai, "Relay protocols for two co-located users," *IEEE Trans. Inform. Theory*,, vol. 52, no. 6, pp. 2329–2344, Jun 2006.

[22] R. Hu and J. Li, "Practical compress-forward in user cooperation: Wyner-Ziv cooperation," in *ISIT, Seattle, USA,* ,, Jul 2006, pp. 489–493.

[23] Y. Qi, R. Hoshyar, M. A. Imran, and R. Tafazolli, "H2-ARQ-Relaying: Spectrum and energy efficiency perspectives," *IEEE Journal on Selected Areas in Communications*, vol. 29, no. 8, pp. 1547–1558, Aug 2011.

[24] R. Fantini, D. Sabella, and M. Caretti, "E3F based assessment of energy efficiency of relay nodes in LTE-Advanced networks," in *22nd annual IEEE international symposium on Personal, Indoor and Mobile Radio Communications (IEEE PIMRC'11)*, 2011.

[25] Y. Qi, R. Hoshyar, M. A. Imran, and R. Tafazolli, "The energy efficiency analysis of HARQ in hybrid relaying systems," in *IEEE Vehicular Technology Conference (VTC)*, Budapest, Hungary, 2011.

[26] I. Törős and P. Fazekas, "An energy efficient cellular mobile network planning algorithm," in *Proceedings of the 2011 IEEE 73rd Vehicular Technology Conference (VTC)*, 2011.

[27] I. Toros and P. Fazekas, "Planning and network management for energy efficiency in wireless systems," in *Proceedings of the 2011 Future Networks Summit (Funems 2011)*, 2011.

[28] S. Gauil, W. Hachem, and P. Ciblat, "Performance analysis of an ofdma transmission system in a multicell environment," *IEEE Transactions on Communications*, vol. 55, no. 4, April 2007.

[29] A. L. Stolyar and H. Viswanathan, "Self-organizing dynamic fractional frequency reuse in ofdma systems," in *27th Conference on Computer Communications*, Phoenix, USA, April 2008.

[30] S. Das and H. Viswanathan, "Interference mitigation through interference avoidance," in *Signals, Systems and Computers, 2006. ACSSC '06. Fortieth Asilomar Conference on*, nov 2006, pp. 1815 –1819.

[31] Ericsson, "Inter-cell interference handling for e-utra," 3GPP TSG-RAN WG1 #42, Tech. Rep., september 2005.

[32] *Evolved Universal Terrestrial Radio Access Network (E-UTRAN); X2 application protocol (X2AP) (Release 8)*, 3GPP Std. TS423-890, 2011.

[33] K. Seong, M. Mohseni, and J. M. Cioffi, "Optimal resource allocation for OFDMA downlink systems," in *Information Theory, 2006 IEEE International Symposium on*, jul 2006, pp. 1394 –1398.

[34] G. Auer, V. Giannini, I. Gódor, and et al., "Cellular energy efficiency evaluation framework," in *Proceedings of the 2011 IEEE 73rd Vehicular Technology Conference (VTC)*, 2011.

[35] A. J. Fehske, P. Marsch, and G. P. Fettweis, "Bit per Joule Efficiency of Cooperating Base Stations in Cellular Networks," in *3rd Workshop on Green Communications*, Miami, 2010.

[36] P. Marsch and G. P. Fettweis, Eds., *Coordinated Multi-Point in Mobile Communications From Theory to Practice*, 2011.

[37] Sandvine, "Mobile internet phenomena report," Sandvine, Tech. Rep., 2010. [Online]. Available: http://www.sandvine.com/downloads/documents/2010%20Global%20Internet%20Phenomena%20Report.pdf

[38] A. Ambrosy, O. Blume, H. Klessig, and W. Wajda, "Energy saving potential of integrated hardware and resource management solutions for wireless base stations," in *Proceedings of W-GREEN Workshop in conjunction with PIMRC 2011*, Toronto, Canada, 09 2011.

[39] L. Hévizi and I. Gódor, "Power savings in mobile networks by dynamic base station sectorization," in *Proceedings of W-GREEN Workshop in conjunction with PIMRC 2011*, Toronto, Canada, 09 2011.

[40] D. Staehle, K. Leibnitz, K. Heck, B. Schroder, A. Weller, and P. Tran-Gia, "Approximating the othercell interference distribution in inhomogeneous umts networks," in *Vehicular Technology Conference, 2002. VTC Spring 2002. IEEE 55th*, vol. 4, 2002, pp. 1640–1644.

[41] M. Shariat, A. U. Quddus, S. A. Ghorashi, and R. Tafazolli, "Scheduling as an important cross-layer operation for emerging broadband wireless systems," *Communications Surveys & Tutorials, IEEE*, 2011.

[42] A. Jalali, R. Padovani, and R. Pankaj, "Data throughput of cdma-hdr a high efficiency-high data rate personal communication wireress system," in *Proc. IEEE VTC00*, Tokio, Japan, May 2000.

[43] M. Shreedhar and G. Varghese, "Efficient fair queueing using deficit round robin," *SIGCOMM Comput. Commun. Rev.*, October 1995.

[44] D. Sabella, M. Caretti, and R. Fantini, "Energy efficiency evaluation of state of the art packet scheduling algorithms for LTE," in *EW2011, European Wireless 2011*, Vienna, Austria (Berlin, Offenbach: VDE VERLAG GMBH), April 2011, pp. 1–4.

[45] F. Chiussi and V. Sivaraman, "Achieving high utilization in guaranteed services networks using early-deadline-first scheduling," in *Proc. Sixth International Workshop on Quality of Service (IWQOS 98)*, May 1998.

[46] E. Calvanese Strinati and P. Greco, "Green resource allocation for OFDMA wireless cellular networks," in *Personal Indoor and Mobile Radio Communications (PIMRC), 2010 IEEE 21st International Symposium on*, sep 2010, pp. 2775–2780.

[47] H. Holtkamp, G. Auer, and H. Haas, "On Minimizing Base Station Power Consumption," in *Proceedings of the IEEE 74th Vehicular Technology Conference*, 2011.

[48] P. Frenger, P. Moberg, J. Malmodin, Y. Jading, and I. Gódor, "Reducing Energy Consumption in LTE with Cell DTX," in *Vehicular Technology Conference Proceedings, 2011. VTC 2011-Spring Budapest. 2011 IEEE 73rd*, 2011.

[49] M. A. Marsan and M. Meo, "Energy efficient management of two cellular access networks," *ACM SIGMETRICS Performance Evaluation Review*, vol. 37, no. 4, March 2010.

[50] M. Ismail and W. Zhuang, "Network cooperation for energy saving in green radio communications," *IEEE Wireless Communications Magazine*, vol. 18, no. 5, October 2011.

[51] J. Lorincz, A. Capone, and D. Begusic, "Heuristic algorithms for optimization of energy consumption in wireless access networks," *KSII Transactions on Internet and INformation Systems*, no. 4, pp. 626–648.

[52] A. Jardosh, K. Papagiannaki, E. Belding, K. Almeroth, G. Iannaccone, and B. Vinnakota, "GreenWLANs: On-Demand WLAN Infrastructures," *Mobile Networks and Applications*, no. 6, pp. 798–814, Dec.

[53] M. A. Marsan, L. Chiaraviglio, D. Ciullo, and M. Meo, "A simple analytical model for the energy-efficient activation of access points in dense wlans," in *Proceedings of the 1st International Conference on Energy-efficient Computing and Networking*, ser. e-Energy '10, 2010, pp. 159–168.

[54] J. Lorincz, A. Capone, and D. Begusic, "Optimized network management for energy savings of wireless access networks," *Computer Networks*, vol. 55, no. 3, pp. 514 – 540, 2011.

[55] D. Ferling, T. Bitzer, T. Bohn, D. Wiegner, and A. Pascht, "Power efficient transceivers to enable energy-efficient mobile radio systems," *Bell Labs Technical Journal*, vol. 15, no. 2, August 2010.

[56] M. J. Gonzalez, D. Ferling, W. Wajda, A. Erdem, and P. Maugars, "Concepts for energy efficient lte transceiver systems in macro base stations," in *Future Network & Mobile Summit*, June 2011, pp. 1–8.

[57] B. Debaillie, A. Giry, M. J. Gonzalez, L. Dussopt, M. Li, D. Ferling, and V. Giannini, "Opportunities for energy savings in pico/femto-cell base-stations," in *Future Network & Mobile Summit*, June 2011, pp. 1–8.

[58] S. Bories, L. Dussopt, A. Giry, and C. Delaveaud, "Duplexer-less rf front-end for lte pico-cell using a dual polarization antenna," in *European Wireless Conference 2011 - Sustainable Wireless Technologies*, vol. 11, April 2011, pp. 1–3.

[59] M. Boldi, S. Petersson, M. Fodrini, A. Orlando, P. Persson, and A. Nilsson, "Multi antenna techniques to improve energy efficiency in lte radio access network," in *Future Network & Mobile Summit*, June 2011, pp. 1–8.

[60] F. Héliot, O. Onireti, and M. A. Imran, "An accurate closed-form approximation of the energy efficiency-spectral efficiency trade-off over the mimo rayleigh fading channel," in *IEEE ICC11, 4th International Workshop on Green Communications*, June 2011, pp. 1–6.

[61] EARTH, "D4.2, green radio technologies," EARTH Project Deliverable, Tech. Rep., dec 2011. [Online]. Available: https://www.ict-earth.eu/publications/deliverables/deliverables.html

[62] L. K. S. Jayasinghe, N. Rajatheva, and M. Latva-aho, "Optimal power allocation for physical layer network coding based MIMO two-way relay system," in *accepted to IEEE WCNC 2012*.

[63] A. Pantelidou, K. Lähetkangas, and M. Latva-aho, "An energy-efficiency comparison of RLNC and ARQ in the presence of FEC," in *Proceedings of the 2011 IEEE 73rd Vehicular Technology Conference (VTC)*, Budapest, Hungary, May 2011.

[64] EARTH, "D6.2b, draft integrated solutions," EARTH Project Deliverable, Tech. Rep., dec 2011. [Online]. Available: https://www.ict-earth.eu/publications/deliverables/deliverables.html

Author Contact Information

István Gódor and Pål Frenger are with Ericsson Research, Budapest, Hungary and Linköping, Sweden, respectively, Email: istvan.Godor@ericsson.com, pal.frenger@ericsson.com. Oliver Blume is with Bell Laboratories, Alcatel-Lucent, Stuttgart, Germany, Email: oliver.blume@alcatel-lucent.com. Hauke Holtkamp is with DOCOMO Euro-Labs, Germany. Muhammad Imran is with University of Surrey, UK, Email: m.imran@surrey.ac.uk. Attila Vidács and Péter Fazekas are with Budapest University of Technology and Economics, Hungary, Email: vidacs@tmit.bme.hu, fazekasp@hit.bme.hu. Dario Sabella is with Telecom Italia, Italy, Email: dario.sabella@telecomitalia.it. Emilio Calvanese Strinati and Rohit Gupta are with CEA-LETI, France, Email: emilio.calvanese-strinati@cea.fr, rohit.gupta@cea.fr. Pekka Pirinen is with University of Oulu, Finland, Email: pekka.pirinen@ee.oulu.fi. Albrecht Fehske is with Technische Universität Dresden, Germany.

18

Energy Efficient Communications in MIMO Wireless Channels

Vineeth S. Varma

CNRS-Supélec-Paris Sud, France

E. Veronica Belmega

ETIS / ENSEA - Université de Cergy-Pontoise - CNRS, France

Samson Lasaulce

CNRS-Supélec-Paris Sud, France

Mérouane Debbah

Supélec, France

CONTENTS

18.1   Introduction

For a long time, the problem of energy consumption mainly concerned autonomous, embarked, or mobile communication terminals. Over the past two decades, designing energy-efficient communication terminals has become a more and more important issue. Concepts such as "green communications" have recently emerged in the literature, e.g., [10], [16] etc. Nowadays, with the existence of large networks involving both fixed and mobile terminals, the energy consumed by the fixed infrastructure has also become a central issue for communications engineers [6]. This chapter presents some of the literature in this framework. More specifically, this is a guide for researchers and engineers on how to devise power control and power allocation schemes in green wireless networks. Among pioneering works on energy-efficient power control there are the works of Goodman [28], [21], [14], [27] and [9] and others like [42]. Therein, the authors define the energy-efficiency of a communication as the ratio between the net data rate (called goodput) and the radiated power; the corresponding quantity is a measure of the average number of bits successfully received per joule of energy consumed at the transmitter. This metric has been used in many works. For example, in [33] it is applied to the problem of distributed power allocation in multi-carrier CDMA (code division multiple access systems) systems. In [34], it is used to model the users delay requirements in energy-efficient systems. In [2], it is re-interpreted as a *capacity per unit cost*[1] measure in MIMO (multiple input multiple output) systems for static and fast fading channels. The use of MIMO in communications brings forth two significant gains characterized as the diversity gain (due to having different channel conditions at different antennas like in MISO) and the multiplexing gain.

---

[1]The capacity per unit cost is an information theoretical notion introduced by Verdu in [42] and a measure of the amount of reliably transmitted information bits over the channel per unit cost.

In this chapter, we analyze the energy-efficiency metric defined as the ratio between the benefit of the transmission, i.e., the information rate, and the cost of the transmission, i.e., the consumed power to achieve this rate. This is measured in bits per Joule.

The information rate can be formulated through an information theoretical approach as studied in [40] and [43] or through a more pragmatic approach as in [42]. However, both of these approaches can be unified via the general concept of the channel capacity per unit cost if the channel is properly defined. The rate defined [28] can be interpreted as the capacity of a binary erasure channel. In a binary channel the signal $X \in \{0, 1\}$ is received as $Y \in \{0, 1, \epsilon\}$. Where ϵ represents the erasure of the signal.

Our analysis is important from both practical and theoretical perspectives. From an engineering point of view, our study helps one to design energy-efficient systems by specifying the optimal transmit power and power allocation policy that maximizes the defined energy-efficiency metric. While from a mathematical point of view, both the information rate and the efficiency can have several interesting properties like concavity, sigmoidal shape or quasi-concavity. This allows the study of interesting optimization problems of the energy-efficiency with respect to the transmit power.

This chapter is structured in two major parts. In the first part, we consider single user MIMO channels, highlighting the impact of imperfect channel state information (CSI). In this case, we study the optimization over not only the transmit power and the power allocation policy, but also over the training sequence length. It turns out that, while using all the available transmit antennas is optimal under perfect CSI, using only a subset of the available antennas is optimal when the channel has to be estimated.

In the second part of the chapter, we focus on the more challenging case of distributed multi-user networks. We analyse the non-cooperative game where several users choose their transmit power allocation policies that maximize their individual energy-efficiency functions. The methodology used to solve games with concave payoffs (e.g. Shannon-rate efficient games [3] [4]) cannot be applied to solve these energy-efficient games.

18.2 On the Design of Energy-Efficient MIMO Single-User Communications

The major focus of this section is on point-to-point MIMO communications. The focus on single user communication systems may be surprising considering that power control is one of the primary problems of interest. However, there are two important reasons for this choice. First of all, the single-user case has most of the main effects we want to emphasize and allows us to describe the proposed approach in a clear and concise manner. Secondly, once the single-user case problem is solved, reasonably more complex multi-user channels can be easily solved (the multiple access channel is one of them). The multi-user scenario becomes tractable provided some conditions are met. One of these conditions being the quasi-concavity of the proposed energy-efficiency metric as we will see in the multi-user section.

The energy-efficiency metric is defined at first in a very general way. Then, we specialize, depending on the specific situation under consideration and study it as a function of various model parameters such as the transmit power.

18.2.1 A General Definition of the Energy-Efficiency Performance Criterion

In what follows, we define and justify the energy-efficiency metric that is suitable to communication systems. Energy-efficiency in general usually refers to how much you gain per unit cost of energy. For example in thermodynamics it refers to the mechanical energy gained per cost of heat energy. In this case, the gain is a measure of how much data was successfully transferred. This leads to the definition of the energy-efficiency metric as the ratio between the data rate and the consumed power. In a MIMO system, the data rate depends on the precoding matrix $\mathbf{Q}$ [20]. The available transmit power p is related to the precoding matrix as $\text{Trace}(\mathbf{Q}) \leq p$.

$$\eta(\mathbf{Q}) = \frac{\text{Rate}(\mathbf{Q})}{\text{NP}[\text{Trace}(\mathbf{Q})]}, \qquad (18.1)$$

where $\text{NP}[\text{Trace}(\mathbf{Q})]$ is the total cost in terms of energy (this may depend on the computation cost, heating, etc.) and the $\text{Rate}(\mathbf{Q})$ represents the corresponding effective data rate. In this paper we assume a linear model where the total power is linearly related to the transmit power as justified by works like [38] and [39].

18.2.2 Optimizing the Total Transmit Power and Power Allocation in MIMO Channels with Perfect CSI

Consider a point-to-point communication with multiple antenna terminals. The signal at the receiver is modeled by:

$$\underline{y}(T) = \mathbf{H}(T)\underline{x}(T) + \underline{z}(T), \qquad (18.2)$$

Where $\mathbf{H}$ is the $n_r \times n_t$ channel transfer matrix and n_t (n_r) the number of transmit (receive) antennas. The vector $\underline{x}$ is the n_t-dimensional column vector of transmitted symbols and $\underline{z}$ is an n_r-dimensional complex white Gaussian noise distributed as $\mathcal{N}(\underline{0}, \sigma^2 \mathbf{I}_{n_r})$. Denoted by $\mathbf{Q} = \mathbb{E}[\underline{x}\underline{x}^H]$ is the input precoding matrix. The corresponding total power constraint is $\text{Trace}(\mathbf{Q}) \leq p$. The channel coherence time is denoted by T and will be omitted in this section for the sake of clarity.

The matrix $\mathbf{H}$ is assumed to be perfectly known at the receiver (coherent communication assumption) whereas only the statistics of $\mathbf{H}$ are available at the transmitter. Three cases will be studied depending on the channel coherence time and on the channel matrix statistics: i) the static links; ii) fast fading links; iii) slow fading links. For the first two cases, the benefit of the transmission will be measured in terms of Shannon transmission rates. For slow fading channels, outage events exist, while makes reliable communication impossible, and the length of the codeword used determines the success rate. Observe that in the first two cases, the solution is trivial and corresponds to the transmitters remaining silent. However, this is no longer the case for slow fading channels. In this case, the solution to the optimization problem is provided only for the particular case of MISO (the receiver is equipped with a single antenna). For the MIMO case, the optimal solution is conjectured and validated through numerical simulations.

18.2.2.1 Static Links

By definition, in the static links case, the frequency at which the channel matrix varies is strictly zero. In other words, $\mathbf{H}$ is a constant matrix. In this particular context, both the transmitter and receiver are assumed to know this matrix. This is the same framework as [20]. Thus, for a given precoding scheme $\mathbf{Q}$, the transmitter can reliably send to the receiver $\log_2 \left| \mathbf{I}_{n_r} + \frac{1}{\sigma^2} \mathbf{HQH}^H \right|$ bits per channel use (bpcu). Let us define the energy-efficiency of this communication by:

$$\eta_{\text{static}}(\mathbf{Q}) = \frac{\log_2 \left| \mathbf{I}_{n_r} + \frac{1}{\sigma^2} \mathbf{HQH}^H \right|}{\text{Trace}(\mathbf{Q})}. \tag{18.3}$$

The energy-efficiency function $\eta_{\text{static}}(\mathbf{Q})$ corresponds to an achievable rate per unit cost for the MIMO channel as defined in [42] under the assumption that the input alphabet does not contain any zero-cost symbols (i.e., silence at the transmitter does not convey information). It turns out that the result obtained in [42] for the single-input single-output channel extends to the MIMO channel.

Proposition 18.1 (Optimal precoding matrix for static MIMO channels). *The energy-efficiency of a MIMO communication over a static channel, measured by η_{static}, is maximized when $\mathbf{Q} = \mathbf{0}$ and this maximum is*

$$\eta_{\text{static}}^* = \frac{1}{ln2} \frac{\lambda_{\max}(\mathbf{HH}^H)}{\sigma^2}, \tag{18.4}$$

where $\lambda_{\max}(\mathbf{HH}^H)$ represents the maximum eigenvalue of the matrix $\mathbf{HH}^H$.

The proof can be found in [2]. It can be seen that, for static MIMO channels, the energy-efficiency defined in Eq. (18.3) is maximized by transmitting at very low powers. This kind of scenario occurs for example, when deploying sensors in the ocean to measure a temperature field (which varies very slowly). In some applications however, the rate obtained by using such a scheme can be insufficient. In these cases, the benefit to cost ratio can turn out to be an irrelevant measure and other performance metrics have to be considered (e.g., minimize the transmit power under a rate constraint).

18.2.2.2 Fast Fading Links

In this section, the channel matrix changes for every transmitted symbol. This means that it is different for each channel use. Therefore, the channel varies over a transmitted codeword (or packet) and, more precisely, each codeword sees as many channel realizations as the number of symbols per codeword. In this framework, let us define energy-efficiency by:

$$\eta_{\text{fast}}(\mathbf{Q}) = \frac{\mathbb{E}_{\mathbf{H}}\left[\log\left|\mathbf{I}_{n_r} + \frac{1}{\sigma^2}\mathbf{H}\mathbf{Q}\mathbf{H}^H\right|\right]}{\text{Trace}(\mathbf{Q})}. \tag{18.5}$$

The proof for the static links case can be applied for any channel realization and thus the trivial solution is obtained irrespective of the channel distribution.

Proposition 18.2 (Optimal precoding matrix for fast fading MIMO channels). *The energy-efficiency of a MIMO communication over a fast fading channel, measured by η_{fast}, is maximized when $\mathbf{Q} = \mathbf{0}$ and this maximum is*

$$\eta_{\text{fast}}^* = \frac{1}{ln2}\frac{\text{Trace}(\mathbb{E}\left[\mathbf{H}\mathbf{H}^H\right])}{n_t\sigma^2}. \tag{18.6}$$

Therefore, similarly to the static MIMO channels, maximizing the energy-efficiency function amounts to transmitting at almost zero power for the fast fading MIMO channels. Interestingly, as opposed to these two cases, in slow fading MIMO channels, where outage events are unavoidable, the answer can be different.

18.2.2.3 Slow Fading Links

In this section, the channel remains constant over a codeword and varies from block to block. As a consequence, the Shannon achievable rate is equal to zero. A suitable performance metric that measures the benefit of the transmission in slow-fading channels is the probability of an outage for a given transmission rate target R given in [1]. This metric allows one to quantify the probability that the rate target R is not reached by using a good channel coding scheme and is defined as follows:

$$P_{\text{out}}(\mathbf{Q}, R) = \Pr\left[\log_2\left|\mathbf{I}_{n_r} + \frac{1}{\sigma^2}\mathbf{H}\mathbf{Q}\mathbf{H}^H\right| < \xi\right], \tag{18.7}$$

where $\xi = R/R_0$, the spectral efficiency and R_0 bpcu (bits per channel use) is used to represent the bandwidth.

For the sake of simplicity, the entries of $\mathbf{H}$ are assumed i.i.d. zero-mean unit-variance complex Gaussian random variables. In terms of information assumptions, here again, it can be checked that only the second-order statistics of $\mathbf{H}$ are required to optimize the precoding matrix $\mathbf{Q}$. In this framework, [2] defines the energy-efficiency as follows:

$$\eta_{\text{slow}}(\mathbf{Q}, R) = \frac{R[1 - \text{P}_{\text{out}}(\mathbf{Q}, R)]}{\text{Trace}(\mathbf{Q})}. \tag{18.8}$$

In other words, the energy-efficiency or goodput-to-power ratio (GPR) is defined as the ratio between the expected throughput (see [28] for details) and the average transmit power. The expected throughput can be seen as the average system throughput over many transmissions. In contrast with static and fast fading channels, energy-efficiency is not necessarily maximized at low transmit powers. Thus, a non-trivial solution may exist to the optimization of GPR.

Finding the optimal covariance matrix is not trivial. Indeed, even the outage probability minimization problem w.r.t. $\mathbf{Q}$ is still an open problem [20], [41]. The general solution is conjectured as follows.

Conjecture 18.1 (Optimal precoding matrix for slow fading MIMO channels). *There exists a power threshold $\bar{p}$ such that:*

- *if $p \leq \bar{p}$ then $\mathbf{Q}^* \in \arg\min_{\mathbf{Q}} P_{\text{out}}(\mathbf{Q}, R) \;\Rightarrow\; \mathbf{Q}^* \in \arg\max_{\mathbf{Q}} \eta_{\text{slow}}(\mathbf{Q}, R)$;*

- *if $p > \bar{p}$ then $\eta(\mathbf{Q}, R)$ has a unique maximum in $\mathbf{Q}^* = \frac{p^*}{n_t}\mathbf{I}_{n_t}$ where $p^* \leq p$.*

This conjecture states that, if the available transmit power is less than a threshold, maximizing the GPR is equivalent to minimizing the outage probability. If it is above the threshold, the uniform power allocation is optimal. However using all the available power is generally suboptimal in terms of energy-efficiency. This conjecture is validated in Fig. 18.1 for the MIMO scenario: $n_t = n_r = 2$, $\xi = 1$, $1/\sigma^2 = 3$ dB. For the exact same threshold $\bar{p} = 0.16$ W(Watt), we have that, for $p \leq \bar{p}$ the beamforming PA structure optimal and above it, UPA structure is optimal.

Regarding the optimization problem associated with (18.8) several comments are in order. First, there is no loss of optimality by restricting the search for optimal precoding matrices to diagonal matrices: for any eigenvalue decomposition $\mathbf{Q} = \mathbf{U}\mathbf{D}\mathbf{U}^H$ with $\mathbf{U}$ unitary and $\mathbf{D} = \text{Diag}(\underline{p})$ with $\underline{p} = (p_1, \ldots, p_{n_t})$, both the outage and trace are invariant w.r.t. the choice of $\mathbf{U}$. The energy-efficiency can be written as:

$$\eta_{\text{slow}}(\mathbf{D}, R) = \frac{R[1 - \text{P}_{\text{out}}(\mathbf{D}, R)]}{\displaystyle\sum_{i=1}^{n_t} p_i}. \tag{18.9}$$

Second, the GPR is generally not quasi-concave w.r.t. $\mathbf{D}$. In [2], a counterexample for which the GPR is proven not to be quasi-concave is provided.

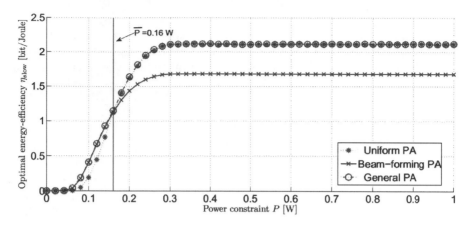

FIGURE 18.1 Optimal energy-efficiency vs. power constraint p, **comparison between beamforming PA, UPA (Uniform power allocation) and General PA.**

Third, the conjecture was validated using Monte-Carlo numerical simulations for the 2×2 case where both the transmitter and receiver are equipped with two antennas. Fourth, the Conjecture 18.1 was rigorously solved for MISO channels where the receiver is equipped with a single antenna (see [2] for details).

Proposition 18.3 (Optimal precoding matrix for slow fading MISO channels). *For all* $\ell \in \{1, ..., n_t - 1\}$, *let* c_ℓ *be the unique solution of the equation (in* x) $\Pr\left[\frac{1}{\ell+1}\sum_{i=1}^{\ell+1}|X_i|^2 \leq x\right] - \Pr\left[\frac{1}{\ell}\sum_{i=1}^{\ell}|X_i|^2 \leq x\right] = 0$ *where* X_i *are i.i.d. zero-mean Gaussian random variables with unit variance. By convention* $c_0 = +\infty$, $c_{n_t} = 0$. *Let* ν_{n_t} *be the unique solution of the equation (in* y) $\frac{y^{n_t}}{(n_t-1)!} - \sum_{i=0}^{n_t-1}\frac{y^i}{i!} = 0$. *Then the optimum precoding matrices have the following form:*

$$
\mathbf{D}^* = \left|
\begin{array}{ll}
\frac{p}{\ell}\boldsymbol{Diag}(\underline{e}_\ell) & \text{if } p \in \left[\frac{c}{c_{\ell-1}}, \frac{c}{c_\ell}\right) \\
\min\left\{\frac{\sigma^2(2^\xi-1)}{\nu_{n_t}}, \frac{p}{n_t}\right\}\mathbf{I}_{n_t} & \text{if } p \geq \frac{c}{c_{n_t-1}}
\end{array}
\right. ,
\qquad (18.10)
$$

where $c = \sigma^2(2^R - 1)$ *and* $\underline{e}_\ell \in \mathcal{S}_\ell$.

Similarly to the optimal precoding scheme that minimizes the outage probability [41], the precoding matrix that maximizes the energy-efficiency function is a diagonal matrix which consists in sharing the available transmit power uniformly among a subset of $\ell \leq n_t$ transmit antennas. As i.i.d entries are assumed for **H**, the choice of these antennas does not matter. What matters is the number of antennas selected, which depends on the available transmit power p: the higher the transmit power, the higher the number of used antennas. The difference between the outage probability minimization and GPR maximization problems appears when the transmit power is greater than the threshold $\frac{c}{c_{n_t}-1}$. In this regime, saturating the power constraint is suboptimal for the GPR optimization. The Conjecture 18.1 has also been solved for the SIMO channel where the transmitter is equipped with a single antenna, and also for the MIMO channel assuming the extreme SNR regimes (low and high SNR regimes).

A special case of interest is the case of uniform power allocation (UPA): $\mathbf{D} = \frac{p}{n_t}\mathbf{I}_{n_t}$ where $p \in [0, P]$ and $\eta_{\text{UPA}}(p, R) \triangleq \eta_{\text{slow}}\left(\frac{p}{n_t}\mathbf{I}_{n_t}, R\right)$. One of the reasons for studying this case is the famous conjecture of Telatar in [20]. This conjecture states that, depending on the channel parameters and target rate (i.e., σ^2, R), the power allocation (PA) policy minimizing the outage probability is to spread all the available power uniformly over a subset of $\ell^* \in \{1, \ldots, n_t\}$ antennas. If this can be proved, then it is straightforward to show that the covariance matrix $\mathbf{D}^*$ that maximizes the GPR is $\frac{p^*}{\ell^*}\mathbf{Diag}(\underline{e}_{\ell^*})$, where $\underline{e}_{\ell^*} \in \mathcal{V}_{\ell^*}{}^2$. Thus, $\mathbf{D}^*$ has the same structure as the covariance matrix minimizing the outage probability except that using all the available power is not necessarily optimal, $p^* \in [0, P]$. In conclusion, solving Conjecture 18.1 reduces to solving Telatar's conjecture and also the UPA case.

The main difficulty in studying the outage probability and/or the energy-efficiency function is the fact that the probability distribution function of the mutual information is generally intractable. In the literature, the outage probability is often studied by assuming an UPA policy over all the antennas and also using the Gaussian approximation of the p.d.f. of the mutual information. This approximation is valid in the asymptotic regime of large number of antennas. However, simulations show that it also quite accurate for reasonable small MIMO systems [13] (for e.g., assuming four-antenna terminals, the approximation is very good and assuming eight-antenna terminals, the error is negligible).

Under the UPA policy assumption, the GPR $\Gamma_{\text{UPA}}(p, R)$ is conjectured to be quasi-concave w.r.t. p.

Conjecture 18.2. *[Quasi-concavity of the energy-efficiency function] Assume that* $\mathbf{D} = \frac{p}{n_t}\mathbf{I}_{n_t}$. *Then* $\eta_{\text{UPA}}(p, R)$ *is quasi-concave w.r.t.* $p \in [0, P]$.

This conjecture was proved for the special cases of MISO and SIMO. Furthermore, it was proved for the general MIMO case assuming the large system approach for three cases: $n_t < +\infty$ and $n_r \to +\infty$; $n_t \to +\infty$ and $n_r < +\infty$; $n_t \to +\infty$, $n_r \to +\infty$ with $\lim_{n_i \to +\infty, i \in \{t,r\}} \frac{n_r}{n_t} = \beta < +\infty$. Numerical simulations were provided to validate the conjecture for finite number of antennas.

---

[2]The set $\mathcal{V}_\ell = \left\{\underline{v} \in \{0,1\}^{n_t} \mid \sum_{i=1}^{n_t} v_i = \ell\right\}$ represents the set of n_t-dimensional vectors containing ℓ ones and $n_t - \ell$ zeros, for all $\ell \in \{1, \ldots, n_t\}$.

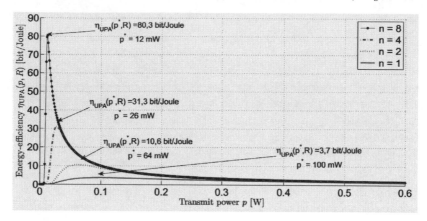

FIGURE 18.2 Energy-efficiency (GPR) vs. transmit power assuming UPA. The energy-efficiency function is a quasi-concave function w.r.t. p. The optimal point p^* is decreasing and $\eta_{\text{UPA}}(p^*, R)$ is increasing with n.

In Fig. 18.2, we plot the energy-efficiency function as a function of power $p \in [0,1]$ W for MIMO channels where $n_r = n_t = n \in \{1, 2, 4, 8\}$, UPA $\mathbf{D} = \frac{p}{n_t} \mathbf{I}_{n_t}$, $\rho = 10$ dB, $R = 1$ bpcu.

Furthermore, the numerical simulations show that the optimal value of the energy-efficiency metric is increasing with the number of antennas. So far, only the case of a MIMO single user system with perfect CSI at the receiver was studied. In practice, the channel has to be estimated at the receiver. In the following, we analyse the effect of channel estimation on the system energy-efficiency. This analysis is best suited for the case of slow fading channels considering that the time to estimate the channel and send data is finite.

18.2.3   Energy Efficient Communication in MIMO Channels with Imperfect CSI

In this section, we consider the effect of constant power consumption by the transmitter which is independent on the radio power transmitted. The main goal of this section is to introduce and justify a definition of energy-efficiency of a communication system with multiple input, multiple output, slow fading links, no CSI at the transmitter, and imperfect CSI at the receiver. Slow fading channels are considered here because if the channel is fast fading then the channel can not be estimated faster than the alteration of the channel making learning a non-viable option. On the other hand if the channel is static, once the channel is estimated, it will never change and so the training does not have a significant cost that has to be optimized in the long run.

Therefore we consider slow fading channels. Since the channel matrix is not known at the transmitter, we assume the UPA, i.e., $\mathbf{Q} = p \frac{\mathbf{I}_{n_t}}{n_t}$. However, similarly to the previous cases, using all the antennas may not always be optimal.

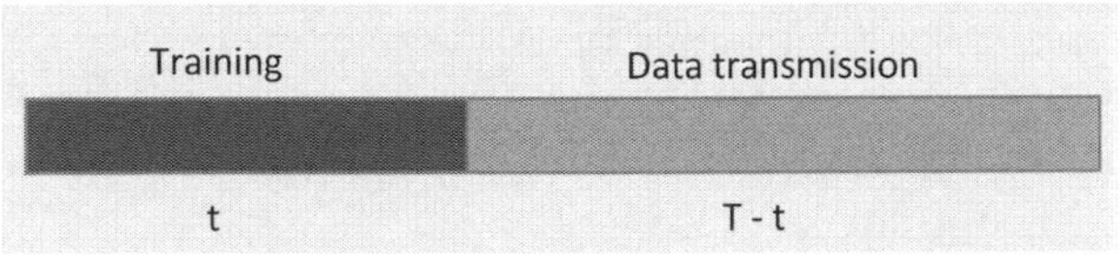

Training	Data transmission
t	T - t

FIGURE 18.3 The coherence time T is divided among training and data transmission

Assume that the channel follows the simple discrete-time block-fading law, where the channel is constant for some time interval, after which it changes to an independent value that it holds for the next interval [17]. This model is appropriate for the slow-fading case where the time with which $\mathbf{H}$ changes is much larger than the symbol duration. Each transmitted block of data is assumed to comprise a training sequence for the receiver to be able to estimate the channel. The training sequence length in symbols is represented by t_s and the block length in symbols by T_s. Continuous counterparts of the latter quantities are defined by $t = t_s \times S_d$ and $T = T_s \times S_d$, where S_d is the symbol duration in seconds. In the training phase, as illustrated in Figure 18.3, all n_t transmitting antennas broadcast orthogonal sequences of known pilot/training symbols of equal power on all antennas. The receiver estimates the channel, based on the observation of the training sequence, denoted by $\widehat{\mathbf{H}}$. The estimation error is denoted by $\delta\mathbf{H} = \mathbf{H} - \widehat{\mathbf{H}}$. Concerning the number of observations needed to estimate the channel, note that typical estimators such as maximum likelihood estimators generally require at least as many measurements as unknowns [18] that is to say:

$$t_s \geq n_t. \tag{18.11}$$

The channel estimate normalized to variance one is denoted by $\widetilde{\mathbf{H}}$. From [17], we have

$$\underline{\widetilde{y}} = \sqrt{\frac{\rho_{eff}(\rho,t)}{n_t}}\,\widetilde{\mathbf{H}}\underline{x} + \underline{\widetilde{z}} \tag{18.12}$$

provided that the effective SNR $\rho_{eff}(\rho,t)$ and equivalent observation noise $\underline{\widetilde{z}}$ are defined properly that is,

$$\begin{cases} \underline{\widetilde{z}} &= \rho \times \delta\mathbf{H} \times \underline{x} + \underline{z} \\ \rho_{eff}(\rho,t) &= \frac{\frac{t_s}{n_t}\rho^2}{1+\rho+\rho\frac{t_s}{n_t}} \end{cases}. \tag{18.13}$$

Note that the above equation does not correspond to any real signal, the observation equation simply denotes a mathematical equivalence of the SNR received due to channel estimation [17], this is also verified in other works like [22]. The worst case scenario for the estimation noise is assumed. Thus, in all formulas derived in the following are lower bounds on the mutual information and success rates. Since the channel matrix is unknown at the transmitter, we assume that $\mathbf{Q} = \mathbf{I}_{n_t}$, meaning that the transmit power is allocated uniformly over all the transmit antennas i.e $p = \frac{\mathbf{I}_{n_t}}{n_t}$. Under this assumption, the energy-efficiency is defined in [25] as:

$$\eta_{n_t}(p,t) = \frac{R \times \left(1 - \frac{t}{T}\right) \times \Pr\left[\log\left|\mathbf{I}_{n_t} + \frac{1}{n_t}\rho_{eff}\left(\frac{Lp}{\sigma^2},t\right)\mathbf{HH}^H\right| \geq \xi\right]}{ap + b} \quad (18.14)$$

where L is a term for the path loss, $a > 0$, $b \geq 0$ are parameters representing the radio-frequency power efficiency and a constant load due to cooling, coding etc., R is the transmission rate in bit/s, R_0 is a parameter which expresses in Hz (e.g., the system bandwidth). The numerator represents the benefit associated with transmitting namely, the net transmission rate (called the goodput in [16]) of the communication and is measured in bits/s. The goodput comprises a term $1 - \frac{t}{T}$ which represents the loss in terms of information rate due to the presence of a training mechanism and a term representing the transmission success probability. Note that a packet is received successfully only if the associated mutual information (which is obtained from the equivalent observation equation (18.12)) is above a certain target. The denominator of (18.14) represents the cost of transmission in terms of power. The proposed form for the denominator of 18.14 is inspired from [19] where the authors propose to relate the average power consumption of a transmitter (base stations in their case) to the average radiated or radio-frequency power by a linear model. Note that, without any loss of generality, we can choose $a = 1$ to simplify the analysis. In what b is concerned, two different regimes are identified:

• The regime where b is small allows one to study not only communication systems where the power consumed by the transmitter is determined by the radiated power but also those which have to been green in terms of electro-magnetic pollution or due to wireless signal restrictions [24].

• The regime where b is large allows one to study not only communication systems where the consumed power is almost independent of the radiated power but also those where the performance criterion is the goodput. Note that when $b = 0$, $t \to +\infty$, $\frac{t}{T} \to 0$, the the framework of [2] can be retrieved as a special case.

Table 18.1 presents a summary of the results available in this framework. In the following, we will present some of these results in detail.

18.2.3.1   Power Control

By inspecting (18.14), we see that using all the available transmit power can be suboptimal. For instance, if the available power is large and all of it is used, then $\eta_{n_t}(p,t)$ tends to zero. Since $\eta_{n_t}(p,t)$ also tends to zero when p

Quasi-concavity of η_{n_t} with respect to p for UPA with imperfect CSI	
SISO, MISO and SIMO	Proven
Large MIMO	
Very low or high SNR	
General MIMO	Conjecture
Concavity of η_{n_t} with respect to t when η_{n_t} is optimized for p	Proven
Quasiconcavity of η_{n_t} with respect to n_t	Conjecture

TABLE 18.1 Summary of known results from the current state of literature on energy efficiency of MIMO systems with imperfect CSI

goes to zero, there must be at least one maximum at which energy-efficiency is maximized, showing the importance not to exploit all the available power in certain regimes. The objective of this section is to study those aspects namely, to show that η_{n_t} has a unique maximum for a fixed training time fraction and provide the equation determining the optimum value of the transmit power.

From [23] we know that a sufficient condition for the function $\frac{f(x)}{x}$ to have a unique maximum is that the function $f(x)$ be sigmoidal/S-shaped and possess some mild properties (which are verified in our setup); a function f is sigmoidal if it is convex up to a point and then becomes concave. To apply this result in our context, [25] defines the function f by

$$f(\rho_{eff}) = \Pr\left[\log\left|\mathbf{I}_{n_t} + \frac{1}{n_t}\rho_{eff}\mathbf{H}\mathbf{H}^H\right| \geq \xi\right]. \tag{18.15}$$

It turns out that proving that f is sigmoidal in the general case of MIMO is a non-trivial problem, as advocated by the current state of relevant literature [2], [41], [8]. This is why we provide here a conjecture and a proposition concerning relevant special cases of MIMO systems [2].

Conjecture 18.3 (Optimization w.r.t. p for general MIMO systems). *For a fixed t and any pair (n_t, n_r), the energy-efficiency function $\eta_{n_t}(p, t)$ is a quasi-concave function with respect to p and has a unique maximum.*

This conjecture becomes a theorem in all classical special cases of interest, which is stated next. It is also supported by an intensive campaign of simulations.

Proposition 18.4 (Optimization w.r.t. p for special cases of MIMO systems). *If one of the following conditions is met:*

(a) $n_t \geq 1, n_r = 1$;

(b) $n_t \to +\infty, n_r < +\infty$;

(c) $n_t < +\infty, n_r \to +\infty$;

(d) $n_t \to +\infty, n_r \to +\infty, \lim\limits_{n_t \to +\infty, n_r \to +\infty} \frac{n_t}{n_r} = \beta < +\infty$;

(e) $p \to 0$;

(f) $p \to +\infty$;

then $\eta_{n_t}(p,t)$ is a quasi-concave function w.r.t. p and has a unique maximum.

This proposition is proved in [25]. Under one of the assumptions of the above proposition, it is relevant to characterize the unique solution of $\frac{\partial \eta_{n_t}}{\partial p}(p,t) = 0$ that is, the root (ρ^*_{eff}) of

$$\frac{L}{\sigma^2}(p+b)\frac{\tau \rho\left[(\tau+1)\rho+2\right]}{\left[(\tau+1)^2+1\right]^2}f'(\rho_{eff}) - f(\rho_{eff}) = 0 \qquad (18.16)$$

with $\tau = \frac{t_s}{n_t}$. Note that p is related to ρ through $p = \frac{\sigma^2 \rho}{L}$ and ρ is related to ρ_{eff} through equation (18.13) and can be expressed as

$$\rho = \frac{1}{2\tau}\rho_{eff}\sqrt{(1+\tau)^2 + \frac{4\tau}{\rho_{eff}}}. \qquad (18.17)$$

Therefore (18.16) can be expressed as a function of ρ_{eff} and solved numerically; once ρ^*_{eff} has been determined, ρ^* follows by 18.17, and $P^* = \rho^* \sigma^2$ follows. A special case is where $b = 0$ and $\tau \to +\infty$; this amounts to finding the unique root of $\rho_{eff}f'(\rho_{eff}) - f(\rho_{eff}) = 0$ which corresponds to the optimal operating SNR in terms of energy-efficiency of a channel with perfect CSI.

Fig. 18.4 illustrates the energy-efficiency as a function of the transmit power (p) for different values of b and illustrates the quasi-concavity of the energy efficiency function w.r.t p, i.e Conjecture 18.3. The parameters used are $R = 1600$, $\xi = \frac{R}{R_0} = 16$, $T_s = 55$ and $n_t = n_r = 4$, $t = 4$ ms.

Now we have presented on how to tune the transmit power in a MIMO system, however there is are additional optimizations than can be performed in this case of imperfect CSI. As we learn the channel within a certain time interval t, this time can be reduced to allow for more time for data transmission but at the cost of a worse channel estimate. This trade-off is studied in the following section.

18.2.3.2   Optimizing Training Sequence Length

The expression of $\eta_{n_t}(p,t)$ shows that only the numerator depends on the fraction of training time. Choosing $t = 0$ maximizes $1 - \frac{t}{T}$ but the packet success rate vanishes. Choosing $t = T$ maximizes the latter but makes the former term go to zero. Again, there is an optimal tradeoff to be found. Interestingly, it is possible to show that the function $\eta_{n_t}(p^*,t)$ is strictly concave w.r.t. t for any MIMO channels in terms of (n_t, n_r), where p^* is a maximum of η_{n_t} w.r.t p. This is stated in the following proposition.

Proposition 18.5 (Optimization w.r.t. t for general MIMO systems). *The energy-efficiency function $\eta_{n_t}(p,t)$ is a strictly concave function with respect*

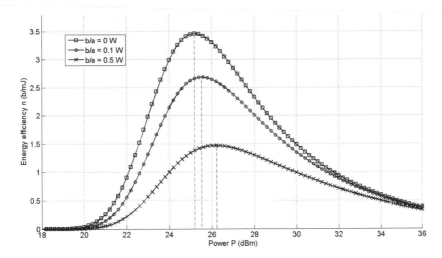

FIGURE 18.4 **Energy-efficiency** (η_{n_t}) **in bits per Joule (bpJ) v.s transmit power**
(*p*) **for a MIMO system with** $t_s = 4$, $n_t = n_r = 4$, $R = 1600$bps **(bits per second)**,
$\xi = \frac{R}{R_0} = 16$ **and** $T_s = 55$ **symbols. Observe that** η_{n_t} **is quasi-concave and has a**
unique maximum for each value of *b*.

to *t* for any *p* satisfying $\frac{\partial \eta_{n_t}}{\partial p}(p, t) = 0$ and $\frac{\partial^2 \eta_{n_t}}{\partial p^2}(p, t) < 0$, *i.e, at the maxima*
of η_{n_t} *w.r.t. p*.

The proof of this proposition is provided in [25]. The proposition means
that once energy-efficiency has been maximized w.r.t. the transmit power, the
uniqueness of the solution to the optimization problem w.r.t. *t*, the optimal
training time follows. Based on this, the optimal fraction of training time is
obtained by setting $\frac{\partial \eta_{n_t}}{\partial t}(p, t)$ to zero which can be written as:

$$\left(\frac{T_s}{n_t} - \tau\right) \frac{\rho^2(\rho + 1)}{[\tau\rho + \rho + 1]^2} f'(\rho_{eff}) - f(\rho_{eff}) = 0 \qquad (18.18)$$

again with $\tau = \frac{t_s}{n_t}$. Here again, following the same reasoning as for optimizing
the η_{n_t} w.r.t. *p*, it is possible to solve numerically the equation w.r.t. ρ_{eff} and
find the optimal t_s, which is denoted by t_s^*.

Fig. 18.5 studies the optimized energy efficiency $\eta_{n_t}^*$ as a function of the
transmit power with various values of t_s. The figure illustrates that beyond a
certain threshold on the available transmit power, there is an optimal training
sequence length that has to be used to maximize the efficiency, when the
optimization w.r.t *p* has been done, which has been proven analytically in
proposition 18.5. The parameters are $R = 1$Mbps, $\xi = 16$, $b = 0$, $n_t = n_r = 4$,
$b = 0$ and $T_s = 55$.

It should be noted that a solution to equation (18.18) necessarily exists
only if η has been optimized w.r.t *p*. However, in many practical situations,

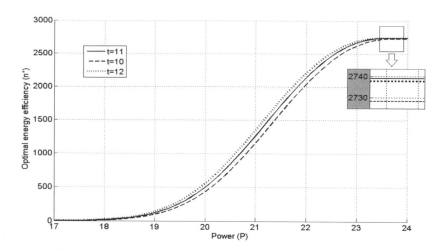

FIGURE 18.5 Optimized efficiency (ν^*) vs. transmit power (p) for a MIMO system with $n_r = 4$, $R = 1$ **Mbps**, $\xi = \frac{R}{R_0}$=16, $n_t = 4$, $T_s = 55$ and $b = 0$**W**. As proved in the proposition, t is concave for optimal p and it can be seen that beyond a certain threshold on the available power, $t_s = 11$ is always optimal.

this optimization might not be possible. Even in such situation, it is favorable to optimize the training time. The following conjecture describes how the optimal training time behaves as the transmit power is varied.

Conjecture 18.4 (Optimal training sequence length). *For a given number of transmit antennas n_t, $\eta_{n_t}(p,t)$ is maximized for $t_s^* = T_s - 1$ in the limit of $p \to 0$. As p increases, t_s^* decreases monotonically until for some P_+, $t_s^* = M$ and then for all $p \geq P_+$, $t_s^* = M$. Where P_+ is simply the smallest p for which $t_s^* = n_t$.*

This shows that the optimal training sequence length clearly depends on the number of antennas used. Note that (18.18) can be easily exploited to prove some parts of the conjecture. This is what the following proposition is about.

Proposition 18.6 (Optimal fraction of training time in extreme SNR regimes). *It can be shown that:* $\lim\limits_{p \to +\infty} t_s^* = n_t$ *for all MIMO systems in general.*

The proof for this can be found in [25].

Fig. 18.6 studies the optimal training sequence length t_s as a function of the transmit power p. Note that in this case, we are not optimizing the efficiency with respect to P and so this figure illustrates conjecture 18.4 and proposition 18.6. With p large enough $t_s = n_t$ becomes the optimal training time and for p small enough $t_s = T_s - 1$ as seen from the figure. The parameters are $R = 1600$,

$b = 0$ W, $\xi = 16$ and $T_s = 10$. (We use $T_s = 10$, as if the coherence time is too large, the outage probabilities for low powers that maximize the training time, such that $t_s^* = T_s - 1$, become too small for any realistic computation.)

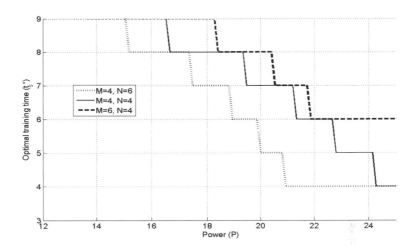

FIGURE 18.6 Optimal training sequence length (t_s) vs. Power (P) MIMO system with $\xi = \frac{R}{R_0} = 16, R = 1$ Mbps $T_s = 10$ symbols. Observe that t_s^* decreases monotonically from $T_s - 1$ to n_t.

18.2.3.3   Optimizing the Number of Transmit Antennas

So far the assumption was that the precoding matrix was chosen to be the identity matrix i.e., $\mathbf{Q} = \mathbf{I}_{n_t}$. Clearly, if nothing is known about the channel, the choice $\mathbf{Q} = \mathbf{I}_{n_t}$ is relevant (and may be shown to be optimal by formulating the problem as an inference problem). On the other hand, if some information about the channel is available (the channel statistics as far as this paper is concerned), it is possible to find a better precoding matrix. As conjectured in [20] and proved in some special cases (see e.g., [41]), the outage probability is minimized by choosing a diagonal precoding matrix and a certain number of 1's on the diagonal. The position of the 1's on the diagonal does not matter since channel matrices with i.i.d. entries are assumed. However, the optimal number of 1's depends on the operating SNR. The knowledge of the channel statistics can be used to compare the operating SNR with some thresholds and lead to this optimal number. Although we consider equation (18.14) as a performance metric instead of the outage probability, we are in a similar situation to [2], meaning that the optimal precoding matrix in terms of energy-efficiency is conjectured to have the same form and that the number of used antennas have to be optimized. In the setting of this paper, as the channel is estimated, an additional constraint has to be taken into account that is, the number of optimal antennas n_t^* cannot exceed the number of training symbols t_s. This leads us to the following conjecture.

Conjecture 18.5 (Optimal number of antennas). *For a given number of training symbols t_s, η_{n_t} is maximized for $n_t^* = 1$ in the limit of $p \to 0$. As p increases, n_t^* also increases monotonically until for some P_+, $n_t^* = t_s$ and then for all $p > P_+$, $n_t^* = t_s$.*

This conjecture can be understood intuitively by noting that the only influence of n_t on η_{n_t} is through the success rate. Therefore, optimizing n_t for any given p and t amounts to minimizing outage. Based on the conjecture in [20], which has been proven for several special cases, we can conclude that the optimal number of antennas is one in the very low SNR regime and that it increments as the SNR increases. However, the effective SNR decreases by increasing n_t, this will result in the optimal n_t for each p with training time lower than or equal to the optimal n_t obtained with perfect CSI. Concerning special cases, it can be easily checked that the optimal number of antennas is 1 at low SNR, and is t_s at high SNR. Work related to proving some special cases of this conjecture can be found in [5]. Finally, a possible refinement of the definition 18.14 regarding to n_t is possible. Indeed, by creating a dependency of the parameter b towards n_t one can better model the energy consumption of a wireless device. For instance, if the transmitter architecture is such that one radio-frequency transmitter is used per antenna, then, each antenna will contribute to a separate fixed cost. In such a situation the total power can written as $ap + n_t b_0$ where b_0 is the fixed energy consumption per antenna. It can be trivially seen that this does not affect the goodput in any manner and only brings in a constant change to the total power as long as n_t is kept a constant. So the optimization w.r.t p and t will not change but it will cause a significant impact on the optimal number of antennas to use.

Fig. 18.7 studies the optimized energy efficiency η^* as a function of the transmit power with various values of n_t. The figure illustrates that beyond a certain threshold on the available transmit power, there is an optimal number of antennas that has to be used to maximize the efficiency. The parameters are $R = 1\text{Mbps}$, $\xi = 16$, $b = 0\text{W}$, $n_r = 4$, $t = 15$ and $T_s = 55$. We observe that for p beyond a certain threshold, the optimal number of antennas is 10. This result is interesting because we are also optimizing the energy efficiency w.r.t p simultaneously and we find that the optimal strategy is to use only a limited number of antennas.

18.2.4 Cross Layer Design in Energy-Efficient Communications

In the previous subsections, the energy-efficiency of systems where data is continuously transmitted has been studied. In practice, this is not the case and the data traffic is typically random and depends on several factors such

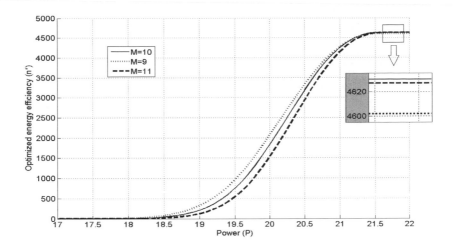

FIGURE 18.7 Optimized efficiency (η^*) vs. transmit power (p) for a MIMO system with n_r=4, R $= 1$ Mbps, $\xi = \frac{R}{R_0}$=16, $t = 15$, $T_s = 55$ and $b = 0$W. It can be seen that beyond a certain threshold, $n_t = 10$ is always optimal while $n_t = 4$ or 9 were optimal for lower powers.

as: the used protocol, the location of the user, the time of the day e.t.c. In the remaining of this section, we focus on the situation where data packets arrive from an upper layer (based on the transfer protocol) randomly into a buffer from which transmission is performed. The energy efficiency of this system is studied in [26].

Consider a buffer of size K at the transmitter. The packets arrival follows a Bernoulli process with probability q for entering the queue, as illustrated in Figure 18.8, (this corresponds to classical ON/OFF sources). All packets in this model are assumed to be of the same size. The throughput (rate) on the radio interface equals to R (bit/s) and this depends on several parameters like modulation and the coding scheme. We consider the case when the transmitter is always active, meaning that it always transmits its packet while the buffer is not empty. Each packet transmitted on the channel is received without any errors with a probability $f(p)$ which depends on the quality of the channel and transmission power p. If the channel fading due to path loss is represented by g, the success probability depends on the SNR $= \frac{gp}{\sigma^2}$. However, based on the block fading channel assumption, we make a slight abuse of notations by using the notation $f(p)$ instead of $f(\text{SNR})$. In some places in this paper, we even remove the variable p for the sake of clarity and use the notation f. We denote by α_t the size of the queue at the transmitter at time slot t. The size of the queue α_t is a Markov process on the state space $\alpha = \{0, \ldots, K\}$. We have the following transition probabilities $\forall i, j \in \alpha$, $P_{i,j} := I\!P(\alpha(t+1) = i|\alpha(t) = j)$ given by:

1. $P_{0,0} = 1 - q + qf$,

2. $P_{K,K} = (1 - q)(1 - f) + q$,

3. for any state $i \in \{0, \ldots, K - 1\}$, $P_{i,i+1} = q(1 - f)$,

4. for any state $i \in \{1, \ldots, K\}$, $P_{i,i-1} = (1 - q)f$,

5. for any state $i \in \{1, \ldots, K - 1\}$, $P_{i,i} = (1 - q)(1 - f) + fq$.

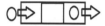

FIGURE 18.8 Packet arrival and transmission from a queue, the boxes represent the available buffers while the circles represent packets of data

A new packet is lost if the queue is full, as illustrated in Figure 18.9, when it comes in and the transmission of the packet currently on the radio interface failed on the same time slot. Indeed, we consider that a packet is in service (occupying the radio interface) until it is transmitted successfully. Thus, a packet in service blocks the queue during $\frac{1}{f(p)}$ time slots on the average. We assume that an arrival of a packet in the queue and a departure (successful transmission) at the same time slot can occur.

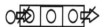

FIGURE 18.9 Packet arrival halted when the buffers are full and the system has failed in transmitting a packet. The boxes represent the buffers while the circles represent packets of data

Given the transition probabilities above, the stationary probability of each state is given by (see e.g., [45]):

$$\forall s \in \alpha, \quad \Pi_s = \frac{\rho^s}{1 + \rho + \ldots + \rho^K}, \tag{18.19}$$

with

$$r = \frac{q(1 - f)}{(1 - q)f}. \tag{18.20}$$

When a packet arrives and finds the buffer full (meaning that the packet currently on the radio interface is not transmitted successfully), it is blocked and this event is considered as a packet loss. The queue is full in the stationary regime with probability Π_K :

$$\Pi_K = \frac{r^K}{1 + r + \ldots + r^K} = \frac{r^K(r - 1)}{r^{K+1} - 1}. \tag{18.21}$$

In order to evaluate the performance of this system, we first determine the expression for the packet loss probability. A packet is lost (blocked) only if a new packet arrives when the queue is full and, on the same time slot, transmission of the packet on the radio interface failed. Note that these two events are independent because the event of "transmit or not" for the current packet on the radio interface, does not impact the current size of the queue, but only the one for the next time slot. This amounts to considering that a packet coming at time slot t, is rejected at the end of time slot t, the packet of the radio interface having not been successfully transmitted. We consider the stationary regime of the queue and then, the fraction of lost packets, Φ, can be expressed as follows:

$$\Phi(p) = [1 - f(p)]\Pi_K(p). \tag{18.22}$$

Thus the average data transmission rate is $q[1 - \Phi(p)]R$. Now, let us consider the cost of transmitting. For each packet successfully transmitted, there have been $\frac{1}{f(p)}$ attempts on an average [28]. $f(p)$ typically depends on the system and for example, in [28] $f(p) = 1 - \exp(-kp)$, where k is a constant. For each time slot, irrespective of whether transmissions occur, we assume that the transmitter consumes energy. A simple model which allows one to relate the radiated power to the total device consumed power is provided in [19] (see also [6] is given by $p_{\text{device}} = ap + b$, where $a \geq 0, b \geq 0$ are some parameters; b precisely represents the consumed power when the transmit power is zero. The average power consumption is in our case $b + \frac{pq(1-\Phi)}{f(p)}$ (we assume without loss of generality that $a = 1$). We are now able to define the energy-efficiency metric $\eta(p)$ as the ratio between the average net data transmission rate and the average power consumption, which gives:

$$\eta(p) = \frac{q[1 - \Phi(p)]R}{b + \frac{pq[1-\Phi(p)]}{f(p)}}. \tag{18.23}$$

The above expression shows that the cross-layer design approach of power control is fully relevant when the transmitter has a cost which is independent on the radiated power; otherwise (when $b = 0$), one falls into the original framework of [28].

In this part, we prove that there exists a unique power where the energy efficiency function is maximized when the transmission rate is a sigmoidal or "S"-shaped function of p. In [23], it was shown that having a sigmoidal success rate $f(p)$ implies quasi-concavity and a unique maximum for $\frac{f(p)}{p}$. This assumption was shown to be highly relevant from a practical viewpoint in [28] as well as from an information theoretical viewpoint in [2].

Theorem 18.1. *The energy efficiency function η_{n_t} is quasi-concave with respective to p and has a unique maximum denoted by $\eta_{n_t}(p^*)$ if the efficiency function $f(p)$ has a sigmoidal shape.*

The proof of this is provided in [26]. We are then able to determine the optimal power p^* which maximize the energy efficiency function, by solving the following equation:

$$0 = \frac{-\mathrm{d}\Phi}{\mathrm{d}p}\{b + \frac{pq(1-\Phi)}{f(p)}\} + (1-\Phi)\{\frac{\mathrm{d}\Phi}{\mathrm{d}p}\frac{p}{f(p)} + \frac{\mathrm{d}(p/f(p))}{\mathrm{d}p}\}. \qquad (18.24)$$

In Fig. 18.10, we study the energy efficiency of a system with $\frac{b}{\sigma^2} = 100$. Here we see that as q decreases p^* decreases. Also seen from the same figure is the quasi-concavity of the energy efficiency function and the asymptotic behavior.

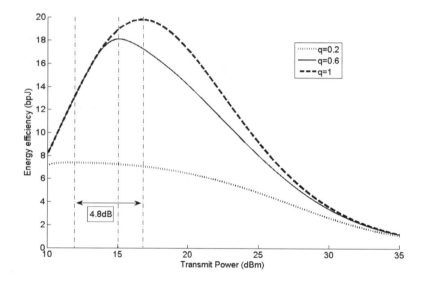

FIGURE 18.10 η_{n_t} in bits per Joule (bpJ)vs p of a system with $\frac{b}{\sigma^2} = 100$ **(20dB)**. Observe that the function is quasi-concave for all q and that p^* decreases as q decreases.

18.3   On the Design of Energy-Efficient MIMO Multi-User Communications

The focus of this section will be on the more practical case of multi-user communications. Furthermore, we will consider distributed multi-user networks, i.e., networks that operate with little or no intervention from a central authority. Other advantages of having a distributed system are the Independence of each user or base station, non-reliance on cooperation, etc.

In reality, there are several factors and layers to be considered while studying the energy efficiency of a network like the MAC layer consisting of random arrival and departure of users, the arrival of packets from each user based on the protocol etc and the physical layer dealing with channel estimation, coding and finally transmission and reception from multiple base stations and antennas. However, research is still in progress in this field and the results presented here are limited to physical layer systems with uniform power allocation and perfect CSI.

18.3.1 A Discussion on the Performance Criterion

In multi-user networks, it is not straightforward to define an energy-efficiency metric. Several different possible notions arise: (i) the ratio of the total utility over the total cost; (ii) the minimal individual energy-efficiency; (iii) individual user energy-efficiency. The first two goals are rather social ones applying to a centralized scenario, whereas the last one applies to distributed multi-user networks where each user tries to optimize its own energy-efficiency. In a network, there are primarily two modes of information transfer i.e the uplink and the downlink. The uplink case is when mobile terminals send information to the corresponding base station and the downlink case is when the base stations send information to the mobile terminals.

For the downlink case, the base station tries to optimize the global efficiency. This problem has been analysed in several works. In [11], the authors consider energy-efficient transmission schemes in cooperative cellular systems with unbalanced traffic between uplink and downlink, and derive the optimal transmission data rate, which minimizes the total energy consumption of battery powered terminals per bit of information. In [12], the quality of service (QoS) constrained radio resource allocation problem at the downlink of multi-user multi-carrier systems is studied based on the trade-off between energy consumption and transmit power within a cross physical and link layer system model, jointly considering power allocation, adaptive modulation and coding and ARQ/HARQ retransmission protocols.

For the uplink, one of the key issues in wireless system design is energy consumption at users terminals. Since in many scenarios, the users' terminals are battery-powered, efficient energy management schemes are required in order to prolong the battery life. Hence, power control plays an even more crucial role in such systems. For the uplink case, for every user, the purpose of power control is to transmit at the optimal power to achieve the required quality of service (QoS) at the uplink receiver without causing unnecessary interference to other users in the system. This motivates the application of game theory, and in this section, we will focus on the uplink case and present some of the novel results in this field.

For the SISO case, several works have analyzed this problem like [30], [31],

[32] and [34]. In [28], the authors study power control algorithms for the multi-user distributed case using a non-cooperative game theoretical framework and propose algorithms that achieve the Nash equilibrium. A Nash equilibrium is defined as a set of strategy choices and the corresponding payoffs in which each player has chosen a strategy and no player can benefit by changing his or her strategy while the other players keep theirs unchanged [29]. The same authors, in [14], propose a pricing scheme for transmit power levels to improve the efficiency of the system, as the Nash equilibrium is not the most optimal scenario in terms of the sum of the utilities.

In the next section we present the results available for the power control game in the case of multi-carrier systems based on [33]. These results can also be interpreted as a power control game for a diagonal MIMO system.

18.3.2   Power Control in Multi-Carrier CDMA Systems

As explained before, due to the competitive nature of the users' interaction, the natural framework for modelling and studying a power control problem in CDMA systems is game theory. Consider a non-cooperative power control game, in which each user seeks to maximize its overall utility by choosing the optimal transmit power over each carrier. The utility function is defined as the ratio of any user's total throughput to its total transmits power over all the carriers. This utility function has units of bits/Joule as before and is suitable for applications where saving power (preserving battery life) is critical. The non-cooperative nature of the proposed game implies that no coordination among the users exists. There are two difficulties to the problem studied in section, being that, firstly, users' strategies in the multi-carrier case are vectors (rather than scalars) and this leads to an exponentially larger strategy set for each user. Secondly, the utility function considered here is not quasi-concave. This means that many of the standard theorems from game theory as well as convex optimization cannot be applied here.

In this section, the Nash equilibrium for the proposed power control game is derived and its existence and uniqueness are studied. Some of the questions answered in this section are the following. The existence of a Nash equilibrium, and the possibility of users reaching it, the kind of carrier allocations and the spread of usage in the carriers at a Nash equilibrium,and the performance of this joint maximization of utility over all the carriers compared with that of an approach where utility is maximized independently over each carrier are some of issues studied in [33] and are presented here.

Consider the uplink of a synchronous multi-carrier DS-CDMA data network with N users, M carriers and processing gain L (for each carrier). The carriers are assumed to be sufficiently far apart so that the (spread-spectrum) signal transmitted over each carrier does not interfere with the signals transmitted over other carriers [11]. It is also assumed that the delay spread and Doppler spread are negligible for each individual carrier. At the transmitter, the incoming bits for user n are divided into M parallel streams and each stream is spread using the spreading code of user n. The M parallel streams are then sent over the M (orthogonal) carriers. For the m-th carrier, the received signal at the uplink receiver (after chip-matched filtering and sampling) can be represented by an $L \times 1$ vector as

$$\underline{y}_m = \Sigma_{n=1}^N \sqrt{P_{n,m}} h_{n,m} x_n + \underline{z}_m \tag{18.25}$$

where b_n, P_n, h_n are the n-th user transmitted bit, transmit power and path gain, respectively, for the m-th frequency channel (carrier); x_n is the spreading sequence for user n which is assumed to be random with unit norm; and $\underline{z}_m$ is the noise vector each element of which is assumed to be Gaussian with mean 0 and covariance σ^2.

The non-cooperative game is studied, in which each user chooses their transmit powers over the D carriers to maximize its overall utility. In other words, each user (selfishly) decides how much power to transmit over each frequency channel (carrier) to achieve the highest overall utility. Let $G_M = [\mathcal{N}, \{A_n\}, \{\eta_n\}]$ denote the proposed non-cooperative game where $\mathcal{N} = \{1, ..., N\}$, and $A_n = [0, P]^M$ is the strategy set for the n-th user. Here, P is the maximum transmit power on each carrier. Each strategy in A_n can be written as $\underline{p}_n = [p_{n,1}, ..., p_{n,M}]$. The utility function (the energy efficiency) for user n is defined as the ratio of the total throughput to the total transmit power for the M carriers, i.e

$$\eta_n(\underline{p}_1, \ldots, \underline{p}_N) = R \frac{\Sigma_{m=1}^M f(\gamma_{n,m})}{\Sigma_{m=1}^M p_{n,m}} \tag{18.26}$$

where R is the target rate (assumed to be the same for all users without any loss in generality), $f()$ is the success rate, $\gamma_{n,m}$ is the SINR of user n on carrier m, hence, the resulting non-cooperative game can be expressed as the following maximization problem:

$$\max_{p_{n,1},\ldots,p_{n,M}} \frac{\Sigma_{m=1}^M f(\gamma_{n,m})}{\Sigma_{m=1}^M p_{n,m}} \tag{18.27}$$

The relationship between $\gamma_{n,m}$ and $p_{n,m}$ is dependent on the uplink receiver. It should be noted that the assumption of equal transmission rates for all users can be made less restrictive. For our analysis, it is sufficient for the users to have equal transmission rates over different carriers but the transmission rate can be different for different users. More generally, the proposed power control game can be extended to allow the users to pick not only their transmit powers but also their transmission rates over the D carriers. While joint power and rate control is important, particularly for data applications, our focus throughout this work is on power control only.

For the non-cooperative power control game, a Nash equilibrium (NE) is a set of power vectors, $(p_1, \ldots, p_N)$, such that no user can unilaterally improve its utility by choosing a different power vector, i.e., $(p_1^*, \ldots, p_N^*)$ is a Nash equilibrium if and only if

$$\eta_n(p_n^*, p_{-n}^*) \geq \eta_n(p_n, p_{-n}^*) \tag{18.28}$$

Here, p_{-n}^* denotes the set of transmit power vectors of all the users except for user n at the NE. This game is particularly difficult to analyze because users' strategies are vectors (rather than scalars) and the utility function is not a quasi-concave function of the user's strategy. For this utility function, it is shown in [33] that at a Nash equilibrium each user transmits only on the carrier that has the best effective channel for that user. Additionally, the conditions required for the existence and uniqueness of the NE are also detailed. The performance of energy efficiency at NE (η^*) versus number of users (K) is illustrated in Figure 18.11.

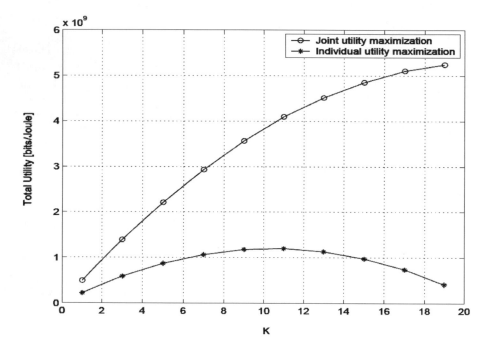

FIGURE 18.11 Energy efficiency at NE (η^*) versus number of users (K) for a system with 2 carriers using matched filter.

Remark: At this point, it is interesting to note that maximizing the spectral efficiency or the rate will lead to a water filling solution over all the carriers [36]. However, it can be seen that when optimizing energy-efficiency, the best strategy is to transmit with all the power on the best carrier [33]. This means that while deciding on a transmission strategy, a choice has to be made on what to maximize.

18.3.3 The Case of Specific Transmitter and Receiver Structures, Multiplexing and Diversity Gain

Using multiple antennas at the transmitter and receiver side increases the transmission performance in terms of rate, due to the multiplexing gain and reliability, due to the diversity gain. Finding the optimal diversity-multiplexing tradeoff that maximizes the energy-efficiency function is a very difficult problem which involves finding the optimal transmitter/receiver structure.

In [40], the authors study two extreme cases: the full diversity mode and the full multiplexing mode. Essentially, to obtain an optimal diversity gain, one can use an adaptive antenna at the transmitter a maximum ratio combiner (MRC) at the receiver. Both techniques amount to weighting each transmit/receive antenna by a complex scalar. This leads to a performance metric under the form $f(\text{SNReq})$. The quantity SNReq is the SNR at the MRC output: the system is equivalent to a SISO channel in which you can partially control the channel gain (by tuning the vector of powers). In the particular case where the channel matrix is diagonal, the energy-efficiency is maximized by implementing channel selection (that is, only one component of the PA vector is not vanishing). This is similar to the case of optimizing the energy efficiency in a multiuser multicarrier system discussed in the previous section. This quantity is maximized by a channel selection policy irrespective of the channel matrix structure. On the other hand, if one wants to maximize the multiplexing gain, one uses a given data flow for each transmit antenna and assume that the receiver decodes each data flow separately so that the performance metric of interest is $\frac{\Sigma_i f(\text{SNR}_i)}{\Sigma_i p_i}$).

Contrarily to the above, the approach that we mostly use in this chapter does not make any assumption on the transmitter/receiver structure. Rather, we assume that the quantity of interest is the mutual information and the connection with the different SNR is possible only if special structures are considered.

18.3.4 Non-Cooperative Resource Allocation in Multi-User MIMO Systems

Finally in this section a power control game for a MIMO system is studied, where the power allocation is chosen by each player trying to maximize their energy-efficiency metric. This section is based on [37].

In this subsection, some results on the uplink of a multi-user MIMO communication system are presented, wherein both the mobile terminals and the common access point (AP) are equipped with multiple antennas. We are interested in the design of non-cooperative resource allocation policies aiming at energy-efficiency maximization, which is defined here as the number of reliably delivered information symbols per unit of consumed energy from the battery. Energy-efficiency maximization is indeed a crucial problem in mobile wireless communications, wherein mobile users are interested in making a careful and smart use of the energy stored in their battery. This section is an extension of [33] to multi-user MIMO wireless systems. We consider three problems depending on the energy-efficiency optimization variables:

1. The transmit power of each user, assuming matched filtering at the receiver;

2. The transmit power and the choice of the uplink linear receiver for each user;

3. The transmit power, the beam forming vector and the choice of the uplink linear receiver for each user;

Consider the uplink of a N-user synchronous, single-cell, MIMO multi-user flat fading channel. Denote by n_t the number of transmit antennas for each user, and by n_r the number of receive antennas. The received signal can be written as

$$y = \sqrt{p_n}\mathbf{H}_n\mathbf{a}_n x_n + z, \qquad (18.29)$$

Where $\mathbf{H}_k$ is the $n_r \times n_t$ channel matrix between the receiver and the nth user, $\mathbf{a}_k$ is the beamforming vector of the n-th user satisfying $\mathbf{a}_n^T\mathbf{a}_n = 1$ and s_k is the symbol transmitted, and p_n is the transmission power of user n.

The energy efficiency function, which is also the utility considered while using a game theoretic approach can be defined for the n-th user as

$$\eta_n = R\frac{f(\gamma_n)}{p_n}, \qquad (18.30)$$

Where γ_n is the SINR of user n which depends on p_n, $\mathbf{a}_k$ and the receiver used.

18.3.4.1 Optimizing Transmit Power with a Matched Filter

For a linear receiver, the SINR can be written as

$$\gamma_n = \frac{p_n(\mathbf{d}_n\mathbf{H}_n\mathbf{a}_n)^2}{\Sigma_{j\neq n}p_j(\mathbf{d}_j\mathbf{H}_j\mathbf{a}_j)^2 + p_n(\mathbf{d}_n\mathbf{H}_n\mathbf{a}_n)^2}, \qquad (18.31)$$

if the linear receiver detects symbols according to $\mathbf{d}_k^T\mathbf{y}$ where $\mathbf{d}_k$ is the L dimensional vector representing the receive filter.

Consider the non-cooperative game in which each user chooses its power so that their utility is maximized. The game can be expressed mathematically as:

$$\max_{p_n\in[0,P]} \frac{f(\gamma_n)}{p_n}, \text{ for } n = 1,\ldots,N, \qquad (18.32)$$

where P is the maximum allowed transmit power for the n-th user, it is proved in [37] that the non-cooperative game defined in 18.32 admits a unique NE point p_n^* for all n which satisfies $f(\gamma_n) = \gamma f'(\gamma_n)$, p_n^* being the power at which γ_n^* is achieved.

18.3.4.2   Optimizing the Choice of Linear Receiver

Now consider the following non-cooperative game where each user chooses its transmit power and its linear receiver competing for the best utility defined mathematically as follows:

$$\max_{p_n \in [0,P], \mathbf{d}_n \in \mathbb{R}^{n_r}} \frac{f(\gamma_n)}{P_n}, \text{ for } n = 1, \ldots, N, \tag{18.33}$$

This is done by at first finding the optimal receiver and then finding the power. As it is known that among linear receivers the MMSE receiver maximizes the SINR, and so, in [37] it is shown that the game described in 18.33 has a unique NE $(P_n^*, \mathbf{d}_n^*)$.

$$\mathbf{d}_p^* = \sqrt{p_n} \mathbf{F}^{-1} \mathbf{H}_n \mathbf{s}_n \tag{18.34}$$

with $\mathbf{F} = I_{n_t} \sigma^2 + \Sigma_{j=1}^N p_j \mathbf{H}_j \mathbf{s}_j \mathbf{s}_j^T \mathbf{H}^T$ and $p_n^* = \min(\bar{P}_n, P)$, such that the n-th user has a SINR of γ_k^* as in the previous section.

18.3.4.3   Optimizing the Beamforming Vector

Now the most general and challenging optimization considered in this chapter performed on a multi-user MIMO system is presented. Here, the users can chose their total transmit power, their linear receiver as well as their beamforming vector. The game that describes this process can be written down as

$$\max_{p_n \in [0,P], \mathbf{d}_k \in \mathbb{R}^N, \mathbf{a}_n \in \mathbb{R}^N} \frac{f(\gamma_n)}{p_n}, \text{ for } n = 1, \ldots, N. \tag{18.35}$$

Given the above equation, first the problem of SINR maximization with respect to the vectors $\mathbf{d}_n$ and $\mathbf{a}_n$ is considered. Again, the SINR-maximizing linear receiver is the MMSE receiver. The n-th user SINR for MMSE detection is already known and the vector that maximizes this for the MMSE receiver is the eigenvector corresponding to the maximum eigen-value of $\mathbf{H}_n^T \mathbf{F}^{-1} \mathbf{H}_n$. Therefore the optimal strategy for each player would be to cyclically update their beam forming vectors satisfying the afore mentioned criteria. The following result holds, from [37] (the proof can be found in the paper).

Theorem 18.2. *Assume that the active users cyclically update their beam forming vectors in order to maximize their own achieved SINR at the output of a linear MMSE receiver. This procedure converges to a fixed point.*

Figure 18.12 (taken from [37]) show the achieved average utility (energy efficiency) at the receiver output versus the number of users, for the considered games, and for a 4 and 4 × 8 MIMO system. Inspecting the curves, it can be observed that a smart resource allocation algorithm brings very remarkable performance improvements.

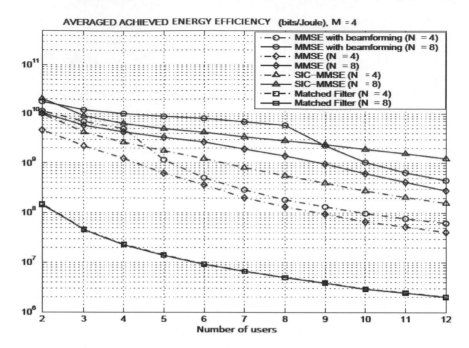

FIGURE 18.12 Energy efficiency at NE (η^*) v.s number of users (N)

18.4   Concluding Remarks and Open Issues

Having studied the energy efficiency in all kinds of situations, it is clear that using zero power is not always optimal. The fraction of power to be used depends on various factors like the number of antennas, the fixed power consumption.

The available results w.r.t. the energy-efficiency of single-user channels are the following:

1. There exists a unique optimal transmit power that achieves the maximum energy-efficiency.

2. There is an optimal number of antennas to be used based on the coherence time.

3. When considering a cross layer approach, the optimal power to use is smaller than when transmission occurs the time, depending on the packet arrival rate.

For the multi-user networks the results are:

1. In a multi-user multi-carrier SISO or diagonal MIMO system, users maxi-

mizing their energy efficiency will converge under specific conditions to an equilibrium where they transmit on their best "effective" carrier.

2. In a multi-user MIMO power allocation game, if users cyclically update their beamforming vectors to maximize their own energy efficiency, this process converges to a fixed point.

We have seen that here are several unsolved and open problems left in this field:

1. Proof of the quasi-concavity of the energy-efficiency function for a MIMO system where an uniform power allocation policy is assumed.

2. Finding the pre-coding matrix, that optimizes the energy-efficiency metric, for an $N \times M$ MIMO system with training.

3. Considering the power consumption of each individual antenna due to additional energy costs of coding, infrastructure.

4. The energy-efficiency study of multi-user MIMO networks with imperfect CSI and finite coherence time.

5. Considering multi-user cross layer models for the energy-efficiency.

18.5 Glossary

SISO: single-input single-output

MIMO: multiple-input multiple-output

CSIT: channel state information at the transmitter

CDIT: channel distribution information at the transmitter

PA: power allocation

RF: radio frequency

SNR: signal-to-noise ratio

SINR: signal-to-interference plus noise ratio

CDMA: code division multiple access

BER: bit error rate

FSK: frequency shift keying

bpcu: bits per channel use

NE: Nash equilibrium

OFDMA: orthogonal frequency-division multiple access

STBC: space-time block coding

AWGN: additive white Gaussian noise

UPA: uniform power allocation

Bibliography

[1] L. H. Ozarow and S. Shamai (Shitz) and A. D. Wyner, "Information theoretic considerations for cellular mobile radio" , IEEE. Journ. VT, 43, 10, p 359-378, May 1994

[2] E.V. Belmega and S. Lasaulce, "Energy-efficient precoding for multiple-antenna terminals", IEEE. Trans. on Signal Processing, 59, 1, Jan. 2011

[3] G. Scutari, D. P. Palomar, and S. Barbarossa, "Competitive design of multiuser MIMO systems based on game theory: A unified view", *IEEE Journal on Sel. Areas in Comm.*, vol. 26, no. 7, pp. 1089–1103, Aug. 2008.

[4] E. V. Belmega, S. Lasaulce, and M. Debbah, "Power allocation games for MIMO multiple access channels with coordination", *IEEE Trans. on Wireless Comm.*, vol. 8, no. 6, pp. 3182–3192, Jun. 2009.

[5] H. Kim,C.B. Chae, G.de Veciana, and R. W. Heath Jr. "Cross layer approach to Energy Efficiency for adaptive MIMO systems exploiting Spare Capacity", IEEE Trans on Wireless Communications, 8(8):4264-4275, August 2009 p.22

[6] L. Saker and S.E. Elayoubi, "Sleep mode implementation issues in green base stations", IEEE PIMRC 2010, Istanbul, September 2010

[7] G. J. Foschini, "Layered space-time architecture for wireless communication in a fading environment when using multi-element antennas", Bell Labs Tech. J., vol. 1, no. 2, pp. 41-59, Autumn 1996

[8] A. Edelman, "Eigenvalues and Condition Numbers of Random Matrices. PhD thesis", Department of Mathematics, Massachusetts Institute of Technology, Cambridge, MA, 1989.

[9] V. Shah, N. B. Mandayam and D. J. Goodman, "Power control for wireless data based on utility and pricing", IEEE Proc. of the 9th Intl. Symp. Personal, Indoor, Mobile Radio Communications (PIMRC), Boston, MA, pp. 427-1432, Sep. 1998.

[10] J. Palicot, C. Roland "On The Use of Cognitive Radio For Decreasing The Electromagnetic Radiations", URSI 05, XXVIII General Assembly, New Delhi, India, October 23-29, 2005.

[11] W.Yang, W.Li, Y.Wang, W.Sun, "Energy-Efficient Transmission Schemes in Cooperative Cellular Systems", Globecom 2010

[12] Q. Bai, J.A. Nossek, "An energy efficient downlink resource allocation strategy for multiuser CP-OFDM and FBMC systems", WiOpt 2010

[13] A. L. Moustakas and S. H. Simon and A. M. Sengupta, "MIMO capacity through correlated channels in the presence of correlated interferers and noise: A (not so) large N analysis", IEEE Journ. IT, 49, 10, p 2545–2561, Oct. 2003

[14] C. U. Saraydar, N. B. Mandayam, and D. J. Goodman, "Efficient power control via pricing in wireless data networks", IEEE Trans. on Communications, vol. 50, No. 2, pp. 291-303, Feb. 2002.

[15] T. M. Cover , J. A. Thomas, "Elements of information theory", Wiley-Interscience, New York, NY, 1991.

[16] F. Meshkati, H. V. Poor, S. C. Schwartz, and N. B. Mandayam, "An energy-efficient approach to power control and receiver design in wireless data networks," IEEE Trans. Commun., vol. 52, pp. 1885-1894, Nov. 2005.

[17] T. L. Marzetta and B. M. Hochwald, "Capacity of a Mobile Multiple-Antenna Communication Link in Rayleigh Flat Fading", *IEEE Trans. on Information Theory*, Vol. 45, No. 1, Jan. 1999

[18] B. Hassibi and B. M. Hochwald, "How Much Training is Needed in Multiple-Antenna Wireless Links?" *IEEE Trans. on Information Theory*, Vol. 49, No. 4, APRIL 2003

[19] F. Richter, A. J. Fehske, G. Fettweis, "Energy Efficiency Aspects of Base Station Deployment Strategies for Cellular Networks", Proceedings of VTC Fall'2009

[20] I. E. Telatar. "Capacity of Multi-antenna Gaussian Channels", *European Trans. Telecommunications*, Vol. 10 (1999), pp. 585-595

[21] I. E. Telatar and R. G. Gallager, "Combining Queueing Theory and Information Theory for Multi-access", IEEE Journal of Selected Areas in Communication, August 1995.

[22] S. Lasaulce and N. Sellami, "On the Impact of using Unreliable Data on the Bootstrap Channel Estimation Performance", *Proc. IEEE SPAWC*, Italy, June 2003, pp. 348-352.

[23] V. Rodriguez, "An Analytical Foundation for Ressource Management in Wireless Communication", IEEE Proc. of Globecom, San Francisco, CA, USA, pp. 898-902, , Dec. 2003.

[24] FCC Sec.15.249, *Operation within the bands 902-928 MHz, 2400-2483.5 MHz, 5725-5875 MHZ, and 24.0-24.25 GHz*

[25] V. Varma and M. Debbah, S. Lasaulce and S.E Ayoubi,"Optimizing Energy-Efficiency in Slow-Fading MIMO Systems with Imperfect Channel State Information", submitted to Computer Communications; Special Issue: Wireless Green, 2011.

[26] V. Varma, S. Lasaulce, Y. Hayel, S.E Ayoubi and M. Debbah ,"Cross-Layer Design for Green Communications: The Case of Power Control", accepted for IEEE International Conference on Communications 2012.

[27] V. Shah, N. B. Mandayam and D. J. Goodman, "Power control for wireless data based on utility and pricing", *IEEE Proc. of Personal, Indoor, Mobile Radio Comm.*, Boston, MA, Sep. 1998.

[28] D. J. Goodman and N. Mandayam, "Power Control for Wireless Data", *IEEE Personal Comm.*, Vol. 7, pp. 48–54, 2000.

[29] D. Fudenberg and J. Tirole, "Game Theory", *MIT Press*, 1991.

[30] S. Lasaulce, M. Debbah, and E. Altman, "Methodologies for analyzing equilibria in wireless games", *IEEE Signal Processing Magazine*, vol. 26, no. 5, pp. 41–52, Sep. 2009.

[31] C. U. Saraydar, N. B. Mandayam and D. J. Goodman, "Efficient power control via pricing in wireless data networks", *IEEE Trans. on Comm.*, vol. 50, no. 2, pp. 291–303, Feb. 2002.

[32] S. Lasaulce, Y. Hayel, R. El Azouzi, and M. Debbah, "Introducing hierarchy in energy games", *IEEE Trans. on Wireless Comm.*, vol. 8, no. 7, pp. 3833–3843, Jul. 2009.

[33] F. Meshkati, A. J. Goldsmith, H. V. Poor and S. C. Schwartz, "A game-theoretic approach to energy-efficient modulation in CDMA networks with delay QoS constraints", *IEEE J. on Sel. Areas in Comm.*, vol. 25, no. 6, Aug. 2007.

[34] F. Meshkati, H. V. Poor, S. C. Schwartz, and R. V. Balan, "Energy-efficient resource allocation in wireless networks with quality-of-service constraints", *IEEE Trans. on Comm.*, vol. 57, no. 11, pp. 3406–3414, Nov. 2009.

[35] F. Meshkati, H. V. Poor, S. C. Schwartz, "An energy-efficient approach to power control and receiver design in wireless data networks", *IEEE Trans. on Comm.*, vol. 53, no. 11, pp. 1885–1894, Nov. 2005.

[36] G. Bacci, and M. Luise, "A noncooperative approach to joint rate and power control for infrastructure wireless networks", *Intl. Conf. on Game Theory for Networks*, Istanbul, Turkey, May 2009.

[37] S. Buzzi, H. V. Poor and D. Saturino, "Energy-efficient resource allocation in multiuser MIMO systems: a game-theoretic framework", EU-SIPCO 2008.

[38] G. Miao, N. Himayat, and G. Li, "Energy-efficient link adaptation in frequency-selective channels", *IEEE Trans. on Commu.*, vol. 58, no. 2, pp. 545–554 , Feb. 2010.

[39] G. Miao, N. Himayat, G. Y. Li, A. T. Koc, and S. Talwar, "Interference-aware energy-efficient power optimization", *IEEE Intl. Conf. on Comm.*, Dresden, Germany, Jun. 2009.

[40] E.V. Belmega, S. Lasaulce, M. Debbah, and A. Hjørungnes, "A New Energy Efficiency Measure for Quasi-Static MIMO Channels" , *International Wireless Communications and Mobile Computing Conference (IWCMC)*, Leipzig, Germany, Jun. 2009.

[41] E. A. Jorswieck, and H. Boche, "Outage probability in multiple antenna systems", European Transactions on Telecommunications, vol. 18, pp. 217-233, 2006.

[42] S. Verdu, "On channel capacity per unit cost", *IEEE Trans. on Inform. Theory*, vol. 36, no. 5, pp. 1019–1030, Sep. 1990.

[43] E. V. Belmega, and S. Lasaulce, "An information-theoretic look at MIMO energy-efficient communications", *ACM Proc. of the Intl. Conf. on Performance Evaluation Methodologies and Tools (VALUETOOLS)*, Pisa, Italy, Oct. 2009.

[44] S. Verdu, "Spectral efficiency in the wideband regime", *IEEE Trans. on Inform. Theory*, vol. 48, no. 6, pp. 1319–1343, Jun. 2002.

[45] R. W. Wolff, "Stochastic Modeling and the Theory of Queues", Englewood Cliffs, NJ: Prentice-Hall, 1989.

Author Contact Information

Vineeth S. Varma and Samson Lasaulce are with Laboratoire des Signaux et Systèmes (LSS), CNRS-Supélec-Paris Sud, Plateau du Moulon, 3 rue Joliot Curie, 91191 Gif-sur-Yvette, France, Email: Vineeth.VARMA@lss.supelec.fr, samson.lasaulce@lss.supelec.fr. E. Veronica Belmega is with ETIS / ENSEA - Université de Cergy-Pontoise - CNRS, 6 Av. du Ponceau, 95014 Cergy-Pontoise, France, Email: belmega@ensea.fr. Mérouane Debbah, Alcatel-Lucent Chair on Flexible Radio, Supélec, Plateau du Moulon, 3 rue Joliot Curie, 91191 Gif-sur-Yvette, France, Email: merouane.debbah@supelec.fr.

19

Minimising Power Consumption to Achieve More Efficient Green Cellular Radio Base Station Designs

John S. Thompson

University of Edinburgh, U.K.

Peter M. Grant

University of Edinburgh, U.K.

Simon Fletcher

NEC Telecom MODUS Ltd., Surrey, U.K.

Tim O'Farrell

University of Sheffield, U.K

CONTENTS

The last ten years have witnessed explosive growth in the number of subscribers for mobile telephony. The technology has evolved from early voice only services to text messaging through to today's mobile wireless (Internet) data delivery. The increasing use of wireless connectivity via smartphones and laptops, predominantly for high bandwidth video streaming, has led to an exponential surge in network traffic. Meeting these mobile traffic demands is causing a significant increase in operator energy cost, as an enlarged network of radio base stations is needed to support mobile broadband effectively and maintain operational competitiveness. This chapter thus explores approaches which will assist in delivering significant energy efficiency gains in future wireless networks. It will investigate several approaches to saving energy in future wireless networks. These include: femtocell and relay deployments; optimizing multiple antenna systems; sleep mode techniques to disable radio transmissions whenever possible; reallocating transmissions under low load; and improving the power amplifier. The impact of these approaches on achieving more energy efficient cellular wireless communication systems is discussed.

19.1   Introduction

It is estimated that 3% of overall global CO_2 emissions come from ICT and the cellular Radio Base Station (RBS) has been identified as the most energy-intensive component in current 3G mobile networks. This *chapter aims to set the scene* on the case for green radio before discussing several approaches to achieving *reduced operating cellular base-station energy consumption* and it extends a previous chapter [1]. The UK has established a legally binding commitment in 2050 to reduce the 1990 CO_2 emission levels by 80%. Although mobile cellular communications accounts for only 2-3% of these emissions there is an enormous growth in mobile subscribers and data traffic.

In 2011 there were 4B(billion) worldwide mobile subscribers, almost 2B Internet subscribers and less than 1.5B fixed phone lines. Previous studies [2] [3] have confirmed that cellular mobile communications consumes more power per user than for fixed systems, such as with optical fibres or for local digital subscriber line (DSL) wireless Internet access, and the RBS is further confirmed the dominant contributing component to the overall cellular CO_2 emissions of the network [4]. Over the 10-15 year lifetime of a RBS, operating energy is the major contributing component in cellular CO_2 emissions and the operators are demanding that the equipment suppliers achieve reductions, as soon as possible. Vodafone [5] is seeking a 50% overall energy reduction over the 2006-2020 period.

In a mobile handset, with a typical 2-year lifetime the manufacturing or embodied energy, i.e. capital expenditure or CAPEX, is a much larger component of the total energy use than the operating energy, Figure 19.1 [6]. The well publicized figure is that the current 3.3B worldwide mobile handsets [4], consume 0.3 GW of power. 2% of the global CO_2 emissions are contributed from the 3M worldwide installed cellular RBS, which consume, in total, 4.5 GW of power. In the RBS, the embodied energy is a much smaller component and the operational energy consumption dominates, Figure 19.1. Reference [7] has also confirmed that the radio access network consumes more than 70% of

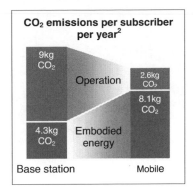

FIGURE 19.1 CO_2 emissions per subscriber per year as derived for the base-station component and mobile handset. Embodied emissions arise from manufacturing process rather than operation [6] ©2011 IEEE

the total energy used by the mobile operator and, sometimes, operator network energy costs equal the personnel costs! Although the RBS emissions per carried traffic data bit will decrease significantly in the future, mobile data traffic increases will inevitably cause a significant overall increase in these emissions!

Hence this chapter concentrates on reducing the operational energy consumption of the RBS in Long Term Evolution (LTE) systems. Figure 19.2 [8] shows that as the standards have progressed through: the GSM General Packet Radio Service (GRPS); Evolved Enhanced Data Rates for GSM Evolution (EDGE); 3rd Generation Partnership Project (3GPP) into High Speed Packet Access (HSPA) and on to $N \times M$ antenna Multiple Input Multiple Output (MIMO) LTE Next Generation Mobile Networks (NGMN), that the spectral efficiency advances, i.e. transmitted bit/s/Hz of available bandwidth, are now progressively reducing.

A typical 3G RBS, which has three 120° sectors, consumes about 300-500 W of AC input operating power to generate an output RF power of 40 W with a modest further allowance for the data backhaul component into the core network. This makes the average annual energy consumption of a 3G RBS around 4.5 MWh (which is already much lower than the earlier GSM RBS designs). The UK 3G mobile network with 12,000 BSs thus consumes over 50 GWh p.a. Operator network energy use has the potential for being much higher in developing markets, e.g. China Mobile operator currently has 500M mobile subscribers, serviced by 500,000 GSM and 200,000 3G CDMA RBS!

The first version of the LTE standard was frozen at the end of 2008 and represents a significant enhancement to existing third generation (3G) wireless systems. It uses orthogonal frequency division multiple access (OFDMA) to transmit data from the RBS to multiple users on different carrier frequencies.

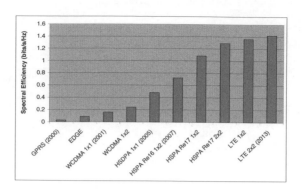

FIGURE 19.2 Progressive improvement in transmission spectral efficiency with new generations of cellular systems, after Qualcomm

It also enables the use of MIMO at both the RBS and on user equipment (UE) terminals to achieve much higher data rates on each link. The sophisticated high data rate LTE RBS, for a 470 m radius cell, has been claimed to consume 3.7 kW of AC power [2], for access plus backhaul, but more realistic industrial estimates are 800 W to 1 kW, which is much larger than for previous 3G networks. However, estimating RBS power consumption per subscriber UE, gives a significant energy advantage to the large cell radius LTE over 3G or HSPA.

Improving base station energy efficiency has thus gained progressively increasing importance over the last 5 to 7 years. In 2006/7 the mobile operators requested a move from the 8% annual improvements in RBS energy consumption seeking a more major step change improvement in overall energy efficiency. This was driven primarily by the increasing critical pressures of energy operating cost on their operating expenditure (OPEX) and thus profitability.

There are several approaches which operators use to provide immediate energy reductions [9]. Many of these relate to operating existing RBSs more efficiently, such as optimizing the power supply; reducing the amount of RBS cooling required through installing more efficient systems or using higher thermostat settings; upgrading older base station equipment, e.g. from GRPS to EDGE system; and moving more quickly to adopt the higher data rate LTE standard. However, longer term enhancements to both RBSs and wireless networks in general still need to be addressed. This is being done through a number of current research projects such as [7]:

Earth: Energy Aware Radio and neTwork tecHnologies: undertaking component and system level energy improvements, an EU Project [10],

OPERA-Net: Optimising Power Efficiency in mobile RAdio Networks: an

EU Project,

Cool BS: Regenerative and Autonomous Femtocell BS,

eWIN: Energy-efficient wireless networking,

GreenTouch: Energy Efficient, sustainable communication networks,

GREENET: Analysis, design and optimization of energy efficient wireless networks: an EU Marie Curie PhD training network [11].

These projects seek to achieve up to a 1,000 fold reduction in energy consumption per transmitted bit. We are partners in the UK Mobile VCE Academic-Industry "Green Radio" programme which is 55 person-year project is addressing a number of new techniques to improve RBS wireless network efficiency [1] [6] [12]. This research programme is monitored by an industry based energy focus group who compare the research outcomes to real world costs, constraints, metrics and expectations. The Mobile VCE works from a "Book of Assumptions" which provides the background reference material which defines and bounds the research programme. As research reports evolve, these are assessed by the energy focus group, to ensure their practical relevance in enhancing next generation system designs. They then use their observations to refine the problem into a new set of targeted questions, which the researchers then address.

19.1.1   Explosive Traffic Growth

Data from [8] shows an exponential growth in cellular downlink HSPA and other **data** traffic of 400% p.a. due to the increased use of smart-phones, video delivery and the use of (Universal Serial Bus) USB dongles for mobile laptop connections. Reference [4] further predicts, in 2020, that smartphones will comprise 33% of user terminal sales and, worldwide mobile data transmissions, will form 99% of the cellular traffic and this will total more than 500 Exabytes p.a. This data traffic growth is completely divergent from the 23% p.a. operator revenue growth curve [8]. This demanding traffic growth implies an increasing future energy consumption as the RBS infrastructure must grow to service this increasing traffic demand. The slow growth in revenues also makes it very difficult to finance a large increase in network size. Thus, there is an urgent need to reduce energy use per data bit carried to minimize the operating power per cell and mitigate the impact of increasing data traffic on the network.

19.1.2   Cellular Scenarios

There are two types of client who require more efficient RBS infrastructure. In the developed world we have: established infrastructure; saturated markets; exponential increase in data traffic; quality of service (QoS) is a key issue; and the drive is mainly to reduce operating costs. In emerging markets have a different set of constraints: less well established infrastructure; rapidly expanding user base; large geographical coverage areas; often no mains power supply;

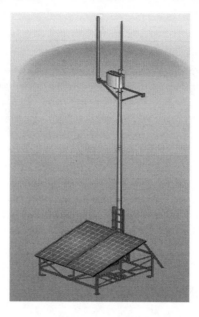

FIGURE 19.3 Example of a solar powered WorldGSM^{TM} village basestation for use in developing countries, courtesy VNL [13].

and thus potential for use of solar and other renewable energy. An interesting observation is that the Vodafone worldwide network currently uses 1,000,000 litres of diesel per day to power geographically remote RBS, predominantly in India! Global diesel RBS consumption is predicted [4] to increase further through to 2020, when it will comprise 13% of the total global RBS energy demand. Figure 19.3 shows an example of a 100W consumption rural Indian solar powered village RBS, designed to alleviate this problem.

19.1.3   Energy Metrics

This fundamental task is to develop metrics to assess the energy efficiency of current and future wireless networks. One major issue is how to trade off the number of base stations, relays, cell sizes and the backhaul locations and bandwidth in order to maximize network overall energy efficiency. Account must be taken of the network data throughput, coverage and the resulting QoS needs to be identified. In addition, the energy consumption of mobile terminals and customer access points/femtocells must be carefully considered in order to avoid simply pushing the burden onto consumers. Results will be obtained initially for some example scenarios to highlight the trade-offs between small and large cells in existing networks.

Our industrial sponsors have recommended the adoption of metrics [6] to quantify the research advances: Energy Consumption Ratio (ECR) is the measurement relating to a single system in Joules per bit and is simply the energy consumed by the system divided by the number of data bits communicated:

$$\text{ECR} = \frac{\text{Energy consumed}}{\text{Data bits communicated}}.$$ (19.1)

The Energy Consumption Gain (ECG), which compares the energy consumed by a proposed system relative to a reference system, where both are communicating the same number of overall data bits:

$$\text{ECG} = \frac{\text{Energy consumed by reference system}}{\text{Energy consumed by system under test}}.$$ (19.2)

Thus an ECG >1 indicates an improvement in efficiency. Directly related to this metric is the Energy Reduction Gain (ERG):

$$\text{ERG} = 1 - \frac{1}{\text{ECG}}.$$ (19.3)

The Earth project uses a similar term Energy Efficiency Gain [14] and Energy Consumption Index [15]. These metrics, such as ECR, are further extended in [4,16] to encompass the coverage of users, in a specific subscriber area and future projections are provided up to 2020, for the likely reductions in ECR and W required to service each user.

19.2   Energy Reduction Techniques for High Traffic Load Scenarios

In the physically larger current wireless cells, often termed macrocells, there is relatively high signal to interference and noise ratio (SINR), close to the RBS. This supports high bit rate modulation schemes such as 64-state Quadrature Amplitude Modulation (64-QAM), which enable a high rate for data transmissions, e.g. up to 5 bits/s/Hz, in the LTE downlink. However, due to signal propagation loss and local cell interference there is much reduced SINR at the more distant cell edge, permitting only relatively unsophisticated Quadrature Phase Shift Keying (QPSK) transmissions, Figure 19.4, which severely limits the achievable cell edge data rates.

Smaller cells, with reduced range to the cell edge, will thus provide the highest possible data rate and an array of such small cells, formed using femtocells or relay devices, may be necessary to secure an overall high network data throughput capability. Although the smaller cells give individually a major improvement in energy efficiency [7], this is offset by the requirement for a 2-D array of such small cells to retain the same overall coverage. Thus the single macrocell and 2-D array of small cells will both give approximately

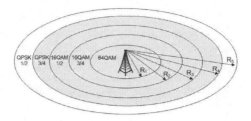

FIGURE 19.4 Shows how, as you approach cell edge where there is reduced SINR, necessitating the adoption of a less spectrally efficient transmission scheme.

the same overall system energy efficiency in transmitted bits per Joule. This observation does not consider the interference between the small cells in the array.

The roll out of femtocells or (HomeNodeB) equipments, now provides ready availability of low power cellular RBS access points for homes or small businesses and picocells can service urban communities, with high traffic demand. These approaches are particularly valuable in servicing known hot spots of traffic, such as offices or shopping complexes. Such femtocell units, which operate in same RF bands as the cellular system, connect to the service network (via DSL or cable) broadband. Informa and the Femto Forum, [17] indicated that 18 operators had launched commercial femtocell services by December 2010, with a total of 30 committed to future deployment. These are predicted to achieve further rapid spectacular future growth [18] with 12M femtocells deployments by 2014 [4]. It is further estimated that, by 2020, there will be even more extensive femtocell deployment but it will consume less than 5% of the total energy in the radio access network.

The aim in deploying femtocells is thus to improve RBS efficiency and achieve high bit/s/Hz/km^2 data transmission throughout the network, combined with low energy consumption. We have modelled a macrocell, where an increasing fraction of the customers are progressively accommodated within femto or pico-cells, and we switch off the unused femtocells, to save energy. This models the relationship between fraction of customers with femtocells and the power consumption per user, for different user densities. The evaluation metric used here is the system power consumption per user; includes CAPEX and OPEX energies. The overall system power consumption in Figure 19.5 [19] decreases with the initial deployment of femtocells. However when the femtocell deployment is greater than 60%, the system energy consumption is found to increase again.

Here system power consumption takes into account not only (OPEX) energy expended in operation but also the embodied (CAPEX) energy used in raw material extraction, transport, manufacture, assembly, installation (for the RBS and femtocell products) including their final disassembly, decon-

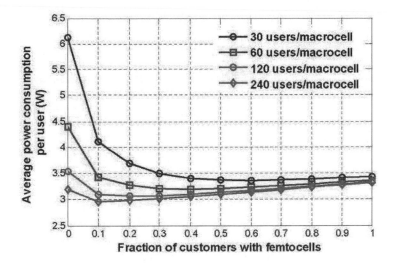

FIGURE 19.5 Power consumption analysis based on the increased penetration of femtocells [19] ©2010 IEEE

struction and decomposition. At the lower user density, of only 30-60 users in total in the macrocell, we achieve an improvement in the power saving. In the Figure 19.5 simulations an overall energy reduction of 3-40% has been achieved.

The macrocell-femtocell two-tiered architecture [20], is attractive for increasing data throughput of users close to the cell edge, and researchers have investigated its QoS performance. The commonly used metrics are system capacity and coverage [21]. During the LTE system design process, system spectrum efficiency was taken as a crucial criterion to evaluate the expected performance [22], thus we have adopted this metric.

An important issue in femtocell deployment is how to control the available transmission power of femtocells to maximize overall efficiency. This is particularly important for the scenario considered for co-channel femtocells [23] which introduce inter-cell interference. We find as more femtocells are deployed, statistically each femtocell consumes less power on average, based on the random distribution of femtocells and the overall range of the average power levels decreases. When deploying the maximum total of 20 femtocells [19] simulated we find that in 80% of these cases the power consumption is reduced to less than 80 mW, which can contribute to a significant overall reduction in consumed energy for the combined network.

It is important to understand the trade-off between energy efficiency and network throughput. We have modeled two scenarios, the first being the variable power femtocell scenario and in the second case each femtocell employs a fixed transmission power level [24]. The higher fixed power level transmissions, does provide an improved spectral efficiency of up to 1 bit/s/Hz compared to the variable power setting. However the fixed transmission power femtocells increase the interference in the macrocell and, the modest gain in spectral efficiency is often offset by the overall increase in energy consumption!

It can thus be concluded from these preliminary results that, to optimize spectral efficiency for femtocells, we should aim to deploy a fixed transmission power allocation strategy and this is particularly effective for small numbers of femtocell deployments. However this will have a relatively high total power requirement and introduce more interference. We find for larger numbers of femtocell deployments, 20 and beyond per macrocell, that a variable power allocation strategy optimizes both the spectral and energy efficiency while, at the same time, minimizing the interference.

The other method of increasing the broadband data throughput, and reducing the cellular networks expenditure, without significantly increasing the overall power drain, is to strategically position relays close to the cell edge [15, 25]. Relays can then forward information between the RBS and cell-edge users to increase the network coverage. An investigation has been undertaken on the relative performance and energy efficiency of data transmission techniques in LTE-Advanced systems [26]. This has focused on replacing the state-of-the-art Hybrid ARQ (HARQ) data transmission protocol with the proposed random network coding (R-NC) in a relay-assisted multi-hop scenario. Such network coding has been shown [27] to improve performance in terms of throughput, reliability, robustness, delay, energy efficiency and security. The R-NC deployment on the end-to-end link between the RBS and the user significantly reduces energy consumption caused by the high HARQ signaling overhead (e.g., the re-transmissions) for users located close to the cell edge. In other words, the deployment of R-NC scheme in a multi-hop scenario with relaying is expected to combine an increased cell edge throughput with improved overall energy efficiency.

The simulation setup is represented by a hexagonal tri-sectored single rural macrocell with a cell radius of 1150 m with twelve outdoor Relay Nodes (RN) deployed close to the cell border at a fixed RBS-RN distance equal to 920 m. The RBS and RN transmit powers are set to 40 W and 7 W respectively. Two communication scenarios are considered between the RBS and the user, a direct or single-hop and a two-hop relay assisted case. Figure 19.6 [24] plots the average energy consumption ratio (ECR), equation (19.1) [6], versus the distance of the mobile UE from the RBS cell centre. The UE is placed on a set of equidistant points along the radial line starting at the RBS location towards the cell edge passing through the central RN in the sector. ECR calculations take into account both the RBS and relay operational power consumption but ignore the small additional power energy loss related to the increased signaling overhead for the RN deployment. This approach assumes that the UE is only attached to one RN and there is no cooperation or coordination between RNs (i.e., only one RN is scheduled per downlink sub-frame). Moreover, RNs are only active when users are present in their coverage areas; otherwise, RNs and their associated signaling overheads are considered as switched off.

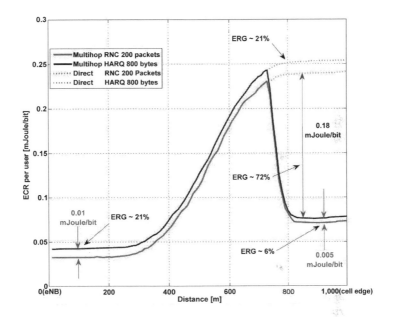

FIGURE 19.6 Plot of Energy Consumption Ratio (ECR) per user across a 1,150 m radius macrocell for direct and multi-hop network transmissions with and without a relay located 230 m from the cell edge [24] ©2011 IEEE

Note in Figure 19.6 how ECR increases rapidly beyond 300 m from the RBS cell centre but it saturates approaching the cell edge. When we replace direct transmission by the relay assisted transmission at 920 m (bolder curves), there is a very significant 72% reduction in required energy for the cell edge transmissions. Note that even though we deploy the relay at 920 m from the RBS, there are significant ECR reductions evident at all distances over 720 m, from the RBS or cell centre. These reductions are more significant than that reported in [28] but, in this latter case, the achieved reductions of up to 50%, are calculated over the entire cell. Also note in Figure 19.6, the energy reduction gain (ERG), equation (19.3) [6], compared to the reference HARQ system, for the deployment of R-NC in preference to the HARQ protocol, particularly for direct transmission. This is especially significant close to the RBS and at the cell edge where a 21% improvement is achieved but a more modest 6% improvement is still evident in the relayed cell edge transmissions.

The signaling overhead will depend on the number of relays simultaneously scheduled in a sub-frame (per cell) and here we have assumed the deployment of low numbers of relays (4 per cell-sector) with only one scheduled relay per sub-frame (i.e., each user is only attached to one relay and there is no coordination or cooperation between relays). Further, relays are only active when users are present in their respective coverage areas. Also note that energy consumption in the core network is very much smaller than in the radio access cellular network, so maximum energy improvements can be expected by optimizing the radio access network.

Energy reduction in other high throughput systems such as those employing multiple antenna transmitters can be obtained by disabling some of the radio antennas as this reduces the required reference signals in the reduced spatial channels. With a smaller control overhead size, the RF energy used for control signaling will potentially decrease. Meanwhile, the operational energy used by each antenna as power amplifier will be reduced significantly if a separate amplifier is used in each radio antenna. In Figure 19.7 [24], we model an operating mode where selection of the optimum MIMO transmission mode is used to minimize the energy consumption.

Similar to results in [14], here we examine a smart MIMO mechanism, to extend the results of [29] from multiple antenna terminals to multiple antenna base stations. We investigate, at the LTE physical layer, switching between different numbers of transmit antennas to improve the overall energy efficiency of the RBS. The LTE standard supports QPSK, 16-QAM and 64-QAM modulation formats and the maximum data rate is achieved (close to the RBS) when 64-QAM is combined with turbo coded and MIMO transmissions. The 64-QAM single antenna transmitter and receiver (1×1 Single Input Single Output (SISO)), 2×2 MIMO and 4×4 MIMO transmission rates then correspond to $R_{max,11} = 5.18$ bit/s/Hz, $R_{max,22} = 10.37$ bit/s/Hz and $R_{max,44} = 20.74$ bit/s/Hz respectively for a single link, to an isolated cell without interference. By averaging over the channel conditions one can subsequently calculate the average or ergodic capacity, C_{NM}, for a specific

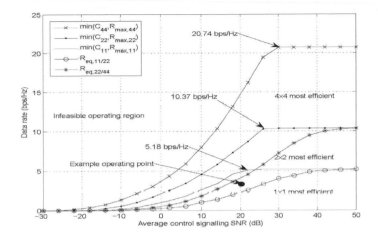

FIGURE 19.7 Plot of achievable data rate versus average control signal SNR for SISO, 2×2 and 4×4 MIMO systems, to assess the optimum energy efficient transmission mode [24] ©2011 IEEE.

$N \times M$ MIMO antenna configuration to determine the feasible region of data rates. In comparison to SISO, 2×2 MIMO can increase the range to the cell edge by 66% for only a <4% energy increase [2].

Control overhead on average, consumes 5-30% of the required resources in each physical radio block in the frame [30]. Under low traffic load, control signaling can possibly be reduced by blanking or disabling the control signal in selected sub-frames to implement a form of sleep mode, see later. Figure 19.7 plots for a 10 MHz LTE system the achievable data rate against the control signaling SNR at the receiver. Increasing the RBS-UE distance will increase the path-loss and generally reduce the control signaling SNR shown in the figure.

The Figure 19.7 plot includes the three graphs of the achievable rate regions for SISO, 2×2 and 4×4 MIMO under unit variance Rayleigh fading channel conditions. Now there is a unique data rate $R_{eq,11/22}$ under which 2×2 MIMO consumes exactly the same amount of energy as the single antenna 1×1 SISO case. With the knowledge of this boundary, if the required data rate $R_d > R_{eq,11/22}$, 2×2 MIMO is preferred, since the energy saving on user data transmission can compensate the extra energy used for signaling the 2×2 MIMO reference signals. In contrast, if $R_d < R_{eq,11/22}$, the single antenna SISO case is more energy-efficient than 2×2 MIMO. Similar methods apply for comparing 2×2 with 4×4 MIMO etc., by considering the $R_{eq,22/44}$ boundary.

Figure 19.7 thus permits one to assess or determine the optimum MIMO configuration. A single antenna transmission consumes less energy for control signaling but the increased spectral efficiency of MIMO makes it more overall energy-efficient for a fixed amount of user data transmissions [31]. The unique boundary data rate curves, R_{eq}, are then plotted to show the energy-efficient transition regions between different numbers of antennas. To evaluate the energy efficiency we require to next super-impose the unique boundary data rate curves, R_{eq}, to model the above transition regions between the different MIMO antenna modes. Thus, if the operating point falls onto the upper left hand side of a $R_{eq,NM/N^\dagger M^\dagger}$ curve, (which defines the boundary where $N \times M$ MIMO and $N^\dagger \times M^\dagger$ MIMO both consume exactly the same energy) then a higher data rate is required than this specific $R_{eq,NM/N^\dagger M^\dagger}$ curve can support, necessitating the deployment of the increased number of MIMO antenna elements. On the other hand, if the operating point falls below, or to the lower right hand side of the $R_{eq,NM/N^\dagger M^\dagger}$ curve, then fewer antenna elements, i.e. reduced complexity MIMO mode may be deployed, to minimise the energy.

Next examine the solid black dot example 3.5 bit/s/Hz operating point in Figure 19.7 at 20 dB SNR. Since this operating point falls below the $R_{max,11}$ value, we can use SISO, 2×2 MIMO or 4×4 MIMO to deliver the required $R_d = 3.5$ bit/s/Hz under an average control signaling SNR = 20 dB channel. Also, as this point is above the corresponding $R_{eq,11/22}$ curve and below the $R_{eq,22/44}$ curve, this implies that 2×2 MIMO is fundamentally more energy-efficient than either SISO or 4×4 MIMO. This demonstrates how the MIMO energy efficiency can be optimized by a detailed assessment of these plots.

Here note that, although the spatial-domain approach is positive from an energy-efficient point of view, noticeable downlink performance degradation may occur. Firstly, decreasing the number of transmit antenna ports may decrease the cell coverage. Secondly, switching from one MIMO mode to another must be handled carefully to avoid incorrect decoding in UE terminals at the point of switching.

A key challenge in the design of radio access networks for cellular mobile systems is how to reduce the energy consumption of the radio-access-network (RAN) without degrading the quality of service provision. In this investigation, which investigates all aspects of RBS energy consumption, a HSPA baseline is compared with a LTE baseline, as well as alternative LTE RAN deployments [32]. The various systems investigated, each employing a 5 MHz band, are:

- **HSPA Baseline:** 3 SISO Sectors with frequency reuse pattern of 1

- **LTE Baseline:** 3 SISO Sectors with frequency reuse pattern of 1

- **LTE Low Green:** 1 SIMO (1×2 Maximal Ratio Combining (MRC)) sector with frequency reuse pattern of 1

- **LTE High Green:** 6 MIMO (2×2 Space Frequency Block Code (SFBC)) sectors with frequency reuse pattern of 3

Figure 19.8 shows in the left hand plot the cell density vs. RAN downlink capacity and offered load in Mbit/s/km^2 and in the right hand plot the cell

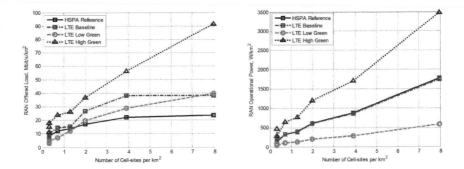

FIGURE 10.8 Cell density vs RAN offered load and cell density vs. RAN power consumption for different deployment solutions

density vs. RAN power consumption in W/km^2 for the above four different deployment solutions The figure models each of the above architecture solutions for different cell-site densities in the same geographic area. The cell-site densities vary with inter-cell-site distances of 300 to 1800 m. The results indicate that, as the number of cell-sites per unit area increases, the offered load the RAN can support increases, but the power consumption, also increases somewhat. For more dense deployment of cells, we achieve lower power consumption per cell (as a picocell consumes less power than a macrocell), however, note that the power consumption does not scale linearly with the increase in cell density.

Figure 19.9 shows in the left hand plot the capacity-power-trade-off between the different architecture solutions by combining the 2 results of Figure 19.8. For each deployment solution, a variety of cell densities are plotted. The figure shows that several solutions can meet the same offered load with different power consumption values. The operational energy reduction is also shown for different solutions. The results at region A show that a dense baseline deployment of HSPA cells can achieve 22 $Mbit/s/km^2$ at a power consumption of 1800 W/km^2. A similar offered load can be satisfied by the baseline LTE solution with fewer cells consuming 500 W/km^2 (region B). This power consumption level can be further reduced by deploying omni-directional cell-sites employing SIMO (Low Green), reducing power consumption to 220 W/km^2 (region C). In order to satisfy higher offered loads of over 60 $Mbit/s/km^2$, 6 sector cell-sites with MIMO can be employed (High Green), but they will consume 1800 W/km^2 (region D).

The key conclusion from this particular investigation is that, compared to a baseline HSPA deployment, approximately 70% energy can be saved by deploying fewer LTE cell-sites, Figure 19.9 right hand part. Furthermore, for the same offered load, a further 55% energy can be saved by replacing the LTE baseline with a low green omni-directional single sector cells with SIMO, i.e. total energy saving compared to the baseline HSPA will be 39%. In order

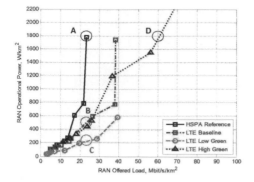 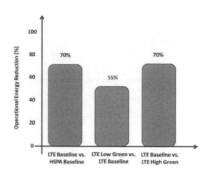

FIGURE 19.9 Capacity-power-trade-off for different architecture solutions.
The left hand graph shows how several distinct solutions can meet the same
offered load for quite different power consumption values. The operational energy reduction is further shown for the different solutions in the right hand
figure.

to increase the system capacity, increasing the number of sectors per cell-site
(from 3 to 6) and employing 2 × 2 SFBC MIMO will increase the system
capacity by 60% but this now causes an increase in energy consumption, as
the LTE baseline consumes only 70% of the high data throughput MIMO
system..

These simulations demonstrate the interesting trade-off between the
achievable capacity and power consumption of a RAN. For a given offered
load, several RAN architectures can meet the load and the trade-off curves
can assist the network designer in identifying the solution that consumes the
least energy. In this modeling, it was found that the radio-head efficiency and
the pathloss model parameters have a large impact on the resultant energy
savings [33].

19.3 Energy Reduction Techniques for Low Traffic Load Scenarios

Currently cellular networks are designed for peak traffic load and hence sleep
mode techniques [34] can reduce energy consumption by switching off either
complete RBS equipments for some of the operators or alternatively parts of an
operator's RBS equipment, whenever this is possible. Past investigations have
indicated potential 8-29% energy savings. There is typical daily 2:1 variation
in cellular data traffic over a 24 hour period, from the lowest load at 7 am to
the highest at 9 pm [35]. They also note major load variations between cells

and further analysis shows that 90% of data traffic is carried in only 40% of the network cells. Thus it is valuable to investigate further the use of sleep modes to achieve better sharing of the scarce spectral resource.

A time domain technique [36] to reduce energy consumption is to switch off the radio transmitter whenever possible. This can lead to significant energy savings as the transmitter power amplifier typically accounts for a significant portion of the overall RBS energy budget. Note also in the earlier Figure 19.5 that selected femtocells were put into sleep mode, when the traffic demand permitted. Further savings may also be possible from other RBS components such as transceiver equipment and baseband digital signal processing (DSPs). The impact on other base station blocks such as cooling systems and power supply circuits may also be reduced through lower levels of heat generation and energy consumption in the base station.

Therefore, in order to optimize the total Radio Frequency (RF) energy saving by time-domain sleep modes, the optimum balance between control and data traffic power consumption must be found. For example, with zero traffic in some cells, we find we can switch off 9 sub-frames per radio frame for these cells. Without sleep mode, the energy consumption would be 600 W. With sleep mode enabled, this reduces to only 262.5 W. Thus our model suggests that the 90% RF ERG contributes to a 56.25% overall energy saving. However these energy savings only apply to low load and become insignificant when the traffic occupies more than 50% of the available bandwidth resource!

Another closely related low load technique reduces the number of active RBS and reallocates mobile subscribers to closely located RBS. This energy saving is achieved by switching off parts of the network at selected frequencies, or reducing the sectorisation. Further if the active RBS is selected to operate in the lower 900 MHz, rather than the higher frequency 1.8 GHz band, then there is also a reduced propagation path loss, requiring less RBS transmission power, with a consequent RF energy saving. Figure 19.10 shows modeling the RF transmission power saving by reallocating links to a lower frequency band, whenever transmission capacity is available in that band, and at low load (i.e. 10-15 high band users), it potentially achieves at "high energy" a considerable saving of 70-80% ERG.

19.4   Other Energy Reduction Techniques

We can also improve efficiency in scenarios such as email transmissions or data download, which can tolerate long delay by adopting store and carry forward relaying techniques. Here we transport (or delay) transmission of the data packets until the transmitter is closer to the RBS to reduce the propagation path loss, required transmitter power level and make the cell overall more efficient. This technique has more limited application to only a few types of

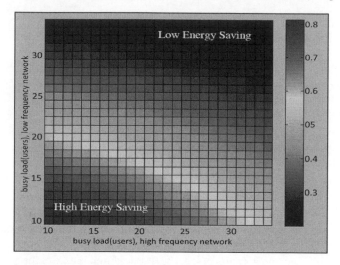

FIGURE 19.10 Energy saving through changing transmission frequencies. RF transmission power reduction by reallocating links to a lower frequency band to facilitate improved propagation, when such capacity is available.

data traffic which are delay insensitive but it does show potential of achieving very significant 80-90% energy reduction.

Techniques across the protocol stack may be employed either to reduce the required radiated power to achieve the required QoS or to achieve the required radiated power using less overall power (noting that the overall power consumption of a device is always greater than the radiated power). Effective radio resource management and signal processing may reduce the radiated power requirement to maintain the required QoS. Power and energy efficient hardware will facilitate an immediate reduction in total energy consumed relative to the radiated power.

Interference reduction is an important factor in securing an energy efficient system. Given a specified radio access technology, a requirement for performance in terms of, e.g., throughput, implies a required ratio of received signal level to received interference level at a receiver. Hence if interference is increased by a given factor, the transmitted power level must be increased by that same factor in order to maintain the required performance, which of course has the effect of reducing energy efficiency for the system. Interference has therefore been carefully addressed in relaying and femtocell deployment.

In all scenarios we must strive to increase component integration, i.e. design more efficient power amplifiers (PAs), produce more power efficient DSP processing or IP cores and minimize the 25% energy loss in the cooling system. With the current move to 4G system deployment with increased use of remote radio head (RRH) masthead electronics we can avoid the 2-3 dB feeder cable losses and better couple the power amplifier to the antenna. Advanced power amplifier techniques, such as Class J design which use a reactive

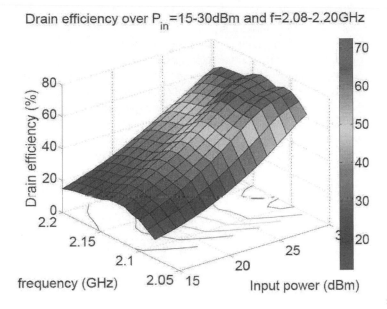

Drain efficiency over P$_{in}$=15-30dBm and f=2.08-2.20GHz

FIGURE 19.11 New Class J Power Amplifier Design which combines the potential of >60% efficiency at maximum output power with a 140 MHz RF bandwidth.

component at the fundamental terminal impedance [37], give Class B like efficiency and achievable output power level. Our current Class J power amplifier design, Figure 19.11 offers 60-70% efficiency, at full output power level, with 140 MHz RF bandwidth at 2.14 GHz nominal operating frequency. Thus by combining such a Class J power amplifier design with techniques, where the supply voltage tracks the RF envelope [38], as pioneered by Nujira in 2008 [12] with recently introduced fast low noise transistors, might offer a considerable overall increase in overall RBS efficiency as the PA is one of the components which contributes significantly to RBS energy consumption.

19.5 Conclusion

This chapter has described the approach being taken in the Mobile VCE green radio project, which started in January 2009, to study novel approaches to reducing the energy consumption of wireless links, particularly in improving the design and operation of RBS. Analysis has shown that when accounting for manufacturing or embodied energy costs, base stations have a much higher operational energy budget than mobile terminals. Proper modeling of the energy consumption of base stations has shown to be an important issue when trying to obtain a clear view of how different radio technologies can reduce energy consumption. A number of methods leading to energy savings for high and low traffic load scenarios have been described and results presented which estimate the performance benefits of these techniques. Our project is being led by industry with the expectation that the most promising research outcomes can feed immediately into future energy efficient wireless standards and products.

RBS overall (RF / AC) efficiency is already increasing year-on-year: from 3% in 2003 to 12% in 2009 and possibly to 25% in 2015? However there is still an urgent need to establish optimum cell size design: macro vs. pico vs. femto taking into account backhaul power requirements, and to ascertain if femtocells or relays or a mixture of these is most energy efficient method of extending the cell range or coverage. MIMO systems do offer higher data rate transmissions, but overall energy consumption must be minimised, to find the optimum transmission scheme.

Our study on the coordination of femtocells and macrocells has been simplified to a user-driven scheme where the users and femtocells are statically distributed and the signaling between femtocells and macrocell has not been considered. In a practical scenario, cooperation must take place between femtocells and macrocell or within femtocells. All femtocells thus require some signaling overhead and this will contribute to an increase in the overall power consumption. The trade-off between reducing RF energy and increasing signaling energy thus needs further consideration.

The move in LTE-Advanced to implement Coordinated Multi-Point (CoMP), transmissions from the individual RBS, to effect better cancellation of interfering signals, particularly at the cell boundaries introduces new issues to consider, such as the energy consumed in achieving coordination and the additional control overhead in operating the femtocells or relays. In all system installations there is also an urgent need to conclude optimization of RBS hardware, increase the integration of the various components and to investigate updates to the network architecture to better facilitate the more efficient delivery of internet based traffic!

Acknowledgments

The work reported here forms part of the Green Radio Core 5 Research Programme of the Virtual Centre of Excellence in Mobile & Personal Communications [12]. This research is funded by the Industrial Companies who are members of Mobile VCE and by the UK EPSRC under Grant Numbers EP/G060584/1, EP/G06041X/1, EP/G062420/1 and EP/G064105/1. The authors express their grateful thanks to industrial leadership team, who steer and review our technical programme. We also thank the academic researchers, from our partner Universities: Bristol; Edinburgh; Kings College London; Southampton and Sheffield, for their research contributions. We specifically thank Vasilis Friderikos of Kings College for Figure 19.10 and Konstantinos Mimis of Bristol for Figure 19.11.

Bibliography

[1] O. Holland, V. Friderikos, and H. Aghvami, "Green Mobile Communications," in *Chapter in Encyclopedia of Wireless and Mobile Communications (2nd Ed.)*. CRC Press, 2011.

[2] W. Vereecken, W. V. Heddeghem, M. Deruyck, B. Puype, B. Lannoo, W. Joseph, D. Colle, L. Martens, and P. Demeester, "Power Consumption in Telecommunication Networks: Overview and Reduction Strategies," *IEEE Communications Magazine*, vol. 49, no. 6, pp. 62–69, June 2011.

[3] J. Baliga, R. Ayre, K. Hinton, and R. Tucker, "Energy Consumption in Wired and Wireless Access Networks," *IEEE Communications Magazine*, vol. 49, no. 6, pp. 70–77, June 2011.

[4] A. J. Fehske, G. Fettweis, J. Malmodin, and G. Biczok, "The Global footprint of Mobile Communications: The Ecological and Economic Perspective," *IEEE Communications Magazine*, vol. 49, no. 8, pp. 55–62, August 2011.

[5] "Vodafone UK Corporate Responsibility 2008." [Online]. Available: http://www.vodafone.com/content/index/about/sustainability.html

[6] C. Han, T. Harrold, I. Krikidis, and et al., "Green Radio: Radio Techniques to Enable Energy Efficient Wireless Networks," *IEEE Communication Magazine*, vol. 49, no. 6, pp. 46–55, June 2011.

[7] Y. Chen, S. Zhang, S. Xu, and G. Li, "Fundamental Trade-offs on Green Wireless Networks," *IEEE Communications Magazine*, vol. 49, no. 6, pp. pp 30–37, June 2011.

[8] L. Collins, "Quart into a pint pot," *IET Engineering and Technology*, vol. 5, no. 13, pp. 60–62, September 2010.

[9] O. Arnold, F. Richter, G. Fettweis, and O. Blume, "Power Consumption Modeling of Different Base Station Types in Heterogeneous Cellular Networks," in *Future Network and Mobile Summit Conference*, 2010.

[10] D. Ferling, T. Bonn, D. Zeller, P. Frenger, I. Godor, Y. Jading, and W. Tomaselli, "Energy Efficiency Approaches for Radio Nodes," in *Future Network and Mobile Summit Conference*, 2010.

[11] M. D. Renzo, L. Alonso, F. Fitzek, and et al., "Greenet - an early stage training network in enabling technologies for green radio," in *2nd Greenet Workshop, colocated with IEEE VTC Spring Conference*, 2011.

[12] Peter Grant, IEEE Globecom 2010 Plenary Talk, http://www.mobilevce.com/frames.htm?core5research.htm.

[13] "**WorldGSM^{TM}** village site," Product Datasheet, Vihaan Networks, Oct. 2010. [Online]. Available: http://www.vnl.in/productsheetsr2/VillageSite_www.pdf

[14] F. Heliot, M. Imran, and R. Tafazolli, "On the Energy Efficiency Gain of MIMO Communication under Various Power Consumption Models," in *Future Network and Mobile Summit Conference*, 2011.

[15] R. Fantini, D. Sabella, and M. Caretti, "Energy efficiency in LTE-Advanced networks with relay nodes," in *73rd IEEE Vehicular Technology Conference Spring*, 2011.

[16] A. Fehske, J. Malmodin, G. Auer, and G. Fettweiss, "Energy Efficiency Metrics for Mobile Communication Networks," in *Future Network and Mobile Summit Conference*, 2011.

[17] Femto Forum White Paper, Femtocell Market Status Report 023, February 2011, http://www.femtoforum.org/femto/pdfs01.php.

[18] Femtocell Forecast 2010, http://www.isuppli.com/Mobile-and-Wireless-Communications/News/Pages/Femtocell-Base-Stations-Poised-for-Spectacular-Growth.aspx.

[19] Y. Hou and D. I. Laurenson, "Energy Efficiency of High QoS Heterogeneous Wireless Communication Network," in *IEEE 72nd Vehicular Technology Conference (VTC 2010-Fall)*, September 2010.

[20] Y. Kim, S. Lee, and D. Hong, "Performance analysis of two-tier femtocell networks with outage constraints," *IEEE Transactions on Wireless Communications*, vol. 9, no. 9, pp. 2695–2700, September 2010.

[21] J. Weitzen and T. Grosch, "Comparing coverage quality for femtocell and macrocell broadband data services," *IEEE Communications Magazine*, vol. 48, no. 1, pp. 40–44, January 2010.

[22] E. Dahlman, S. Parkvall, J. Skold, and P. Beming, *3G Evolution: HSPA and LTE for Mobile Broadband*, 2nd ed. Academic Press, 2008.

[23] P. Pirinen, "Co-channel co-existence study of outdoor macrocell and indoor femtocell users," in *IEEE European Wireless (EW) Conference*, April 2010.

[24] S. McLaughlin, P. Grant, J. Thompson, H. Haas, D. Laurenson, C. Khirallah, Y. Hou, and R. Wang, "Techniques for improving cellular radio base station energy efficiency," *IEEE Wireless Communications Magazine*, vol. 18, no. 5, pp. 10–17, October 2011.

[25] C. Khirallah, J. Thompson, and H. Rashvand, "Energy and Cost Impacts of Relay and Femtocell deployments in LTE-Advanced," *IET Communications, Special Issue on Green Technologies for Wireless Communications and Mobile Computing, to appear,* 2011.

[26] T. Beniero, S. Redana, J. Hamalainen, and B. Raaf, "Effect of Relaying on Coverage in 3GPP LTE-Advanced," in *IEEE 69th Vehicular Technology Conference Spring,* April 2009.

[27] P. Chou, Y. Wu, and K. Jain, "Practical Network Coding," in *Proc. 41st Allerton Conference on Communications, Control and Computing,* October 2003.

[28] Z. Niu, Y. Wu, J. Gong, and Z. Yang, "Cell Zooming for Cost-efficient Green Cellular Networks," *IEEE Communications Magazine,* vol. 48, no. 11, pp. 74–79, November 2010.

[29] B. Bougard, G. Lenoir, A. Dejonghe, L. Perre, F. Catthoor, and W. Dehaene, "Smart MIMO: An energy aware adaptive MIMO-OFDM radio link control for next generation wireless local area networks," *EURASIP Journal on Wireless Communications and Networking,* vol. 2007, no. 3, pp. 1–15, 2007.

[30] P. Mogensen, W. Na, I. Kovacs, and et al., "LTE Capacity Compared to the Shannon Bound," in *IEEE 65th Vehicular Technology Conference,* April 2007.

[31] D. Laselva, F. Capozzi, F. Frederiksen, K. Pedersen, J. Wigard, and I. Kovacs, "On the impact of realistic control channel constraints on QoS provisioning in UTRAN LTE," in *IEEE 70th Vehicular Technology Conference Fall,* September 2009.

[32] W. Guo, C. Turyagyenda, H. Hamdoun, S. Wang, P. Loskot, and T. O'Farrell, "Towards A Low Energy LTE Cellular Network: Architectures," in *EURASIP European Signal Processing Conference (EUSIPCO),* August 2011.

[33] W. Guo and T. O'Farrell, "Green cellular network: Deployment solutions, sensitivity and tradeoffs," in *IEEE International Conference on Wireless Advanced (Wi-Ad),* June 2011.

[34] G. Fischer, "Next-Generation Base Station Radio Frequency Architecture," *Bell Labs Technical Journal,* vol. 12, no. 2, pp. 3–18, 2007.

[35] H. Holma and A. Toskala, *LTE for UMTS - OFDMA and SC-FDMA Based Radio Access.* John Wiley and Sons, 2009.

[36] R. Wang, J. Thompson, and H. Haas, "A Novel Time-domain Sleep Mode Design for Energy Efficient LTE," in *4th International Symposium on Communications, Control and Signal Processing, (ISCCSP),* March 2010.

[37] K. Mimis, K. Morris, and J. McGeehan, "A 2GHz GaN Class-J power amplifier for base station applications," in *IEEE Topical Conference on*

Power Amplifiers for Wireless and Radio Applications (PAWR), January 2011.

[38] V. Mancuso and S. Alouf, "Reducing costs and pollution in cellular networks," *IEEE Communications Magazine*, vol. 49, no. 8, pp. 63–71, August 2011.

Author Contact Information

John S. Thompson and Peter M. Grant are with Institute for Digital Communications, School of Engineering, University of Edinburgh, Kings Buildings, Edinburgh, EH9 3JL, U.K., Email: john.thompson@ed.ac.uk, Peter.Grant@ed.ac.uk. Simon Fletcher is with NEC Telecom MODUS Ltd., Olympus House, Business Park 5, Cleeve Road, Leatherhead, Surrey KT22 7SA, U.K, Email: simon.fletcher@eu.nec.com. Tim O'Farrell is with the Department of Electronic and Electrical Engineering, University of Sheffield, Mappin Street, Sheffield, S1 3JD, U.K, Email: T.OFarrell@sheffield.ac.uk.

20

Energy Conservation of Mobile Terminals in Multi-cell TDMA Networks

Liqun Fu

The Chinese University of Hong Kong, Hong Kong,China

Hongseok Kim

Sogang University, Seoul, Korea

Jianwei Huang

The Chinese University of Hong Kong, Hong Kong, China

Soung Chang Liew

The Chinese University of Hong Kong, Hong Kong, China

Mung Chiang

Princeton University, NJ, USA

CONTENTS

20.1   Introduction

The main objective of green wireless research is to reduce the carbon footprint and energy consumption of information technology (IT) industry. There are more than 4 billion cell-phones in the world [1], and wireless devices and equipments consume 9% of the total energy of IT, i.e., as much as 6.1 TWh/year [2]. Future wireless systems such as 3GPP-LTE or WiMAX2 are evolving to support broadband services that demand a higher capacity than can be provided by today's wireless networks. In most cases, this is achieved at the expense of higher energy consumption and severe impact on the environment.

Cellular networks are not likely to be fully utilized all the time [3]. That is, cellular networks are often designed to support peak traffic load rather than the average traffic load. A substantial amount of bandwidth is reserved for time varying, non-stationary loads, and to facilitate handoffs in cellular system. As a result, cellular networks are often under-utilized. As will be seen, exploiting this fact can reduce energy consumption without compromising the users' perceived quality of service (QoS).

Like other under-utilized network elements such as servers in the data centers or switches/routers in the Internet, one of energy saving techniques in cellular networks is to turn off the under-utilized base stations when necessary, i.e., during the night or for the area where traffic is low. There have been extensive studies on base station energy conservation [4–8]. In this chapter, however, we will focus on reducing the energy consumption of mobile users, which not only addresses the environmental concern, but can also lengthen the battery lifetime of devices and improve users' experiences. In particular, we will consider a time-division-multiple-access (TDMA) cellular network. In each cell, a base station serves a number of users. The transmissions of these users do not overlap in time. However, the transmissions of users in different cells may overlap and interfere with one another. Each user has a certain traffic requirement. We want to answer the following question: how do we schedule the uplink transmissions so as to minimize the total energy consumption while satisfying the traffic requirements of all users?

The gist of the problem is as follows. In the absence of interference, for a transmission, Shannon's capacity formula states that $x = w\log\left(1 + \frac{pG}{\sigma^2}\right)$, where x is the data rate, w is the bandwidth, p is the transmit power, G is the channel gain, and σ^2 is the noise power. Suppose that the transmission is turned on for T seconds within a frame. Then, the number of nats delivered per frame is $b = xT = wT\log\left(1 + \frac{EG}{T\sigma^2}\right)$, where E is the energy consumption per b nats. From this expression, we immediately see a tradeoff between the transmission time T and the energy E when delivering b nats: increasing the transmission time T makes E smaller.

20.1.1 Challenges and Our Solution

The energy conservation problem in multi-cell networks is complicated in two ways:

1. *Intra-cell interaction:* Each TDMA frame has a finite amount of time resource. Within each cell, a longer transmission time of one terminal means a less transmission time for other terminals. Thus, their transmit energies trade off against each other.

2. *Inter-cell interaction:* Across cells, the interference received by a base station depends on simultaneous transmissions in other cells. If simultaneous transmissions can be properly scheduled, mutual interferences can be reduced, which in turn can reduce the total energy consumption. This can be intuitively seen from $b = wT\log\left(1 + \frac{EG}{T(\sigma^2+q)}\right)$, where q is the interference; that is, all things being equal, a smaller E is required if the interference q can be reduced.

Thus, in general the energy conservation and the inter-cell interference are coupled. To minimize the total energy consumption, we need to jointly consider the time fraction allocated to each transmission within each cell and the scheduling of simultaneously transmissions across cells. Besides the transmit energy E, wireless devices also consume circuit energy when they transmit, and "idle" energy when they do not. The relative magnitudes of these energies have a subtle but important effect on the solution to our problem.

Finding an overall optimal solution to the energy minimization problem is non-trivial, as elaborated in Section 20.3. In this chapter, we propose a method that decomposes the overall problem into two sub-problems along the line of 1 and 2 above. That is, we first consider the sub-problem of intra-cell time fraction allocation, assuming interference is constant throughout a frame (this assumption is to a large extent valid according to our simulation experiments – see Section 20.5.2). After the transmission time fractions (and target SINRs) in each cell are fixed, we then consider the transmission scheduling across cells and set the transmit powers of the terminals to fulfill the target SINRs. Based on the solution to the second sub-problem, we then adjust the inter-cell interferences and solve the first sub-problem again. The process is iterated, if necessary, by alternating between these two modules.

The solution found by this decomposition method is guaranteed to be feasible, albeit not necessarily optimal. Simulations indicate that this decomposition method can achieve energy reduction of more than 70% and inter-cell interference power reduction of more than 35% compared with the simplistic scheme of maximum power transmission. We also derive an interesting *decoupling property* under the assumption that the inter-cell interference power stays constant over a TDMA frame: if the idle power consumption of terminals is no less than their circuit power consumption, or when both are negligible, then the energy-optimal transmission rates of the users are *independent* of the inter-cell interference power.

20.1.2   Related Work

The key focus of early research on cellular networks was on interference control instead of energy saving. The network performance maximization problem was often posed as maximizing the system throughput while meeting some required signal to interference-and-noise ratio (SINR) to achieve a target data rate, e.g., for reliable voice connections [9–16].

Energy-efficient transmission schemes were first explored in the context of sensor networks [17–20]. In [20], each sensor node transmits packets as slowly as is allowed by some delay constraint, using the so-called lazy scheduling. Lazy scheduling reduces energy consumption by making use of all available time resource before the deadline, thus make the bursty arrival packets as smooth as possible at the output.

The energy-delay tradeoffs in wireless networks have been explored under various channel models [21,22]. Ref. [21] studied the problem of minimizing the average transmit power with delay constraint under fading channels. Ref. [22] studied the energy-delay tradeoffs under the additive white Gaussian noise (AWGN) channels with bursty traffic. Under the fading channels, [23,24] considered the use of opportunistic scheduling to reduce energy consumption. The key idea is opportunistic transmission, where terminals transmit only when the channel conditions are good enough so that the same traffic requirement can be satisfied with smaller energy consumption.

The key results related to energy-delay tradeoff in sensor networks are built upon the Shannon's capacity formula, which are applicable to general wireless systems. Therefore, the extensive energy-delay tradeoff results in sensor networks can also shed some light on the energy saving in cellular networks. However, the QoS requirements for sensor networks and for cellular networks are quite different. In sensor networks, the system traffic load is usually small, and often there is no strict rate requirement for each user. The transmission requirement is usually characterized by a delay constraint. However, for cellular networks, the system traffics are usually voice/video connections or file transfers, which are much heavier than those in sensor networks. Furthermore, the QoS requirement for voice/video connections is usually characterized by a strict target-rate requirement, which is more stringent than those in sensor networks. Therefore, the energy saving in cellular networks requires different formulations and solution techniques.

Recent research results show that if the circuit power (e.g., the power consumption of the circuit blocks, e.g., mixers, filters, D/A converters) is taken into account, slow transmission is not always energy efficient [25–27]. This is because that although the transmission energy consumption decreases as the transmission time grows, the circuit energy consumption increases as the transmission time grows. Thus, sometimes fast transmissions may be beneficial to energy conservation. Indeed, there is an energy optimal transmission rate when the circuit power consumption is taken into consideration. Ref. [28] studied the problem of minimizing the total energy consumption, including both the transmit and circuit energy consumption, in a multiple-input multiple-output (MIMO) cellular network. It was shown that by swiching the transmission mode between MIMO and SIMO (single-input multiple-output), significant energy-saving can be achieved. Optimizing the total transmission energy including the circuit power was also considered in IEEE 802.16m WiMAX2 [29, 30]. In optimizing the energy efficiency, there are mainly two kinds of metrics. The first one is to minimize the total energy consumption per bit (or per flow) [28, 31, 32], and the other is to maximize the energy utility, which is defined as the total bits that can be delivered per Joule [33–36]. The idea of leveraging spare capacity of TDMA cellular systems to save mobile terminals' total energy consumptions under stochastic traffic loads has been explored in [32]. Similar idea can be easily applied to the frequency division multiple access (FDMA) cellular system. It was shown that, by properly choosing the transmit powers, as well as the instantaneous rates and the time fractions of the users within a cell, average energy consumption per real-time session can be minimized. In addition, it was demonstrated that energy saving ratio is substantial, e.g., more than 50% when the network is under-utilized.

However, most of these works [28, 29, 32] focused on the saving the energy consumption in the single-cell case (interference-free environment). The *multi-cell* case is of much interest because practical deployments of wireless networks contain multiple cells. In a multi-cell network, the inter-cell interference power not only affects the energy consumptions of the users but also affects the perceived QoS by the users in the system. In general, the energy conservation and the inter-cell interference are tightly coupled. A very recent paper [36] studied the energy-efficient power control in OFDMA based multi-cell networks. The authors proposed a distributed non-cooperative game approach to maximize the overall network energy efficiency, which achieves a trade-off between system throughput and energy consumption. However, no QoS guarantees are provided to the users in the system. In this chapter, our focus is to conserve the total energy consumptions in a multi-cell TDMA network, while satisfying the QoS requirement of each user in the system. We find that combining intra-cell time fraction allocation and inter-cell scheduling/power control can potentially be more energy-efficient. Extensive simulations by us verify that combining energy-optimal transmission with inter-cell power control could improve the energy efficiency by 50% compared with the case when only intra-cell energy optimal transmission, as in [32], is performed.

The remainder of this chapter is organized as follows. In Section 20.2, we describe our system model and assumptions. Section 20.3 is devoted to the problem formulation. The proposed energy-efficient policy is provided in Section 20.4. We provide the simulation results in Section 20.5. In Section 20.6, we discuss possible future works, followed by the conclusion in Section 20.7.

20.2 System Model

We consider energy efficient uplink communications in wireless cellular networks. Within each cell, the users send traffic to the same base station (BS) via Time Division Multiple Access (TDMA). The time is divided into fixed length frames. Within a frame, each user is allocated a dedicated time period, during which it is the only uplink transmitter within the cell. There is no interferences among users in the same cell. The concurrent transmissions of different users at different cells, however, lead to inter-cell interferences. We would like to choose the proper time allocations and transmission powers for users in multiple cells, such that the total energy consumption is minimized while satisfying the QoS requirements.

20.2.1 Power Consumption Model

We consider a comprehensive terminal power consumption model, which includes the transmit power, the circuit power, and the idling power [17, 28, 32, 37].

A terminal's transmission rate x depends on the transmit power p according to Shannon's capacity formula:

$$x = w \log \left(1 + \frac{pG}{\sigma^2 + q}\right) \Leftrightarrow p = \left(\exp\left(\frac{x}{w}\right) - 1\right) \frac{\sigma^2 + q}{G}, \qquad (20.1)$$

where w is the bandwidth, G is the channel gain, σ^2 is the noise power, and q is the inter-cell interference. There is drain efficiency of the RF power amplifier at a transmitter, denoted by $\theta \in (0, 1)$, which is defined as the ratio of the output power and the power consumed in the power amplifier. Therefore, given an output power of p, the power consumption at the RF amplifier of a transmitter is p/θ[1].

---

[1] In practical wireless systems, different modulation schemes and forward error correction (FEC) codes may be used. Compared with the Shannon's capacity formula, the impact of adaptive modulation and coding (AMC) schemes results in a constant SINR gap [38]. This constant factor can be absorbed by the parameter θ, which denotes the cumulative effect of the drain efficiency, modulation and FEC.

Besides the transmit power, an active terminal also consumes non-negligible circuit power [17, 28], which is the power of the circuit blocks in the transmission chain, e.g., mixers, filters, local oscillators, and D/A converters. When a transmitter is idle, there is also power consumption due to leakage currents [37]. Therefore, the total power consumption $f(x)$ of a terminal with transmission rate x is given as

$$
f(x) = \begin{cases} \left(\exp\left(\frac{x}{w}\right) - 1\right) \frac{\sigma^2 + q}{\theta G} + \alpha, & \text{if } x > 0 \text{ (active)}, \\ \beta, & \text{if } x = 0 \text{ (idling)}, \end{cases} \tag{20.2}
$$

where α is the circuit power when a terminal is active, and β is the power consumed in idle state. In Section 20.3.3 and Section 20.4, we will show that the circuit power and the idling power have a substantial impact on the time and power solutions of energy efficient transmissions.

Main notations of this chapter are summarized in Table 20.1. We use lower boldface symbols (e.g., **p**) to denote vectors and uppercase boldface symbols (e.g., **B**) to denote matrices. We use calligraphic symbols (e.g., $\mathcal{A}$) to denote sets. The vector inequalities denoted by $\succeq$ and $\preceq$ are component-wise.

20.2.2 Inter-Cell Interference

Consider a system with a set of M cells: $\{C(m), 1 \leq m \leq M\}$. Each cell $C(m)$ contains a set of users (terminals) $\mathcal{A}(m)$. The users within the same cell are allocated different time fractions for uplink transmissions. However, users in different cells may transmit simultaneously and cause interference to each other. As can be seen from (20.1), the transmit power consumption is closely related to the interference power level. Given a fixed transmission rate x, a larger inter-cell interference power q leads to a larger transmit power p. Next we calculate the minimum transmit power vector and the minimum interference power vector that can support the rate requirements of several simultaneous transmissions.

Let $\mathcal{S}$ denote the set of users that are active simultaneously in the multi-cell network at a particular instant. Since TDMA is considered within each cell, the size of set $\mathcal{S}$ is no larger than the number of cells M, i.e., $|\mathcal{S}| \leq M$. Without loss of generality, we only need consider the $|\mathcal{S}|$ cells with active users. Let us define an $|\mathcal{S}| \times |\mathcal{S}|$ nonnegative cross channel gain matrix $\mathbf{G}_{\mathcal{S}} = [g_{mn}]$, with entries as follows:

$$
g_{mn} = \begin{cases} 0, & \text{if } m = n, \\ G_{i(n), C(m)}, & \text{if } m \neq n, \end{cases} \tag{20.3}
$$

where $G_{i(n), C(m)}$ is the channel gain from user $i(n)$ in cell $C(n)$ to the BS of cell $C(m)$. We further define an $|\mathcal{S}| \times |\mathcal{S}|$ nonnegative relative-channel-gain matrix $\mathbf{B}_{\mathcal{S}}$ of set $\mathcal{S}$, which is the cross channel gain matrix $\mathbf{G}_{\mathcal{S}}$ normalized by

TABLE 20.1 Notation Summary [39] © 2011 IEEE

Notation	Physical Meaning
m, n	the indices of cell
i, j	the indices of user
k	the index of concurrent transmission set
$\mathcal{A}$	the set of all users in one cell
$\mathcal{S}$	the concurrent transmission set
M	the number of cells
K	the number of concurrent transmission sets in one frame
λ	the arrival rate of users to the multi-cell network
r	session rate requirement
x	instantaneous transmission rate
p	transmit power
q	inter-cell interference power
σ^2	noise power
γ	target SINR requirement
w	spectral bandwidth
α	circuit power
β	idling power
δ	$\alpha - \beta$
θ	drain efficiency
$G_{i(m)C(m)}$	the channel gain of user $i(m)$ in cell $C(m)$
$G_{i(n)C(m)}$	the cross channel gain of user $i(n)$ in cell $C(n)$ to the base station of cell $C(m)$
$\mathbf{B}$	relative channel gain matrix
φ	Lagrange multiplier

the direct channel gains. The elements in matrix $\mathbf{B}_\mathcal{S} = [b_{mn}]$ are as follows:

$$b_{mn} = \begin{cases} 0, & \text{if } m = n, \\ \dfrac{G_{i(n),C(m)}}{G_{i(m),C(m)}}, & \text{if } m \neq n, \end{cases} \tag{20.4}$$

where $G_{i(m),C(m)}$ is the channel gain from user $i(m)$ in cell $C(m)$ to the BS of cell $C(m)$. Let $\gamma_\mathcal{S} = \left(\gamma_{i(m)} : i(m) \in \mathcal{S} \right)$ denote the target SINR vector of the users in set $\mathcal{S}$. Let $\mathbf{D}\left(\gamma_\mathcal{S} \right)$ be the $|\mathcal{S}| \times |\mathcal{S}|$ diagonal matrix whose diagonal entries are the elements in $\gamma_\mathcal{S}$. The SINR requirements of the users in set $\mathcal{S}$ can be written in matrix form as

$$\left(\mathbf{I} - \mathbf{D}\left(\gamma_\mathcal{S} \right) \mathbf{B}_\mathcal{S} \right) \mathbf{p}_\mathcal{S} \succ \mathbf{D}\left(\gamma_\mathcal{S} \right) \mathbf{v}_\mathcal{S}, \tag{20.5}$$

where $\mathbf{I}$ is an $|\mathcal{S}| \times |\mathcal{S}|$ identity matrix, and vector $\mathbf{v}_\mathcal{S} = \left(\dfrac{\sigma^2}{G_{i(m),C(m)}} : i(m) \in \mathcal{S} \right)^T$ is the noise power vector normalized by the channel gain.

Let $\rho\left(\mathbf{D}\left(\gamma_\mathcal{S} \right) \mathbf{B}_\mathcal{S} \right)$ denote the largest real eigenvalue (also called the Perron-Frobenius eigenvalue or the spectral radius) of matrix $\mathbf{D}\left(\gamma_\mathcal{S} \right) \mathbf{B}_\mathcal{S}$. The following well-known proposition gives the necessary and sufficient condition of checking the feasibility of a target SINR vector $\gamma_\mathcal{S}$ and computing the minimum transmit power solutions that achieves $\gamma_\mathcal{S}$.

Proposition 20.1 ([40–42]). *The necessary and sufficient condition for a target SINR vector $\gamma_\mathcal{S}$ to be feasible is*

$$\rho\left(\mathbf{D}\left(\gamma_\mathcal{S} \right) \mathbf{B}_\mathcal{S} \right) < 1. \tag{20.6}$$

If $\gamma_\mathcal{S}$ is feasible, the component-wise minimum transmit power to achieve $\gamma_\mathcal{S}$ is

$$\mathbf{p}_\mathcal{S}(\gamma_\mathcal{S}) = \left(\mathbf{I} - \mathbf{D}\left(\gamma_\mathcal{S} \right) \mathbf{B}_\mathcal{S} \right)^{-1} \mathbf{D}\left(\gamma_\mathcal{S} \right) \mathbf{v}_\mathcal{S}. \tag{20.7}$$

Proof sketch. By the Perron-Frobenius theorem [40], we know that $\rho\left(\mathbf{D}\left(\gamma_\mathcal{S} \right) \mathbf{B}_\mathcal{S} \right)$ is a positive, simple eigenvalue of matrix $\mathbf{D}\left(\gamma_\mathcal{S} \right) \mathbf{B}_\mathcal{S}$, and its corresponding eigenvector is positive componentwise. From matrix theory, we know that $\rho\left(\mathbf{D}\left(\gamma_\mathcal{S} \right) \mathbf{B}_\mathcal{S} \right) < 1$ is a necessary and sufficient condition for $\left(\mathbf{I} - \mathbf{D}\left(\gamma_\mathcal{S} \right) \mathbf{B}_\mathcal{S} \right)^{-1}$ to exist [41]. Furthermore, [42] shows that (20.7) is a Pareto-optimal solution to (20.5). That is, any transmit power $\mathbf{p}$ that satisfies (20.5) is component-wise no smaller than $\mathbf{p}_\mathcal{S}(\gamma_\mathcal{S})$, i.e., $\mathbf{p} \succeq \mathbf{p}_\mathcal{S}(\gamma_\mathcal{S})$. $\quad \Box$

The total interference and noise power at the BS of cell $C(m)$ is given by

$$q_{C(m)} = \sum_{i(n) \in \mathcal{S}, n \neq m} G_{i(n),C(m)} \cdot p_{i(n)} + \sigma^2,$$

which can be written in matrix form as

$$\mathbf{q} = \mathbf{G}_\mathcal{S} \cdot \mathbf{p} + \eta_\mathcal{S}. \tag{20.8}$$

Proposition 20.2 ([13]). *The interference power vector of set $\mathcal{S}$ correspond-ing to the minimum transmit power solution in (20.7) is given by*

$$\mathbf{q}_{\mathcal{S}}(\boldsymbol{\gamma}_{\mathcal{S}}) = (\mathbf{I} - \mathbf{B}_{\mathcal{S}}\mathbf{D}\,(\boldsymbol{\gamma}_{\mathcal{S}}))^{-1}\,\boldsymbol{\eta}_{\mathcal{S}}, \qquad (20.9)$$

where $\boldsymbol{\eta}_{\mathcal{S}} = \left(\sigma^2, \sigma^2, \cdots, \sigma^2\right)^T$ *is the noise power vector. Each element in* $\mathbf{q}_{\mathcal{S}}(\boldsymbol{\gamma}_{\mathcal{S}})$ *denotes the interference power received by the corresponding base sta-tion. Furthermore,* $\mathbf{q}_{\mathcal{S}}(\boldsymbol{\gamma}_{\mathcal{S}})$ *is the component-wise minimum interference power vector with the target SINR vector* $\boldsymbol{\gamma}_{\mathcal{S}}$. *That is, for any transmit power solution* $\mathbf{p}$ *that achieves an SINR vector no less than* $\boldsymbol{\gamma}_{\mathcal{S}}$, *its corresponding interference power vector* $\mathbf{q}$ *satisfies*

$$\mathbf{q} \succeq \mathbf{q}_{\mathcal{S}}(\boldsymbol{\gamma}_{\mathcal{S}}).$$

Proof. The interference power vector corresponding to the transmit power solution $\mathbf{p}_{\mathcal{S}}(\boldsymbol{\gamma}_{\mathcal{S}})$ in (20.7) is

$$
\begin{aligned}
\mathbf{q}_{\mathcal{S}}(\boldsymbol{\gamma}_{\mathcal{S}}) &= \mathbf{G}_{\mathcal{S}} \cdot \mathbf{p}_{\mathcal{S}}(\boldsymbol{\gamma}_{\mathcal{S}}) + \boldsymbol{\eta}_{\mathcal{S}} \\
&= \mathbf{G}_{\mathcal{S}}\left(\mathbf{I} - \mathbf{D}\,(\boldsymbol{\gamma}_{\mathcal{S}})\,\mathbf{B}_{\mathcal{S}}\right)^{-1}\mathbf{D}\,(\boldsymbol{\gamma}_{\mathcal{S}})\,\mathbf{v}_{\mathcal{S}} + \boldsymbol{\eta}_{\mathcal{S}} \\
&= \mathbf{B}_{\mathcal{S}}\left(\mathbf{I} - \mathbf{D}\,(\boldsymbol{\gamma}_{\mathcal{S}})\,\mathbf{B}_{\mathcal{S}}\right)^{-1}\mathbf{D}\,(\boldsymbol{\gamma}_{\mathcal{S}})\,\boldsymbol{\eta}_{\mathcal{S}} + \boldsymbol{\eta}_{\mathcal{S}} \\
&= \left(\mathbf{B}_{\mathcal{S}}\left(\mathbf{I} - \mathbf{D}\,(\boldsymbol{\gamma}_{\mathcal{S}})\,\mathbf{B}_{\mathcal{S}}\right)^{-1}\mathbf{D}\,(\boldsymbol{\gamma}_{\mathcal{S}}) + \mathbf{I}\right)\boldsymbol{\eta}_{\mathcal{S}} \\
&= (\mathbf{I} - \mathbf{B}_{\mathcal{S}}\mathbf{D}\,(\boldsymbol{\gamma}_{\mathcal{S}}))^{-1}\,\boldsymbol{\eta}_{\mathcal{S}}.
\end{aligned}
$$

For any transmit power solution $\mathbf{p}$ that achieves an SINR vector no less than $\boldsymbol{\gamma}_{\mathcal{S}}$, we have $\mathbf{p} \succeq \mathbf{p}_{\mathcal{S}}(\boldsymbol{\gamma}_{\mathcal{S}})$. Furthermore the cross channel gain matrix $\mathbf{G}_{\mathcal{S}}$ is non-negative. According to (20.8), the interference power vector corre-sponding to $\mathbf{p}$ satisfies $\mathbf{q} \succeq \mathbf{q}_{\mathcal{S}}(\boldsymbol{\gamma}_{\mathcal{S}})$. $\qquad\square$

20.2.3   Dynamic User Sessions

We study a dynamic system with real-time application sessions (e.g., video/voice sessions). Our target is to minimize the average energy consump-tion per session in a stationary system. We assume that the users' arrival to each cell $C(m)$ follows a Poisson process with rate $\lambda_{C(m)}$. Then the arrival rate to all the cells is $\lambda = \sum\limits_{m=1}^{M} \lambda_{C(m)}$. Let J be a random variable denoting the energy consumption per *session* and P be a random variable denoting the total power consumption in the system. The following proposition shows the relation between $\mathrm{E}[P]$ and $\mathrm{E}[J]$ in a stationary system:

Proposition 20.3 (*[32]*). *In a stationary system with user arrival rate* λ, *we have* $\mathrm{E}[P] = \lambda\mathrm{E}[J]$.

According to Proposition 20.3, minimizing the average energy consumption per session is equivalent to minimizing the average power consumption of all the users in the system. Furthermore, there is a special feature for real-time sessions: the connection duration of a real-time session is independent of the allocated transmission rate. For example, allocating a higher transmission rate to a voice session cannot make the phone call end earlier, and the stationary distribution of the number of users in the TDMA system is independent of the transmit powers as long as the rate requirements are satisfied [32]. Therefore, minimizing the energy consumption in a *dynamic* system that supports real-time sessions is equivalent to minimizing the energy consumption with a *static* number of users in the TDMA system[2]. In the rest of this chapter, we will focus on the average power minimization problem in the multi-cell system with a static number of users.

20.3 Problem Formulation and Decoupling Property

In this section, we will show that the energy conservation of mobile users in a multi-cell TDMA network can be formulated as a joint scheduling and power control optimization problem, which is quite challenging to solve in general. We propose a decomposition method to tackle this problem based on one key assumption: the interference power at the base station remains constant within a time frame. This assumption is verified reasonable with simulations results for our problem. Furthermore, we derive an interesting decoupling property: if the idle power consumption of terminals is no less than their circuit power consumption, or when both are negligible, then the energy-optimal transmission rates of the users are independent of the inter-cell interference power.

20.3.1 Power Minimization in Multi-Cell Networks

We assume that the frames are synchronized across all cells in the multi-cell network. Without loss of generality, the frame duration is normalized to be 1. Since different users are active at different times in different cells, we will have different concurrent transmission sets in the multi-cell network. Suppose there are a total K concurrent transmission sets, denoted by $\{\mathcal{S}_k, 1 \leq k \leq K\}$. Each set $\mathcal{S}_k$ is active for a time fraction of t_k ($0 \leq t_k \leq 1$) within a frame. If we consider all possible combinations of simultaneous active users, then K can be as large as $\prod_{m=1}^{M} (|\mathcal{A}(m)| + 1)$. For example, in a multi-cell network

---

[2]This only holds for dynamic systems that support real-time sessions, but does not hold for other non-real-time sessions such as file transfer. For delay-tolerant non-real-time sessions, the stationary distribution of the number of users heavily depends on the rate and power control allocations of the users. For example, allocating a lower transmission rate to a file transfer session will keep the corresponding user staying longer in the system.

with 19 cells with each cell having 9 users, we have $K = 10^{19}$. Let $\mathbf{x}_{\mathcal{S}_k} = \left(x_{i(m)}(k) : i(m) \in \mathcal{S}_k\right)$ denote the instantaneous transmission rate vector of set $\mathcal{S}_k$. According to Shannon's capacity formula, the relation between the instantaneous transmission rate vector $\mathbf{x}_{\mathcal{S}_k}$ and the corresponding SINR vector $\gamma_{\mathcal{S}_k}$ is

$$\mathbf{x}_{\mathcal{S}_k} = w\log\left(1 + \gamma_{\mathcal{S}_k}\right) \Leftrightarrow \gamma_{\mathcal{S}_k} = \exp\left(\frac{\mathbf{x}_{\mathcal{S}_k}}{w}\right) - 1. \qquad (20.10)$$

Substituting (20.10) into (20.7), then the minimal power vector $\mathbf{p}_{\mathcal{S}_k}$ that supports $\mathbf{x}_{\mathcal{S}_k}$ is

$$\mathbf{p}_{\mathcal{S}_k}(\mathbf{x}_{\mathcal{S}_k}) = \left(\mathbf{I} - \mathbf{D}\left(\exp\left(\frac{\mathbf{x}_{\mathcal{S}_k}}{w}\right) - 1\right)\mathbf{B}_{\mathcal{S}}\right)^{-1} \mathbf{D}\left(\exp\left(\frac{\mathbf{x}_{\mathcal{S}_k}}{w}\right) - 1\right)\mathbf{v}_{\mathcal{S}}. \qquad (20.11)$$

Recall that $\mathcal{A}(m)$ is the set of users in cell $C(m)$. For a user $i(m) \in \mathcal{A}(m)$ with real-time sessions, its QoS requirement is measured as its session rate requirement $r_{i(m)}$. We assume that there is call admission control that guarantees that the system load is no larger than the system capacity. This guarantees that the rate requirements of all the users admitted to system can be satisfied. As shown in Section 20.2.3, under Proposition 20.3, given an arrival rate λ to the system, the average energy consumption per session is proportional to the expected power usage of all users at a moment in time in a stationary system. Thus minimizing the average energy per session is equivalent to minimizing the expected power usage of the system in a multi-cell system. To represent this problem mathematically, we define the following binary coefficients for each user $i(m) \in \mathcal{A}(m)$, $1 \leq m \leq M$, and $1 \leq k \leq K$,

$$z_{i(m)}(k) = \begin{cases} 1, & \text{if } i(m) \in \mathcal{S}_k, \\ 0, & \text{if } i(m) \notin \mathcal{S}_k. \end{cases} \qquad (20.12)$$

Problem: average power minimization in a multi-cell network

$$\text{minimize} \quad \sum_{k=1}^{K} t_k \left(\sum_{m=1}^{M} \left(\sum_{i(m) \in \mathcal{A}(m)} ((1 - z_{i(m)}(k))\beta \right.\right.$$
$$\left.\left. + z_{i(m)}(k)\left(\alpha + \frac{p_{i(m)}(k)}{\theta}\right)\right)\right)\right)$$

$$\text{subject to} \quad \sum_{k=1}^{K} t_k = 1, \qquad (20.13)$$

$$\sum_{k=1}^{K} z_{i(m)}(k) \cdot x_{i(m)}(k) \cdot t_k = r_{i(m)}, \forall i(m), \forall m,$$

$$\text{variables} \quad x_{i(m)}(k) \geq 0, \quad \forall k, \forall i(m), \forall m,$$
$$t_k \geq 0, \quad \forall k.$$

The objective function in (20.13) is the total average power consumption of all the users in the system and consists of two parts. The first part is the power consumption when the users are idle. The second part is the power consumption when the users are active in transmissions, where $p_{i(m)}(k)$ is computed according to (20.11) as a function of $\mathbf{x}_{\mathcal{S}_k}$. The first constraint in (20.13) states that the total time allocated to all the concurrent transmission sets equals the frame length, which is normalized to be 1. Here, we treat the case where no user is active in any cell as a special concurrent transmission set of $\mathcal{S}_k = \emptyset$. The second constraint in (20.13) states that each user's session rate requirement is satisfied. The variables in (20.13) are the time fraction variables t_k and the instantaneous rate variables $x_{i(m)}(k)$.

It is challenging to solve Problem (20.13) directly and optimally. First, if we consider all possible combinations of simultaneous active users, then the total number of concurrent transmission sets K increases exponentially with the cell number M. Second, the transmit power $p_{i(m)}(k)$ in the objective function of (20.13) is a complicated function of the instantaneous rate variables $x_{i(m)}(k)$'s. The transmit power is different for each user $i(m)$ and each different concurrent transmission set $\mathcal{S}_k$.

In this chapter, we focus on designing a heuristic algorithm to solve Problem (20.13) based on one key assumption:

Assumption 20.1. *For each cell $C(m)$, we assume the interference experienced by the BS, $q(m)$, remains constant within a time frame.*

Assumption 20.1 is later verified reasonable with the simulation results in Section 20.5.2. With this assumption, the users' transmission schedule in one cell does not affect the transmissions in other cells. Without loss of generality, we will simply assume that the transmission order of the users in each cell is fixed based on the arrival order of the corresponding sessions. We will tackle Problem (20.13) by solving intra-cell average power minimization and inter-cell power control separately.

20.3.2 Intra-Cell Average Power Minimization

Based on Assumption 20.1, the average power minimization problem of a given cell turns out to be a convex optimization problem. Let us consider cell $C(m)$. The session rate requirement of user $i(m) \in \mathcal{A}_m$ is $r_{i(m)}$. If the instantaneous transmission rate of $i(m)$ is $x_{i(m)}$, then the time fraction that user $i(m)$ needs to satisfy its session rate requirement is $t_{i(m)} = \frac{r_{i(m)}}{x_{i(m)}}$. During the time fraction $t_{i(m)}$, the power consumption of the active user $i(m)$ is $\frac{\exp\left(\frac{x_{i(m)}}{w}\right) - 1}{\theta G_{i(m)C(m)}} \left(\sigma^2 + q(m)\right) + \alpha$. All other users in cell $C(m)$ remain in idle state during $t_{i(m)}$. The power consumption of all idle users during the time fraction $t_{i(m)}$ is $(|\mathcal{A}(m)| - 1)\beta$. If $1 - \sum_{i \in \mathcal{A}(m)} \frac{r_{i(m)}}{x_{i(m)}} > 0$, then all users will remain idle during the time fraction of $1 - \sum_{i \in \mathcal{A}(m)} \frac{r_{i(m)}}{x_{i(m)}}$, with the total power consumption of $|\mathcal{A}(m)| \beta$. The intra-cell average power minimization problem can be formulated as follows:

Problem: intra-cell average power minimization:

$$
\text{minimize} \quad \sum_{i(m) \in \mathcal{A}(m)} \frac{r_{i(m)}}{x_{i(m)}} \left(\frac{\exp\left(\frac{x_{i(m)}}{w}\right) - 1}{\theta G_{i(m)C(m)}} \left(\sigma^2 + q(m)\right) + \alpha \right.
$$

$$
\left. + \left(|\mathcal{A}(m)| - 1\right)\beta \right) + \left(1 - \sum_{i \in \mathcal{A}(m)} \frac{r_{i(m)}}{x_{i(m)}}\right) |\mathcal{A}(m)| \beta \quad (20.14)
$$

$$
\text{subject to} \quad \sum_{i(m) \in \mathcal{A}(m)} \frac{r_{i(m)}}{x_{i(m)}} \le 1,
$$

$$
\text{variables} \quad x_{i(m)} \ge 0, \forall i(m) \in \mathcal{A}(m).
$$

The objective in (20.14) is to minimize the total average power consumptions of all users in cell $C(m)$ during the unit time frame. Since we consider uplink transmissions, the base station is the common receiver for all the users in $\mathcal{A}(m)$. Thus, the inter-cell interference power at the base station (i.e., $q(m)$) is the same for every user. The constraint in (20.14) states that the total active time fraction is no larger than the frame length.

Problem (20.14) can be shown to be equivalent to,

$$
\text{minimize} \quad \sum_{i(m) \in \mathcal{A}(m)} \frac{r_{i(m)}}{x_{i(m)}} \left(\frac{\exp\left(\frac{x_{i(m)}}{w}\right) - 1}{\theta G_{i(m)C(m)}} \left(\sigma^2 + q(m)\right) + \alpha - \beta \right)
$$

$$
\text{subject to} \quad \sum_{i(m) \in \mathcal{A}(m)} \frac{r_{i(m)}}{x_{i(m)}} \le 1,
$$

$$
\text{variables} \quad x_{i(m)} \ge 0, \forall i(m) \in \mathcal{A}(m).
$$

$$(20.15)$$

If we change the variable $x_{i(m)}$ to the time fraction variable $t_{i(m)} = \frac{r_{i(m)}}{x_{i(m)}}$, Problem (20.15) is further equivalent to,

$$
\text{minimize} \quad \sum_{i(m) \in \mathcal{A}(m)} t_{i(m)} \left(\frac{\exp\left(\frac{r_{i(m)}}{wt_{i(m)}}\right) - 1}{\theta G_{i(m)C(m)}} \left(\sigma^2 + q(m)\right) + \alpha - \beta \right)
$$

$$
\text{subject to} \quad \sum_{i(m) \in \mathcal{A}(m)} t_{i(m)} \le 1,
$$

$$
\text{variables} \quad t_{i(m)} \ge 0, \forall i(m) \in \mathcal{A}(m).
$$

$$(20.16)$$

The second derivative of the objective function in (20.16) with respect to variable $t_{i(m)}$ is

$$\frac{\left(\sigma^2 + q(m)\right) r_{i(m)}^2}{\theta G_{i(m)C(m)} w^2 t_{i(m)}^3} \exp\left(\frac{r_{i(m)}}{w t_{i(m)}}\right),$$

which is always positive. So the objective function in (20.16) is convex. The constraints in (20.16) are linear constraints. Therefore, Problem (20.16) is a convex optimization problem. The optimal instantaneous rate $x_{i(m)}^*$ (or equivalently the optimal time fraction $t_{i(m)}^*$) of the intra-cell power minimization problem in general depends on the inter-cell interference power $q(m)$. To simplify notation, let $\delta = \alpha - \beta$.

Next we show that the optimal solutions to the intra-cell power minimization problem and the inter-cell interference power can be decoupled if $\delta \leq 0$.

20.3.3 Decoupling Property

If $\delta \leq 0$, the idling power β is no smaller than the circuit power α. Then we have the following theorem. In addition, the theorem is also valid when both the circuit power and the idling power are negligible (i.e., $\beta \approx \alpha \approx 0$).

Theorem 20.1. *If $\delta \leq 0$, the optimal instantaneous transmission rate solutions, the optimal time fractions, and the optimal target SINRs of the intra-cell power minimization problem* (20.15) *(i.e., $x_{i(m)}^*$, $t_{i(m)}^*$, and $\gamma_{i(m)}^*$ for all $i(m) \in \mathcal{A}(m)$) are independent of the inter-cell interference power level, the circuit power, and the idling power.*

Proof. The first order derivative of the objective function in (20.16) with respect to variable $t_{i(m)}$ is

$$\frac{\sigma^2 + q(m)}{\theta G_{i(m)C(m)}} \left(-\frac{r_{i(m)}}{w t_{i(m)}} \exp\left(\frac{r_{i(m)}}{w t_{i(m)}}\right) + \exp\left(\frac{r_{i(m)}}{w t_{i(m)}}\right) - 1\right) + \delta. \quad (20.17)$$

The first part of (20.17) (except δ) is always negative when $0 \leq t_{i(m)} \leq 1$. This can be easily shown if we let $u_{i(m)} = \frac{r_{i(m)}}{w t_{i(m)}}$. The first part of (20.17) then becomes

$$\frac{\sigma^2 + q(m)}{\theta G_{i(m)C(m)}} \left(-u_{i(m)} \exp\left(u_{i(m)}\right) + \exp\left(u_{i(m)}\right) - 1\right). \quad (20.18)$$

The first order derivative of (20.18) with respect to $u_{i(m)}$ is $\frac{\sigma^2 + q(m)}{\theta G_{i(m)C(m)}}$ $\left(-u_{i(m)} \exp\left(u_{i(m)}\right)\right)$, which is negative for any positive $u_{i(m)}$. So (20.18) is a monotonically decreasing function of $u_{i(m)}$. When $u_{i(m)} = 0$, (20.18) equals zero. So (20.18) is negative for any positive $u_{i(m)}$. When $0 \leq t_{i(m)} \leq 1$, we have $u_{i(m)} \geq \frac{r_{i(m)}}{w}$. So the first part of (20.17) is always negative when $0 \leq t_{i(m)} \leq 1$.

Therefore, when $\delta \leq 0$, (20.17) is always negative. So the object function in (20.16) is a monotonically decreasing function of the transmission time fraction $t_{i(m)}$. As a result, the optimal solution to Problem (20.16) is achieved when the inequality constraint is tight, i.e., $\sum_{i(m)\in\mathcal{A}(m)} t_{i(m)} = 1$. In this case, minimizing

$$\sum_{i(m)\in\mathcal{A}(m)} t_{i(m)} \left(\frac{\exp\left(\frac{r_{i(m)}}{w\cdot t_{i(m)}}\right) - 1}{\theta G_{i(m)}C(m)} \left(\sigma^2 + q(m)\right) + \delta \right)$$

is equivalent to minimizing

$$\sum_{i(m)\in\mathcal{A}(m)} t_{i(m)} \left(\frac{\exp\left(\frac{r_{i(m)}}{w\cdot t_{i(m)}}\right) - 1}{\theta G_{i(m)}C(m)} \left(\sigma^2 + q(m)\right) \right).$$

Furthermore, $\sigma^2 + q(m)$ becomes a common scaling factor in the objective function and thus can be removed. Therefore, Problem (20.16) is equivalent to a simplified formulation where $q(m)$ and δ can be removed:

$$\begin{aligned}
\text{minimize} \quad & \sum_{i(m)\in\mathcal{A}(m)} t_{i(m)} \left(\frac{\exp\left(\frac{r_{i(m)}}{w\cdot t_{i(m)}}\right) - 1}{G_{i(m)}C(m)} \right) \\
\text{subject to} \quad & \sum_{i(m)\in\mathcal{A}(m)} t_{i(m)} = 1, \\
\text{variables} \quad & t_{i(m)} \geq 0.
\end{aligned} \tag{20.19}$$

This completes the proof. $\qquad\qquad\square$

The physical meaning of Theorem 20.1 is that if $\delta \leq 0$ (i.e., the idle power consumption is no less than the circuit power consumption), the users in the system will make use of all the time resource for transmissions in order to minimize the system power consumption. When the whole time frame is utilized, the interference power at the base station is a common influence that affects all the users in the cell, which does not affect the time fraction allocation among the users in the system. Theorem 20.1 will be referred to the "decoupling property" for $\delta \leq 0$, which decouples the intra-cell average power optimization from the inter-cell power control.

20.4 The DSP Algorithm

Theorem 20.1 motivates us to propose an algorithm, called Decomposed Scheduling and Power control (DSP), to achieve energy-efficient transmissions in a multi-cell system. Different values of δ will lead to different executions in the algorithm.

20.4.1 DSP Algorithm When $\delta \leq 0$

Because of the decoupling property when $\delta \leq 0$, we will optimize the average power consumption in two separate steps:

- Step 1 (intra-cell average power minimization): Each cell $C(m)$ solves Problem (20.19) to determine the optimal time fraction, the optimal instantaneous rate, and the optimal target SINR of each user in $\mathcal{A}(m)$.

- Step 2 (inter cell power control): Given the optimal target SINRs of the users in each cell, we can get the optimal target SINR vector for the users that are active simultaneously (i.e., in each set $\mathcal{S}_k$). Then we will compute the component-wise minimum power solution that satisfies the target SINR vector.

The flowchart of the DSP algorithm for the case $\delta \leq 0$ is shown in Fig. 20.1.

In Step 1, each cell $C(m)$ solves the convex optimization problem (20.19) using the Lagrangian method. Let φ denote the Lagrangian multiplier of the constraint in (20.19). The Lagrangian function is

$$L\left(\mathbf{t}, \varphi\right) = \sum_{i(m) \in \mathcal{A}(m)} t_{i(m)} \left(\frac{\exp\left(\frac{r_{i(m)}}{w \cdot t_{i(m)}}\right) - 1}{G_{i(m)C(m)}}\right) + \varphi \left(\sum_{i(m) \in \mathcal{A}(m)} t_{i(m)} - 1\right).$$

Since Problem (20.19) is convex, the necessary and sufficient conditions for an optimal solution are the KKT conditions:

$$\nabla_{\mathbf{t}} L\left(\mathbf{t}, \varphi\right) = 0 \quad \text{and} \quad \varphi \left(\sum_{i(m) \in \mathcal{A}(m)} t_{i(m)} - 1\right) = 0.$$

From $\nabla_{\mathbf{t}} L\left(\mathbf{t}, \varphi\right) = 0$, we have

$$\varphi^* = \frac{1}{G_{i(m)C(m)}} \left(\exp\left(\frac{r_{i(m)}}{wt^*_{i(m)}}\right)\left(\frac{r_{i(m)}}{wt^*_{i(m)}} - 1\right) + 1\right), \tag{20.20}$$

where φ^* is the optimal Lagrange multiplier and $t^*_{i(m)}$ is the optimal time fraction solution to (20.19). Given the parameters of $r_{i(m)}$, $G_{i(m)C(m)}$, and w, the optimal Lagrange multiplier φ^* can be computed by the Newton's method, which guarantees superlinear convergence (faster than exponential) [43]. After obtaining φ^*, the optimal time fraction $t^*_{i(m)}$ can be calculated by

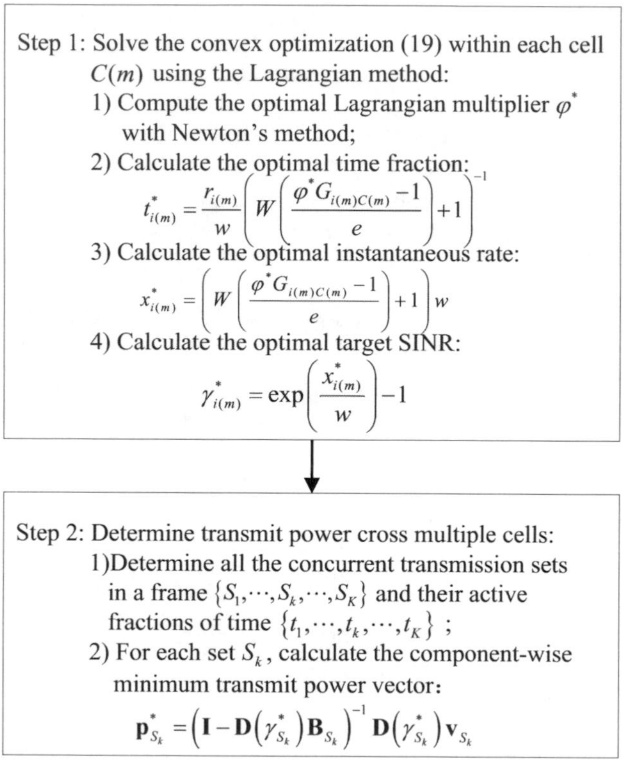

Step 1: Solve the convex optimization (19) within each cell
$C(m)$ using the Lagrangian method:
1) Compute the optimal Lagrangian multiplier φ^*
with Newton's method;
2) Calculate the optimal time fraction:

$$t^*_{i(m)} = \frac{r_{i(m)}}{w}\left(W\left(\frac{\varphi^* G_{i(m)C(m)} - 1}{e} \right) + 1 \right)^{-1}$$

3) Calculate the optimal instantaneous rate:

$$x^*_{i(m)} = \left(W\left(\frac{\varphi^* G_{i(m)C(m)} - 1}{e} \right) + 1 \right) w$$

4) Calculate the optimal target SINR:

$$\gamma^*_{i(m)} = \exp\left(\frac{x^*_{i(m)}}{w} \right) - 1$$

Step 2: Determine transmit power cross multiple cells:
1) Determine all the concurrent transmission sets
in a frame $\{S_1, \cdots, S_k, \cdots, S_K\}$ and their active
fractions of time $\{t_1, \cdots, t_k, \cdots, t_K\}$;
2) For each set S_k, calculate the component-wise
minimum transmit power vector:

$$\mathbf{p}^*_{S_k} = \left(\mathbf{I} - \mathbf{D}\left(\gamma^*_{S_k}\right) \mathbf{B}_{S_k} \right)^{-1} \mathbf{D}\left(\gamma^*_{S_k}\right) \mathbf{v}_{S_k}$$

FIGURE 20.1 Flowchart of the DSP method for the case $\delta \leq 0$ [39] © 2011 IEEE.

solving (20.20). An efficient way to solve (20.20) is to tabulate the Lambert W function [44], which is defined as

$$W(y)\exp\left(W(y)\right) = y.$$

Then $t^*_{i(m)}$ is given by

$$t^*_{i(m)} = \frac{r_{i(m)}}{w}\left(W\left(\frac{\varphi^* G_{i(m)C(m)} - 1}{e}\right) + 1\right)^{-1}. \tag{20.21}$$

The optimal instantaneous rate solution $x^*_{i(m)}$ is:

$$x^*_{i(m)} = \frac{r_{i(m)}}{t^*_{i(m)}} = \left(W\left(\frac{\varphi^* G_{i(m)C(m)} - 1}{e}\right) + 1\right)w. \tag{20.22}$$

Given the instantaneous rate solution $x^*_{i(m)}$, the target SINR $\gamma^*_{i(m)}$ then can be determined by equation (20.10).

In Step 2, optimal power control is performed across multiple cells to determine the optimal transmit powers for the users in each cell. We have obtained the active time fraction $t^*_{i(m)}$, the instantaneous rate $x^*_{i(m)}$, and the target SINR $\gamma^*_{i(m)}$ of each user in each cell. Because the scheduling order in each cell is determined by its arrival order, we can determine all the concurrent transmission sets $\{\mathcal{S}_k, 1 \leq k \leq K\}$ and their active fractions of time $\{t_k, 1 \leq k \leq K\}$ in the frame. According to Proposition 20.1, we can compute the component-wise minimum transmit power solutions of each set $\mathcal{S}_k$ that achieve the target SINR vector $\gamma^*_{\mathcal{S}_k}$ as in (20.7).

20.4.2 DSP Algorithm When $\delta > 0$

When $\delta > 0$, the circuit power is greater than the idling power, which is more likely to happen in practice [17]. The intra-cell power minimization problem for $\delta > 0$ is given in (20.16). The optimal time fraction and the optimal instantaneous rate solution to (20.16) are *dependent* on the inter-cell interference power $q(m)$. This motivates us to use an iterative method to minimize the energy consumption in the multi-cell network. At the beginning of each iteration, we replace $q(m)$ with the average interference power $\hat{q}(m)$ obtained from the previous iteration for every cell $C(m)$. For the first iteration, the estimated interference power $\hat{q}(m)$ is the averaged interference power of the previous frame.

The flowchart of the DSP algorithm for the case of $\delta > 0$ is shown in Fig. 20.2. It involves an iteration between two steps. In Step 1, each cell $C(m)$ solves Problem (20.16) using the Lagrangian method, where $q(m)$ is replaced

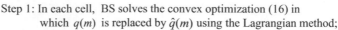

Estimate $\hat{q}(m)$ with the averaged
interference power of the previous frame

Step 1: In each cell, BS solves the convex optimization (16) in
which $q(m)$ is replaced by $\hat{q}(m)$ using the Lagrangian method;
1) Compute the optimal Lagrangian multiplier φ^* with
Newton's method;
2) Calculate the optimal time fraction:

$$t^*_{i(m)} = \frac{r_{i(m)}}{w}\left(W\left(\frac{(\varphi^* + \delta)\theta G_{i(m)C(m)} - (\sigma^2 + \hat{q}(m))}{e(\sigma^2 + \hat{q}(m))}\right) + 1\right)^{-1}$$

2) Calculate the optimal instantaneous rate:

$$x^*_{i(m)} = \left(W\left(\frac{(\varphi^* + \delta)\theta G_{i(m)C(m)} - (\sigma^2 + \hat{q}(m))}{e(\sigma^2 + \hat{q}(m))}\right) + 1\right) w$$

3) Calculate the optimal target SINR:

$$\gamma^*_{i(m)} = \exp\left(\frac{x^*_{i(m)}}{w}\right) - 1$$

Step 2: Determine transmit power cross multiple cells:
1) Determine all the concurrent transmission sets in a frame
$\{S_1, \cdots, S_K\}$, and their active fractions of time $\{t_1, \cdots, t_K\}$;
2) For each set S_k, calculate the component-wise minimum
transmit power vector:

$$\mathbf{p}_{S_k} = \left(\mathbf{I} - \mathbf{D}(\gamma^*_{S_k})\mathbf{B}_{S_k}\right)^{-1}\mathbf{D}(\gamma^*_{S_k})\mathbf{v}_{S_k}$$

3) Calculate the interference power vector

$$\mathbf{q}_{S_k} = \left(\mathbf{I} - \mathbf{B}_{S_k}\mathbf{D}(\gamma^*_{S_k})\right)^{-1}\boldsymbol{\eta}_{S_k}$$

4) Calculate the total power consumption in the current iteration:

$$\sum_{k=1}^{K} t_k \left(\sum_{m=1}^{M}\left(\sum_{i(m)\in A(m)} \left((1 - z_{i(m)}(k))\beta + z_{i(m)}(k)\left(\alpha + \frac{p_{i(m)}(k)}{\theta}\right)\right)\right)\right)$$

5) Update the estimated average interference power:

$$\hat{\mathbf{q}} = \sum_{k=1}^{K} t_k \mathbf{q}_{S_k}$$

Yes

Is the total power consumption
reduced by more than or equal to ε?

No

Terminate

FIGURE 20.2 Flowchart of the DSP method when $\delta > 0$ [39] © 2011 IEEE.

by $\hat{q}(m)$. The Lagrangian function of (20.16) is given by

$$L(\mathbf{t}, \varphi) = \sum_{i(m) \in \mathcal{A}(m)} t_{i(m)} \left(\frac{\exp\left(\frac{r_{i(m)}}{w \cdot t_{i(m)}}\right) - 1}{\theta G_{i(m)C(m)}} \left(\sigma^2 + \hat{q}(m)\right) + \delta \right)$$

$$+ \varphi \left(\sum_{i(m) \in \mathcal{A}(m)} t_{i(m)} - 1 \right).$$

Similarly, we use the KKT conditions to solve formulation (20.16). Compared with (20.20), (20.21), and (20.22), the optimal Lagrange multiplier φ^*, the optimal time fraction $t^*_{i(m)}$, and the optimal instantaneous rate $x^*_{i(m)}$ under the case of $\delta > 0$ are modified to

$$\varphi^* = \frac{\sigma^2 + \hat{q}(m)}{\theta G_{i(m)C(m)}} \left(\exp\left(\frac{r_{i(m)}}{w t^*_{i(m)}}\right) \left(\frac{r_{i(m)}}{w t^*_{i(m)}} - 1\right) + 1 \right) - \delta, \qquad (20.23)$$

$$t^*_{i(m)} = \frac{r_{i(m)}}{w} \left(W \left(\frac{(\varphi^* + \delta) \theta G_{i(m)C(m)} - (\sigma^2 + \hat{q}(m))}{e(\sigma^2 + \hat{q}(m))} \right) + 1 \right)^{-1}, \qquad (20.24)$$

and

$$x^*_{i(m)} = \frac{r_{i(m)}}{t^*_{i(m)}} = \left(W \left(\frac{(\varphi^* + \delta) \theta G_{i(m)C(m)} - (\sigma^2 + \hat{q}(m))}{e(\sigma^2 + \hat{q}(m))} \right) + 1 \right) w. \qquad (20.25)$$

In Step 2, given the active time fraction $t^*_{i(m)}$, the instantaneous rate $x^*_{i(m)}$, and the target SINR $\gamma^*_{i(m)}$ obtained in step 1, the concurrent transmission sets $\{\mathcal{S}_k, 1 \leq k \leq K\}$ and their active fractions of time $\{t_k, 1 \leq k \leq K\}$ are determined. The transmit power vector $\mathbf{p}_{\mathcal{S}_k}$ and the interference power vector $\mathbf{q}_{\mathcal{S}_k}$ for each set $\mathcal{S}_k$ can be determined according to equations (20.7) and (20.9), respectively. The total power consumption in the current iteration is computed by

$$\sum_{k=1}^{K} t_k \left(\sum_{m=1}^{M} \left(\sum_{i(m) \in \mathcal{A}(m)} \left((1 - z_{i(m)}(k))\beta + z_{i(m)}(k) \left(\alpha + \frac{p_{i(m)}(k)}{\theta} \right) \right) \right) \right), \qquad (20.26)$$

where $z_{i(m)}(k)$ (defined in (20.12)) denotes whether user $i(m)$ is active in set $\mathcal{S}_k$, and $p_{i(m)}(k)$ is the mth element in the transmit power vector $\mathbf{p}_{\mathcal{S}_k}$.

We use the averaged interference power vector in the current frame to serve as the estimate interference power in the next iteration, which is given by

$$\hat{\mathbf{q}} = \sum_{k=1}^{K} t_k \mathbf{q}_{\mathcal{S}_k}. \qquad (20.27)$$

The mth element in vector $\hat{\mathbf{q}}$ is the averaged interference power experienced by the BS in cell $C(m)$, $\hat{q}(m)$. Notice that in each iteration of the DSP algorithm, the total power consumption is compared with last iteration, and the next iteration starts if the total power consumption is reduced by more than or equal to a percentage threshold $\varepsilon \in (0,1)$. If the improvement of the total power consumptions is less than ε, the DSP algorithm terminates. The total power consumption is monotonically decreasing and the DSP algorithm is guaranteed to converge in a finite number of iterations [3].

20.5   Simulation results

We carry out extensive simulations to evaluate the performance of the proposed DSP algorithm. We simulate a multi-cell network with a frequency reuse factor of 3, i.e., one of every 3 cells use the same channel. The network topology is shown in Fig. 20.3. There are a total of 7 cells using the same channel, and the radius of each cell is 300 m. The users are uniformly distributed in each cell. For a given number of users, we investigate 100 sets of random user positions and present the averaged results. The session rate requirement of each user is 70 kbps (48.52 knats/second). The bandwidth is 1 MHz. The frame length is normalized to be 1 second. The maximum output power is 27.5 dBm. The drain efficiency is 0.2. The noise power density is -174 dBm/Hz. The power related parameters are cited from [32, 37]. We adopt the distance-based path loss model with a path loss exponent of 4.

20.5.1   Power Consumption Improvement

We evaluate the performance of the DSP algorithm proposed for both the two cases where $\delta \leq 0$ and $\delta > 0$. For $\delta \leq 0$, we only consider the transmit power consumption and neglect the circuit power and the idling power consumption. Then the algorithm in Section 20.4.1 is used. For $\delta > 0$, the idling power and the circuit power are set as 25 mW and 30 mW, respectively, and therefore the algorithm in Section 20.4.2 is used. The improvement threshold ε is set as 0.001%.

We compare the power consumption performances of the following three transmission policies:

1. Maximum power transmission: each user transmits with the same maximum transmit power.

---

[3]The maximum number of iterations is upper bounded by $\log_\varepsilon \left(\frac{P_{\min}}{P_1} \right)$, where P_1 is the total power consumption in the first iteration and $P_{\min}$ is the minimum total power consumption in the system.

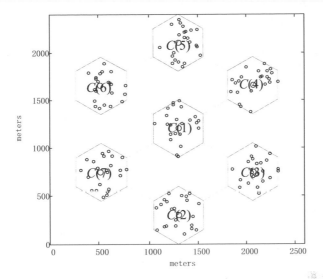

FIGURE 20.3 A multi-cell network with 7 cells operated on the same channel (the frequency reuse factor is 3), and there are 23 users uniformly distributed in each cell. The big circles are the base stations and the small circles are the users. Here we only show the users which transmit on one particular channel [39] © 2011 IEEE.

2. Single-EOT: the Single-cell Energy Optimal Transmission policy proposed in [32] [4].

3. DSP: Decomposed Scheduling and Power control proposed in this chapter.

Figure 20.4 shows the system power consumptions of the above three algorithms as a function of the number of users in each cell when only the transmit power consumption is considered. Figure 20.5 shows the system total power consumptions including the transmit power, the circuit power, and the idling power. As expected, DSP outperforms single-EOT, which in turn outperforms the maximum transmit power policy in both Fig. 20.4 and Fig. 20.5. The system power consumptions of the Single-EOT and DSP algorithms increase more slowly as the number of users increases. Because the connection duration of a real time session is the same among these three algorithms, so

---

[4]Reference [32] considered an isolated single cell network, where the inter-cell interference power is 0. Here we consider multi-cell network extension. In order to make sure the target transmission rate can be achieved when the actual interference power is unknown, we assume the worst case inter-cell interference power. In this case, the BS assumes that the users in the adjacent cells use maximum transmit power, and the worst case interference distance is twice of the cell radius.

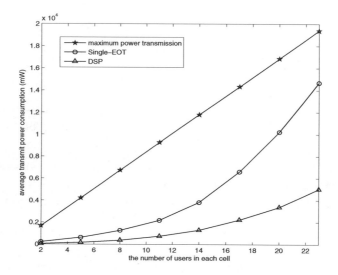

FIGURE 20.4 Transmit power consumptions, $\delta = 0$ and the algorithm in Section 20.4.1 is used. The number of users in each cell ranges from 2 to 23 [39] © 2011 IEEE.

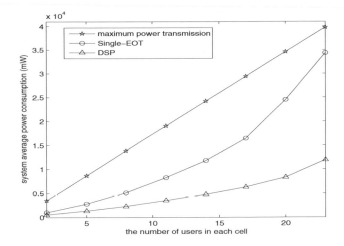

FIGURE 20.5 System total power consumptions, $\delta > 0$ and the algorithm in Section 20.4.2 is used. The number of users in each cell ranges from 2 to 23 [39] © 2011 IEEE.

the system power reduction ratio is equivalent to the system energy reduction ratio. For all simulation settings (i.e., the number of users per cell ranges from 2 to 23), compared with the maximum transmit power policy, DSP achieves a power/energy reduction of more than 74% and 70% in Fig. 20.4 and Fig. 20.5, respectively. The energy saving benefits become more significant when only the transmit power consumption is considered.

In single-EOT, the BS trades off energy consumption and transmission time from a single cell's perspective. However, since BSs of different cells do not cooperate in single-EOT, the power saving is still limited due to conservative estimation of the inter-cell interferences. The DSP algorithm combines the intra-cell average power minimization with inter-cell power control. As a result, the system power/energy consumption reduction ratio can be further improved compared with the Single-EOT algorithm: for all the simulated numbers of users per cell, DSP algorithm achieves a further system power/energy reduction of more than 65% and 50% in Fig. 20.4 and Fig. 20.5, respectively.

20.5.2   Reduction of the Inter-Cell Interference Power Level

We next investigate the interference power levels of the DSP algorithm when $\delta > 0$. Specifically, we focus on the interference power at the base station of the central cell $C(1)$ in the network topology in Fig. 20.3. Figure 20.6 shows the average interference power as a function of the number of users in each cell. It is clear that DSP outperforms single-EOT, which in turn outperforms the

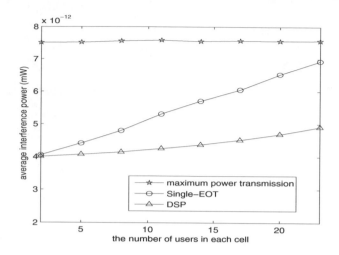

FIGURE 20.6 Averaged interference power at the base station of the central cell $C(1)$, where the number of users in each cell ranges from 2 to 23 [39] © 2011 IEEE.

maximum transmit power policy. The maximum transmit power policy not only consumes a large system power consumption but also generates a large interference power at the base station. Compared with the maximum transmit power policy, DSP achieves an interference power reduction of more than 35% for all the simulated number of users per cell. DSP leads to a "win-win" situation: it reduces both the transmit power and the inter-cell interference. Furthermore, we find that there is a tradeoff between the interference power levels and the system work load in both the Single-EOT and DSP algorithms. The interference power levels of the Single-EOT and DSP increase as the number of users increases. When each cell has a small number of users, each user has more time to transmit and thus the inter-cell interference powers can be reduced significantly. However, in the maximum transmit power policy, the interference power levels are similar as the number of users changes.

We further investigate how the interference power changes over time. Figure 20.7 exhibits the interference power levels of a sample random network with 23 users uniformly distributed in each cell under the maximum transmit power policy and the DSP algorithm. The x-axis represents the time within a single frame. The y-axis is the interference power at the base station of cell $C(1)$. Figure 20.7 shows that the interference power at the base station fluctuates a lot in the maximum transmit power policy; however the interference power remains roughly constant within a time frame in the DSP algorithm.

Specifically, to measure the fluctuation of the interference power, we examine the coefficient of variation. Given the interference power vector that

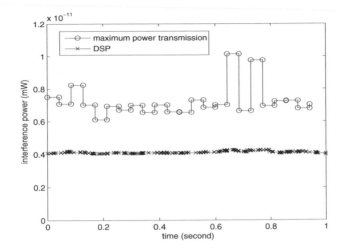

FIGURE 20.7 The fluctuation of the interference power at the base station of the central cell $C(1)$ within one time frame [39] © 2011 IEEE.

contains all the interference powers at the base station of $C(1)$ within a time duration of one frame, the coefficient of variation is defined by the ratio between the standard deviation and the mean of the interference. A large coefficient of variation indicates a large fluctuation of the interference power within the frame. Quantitatively, the coefficient of variation of the interference power in Fig. 20.7 under the maximum transmit power policy is 0.1316. Under the DSP algorithm, the coefficient of variation is reduced to 0.0146. This is because under the maximum transmit power policy, each user in the adjacent cells $(C(2)$ to $C(7))$ uses the same transmit power. The interference power at base station of $C(1)$ heavily depends on the locations of the active users in cells $C(2)$ to $C(7)$. If a user is at the cell boundary that is close to the base station of $C(1)$, it will generate a large interference. In the DSP algorithm, after doing single-cell optimization, the user at the cell boundary is allocated a larger fraction of time resource so that its instantaneous transmission rate requirement can be reduced. Therefore, the transmit power of the cell-boundary user can be reduced, which causes less interference to the base station of cell $C(1)$.

Table 20.2 shows the averaged coefficient of variation when the number of users in each cell changes ranges from 2 to 23. We find that for all the simulated numbers of users per cell, the averaged coefficients of variation of the DSP algorithm is very small, i.e., the interference power fluctuates very little. These results verify our constant interference assumption, which was the basis for the decomposition method proposed in this chapter. The DSP algorithm has the effect of smoothing out the interference power received at

TABLE 20.2 The Averaged Coefficient of Variation of the Interference Power at the Base Station of Cell $C(1)$ [39] © 2011 IEEE

the number of users in each cell	2	5	8	11	14	17	20	23
maximum power transmission	0.107	0.111	0.120	0.130	0.133	0.125	0.128	0.125
DSP algorithm	0.001	0.002	0.004	0.006	0.008	0.010	0.013	0.016

the base stations. This observation further indicates that the scheduling order of the users in each cell is not important in the DSP algorithm. Our DSP algorithm can alleviate the combinatorial part in formulation (20.13), which is the most challenging part in solving the joint power control, rate control, and scheduling problem.

20.5.3   Convergence Performance

When $\delta > 0$, the DSP algorithm involves iterations between two alternative steps. The total power consumption is reduced in each iteration. The DSP algorithm terminates if the improvement in the current iteration is less than a percentage threshold. Figure 20.8 shows the number of iterations that the DSP algorithm needs for convergence. For each given number of links, we investigate 200 random networks and present both the maximum numbers and the average numbers of iterations of the DSP algorithm. We find that for all the simulated networks with different number of users per cell, the average numbers of iterations for DSP to converge are around 3. The maximum number of iterations of the DSP algorithm is no larger than 8. In Section 20.4.2, we show that the DSP algorithm is guaranteed to converge. Figure 20.8 further indicates that the DSP algorithm converges very fast.

20.6   Potential Research Directions

In this section, we will discuss two possible future research directions: the first one is the energy conservation in cellular networks that support *mobility*; the second one is the energy conservation problem in the wireless networks that support *non-real-time* applications.

When users are moving, their channels are often fast time-varying. The power solutions of the DSP algorithm may not satisfy the users' target SINR requirements, since the channel gains may have been changed before the algorithm converges. One possible solution is to set an SINR margin to combat the negative impact of mobility [45], i.e., increase the target SINR by a certain

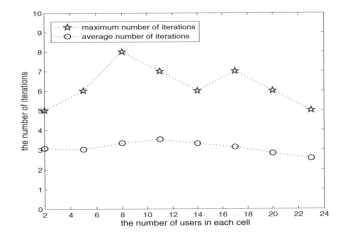

FIGURE 20.8 The maximum and average numbers of iterations for the DSP algorithm ($\delta > 0$) to converge [39] © 2011 IEEE.

amount. As a result, although the channel gains may have been changed, the users transmission rate requirements can still be satisfied if the SINR margin is sufficiently large. For example, a margin of 3 dB is reserved for up-link transmissions in *mobile* WiMAX assuming a frequency reuse factor of 3 [46]. It is clear that there is a trade-off between the SINR margin and the energy efficiency: a small SINR margin may not guarantee the mobile users' QoS requirements; a large SINR margin may lead to unnecessary waste of energy consumptions. Furthermore, the optimization of the SINR margin is affected by several other factors, e.g., the moving speeds of the mobile users, the frequency reuse factor, and the frame length. The energy-efficient transmission in mobile multi-cell networks while providing QoS guarantees is an interesting topic for further study.

In this chapter, we focus on the cellular networks that support *real-time* application sessions (video/ voice sessions). The extension to the dynamic systems that support *non-real-time* sessions (e.g., file transfers) is an interesting yet challenging topic. The real-time sessions have a special feature: the connection duration of a real-time session is independent of the allocated network resources as long as its target rate requirement is satisfied; otherwise, the session may be dropped. However, *non-real-time* sessions are delay-tolerant. The holding time of a non-real-time session depends on the rate and power allocation policy. For example, allocating a lower transmission rate to a file transfer session will keep the corresponding user staying longer in the system. The stationary distribution of the number of users depends heavily on the rate and power control allocations. In addition, the QoS metric of non-real-time sessions is less stringent than the real-time sessions, and thus the system constraints are different. Therefore, the energy-conservation problem for non-real-time sessions requires different formulation and solution techniques.

20.7   Conclusion

In this chapter, we study the problem of energy conservation of terminals in a multi-cell TDMA network supporting bursty real-time sessions. The associated optimization problem involves joint scheduling, rate control, and power control.

We propose a method that decomposes the overall problem into two sub-problems: intra-cell energy optimization and inter-cell power control. This decomposition method is guaranteed to find a feasible solution, albeit not an optimal one. The decomposition is motivated and made simple by the following observations:

1. The original optimization problem is too complicated to solve directly online.

2. In cellular networks, the cells using the same frequency band are usually geographically separated by a distance. Interference is a strong function of distance when the distance is small, but a weak function of distance when the distance is large. Furthermore, after doing intra-cell averaged power minimization, the base station trades off energy consumption with transmission time. This will reduce the interference power generated by the cell-boundary users. Thus, we could make the approximation that the interference is constant when we make intra-cell time fraction allocations to the users within a cell.

3. If the idle power is no less than the circuit power, or both are negligible, then there is a "decoupling property": the energy-optimal time allocations to individual users within each cell are independent of the inter-cell interference (under the assumption that the interference stays constant throughout a frame).

4. If the idle power is less than the circuit power, the sub-problems are coupled. We then need to iteratively solve the two sub-problems until convergence.

Acknowledgments

This work is supported by AoE grant E-02/08 from the UGC of the Hong Kong SAR, China, the General Research Funds (Project Number 412511, 412710, and 414911) established under the University Grant Committee of the Hong Kong Special Administrative Region, China, the Sogang University Research Grant of 2011, US NSF CNS-1011962, a Princeton Grand Challenge grant, and a Google grant. The key results of this chapter are drawn from [39].

Bibliography

[1] "Nice talking to you ... mobile phone use passes milestone," March 2009. [Online]. Available: http://www.guardian.co.uk/technology/2009/mar/03/mobile-phones1

[2] J. Mogul, "Improving energy efficiency for networked applications," in *Architectures for Networking and Communications Systems (ANCS) Keynote Speech*, 2007.

[3] B. Rengarajan and G. de Veciana, "Architecture and abstractions for environment and traffic aware system-level coordination of wireless networks: the downlink case," in *Proc. IEEE INFOCOM*, 2008.

[4] E. Oh, B. Krishnamachari, X. Liu, and Z. Niu, "Towards dynamic energy-efficient operation of cellular network infrastructure," pp. 56–61, Jun. 2011.

[5] C. Peng, S.-B. Lee, S. Lu, H. Luo, and H. Li, "Traffic-driven power saving in operational 3G cellular networks," in *Proc. ACM MOBICOM*, Las Vegas, NV, USA, Sep. 2011.

[6] K. Son, H. Kim, Y. Yi, and B. Krishnamachari, "Base station operation and user association mechanisms for energy-delay tradeoffs in green cellular networks," *IEEE Jour. Select. Areas in Commun.*, vol. 29, no. 8, pp. 1525–1536, Aug. 2011.

[7] K. Dufkova, M. Popovic, R. Khalili, J.-Y. L. Boudec, M. Bjelica, and L. Kencl, "Energy consumption comparison between macro-micro and public femto deployment in a plausible LTE network," in *Proc. e-Energy 2011: 2nd International Conference on Energy-Efficient Computing and Networking*, New York, NY, USA, Jun. 2011.

[8] K. Son, E. Oh, and B. Krishnamachari, "Energy-aware hierarchical cell configuration: from deployment to operation," in *Proc. IEEE INFOCOM 2011 Workshop on Green Communications and Networking*, Apr. 2011.

[9] G. J. Foschini and Z. Miljanic, "A simple distributed autonomous power control algorithm and its convergence," *IEEE Trans. Veh. Technol.*, vol. 42, no. 4, pp. 641–646, Nov. 1993.

[10] R. Yates, "A framework for uplink power control in cellular radio systems," *IEEE Jour. Select. Areas in Commun.*, vol. 13, pp. 1341–1347, 1995.

[11] D. Goodman and N. Mandayam, "Power control for wireless data," *IEEE Personal Commun. Mag.*, vol. 7, no. 2, pp. 48–54, Apr. 2000.

[12] D. Gesbert, S. G. Kiani, A. Gjendemsjo, and G. E. Oien, "Adaptation, coordination, and distributed resource allocation in interference-limited wireless networks," *Proc. IEEE*, vol. 95, no. 12, pp. 2393–2409, Dec. 2007.

[13] M. Chiang, P. Hande, T. Lan, and C. W. Tan, *Power Control in Wireless Cellular Networks, Foundation and Trends in Networking*, July 2008.

[14] M. Johansson and L. Xiao, "Cross-layer optimization of wireless networks using nonlinear column generation," *IEEE Trans. Wireless Commun.*, vol. 5, no. 2, pp. 435–445, Feb. 2006.

[15] A. Gjendemsjo, D. Gesbert, G. E. Oien, and S. G. Kiani, "Binary power control for sum rate maximization over multiple interfering links," *IEEE Trans. Wireless Commun.*, vol. 7, pp. 3164–3173, Aug. 2008.

[16] S. Singh, N. B. Mehta, A. F. Molisch, and A. Mukopadhyay, "Moment-matched lognormal modeling of uplink interference with power control and cell selection," *IEEE Trans. Wireless Commun.*, vol. 9, no. 3, pp. 932–938, Mar. 2010.

[17] S. Cui, A. J. Goldsmith, and A. Bahai, "Energy-constrained modulation optimization," *IEEE Trans. Wireless Commun.*, vol. 4, pp. 2349–2360, Sep. 2005.

[18] S. Cui, R. Madan, A. J. Goldsmith, and S. Lall, "Cross-layer energy and delay optimization in small-scale sensor networks," *IEEE Trans. Wireless Commun.*, vol. 6, pp. 3688–3699, Oct. 2007.

[19] U. C. Kozat, I. Koutsopoulos, and L. Tassiulas, "Cross-layer design for power efficiency and QoS provisioning in multi-hop wireless networks," *IEEE Trans. Wireless Commun.*, vol. 5, no. 11, pp. 3306–3315, Nov. 2006.

[20] B. Prabhakar, E. Uysal Biyikoglu, and A. El Gamal, "Energy-efficient transmission over a wireless link via lazy packet scheduling," in *Proc. IEEE INFOCOM*, vol. 1, 2001, pp. 386–394.

[21] R. A. Berry and R. G. Gallager, "Communication over fading channels with delay constraint," *IEEE Trans. Inform. Theory*, vol. 48, no. 5, pp. 1135–1149, May 2002.

[22] D. Rajan, A. Sabharwal, and B. Aazhang, "Delay-bounded packet scheduling of bursty traffic over wireless channels," *IEEE Trans. Inform. Theory*, vol. 50, no. 1, pp. 125–144, Jan. 2004.

[23] H. Wang and N. B. Mandayam, "Opportunistic file transfer over a fading channel under energy and delay constraints," *IEEE Trans. Commun.*, vol. 53, no. 4, pp. 632–644, Apr. 2005.

[24] K.-K. Leung and C. W. Sung, "An opportunistic power control algorithm for cellular network," *IEEE/ACM Trans. Networking*, vol. 14, no. 3, pp. 470–478, Jun. 2006.

[25] S. Pollin, R. Mangharam, B. Bougard, L. V. der Perre, I. Moerman, R. Rajkumar, and F. Catthoor, "MEERA: Cross-layer methodology for energy efficient resource allocation in wireless networks," *IEEE Trans. Wireless Commun.*, vol. 6, no. 2, pp. 617–628, Feb. 2007.

[26] S. Cui, A. J. Goldsmith, and A. Bahai, "Energy-efficiency of MIMO and cooperative MIMO techniques in sensor networks," *IEEE Jour. Select. Areas in Commun.*, vol. 22, no. 6, pp. 1089–1098, Aug. 2004.

[27] R. Madan, S. Cui, S. Lall, and A. J. Goldsmith, "Modeling and optimization of transmission schemes in energy-constrained wireless sensor networks," *IEEE/ACM Trans. Networking*, vol. 15, no. 6, pp. 1359–1372, 2007.

[28] H. Kim, C.-B. Chae, G. de Veciana, and R. W. H. Jr., "A cross-layer approach to energy efficiency for adaptive MIMO systems exploiting spare capacity," *IEEE Trans. Wireless Commun.*, vol. 8, no. 8, pp. 4264–4275, Aug. 2009.

[29] N. H. et al., "Improving client energy consumption in 802.16m," *C80216m-09_0107r2, IEEE 802.16m WiMAX2*, Jan. 2009.

[30] H. Kim, X. Yang, and M. Venkatachalam, "Dual access points association in relay networks to conserve mobile terminalsenergy," *under revision for IET Communications, Special Issue on Green Technologies for Wireless Communications and Mobile Computing*, 2011.

[31] M. J. Neely, "Energy optimal control for time varying wireless networks, journal = IEEE Trans. Inform. Theory, year = 2006, volume = 52, number = 7, pages = 1-18, month = Jul."

[32] H. Kim and G. de Veciana, "Leveraging dynamic spare capacity in wireless systems to conserve mobile terminals' energy," *IEEE/ACM Trans. Networking*, vol. 18, no. 3, pp. 802–815, June 2010.

[33] F. Meshkati, H. V. Poor, S. C. Schwartz, and N. B. Mandayam, "An energy-efficient approach to power control and receiver design in wireless data networks," *IEEE Trans. Commun.*, vol. 53, no. 11, pp. 1885–1894, Nov. 2005.

[34] F. Meshkati, H. V. Poor, and S. C. Schwartz, "Energy-efficient resource allocation in wireless networks," *IEEE Sig. Proc. Mag.*, vol. 23, pp. 58–68, May 2007.

[35] G. Miao, N. Himayat, Y. Li, and A. Swami, "Cross-layer optimization for energy-efficient wireless communications: A survey," *Wiley Journal Wireless Commun. and Mobile Computing*, vol. 9, Apr. 2009.

[36] G. Miao, N. Himayat, G. Y. Li, and S. Talwar, "Distributed interference-aware energy-efficient power optimization," *IEEE Trans. Wireless Commun.*, vol. 10, no. 4, pp. 1323–1333, Apr. 2011.

[37] *WiMAX power amplifier ADL5570 and 5571*, "*http://www.analog.com/uploadedfiless/data_sheets/adl5570.pdf*, *http://www.analog.com/uploadedfiless/data_sheets/adl5571.pdf*", Std. Analog Device, Sep. 2007.

[38] A. J. Goldsmith and P. P. Varaiya, "Capacity of fading channels with channel side information," *IEEE Trans. Inform. Theory*, vol. 43, no. 6, pp. 1986–1992, Nov. 1997.

[39] L. Fu, H. Kim, J. Huang, S. C. Liew, and M. Chiang, "Energy conservation and interference mitigation: From decoupling property to win-win strategy," *IEEE Trans. Wireless Commun.*, vol. 10, no. 11, pp. 3943–3955, Nov. 2011.

[40] R. A. Horn and C. R. Johnson, *Matrix Analysis*. New York: Cambridge Univ. Press, 1991.

[41] N. Bambos, C. Chen, and G. Pottie, "Channel access algorithms with active link protection for wireless communication networks with power control," *IEEE/ACM Trans. Networking*, vol. 8, no. 5, pp. 583–597, Oct. 2000.

[42] D. Mitra, "An asynchronous distributed algorithm for power control in cellular radio systems," in *Proc. 4th WINLAB Workshop*, Rutgers University, New Brunswick, NJ, 1993.

[43] S. Boyd and L. Vandenberghe, *Convex Optimization*. Cambridge University Press, 2004.

[44] R. M. Corless, G. H. Gonnet, D. E. G. Hare, D. J. Jeffrey, and D. E. Knuth, "On the lambert W function," *Advances in Computational Mathematics*, vol. 5, pp. 329–359, 1996.

[45] C. W. Tan, D. P. Palomar, and M. Chiang, "Energy-robustness tradeoff in cellular network power control," *IEEE/ACM Trans. Networking*, vol. 17, no. 3, pp. 912–925, June 2009.

[46] "Mobile WiMAX part I: A technical overview and performance evaluation," Aug. 2006. [Online]. Available: http://www.wimaxforum.org/technology/downloads/

Author Contact Information

Liqun Fu is with The Institute of Network Coding, The Chinese University of Hong Kong, Shatin, New Territories, Hong Kong, China, Email: lqfu@inc.cuhk.edu.hk. Hongseok Kim is with The Department of Electronic Engineering, Sogang University, Seoul, Korea, Email: hongseok@sogang.ac.kr. Jianwei Huang and Soung Chang Liew are with The Department of Information Engineering, The Chinese University of Hong Kong, Shatin, New Territories, Hong Kong, China, Email: jwhuang@ie.cuhk.edu.hk and soung@ie.cuhk.edu.hk. Mung Chiang is with The Department of Electrical Engineering, Princeton University, NJ 08544, USA, Email: chiangm@princeton.edu.

21

Energy Efficiency for Wireless Relay Systems

Jinho Choi

Swansea University, UK

Duc To

Aeroflex Ltd., UK

Weixi Xing

Swansea University, UK

Ye Wu

Huawei Technologies Co. Ltd., China

Shugong Xu

Huawei Technologies Co. Ltd., China

CONTENTS

21.1   Introduction

Climate change has been an important international agenda these days and will be more important in the future. A number of climate groups have been working on to reduce CO_2 emissions. A climate group (called GeSI) has identified that information and communications technology (ICT) sectors can play a key role in reducing CO_2 emissions by 15% in Smart 2020 report [1]. To reduce CO_2 emissions, it is also important to provide energy efficient solutions in wireless communications. In this chapter, we discuss energy efficiency for wireless relay systems. Energy efficiency can be measured by the energy consumed to transmit one bit over wireless channels. Depending on transmission schemes and system configurations, the energy efficiency can be different and this measure helps choose a better scheme or system for information transmissions. We will focus on wireless relay systems and analyze their energy efficiency in terms of energy and delay tradeoff (EDT). The notion of EDT plays a crucial role in deciding energy efficient transmission schemes or systems with delay constraints as quality of service (QoS) constraints.

Wireless relay systems can provide reliable transmissions using relay nodes (RNs) located between a source node (SN) and destination node (DN). In [2], it is shown that the performance can be improved by exploiting cooperative diversity and a few relay protocols are proposed including amplify-and-forward (AF) and decode-and-forward (DF) protocols. One-way relay systems in [2] have been extended to two-way relay systems in [3,4]. The notion of physical-layer network coding (PNC) can be applied to two-way relay systems [5,6]. In [7,8], the capacity of two-way relay channels are derived using nested lattice coding.

Hybrid automatic repeat request (HARQ) protocols can be employed for reliable packet transmissions over fading channels in wireless communications [9]. In [10], HARQ protocols are analyzed by information-theoretic approaches and it is shown that the HARQ protocol with incremental redundancy (HARQ-IR) can achieve the ergodic channel capacity over block fading channels. Various coding approaches are proposed for HARQ-IR [11–13]. The application of HARQ protocols to wireless relay systems has been considered in [2,14].

Although the role of RNs in relay-aided transmissions (throughout this chapter, as opposed to direct transmissions, signal transmissions that are carried out with the aid of RNs to transmit packets are called relay-aided transmission schemes) is to provide spatial diversity or cooperative diversity in [2], RNs can effectively increase the coverage and improve the connectivity for wireless networks of multiple nodes when the transmission power of each node is limited (in this case, other nodes become RNs) [15]. As shown in [14], relay-aided transmissions can be more energy efficient than direct transmissions which require a high transmission power in terms of the tradeoff between the energy efficiency and delay.

The energy efficiency of relay-aided transmission schemes in wireless relay systems has been studied in the literature. With a bit error rate (BER) constraint, the energy efficiency of relay-aided transmission schemes for 3-node cooperative relay systems has been studied in [16]. It is shown that the energy efficiency of relay protocols depends on the location of RN and constellation size. In [17], with multiple RNs, the energy efficiency is investigated with the overhead energy consumption that is required by CSI acquisition for distributed beamforming. Information theoretic approaches are studied for the energy efficiency of relay systems in [18]. In [19], the energy consumption for both signal transmissions and operations of electronic circuits [20] is considered to see the energy efficiency of various HARQ protocols with and without cooperative diversity. Since RN is to decode coded packets from SN with cooperative diversity, it can be seen that HARQ protocols are applied with the DF protocol in [19].

In this chapter, we focus on the HARQ-IR protocol for reliable packet transmissions in wireless relay systems and study its EDT. In wireless relay systems, two different protocols should be employed for relaying and reliable packet transmissions. For example, in one-way relay systems, we can have two different combinations of protocols resulting in different relay-aided transmission schemes: *i)* HARQ-IR with AF and *ii)* HARQ-IR with DF. In two-way relay systems, we also have different relay-aided transmission schemes depending on combinations of protocols. While a similar issue has been studied in [14][1] for one-way relay systems, we attempt to define the EDT curve explicitly to understand the tradeoff between energy efficiency and delay and derive EDT curves of the HARQ-IR protocol with different relay protocols in both one-way and two-way relay systems.

It is shown that direct transmissions without an RN can be more energy efficient under a short delay constraint (or in the high power regime). On the other hand, in the low power regime (or under a long delay constraint), we can show that decoding at an RN plays a crucial role in improving energy efficiency. In particular, if decoding is performed at an RN in both one-way and two-way relay systems, relay-aided transmissions are more energy efficient than direct transmissions by a factor of greater than or equal to 8 (with a path loss exponent of 3) as the transmission power approaches 0.

Throughout this chapter, we do not consider the energy consumption by electronic circuits [20], while it is taken into account in [16, 19]. Since the modeling for energy consumption by electronic circuits depends on hardware implementation technologies, it is difficult to have a proper model into the analysis of energy efficiency for wireless relay systems. Thus, although we may obtain optimistic results when the transmission delay is long (i.e., in the

---

[1]To the best of our knowledge, the EDT of HARQ protocols in wireless relay systems has been studied for the first time in [14].

low power regime), the energy consumption by electronic circuits is ignored and it helps us to focus on the fundamental tradeoff between energy efficiency and delay.

21.2   Background

In this section, we present a brief description of the HARQ-IR protocol and its tradeoff between energy efficiency and delay. In addition, two different wireless relay systems, one-way and two-way relay systems, are presented.

21.2.1   HARQ and Energy-Delay Tradeoff

Suppose that a message sequence is to be transmitted over a fading channel. Provided that the channel signal-to-noise ratio (SNR) is not known prior to transmissions, we can use HARQ protocols for reliable communications. Among HARQ protocols, we will focus on the HARQ-IR protocol as it can achieve the channel capacity.

For HARQ-IR, we need to have a channel encoder that can generate a codeword of an arbitrary length or a sequence of subcodewords of a given message sequence. A channel decoder should also be capable of decoding with a sequence of corrupted subcodewords. A detailed procedure for the HARQ-IR protocol is given below.

1. For a given message block, $\mathbf{m}$, a sequence of coded subblocks is generated as $\mathbf{c}_1, \mathbf{c}_2, \ldots$, where $\mathbf{c}_k = \text{ENC}_k(\mathbf{m})$. We will denote by $\text{ENC}_k(\cdot)$ the kth encoding operation and by $\text{DEC}_k(\cdot)$ the kth decoding operation. Let $k = 1$.

2. A transmitter transmits $\mathbf{c}_k$ and a receiver attempts to decode to recover $\mathbf{m}$ with $\{\mathbf{r}_1, \ldots, \mathbf{r}_k\}$ as $\hat{\mathbf{m}}_k = \text{DEC}_k(\mathbf{r}_1, \ldots, \mathbf{r}_k)$, where $\mathbf{r}_k$ is the received signal subblock and $\hat{\mathbf{m}}_k$ denotes the decoded message block obtained with k retransmissions.

3. If decoding is successful, the receiver sends positive acknowledgment (ACK). Otherwise, a retransmission request (RQ) or negative acknowledgment (NACK) is sent to the transmitter. In the former case, the transmitter finishes the transmission and retransmissions of the current message block and starts a new transmission for the next message block. Let $k \leftarrow k + 1$ and move to Step 2 in the latter case.

For point-to-point communications over a block-fading channel, the received signal block is given by

$$\mathbf{r} = \sqrt{\alpha_k}\mathbf{x} + \mathbf{n}, \tag{21.1}$$

where $\mathbf{x}$ is a modulated signal of a coded block, $\sqrt{\alpha_k}$ is the channel gain over the kth block, and $\mathbf{n} \sim \mathcal{CN}(0, \mathbf{I})$. Let

$$Z_k = \log(1 + P\alpha_k).$$

Assuming that a capacity achieving code of rate R, which is also called the initial rate, is used, the number of retransmissions of HARQ-IR is given by [10]

$$K(R, P) = \min\left\{ k \mid \sum_{i=1}^{k} Z_i > R \right\}. \tag{21.2}$$

The number of retransmissions is a random variable that depends on R and P. In general, $K(R, P)$ is a non-decreasing function of R and a non-increasing function of P.

Define the average number of retransmissions of HARQ-IR as

$$\bar{K}(R, P) = \mathbb{E}[K(R, P)], \tag{21.3}$$

where $\mathbb{E}[\cdot]$ denotes the statistical expectation and this expectation is carried out over random channel gains, $\{\alpha_i\}$. The *effective delay* is defined as the average number of retransmissions, $\bar{K}(R, P)$, to the transmission rate as follows:

$$\mathsf{D}(R, P) \triangleq \frac{\bar{K}(R, P)}{R}. \tag{21.4}$$

Let T denote the time for one packet transmission. Then, the average transmission time for a successful packet delivery by HARQ-IR is $\bar{K}(R, P)T$, while the number of bits per packet is BTR. From this, the transmission time per bit becomes $\frac{\bar{K}(R,P)T}{BTR}$ (sec/bit). Therefore, $\mathsf{D}(R, P)$ can be interpreted as the average duration (in seconds) required to deliver one bit scaled by the system bandwidth. The unit of $\mathsf{D}(R, P)$ becomes sec/(bit/Hz). For a fixed B, $\mathsf{D}(R, P)$ is the reciprocal of the the average transmission rate or spectral efficiency.

We can now define the energy per bit, EB, normalized by the system bandwidth as follows:

$$\mathsf{EB}(R, P) \triangleq \frac{P\bar{K}(R, P)}{R} = P\mathsf{D}(R, P). \tag{21.5}$$

The average energy consumed for a successful packet delivery is $P\bar{K}(R, P)T$, while the number of bits per packet is BTR. The unit of the EB in (21.5) is Joules/bit/Hz. From (21.4) and (21.5), we can see the relationship between EB and delay. Using the pair of $\{\mathsf{EB}(R, P), \mathsf{D}(R, P)\}$, we can characterize the energy and delay tradeoff (EDT) of HARQ-IR for wireless communication systems.

In order to see the energy efficiency through EDT, we can consider an example for point-to-point communications. Suppose that the channel gains are given by $\{\alpha_1, \ldots, \alpha_6\} = \{1, 3, 0, 1, 2, 1\}$. Suppose that two different power levels can be used: $P_1 = 10$ and $P_2 = 3$. If $R = 10$, with P_1, the required number of retransmissions is $K(R, P_1) = 4$ as

$$\sum_{i=1}^{3} \log_2(1 + P_1\alpha_i) = 8.41 < R = 10 < \sum_{i=1}^{4} \log_2(1 + P_1\alpha_i) = 11.87.$$

For the case of $R = 10$ and $P_2 = 3$, the required number of retransmissions is $K(R, P_2) = 5$ as

$$\sum_{i=1}^{4} \log_2(1 + P_2\alpha_i) = 7.32 < R = 10 < \sum_{i=1}^{5} \log_2(1 + P_2\alpha_i) = 10.12.$$

Although P_2 results in a longer delay, its EB is $\mathsf{EB} = \frac{P_2 5}{R} = 1.5$ that is lower than the EB with P_1 which is $\mathsf{EB} = \frac{P_1 4}{R} = 4$.

In general, EB decreases with D. That is, the longer the transmission delay, the better the energy efficiency can be achieved, which is also observed in [10] in terms of throughput. Furthermore, as R increases, the EDT of HARQ-IR can be improved. This improvement results from the increasing throughput of HARQ-IR when R increases [10]. In general, a large R requires a higher order modulation and a channel code of a high code rate, which are not practical yet in most wireless communication systems. However, as the effective delay increases, the difference of EB's between large and small values of R vanishes. Thus, for applications where delay constraints are not stringent, R can be low without degrading EDT significantly.

In Fig. 21.1, a typical EDT curve is illustrated. In the high power regime, the EB is high and the effective delay is short. As $P \to \infty$, we have $(\mathsf{EB}, \mathsf{D}) \to (\infty, \frac{1}{R})$ asymptotically. On the other hand, in the low power regime, the EB is low and the effective delay is long. As $P \to 0$, the EDT curve will approach $(\mathsf{EB}, \mathsf{D}) \to (\mathsf{EB}_{\min}, \infty)$, where $\mathsf{EB}_{\min}$ denotes the minimum EB when the average effective delay approaches infinity. It is desirable to operate a system in the low power regime to be more energy efficient, while the effective delay becomes long. A practical system of a finite R will have a pair of certain EB and D and this operating point is above the EDT curve as the performance of coding may not be ideal.

21.2.2   Wireless Relay Systems

We consider two different relay systems: one-way and two-way relay systems. In each system, we assume that there is only one RN.

A one-way relay system consists of three nodes: SN, RN, and DN as shown in Fig. 21.2. In order to transmit a packet, two time slots are required. During the first time slot, SN transmits a packet to RN. RN forwards the packet to DN during the second time slot. There could be various ways to forward the received signal at RN to DN [2]. If RN forwards the received signal without any detection or decoding, the resulting approach is referred to as the AF

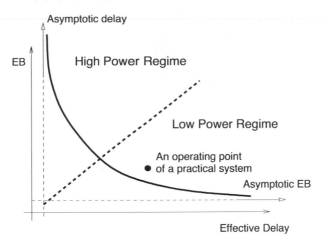

FIGURE 21.1 A typical EDT curve and its asymptotic behavior.

protocol. RN can attempt to decode the packet and re-encode it to forward to DN. This approach is referred to as the DF protocol. Multiple relay nodes can be used for reliable transmissions. However, we do not consider multiple relay nodes to focus on more fundamental issues of the EDT of wireless relay systems. In addition, unlike [2], we assume that the direct link from SN to DN is not available as the distance is sufficiently long.

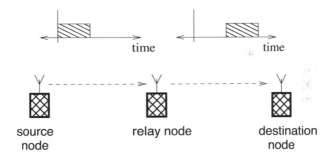

FIGURE 21.2 One-way relay system.

If two nodes attempt to exchange information, a certain node between two nodes can help information exchange. A two-way relay system consists of two SNs and one RN. For information exchange, two time slots are required. During the first time slot, two SNs transmit their signals, s_1 and s_2, simultaneously, and RN receives the superposition of two signals. Then, RN broadcasts the received signal to both two SNs during the second time slot. The resulting exchange protocol consists of multiple access (MAC) (for the first time slot) and broadcast (BC) (for the second time slot) phases as illustrated in Fig. 21.3. This protocol consisting of the MAC and BC phases can achieve a higher spectral efficiency than conventional protocols using four or three time slots [3,4]. As in one-way relay systems, we assume that the direct link between two SNs is not available in two-way relay systems. Thus, no cooperative diversity is exploited. Furthermore, it is assumed that the lengths of two time slots are equal throughout the chapter.

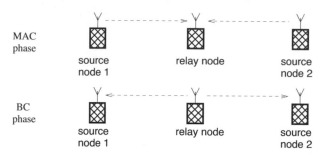

FIGURE 21.3 **Two-way relay system with MAC and BC phases.**

21.3   One-Way Relay Systems

In this section, we study the EDT of one-way relay systems and compare the energy efficiencies of two relay protocols, AF and DF relay protocols, through EDT curves. The energy efficiency in the low power regime is studied to see the advantage of relay-aided transmissions over direct transmissions.

21.3.1   EDT with Amplify-and-Forward Protocol

A relay-aided transmission scheme can employ the HARQ-IR protocol for reliable packet transmissions and the AF protocol for relaying. In this case, RN does not provide any feedback to SN. In the first time slot, SN sends a packet to RN and RN forwards this packet to DN with a proper amplification in the next time slot. Then, DN attempts to decode the packet. For the HARQ-IR protocol, we assume that DN can directly send its binary signal (RQ or ACK) to SN.

Let h_{SR} and h_{RD} denote the channel coefficients from SN to RN and from RN to DN, respectively. Then, the received signals at RN and DN, respectively, are given by

$$r_R = h_{SR}\sqrt{P_S}s + n_R;$$
$$r_D = h_{RD}\sqrt{P_R}\alpha_R r_R + n_D, \qquad (21.6)$$

where s denotes the signal from SN, P_S and P_R represent the signal transmission powers from SN and RN, respectively, and $n_R \sim \mathcal{CN}(0, \sigma_R^2)$ and $n_D \sim \mathcal{CN}(0, \sigma_D^2)$ denote the noise terms at RN and DN, respectively. In (21.6), α_R is the power normalization factor that is given by

$$\alpha_R = \sqrt{\frac{1}{\mathbb{E}[|r_R|^2]}} = \sqrt{\frac{1}{|h_{SR}|^2 P_S + \sigma_R^2}}, \tag{21.7}$$

where $\mathbb{E}[|s|^2] = 1$. The received signal at DN can be rewritten as

$$r_D = h_{RD} h_{SR} \sqrt{P_R} \sqrt{P_S} \alpha_R s + h_{RD} \sqrt{P_R} \alpha_R n_R + n_D \tag{21.8}$$

and, assuming that $\sigma_R^2 = \sigma_D^2 = 1$, the SNR is given by

$$\mathsf{SNR}(P_S, P_R) = \frac{P_S P_R |h_{SR}|^2 |h_{RD}|^2}{P_S |h_{SR}|^2 + P_R |h_{RD}|^2 + 1}. \tag{21.9}$$

If $P_S |h_{SR}|^2$ and $P_R |h_{RD}|^2$ are sufficiently high (i.e. high SNR), we have the following approximation:

$$\mathsf{SNR}(P_S, P_R) \approx \frac{P_S P_R |h_{SR}|^2 |h_{RD}|^2}{P_S |h_{SR}|^2 + P_R |h_{RD}|^2} \approx \min\left\{ P_S |h_{SR}|^2, P_R |h_{RD}|^2 \right\}.$$

This shows that the weakest link can decide the SNR at DN.

For the HARQ-IR protocol, we assume that DN can directly send its binary signal (RQ or ACK) to SN. We assume that the channels are modeled as block fading channels, where each channel coefficient remains constant for the duration of a time-slot, but independent for each time-slot. Denote by $Z_q = \log_2\left(1 + \mathsf{SNR}(P_S, P_R)\right)$ the qth realization of the instantaneous channel capacity, where $\mathsf{SNR}(P_S, P_R)$ is considered as a random variable. The number of retransmissions from SN to DN is given by

$$K_{AF}(R, P_S, P_R) = \min\{k \mid \sum_{q=1}^{k} Z_q > R\}, \tag{21.10}$$

where the total number of retransmissions becomes doubled as one (re)transmission requires two time slots. Provided that the total power is fixed, P_S and P_R can be optimized as follows:

$$\{P_S^*, P_R^*\} = \arg \min_{P_S + P_R \leq P} \bar{K}_{AF}(R, P_S, P_R), \tag{21.11}$$

where $\bar{K}_{AF}(R, P_S, P_R) = \mathbb{E}[K_{AF}(R, P_S, P_R)]$. The delay efficiency becomes

$$\mathsf{D}_{AF}(R, P) = \frac{2\bar{K}_{AF}(R, P_S^*, P_R^*)}{R}, \tag{21.12}$$

while the EB for the AF protocol is given by

$$\mathsf{EB}_{AF}(R,P) = \frac{P}{2}\mathsf{D}_{AF}(R,P) = \frac{P\bar{K}_{AF}(R,P_\mathsf{S}^*,P_\mathsf{R}^*)}{R}. \tag{21.13}$$

In one-way relay systems, since only one link out of two links, SN-to-RN and RN-to-DN, is active for each retransmission, the transmission time is halved and this is taken into account in the EB in (21.13).

Note that, in (21.11), the minimization of the average number of retransmissions is considered for the power allocation. This is equivalent to minimizing the effective delay as shown in (21.12). For the power allocation, we can also consider the minimization of EB. However, in this case, since the EB can be minimized when P_S and P_R become zero, the effective delay can be infinity, which is not desirable. Therefore, throughout the chapter, we consider the minimization of the average number of retransmissions for the power allocation.

21.3.2   EDT with Decode-and-Forward Protocol

When the DF protocol is employed for relaying in conjunction with the HARQ-IR protocol, we assume that there are two hops with two independent HARQ-IR operations. That is, there are one HARQ-IR operation for the link of SN-to-RN and the other for the link of RN-to-DN. In this case, unlike the AF protocol, RN attempts to decode packets and sends a feedback signal to SN. Once RN succeeds to decode a packet, it encodes the packet and forwards it to DN using another HARQ-IR protocol.

Let

$$\mathsf{SNR}_\mathsf{R} = P_\mathsf{S}|h_\mathsf{SR}|^2 \text{ and } \mathsf{SNR}_\mathsf{D} = P_\mathsf{R}|h_\mathsf{RD}|^2. \tag{21.14}$$

Then, the total number of retransmissions is given by

$$K_{DF}(R,P_\mathsf{S},P_\mathsf{R}) = K_\mathsf{SR}(R,P_\mathsf{S}) + K_\mathsf{RD}(R,P_\mathsf{R}), \tag{21.15}$$

where

$$K_\mathsf{SR}(R,P_\mathsf{S}) = \min\{k \mid \sum_{q=1}^{k} Z_{\mathsf{SR},q} > R\};$$

$$K_\mathsf{RD}(R,P_\mathsf{R}) = \min\{k \mid \sum_{q=1}^{k} Z_{\mathsf{RD},q} > R\}.$$

Here, $Z_{\mathsf{SR},q}$ and RD,q are the qth realizations of $\log_2(1+\mathsf{SNR}_\mathsf{R})$ and $\log_2(1+\mathsf{SNR}_\mathsf{D})$, respectively. Let P_S^* and P_R^* denote the optimal power allocation that minimizes the effective delay. Then, the effective delay is given by

$$\mathsf{D}_{DF}(R,P) = \frac{\bar{K}_{DF}(R,P_\mathsf{S}^*,P_\mathsf{R}^*)}{R}, \tag{21.16}$$

where $\bar{K}_{DF}(R, P_{\mathsf{S}}, P_{\mathsf{R}}) = \mathbb{E}[K_{DF}(R, P_{\mathsf{S}}, P_{\mathsf{R}})]$, while the EB becomes

$$\mathsf{EB}_{DF}(R, P) = \frac{P_{\mathsf{S}}^* \bar{K}_{\mathsf{SR}}(R, P_{\mathsf{S}}^*) + P_{\mathsf{R}}^* \bar{K}_{\mathsf{RD}}(R, P_{\mathsf{R}}^*)}{R}, \tag{21.17}$$

where $\bar{K}_{\mathsf{SR}}(R, P_{\mathsf{S}}) = \mathbb{E}[K_{\mathsf{SR}}(R, P_{\mathsf{S}})]$ and $\bar{K}_{\mathsf{RD}}(R, P_{\mathsf{R}}) = \mathbb{E}[K_{\mathsf{RD}}(R, P_{\mathsf{R}})]$.

21.3.3 Analysis and Numerical Results

If $\mathbb{E}[|h_{\mathsf{SR}}|^2] = \mathbb{E}[|h_{\mathsf{RD}}|^2]$, we have an equal power allocation. We can consider the asymptotic performance for this case. That is, under the assumption that the distance between SN and DN is normalized to be unity, we consider the following conditions: $\mathbb{E}[|h_{\mathsf{SR}}|^2] = \mathbb{E}[|h_{\mathsf{RD}}|^2] = \left(\frac{1}{2}\right)^{-\eta}$ and $P_{\mathsf{S}} = P_{\mathsf{R}} = \frac{P}{2}$, where η denotes the path loss exponent. For comparison, we consider the HARQ-IR protocol without RN. Without RN (i.e. direct transmissions), if P is very small,

$$\sum_{i=1}^{k} Z_i = \sum_{i=1}^{k} \log_2(1 + \alpha_i P) \approx \frac{\sum_{i=1}^{k} \alpha_i P}{\log 2},$$

where α_i is the channel gain for direct transmissions and $\mathbb{E}[\alpha_i] = 1$. Thus, the number of retransmissions can be approximated as $\bar{K}(R, P) \approx \frac{R}{P} \log 2$ based on the law of large numbers. This results in the following asymptotic EB for direct transmissions:

$$\lim_{P \to 0} \mathsf{EB}_{direct}(R, P) = \log 2 = -1.59 \text{ (dB)}, \tag{21.18}$$

which is also the minimum EB, $\mathsf{EB}_{\min}$ for direct transmissions. Under the channel gain normalization, $\mathbb{E}[\alpha_i] = 1$, this limit corresponds to the minimum energy per bit for reliable communications in [21]. For the HARQ-IR protocol with DF, each link has a half power, $P/2$. Thus, the total number of retransmissions for a low P, $\bar{K}_{DF}(R, P) \approx 2 \frac{R}{\frac{P}{2}\left(\frac{1}{2}\right)^{-\eta}} \log 2 = \frac{R}{P2^{\eta-2}} \log 2$. From this, the asymptotic EB for the HARQ-IR protocol with DF becomes

$$\lim_{P \to 0} \mathsf{EB}_{DF}(R, P) = \frac{\log 2}{2^{\eta-1}}. \tag{21.19}$$

This shows that relay-aided transmissions using the HARQ-IR protocol with DF can be more energy efficienct than direct transmissions with the HARQ-IR protocol when $\eta > 1$ in the low power regime ($P \to 0$). If $\eta = 3$, it can be shown that the EB can be quartered. On the other hand, if $P \to \infty$ (i.e. high power regime), direct transmissions are preferable to relay-aided transmissions, because one transmission would be sufficient in direct transmissions, while at least two transmissions (from SN to RN and from RN to DN) are required in relay-aided transmissions.

If M-hop relay-aided transmissions with $M - 1$ RNs are considered using the HARQ-IR protocol with DF for the case of equal distance between nodes and equal power allocation, it can be shown that

$$\lim_{P \to 0} \mathsf{EB}_{DF}(R, P) = \frac{\log 2}{M^{\eta-1}}; \tag{21.20}$$

$$\lim_{P \to \infty} \mathsf{D}_{DF}(R, P) = \frac{M}{R}. \tag{21.21}$$

Note that the energy efficiency gain in (21.20) is the relative gain with respect to the energy efficiency of direct transmissions. In the low power regime, relay-aided transmissions (using the HARQ-IR protocol with DF) can be more energy efficient than direction transmissions by a factor of $M^{\eta-1}$. However, this energy efficiency could be degraded once the circuit power consumption of RNs is taken into account.

Unfortunately, in the low power regime (as well as in the high power regime), relay-aided transmissions using the HARQ-IR protocol with AF do not provide better EDT than direct transmissions. This can be shown by the SNR expression in (21.9). For $P \ll 1$ (i.e. low power regime), we have

$$\mathsf{SNR}(P_\mathsf{S}, P_\mathsf{R}) \approx \frac{P^2}{4} |h_\mathsf{SR}|^2 |h_\mathsf{RD}|^2.$$

Thus, the average number of retransmissions is proportional to $\frac{1}{P^2}$. This means that the EB can increase as $P \to 0$ when the HARQ-IR protocol with AF is employed. Therefore, the AF protocol does not suit for relaying when the HARQ-IR protocol is employed for reliable packet transmissions in one-way relay systems. This is an interesting observation as the AF protocol usually performs well and better than the DF protocol for delay-limited applications (i.e. without HARQ protocols) [2].

We carry out simulations to see the EDT of the HARQ-IR protocols with AF and DF. In our simulations, the distance between SN and DN is normalized to be unity and the distance between SN and RN is denoted by β, where $0 < \beta < 1$. Furthermore, we have assumed that $\mathbb{E}[|h_\mathsf{SR}|^2] = \frac{1}{\beta^\eta}$ and $\mathbb{E}[|h_\mathsf{RD}|^2] = \frac{1}{(1-\beta)^\eta}$.

With $\eta = 3$, Figs. 21.4 and 21.5 show the EDT curves when $\beta = 0.5$ and 0.1, respectively. They both exhibit that the HARQ-IR protocol with DF is better than that with AF in terms of EDT. In the DF protocol, since RN attempts to decode packets, if the received packets are not reliable at RN, it does not waste the second time slot to forward unreliable packets to DN. Thus, the DF protocol requires less number of retransmissions than the AF protocol in HARQ-IR, which results in a better EDT.

As shown earlier, relay-aided transmissions using the HARQ-IR protocol with DF can be more energy efficient than direct transmissions in the low power regime. With $\eta = 3$, Fig. 21.6 shows the EDT curves of relay-aided and direct transmissions. We can confirm that the EB of relay-aided transmissions using the HARQ-IR protocol with DF is a quarter of that of direct

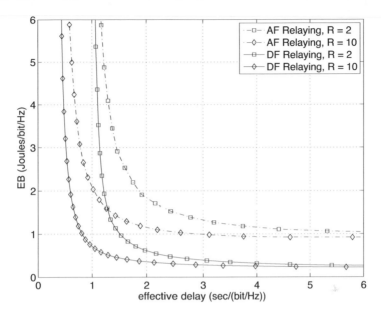

FIGURE 21.4 EDT of HARQ-IR with AF and DF when $\beta = 0.5$.

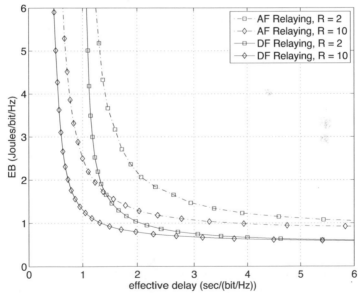

FIGURE 21.5 EDT of HARQ-IR with AF and DF when $\beta = 0.1$.

transmissions in the low power regime, where the transmission delay is long. On the other hand, in the high power regime, the effective delay of relay-aided

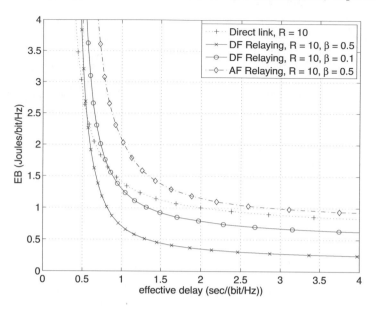

FIGURE 21.6 Comparison between DF relaying and direct link transmission.

transmissions is doubled compared to that of direct transmissions as expected in (21.21). For the case where RN is close to one of two terminals (SN or DN), i.e., β is close to 0 or 1, we still can see the advantage of relay-aided transmissions over direct transmissions in the low power regime, which is observed in Fig. 21.6.

21.4   Two-Way Relay systems

In two-way relay systems shown in Fig. 21.3, the notion of physical-layer network coding (PNC) [5, 6] plays a critical role in improving the spectral efficiency. In this section, the HARQ-IR protocol is applied to two-way relay systems with analog network coding (ANC) and PNC protocols.

21.4.1   EDT with Analog Network Coding

The ANC protocol can be considered as a generalization of the AF protocol in two-way relay systems. [3,4]. In the MAC phase, two SNs, denoted by SN_1 and SN_2, transmit their signals to RN and the received signal at RN is given by

$$r_R = \sqrt{P_1}h_1s_1 + \sqrt{P_2}h_2s_2 + n_R, \qquad (21.22)$$

where P_m, h_m, s_m, and $n_R \sim \mathcal{CN}(0,1)$ denote the signal power, the channel coefficient from SN_m to RN, the transmitted signal from SN_m, and the background noise at RN, respectively. In the BC phase, RN broadcasts the received signal, r_R, and the received signal at SN_m is given by

$$r_m = g_m \sqrt{P_R} \alpha_R r_R + n_m, \qquad (21.23)$$

where g_m, P_R, and $n_m \sim \mathcal{CN}(0,1)$ are the channel coefficient from RN to SN_m, the transmission power from RN, and the background noise at SN_m, respectively. Here, α_R denotes the power normalization factor that is given by

$$\alpha_R = \sqrt{\frac{1}{\mathbb{E}[|r_R|^2]}} = \sqrt{\frac{1}{|h_1|^2 P_1 + |h_2|^2 P_2 + 1}}.$$

As its own signal is available at each SN, this signal can be canceled. The resulting SNR at each SN is given by

$$SNR_1(P_1, P_2, P_R) = \frac{P_1 P_R |h_1|^2 |g_2|^2}{|g_2|^2 P_R + |h_1|^2 P_1 + |h_2|^2 P_2 + 1};$$

$$SNR_2(P_1, P_2, P_R) = \frac{P_2 P_R |h_2|^2 |g_1|^2}{|g_1|^2 P_R + |h_1|^2 P_1 + |h_2|^2 P_2 + 1}, \qquad (21.24)$$

where $SNR_1(P_1, P_2, P_R)$ and $SNR_2(P_1, P_2, P_R)$ denote the SNR at SN_2 and SN_1, respectively. Let $SNR_{m,q}$ denote the qth realization of $SNR_m(P_1, P_2, P_R)$ and let $Z_{m,q} = \log_2(1 + SNR_{m,q})$. Then, in HARQ-IR, the number of retransmissions for the packets transmitted by SN_m at rate R_m is given by

$$K_{ANC,m}(R_m, P_1, P_2, P_R) = \min\{k \mid \sum_{q=1}^{k} Z_{m,q} > R_m\}. \qquad (21.25)$$

Let $\bar{K}_{ANC,m}(R_m, P_1, P_2, P_R) = \mathbb{E}[K_{ANC,m}(R_m, P_1, P_2, P_R)]$. The optimal power allocation can be performed with a total power constraint to minimize the total number of retransmissions follows:

$$\{P_1^*, P_2^*, P_R^*\} = \arg \min_{P_1 + P_2 + P_R \leq P} \sum_{m=1}^{2} \bar{K}_{ANC,m}(R_m, P_1, P_2, P_R), \qquad (21.26)$$

where P denotes the total power.

The resulting effective delay for two-way relay systems is given by

$$D_{ANC}(R_1, R_2, P) = \frac{\sum_{m=1}^{2} 2\bar{K}_{ANC,m}(R_m, P_1^*, P_2^*, P_R^*)}{R_1 + R_2}. \qquad (21.27)$$

The average energy consumption for signal transmissions during the MAC phase is given by $\sum_{m=1}^{2} P_m^* \bar{K}_{ANC,m}(R_m, P_1^*, P_2^*, P_R^*)$ and during the BC phase is given by $\frac{P_R^*}{2} \sum_{m=1}^{2} \bar{K}_{ANC,m}(R_m, P_1^*, P_2^*, P_R^*)$. The resulting EB is given by

$$\mathsf{EB}_{ANC}(R_1, R_2, P) = \frac{\sum_{m=1}^{2} \left(P_m^* + \frac{P_R^*}{2}\right) \bar{K}_{ANC,m}(R_m, P_1^*, P_2^*, P_R^*)}{R_1 + R_2}. \quad (21.28)$$

21.4.2   EDT with Physical-Layer Network Coding

When PNC is employed with the HARQ-IR protocol for two-way relay systems, in the MAC phase, RN attempts to jointly decode the signals transmitted by SN_1 and SN_2 [7, 22, 23]. Retransmissions will take place until RN succeeds to decode both the two signals without errors. Once RN can decode the two signals, base on PNC, it can re-encode the two signals into a single network coded message. This network coded message is repeatedly transmitted in the BC phase until both the two SNs can decode it without errors. Once the network coded message is successfully decoded at both SNs, each source node can use it to decode the message sent by the other. For example, if the network coded message is formed by taking the bitwise exclusive OR of the two messages generated at the two source nodes. Each source node decodes the message sent by the other by simply taking the bitwise exclusive OR of its own message and the network coded message. Thus, for one message exchange between SN_1 and SN_2, there could be multiple time slots for each phase.

In the MAC phase, RN performs joint decoding of two signals transmitted by SN_1 and SN_2. Let $Z_{R,1,q}$, $Z_{R,2,q}$, and $Z_{R,q}$ represent the qth realizations of $\log_2(1 + \mathsf{SNR}_{R,1})$, $\log_2(1 + \mathsf{SNR}_{R,2})$, and $\log_2(1 + \mathsf{SNR}_{R,1} + \mathsf{SNR}_{R,2})$, respectively, where $\mathsf{SNR}_{R,1} = |h_1|^2 P_1$ and $\mathsf{SNR}_{R,2} = |h_2|^2 P_2$. Using the MAC capacity bound [24], for given R_1 and R_2, the number of retransmissions is the minimum of k that satisfies the following inequalities:

$$\sum_{q=1}^{k} Z_{R,1} > R_1; \quad \sum_{q=1}^{k} Z_{R,2} > R_2; \text{ and } \sum_{q=1}^{k} Z_{R,q} > R_1 + R_2.$$

In other words, the number of retransmissions is given by

$$K_{MAC}(R_1, R_2, P_1, P_2) = \min\left\{k \;\middle|\; \left\{\sum_{q=1}^{k} Z_{R,1,q} > R_1\right\} \cap \left\{\sum_{q=1}^{k} Z_{R,2,q} > R_2\right\}\right.$$

$$\left. \cap \left\{\sum_{q=1}^{k} Z_{R,q} > R_1 + R_2\right\}\right\}.$$

$$(21.29)$$

Clearly, the number of retransmissions in (21.29) is optimistic as we employ the MAC capacity bound.

It is noteworthy that in (21.29), it is assumed that RN decode two signals,

s_1 and s_2, jointly. However, as in [22, 23], RN only needs to decode an XOR-like version of s_1 and s_2. For example, if $s_1, s_2 \in \{-1, +1\}$, RN needs to detect whether $s_1 = s_2$ or $s_1 \neq s_2$. In this case, the number of retransmissions can be smaller than that in (21.29).

If RN succeeds to decode signals, it can broadcast the decoded signals after encoding to both SN_1 and SN_2. In the BC phase, the SNR to SN_m is given by

$$\mathsf{SNR}_{\mathsf{S},m} = |g_m|^2 P_{\mathsf{R}}. \tag{21.30}$$

Then, the number of retransmissions to SN_m is given by

$$K_{BC,m}(R_m, P_{\mathsf{R}}) = \min\{k \mid \sum_{q=1}^{k} Z_{m,q} > R_{v(m)}\}, \tag{21.31}$$

where $Z_{m,q}$ denotes the qth realization of $\log_2(1 + \mathsf{SNR}_{\mathsf{S},m})$ and $v(m) = 1$ if $m = 2$ and 2 if $m = 1$. Since jointly encoded XOR-like version of a pair of transmitted signals from SN_1 and SN_2 for PNC is retransmitted by RN until two SNs can successfully decode, the number of retransmissions in the BC phase is given by

$$K_{BC}(R_1, R_2, P_{\mathsf{R}}) = \max\{K_{BC,1}(R_1, P_{\mathsf{R}}), K_{BC,2}(R_2, P_{\mathsf{R}})\}. \tag{21.32}$$

Let

$$\bar{K}_{MAC}(R_1, R_2, P_1, P_2) = \mathbb{E}[K_{MAC}(R_1, R_2, P_1, P_2)];$$
$$\bar{K}_{BC}(R_1, R_2, P_{\mathsf{R}}) = \mathbb{E}[K_{BC}(R_1, R_2, P_{\mathsf{R}})].$$

Denote by P_1^*, P_2^*, and P_{R}^* the optimal power allocation for P_1, P_2, and P_{R}, respectively, under a total power constraint with P. The resulting effective delay becomes

$$\mathsf{D}_{PNC}(R_1, R_2, P) = \frac{\bar{K}_{MAC}(R_1, R_2, P_1^*, P_2^*) + \bar{K}_{BC}(R_1, R_2, P_{\mathsf{R}}^*)}{R_1 + R_2}, \tag{21.33}$$

while the EB is given by

$$\mathsf{EB}_{PNC}(R_1, R_2, P) = \frac{(P_1^* + P_2^*)\,\bar{K}_{MAC}(R_1, R_2, P_1^*, P_2^*) + P_{\mathsf{R}}^*\bar{K}_{BC}(R_1, R_2, P_{\mathsf{R}}^*)}{R_1 + R_2}. \tag{21.34}$$

21.4.3   Analysis And Numerical Results

Throughout this subsection, for the sake of simplicity, we only consider the symmetric case, where $\mathbb{E}[|h_m|^2] = \mathbb{E}[|g_m|^2] = \frac{1}{(1/2)^\eta} = 2^\eta$, $m = 1, 2$. Let λ and $1 - \lambda$ be the fractions of power allocated to two SNs and RN, respectively. Then, we have $P_1 = P_2 = \frac{\lambda P}{2}$ and $P_{\mathsf{R}} = (1 - \lambda)P$. Using numerical techniques, we can find λ^* that minimizes the average number of retransmissions for the HARQ-IR protocols with PNC and ANC.

In the high power regime, both the HARQ-IR protocols with ANC and PNC require two time slots for the exchange of a pair of packets and no additional retransmission is required. In this case, i.e. as $P \to \infty$, the two EDT curves approach $\left(\infty, \frac{2}{R_1+R_2}\right)$, which is the same as in direct transmissions. Note that, in the high power regime, the advantage of relay-aided transmissions using wireless relay systems over direct transmissions vanishes as mentioned in Subsection 21.3.3.

In the low power regime, suppose that the HARQ-IR protocol with PNC has allocated an equal power for all transmissions, $\bar{P}$. Thus, the total power P is $3\bar{P}$. Furthermore, assume that $R_1 = R_2 = R$. As in Subsection 21.3.3, from (21.29) and (21.32), it can be shown that

$$\bar{K}_{MAC}(R, R, \bar{P}, \bar{P}) \approx \frac{2R\log 2}{2\bar{P}2^\eta}$$

and

$$\bar{K}_{BC}(R, R, \bar{P}) \approx \frac{R\log 2}{\bar{P}2^\eta}$$

as $\bar{P} \to 0$. Thus, from (21.34), we have

$$\lim_{\bar{P}\to 0} \mathsf{EB}_{PNC}(R, R, 3\bar{P}) = \frac{2\bar{P}\frac{R\log 2}{\bar{P}2^\eta} + \bar{P}\frac{R\log 2}{\bar{P}2^\eta}}{2R} = \frac{3}{4}\frac{\log 2}{2^{\eta-1}}. \qquad (21.35)$$

From this result and (21.19), we can see that the asymptotic EB of two-way relay systems can be lower than that of one-way relay systems by a factor of 3/4 in the low power regime.

Fig. 21.7 presents the EDT curves of the HARQ-IR protocols with ANC and PNC for two-way relay systems. We also include the EDT curve for the direct transmission for comparison purposes. In the low power regime, we can see clearly that relay-aided transmissions using two-way relay systems can be more energy efficient than direct transmissions when the HARQ-IR protocol with PNC is employed. Furthermore, from Fig. 21.8, we can confirm that *i)* relay-aided transmissions are more energy efficient than direct transmissions in the low power regime when an RN performs decoding; *ii)* two-way relay systems asymptotically have lower EB than one-way relay systems by a factor of 3/4, which is predicted in (21.35) when the DF and PNC protcols are employed for relaying in one-way and two-way relay systems, respectively.

21.5   Conclusions

In this chapter, we studied the energy efficiency of wireless relay systems when the HARQ-IR protocol is employed for reliable packet transmissions. In order

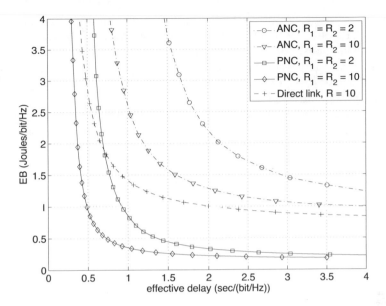

FIGURE 21.7 EDT for two-way relay system with HARQ-IR.

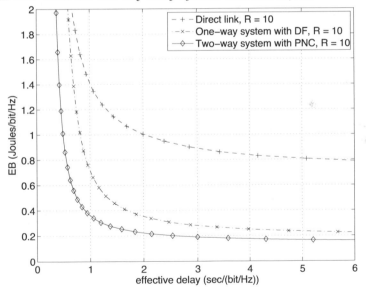

FIGURE 21.8 EDT curves of direct transmissions and relay-aided transmissions in one-way and two-way relay systems.

to see the tradeoff between energy efficiency and delay, the EDT curve of the HARQ-IR protocol was derived in both one-way and two-way relay systems with different relay protocols. In the high power regime, it was shown that direct transmissions without an RN can be more energy efficient than relay-aided transmissions. In the low power regime, which is more important in terms of energy efficiency, relay-aided transmissions can have much higher energy efficiency than direct transmissions if the DF and PNC protocols are used in one-way and two-way relay systems, respectively.

Unfortunately, this was not true if the AF and ANC protocols are employed. As in [2], for the case of unreliable transmissions where packets can be lost, the AF protocol can perform better than the DF protocol in general. However, as in this chapter, if HARQ protocols are employed for reliable packet transmissions, RN needs to perform decoding for better performance.

There are issues to be addressed in the future. The energy consumption by electronic circuits [20] needs to be taken into account to see an overall energy efficiency. This issue will be more serious if more RNs are used for wireless relay systems and the transmission power is low (i.e., in the low power regime) where the energy consumption by electronic circuits could be dominant. In this chapter, we do not consider the rate control for the energy efficiency. Under various QoS constraints, the rate control could be studied for better energy efficiency. Furthermore, we only consider simple wireless relay systems (one-way and two-way relay systems with 3 nodes). It could be extended to more complicated wireless relay networks with a number of nodes.

Bibliography

[1] *Smart 2020: Enabling the low carbon economy in the information age*, The Climate Group. [Online]. Available: http://www.smart2020.org/_assets/files/02_Smart2020Report.pdf

[2] J. Laneman, D. Tse, and G. Wornell, "Cooperative diversity in wireless networks: Efficient protocols and outage behavior," *IEEE Trans. Inform. Theory*, vol. 50, no. 12, pp. 3062–3080, Dec. 2004.

[3] P. Popovski and H. Yomo, "Bi-directional amplification of throughput in wireless multi-hop network," in *Proc. IEEE VTC*, May 2006, pp. 588–593.

[4] R. Rankov and A. Wittneben, "Spectral efficient protocols for half-duplex fading relay channels," *IEEE J. Selected Areas Commun.*, vol. 25, no. 2, pp. 379–289, Feb. 2007.

[5] S. Zhang, S. C. Liew, and P. Lam, "Hot topic: physical-layer network coding," in *Proc. ACM MOBICOM*, Sep. 2006, pp. 358–365.

[6] S. Zhang, S. C. Liew, and L. Lu, "Physical-layer network coding schemes over finite and infinite fields," in *Proc. IEEE Globecom*, Nov. 2008, pp. 3784–3789.

[7] M. P. Wilson, K. Narayanan, H. Pfister, and A. Sprintson, "Joint physical layer coding and network coding for bidirectional relaying," *IEEE Trans. Inform. Theory*, vol. 56, no. 11, pp. 5641–5654, Nov. 2010.

[8] S.-Y. Nam, W. Chung and Y. H. Lee, "Capacity of the gaussian two-way relay channel within 1/2 bit," *IEEE Trans. Inform. Theory*, vol. 56, no. 11, pp. 5488–5494, Nov. 2010.

[9] S. B. Wicker, *Error Control Systems for Digital Communication and Storage*. Prentice Hall, 1995.

[10] G. Caire and D. Tuninetti, "The throughput of hybrid-ARQ protocols for the Gaussian collision channel," *IEEE Trans. Inform. Theory*, vol. 47, no. 5, pp. 1971–1988, Jul. 2001.

[11] S. Sesia, G. Caire, and G. Vivier, "Incremental redundancy hybrid ARQ schemes based on low-density parity-check codes," *IEEE Trans. Commun.*, vol. 52, no. 8, pp. 1311–1321, Aug. 2004.

[12] E. Soljanin, N. Varnica, and P. Whiting, "Punctured vs rateless codes for hybrid ARQ," in *Proc. IEEE ITW*, Mar. 2006, pp. 155–159.

[13] J.-F. Cheng, "Coding performance of HARQ with BICM – Part I: unified performance analysis," in *Proc. IEEE PIMRC*, Sep. 2010, pp. 976–981.

[14] B. Zhao and M. C. Valenti, "Practical relay networks: a generalization of hybrid-ARQ," *IEEE J. Selected Areas Commun.*, vol. 23, no. 1, pp. 7–18, Jan. 2005.

[15] P. Gupta and P. R. Kumar, "Toward an information theory of large networks: an achievable rate region," *IEEE Trans. Inform. Theory*, vol. 53, pp. 1877–1894, Aug. 2003.

[16] Q. Chen and M. C. Gursoy, "Energy efficiency analysis in amplify-and-forward and decode-and-forward cooperative networks," in *Proc. IEEE WCNC*, Apr. 2010, pp. 1–6.

[17] R. Madan, N. B. Mehta, A. F. Molisch, and J. Zhang, "Energy-efficient cooperative relaying over fading channels with simple relay selection," *IEEE Trans. Wireless Commun.*, vol. 7, no. 8, pp. 3013–3025, Aug. 2008.

[18] A. El Gamal, M. Mohseni, and S. Zahedi, "Bounds on capacity and minimum energy-per-bit for awgn relay channels," *IEEE Trans. Inform. Theory*, vol. 52, no. 4, pp. 1545–1561, Apr. 2006.

[19] I. Stanojev, O. Simeone, Y. Bar-Ness, and D. H. Kim, "Energy efficiency of non-collaborative and collaborative hybrid-ARQ protocols," *IEEE Trans. Wireless Commun.*, vol. 8, no. 1, pp. 326–335, Jan. 2009.

[20] S. Cui, A. Goldsmith, and A. Bahai, "Energy-constrained modulation optimization," *IEEE Trans. Wireless Commun.*, vol. 4, no. 5, pp. 2349–2360, Sep. 2005.

[21] C. E. Shannon, "Communication in the presence of noise," *Proc. IRE*, vol. 37, pp. 10–21, Jan. 1949.

[22] S. Zhang and S.-C. Liew, "Channel coding and decoding in a relay system operating with physical-layer network coding," *IEEE J. Selected Areas Commun.*, vol. 27, no. 5, pp. 788–796, Jun. 2009.

[23] D. To and J. Choi, "Convolutional codes in two-way relay networks with physical layer network coding," *IEEE Trans. Wireless Commun.*, vol. 9, no. 9, pp. 2724–2729, Sep. 2010.

[24] T. M. Cover and J. A. Thomas, *Elements of Information Theory*, 2nd ed. NJ: John Wiley, 2006.

Author Contact Information

Jinho Choi and Weixi Xing are with Swansea University, UK, Email: J.Choi@swansea.ac.uk. Duc To was with Swansea University and is currently with Aeroflex Ltd., UK, Email: duc.to@aeroflex.com. Ye Wu and Shugong Xu are with Huawei Technologies Co. Ltd., China, Email: xushugong@huawei.com.

22

Holistic Approach to Green Wireless Communications Based on Multicarrier Technologies

Hanna Bogucka

Poznan University of Technology, Poznan, Poland

Adrian Kliks

Poznan University of Technology, Poznan, Poland

Paweł Kryszkiewicz

Poznan University of Technology, Poznan, Poland

Oliver Holland

King's College London, London, UK

CONTENTS

Sustainable development has been an issue and the aspiration of modern societies already for over two decades, i.e. since the United Nations General

Assembly in December 1987, and its Resolution 42/187. It is also well known that 3% of the world-wide energy is consumed by the Information and Communication Technology (ICT) infrastructure (out of which 80% is consumed by the base stations of mobile communication systems) which causes about 2% of the world-wide CO_2 emissions. Moreover, wireless communication sector of the ICT domain that revolutionized communication between people by providing *anywhere, anytime* and *anyhow* linking is now moving to mass connections between machines. Lowering the energy consumption and related CO_2 emission caused by this ubiquitous wireless communication seems to be a major challenge for the new system designers.

A holistic approach to any wireless-technology eco-sustainability is based on the idea that all the system components should contribute to the overall sustainable development of this technology, and that the properties of that technology related to the impact on the eco-system cannot be determined only by its components alone. (The general principle of holism was concisely summarized by Aristotle in the Metaphysics: "The whole is more than the sum of its parts".)

This chapter addresses a number of issues related to green wireless communications based on multicarrier technologies considered for 4G systems and for the future flexible and opportunistic access to spectrum resources. These issues encompass the efficient use of natural resources, namely energy and spectrum, for a given the transmission scheme (minimization of the transmit power, efficient usage of the spectrum-energy resources for the required transmission quality, or minimization of the power "wasted" on pilots), lowering of the electromagnetic pollution (minimization of the out-of-band radiation), and spectrum recycling (opportunistic use of the spectrum).

22.1   Introduction

The Orthogonal Frequency Division Multiplexing (OFDM), a well-known modulation scheme, and related Orthogonal Frequency Division Multiple Access (OFDMA) technique are the most well known examples of multicarrier transmission technologies. Both of them make use of orthogonal SubCarriers (SCs) for the parallel transmission of data streams. This results in a relatively long symbol-period, noticably exceeding the duration of a channel impulse response. Moreover, insertion of the so-called Cyclic Prefix (CP) and its removal at the receiver can compensate for the Inter-Symbol Interference (ISI). The OFDM equalizer is thus very simple, and consists in a number of single-tap filters, one for each subcarrier-frequency channel. Application of low complex Inverse Fast Fourier Transform (IFFT) at the transmitter and the Fast Fourier Transform (FFT) at the receiver for the OFDM modulation and demodulation is another advantage of the OFDM/OFDMA techniques. OFDM transmission has also some disadvantages: it is sensitive to the phase noise and frequency offsets, has high Peak-to-Average Power Ratio (PAPR) and relatively high Out-of-Band (OOB) power radiation. The OFDM has been investigated extensively, and a number of books and papers have been published detailing research results on this topic [1] - [6].

In general, OFDM and OFDMA are considered as flexible, scalable and easily optimizable technologies, suited for future wireless communication systems. They have already been applied in a number of existing wireless communication standards or recommended by relevant working groups for use in indoor, outdoor and mobile radio environment, i.e. in IEEE 802.15.3.a (WPANs), IEEE 802.11.a/g/n/e (WLAN), 802.16 (WMAN), IEEE 802.22 (WRAN), or in the 3GPP Long-Term Evolution (LTE) mobile telecommunication system. OFDM is also used by the broadcasting systems, such as Digital Audio Broadcasting (DAB), Digital Radio Mondiale (DRM, DRM+) and Digital Video Broadcasting Terrestrial (DVB-T, DVB-T2) and Handheld (DVB-H) standards. Finally, OFDM and its version that uses non-contiguous subcarriers, namely Non-Contiguous Orthogonal Frequency Division Multiplexing (NC-OFDM) is proposed for future Cognitive Radio (CR) technology [7,8], that can make use of the available spectrum holes (the spectrum that is temporarily unoccupied by a licensed Primary User - PU) appearing at a particular time and location. The available spectrum could be fragmented with a number of narrow frequency bands. Using OFDM has the advantage that these bands can be relatively easily aggregated to convey the Secondary User's (SU's) (unlicensed user's) traffic, and that the spectrum utilization increases significantly [9].

The cognitive radios are expected to utilize the location information for multiple purposes, including the environment characterization, optimization of the transmission and reception of a single link, optimization of the network and provision of the location-based services [10]. OFDM transmission commonly applies pilot sequences and cyclic prefix, which can be used for positioning procedures and algorithms [11], [12]. These pilots and the CP do not bear any information data, and although necessary in the mobile-communication scenarios to assist the channel estimation and combat ISI, they are usually considered as a "waste" of system resources (including the transmit power).

We may thus summarize, that OFDM (and related NC-OFDM) and OFDMA are technologies that will be widely used for wireless communication due to their inherent interoperability, flexibility in the transmission parameters adjustment, efficient spectrum utilization, and spectrum aggregation capabilities. Thus, focusing on this technology for the future green wireless communication seems justified. Below, we address the following major issues of green OFDM-based wireless communication: its energy- and spectrum efficiency (that would include minimization of the transmit power, maximization of the transmission rate given low power limit, and minimization of the power "wasted" on pilots), lowering of the electromagnetic pollution that includes minimization of the OFDM OOB radiation, and spectrum recycling meaning efficient reuse of the spectrum temporarily not used by the PUs.

22.2   Efficient Use of Natural Resources in Wireless Multicarrier Transmission

Let us first consider, what natural resources are required and utilized for the wireless communication between the peers in a network. The energy consumption by the applied wireless transmission technology can be translated to the usage of the energy-providing resources: carbon, crude oil, or other conventional deposits or modern non-conventional resources. Moreover, the electromagnetic spectrum constitutes a natural and limited resource. The goal in the green wireless-communication design should be to possibly modestly but also very efficiently make use of these resources, i.e. of the available spectrum and energy.

OFDM technique can add flexibility to the communication system layers to reduce the energy consumption. Link-adaptation techniques aim at making the best use of available resources to achieve a target Quality of Service (QoS) while optimizing one of the transmission parameters. Based on the target QoS and on the channel state information (CSI), the transmission parameters (usually the modulation constellation and the coding scheme at distinct subcarriers) are adopted to optimize the goal function given the constraints imposed in the optimization algorithm. The OFDM link-adaptation performance, strongly depends on the accuracy of the CSI, affected by the channel estimation errors and the CSI feedback delay.

A number of variants of the adaptation strategies can be considered. In the throughput-oriented (rate-adaptive) strategy an adaptation algorithm aims at provisioning of the highest bit rate (or spectral efficiency) for a required bit error rate (BER) with a given radiated power limit. In the power-oriented (margin-adaptive) strategy, the algorithm would aim at minimization of the transmit power for a required BER and bit-rate. Finally, in less popular approaches, the strategy is to minimize the BER given the total-power and required bit-rate.

22.2.1   Energy-and-Spectrum Efficient OFDM Adaptive Transmission

As mentioned before, the energy-efficiency in adaptive OFDM (e.g. bit and power loading) is typically addressed by considering the margin-adaptive (MA) schemes, where the required QoS parameters, such as the data-rate and BER are constraints, and the power (or energy) allocated to the OFDM-symbol transmission is the subject of the minimization. The rate-adaptive

(RA) schemes maximize the user's data rate, and result in better utilization of the spectrum and thus, also in lower transmit-power-level requirements. The RA scheme has the following problem formulation: to find power levels at data subcarriers to maximize the total transmission data-rate under the power limitation, i.e. to find:

$$\max_{P(f_n)} \sum_{f_n \in \mathbf{S_D}} R(f_n) = \max_{P(f_n)} \sum_{f_n \in \mathbf{S_D}} \Delta f \log_2[1 + \delta P(f_n)\gamma(f_n)] \tag{22.1}$$

subject to:

$$\sum_{f_n \in \mathbf{S_D}} P(f_n) \leq P_{\text{req}}.$$

In the above formulas, $\mathbf{S_D}$ is the (finite and countable) subset of the OFDM subcarriers assigned to data symbols only (not to pilots), Δf is the subcarriers spacing, $R(f_n)$ is the data bit-rate at subcarrier n (at the frequency f_n), $\gamma(f_n) = |H(f_n)|^2 / \mathcal{N}_0$ is the Carrier-to-Noise Ratio (CNR) measured at the frequency f_n, $H(f_n)$ and $P(f_n)$ are the user's channel characteristic and the power spectral density allocated to this frequency respectively, and $\mathcal{N}_0$ is the noise power spectral density. Moreover, in (22.1), δ is the factor (often called the Signal-to-Noise Ratio (SNR)-gap) depending on the assumed user's Bit Error Probability (BEP) P_e, while P_{req} is the total data-power constraint, i.e. the power that can be assigned to data at the given OFDM-symbol period (the total OFDM symbol energy constraint minus the pilots energy over the OFDM period).

The MA bit-and-power loading algorithm is formulated such as to find subcarriers power assignments to minimize the total transmit power under the minimal data-rate constraint:

$$\min_{P(f_n)} \sum_{f_n \in \mathbf{S_D}} P(f_n) \tag{22.2}$$

subject to:

$$\sum_{f_n \in \mathbf{S_D}} R_n = \sum_{f_n \in \mathbf{S_D}} \Delta f \log_2[1 + \delta P(f_n)\gamma(f_n)] \geq R_{\text{req}},$$

where R_{req} is the target required data bit-rate that must be satisfied every OFDM-symbol period. Note, that although the problem formulation could also include optimal pilots energy allocation, it is not useful due to the additional data overhead involved. Therefore the above optimization problems are formulated in relation to the data subcarriers only and introduce parameterized pilots energy value.

The solutions to the above problems are obtained by the well known water-filling algorithm providing real-valued numbers of data rates at each subcarrier. The optimal algorithms resulting in the required integer granularity reflecting digital modulation constellations applied are: Hughes-Hartogs algorithm [13], Chow algorithm [14], Fisher-Huber algorithm [15] or Levin-Campello algorithm [16, 17]. As can be seen in (22.1) and (22.2), the optimization bases on the estimated channel characteristic, target BER, and the posed constraints.

Note, that in an OFDM transmitter, the consumed power consists of the transmit (radiated) power and the power needed for the signal processing. The physical layer techniques that result in the more error-robust transmission, and thus allow for lower transmit power, are usually computationally complex. This computation complexity in turn requires more processing power. Thus, the energy-oriented adaptation-algorithm design has to trade-off the optimization criteria, e.g. to minimize the radiated power given a limit for the baseband processing. Moreover, advantageous simplifications in the CSI representation can be made and suboptimal adaptation-algorithm versions can be applied that allow for less complex processing toward the QoS-acceptable solution.

The above considerations are related mostly to the single-link and single-user case. In literature, we find some interesting multi-user OFDM adaptive schemes. An optimal assignment of subcarriers, bits and powers to multiple users is based on the multiuser water-filling theorem [18], maximizing the total bit-rate given a maximum transmit power constraint for each user. Some practical approaches to this problem have been proposed, which aim at the maximization of the spectral efficiency while respecting transmit power constraints ([19, 20]), or at minimization of the total transmit power while guaranteeing a minimum bit-rate for each user ([21]). Some practical suboptimal solutions are also proposed, e.g. in [22, 23].

As mentioned above, in green approach to the utilization of natural resources two challenges are addressed: to minimize the usage of resources and to maximize the efficiency of this usage. Thus, both MA and RA schemes are useful for this green approach either for the minimization of the energy, or for the efficiency of the spectrum usage.

22.2.2   OFDM Energy-Wastage Minimization

In practical adaptive OFDM schemes, imperfect CSI has to be dealt with, what requires some energy for the transmission of pilots. To assist the accurate channel estimation necessary for the MA and RA schemes and for the successful data detection, a number of pilot symbols are usually transmitted over a wireless link, usually once within the channel coherence time. This energy *wasted* on pilots should be minimized because the pilot symbols do not convey any information. In some papers dealing with the problem of the channel-estimation quality and with the adaptive modulation, minimization of the estimation error has been the main goal to come up with the maximum channel capacity. In [24] the adaptive pilot allocation has been considered with a predetermined density in an OFDM symbol. In [25], the optimal pilots-to-data power ratio has been found for adaptive MIMO-OFDM system, optimizing the symbol-error-rate. As shown in [26] and [27], the pilots density and their energy in an OFDM symbol can be adaptively allocated depending on the channel estimation quality necessary for the signal reception and on the assumed OFDM-link-adaptation quality. There, the trade-off for the proper number of pilots in an OFDM symbol versus the single-pilot energy is considered for two cases: the *constant single-pilot energy* and the *constant total pilots-energy*.

In the *constant single-pilot energy* (CSPE) case, each single pilot energy is kept on the same level, independent of the pilot-symbols number. If the power limit is set for an OFDM symbols, an increase of the pilots number results in the decrease of the energy assigned to data symbols. In the *constant total pilots-energy* (CTPE) case, the energy assigned to all pilots is constant, and so is the energy assigned to data. Therefore, if the number of pilots increases, the energy assigned to a single pilot must decrease, because in both cases (CSPE and CTPE) the total energy of the OFDM-symbol is a subject to the constraint.

In [26, 27] results are presented for the rate-adaptive and margin-adaptive M-QAM-based OFDM systems (where $M \in \{2, 4, 16, 64, 256\}$) with pilots for both the CSPE and the CTPE cases. There, the energy consumption due to the pilots usage relative to the OFDM data symbols energy is shown versus the average SNR. The number of subcarriers $N = N_\mathrm{p} + N_\mathrm{s} = 256$ (where N_p and N_s are the number of pilot symbols and the number of information data-symbols respectively in one OFDM symbol). The following channel models have been onsidered: a two-paths Rayleigh frequency-selective channel with the delay-spread of $1/4$ of the OFDM-symbol period and the second path power 6 dB below the first one (in [27]) and LTE-conform Extended Pedestrian type A and Extended Vehicular channel models (in [26]). The assumed target BER has been 10^{-3}. The pilots density and their energy have been the subject of adaptation in both CSPE and CTPE cases.

It has been shown that as the average SNR increases, in the CSPE case, we can use pilots of lower density and lower average energy, what results in more subcarriers available for the data transmission or more energy that can be flexibly assigned to data. Consequently, the pilots energy decreases. In the CTPE case, the ratio between the pilots-energy and data-energy is fixed, so an increase of the average SNR results in better channel estimation but the total transmission energy per an OFDM-symbol as well as the minimum pilots density is fixed. For low SNRs transmission, outage is observed more often, so we obtain some unwanted energy-savings. The choice of the transmission parameters in adaptive OFDM should result in a green-transmission model with low energy "wasted" on pilots, but with the desirable spectral- and energy-efficiency.

22.3 Reducing the Electromagnetic Pollution Caused by Multicarrier Technologies

One of the issues of the OFDM and related NC-OFDM techniques is the shape of the spectrum outside of the intended transmission bandwidth. The NC-OFDM spectrum sidelobes have relatively high power due to the shape of an individual data-bearing subcarrier spectrum (*Sinc* function). Consequently, for the transmissions closely located (on the frequency axis) to a collection of data-bearing subcarriers of the considered NC-OFDM signal, the experienced interference may not be acceptable. Such an out-of-band (OOB) radiation can be treated as the electromagnetic *pollution* that taints the neighboring frequency bands. It is essential to limit this unwanted radiation that distorts (pollutes) other transmissions. For the OFDM and NC-OFDM signals it means that their spectrum shapes have to be treated such that the OOB radiation is minimized.

Moreover, the OFDM-based transmission is generally characterized by relatively high Peak-To-Average-Power Ratio (PAPR), and may suffer from non-linear distortions particularly occurring in High-Power Amplifiers (HPA). As a result the transmission signal spectrum broadens, what becomes the source of interference in the adjacent frequency bands. Thus, it is important to investigate methods for reducing both the OOB radiation and the PAPR of an NC-OFDM transmission signal. These methods should aim at achieving reasonable computational complexity, and low energy cost.

22.3.1 OFDM Out-of-Band Radiation Reduction

There have been a number of methods elaborated and presented in the literature for the OOB interference reduction in multicarrier systems (see [28] and the references therein). Here, we show some results of a very promising method that combines Windowing (W) and the so-called Cancellation Carriers (CC) method with some additional proposed improvements. Windowing is usually applied to the time-domain OFDM symbol with a cyclic prefix (CP). In order to use the window while simultaneously not negatively impact the bit-error rate (BER), the CP must be increased in length (to N_{CP}) and a cyclic suffix (CS) must also be added, which is a cyclic extension of β samples at the end of the considered symbol. The OFDM-symbol time-domain samples result from the multiplication of the samples vector with CP and CS at the output of IFFT by the window shape. In [29], it has been shown that high sidelobe suppression is achieved with the Hanning window. Windowing results in a relatively high throughput decrease, because the consecutive symbols overlapping each other by β samples yield to longer effective OFDM symbol duration. The windowing method is however low computationally complex, and its application does not depend on the modulated data.

The Cancellation Carriers (CCs) method [30] takes advantage of the spectrum shape of each subcarrier in order to reduce the resulting OOB interference level. There, a small subset of active subcarriers is selected to compensate the sidelobes of the adjacent subcarriers. As the subcarriers closest to the spectrum edge have the strongest influence on the OOB radiation, they are usually chosen to carry the canceling signal. The values of the CCs have to be calculated for each OFDM symbol separately, since the independent modulated data symbols cause different OOB interference levels. For this purpose, frequency-sampling points are defined in order to determine the values of the OOB signal spectrum in the frequency range that is targeted for the OOB reduction. Then, calculation of the values to be transmitted on the CCs is based on the optimization algorithm. This optimization problem has high computational complexity requiring solution of the Lagrange inequality for each OFDM symbol. A drawback of the CCs method is the BER-performance deterioration due to the fact that an OFDM system usually operates under the total power constraint. If part of an OFDM symbol energy is sacrificed to the cancellation carriers, the remaining energy that can be used for data transmission is reduced.

The CCs method has been extensively investigated, and a number of modifications and combinations of the CC algorithm with other methods has been presented, e.g. in [31–33]. The CCs method is very flexible, i.e. its key parameters such as the number of cancellation subcarriers and their power can be flexibly adjusted to the changing channel conditions to meet the required transmission quality. Some of its shortcomings can be efficiently equalized when it is combined with the windowing method. While windowing provides higher OOB radiation reduction for spectrum components more distant from the occupied OFDM band on the frequency axis, the CCs method performes better for components closer to the OFDM nominal band as shown in [34]. Thus, the combination of both methods provides additional degrees of freedom as the number of cancellation carriers and window shapes can be altered to fulfill the transmission requirements. In [28] several enhancements to this combined approach have been proposed, which reduce both the computational complexity of the method and the energy-loss (energy-wastage) due to the use of the CCs. An improvement of the BER performance is also observed.

As shown in [28], the CCs optimization formula is designed for the time domain windowed signal. Let us denote the input of the N-order IFFT as the vector $\mathbf{s} = \{s_{-N/2}, \ldots, s_{N/2-1}\}$ of N elements, which contains α data symbols at inputs indexed as: $\mathbf{d} = \{d_1, ..., d_\alpha\}$, φ cancellation symbols at the CCs: $\mathbf{c} = \{c_1, \ldots, c_\varphi\}$ and zeros for deactivated subcarriers. The optimization of cancellation symbols is based on the estimation of the spectrum values resulting from the superposition of the spectra of each cancellation and data carrier. For a set of ν frequency-sampling points $\mathbf{l} = \{l_1, \ldots, l_\nu\}$ in the optimization OOB region, the values of the cancellation subcarriers are calculated for each OFDM symbol separately since the independent modulated data symbols cause different OOB interference levels. These frequency-sampling points describe the optimization region, in which the estimates of the spectrum values resulting from the spectral superposition of the data subcarriers (DCs) and the CCs have to be calculated.

The optimization problem defined in [28] for the improved OOB-power reduction method is expressed by the following formula:

$$\min_{\mathbf{s_c}}\{\|\mathbf{P}_{\mathrm{CC}}^{(\nu\times\varphi)}\mathbf{s_c} + \mathbf{P}_{\mathrm{DC}}^{(\nu\times\alpha)}\mathbf{s_d}\|^2 + \mu\|\mathbf{s_c}\|^2\}, \tag{22.3}$$

where $\mathbf{s_d}$ and $\mathbf{s_c}$ contain complex values modulating data-carriers and cancellation-carriers respectively, and form the subvectors of vector $\mathbf{s}$ created by the $\mathbf{d}$ and $\mathbf{c}$ indexed cells, respectively. Moreover, $\mathbf{P}_{\mathrm{CC}}^{(\nu\times\varphi)}$ is the matrix of dimensions $(\nu\times\varphi)$ transforming the vector of the CCs values $\mathbf{s_c}$ of length φ to the spectrum estimates. For the DCs, the matrix $\mathbf{P}_{\mathrm{DC}}^{(\nu\times\alpha)}$ of dimensions $(\nu\times\alpha)$ and vector $\mathbf{s_d}$ of length α play the same role. Finally, μ factor is used to balance between the CCs power and resulting OOB power reduction to overcome the problem of an unacceptable power assigned to the CCs. For $n\in\mathbf{c}$, the coefficients $p_{n,l}$ are the elements of the matrix $\mathbf{P}_{\mathrm{CC}}^{(\nu\times\varphi)}$, and can be pre-calculated. Similarly, for the data carriers, when $n\in\mathbf{d}$, $p_{n,l}$ defines the matrix $\mathbf{P}_{\mathrm{DC}}^{(\nu\times\alpha)}$, and can be calculated off-line.

The solution of this problem is relatively low computationally complex, since the multiplication of vector $\mathbf{s_d}$ by a precalculated matrix is performed for each OFDM symbol. Moreover, the optimization procedure described above significantly reduces the SNR loss typical for a CCs method, because it limits the power assigned to CCs while increasing the power reserved for the DCs. The system BER performance can be additionally enhanced when the CCs are also used for the improved detection. This is because the CCs are correlated with the DCs and can be used as redundancy-tones. Thus, the CCs can be used to regain the power devoted to these subcarriers in the first place, but also make use of the frequency diversity for achieving a higher degree of robustness with respect to the frequency-selective fading [28].

Finally, a metric that indicates the potential throughput loss caused by the introduction of CCs, windowing or the combination of CCs and W can be derived. This throughput loss can be assessed in comparison with a system not employing any OOB interference reduction method, in which all subcarriers are occupied by the DCs. It is described by the following expression [28]:

$$R_{\mathrm{loss}} = \left(1 - \frac{1 - \frac{\varphi}{\varphi+\alpha}}{1 + \frac{\beta}{N+N_{\mathrm{CP}}}}\right) \cdot 100\%. \tag{22.4}$$

Below, simulation results are presented of the combined CCs and W methods improved as discussed above for the NC-OFDM system. In our experiments $N = 256$, where the subcarriers indexed as $\{-96, .., -1\} \cup \{1, .., 28\} \cup \{53, .., 96\}$ are occupied by the data symbols, and there are four cancellation carriers placed on each side of data carriers blocks. The duration of the cyclic prefix equals $N_{\mathrm{CP}} = 16$ samples, but $\beta = 30$ samples of the Hanning window extension are also used. The number of CCs and shaping window duration was chosen in such a way that the mean OOB interference power level is achieved at least 50 dB below the mean in-band power level for a reasonable value of μ, i.e., $\mu = 0.002$.

In Fig. 22.1, we can see the right-hand side of normalized Power Spectral Density (PSD) of the considered example NC-OFDM signal and its OOB power reduction obtained for three methods: CC method, windowing, and combined CC and W scheme. For comparison also the PSD of the reference system (an NC-OFDM system, in which all subcarriers are used for data carriers, without any spectrum-shaping technique applied) is shown. The comparison has been performed for the schemes that present the same potential throughput loss metric, which for our evaluation system equals $R_{\mathrm{loss}} = 17.77\%$. Similar throughput loss is obtained either from the CC method with $\phi_e = 8$ CCs per edge (a subset of the total of φ CCs) of the DCs band (data subcarriers are now indexed as: $\{-92, .., -1\} \cup \{1, .., 24\} \cup \{57, .., 92\}$), from the W method with Hanning window extension of $\beta = 65$ samples (DC subcarriers are now $\{-100, .., -1\} \cup \{1, .., 32\} \cup \{49, .., 100\}$), or from the described above combined CCs and W method with $\phi_e = 4$ and $\beta = 30$. Let us note, that the combination of both methods results in high and steep OOB power attenuation, thus confirming that the combination of these methods possesses the potential for protecting both wideband and narrowband neighboring PU's signals and significantly lowering the electromagnetic pollution in these neighboring bands.

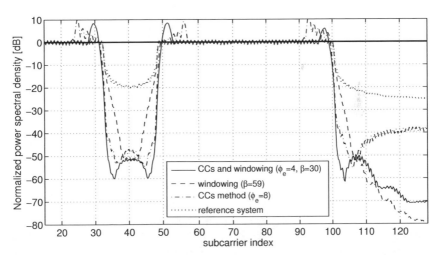

FIGURE 22.1 Normalized PSD of the NC-OFDM transmission signal with one narrow-band spectrum notch in case of the application of CC, W and combined CC and W methods.

22.3.2   Holistic View On Reduction of Non-Linear Distortions In OFDM

As mentioned above, the presence of the nonlinear element such as HPA in the processing chain could have additional significant influence on the OOB emission, in particular when the variation of the amplitude of the time-domain signal samples is high. In practice, the above mentioned signal's amplitude variation is measured by the means of PAPR metric. Various PAPR definitions can be found in the literature depending on the considered signal nature, i.e whether it is finite or infinite, continuous or discrete [35]. Here, we adopt the following PAPR definition:

$$\text{PAPR} = \frac{\|\mathbf{r}\|_\infty^2}{\text{E}\{\|\mathbf{r}\|_2^2\}}, \tag{22.5}$$

where $\mathbf{r}$ is the vector of time-domain samples (at the output of the IFFT after cyclic extension), $\text{E}\{\cdot\}$ denotes the expected value of a random variable, the $\|\mathbf{r}\|_\infty^2$ is the infinite norm denoting the maximum absolute value of vector $\mathbf{r}$ and $\|\mathbf{r}\|_2^2$ is the second order (Euclidean) norm denoting the power of the signal samples in vector $\mathbf{r}$.

A great number of various PAPR reduction methods can be found in the literature [36–40]. From all of the well-known PAPR reduction methods some of these techniques can be considered as preferable ones, because they do not require transmission of the side information necessary to restore original data signal intentionally distorted at the transmitter by the PAPR reducing algorithm. Another preference for a PAPR reduction method could be to choose a scheme that should not require any modifications in the reception algorithm. In this view, the Active Constellation Extension (ACE) method is considered to be very practical in the OFDM-based wireless transmission [41] and has been proposed for the new DVB-T2 standard, released in 2009 [42]. The ACE method is based on the amplitude predistortion of some selected data symbols at the input of IFFT to decrease the PAPR at its output. In this method, only the outer constellation points can be predistorted in order to maintain fixed minimum distance between the constellation points. The optimization problem defined for ACE is to find the set of data symbols to be predistorted to minimize the infinite norm of the IFFT-output signal samples. Detailed analysis of this optimization problem for the ACE method can be found in [43,44], where the problem has been defined as the quadratically-constrained quadratic program [45].

Interestingly, when PAPR is small the main source of the electromagnetic pollution in the adjacent frequency bands is the shape of the OFDM subcarrier spectrum, while when PAPR is relatively high, reduction of the sidelobes is not sufficient, and the PAPR reduction is of higher importance. Below, we try to provide the holistic view on the OOB power and PAPR reduction. One can find various approaches to this problem in the literature [44, 46], which are the combination of the formulation of particular problems presented earlier in this section. However, as it has been stated in [44], in most cases there does not exist an optimal value (solution) that optimizes jointly all cost functions targeting optimal bit and power loading, OOB power reduction and PAPR reduction. Thus, in [44] the authors have proposed the weighted joint PAPR-OOB optimization problem, where the weighted impact of the specific criterion (e.g. PAPR or OOB power) on the final optimization function is considered. In practice, however, the step-wise application of the OOB and PAPR reduction algorithm is adopted. This means that these respective PAPR and OOB power reduction procedures are applied in a consecutive manner.

Let us examine the effectiveness of the ACE method (used for PAPR reduction) associated with CCs algorithm described before. Extensive simulations have been carried out to verify the efficiency of these chosen methods for QPSK NC-OFDM transmission. In our investigations, the number of possible subcarriers (IFFT order) has been again set to $N = 256$, where the system effectively utilizes subcarriers indexed from -100 to 100, with zero-frequency carrier turned off and the frequency notch of 16 subcarriers required for a narrow-band transmission of a primary user. The ACE has been configured in the following way: the scaling (predistorting) parameter is assumed to be flexible, and can be modified in the range from 1 to 3; the number of predistorted symbols is set to 25% of all data subcarriers; the samples to be attenuated must have their amplitude higher than 1.4 times the mean amplitude of all time-domain samples in one OFDM symbol. The number of the cancellation carriers on each edge of the frequency band used for data transmission has been set to 4. It has been also assumed that the ACE method cannot modify the values carried on the CCs. Finally, the Rapp model with parameter $p = 4$ has been examined for the HPA.

The Complementary Cumulative Distribution Function (CCDF) of PAPR for five cases have been investigated. First, the PAPR characteristic has been obtained for the original unmodified signal. In the next two cases, the application of CCs without ACE and the application of ACE without the CCs method have been considered, respectively. Finally, both methods have been implemented jointly in the following configurations: the ACE method performed before the CCs method, the CCs algorithm proceeding the ACE. The obtained results are presented in the Fig. 22.2.

One can note that a small PAPR degradation is observed regarding the reference signal when only the CC method is applied. It means that the power back-off of the HPA in the processing chain has to be increased in order to keep the achieved OOB level unchanged. This, however, leads toward worse energy efficiency of the power amplifier. Significant PAPR improvement (more than 1dB at the CCDF level of 10^{-4}) can be observed when OOB and ACE reduction methods are applied jointly, however better performance is achieved for the case when the OOB method is applied first. In such a case, the PAPR characteristic is almost the same as observed for the reference signal. It means that the best option is to apply the CC method first and follow with the ACE procedure.

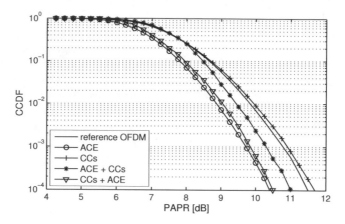

FIGURE 22.2 PAPR characteristics (CCDF of PAPR) for the considered scenarios of joint approach to PAPR and OOB power reduction.

The influence of subcarrier spectrum sidelobes and non-linearities of HPA on the signal spectrum shape, particularly in the adjacent bands can be observed in Fig. 22.3 for the HPA Input-Backoff (IBO) equal to 7 dB. For this IBO parameter value, the most advantageous in terms of the spectrum shape is the combination of PAPR reduction using ACE method followed by the insertion of CCs. Either the use of PAPR reduction or the sidelobes suppression alone provide worse results.

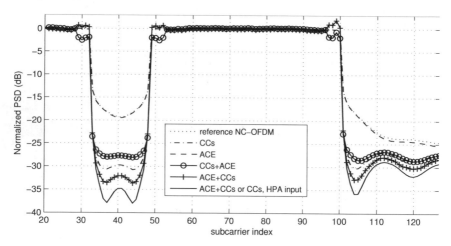

FIGURE 22.3 PSDs observed after and before the HPA (Rapp model with $p = 4$) for IBO=7dB

22.4 Spectrum Recycling in Future Multicarrier-Based Networks

Electromagnetic spectrum is a limited and scarce natural resource, and its efficient usage for mobile telecommunication services has been in focus of research in the last decades. Intelligent and flexible spectrum access procedures and resource allocation methods are continuously developed and improved to increase this efficiency. The ambition of the modern radio network design is to *recycle* the spectrum that is temporarily unused, and to rationalize its distribution through the application of the concept of dynamic and cognitive spectrum access. It has been observed that the spectrum in many licensed frequency bands is under-utilized, i.e. the duty cycle is quite low even in densely populated areas. The Dynamic Spectrum Access (DSA) procedures are often implemented in the centralized manner and require the Channel State Information (CSI) of all links in the network. However, this necessitates a volume of overhead traffic. The cognitive radio (CR) concept, on the other hand, involves more distributed operation of intelligent wireless devices as autonomous as possible.

Below, we focus on the DSA methods in the OFDMA-based network, i.e. on the allocation of subcarriers aiming at rational and efficient utilization of the spectrum recycled from the licensed users who temporarily do not use it. Rationality in our case means that apart from maximizing the spectral efficiency the network and each individual node aim at lowering the energy-cost of this efficiency, as well as at increasing the number of served users. This can be done based on the concept of *coercive taxation* [47]. The actors in such a network are intelligent (cognitive) radio devices using NC-OFDM technique, trying to acquire subcarriers in the available transmission band, possibly many of them to maximize their throughput. Naturally, a selfish node would occupy the whole available spectrum. However, from the green communication network perspective, this behavior would decrease its capacity in terms of the number of users served. The problem is related to the classical resource utilization dilemma known as the *Tragedy of Commons* and described in [47]. The problem is that utilization of limited resources is inefficient when the users expose selfish behavior.

22.4.1 NC-OFDM Spectrum Access with Taxation

In [48], resource taxation method has been presented for the OFDMA-based network area that prevents the selfish behavior of the nodes in the network. In this framework, a cognitive NC-OFDM radio terminal senses the spectrum for the available subcarriers. There is a collision-avoidance mechanism to handle two nodes trying to access the spectrum at the same time. After sensing of the available spectrum resources, each actor (an NC-OFDM radio node) can occupy a number of available SCs. Each node is allowed take decisions independently while treating the rest of the actors as a whole (the Network Nodes Community - NNC). In the considered case, when the node wants to use a number of i subcarriers, for sure these should be its i strongest SCs in terms of the highest Carrier-to-Interference and Noise Ratio (CINR), because from both the individual node perspective and from the whole network perspective, making use of the strongest SCs results in higher spectral efficiency. The idea is first to limit the nodes in the maximum number of subcarriers a node can take at the time, and then to apply taxation to prevent them from excessive (and selfish) usage of the resources. Note that limiting the nodes in the number of accessed subcarriers has been proved beneficial from the network efficiency standpoint [49].

In [50], two taxation-based models for the scenario described above have been presented: the *Selfish-Behavior Model* (SelBM) and the *Social-Behavior Model* (SocBM) with different utility functions used by the nodes in the network with orthogonal channels, e.g. OFDM subcarriers. Moreover, the nodes use only their local information on their own CSI, the number of sensed available channels and the number of other actors in the network (their competitors). In the SelBM, the utility for the kth user is defined as:

$$\zeta_k(b_k) = \int_{f \in \mathbf{S}_k} \log_2 \left[1 + \delta_k P_k(f)\gamma_k(f)\right] df - r_{\text{tax}} b_k , \qquad (22.6)$$

where $\mathbf{S}_k$ is the (non-countable) set of frequencies potentially available for user k, b_k is the amount of bandwidth the user acquires, r_{tax} is the tax-rate parameter, $\gamma_k(f) = |H_k(f)|^2 / \mathcal{N}_0$ is the Carrier-to-Noise Ratio (CNR) measured at the frequency f, $H_k(f)$ and $P_k(f)$ are the kth user's channel characteristic and the power spectral density allocated to this frequency respectively, and $\mathcal{N}_0$ is the noise power spectral density. Let us note, that for the case of orthogonal channels, interference that is usually added to noise equals zero. Moreover, in (22.6), δ_k is the SNR-gap for the assumed BEP.

In the SocBM, the utility is defined as:

$$\xi_k(b_k, c_k) = \left\{ \frac{1}{B} \int_{f \in \mathbf{S}_k} \log_2 \left[1 + \delta_k P_k(f)\gamma_k(f)\right] df \right\} \cdot \{B_k - b_k - c_k\} - r_{\text{tax}} b_k ,$$
$$(22.7)$$

where c_k is the amount of bandwidth that will be occupied by NNC, and B_k is the amount of bandwidth available for the kth node at its turn. One may interpret formula (22.7) as the total normalized throughput (throughput per frequency unit of the total available bandwidth B) which could be obtained by the new incoming users in case they occupied the remaining bandwidth and had the same average spectral efficiency as the considered user (node). This way, in the decision-making on how much of the spectrum to occupy, the network nodes factor the social aspect of the network (to serve multiple nodes) and not just their own benefit.

Properly established taxes aim at extorting the desirable actors' (network nodes or users) behavior. The tax rate can be fixed (linear tax) and not dependent on the chosen strategy, or contrary, it may depend on the number of chosen SCs (progressive tax). The tax-rate should be carefully adopted to the number of resources, the number of competing nodes and their anticipated behavior. In [50], a reduced-complexity algorithm searching the optimal tax-rate r_{tax} has been proposed to maximize the total network throughput for both the SelBM and SocBM.

The average network throughput per available subcarrier resulting from the optimal taxation is presented in Figure 22.4 for $\Delta f = B/256$. The power-control mechanism has been assumed to control the transmitted power for each node, with a tolerance of 3 dB. The channel model is the six-path channel, with paths having the same power, and delays uniformly spread between 0 to $1/B$. (This is a test-channel model often used for the test of equalizers, that reflects particularly hostile environment with very small coherence bandwidth and very deep fading.) The target BEP $P_e = 10^{-3}$ for all links. The results are compared with the greedy algorithm (that assigns the frequencies to the nodes with the highest CNR value at these frequencies) and Round-Robin algorithm. Although both of these algorithms can be only implemented in a centralized manner, they give the two-opposite extremes: either maximum spectral efficiency or maximum fairness for the case of the whole used bandwidth. The simulation results show that it is highly beneficial for the network and for the individual nodes to use taxation with the tax-rate maximizing the network sum-throughput. It is also beneficial due to better utilization of the spectrum resources and higher percentage of served nodes. Simulation results also show that in the considered scenario, when the optimal tax-rate is applied, the achievable sum-throughput per frequency unit is as high as 5.5 [bits/s/Hz] for sufficiently high number of nodes. Moreover, in such a case, $99 - 100\%$ of nodes are served in the network, i.e. are able to acquire some resources satisfying their target BEP.

22.5   Concluding Remarks

Multicarrier technologies are considered for future wireless communication systems due to their interoperability, flexibility in the transmission parameters adjustment, efficient spectrum utilization and the spectrum aggregation and shaping capabilities. The challenges posed by the green OFDM-based radio include the development of methods for the energy-aware, efficient and rational use of natural resources (energy and spectrum) under multiple limitations posed by the network. We have presented methods that jointly treat the energy-efficiency, electromagnetic pollution reduction and efficient spectrum reuse (recycling) in the OFDM and related NC-OFDM transmission.

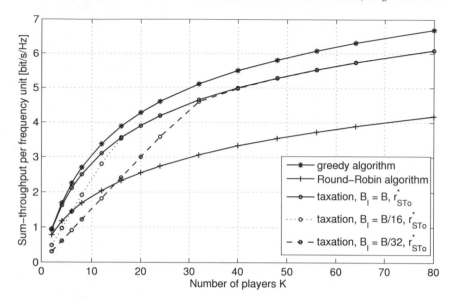

FIGURE 22.4 Average sum throughput for the optimal tax-rate r^*_{tax} for both SelBM and SocBM; six-path channel model.

Although presented results are promising, further integration of these techniques is needed to account for cross-layer issues in future green radio.

Acknowledgements

The work presented in this chapter has been supported by the European 7th Framework project ACROPOLIS (no. 257626) on *Advanced coexistence technologies for radio optimisation in licenced and unlicensed spectrum.*

Bibliography

[1] L. Hanzo, M. Munster, B.J. Choi, T. Keller, *OFDM and MC-CDMA for Broadband Multi-user Communications, WLANs and Broadcasting*, John Wiley & Sons, July 2003.

[2] R. Van Nee, R. Prasad, *OFDM for Wireless Multimedia Communications*, Artech House Publishers, Boston, London, 2000.

[3] T. Hwang, C. Yang, G. Wu, S. Li, G.Y. Li, "OFDM and Its Wireless Applications: A Survey", *IEEE Transactions on Vehicular Technology*,

Vol. 58, Issue 4, May 2009, pp.1673-1694.

[4] Y. (G.) Li and G. Stuber, *Orthogonal Frequency Division Multiplexing for Wireless Communications*, Boston, MA: Springer-Verlag, Jan. 2006.

[5] T. Keller, L. Hanzo, "Adaptive Modulation Techniques for Duplex OFDM Transmission", *IEEE Transactions On Vehicular Technology*, Vol. 49, No. 5, September 2000.

[6] L. Hanzo, C.H. Wong, M.S. Yee, *Adaptive Wireless Transceivers: Turbo-Coded, Turbo-Equalised and Space-Time Coded TDMA, CDMA and OFDM systems*, John Wiley, March 2002.

[7] J. Mitola, "Cognitive Radio: an integrated agent architecture for software defined radio", Dissertation, Doctor of Technology, Royal Institute of Technology (KTH), Sweden, 2000.

[8] S. Haykin, "Cognitive Radio: Brain-Empowered Wireless Communications", *IEEE Journal Select. Areas In Commun.*, Vol. 23, No. 2, Feb. 2005, pp. 201-220.

[9] R. Rajbanshi, A.M. Wyglinski, G.J. Minden, "VOFDM-Based Cognitive Radios for Dynamic Spectrum Access Networks", Chapter 6 in E. Hossain, V.K. Bhargava, *Cognitive Wireless Communication Networks*, Springer Science and Business Media, 2007.

[10] H. Colebi, H. Arslan, "Utilization of Location Information in Cognitive Wireless Networks", *IEEE Wireless Communications*, vol. 14, no. 4, 2007, pp 6-13.

[11] X. Li, K. Pahlavan, M. Latva-aho, M. Ylianttila, "Indoor geolocation using OFDM signal in HIPERLAN/2 wireless LANs", Personal Indoor and Mobile Radio Communication Conference, PIMRC 2000, vol. 2, pp. 1449-1453.

[12] H. Ni, G. Ren, Y. Chang, "A TDOA Location Scheme in OFDM Based WMANs", *IEEE Transactions on Consumer Electronics*, Vol. 54, No. 3, August 2008, pp. 1017-1021.

[13] D. Hughes-Hartogs, "Ensemble Modem Structure for Imperfect Transmission Media,"U.S. Patents Nos. 4,679,227 (July 1987), 4,731,816 (Mar. 1988), and 4,833,706 (May 1989).

[14] P. S. Chow, J. M. Cioffi, and J. A. C. Bingham, "A practical Discrete Multitone Transceiver Loading Algorithm for Data Transmission over Spectrally Shaped Channels," *IEEE Trans. Commun.*, Vol. 43, February/March/April 1995, pp. 773 - 775.

[15] R. F. H. Fischer and J. B. Huber. "A new loading Algorithm for Discrete Multitone Transmission". in Proc. IEEE Globecom '96 pp 724-728, London, November 1996.

[16] J.C. de Souza, "Discrete Bit Loading for Multicarrier Modulation Systems," PhD dissertation, Stanford University, June 1999.

[17] H. Levin, "A complete and optimal data allocation method for practical discrete multitone systems," in Proc. of 2001 IEEE Global Telecommun. Conf. , vol. 1, pp. 369-374, Nov. 2001.

[18] R.S. Cheng, S. Verdu, "Gaussian Multiaccess Channels with ISI: Capacity Region and Multiuser Water-Filling", *IEEE Transactions on Information Theory*, Vol. 39, No. 3, May 1993, pp. 773-785.

[19] C. Zeng, M. Luise, C. Hoo, J.M. Cioffi, "Efficient water-filling algorithms for a Gaussian multiaccess channel with ISI", Vechicular Technology Conference VTC-Fall 2000, Boston, Sept. 2000.

[20] G. Munz, S. Pfletschinger, J. Speidel, "An efficient waterfilling algorithm for multiple access OFDM", IEEE Global Telecommunications Conference, 2002,GLOBECOM'02, Vol. 1, 17-21 Nov. 2002, pp. 681-685.

[21] C.Y. Wong, R.S. Cheng, K.B. Letaief, D. Mursh, "Multiuser OFDM with adaptive subcarrier, bit, and power allocation", *IEEE Journal on Selected Areas in Communications*, Vol. 17, No. l0, pp. 1747-1757, Oct. 1999.

[22] S.H. Ali, K.-D. Lee, V.C.M. Leung, "Dynamic Resource Allocation in OFDMA Wireless Metropolitan Area Networks", *IEEE Wireless Communications*, Feb. 2007, pp. 6-13

[23] 65. A. Kliks, H. Bogucka, "Computationally-efficient bit-and-power allocation for multicarrier transmission", Future Networks and Mobile Summit, FuNeMS 2010, 16-18 June, 2010, Florence, Italy

[24] G. Jiann-Ching, A. Osseiran, "Adaptive Pilot Allocation in Downlink OFDM", IEEE Wireless Communications and Networking Conference, WCNC 2008, pp. 840-845.

[25] T. Kim, J.G. Andrews, "Balancing pilot and data power for adaptive MIMO-OFDM systems", IEEE Global Communications Conference, GLOBECOM 2006, Nov. 2006, pp. 1-5.

[26] L. Koschel, H. Bogucka, "Energy-Efficient Bit and Power Loading Options for Practical OFDM Systems with Pilot-Aided Channel Estimation," IEEE International Symposium on Personal Indoor and Mobile Radio Communications, PIMRC'10, Istanbul, Turkey, 26-29 Sept, 2010,

[27] H. Bogucka, A. Conti, "Degrees of Freedom for Energy Savings in Practical Adaptive Wireless Systems," *IEEE Communications Magazine*, June 2011, Vol. 49, No. 6, pp. 38-45.

[28] P. Kryszkiewicz, H. Bogucka, A.M. Wyglinski, "Protection of Primary Users in Dynamically Varying RadioEnvironment: Practical Solutions and Challenges," *EURASIP Journal on Wireless Communications and Networking*, special issue on "Ten Years of Cognitive Radio: State of the Art and Perspectives," in press.

[29] M.S. El-Saadany, A.F. Shalash, M. Abdallah, "Revisiting active cancellation carriers for shaping the spectrum of OFDM-based Cognitive Radios",

Sarnoff Symposium, 2009. IEEE SARNOFF '09, vol., no., pp.1-5, March 30 2009-April 1 2009.

[30] S. Brandes, I. Cosovic, M. Schnell, "Reduction of out-of-band radiation in OFDM systems by insertion of cancellation carriers", *IEEE Communications Letters*, vol.10, no.6, pp.420-422, June 2006.

[31] S. Brandes, I. Cosovic, M. Schnell, "Reduction of out-of-band radiation in OFDM based overlay systems", IEEE DySPAN 2005, pp.662-665, 8-11 Nov. 2005.

[32] N. Sokhandan, S.M. Safavi, "Sidelobe Suppression in OFDM-based cognitive radio systems "10th International Conference on Information Sciences Signal Processing and their Applications (ISSPA) 2010, pp.413-417, 10-13 May 2010.

[33] Shih-Gu Huang, Chien-Hwa Hwang, "Improvement of active interference cancellation: avoidance technique for OFDM cognitive radio " *IEEE Transactions on Wireless Communications*, vol.8, no.12, pp.5928-5937, December 2009.

[34] Mahmoud, H.A.; Arslan, H.; "Spectrum shaping of OFDM-based cognitive radio signals", *IEEE Radio and Wireless Symposium*, pp.113 - 116 , January 2008.

[35] M. Sharif, M. Gharavi-Alkhansari, B.H. Khalaj, "On the peak-to-average power of OFDM signals based on oversampling," *IEEE Transactions on Communications*, Vol. 51, No. 1, Jan. 2003, pp. 72 - 78.

[36] H. Bogucka, H., "Directions and Recent Advances in PAPR Reduction Methods," IEEE International Symposium on Signal Processing and Information Technology, Aug. 2006, pp. 821 -827.

[37] H.S. Han, J.H. Lee, "An overview of peak-to-average power ratio reduction techniques for multicarrier transmission," *IEEE Wireless Communications*, Vol. 12, No. 2, April 2005, pp. 56 - 65.

[38] T. Jiang, T. Wu, "An Overview: Peak-to-Average Power Ratio Reduction Techniques for OFDM Signals," IEEE Transactions on Broadcasting, Vol. 54, No. 2, June 2008, pp. 257 -268.

[39] Y. Louët, J. Palicot, "A classification of methods for efficient power amplification of signals," *Annals of Telecommunications*, Vol. 63, No. 7, 2008, pp. 351-368.

[40] L. Wang, C. Tellambura, "An Overview of Peak-to-Average Power Ratio Reduction Techniques for OFDM Systems," IEEE International Symposium on Signal Processing and Information Technology, Aug. 2006, pp. 840 -845.

[41] S. Sezginer, H. Sari, "OFDM peak power reduction with simple amplitude predistortion," *IEEE Communications Letters*, Vol. 10, No. 2, Feb. 2006, pp. 65 -67.

[42] ETSI EN 302 755 V1.1.1 (2009-09),"ETSI Digital Video Broadcasting (DVB); Frame structure channel coding and modulation for a second generation digital terrestrial television broadcasting system (DVB-T2)," Sep. 2009.

[43] B.S. Krongold, D.L. Jones, "PAR reduction in OFDM via active constellation extension," *IEEE Transactions on Broadcasting*, Vol. 49, No. 3, Sept. 2003, pp. 258 - 268.

[44] M. Senst, M. Jordan, M. Dorpinghaus, M. Farber, G. Ascheid, H. Meyr, "Joint Reduction of Peak-to-Average Power Ratio and Out-of-Band Power in OFDM Systems," IEEE Global Telecommunications Conference, GLOBECOM'07, Nov. 2007, pp. 3812 -3816.

[45] D. Bertsekas, A. Nedic, A. Ozdaglar,"Convex Analysis and Optimization," *Athena Scientific*, April 2003.

[46] A. Ghassemi, L. Lampe, A. Attar, T.A. Gulliver, "Joint Sidelobe and Peak Power Reduction in OFDM-Based Cognitive Radio," IEEE 72nd Vehicular Technology Conference Fall (VTC 2010-Fall), Sept. 2010, pp. 1 -5.

[47] G. Hardin, "The Tragedy of the Commons," Science, no. 162, 1968, pp. 1243-1248.

[48] H. Bogucka, "Efficient and Rational Spectrum Utilization in Opportunistic OFDMA Networks with Imperfect CSI: a Utility-Based Top-Down Approach," *Wireless Communications and Mobile Computing Journal*, Wiley InterScience, 2010, DOI: 10.1002/wcm.973.

[49] S.M. Perlaza, M. Debbah, S. Lasaulce, H. Bogucka, "On the Benefits of Bandwidth Limiting in Decentralized Vector Multiple Access Channels," 4th International Conference on Cognitive Radio Oriented Wireless Networks and Communications, Hannover, Germany, June, 22-24 2009.

[50] H. Bogucka, "Optimal Resource Pricing Coercing Social Behavior in Wireless Networks," IEEE International Communications Conference ICC'11, 5-9 June, 2011, Kyoto, Japan.

Author Contact Information

Hanna Bogucka, Adrian Kliks, and Paweł Kryszkiewicz are with Poznan University of Technology, Poznan, Poland, Email: hbogucka@et.put.poznan.pl. Oliver Holland is with King's College London, London, UK, Email: oliver.holland@kcl.ac.uk.

23

Green Video Streaming over Cellular Networks

Yingsong Huang

Auburn University, USA

Shiwen Mao

Auburn University, USA

Yihan Li

Auburn University, USA

CONTENTS

659

23.1   Introduction

According to a recent study by Cisco, mobile data traffic is expected to grow to 6.3 Exabytes per month by 2015, a 26—fold increase over 2010 [1]. In addition, mobile video will generate much of the mobile traffic growth by 2015. Of the 6.3 Exabytes per month wireless data traversing mobile wireless networks by 2015, 4.2 Exabytes will be related to video. This trend is driven by the overwhelming proliferation of intelligent handheld devices, which brings about the compelling need for ubiquitous access to video content over wireless access networks. This trend will significantly influence the design and operation of future wireless networks. Wireless video applications not only are bandwidth intensive, but also involve stringent user quality of experience (QoE) requirements. Under this context, the problems of green communications and power efficiency become ultimate important: it not only determines how long a wireless handheld device can operate without recharging battery, but also has significant implications on environmental issues.

To meet the explosively increasing wireless data (and in particular, wireless multimedia) demands, more and more wireless access networks are being deployed. It is reported that every year, 120,000 new base stations (BS) are added, catering to the 300 Million to 400 Million new mobile phone users adopting mobile services around the world [2]. Furthermore, there is a rapid growth in the deployment of femtocells [3–6]. A femtocell is a small cellular BS, typically used for serving approved users within a small coverage such as a house. Femtocells can be used to extend coverage, improve capacity, and reduce both power consumption and interference, by reducing the distance of wireless transmissions [3]. Many wireless operators have launched femtocell service recently, such as AT&T, Sprint, Verizon, and Vodafone. The incipient wide adoption of femtocells will greatly intensify the proliferation of BS's.

Given the intensive concerns on carbon dioxide emission and global warming, the concept of "green" communications has attracted considerable interest and effort from the research community. Among various green communication technologies, we focus on the energy efficiency of base stations for downlink video streaming. This is due to the expected surge in wireless video data, as well as the drastic increase in the deployment of BS's. It is reported that, in a typical cellular network, more than 50% of the total power consumption is directly attributed to BS equipment [7]. Therefore, any small improvement in the energy efficiency of video coding or wireless transmission systems will be amplified by the huge volume of wireless video data and large number of deployed BS's, and will result in considerable environmental impact as given by the Amdahl's law. Considerable savings on electrical bills could be achieved for wireless operators when the power of BS's is minimized for video streaming.

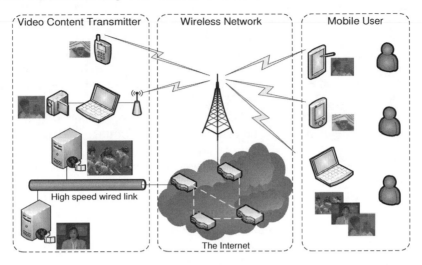

FIGURE 23.1 Overview of the wireless video networking system.

The reduced electricity consumption will also bring about important improvement in the overall carbon footprint of the wireless industry and achieve the goal of "green" communications.

In this book chapter, we will focus on the energy consumption of BS's for downlink video streaming. In particular, we first provide an overview of the general background of green wireless multimedia communications in Section 23.2. Next, we consider the problem of downlink power allocation for variable-bit-rate (VBR) video streaming in a cellular network in Section 23.3, as a case study of green wireless video communications. We present a cross-layer optimization and scheduling framework with the object to minimize the BS's power consumption while maximizing the QoE of video users. We conclude the chapter with a brief discussion of open problems and future directions in Section 23.4.

23.2 Green Video Communications

23.2.1 Video Communication System Overview

A typical wireless video system is illustrated in Fig. 23.1, which generally consists of video content transmitters, wireless/wireline network systems, and mobile receiving/playout devices. To achieve energy efficiency in wireless multimedia systems, various schemes have been developed in prior work, each focusing on one (or several) of the system components.

The video content provided by commercial video providers or individuals is first encoded into compressed videos using different codecs. Generally constant-bit-rate (CBR) videos are coded for streaming videos, since the relatively stable bit rate of the CBR video makes it easier for network scheduling and resource allocation [8–10]. However, there may be relatively large fluctuations in the quality of different frames. VBR videos are usually used instead for stored videos since they can provide stable quality for the video frames, but having bit rates with extremely high variations [11–13]. Scalable videos are particularly appealing for video streaming since they allow the trade-off between video quality and the amount of network resource required (or available) [4, 6, 14, 15].

At the video content transmitter, the encoded frames are then packetized by the network transmission protocols. The video packets are streamed towards the destination through wired and/or wireless networks. After frames arrive at the destination, the codec decodes the received frames. Since the video packets could be corrupted or lost during transmission, error control and concealment techniques may be applied to mitigate the impact of transmission errors when decoding the video. Finally the reconstructed video frames are played out on screens at the receiving device.

During the video streaming process, energy are mainly consumed by video codec encoding, network transmissions, receiver decoding, error mitigation, and playout. The energy consumption incurred at the encoder and decoder are mainly due to the processing of video data at the end nodes, while encoding is usually more computation and energy intensive than decoding. The codec computation and network transmission use up the largest part of the overall energy, and are also critical components for the achievable QoE of video service [16]. It is important to address the "green" communication problem in video streaming by exploring the energy savings in both video codec and video transmission.

23.2.2   Energy Efficiency in Video Coding

In the past decades, the advances of wireless communications and networking technologies are much significant than that of the battery technology. Consequently, how to prolong the battery life of mobile devices becomes one of the major environmental and economical concerns. We focus on power efficient codec design in this section and will explore the battery-aware transmission in the next subsection.

As the increasing demand of high quality video, high compression efficiency codecs are designed to enable higher resolution, which significantly increases the complexity of encoding algorithms. Moreover, the stringent delay requirements of video service usually keep the video device processor constantly busy for managing the high computation tasks. The processor may consumes as much as 2/3 of the total power of a mobile device [17]. Thus it is important to balance video quality and the computational complexity to achieve power-aware video coding.

Typical encoder and decoder block diagrams are illustrated in Fig. 23.2 and

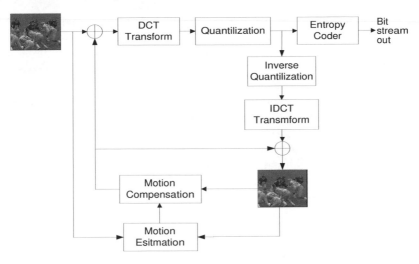

FIGURE 23.2 Block diagram of a typical video encoder.

FIGURE 23.3 Block diagram of a typical video decoder.

Fig. 23.3, respectively. Such approaches have been adopted in many video coding standards, such as H.264, and MPEG-4. The framework consists of motion estimation and compensation, discrete cosine transform (DCT), quantization, entropy coding, inverse quantization, and inverse DCT (IDCT). According to recent research [18], motion estimation/compensation constitutes more than 40% of the CPU workload, DCT/IDCT and quantization/inverse-quantization makes up over 16% of the CPU workload, and the entropy encoder, whose computational complexity largely depends on the coding bit rate, composes less than 10% of the CPU workload. Thus, it is important to explore the energy efficiency of these components that consume the most part of the processing power at a video codec.

There are many power-aware codec designs to balance power consumption and video quality. The main idea behind these techniques is that the diverse complexity of the video content may require different levels of compression. To achieve a certain video quality, slow motion and simple scenes require much less computation than high motion sports and movie streams. Thus it is theoretically feasible to obtain the optimal power efficiency by dynamically adjusting the computation complexity of the codec components for different videos, frames, macro blocks (MBs) and blocks, while keeping the video quality relatively constant at a certain level.

In [19], the authors present a configurable coding scheme, which adjusts the codec control parameters to achieve an optimal operation point on complexity-distortion curves based on exhaustive search and the Lagrangian multiplier method. In [20], a power-aware motion estimation algorithm is presented, which is adaptive to the battery status by a content-based subsample algorithm. When the battery is in the full capacity, all the processing elements in the motion estimation function are turned on to provide the best quality. On the other hand, when the battery capacity is decreased, some processing elements are disabled to extend the battery life with little quality degradation. In [21], the authors extend the functions of DCT/IDCT in a framework to decrease the power consumption by skipping the low energy MBs in DCT and all zero coefficients input data in IDCT. The combined method reduce, on average, 94% of power dissipation.

Another class of power-aware video codec design aims to dynamic adjust the voltage and frequency of the CPU for energy conservation. Various dynamic voltage scaling (DVS) algorithms are provided to determine the minimum energy consumption for processing video tasks under stringent delay requirements. With a DVS enabled processor, the voltage level and associated clock frequency are adapted to the time-varying video processing workload to save energy. The trade-off between reducing voltage level/clock frequency and increasing processing time is the core in the DVS-based design.

In [22], the authors derive the optimal voltage scheduling with linear programming. The algorithm calculates the optimal scheduling offline with knowledge of the precise complexity and arrival time of each decoding job, which may not be easy to acquire in real time. A heuristic algorithm is then introduced by predicting the stochastic complexity of the workloads. In [23], a DVS algorithm is presented that adjusts both the clock frequency and the voltage level of the CPU to achieve energy efficiency for video content processing, while maintaining the QoS of the video. A comprehensive statistical analysis of the CPU workload is presented in [24] for multimedia applications. The statistical results show that there is a large room for DVS to reduce energy consumption for multimedia streaming and the processor workload can also be accurately predicted with a moderate effort. The DVS system is based on the control theoretic framework. A PID-based DVS controller is developed to achieve a penalty controllable energy reduction, which can be incorporated into an online algorithm.

In summary, the power-aware codec design focuses on the tension between video quality and power consumption based on content diversity. The existing power-aware schemes extend the traditional video codec functions by jointly considering the video content and power constraint. The algorithms aim to adjust the codec parameters to minimize the power consumption while preserving good video quality. In addition, the hardware support for DVS technologies enables adaptive adjustment of the clock frequency and operating voltage level of the CPU, to accommodate varying codec workload. It should be noted that the power-aware codec design needs to jointly adjust large number of configurable parameters, which provides the context for applying effective global optimal techniques and algorithms.

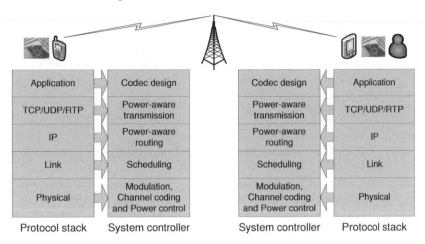

Protocol stack | System controller | System controller | Protocol stack

FIGURE 23.4 Protocol layers and system controls involved in wireless video transmission.

23.2.3 Energy Efficiency in Video Transmission

A typical video transmission path is shown in Fig. 23.4. The video frames generated by the codec are packetized and delivered through the network protocol stack (UDP/RTP/IP). The link layer schedules the packets with a MAC protocol (e.g., TDMA, FDMA, CDMA, or CSMA) and passes the frames down to the physical layer, where channel coding, modulation and power allocation may be applied to overcome the time-varying and unreliable wireless channels. At the receiver side, received video packets are decoded. The video bit stream is restored and then decompressed by the decoder. Error concealment techniques may be applied to mitigate the impact of delayed and corrupted video frames.

The key challenge of video over wireless networks is the time-varying wireless channel, which has a gain that varies over time due to channel fading, shadowing, and inter channel interference [25]. This causes random packet losses and delays. Thus, fixed resource allocation or scheduling schemes may not be sufficient to achieve the best video quality or energy efficiency. An adaptive, video content-aware resource allocation scheme is necessary for supporting energy efficient video streaming over wireless networks.

It is reported that BS equipment consumes more than 50% of the total power in a typical cellular network. The energy efficiency and power control in cellular networks thus demand careful reexamination to achieve the goal of green communications. Power control in cellular networks has been widely studied for more than 15 years for voice or data applications. Many effective power control algorithms are proposed in the literature (see [26–31]), and the closely related admission control problems are also investigated [32–34]. The problem of video communications in cellular networks brings about many new challenges to the BS power control problem, since the power allocations need to achieve not only the target Signal to Interference-plus-Noise Ratio (SINR), but also the QoS or QoE of the streamed videos [12, 13, 35, 36].

23.2.4   Joint Video Coding and Transmission Design

As shown in Fig. 23.4, the wireless video system is a complex system with many closely coupled control knobs and parameters. Clearly, a cross-layer design that jointly optimizes multiple parameters in different layers has the potential of achieving better energy efficiency and video performance, comparing to the traditional layered approach.

Consider power allocation as an example. Normally, a lower transmission rate requires a smaller transmit power, as well as high compression ratio at the video codec. However, as discussed in the previous subsection, a higher compression ratio incurs more intensive computations and consumes more power at the codec. Apparently, a cross-layer design framework would be useful in this case, where video coding in the application layer and the transmission schemes in the lower network layers are jointly considered and optimized [13, 37]. Specifically, the lower layer network protocols now have the opportunity to exploit the information from the video content and source coding parameters to optimize the transmission strategy. On the other hand, video coding in the application layer may also take advantage of the channel and network information, and thus can select the coding parameters to provide the best coding quality and be adaptive to the status of wireless networks and channels. This approach also provides future support for the "content-centric" multimedia network design [38–41].

A large number of designs that jointly consider video coding and transmission have been proposed in the literature. We review some selected ones in the following due to space limit. In [42], the authors adopt joint source coding and channel coding to minimize the total power consumption, while keeping the end-to-end video quality at a fixed level. A framework of joint source-channel coding and power adaptation is presented in [43], where error resilient source coding , channel coding and transmission power adaptation are jointly designed to optimize video quality, given constraints on the total transmission energy and delay. An algorithm is introduced to find the minimal energy source coding and power allocation by adaptively allocating resources to different video segments based on their relative importance. In [44], the RF front-end circuit energy is controlled for wireless video transmission by adjusting parameters in physical layer and MAC layer. In [45], the authors investigate the transmission over bandwidth-limited multi-access wireless uplink channel. The energy-efficient video communication is obtained by jointly adapting video summarization, coding schemes, modulation schemes, and packet transmission. In [46],the authors present a framework for joint network optimization, source adaption and deadline-driven scheduling for multi-user video streaming over wireless networks. Both the physical layer and application layer are jointly considered to maximize the total users' reception quality under the power consumption constraint. In [47], the authors investigate the joint optimization among source coding at application layers, ARQ scheme at data link layers and adaptive modulation and channel coding at physical layer. Within the delay-distortion framework, the parameters of above layers are jointly optimized to achieve the best quality of the received videos.

23.2.5 Video Over Emerging Wireless Networks

The video streaming problem has also been considered for several emerging wireless networking paradigms. Visual sensor networks (VSN) also represent an important application of energy efficient multimedia networks. VSN consists of a large number of low-power camera nodes, which integrate the image sensor, embedded processor, and wireless transceiver. The development of VSN has brought about many potential applications, such as surveillance, environmental monitoring, smart homes/cities, and visual reality [48], to name a few. Due to the battery limitation, the life time of VSN camera nodes is limited by their energy consumption in wireless channel sensing, transmission, and video and image data processing. Energy efficiency is a critical issue in the design of VSN nodes, since they may not be recharged as often as smart phones, and are expected to operate over extended periods of time (e.g., on two AA batteries for one year [49]). Therefore power efficient designs are highly preferable at all the protocol layers in VSNs.

Comprehensive surveys of VSNs can be found in [48,49]. It has been shown that power-aware routing is highly effective in prolonging the lifetime of wireless sensor networks [50]. In [51, 52], the authors investigate the directional-control data fusion scheme to reduce the amount of sensory data transmission in sensor networks. When processing video data is allowed within the network, data fusion can be employed to reduce the redundancy among multiple video streams along the routing path, thus reducing the volume of transmitted video data and saving energy at the intermediate nodes. Power-aware transport layer designs are mainly based on de facto standard of TCP. In [53,54], the authors incorporate a new error-recovery mechanism into TCP to avoid unnecessary retransmissions caused by AIMD, especially when the network is disconnected or there are losses due to high bit error. This scheme is shown to prolong the lifetime of wireless sensor networks.

In [14,55,56], the authors investigate the problem of video transmission over cognitive radio (CR) networks, where secondary users sense the licensed channels and aim to exploit the transmission opportunities in the spectrum holes. The uncertain channel availability condition brings about many unique challenges. The problem of supporting video communications in two-tier femtocell networks is investigated in [4, 6], which consists of a macro BS that serves the entire network area and multiple femtocell BS's that serves approved users with a small area. Effective algorithms are developed to allocate network resource to the video sessions with optimal or sub-optimal performance and proved performance bounds. In a recent work [57], we investigate the problem of combing cooperative relay with CR for multiuser downlink video streaming, where interference alignment is incorporated to facilitate concurrent transmissions of multiple video packets.

23.3   Energy Efficient Downlink VBR Video Streaming

23.3.1   Overview

As a case study, we investigate the challenging problem of energy efficient downlink VBR video streaming in a cellular network in this section. This study is presented in part in [36]. This research distinguish itself from prior work in the following aspects: First, instead of power-aware mobile video devices, we focus on the BS power efficiency when transmitting multiuser videos. As mentioned before, the BS equipment consumes more than 50% of the total power in a typical cellular network. Thus, it is important to improve the BS energy efficiency to achieve the goal of green communications. Second, we explicitly investigate streaming of multiple VBR videos. VBR videos can offer constant and better QoS over CBR videos with the same bit budget. However, VBR videos are notoriously difficult to schedule and control in wireless networks, due to the high variability and the complex autocorrelation structure [11,58,59]. Third, we adopt a stochastic majorization theoretic approach to minimize the total BS energy consumption, by jointly considering power control, wireless channel condition, playout buffer constraints, and playout deadlines.

In particular, we consider the problem of optimal power allocation for multiuser VBR video streaming in the downlink of a cellular network with orthogonal channels. We assume the wireline segment of a video session path is reliable with sufficient bandwidth, while the last-hop wireless link is the bottleneck. Thus the corresponding video data is always available at the BS before the scheduled transmission time. We adopt a deterministic model for VBR video traffic that incorporates video frame and playout buffer characteristics. The BS allocates a transmit power to each user in each time slot. The problem is to find the optimal power control schedule to stream the requested VBR video data to users, such that the total transmit power consumption can be minimized, while minimizing the buffer underflow and overflow events.

The problem is formulated as a constrained stochastic optimization problem. We show that the problem fits well with majorization theory, which concerns with partial ordering of real vectors and order-preserving functions. It answers the question of how to order vectors with nonnegative real components and its order-preserving functions [60]. A majorization-based solution framework is developed to tackle the problem. First, we prove that the objective function of the formulated problem is Schur-convex with the order-preserving

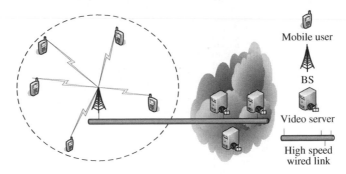

FIGURE 23.5 Cellular network and video streaming system model [36] © 2011 IEEE.

property [60]. Second, we investigate the case of a single VBR video session with relaxed peak power constraint. We develop a majorization-based power optimal algorithm with low complexity, and prove the power optimality of the proposed algorithm and the uniqueness of the global optimum. We also demonstrate that the proposed algorithm is smoothness optimal as well. Third, we investigate the case of multiuser VBR streaming, where power allocations for the users are coupled with the BS peak power constraint. We develop a heuristic algorithm that selectively suspends some video sessions, which will not incur underflow in the next time slot, when the peak power constraint is violated. Finally, the proposed algorithms are evaluated with trace-driven simulations [61], and are shown to achieve considerable power savings and improved video quality over a conventional "lazy" scheme [62].

23.3.2 System Model

23.3.2.1 Network and Video Source Model

We consider the downlink in a cellular network, as shown in Fig. 23.5. There are N active mobile users in a set $\mathcal{U} = \{1, 2, \cdots, N\}$ in the cell that subscribe to the video service. A BS transmits multiple VBR videos to the mobile users. Each user occupies a downlink channel, which is a spectral/time resource slot, the nature of which depends on the specific multiple access technique adopted. We assume that the downlink channels within a cell are orthogonal, due to perfect synchronization of the spreading codes or the use of guard times or frequencies. We further assume the wireline segment of a video session path is reliable with sufficient bandwidth, while the last-hop wireless link is the bottleneck. Thus the corresponding video data is always available at the BS before the scheduled transmission time.

It is non-trivial to accurately model VBR video traffic, which exhibits both strong asymptotic self-similarity and short-range correlation [58]. In this work, we adopt a *deterministic model* that considers frame sizes, frame intervals, and playout buffers [63]. Let $D_n(t)$ denote the *cumulative consumption curve* of

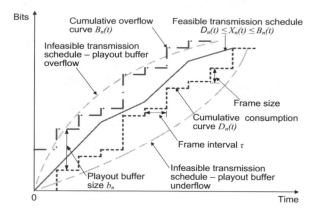

FIGURE 23.6 **Transmission schedules/curves for VBR video session** n **[36], ©**
2011 IEEE.

the n-th user, representing the cumulative amount of bits consumed by the
decoder from the beginning to time t. The cumulative consumption curve
is determined by video characteristics such as frame sizes and frame rates.
The frame playout deadline is implicitly marked by the stairs in $D_n(t)$. If
the required frame is not available at the corresponding stairs, the playout
process will be stalled. User n has a playout buffer of b_n bits. Its video has
T_n frames, which translates to Ω_n bits in total. We can derive a *cumulative
overflow curve* for user n as

$$B_n(t) = \min\{D_n(t-1) + b_n, D_n(T_n)\},\ 0 \le t \le T_n. \qquad (23.1)$$

$B_n(t)$ is the maximum number of bits that the receiver can receive from the
beginning to time t without overflowing user n's playout buffer (i.e., part of
$B_n(t)$ is already consumed by the decoder, while the remaining part, up to b_n
bits, is stored in the playout buffer). Finally we define *cumulative transmission
curve* $X_n(t)$ as the total amount of bits that are actually transmitted to user
n at time t.

The three curves are illustrated in Fig. 23.6. We assume time is divided
into time slots, and power control is performed on time slot units. Note that
the frame intervals of the VBR videos do not have to be aligned, and time
slots do not have to be aligned with the frame intervals. To simplify notation,
we assume the frames are aligned and the duration of a time slot is τ in the
rest of this paper. A feasible transmission schedule will produce a cumulative
transmission curve $X_n(t)$ that lies in between $D_n(t)$ and $B_n(t)$, i.e., causing
neither underflow nor overflow at the playout buffer. In practice, $D_n(t)$'s are
known for stored videos and can be delivered to the BS as metadata during
the session setup phase, and $B_n(t)$'s are then derived as in (23.1).

The BS allocates a transmit power to each user in each time slot. Let
$\vec{P}(t) = [P_1(t), \cdots, P_N(t)]$ be the power allocation in time slot t. The Signal
to Interference-plus-Noise Ratio (SINR) at user n in time slot t can be written

as

$$\gamma_n(t) = G_n(t)P_n(t)/\eta_n(t), \tag{23.2}$$

where $G_n(t)$ is the path gain from BS to user n and $\eta_n(t)$ is the the noise power at user n in time slot t. We assume block fading channels, where the $G_n(t)$'s are i.i.d. random variables with a certain distribution, for $t = 1, \cdots, T_n$ [30]. The downlink data rate can be written as $c_n(t) = B_w\log\left(1 + \kappa\gamma_n(P_n(t))\right)$, where B_w is the channel bandwidth and κ depends on the transceiver design, such as modulation and channel coding. Without loss of generality, we use the Shannon capacity as an upper bounding approximation:

$$c_n(t) = B_w\log\left(1 + \gamma_n(P_n(t))\right). \tag{23.3}$$

Once the link capacity is determined, $c_n(t)\tau$ bits of video data will be delivered to user n in that time slot. The cumulative transmission curve $X_n(t)$ can be written as

$$X_n(0) = 0; \quad X_n(t) = X_n(t-1) + c_n(t)\tau. \tag{23.4}$$

A feasible transmission schedule should cause neither playout buffer underflow nor overflow, i.e., satisfying

$$D_n(t) \leq X_n(t) \leq B_n(t), \text{ for all } t, n. \tag{23.5}$$

23.3.2.2 Power-Aware Transmission Scheduling

As discussed, we jointly consider the traffic source model in the application layer and power allocation in the physical layer. We adopt cross-layer design to compute the optimal feasible transmission schedule $\{X_n(t), 0 < t \leq T_n\}$, for all users $n \in \mathcal{U}$, such that the total transmit power consumption can be minimized. From (23.3), the required transmit power for user n is

$$P_n(t) = (2^{c_n(t)/B_w} - 1)\eta_n(t)/G_n(t). \tag{23.6}$$

A peak power constraint may be applied at the BS, i.e., $\sum_{n \in \mathcal{U}} P_n(t) \leq \bar{P}$, for all t. We then formulate the following *constrained stochastic optimization problem*, aiming to minimize the expected total transmit power.

$$\text{minimize} \sum_{n \in \mathcal{U}} \sum_{t=1}^{T_n} \mathbb{E}[P_n(t)] \tag{23.7}$$

$$\text{subject to: } D_n(t) \leq X_n(t) \leq B_n(t), \text{for all } n, t \tag{23.8}$$

$$\sum_{n \in \mathcal{U}} P_n(t) \leq \bar{P}, \text{for all } t. \tag{23.9}$$

Due to orthogonal channels, transmission in one channel does not interfere with those in other channels. We first relax the peak power constraint (23.9) (i.e., the case when $\bar{P}$ is large). Then, Problem (23.7) can be decomposed into N sub-problems, each minimizing the transmit power of a video session.

$$\text{minimize} \sum_{t=1}^{T_n} \mathbb{E}[P_n(t)], \text{for all } n \in \mathcal{U} \tag{23.10}$$

$$\text{subject to: } D_n(t) \leq X_n(t) \leq B_n(t), \text{ for all } n, t. \tag{23.11}$$

For given $B_n(t)$ and $D_n(t)$, the feasible transmission schedule satisfying (23.11) is not unique. The i-th feasible transmission schedule is a piece-wise linear curve that can be represented as a vector $\overrightarrow{C}_n^i = [c_n^i(1), \cdots, c_n^i(T_n)]$, where $c_n^i(t) \geq 0$ is the data rate in time slot t, for all t. Let $\overrightarrow{C}_n^* = [c_n^*(1), \cdots, c_n^*(T_n)]$ be the optimal solution to (23.10). For a given VBR video, all the feasible transmission schedules transmit the same amount of video data, i.e.,

$$\sum_{t=1}^{T_n} c_n^i(t) = \sum_{t=1}^{T_n} c_n^*(t) = \Omega_n, \text{ for all } i, n. \tag{23.12}$$

Furthermore, the total transmit power for a feasible schedule can be viewed as a mapping function $\Phi : \mathcal{R}^{T_n} \to \mathcal{R}$ with

$$\Phi(\overrightarrow{C}_n^i) = \sum_{t=1}^{T_n} (2^{c_n^i(t)/B_w} - 1)\eta_n(t)/G_n(t). \tag{23.13}$$

Given such an interpretation of the relaxed Problem (23.10), the objective is to find an optimal feasible vector $\overrightarrow{C}_n^*$, such that its total power P_n^*, obtained through the mapping $\Phi(\cdot)$, is the minimum among all feasible vectors $\overrightarrow{C}_n^i$. This interpretation fits well with the *majorization* theory, which provides useful order preserving results for inequality problems [60]. Applying these results, we design an optimal algorithm for solving the decomposed sub-problem (23.10) in Section 23.3.4.1. Then we will examine the case of multiuser VBR video streaming coupled with the peak power constraint in Section 23.3.4.4.

23.3.3   Inequality and Theory of Majorization

23.3.3.1   Majorization Preliminaries

Majorization theory formalizes the intuitive notion of inequality and describes when the components of a vector are "less spread out" or "more nearly equal" than the components of another vector [60]. It concerns with how to order vectors with nonnegative real components and order-preserving functions. Majorization theory has been used to solve communications and networking problems in the literature [63–65]. In this paper, we apply majorization to the problem of optimal power control for streaming VBR videos. For simplicity, all the vectors in this section are row vectors.

Definition 23.1. *Consider two n-dimensional vectors $\vec{X}$ and $\vec{Y}$ with non-negative real elements. Let the elements be ordered non-increasingly and re-indexed as $x_1 \geq \cdots \geq x_n \geq 0$ for $\vec{X}$ and $y_1 \geq \cdots \geq y_n \geq 0$ for $\vec{Y}$. $\vec{X}$ is said to be majorized by $\vec{Y}$, denoted as $\vec{X} \prec \vec{Y}$, if $\sum_{i=1}^{t} x_i \leq \sum_{i=1}^{t} y_i$, $t = 1, \cdots, n-1$ and $\sum_{i=1}^{n} x_i = \sum_{i=1}^{n} y_i$ [60].*

For example, we have $[1/n, \cdots, 1/n] \prec [1/(n-1), \cdots, 1/(n-1), 0] \prec [1/2, 1/2, 0, \cdots, 0] \prec [1, 0, \cdots, 0]$. It can be seen that the total weight of 1 becomes more and more concentrated into fewer and fewer elements in the above majorized vector sequence. When such partial ordering is obtained, an *order-preserving function* can be applied to the vectors and the outcomes remain monotonic with respect to the partial order of the vectors.

Definition 23.2. *A real-valued function $\phi(\cdot)$ defined on a set $\mathcal{S} \subset \mathcal{R}^n$ is said to be Schur-convex on $\mathcal{S}$, if $\vec{X} \prec \vec{Y}$ implies $\phi(\vec{X}) \leq \phi(\vec{Y})$, for all $\vec{X}, \vec{Y} \in \mathcal{S}$ [60].*

Schur-convex functions have the order-preserving property, which makes majorization useful for solving optimization problems. The following facts can be used to verify whether a function is Schur-convex.

Fact 1. *If $\phi(\cdot)$ is symmetric and convex, then ϕ is Schur-convex. Consequently, $\vec{X} \prec \vec{Y}$ implies $\phi(\vec{X}) \leq \phi(\vec{Y})$ [60].*

If we choose $\phi = \sum_i f(x_i)$ and $f(\cdot)$ is continuous and convex, then we have the following strong fact.

Fact 2. *$\sum_i f(x_i) \leq \sum_i f(y_i) \Leftrightarrow \vec{X} \prec \vec{Y}$ holds true for all continuous convex functions $f : \mathcal{R} \to \mathcal{R}$ [60].*

We can extend the concept of majorization to the case of a set of vectors or multiple segments of a long vector.

Lemma 23.1. *Consider $\vec{X} = [\vec{X}_1, \cdots, \vec{X}_K]$ and $\vec{Y} = [\vec{Y}_1, \cdots, \vec{Y}_K]$, where each element vector $\vec{X}_i$ and $\vec{Y}_i$ has dimension J_i and satisfying $\vec{X}_i \prec \vec{Y}_i$, $i = 1, \cdots, K$. Then we have $\vec{X} \prec \vec{Y}$.*

Proof. Let $f(\cdot)$ be a continuous and convex function. From Fact 2, we have $\vec{X}_i \prec \vec{Y}_i \Leftrightarrow \sum_{j=1}^{J_i} f(x_i^j) \leq \sum_{j=1}^{J_i} f(y_i^j)$. It follows that $\sum_{i=1}^{K} \sum_{j=1}^{J_i} f(x_i^j) \leq \sum_{i=1}^{K} \sum_{j=1}^{J_i} f(y_i^j)$, i.e., $\vec{X} \prec \vec{Y}$. $\qquad \square$

Consider three real vectors $\vec{X} = [\bar{x}, \cdots, \bar{x}]$, $\vec{Y} = [y_1, \cdots, y_n]$, and $\vec{Z} = [z_1, \cdots, z_n]$, where $\sum_{i=1}^{n} y_i = \sum_{i=1}^{n} z_i = n\bar{x}$. If the elements in each vector are non-increasing, we can plot the cumulative curves for the normalized vectors $\vec{X}/(n\bar{x})$, $\vec{Y}/(n\bar{x})$ and $\vec{Z}/(n\bar{x})$, which are piece-wise linear curves A, B, and C in Fig. 23.7. The i-th point on the cumulative curve is the sum of the first

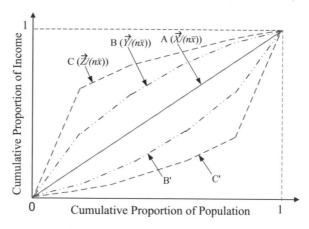

FIGURE 23.7 Lorenz curves [36] © 2011 IEEE.

i elements of the corresponding vector. If each element of the vectors is the income of an individual, these are known as *Lorenz Curves* that evaluate social income inequality [66]; each point $(a\%, b\%)$ on the curve indicates the top $a\%$ of the population earn $b\%$ of the income.

In Fig. 23.7, the straight line A is for vector $\vec{X}/(n\bar{x})$, i.e., equal distribution. Unequal distributions $\vec{Y}/(n\bar{x})$ and $\vec{Z}/(n\bar{x})$ produce curves B and C, respectively, which are bent in the middle. As the bow-shaped curves are bent, concentration increases. We call B and C *concave bow curves*. The bow curve closer to A represents more even distribution [60], which leads to $\vec{X} \prec \vec{Y} \prec \vec{Z}$. Furthermore, we can permute $\vec{Y}$ and $\vec{Z}$ by reordering their elements, to obtain new vectors $\vec{Y}'$ and $\vec{Z}'$. We call the resulting curves B' and C' *convex bow curves*. Since the order of elements in a vector plays no role in majorization (see Definition 23.1), the same majorization relationship $\vec{X} \prec \vec{Y}' \prec \vec{Z}'$ still holds true in this convex bow curve case.

23.3.3.2   Schur-convexity of Problem (23.10)

As discussed, Problem (23.10) fits well with majorization theory with a mapping function (23.13). To solve the problem, we need to find the optimal rate vector $\vec{C}_n^*$ that is majorized by all other feasible transmission rate vectors $\vec{C}_n^i$, as $\vec{C}_n^* \prec \vec{C}_n^i$, for all i. If the mapping (23.13) is Schur-convex, then the total transmit power to achieve $\vec{C}_n^*$ will also be dominated by those of other feasible transmission rate vectors. That is, the minimum power is found for Problem (23.10). Due to random path gains and noise powers, stochastic majorization (rather than ordinary majorization) should be used, which investigates the inequality properties related to random variables [60]. We have the following theorem for the mapping (23.13) in Problem (23.10).

Theorem 23.1. *The objective function of (23.10) is an increasing Schur-convex function.*

Proof. Due to i.i.d. channel gains and noise powers, the random variables $\eta_n(t)/G_n(t)$'s are *exchangeable*, for all t. Define $w(g, \eta, c) = (2^{c/B_w} - 1)\eta/g$, which is convex and increasing with c, for all $g \geq 0$ and $\eta \geq 0$. Let $\psi(\overrightarrow{C})$ $= \mathbb{E}[\Phi(\overrightarrow{C})] = \mathbb{E}[\sum_{t=1}^{T} w(g(t), \eta(t), c(t))]$. $\Phi(\overrightarrow{C})$ is a symmetric, convex and increasing function in $\overrightarrow{C}$ for each fixed $\overrightarrow{G}$ and $\overrightarrow{\eta}$. According to Proposition 11.B.5 in [60], $\psi(\overrightarrow{C})$ is symmetric, convex and increasing. Following Fact 1, the objective function (23.10) is Schur-convex and increasing. $\qquad \square$

With Theorem 23.1, solving Problem (23.10) is equivalent to finding the optimal rate vector $\overrightarrow{C}_n^*$, such that $\overrightarrow{C}_n^* \prec \overrightarrow{C}_n^i$, for all i. Then the total power associated with $\overrightarrow{C}_n^*$ is the minimum since the mapping (23.13) is order-conserving. As discussed in Section 23.3.3.1, the feasible rate vector that is closest to equal distribution (i.e., curve A in Fig. 23.7) will be majorized by all other feasible rate vectors. Therefore, we transform Problem (23.10) to finding a transmission schedule with the most evenly distributed rates for all the time slots.

23.3.4 Power Allocation Algorithms

Based on the stochastic majorization interpretation of Problem (23.10) and the Schur-convex property of its objective function, we first develop a power optimal algorithm (PMA) for the case of relaxed peak power constraint. We prove its optimality and the uniqueness of the global optimal, as well as the equivalence of power optimal and smoothness optimal. We then describe a heuristic algorithm for the case of multiple videos coupled with the peak power constraint.

23.3.4.1 Optimal Algorithm for Problem (23.10)

23.3.4.2 Power Minimization Algorithm

From Section 23.3.3.1, an evenly distributed rate vector $\overrightarrow{C}_n^{opt}$ $=$ $[\Omega_n/(T_n\tau), \cdots ,$ $\Omega_n/(T_n\tau)]$ is majorized by all feasible schedules, i.e., $\overrightarrow{C}_n^{opt} \prec \overrightarrow{C}_n^i$, for all i. However, due to the high variability of VBR video frames, limited playout buffer size, random path gains and noise powers, $\overrightarrow{C}_n^{opt}$ may not always be feasible. In general, each feasible schedule is piece-wise linear with a set of rate change points, where the rate is increased or decreased to prevent buffer underflow or overflow. $\overrightarrow{C}_n^{opt}$ is a special case with no such rate change points.

The algorithm in Table 23.1, termed PMA, can generate a piece-wise linear schedule, while keeping each piece as long as possible and rate variation as small as possible. The operation of the algorithm is illustrated in Fig. 23.8. Starting from t_{start} (e.g., h_1 in Fig. 23.8), PMA first computes two probe lines:

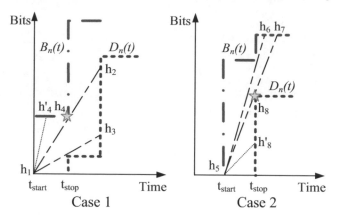

FIGURE 23.8 Two cases for determine the next transmission rate [36] © 2011 IEEE.

- One through the starting point and one of the future corner points of $B_n(t)$, which can go the furthest into the future without causing buffer underflow or overflow (e.g., lines h_1h_2 in Case 1 and h_5h_6 in Case 2 of Fig. 23.8). The rate of this probe line is $C_{max}(t) = \frac{B_n(t) - X_n(t_{start})}{t - t_{start}}$.

- The other through the starting point and one of the future corner points of $D_n(t)$, which can go the furthest into the future without causing buffer underflow or overflow (e.g., lines h_1h_3 in Case 1 and h_5h_7 in Case 2 of Fig. 23.8). The rate of this probe line is $C_{min}(t) = \frac{D_n(t) - X_n(t_{start})}{t - t_{start}}$.

All feasible transmission curves should lie in between these two probe lines in order to go that far. Furthermore, when the probe lines end, they hit *both* on either $B_n(t)$ or $D_n(t)$. Otherwise, we can always adjust one of the probe lines to make them go even further into the future. For example, see lines h_1h_3 and $h_1h'_4$ in Case 1 of Fig. 23.8. We can use line h_1h_2, which goes further into the future, to replace line $h_1h'_4$, and both probe lines hit $D_n(t)$ eventually (also see lines h_5h_6 and $h_5h'_8$ in Case 2).

If both probe lines hit $D_n(t)$ (i.e., Case 1 in Fig. 23.8), any feasible schedule for this interval will also hit $D_n(t)$, since they must lie in between the two probe lines. We then trace back the upper probe line (i.e., line h_1h_2) to find the latest time when the buffer is full (i.e., point h_4 at time t_{stop}). Then segment h_1h_4 will be chosen as the transmission schedule for this interval, with rate $\frac{B_n(t_{stop}) - X_n(t_{start})}{t_{stop} - t_{start}}$.

If both probe lines hit $B_n(t)$ (i.e., Case 2 in Fig. 23.8), any feasible schedule for this interval will also hit $B_n(t)$. We then trace back the lower probe line (i.e., line h_5h_7) to find the latest time when the buffer is empty (i.e., point h_8 at time t_{stop}). Then segment h_5h_8 will be chosen as the transmission schedule for this interval, with rate $\frac{D_n(t_{stop}) - X_n(t_{start})}{t_{stop} - t_{start}}$.

After the transmission schedule for $[t_{start}, t_{stop})$ is determined, we set

TABLE 23.1 Power Minimization Algorithm (PMA-1) ([36], © 2011 IEEE)

1	BS obtains b_n, D_n, B_n, and B_w for all user n;
2	$t = 1, t_{start} = 0, t_{stop} = t_{c_1} = t_{c_2} = 1, C_{min} = 0,$ $C_{max} = \infty$;
3	DO:
4	Calculate $C_{max}(t)$ and $C_{min}(t)$ over interval $[t_{start}, t]$;
5	IF ($C_{min} \leq C_{min}(t)$ & $C_{min}(t) \leq \min\{C_{max}, C_{max}(t)\}$)
6	$C_{min} = C_{min}(t)$ and $t_{c_1} = t$;
7	END IF
8	IF ($C_{max} \geq C_{max}(t)$ & $C_{max}(t) \geq \max\{C_{min}, C_{min}(t)\}$)
9	$C_{max} = C_{max}(t)$ and $t_{c_2} = t$;
10	END IF
11	IF ($C_{min} > \min\{C_{max}, C_{max}(t)\}$)
12	Select C_{min} from t_{start} to $t_{stop} = t_{c_1}$;
13	ELSEIF ($C_{max} < \max\{C_{min}, C_{min}(t)\}$)
14	Select C_{max} from t_{start} to $t_{stop} = t_{c_2}$;
15	ELSE
16	$t++$;
17	CONTINUE;
18	END IF
19	$t_{start} = t_{stop}, t_{stop} = t_{c_1} = t_{c_2} = t_{start} + 1,$ $t = t_{start} + 1, C_{min} = 0, C_{max} = \infty$;
20	WHILE (some time slots are not assigned a rate)
21	DO:
22	Measure the channel gain of the time slot, calculate power using (24.22), and transmit the video data;
23	WHILE (more video frames to transmit)

$t_{start} = t_{stop}$ and repeat the above procedure to find the schedule for the next time interval. In Table 23.1, the algorithm probes for the longest feasible rate starting from t_{start} in Steps 4–10. In Steps 11–14, the transmission rate for the interval $[t_{start}, t_{stop})$ is determined depending on which of the two cases it is as illustrated in Fig. 23.8. Steps 16–17 are for the case that the rate does not change in the time slot. Step 19 resets the variables to start the computation for the next segment of $X_n(t)$. Finally, Steps 21–23 transmit the frames following the computed schedule.

23.3.4.3 Optimality Proof

We next show that the algorithm given in Table 23.1 computes the optimal solution to Problem (23.10).

Theorem 23.2. *The power minimization algorithm PMA is optimal to Problem (23.10).*

Proof. The proof is similar to that in [63] for smoothing a VBR video over a constant capacity channel. The schedule computed by PMA, $\overrightarrow{C}_n^*$, could be a

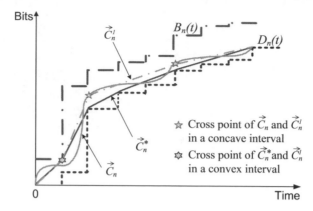

FIGURE 23.9 Relationship among the three curves: C^*, C_1 and C.

straight line or in the general case, consist of one or more convex and concave segments. If $\vec{C}_n^*$ is a straight line, it is obvious that $\vec{C}_n^* \prec \vec{C}_n^i$ for any other $\vec{C}_n^i$ (see Fig. 23.7) and it is power optimal. In the general case, we need to show $\vec{C}_n^* \prec \vec{C}_n^i$, for all i in every convex or concave segment. Then according to Lemma 23.1, we have $\vec{C}_n^* \prec \vec{C}_n^i$ for all i and it is optimal.

Let $\vec{C}_n$ denote an arbitrary feasible schedule. We introduce an auxiliary schedule $\vec{C}_n^1$, which intersects with $\vec{C}_n^*$ at all its rate change points in every convex segment, and with $\vec{C}_n$ at all its rate change points in every concave segment, as shown in Fig. 23.9.

First, we prove that $\vec{C}_n^* \prec \vec{C}_n^1$. For a convex segment of $\vec{C}_n^*$, because $\vec{C}_n^1$ intersects with $\vec{C}_n^*$ at all the rate change points of $\vec{C}_n^*$, we have $\vec{C}_n^* = \vec{C}_n^1$ in all the convex segments. For a concave segment of $\vec{C}_n^*$, the endpoints of the concave segment should be the last (first) rate changing point of the previous (next) convex segment, where $\vec{C}_n^*$ intersects with $\vec{C}_n^1$. The rate change points within the concave segment are all on $D_n(t)$, as in PMA. Therefore, $\vec{C}_n^1$ is an outer concave curve above $\vec{C}_n^*$ (or, it is farther away from the straight line A in Fig. 23.7) in this segment. From the discussion of Fig. 23.7, we have $\vec{C}_n^* \prec \vec{C}_n^1$ for all the concave segments. It follows that $\vec{C}_n^* \prec \vec{C}_n^1$ according to Lemma 23.1.

We next prove that $\vec{C}_n^1 \prec \vec{C}_n$. For a convex segment of $\vec{C}_n^*$, the endpoints of the convex segment should be the last (first) rate changing point of the previous (next) concave segment, where $\vec{C}_n$ intersects with $\vec{C}_n^1$. The rate change points of $\vec{C}_n^1$ (or, of $\vec{C}_n^*$) in the convex segment are all on $B_n(t)$. Therefore, $\vec{C}_n$ is an outer convex curve below $\vec{C}_n^1$ in this segment. From the discussion of Fig. 23.7, it follows that $\vec{C}_n^1 \prec \vec{C}_n$ in all the convex segments. In a concave segment, we have either $\vec{C}_n = \vec{C}_n^1$ or $\vec{C}_n \prec \vec{C}_n^1$, because $\vec{C}_n^1$

intersects with $\overrightarrow{C}_n$ at each rate changing point. Thus, we obtain $\overrightarrow{C}_n^1 \prec \overrightarrow{C}_n$ for all the concave segments, and $\overrightarrow{C}_n^1 \prec \overrightarrow{C}_n$ according to Lemma 23.1.

Finally we have $\overrightarrow{C}_n^* \prec \overrightarrow{C}_n^1 \prec \overrightarrow{C}_n$. Theorem 23.1 states that Problem (23.10) is Schur-convex and order preserving. It follows from Fact 1 that $\overrightarrow{C}_n^*$ is optimal to Problem (23.10). $\square$

Corollary 23.1. *The power optimal transmission scheme $\overrightarrow{C}_n^*$ is unique for given $B_n(t)$ and $D_n(t)$.*

Proof. Suppose $\overrightarrow{C}_n^*$ is not unique. Then there exists $\overrightarrow{C}_n' \prec \overrightarrow{C}_n^i$, for all i, and $\overrightarrow{C}_n' \neq \overrightarrow{C}_n^*$. $\overrightarrow{C}_n'$ must have a different set of rate change points from that of $\overrightarrow{C}_n^*$. According to the proof of Theorem 23.2, we can construct an auxiliary schedule $\overrightarrow{C}_n^1$, such that $\overrightarrow{C}_n^* \prec \overrightarrow{C}_n^1 \prec \overrightarrow{C}_n'$, which contradicts the assumption that $\overrightarrow{C}_n'$ is optimal. $\square$

Corollary 23.2. *The computational complexity of Algorithm PMA is $\mathcal{O}(T_n^2)$.*

Proof. In the worst case, the PMA algorithm computes the optimal schedule for each time slot by probing the full length of the remaining video sequence (as in Steps 4–12 in Table 23.1). The worst case execution time is $\sum_{i=T_n}^1 i = \frac{T_n(T_n+1)}{2} \Rightarrow \mathcal{O}(T_n^2)$. $\square$

Note that Algorithm PMA is executed during the session setup time. It only incurs a small initialization delay. In our simulations with VBR video traces, we find the execution time is usually negligible. When the channel statistics are changed (i.e., due to handoff), the schedule will be recomputed for the remaining video frames.

Corollary 23.3. *The power optimal transmission schedule $\overrightarrow{C}_n^*$ is also the smoothest one among all feasible schedules.*

Proof. To evaluate the smoothness of a transmission schedule $\overrightarrow{C}$, the following smoothness utility function can be used:

$$U(\overrightarrow{C}) = \sum_{t=1}^{T_n}([c(t) - \bar{c}]/T_n), \tag{23.14}$$

where $\bar{c} = \sum_{t=1}^{T_n} c(t)/T_n$ is the average rate. This is a continuous symmetric convex function $U : \mathcal{R}^{T_n} \to \mathcal{R}$. From Fact 1, U is Schur-convex and order preserving. The optimal power transmission schedule $\overrightarrow{C}_n^*$ satisfies $\overrightarrow{C}_n^* \prec \overrightarrow{C}_n^i$ for all i. Therefore, it also achieves the minimum value for $U(\cdot)$. $\square$

TABLE 23.2 Power Minimization Algorithm for Multiuser Videos (PMA-m)
([36], © 2011 IEEE)

1	Execute power minimization algorithm PMA to compute transmission schedules for all active users;
2	DO:
3	Measure channel gains of the current time slot and calculate the transmit powers using (24.22);
4	IF (peak power constraint is violated)
5	Select the users who won't have underflow even without transmission in this time slot;
6	Sort the selected users in decreasing order of powers;
7	DO:
8	Decrease the power of the selected users by the order;
9	WHILE (peak power constraint is not satisfied)
10	END IF
11	Transmit the videos and recalculate the optimal transmission scheme for the paused mobile users for the next time interval;
12	WHILE (there are more video frames to transmit)

23.3.4.4   Multiuser Video Transmissions

We now consider Problem (23.7) to compute transmission schedules for N VBR video sessions, which are coupled by the peak power constraint (23.9). Due to the peak power constraint and random channel gains, the individually calculated transmit powers may violate (23.9) in some time slots. The problem is further complicated because of the random channel gains, which is not available a priori (except for the statistics of the channels).

To solve Problem (23.7), we develop a heuristic algorithm, termed PMA-m, as presented in Table 23.2. The PMA-m algorithm uses PMA to compute transmission schedules for all active users. Then based on current channel state information, it computes the power needed to achieved the rate for each user, and checks the peak power constraint $\sum_{n \in \mathcal{U}_n} P_n(t) \leq \bar{P}$. If the constraint is not violated, each user's video data will be transmitted at the computed power. Otherwise, as in Steps 4–10, PMA-m selects those users who will not have buffer underflow if their transmissions are suspended in the following time slot, and sort them in the deceasing order of their required powers. Starting with the first user, PMA-m decreases the powers of the users in the list; if the first user's power reaches 0 W but the peak power constraint is till not satisfied, PMA-m starts to reduce the power of the second user in the list; and so forth until the peak power constraint is satisfied.

In some extremely severe channel conditions, the total power $\bar{P}$ cannot even support the minimum required bit rate for all the users. Some users have to be paused and the current frames be discarded. The corresponding playout of such a user will be frozen until the next time slot. Finally, the transmission

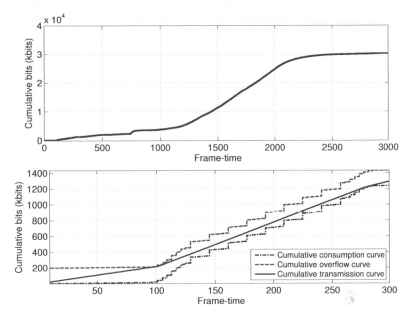

FIGURE 23.10 Simulation results: transmission curves of *Star wars*.

schedules for the suspended users will be recomputed using PMA as in Step 11 and the above procedure is repeated.

23.3.5 Performance Evaluation

We demonstrate the performance of the proposed optimal power control algorithm through trace-driven simulations. We simulate the downlink of a cell with 1 mile radius. The channels are assumed to be orthogonal, each with $B_w = 1$ MHz bandwidth. We assume that bit errors can be corrected by error correction codes. The path gain averages are $\bar{G}_n = d_n^{-4}$, where d_n is the physical distance from the BS to user n. We assume log-normal fading with zero mean and 8 dB standard deviation. The device temperature is 290 Kelvin and the equivalent noise bandwidth is $B_w = 1$ MHz. The BS streams three movies *Star Wars*, *NBC News*, and *Tokyo Olympics* to active users. The video traces are obtained from the Video Trace Library at Arizona State University [61].

We first investigate the performance of the power optimal algorithm. In the simulation, the BS streams 3,000 frames of a video sequence to each mobile user located at different distances to the BS. The cumulative consumption, overflow and transmission curves of the *Star wars* video session are plotted in Fig. 23.10. It can be seen that the transmission schedule always lie in between the cumulative consumption and overflow curves, indicating that there is no playout buffer underflow or overflow events in this simulation.

We next compare the optimal power algorithm with a conventional transmission scheme with respect to the average power consumption at the BS. In

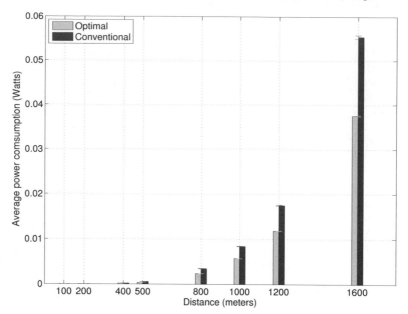

FIGURE 23.11 Simulation results: average power consumption.

each time slot, the conventional scheme only transmits the video data that is needed by the decoder at the end of the time slot. It achieves a cumulative transmission curve that connects all the corner points of $D_n(t)$ (also called the "*lazy*" scheme). Intuitively, such lazy approach should be energy efficient since it always transmits the minimal amount of data as needed. However, we will see that the proposed algorithm outperform this lazy approach in the simulations.

In Fig. 23.11, we plot the average power consumption achieved by the two schemes for increased distance to the BS. Each point in the figure is the average of 10,000 simulation runs. The 95% confidence intervals are plotted as vertical bars in the figure, which are all very small.

It can be seen that the proposed algorithm outperforms the conventional scheme for the entire range of distances examined. When the distance to BS is small, both schemes use small transmit powers and the power savings are not very big. However, when the distance is increased, channel fading has a bigger impact on interference and channel capacity. The proposed algorithm achieves considerable power savings than the conventional scheme. When the distance is 1,600 m, the total power of the proposed scheme is only 46.62% of that of the conventional approach, corresponding to a 54.34% normalized improvement.

We also obtain the average execution time of the proposed algorithm, under the same setting but for 20,000 *Star Wars* frames. We find that the average execution time is about 0.06 s on an IBM Laptop with Intel T2400 1.83 GHz processor and 2 GB RAM.

TABLE 23.3 Simulation Results: Playout Buffer Underflow Rates

	Scenario 1	Scenario 2	Scenario 3
PMA-m	0.0005%	1.8%	1.66%
Conventional	0.89%	13.24%	11.64%

Finally, we examine the buffer underflow events. The following scenarios are simulated:

- *Scenario 1*: $\bar{P} = 1$ W; the movies are *Star Wars*, *NBC News*, and *Tokyo Olympics*; 50 mobile users; $B_w = 1$ MHz;

- *Scenario 2*: The same setting as i) except that $B_w = 125$ KHz;

- *Scenario 3*: $\bar{P} = 10$ W; the HD movies are *Terminator 2*, *From Mars to China*, and *Sony Demo*; 20 mobile users; $B_w = 1$ MHz.

The HD movies have larger frame sizes and higher variability in frame sizes. The buffer underflow rates are presented in Table 23.3, each being the ratio of the number of underflow frames over the total number of frames. PMA-m achieves considerably lower underflow rates for all the three scenarios. The PMA-m underflow rates are 0.056%, 13.60%, and 14.26% of that of the conventional scheme. Therefore, PMA-m achieves not only considerable energy savings, but also much better video quality for the mobile users.

23.4 Conclusion and Future Directions

Given the foreseeable drastic growth in wireless video data volume and BS deployment, the energy saving for BS in cellular networks is an important problem in "green" communications. In this chapter, we addressed the problem of energy efficient wireless video communications. We first provide a brief survey of related work in this important area, including energy efficient video coding and energy efficient network streaming systems. We then provided a case study on energy efficient downlink multiuser VBR video streaming. Our formulation took into account the interactions among power control, fading channels, VBR video traffic, and playout characteristics. A constrained stochastic optimization problem is formulated aiming to minimize the BS power consumption and to avoid playout buffer overflow or underflow. We then developed majorization-based algorithms to the formulated problem for highly competitive solutions.

Although there have been considerable advances made in green wireless video communications, many interesting problems still remain open in this interesting problem area. Due to the intrinsic complexity of video sources and the dynamics and uncertainty of wireless systems, we conjecture that a holistic approach that encompasses the parameter space would be necessary and the trade-off between complexity and efficiency of a solution algorithm should be carefully investigated. In particular, we list some interesting problems that may worth further investigation in the following.

- Complexity-distortion analysis and the design of energy-efficient video codecs require accurate models of video codecs. However, the video codec is a combination of complex functional blocks, which makes accurate mathematic modelling extremely difficult. In addition, the quality of compressed video and the power consumption of the video codec depend on a large number of parameters. A content-based power-aware design may encounter a large search space for optimal solutions. It would be useful to develop an accurate and effective model for video codec that can be incorporated into mathematical optimization frameworks for both ernergy efficient codec design and wireless video system design.

- Cross-layer design and optimization has been widely adopted for video networking problems. It has been shown that an adaptive strategy with cooperation of several layers can achieve optimal power efficiency for video streaming. Most prior work assume that the wireless channel information and network status are known a priori (e.g., by accurate estimation, measurement, and timely feedback). In practice, this assumption may not be true, because of channel/network uncertainty and dynamics, and delay and congestion in the network. Thus how to balance the achievable performance and the control overhead of the design is still an open problem. Effective schemes that are robust to the channel/network uncertainties would be highly appealing.

- Many wireless video systems exploit scalable video coding (SVC) to adapt to wireless channel and network variabilities [67]. However, given the implicit limit on the lowest video quality that can be tolerated, and the limited amount of wireless network resources, the number of active video sessions must be kept below a maximum number. An admission control mechanism is necessary for guaranteeing the minimum performance of video sessions. Based on the present and predicted network status, the admission control mechanism decides whether a new video streaming request could be accepted, such that the required minimum performance of the new session and the existing sessions could all be satisfied. Given the realtime nature of video applications, a fast algorithm that does not require excess and accurate network state information, yet is able to achieve high utilization of the wireless network resources, would be very useful.

- We consider power efficient downlink VBR streaming in cellular networks with a stochastic optimization approach. Assuming that the channels are orthogonal, the formulated problem can be solved by applying stochastic majorization theory. It would be more challenging to consider the non-orthogonal scenario, where the inter-channel interference further complicates the power allocations for multiple mobile users. The majorization theory may not be directly applicable to such a situation. New mathematical theory and tools are needed to tackle this more challenging problem, such as dynamic programming, nonlinear stochastic optimization, and game theory.

Acknowledgment

This work is supported in part by the US National Science Foundation (NSF) under Grants CNS-0953513, ECCS-0802113, DUE-1044021, IIP-1127952, and CNS-1145446, and through the NSF Wireless Internet Center for Advanced Technology (WICAT) at Auburn University (under Grant IIP-0738088). Any opinions, findings, and conclusions or recommendations expressed in this material are those of the author(s) and do not necessarily reflect the views of the Foundation.

Bibliography

[1] Cisco, "Cisco visual networking index: Global mobile data traffic forecast update, 2010-2015," Feb. 2011. [Online]. Available: http://www.cisco.com/en/US/solutions/collateral/ns341/ns525/ns537/\\ns705/ns827/white_paper_c11-520862.html

[2] H. Sistek, "Green-tech base stations cut diesel usage by 80 percent," Apr. 2008. [Online]. Available: http://news.cnet.com/8301-11128_3-9912124-54.html

[3] V. Chandrasekhar, J. G. Andrews, and A. Gatherer, "Femtocell networks: A survey," *IEEE Commun. Mag.*, vol. 46, no. 9, pp. 59–67, 2008.

[4] D. Hu and S. Mao, "Resource allocation for medium grain scalable videos over femtocell cognitive radio networks," in *IEEE ICDCS'11*, Minneapolis, MN, June 2011, pp. 258–2672.

[5] ——, "Multicast in femtocell networks: A successive interference cancellation approach," in *IEEE GLOBECOM'11*, Huston, TX, Dec. 2011, pp. 1–6.

[6] ——, "On medium grain scalable video streaming over cognitive radio femtocell networks," *IEEE J. Sel. Areas Commun.*, vol. 30, no. 4, Apr. 2012.

[7] S. Vadgama, "Trends in green wireless access," *FUJITSU Scientific Technical Journal*, vol. 45, no. 4, 2009.

[8] S. Mao, S. Kompella, Y. Hou, H. Sherali, and S. Midkiff, "Routing for concurrent video sessions in ad hoc networks," *IEEE Trans. Veh. Technol.*, vol. 55, no. 1, pp. 317–327, Jan. 2006.

[9] S. Kompella, S. Mao, Y. Hou, and H. Sherali, "Cross-layer optimized multipath routing for video communications in wireless networks," *IEEE J. Sel. Areas Commun.*, vol. 25, no. 4, pp. 831–840, May 2007.

[10] ——, "On path selection and rate allocation for video in wireless mesh networks," *IEEE/ACM Trans. Networking*, vol. 17, no. 1, pp. 212–224, Feb. 2009.

[11] D. P. Heyman and T. V. Lakshmanr, "What are the implications of long-range dependence for VBR-video traffic engineering?" *IEEE/ACM Trans. Networking*, vol. 4, no. 3, pp. 301–317, June 1996.

[12] Y. Huang, S. Mao, and Y. Li, "Downlink power allocation for stored variable-bit-rate videos," in *Proc. ICST QShine'10*, Houston, TX, Nov. 2010, pp. 423–438.

[13] Y. Huang and S. Mao, "Downlink power control for variable bit rate video over multicell wireless networks," in *Proc. IEEE INFOCOM'11*, Shanghai, China, Apr. 2011, pp. 2561–2569.

[14] D. Hu, S. Mao, Y. T. Hou, and J. H. Reed, "Fine grained scalability video multicast in cognitive radio networks," *IEEE J. Sel. Areas Commun.*, vol. 28, no. 3, pp. 334–34, Apr. 2010.

[15] D. Hu and S. Mao, "Streaming scalable videos over multi-hop cognitive radio networks," *IEEE Trans. Wireless Communications*, vol. 9, no. 11, pp. 3501–3511, Nov. 2010.

[16] A. K. Katsaggelos, F. Zhai, Y. Eisenberg, and R. Berry, "Energy-efficient wireless video coding and delivery," *IEEE Trans. Wireless Commun.*, vol. 12, no. 4, pp. 24–30, Aug. 2005.

[17] P. Agrawal, J. Chen, S. Kishore, P. Ramanathan, and K. Sivalingam, "Battery power sensitive video processing in wireless networks," in *Proc. IEEE Personal, Indoor and Mobile Radio Communications'98*, Sep. 1998, pp. 116–120.

[18] Z. He, Y. Liang, L. Chen, I. Ahmad, and D. Wu, "Power-rate-distortion analysis for wireless video communication under energy constraints," *IEEE Trans. Circuits Syst. Video Technol.*, vol. 15, no. 5, pp. 645–658, May 2005.

[19] D. N. Kwon, P. F. Driessen, A. Basso, and P. Agathoklis, "Performance and computational complexity optimization in configurable hybrid video coding system," *IEEE Trans. Circuits Syst. Video Technol.*, vol. 16, no. 1, pp. 31–42, Jan. 2006.

[20] H. Cheng and L. Dung, "A content-based methodology for power-aware motion estimation architecture," *IEEE Trans. Circuits Syst. Video Technol.*, vol. 52, no. 10, pp. 631–635, Oct. 2005.

[21] N. J. August and D. S. Ha, "Low power design of dct and idct for low bit rate video codecs," *IEEE Trans. Circuits Syst. Video Technol.*, vol. 6, no. 3, pp. 414–422, Jun. 2004.

[22] Z. Cao, B. Foo, L. He, and M. van der Schaar, "Optimality and improvement of dynamic voltage scaling algorithms for multimedia applications," *IEEE Trans. Circuits Syst. Video Technol.*, vol. 57, no. 3, pp. 681–690, Mar. 2010.

[23] H. M. Wang, H. S. Choi, and J. T. Kim, "Workload-based dynamic voltage scaling with the qos for streaming video," in *Proc. IEEE Electronic Design, Test and Applications'08*, Jan. 2008, pp. 236–239.

[24] M. Li, Z. Guo, R. Y. Yao, and W. Zhu, "A novel penalty controllable dynamic voltage scaling scheme for mobile multimedia applications," *IEEE Trans. Mobile Computing*, vol. 5, no. 12, pp. 1719–1733, Dec. 2006.

[25] A. Goldsmith, *Wireless Communications*. New York, NY: Cambridge University Press, 2005.

[26] G. J. Foschini and Z. Miljanic, "A simple distributed autonomous power control algorithm and its convergence," *IEEE Trans. Veh. Technol.*, vol. 42, no. 4, pp. 641–646, Nov. 1993.

[27] D. Mitra, "An asynchronous distributed algorithm for power control in cellular radio system," in *Proc. 4th Winlab Workshop on Third Generation Wireless Information Networks*, New Brunswick, NJ, Oct. 1993, pp. 249–257.

[28] S. A. Grandhi, J. Zander, and R. Yates, "Constrained power control," *Int. J. Wireless Personal Commun.*, vol. 1, no. 4, pp. 257–270, Apr. 1995.

[29] P. Hande, S. Rangan, M. Chiang, and X. Wu, "Distributed uplink power control for optimal sir assignment in cellular data networks," *IEEE/ACM Trans. Networking.*, vol. 16, no. 6, pp. 1420–1433, Dec. 2008.

[30] J. Lee, R. Mazumdar, and N. Shroff, "Downlink power allocation for multi-class wireless systems," *IEEE/ACM Trans. Networking*, vol. 13, no. 4, pp. 854–867, Aug. 2005.

[31] C. W. Tan, D. P. Palomar, and M. Chiang, "Robustness tradeoff in cellular network power control," *IEEE/ACM Trans. Networking*, vol. 17, no. 3, pp. 912–925, Jun. 2009.

[32] N. Bambos, S. C. Chen, and G. J. Pottie, "Radio link admission algorithm for wireless networks with power control and active link quality protection," in *Proc. IEEE INFOCOM'95*, Boston, MA, Apr. 1995, pp. 97–104.

[33] M. Xiao, N. B. Shroff, and E. K. P. Chong, "Distributed admission control for power-controlled cellular wireless systems," *IEEE/ACM Trans. Networking*, vol. 9, no. 6, pp. 790–800, Dec. 2001.

[34] C. Comaniciu and H. V. Poor, "Jointly optimal power and admission control for delay sensitive traffic in cdma networks with lmmse receivers," *IEEE Trans. Signal Processing*, vol. 51, no. 8, pp. 2031–2042, Aug. 2003.

[35] G. Liang and B. Liang, "Balancing interruption frequency and buffering penalties in VBR video streaming," in *Proc. IEEE INFOCOM'07*, Anchorage, AK, May 2007, pp. 1406–1414.

[36] Y. Huang, S. Mao, and Y. Li, "Downlink power control for vbr video streaming in cellular networks: a majorization approach," in *Proc. IEEE GLOBECOM'11*, Dec. 2011.

[37] M. Chiang, "Balancing transport and physical layers in wireless multihop networks: jointly optimal congestion control and power control," *IEEE J. Sel. Areas Commun.*, vol. 23, no. 1, pp. 104–116, Jan. 2005.

[38] S. Mao, S. Lin, Y. Wang, S. S. Panwar, and Y. Li, "Multipath video transport over wireless ad hoc networks," *IEEE Wireless Commun.*, vol. 12, no. 4, pp. 42–49, Aug. 2005.

[39] S. Mao, Y. T. Hou, H. D. Sherali, and S. F. Midkiff, "Multimedia-centric routing for multiple description video in wireless mesh networks," *IEEE Network*, vol. 22, no. 1, pp. 19–24, Jan./Feb. 2008.

[40] H. Y. Shutoy, D. Gunduz, E. Erkip, and Y. Wang, "Cooperative source and channel coding for wireless multimedia communications," *IEEE J. Sel. Signal Process.*, vol. 1, no. 2, pp. 295–307, Aug. 2007.

[41] T. Koponen, T. Koponen, B. Chun, A. Ermolinskiy, K. H. Kim, and S. Shenker, "A data-oriented (and beyond) network architecture," in *ACM SIGCOMM'07*, Kyoto, Japan, Agu. 2007, pp. 181–192.

[42] X. Lu, E. Erkip, Y. Wang, and D. Goodman, "Power efficient multimedia communication over wireless channels," *IEEE J. Sel. Areas Commun.*, vol. 21, no. 10, pp. 1738–1751, Dec. 2003.

[43] Y. Eisenberg, C. E. Luna, T. N. Pappas, R. Berry, and A. K. Katsaggelos, "Joint source coding and transmission power management for energy efficient wireless video communications," *IEEE Trans. Circuits Syst. Video Technol.*, vol. 12, no. 6, pp. 411–424, Jun. 2002.

[44] Y. Li, M. Reisslein, and C. Chakrabarti, "Energy-efficient video transmission over a wireless link," *IEEE Trans. Veh. Technol.*, vol. 58, no. 3, pp. 1229–1244, Mar. 2009.

[45] Z. Li, F. Zhai, and A. K. Katsaggelos, "Joint video summarization and transmission adaptation for energy-efficient wireless video streaming," *EURASIP J. on Advances in Signal Processing*, vol. 2008, pp. 1–11, Jan. 2008.

[46] J. Huang, Z. Li, M. Chiang, and A. Katsaggelos, "Joint source adaptation and resource allocation for multi-user wireless video streaming," *IEEE Trans. Circuits Syst. Video Technol.*, vol. 18, no. 5, pp. 582–595, May. 2008.

[47] D. Wu, S. Ci, and H. Wang, "Cross-layer optimization for video summary transmission over wireless networks," *IEEE J. Sel. Areas Commun.*, vol. 25, no. 4, pp. 841–850, May 2007.

[48] S. Soro and W. Heinzelman, "A survey of visual sensor networks," *Advances in Multimedia*, vol. 2009, 2009, article ID 640386, 21 pages, doi:10.1155/2009/640386.

[49] A. Seema and M. Reisslein, "Towards efficient wireless video sensor networks: A survey of existing node architectures and proposal for a flexiwvsnp design," *IEEE Commun. Surveys Tutorials*, vol. 13, no. 3, pp. 462–486, Quarter 2011.

[50] M. Chen, T. Kwon, S. Mao, Y. Yuan, and V. C. M. Leung, "Reliable and energy-efficient routing protocol in dense wireless sensor networks," *International Journal of Sensor Networks*, vol. 4, no. 1/2, pp. 104–117, 2008.

[51] M. Chen, V. C. M. Leung, and S. Mao, "Directional controlled fusion in wireless sensor networks," *ACM/Springer MONET*, vol. 14, no. 2, pp. 220–229, Apr. 2009.

[52] ——, "Directional controlled fusion in wireless sensor networks," in *Proc. QShine'08*, Hong Kong, P.R. China, Jul. 2008, pp. 1–7.

[53] J. Liu and S. Singh, "Atcp: Tcp for mobile ad hoc networks," *IEEE J. Sel. Areas Commun.*, vol. 19, no. 7, pp. 1300–1315, Jul. 2001.

[54] V. Tsaoussidis and H. Badr, "Tcp-probing: towards an error control schema with energy and throughput performance gains," in *Proc. Network Protocols'00*, 2000, pp. 12–21.

[55] D. Hu and S. Mao, "Streaming scalable videos over multi-hop cognitive radio networks," *IEEE Trans. Wireless Commun.*, vol. 9, no. 11, pp. 3501–3511, Nov. 2010.

[56] H. Luo, S. Ci, and D. Wu, "A cross-layer design for the performance improvement of real-time video transmission of secondary users over cognitive radio networks," *IEEE Trans. Circuits Syst. Video Technol.*, vol. 21, no. 8, pp. 1040–1048, Aug. 2011.

[57] D. Hu and S. Mao, "Cooperative relay with interference alignment for video over cognitive radio networks," in *Proc. IEEE INFOCOM'12*, Orlando, FL, Mar. 2012.

[58] M. W. Garrett and W. Willinger, "Analysis, modeling and generation of self-similar VBR video traffic," *ACM SIGCOMM Comput. Commun. Rev.*, vol. 24, no. 4, pp. 269–280, 1994.

[59] J. Beran, R. Sherman, M. S. Taqqu, and W. Willinger, "Long-range dependence in variable-bit-rate video traffic," *IEEE Trans. Commun.*, vol. 43, no. 2/3/4, pp. 1566–1579, Feb./Mar./Apr. 1995.

[60] A. W. Marshall and I. Olkin, *Inequalities: Theory of Majorization and Its Applications.* New York, NY: Academic Press, 1979.

[61] M. Reisslein, "Video trace library," Arizona State University, [online] Available: http://trace.eas.asu.edu/.

[62] S. Sen, D. Towsley, Z. Zhang, and J. K. Dey, "Optimal multicast smoothing of streaming video over the internet," *IEEE J. Sel. Areas Commun.*, vol. 20, no. 7, pp. 1345–1359, Sep. 2002.

[63] J. D. Salehi, Z.-L. Zhang, J. Kurose, and D. Towsley, "Supporting stored video: reducing rate variability and end-to-end resource requirements through optimal smoothing," *IEEE/ACM Trans. Networking.*, vol. 6, no. 4, pp. 397–410, Aug. 1998.

[64] S. Ulukus and A. Yener, "Iterative transmitter and receiver optimization for CDMA networks," *IEEE Trans. Wireless Commun.*, vol. 3, no. 6, pp. 1879–1884, Nov. 2004.

[65] E. A. Jorswieck and H. Boche, "Optimal transmission strategies and impact of correlation in multiantenna systems with different types of channel state information," *IEEE Trans. Signal Processing*, vol. 52, no. 12, pp. 3440–3453, Dec. 2004.

[66] B. C. Arnold, *Majorization and the Lorenz Order: A Brief Introduction.* New York, NY: Springer-Verlag, 1987.

[67] M. Wien, H. Schwarz, and T. Oelbaum, "Performance analysis of SVC," *IEEE Trans. Circuits Syst. Video Technol.*, vol. 17, no. 9, pp. 1194–1203, 2007.

Author Contact Information

Yingsong Huang, Shiwen Mao, and Yihan Li are with Department of Electrical & Computer Engineering Auburn University, Auburn, AL 36849-5201, USA, Email: huangys@auburn.edu, smao@ieee.org, yli@ieee.org.

Part III

Focus on Wireline Communications

24

Trading off Energy and Forwarding Performance in Next-Generation Network Devices

Raffaele Bolla

DITEN-University of Genoa / CNIT - University of Genoa Research Unit, Genoa, Italy

Roberto Bruschi

CNIT - University of Genoa Research Unit, Genoa, Italy

Franco Davoli

DITEN-University of Genoa / CNIT - University of Genoa Research Unit, Genoa, Italy

Paolo Lago

DITEN-University of Genoa / CNIT - University of Genoa Research Unit, Genoa, Italy

CONTENTS

24.1   Introduction

In the last few years, the research field of "green" and energy-efficient networking has gained great interest on the part of network providers and equipment manufacturers. Such interest springs from heavy and critical economical needs, since both energy cost and network electrical requirements exhibit an almost steadily growing trend. For example, as shown in [1], energy consumption of the Telecom Italia network had reached more than 2 TWh (about 1% of the total Italian energy demand) in 2006, increasing by 7.95% with respect to the previous year. Similar trends can be generalized to a large part of the other telecoms and service providers; recent studies done by the Global e-Sustainability Initiative (GeSI) [2] foresee a jump in the overall network energy requirement of European Telcos from about 21.4 TWh in 2010 to 35.8 TWh in 2020 if no Green Network Technologies (GNTs) would be adopted. This alarming growth in energy network requirements is essentially a consequence of the increase in data traffic volume (which follows Moore's law, by doubling every 18 months [3]), causing an even larger increase in the number and capacity of deployed network devices. For instance, high-end IP routers are even more based on complex multi-rack architectures, which provide more and more network functionalities and continue to increase their capacities with a factor of 2.5 every 18 months [4]. At the same time, as shown in Figure 24.1, based on the data from [4], and as suggested by Dennard's scaling law [5], silicon technologies (e.g., CMOS) improve their energy efficiency at a lower rate with respect to routers' capacities and traffic volumes, by increasing of a factor of 1.65 every 18 months. Though core routers represent a minority of deployed

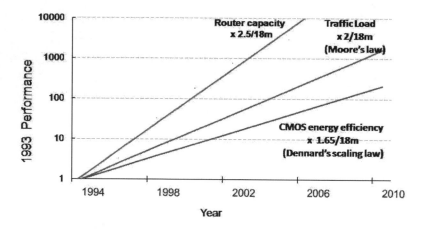

FIGURE 24.1 Evolution from 1993 to 2008 of high-end IP routers' capacity (per rack) vs. traffic volumes (Moore' s law) and energy efficiency in silicon technologies.

devices, similar considerations could be done also for the metro/transport and access network segments, where less energy-hungry equipment is present, but in numbers that are larger by orders of magnitude. In this respect, it is interesting to note how different assumptions on equipment evolution may lead to widely different forecasts. If no technology improvements and no green networking techniques were considered, the increasing trend in user access speed would lead to a predominance of the energy consumed in the core with respect to other network segments [6]. On the other hand, the situation would be reversed, with a clear predominance of the access, by only taking into account a 10% rate of improvement in the energy efficiency of the deployed technology [7]. In any case, the sole introduction of novel low consumption silicon technologies cannot clearly cope with such trends, and be enough for drawing ahead current network equipment towards a greener Future Internet. In such an environment, the challenge for network equipment manufacturers is, nowadays, to design new architectures able to scale their performance and functionalities, while preserving a high level of energy efficiency. It is well known that network links and devices are provisioned for busy or rush hour load, which typically exceeds their average utilization by a wide margin [8]. While this margin is seldom reached, nevertheless the power consumption is determined by it and remains more or less constant even in the presence of fluctuating traffic loads. Thus, the key of any advanced power saving criteria resides in dynamically adapting resources, provided at network, link or equipment levels, to current traffic requirements and loads [9] [10].

As a first step, we will try to identify and to characterize the main sources of energy consumption in a heterogeneous set of wire-line network devices. This analysis will give us the possibility of deeply understanding the most energy-hungry components and elements of current network devices. Then, we will move our attention on analysing and evaluating the impact of power scaling green networking technologies (GNTs) on next-generation network devices. We will focus on two basic techniques: Adaptive Rate (AR) and Low Power Idle (LPI). The former allows dynamically modulating the capacity of a link, or of a processing engine, in order to meet traffic loads and service requirements; the latter forces links or processing engines to enter low power states when not sending/processing packets and to quickly switch to a high power state when sending one or more packets. However, these techniques are not exclusive, and can be jointly adopted (as for example in general-purpose processors), in order to adapt system performance to current workload requirements. For instance, the IEEE 802.3az task force [11] considered and evaluated both techniques, and decided to base the new Green Ethernet standard only on the LPI primitive. This decision stemmed from the need for maintaining the implementation complexity and cost as low as possible. In other network contexts, the evaluations and the resulting decisions may be very different.

In such scenario, in order to give the reader the opportunity of understanding the impact of such techniques on network performance indexes (mainly packet delay and jitter), we will summarize a simple recently introduced analytical model based on classical concepts of queuing theory and able to capture the trade-off between energy- and network-aware performance metrics, when AR and/or LPI techniques are adopted in a network device [12]. In order to validate the proposed model, we performed several tests by using real-world traffic traces, and compared the estimated performance indexes with experimental measurements, obtained with Component Off-The-Shelf (COTS) SW routers [13].

The chapter is organized as follows. We briefly examine the energy consumption of some network devices in the next section. Section 24.3 describes power saving techniques in more detail, and introduces AR and LPI. Section 24.4 addresses some details of the analytical model and its validation. Section 24.5 contains the conclusions.

24.2   The Devices' Internal Sources of Energy Consumption

In order to face the energy efficiency issue in wire-line networks, we have to firstly understand and accurately characterize the real sources of power consumption in network equipment. An interesting contribution in this direction can be found in [6], where Tucker *et al.* present a stimulating perspective on network design by taking into account the energy efficiency aspects. As pointed out by Tucker *et al.*, network devices working in the different network portions play a central role, since the overall energy consumption in networks arises from their operational power requirements and their density. In more detail, operational power requirements arise from all the hardware (HW) elements realizing network-specific functionalities, like the ones related to data- and control-planes, as well as from elements devoted to auxiliary functionalities (e.g., air cooling, power supply, etc.). In this respect, the data-plane certainly represents the most energy-starving and critical element in the largest part of network device architectures, since it is generally composed by special purpose HW elements (packet processing engines, network interfaces, etc.) that have to perform per-packet forwarding operations at very high speeds. From a general point of view, IP routers have similar architectures with respect to high-end switching systems, i.e., highly modular and hierarchical architectures. Focusing on high-end IP routers, Tucker et al. [6] estimated that the data-plane power consumption weighs for 54% of the total, vs. 11% for the control plane and 35% for power and heat management (see Fig. 24.2). The same authors further broke out energy consumption sources at the data-plane on a per-functionality basis. Internal packet processing engines require about 60% of the power at the data-plane of a high-end router, network interfaces weigh for 13%, switching fabric for 18.5% and buffer management for 8.5%.

Notwithstanding that this study specifically refers to high-end router plat-

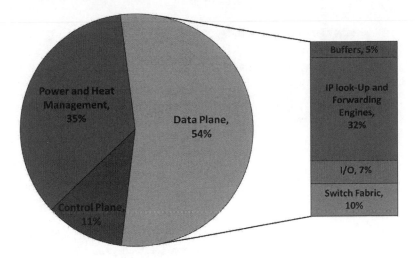

FIGURE 24.2 Estimate of power consumption sources in a generic platform of high-end IP router.

forms, and the same internal distribution of power cannot be obviously maintained for all the typologies and the architectures of network devices, the resulting estimations provide a relevant and clear indication on how and where future research efforts need to be focused in order to build next-generation green devices. Due to the high processing capacity, high-end IP routers based on multi-chassis platforms can be certainly considered as the network node typology with the highest complexity level. Traffic processing engines have generally to support complex forwarding and lookup functionalities, and internal HW elements are generally dimensioned for processing enormous traffic volumes. For these reasons, in order to effectively reduce the carbon footprint of such kind of devices, green technologies will have to mainly address the energy efficiency in packet processing engines that nowadays do not include any power management capabilities like low power idle and/or adaptive rate. On the other hand, Digital Subscriber Line Access Multiplexers (DSLAMs) generally include less and simpler packet processing functionalities with respect to routers, but also present a much larger number of network interfaces. In detail, an IP-DSLAM supports several hundreds of DSL customers, housing a corresponding number of line cards. Every line card (LC) is connected to the Up-link card (UC), where the Ethernet switch performs multiplexing operations in order to direct the aggregated traffic to the transport network (Figure 24.3).

The power decomposition among the different functional blocks of a fully equipped ADSL2+ and VDSL2+ IP DSLAM node is shown in Figure 24.4. As can be observed in the pie charts, line cards contribute to about 80% of the global power consumption. In detail, line drivers (LD), which perform the physical layer operations, require about 36-46% of the power share, while the

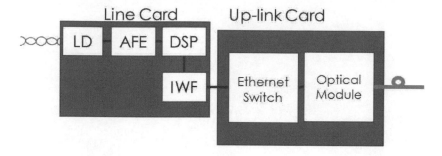

FIGURE 24.3 IP-DSLAM architecture

DSL modem chipsets (DSP+AFE) and the Inter-working functions (IWF) weigh 15-26% and 19%, respectively. Since an Up-link card can serve a number of LCs, its energy weight turns out to be negligible with respect to the global energy requirements of the equipment. Certainly, since the power con-

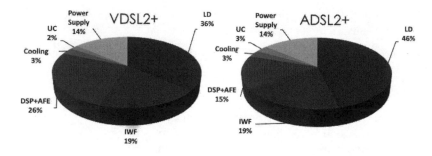

FIGURE 24.4 Power decomposition of ADSL2+ and VDSL2+ IP DSLAM.

sumption of today's DSLAMs mainly stems from link interfaces, future green technologies and solutions for access networks will have especially to focus on energy-efficiency at the link/network interface layer. Following these basic ideas, the largest part of current research contributions generally focused on introducing novel extensions and solutions for reducing the carbon footprint of particular devices, by working at the level of internal and transport network nodes, and that of network interfaces and/or link protocols for access and processing engines of core home equipment.

24.3   Power Scaling Techniques

Power scaling techniques are aimed at modulating capacities of network device resources (e.g., link bandwidths, computational capacities of packet processing engines, etc.) according to current traffic loads and service requirements.

Such approaches are generally founded on the two main kinds of power saving capabilities provided by the HW level, namely, AR and LPI. As mentioned in the previous section, the largest part of today's network equipment does not include such HW capabilities, but power management is a key feature in today's processors across all market segments. In detail, AR capabilities allow dynamically reducing the working rate of processing engines or of link interfaces. This is usually accomplished by scaling the clock frequency or by throttling the CPU clock (i.e., the clock signal is gated or disabled for a certain number of cycles at regular intervals). For instance, the power consumption of CMOS based silicon can be roughly characterized as follows:

$$P = CV^2 f, \tag{24.1}$$

where P is the active power absorbed, C the CMOS capacitance, and V and f are the operating voltage and frequency values, respectively. It is worth noting that V and f need to be directly proportional for a correct working of the CMOS silicon. Decreasing the operating frequency and the voltage of a processor, or throttling its clock, obviously allows the reduction of the power consumption and of heat dissipation at the price of slower performance. On the other hand, LPI allows reducing power consumption by rapidly turning off sub-components when no operations are performed, and by re-waking them up when the system has to resume activity. Wake-up instants may be triggered by external events in a pre-emptive mode. In a network device this can be performed by leaving the circuitry that senses packets on, so that the wake-up event is triggered by the arrival of a packet or burst of packets. Timer-driven sleeping is another way for entering and exiting sleep states; in this case, the network element wakes up every certain time periods, and controls if there are new activities to process. In COTS CPUs, the HW implementation of both LPI and AR solutions is generally performed by pre-selecting a set of feasible and stable HW configurations, which provide different trade-offs between energy consumption and performance. In more detail, in general purpose computing systems, the Advanced Configuration and Power Interface (ACPI) provides a standardized interface between the hardware and software layers by modelling the LPI and AR primitives as a set of states (C-states and P-states, respectively). Regarding the C-states, C_0 is an active power state where the CPU executes instructions, while power states C_1 through C_n are processor sleeping or idle states, where the processor consumes less power and dissipates less heat. On the other hand, as the sleeping power state $(C_1, ..., C_n)$ becomes deeper, the transition between the active and the sleeping state (and vice versa) requires longer time. For example, as outlined in Table 24.1, the transition between states C_0 and C_1 needs just only few nano-seconds, while 50 μs are required for entering the C_3 state.

While in the C_0 state, ACPI allows the performance of the processor's core to be tuned through P-state transitions. In this way, a CPU/Core can consume

TABLE 24.1 Indicative energy saving and transition times for COTS processors' C-states

C-State	Energy Saving with respect to the C_0 state	Transition Times
C_0	0%	–
C_1	70%	10 ns
C_2	75%	100 ns
C_3	80%	50 μs
C_4	98%	160 μs
C_5	99%	200 μs
C_6	99.9%	unknown

different amounts of power while providing different performance at the C_0 (running) state. In general, the higher the index of P and C states is, the less will be the power consumed and the heat dissipated. Both these power saving

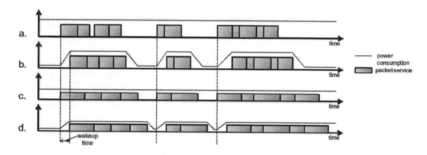

FIGURE 24.5 Packet service times and power consumptions in the following cases: (a) no power-aware optimizations, (b) only low power idle, (c) only adaptive rate, (d) both low power idle and adaptive rate.

capabilities can be jointly adopted in order to adapt system performance to current workload requirements. As shown in Figure 24.5, rate scaling (Fig. 24.5-c) obviously causes a stretching of packet service times (i.e., header processing time in a processing engine, or packet transmission time in a link interface), while the sole adoption of low power idle (Fig. 24.5-b) introduces an additional delay in packet service, due to the wake-up times. Finally, as outlined in Fig. 24.5-d, the joint adoption of both LPI and AR capabilities may not lead to outstanding energy gains, since performance scaling causes larger packet service times, and consequently shorter idle periods. However, the energy- and network-aware effectiveness of LPI and AR (and their possible joint adoption) must be accurately evaluated by taking HW and traffic features and requirements into account. In this respect, it is worth noting that the overall energy saving and the network performance strictly depend on incoming traffic volumes and statistical features (i.e., inter-arrival times, burstiness levels, etc.). For these reasons an optimization policy is generally needed to configure and control the usage of LPI and AR primitives with respect the to the estimated workload and service requirements. For instance, LPI techniques provide more energy gain and network performance when incoming traffic has a high burstiness level. This is because bursty traffic is characterized by long inactivity periods that lead the hardware to remain longer in low power consumption; moreover, less active-idle transitions are needed and the additional delay due to the wake-up time becomes negligible. Starting from these considerations, Nedevschi et al. proposed in [8] to shape the traffic into small bursts at the edge of the network in order to explicitly control the trade-off between network performance and energy saving. In particular, the traffic shaping policy proposed introduces a buffer interval B in the ingress router of the network. The ingress router buffers incoming traffic for up to B ms and, once every B ms, forwards buffered traffic in a burst. To ensure that bursts created at the ingress are retained as they traverse the network, an ingress router arranges packets within the burst such that all packets destined for the same egress router are contiguous within the burst. By doing so, the LPI-capable devices in the network can exploit the periodic behaviour of the traffic, leading to fewer transitions and amortizing the transition penalty over multiple packets.

An example of application of LPI mechanisms in network equipment can be found in the novel IEEE 802.11az standard [11] for network interfaces. IEEE 802.11az defines a simple protocol that enables the transmitter to send LPI signaling to indicate that the link can go to sleep. After a period T_s, the transmitter can stop signaling altogether, so that the link becomes quiescent and consumes less power. Periodically, the transmitter sends some signals so that the link does not remain quiescent for too long without a refresh. Finally, when the transmitter wishes to resume the fully functional link, it sends a wakeup signal and after a pre-determined time T_w the link is active and data can be sent. The refresh signal that is sent periodically while the link is idle is important for multiple reasons. The heartbeat of the refresh signal helps ensure that both connected interfaces know that the link is present and allows for immediate notification following a disconnection. Second, the refresh signal can be used to test the channel and create an opportunity for the receiver to adapt to changes in the channel characteristics. Moreover, this is vital to support the rapid wake-up to the full speed data transfer without sacrificing forward performance and/or data integrity.

For what concerns the adaptive rate techniques, Gunaratne *et. al.* [14] proposed an interesting AR policy, called Adaptive Link Rate (ALR), specifically targeted for Network Interface Cards (NIC). ALR is fundamentally conceived in order to adjust NIC speed to effective traffic levels. In fact, higher rates require more energy, which is mostly used to send small amounts of data. Moreover, within a few years, 10 Gbps NICs may become standard in desktop PCs. For instance, a 100 Mbps Ethernet NIC consumes in the order of 1 W, while a 10 Gbps NIC consumes few tens of Watts. The Ethernet standard already includes auto-negotiation handshake mechanisms, which allow the link data rate to change, but their behaviour in today's HW is far too slow to be used for energy efficiency purposes. Therefore, new methods are needed in the

Ethernet standard to effectively support ALR, and to quickly modulate link data rates with traffic levels to scale energy consumption with actual service demand. As proposed by Gunaratne *et. al.*, a faster two-way handshake can be implemented using Ethernet MAC frames, as follow:

- When the end of the link determines a need to increase or decrease its data rate, an ALR REQUEST MAC frame is sent. The request could be "go to low" or "go to high" data rate.

- The receiving-end link acknowledges the data rate change request with either an ALR ACK (agrees to change the data rate) or ALR NACK (does not agree) MAC frame.

The ALR ACK response will trigger a link data rate switch (rate transition) and link resynchronization. The total time for the handshake plus resynchronization, T_{switch} , could likely be less than 100 microseconds for 1 Gbps Ethernet. This is based on a reasonable 10,000 clock cycles needed for link resynchronization.

The same authors proposed two optimization policies able to scale power consumption of network interfaces with respect to the traffic level. The first one is based on a thresholding policy, by monitoring the output buffer occupancy; two thresholds (qHigh and qLow) are used to introduce hysteresis into the system and prevent a trivial oscillation between rates. If the output buffer queue length exceeds qHigh, then the data rate must be transitioned to high and a request "go to high" is sent. The only allowable response from the other side is an ACK. That is, the other side cannot disagree with a request to increase the link data rate. If the output queue decreases below qLow, then the data rate can be reduced to low if the other side of the link agrees, by checking its output queue length. Since the network traffic could be characterized by a high burstiness coefficient, the previous dual-threshold policy can cause an unsustainable link rate oscillation that increases the packet delay due to the service interruption T_{switch}. The high occurrence of the data rate oscillations can be moderated by explicitly monitoring the link utilization and using it in the decision to transition between data rates. The link utilization can be simply obtained by counting the number of bytes transmitted during a time period. Thus, by taking into account also the link utilization in the dual-threshold policy, trivial oscillations between rates can be reduced at the price of increasing the complexity of the network interface. The previous example of AR policy for a network interface raises some major issues to deal with in order to introduce power saving algorithms in network devices. The first one is the time spent to switch components to different speed, which introduces an additional delay in the packet service time; moreover, if T_{switch} is exceedingly greater than the packet time scale, an optimization procedure cannot quickly react to sudden increases in the traffic level. However, even considering a perfect device with instantaneous speed scaling, the selection of the optimal rate comes from the monitoring of the traffic statistical features during a time window: this procedure could place a significant weight in the total power consumption due to the increase in the complexity of an AR-capable device. Another critical variable in this problem is the choice of the time window. Obviously, decreasing the traffic monitoring time allows fitting well the traffic level profile; but, on the other hand, it increases the number of speed selections and, consequently, the rate oscillations.

24.4   A Theoretical Model of the Trade-Off Between Energy Consumption and Quality of Service

In this section, our aim is to characterize the impact of the previous power saving mechanisms (AR and LPI) on network performance indexes and how they can affect the overall device behaviour. To this purpose, we will summarize a simple analytical model based on classical concepts of queuing theory able to capture the trade-off between energy- and network-aware performance metrics, when AR and/or LPI techniques are adopted in a network device [12]. The proposed analytical model can represent a large set of network switching architectures, technologies and components (from green link interfaces to packet processing engines) and turns out to be an interesting estimation tool suitable to be effectively adopted inside optimization procedures for the design and the dynamic control of next-generation green networking equipment.

Starting from the estimated incoming traffic load and from a fixed AR and LPI device configuration, the main objective of the model is twofold: it has to accurately represent the energy consumption, as well as to estimate the packet forwarding performance in terms of packet latency times and loss rates. Our approach considers a network device substantially as a traffic-forwarding engine, which includes LPI and AR primitives and processes incoming packet headers at a finite capacity μ. For the sake of simplicity, we refer to LPI and AR primitives by using the ACPI terminology, which treats LPI and AR primitives in terms of C- and P-states, respectively. The selection of different P- and C-states is supposed to impact on the forwarding engine performance in terms of both the packet service (processing) capacity and wakeup times. Moreover, the model assumes that all packet headers require a constant service time. This hypothesis represents a reasonable approximation for a large part of current routing and switching devices. A finite buffer, with a size equal to N packets, is assumed to be bound to the server for backlogging incoming traffic packets. The rest of this section is organized as follows. Sub-section 24.4.1 introduces the main parameters to be considered in a device with AR and LPI capabilities. Sub-section 24.4.2 shows the model for representing the traffic incoming to the energy-aware device. The proposed analytical model is described in Sub-section 24.4.3. Finally, Sub-section 24.4.4 reports some numerical results in order to validate the proposed model.

24.4.1   Introducing Energy-Aware Parameters

Let $\{ C_0, C_1, ..., C_X \}$ and $\{P_0, P_1, ..., P_Y\}$ be the set of sleeping and performance states, respectively, available in the device. Each sleeping state is bound to distinct values of idle power consumption $\Phi_{idle}(C_x)$ and of transition times $\tau_{off}(C_x)$ and $\tau_{on}(C_x)$, needed to enter the idle state and to wake-up from it. As mentioned in section 24.2, a deeper sleeping state is characterized both by lower power consumption and by a larger transition period. Similarly, each P state can be related to a certain active power consumption $\Phi_a(P_y)$, as well as a packet processing capacity $\mu(P_y)$. As the y index is higher, both $\Phi_a(P_y)$ and $\mu(P_y)$ values decrease. Once fixed the state pair $\{C_x, P_y\}$, the system works with the renewal process representation shown in Figure 24.6. The engine has infinitely many alternating busy $T_B^{(n)}$ and idle $T_I^{(n)}$ periods, where the index n denotes the order of the interval. During a generic $T_B^{(n)}$, the server is active and performing packet forwarding activities, and then it has instantaneous power consumption equal to $\Phi_a(P_y)$. Afterwards, when it serves the last backlogged packet, it enters the $T_I^{(n)}$ period corresponding to the low-consumption C_x state. However, transitions from the active state C_0 to the C_x state are not

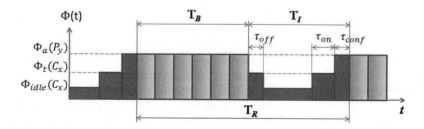

FIGURE 24.6 Power consumption during a renewal busy-idle cycle.

instantaneous, and a transition time τ_{off} is required. When new packets are received, the device has to wake-up by exiting the C_x state and returning to the active one (this requires an additional τ_{on} period). Furthermore, depending on the specific device architecture and implementation, an additional time τ_{conf} is required to setup and to suitably configure the packet elaboration process. It is worth noting that, while τ_{on} and τ_{off} depend on the sleeping C_x state, the τ_{conf} parameter depends on the P_y state, since it represents a certain number of operations that have to be performed by the server, before starting packet-forwarding. Therefore, the instantaneous power requirements can be expressed as follows:

$$\Phi(t) = \begin{cases} \Phi_{idle}(C_x), & \text{if the server is in the } C_x \text{ state.} \\ \Phi_a(P_y), & \text{if the server is in the } C_0 \text{ state and working} \\ & \text{in the } P_y \text{ power state.} \\ \Phi_t(C_x), & \text{if the server is moving from } C_x \\ & \text{to the } C_0 \text{ state and vice versa.} \end{cases} \quad (24.2)$$

As in most COTS platforms $\tau_{off} << \tau_{on}$, we have decided to neglect τ_{off} in the model derivation.

24.4.2   The Traffic Arrival Process Model

The modeling and the statistical characterization of packet inter-arrival times are well known to have Long Range Dependency (LRD) and multi-fractal statistical features. However, we decided to adopt the more recent Batch Markov Arrival Process (BMAP) model with Long Range Dependent (LRD) batch sizes [15], which can effectively estimate the network traffic behaviour. This model assumes to receive groups (or batches, or bursts) of j packets at exponential inter-arrival times with average value equal to $\frac{1}{\lambda}$. The probability distribution of the number of packets in the batches is supposed to follow Zipf's law (which can be regarded as the discrete version of a continuous Pareto probability distribution). In more detail, we assume that incoming packet batches have the following probability mass function:

$$\beta_j = \begin{cases} \frac{1}{j^\nu \sum_{i=1}^{j_{max}} \frac{1}{i^\nu}}, & 1 \le j \le j_{max} \\ 0, & j > j_{max}, \end{cases} \quad (24.3)$$

where β_j represents the probability that an incoming burst contains j packets, with $j \in [1, j_{max}]$. The average number of packets in a batch, β, is then obtained as:

$$\beta = \frac{\sum_{i=1}^{j_{max}} \frac{1}{i^{\nu-1}}}{\sum_{i=1}^{j_{max}} \frac{1}{i^\nu}}. \quad (24.4)$$

The Probability Generating Function (PGF) of batch sizes can be expressed as:

$$X(z) = \sum_{j=1}^{\infty} \beta_j z^j = \sum_{j=1}^{j_{max}} \frac{z^j}{j^\nu (\sum_{i=1}^{j_{max}} \frac{1}{i^\nu})}. \quad (24.5)$$

24.4.3   The Queuing System Model

Starting from the assumed traffic arrival process and the hypothesis of deterministic service times, the proposed model corresponds to a $M^x/D/1$ queuing system with a server SET-up period ($M^x/D/1/SET$) [16], where customers arrive in batches at Markov inter-arrival times with average rate λ, and are served by a single server at a fixed rate μ. In order to take the LPI transition periods into account, the model considers deterministic server setup times. When the system becomes empty, the server is turned off. The system returns operative only when a batch of packets arrives. At this point in time service can begin only after an interval $\tau_{setup} = \tau_{on} + \tau_{conf}$ has elapsed.

The stationary probability P_n, of having n packets in the queuing system can be obtained by exploiting the following probability generating function for the $M^x/G/1$:

$$P(z; M^x/G/1) = (1 - \rho)\frac{(1 - z)B(\lambda - \lambda X(z))}{B(\lambda - \lambda X(z)) - z} \tag{24.6}$$

where $B(.)$ is the Laplace transform of service times, which, under the assumption that service times are deterministic, can be expressed as:

$$B(s) = e^{-\frac{s}{\mu}} \tag{24.7}$$

In order to find the PGF of the $M^x/D/1/SET$ queuing system, we can exploit the PGF for the $M^x/D/1$, which by using Eqs. 24.6 and 24.7, turns out to be:

$$P(z, M^x/D/1) = (1 - \rho)\frac{(1 - z)e^{-\frac{\lambda}{\mu}(1 - X(z))}}{(e^{-\frac{\lambda}{\mu}(1 - X(z))} - z)} \tag{24.8}$$

Since the period τ can be considered as server vacation, we can exploit the stochastic decomposition results of Doshi [17] for the single unit arrival case for bulk arrivals; thus, the PGF of the $M^x/D/1$ queue with setup times turns out to be:

$$P(z, M^x/D/1/SET) = \zeta(z)P(z, M^x/D/1), \tag{24.9}$$

where

$$\zeta(z) = \frac{1 - zV(\lambda - \lambda X(z))}{(1/\beta + \lambda\tau)(1 - X(z))} \tag{24.10}$$

is the PGF of the number of arrivals during the residual life of the vacation period, defined as an idle period plus an interval τ, and $V(s)$ is the Laplace transform of the setup time. Since server setup times have constant durations equal to τ, we can express $V(s)$ as:

$$V(s) = e^{-\tau s}. \tag{24.11}$$

By using Eqs. 24.10 and 24.11 in Eq. 24.9, we can obtain the PGF of our $M^x/D/1/SET$ system:

$$P(z, M^x/D/1/SET) = (1 - \rho)\frac{1 - ze^{-\lambda\tau[1 - X(z)]}(1 - z)e^{-\frac{\lambda}{\mu}[1 - X(z)]}}{(\frac{1}{\beta} + \lambda\tau)[1 - X(z)]e^{-\frac{\lambda}{\mu}[1 - X(z)]} - z}. \tag{24.12}$$

Remembering that the PGF is defined as:

$$P(z, \cdot) = \sum_{n=0}^{\infty} P_n z^n, \qquad (24.13)$$

we can obtain the state probabilities P_n by calculating the Taylor series' coefficients of the $P(z)$ function:

$$P_n = \frac{1}{n!} \frac{\partial^n}{\partial z^n} P(z, \cdot) \bigg|_{z=0}. \qquad (24.14)$$

In order to evaluate the energy saving provided by the LPI and AR mechanisms, we have to statistically characterize the server idle and busy times. As mentioned in sub-section 24.4.1, when the server utilization $\rho < 1$, a single-server queuing system is known to empty infinitely often; this obviously remains true also for our $M^x/D/1/SET$ model. Hence, using classical principles of renewal theory, we can identify independent and identically distributed (iid) "cycles" of the form:

$$T_R^{(n)} = T_B^{(n)} + T_I^{(n)}, \qquad (24.15)$$

where $T_B^{(n)}$ is the nth busy period, and $T_I^{(n)}$ is the nth idle period. The average durations of idle and busy periods are given by:

$$T_I = E\{T_I^{(n)}\} = \frac{1}{\lambda}, \qquad (24.16)$$

$$T_B = \frac{1}{\lambda} \frac{\rho}{(1-\rho)} + \frac{\beta\tau}{1-\rho}. \qquad (24.17)$$

We can obtain T_R as follows:

$$T_R = T_I + T_B = \frac{\frac{1}{\lambda} + \beta\tau}{1-\rho}. \qquad (24.18)$$

Starting from the stationary probabilities P_n, as well as the idle and busy periods, we can easily derive a large set of network performance indexes. The mean value $\bar{L}$ of packets in the queuing system can be obtained by specializing the general expressions in [16] to our case of deterministic service time and Zipf-distributed packet batches:

$$\bar{L} = lim_{z \to 1} P'(z, M^x/D/1/SET) = \frac{2\lambda\beta\tau + \lambda^2\beta^2\tau^2 - \beta + \sum_{j=1}^{j_{max}} \beta_j j^2}{2(1 + \lambda\beta\tau)} +$$

$$+ \frac{\rho^2 - \beta + \sum_{j=1}^{j_{max}} \beta_j j^2}{2(1-\rho)}. \qquad (24.19)$$

Using Little's law, the average waiting time $\bar{W}$ is:

$$\bar{W} = \frac{\bar{L}}{\lambda\beta} = \frac{2\tau + \lambda\beta\tau^2 - \frac{1}{\lambda} + \frac{1}{\lambda\beta}\sum_{j=1}^{j_{max}}\beta_j j^2}{2(1+\lambda\beta\tau)} + $$
$$+ \frac{\rho^2 - \beta + \sum_{j=1}^{j_{max}}\beta_j j^2}{2\lambda\beta(1-\rho)}. \tag{24.20}$$

It is worth noting that both the $P(z)$ function in Eq. 24.13 and the stationary probabilities P_n in Eq. 24.14 are referred to the $M^x/D/1/SET$ queue with an infinite buffer. However, by assuming a low value of loss probability, we can approximate the stationary probabilities of the finite buffer queuing system with the $\{P_0, P_1, ..., P_N\}$ probabilities of the $M^x/D/1/SET$ queue. In more detail, the average value of packet loss probability can be expressed through the following approximation:

$$P_{loss} = 1 - \sum_{n=0}^{N} P_n. \tag{24.21}$$

The approximation might be used also to re-compute $\bar{L}$ and $\bar{W}$ for the finite buffer case. However, if P_{loss} is minute (as it actually turns out to be in most practical cases), Eqs. 24.19 and 24.20 already provide a good approximation.

Recalling Eq. 24.6 and Eq. 24.2, we can express the average power consumption in a renewal cycle for a fixed pair of $\{C_x, P_y\}$ as follows:

$$\bar{\Phi} = \frac{[\Phi_a \cdot (T_B - \tau_{on}) + \Phi_t \cdot \tau_{on} + \Phi_{idle} \cdot T_I]}{T_R}, \tag{24.22}$$

and, by using Eqs. 24.16 and 24.17 in Eq. 24.22:

$$\bar{\Phi} = \frac{[\Phi_a \cdot (\frac{1}{\lambda}\rho + \beta\tau - (1-\rho)\tau_{on}) + (1-\rho)(\tau_{on} \cdot \Phi_t + \frac{1}{\lambda} \cdot \Phi_{idle})]}{\frac{1}{\lambda} + \beta\tau}. \tag{24.23}$$

24.4.4   Model Validation

The proposed analytical model is able to capture the impact of power management capabilities on network performance metrics and explicitly takes a large set of features and parameters into account, like, for example, the transition times to exit low-consumption sleeping states, or packet latencies and loss probabilities. For this reason the proposed model is suitable to be adopted inside optimization procedures for controlling next-generation energy-aware devices. For the sake of completeness, we report in the following the validation results that were obtained in [12].

In order to validate the proposed model, we decided to use a large data set from real-world, and experimental devices as terms of comparison. Regarding the experimental devices, we used a multi-core Linux Software (SW) router, based on general purpose PC architecture equipped with two Xeon X5550 Quad Core, since the HW already provides advanced power management capabilities by means of the ACPI technology. The considered SW Router is equipped a Gigabit Ethernet adapter, which support multiple Tx/Rx buffer and multiple HW interrupts per network interface. Using such innovative network board features, we deployed a new SW architecture/configuration that allows us to optimize SR data-plane performance by reducing memory sharing among cores and avoiding the contention on shared memory [18]. In detail, each Rx buffer is directly bound to a single core. Regarding the Tx buffer, each link provides one separate buffer for each core; thus, all cores perform packet-forwarding operations in a fully parallel and independent way. Owing to such details, and as already discussed in [19], the operating behaviour of each core can be represented by means of an independent $M^x/D/1/SET$ queueing system. Each processor core includes AR and LPI capabilities in terms of 4 available P-states, and 3 C-states (including the C_0 one), respectively. Tables 24.2 and 24.3 report the power consumption and the network performance indexes (e.g., τ_{on}, and μ) for the available C- and P-states. Previous experimentations on SW router architectures [13] suggest to use the values for the τ_{on} parameter indicated in Table 24.2, and to fix $\tau_{setup} = \mu^{-1}$.

TABLE 24.2 Power consumption and transition times of the device's C-states

C_x state	$\Phi_{idle}(C_x)$	τ_{on}
C_0	active	active
C_1	10 Watt	10 ns
C_2	8 Watt	100 ns

TABLE 24.3 Power consumption and forwarding capacities of the device's P-states

P_y state	$\Phi_a(P_y)$	μ
P_3	50 Watt	650 kpkts/s
P_2	60 Watt	770 kpkts/s
P_1	70 Watt	890 kpkts/s
P_0	80 Watt	1010 kpkts/s

For the sake of simplicity, since each core, serves packets from its reception buffer, we decided to show the validation results for a single processor core, receiving traffic from a single Gigabit Ethernet interface with reception and transmission buffer sizes equal to 512 packets, and forwarding it towards another Gigabit Ethernet link. Regarding incoming traffic, we performed the

SW router experimentations and the proposed model estimation by using real-world traffic traces that are publicly available [20]. In detail, a 96-hour-long traffic trace divided into sequential time windows of 15 minutes was used. Thus, for each time window, we obtained energy- and network-aware performance indexes both with the SW router and with the proposed model. As far as the proposed model is concerned, for each time window, we used the λ, β, ν, and j_{max} values calculated from the traffic trace. In detail, parameters ν and j_{max} were obtained by least squares fitting of the Zipf distribution in Eq. 24.3 with the trace sample. The evolution of the traffic offered load over the time of the reference traffic trace is reported in Figure 24.7 in terms of burst arrival rates and burst sizes. The minimum value of traffic loads is from 3:00 to 6:00, while rush hours occur at 11:00 and 14:00. It is interesting to underline how an increase in incoming traffic volume is due to the rise of both burst arrival rate and burst sizes. For the sake of clarity, in the rest of this sub-section we will show the estimation results and measurements for only a 24-hour-long traffic trace and for a subset of P- and C-states combinations. Figure 24.8 reports the power consumption values estimated by the analytical model (AM), the values measured, and the maximum estimation error in each time window. The AM estimation was obtained with Eq. 24.23. The results in Figure 24.8 outline the

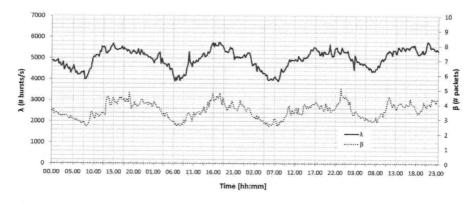

FIGURE 24.7 Average values of λ and β measured in the traffic trace in [20]

good accuracy level provided by the model. When the probability that burst inter-arrival time is larger than τ drops, the device enters low-power sleeping states more and more rarely and for shorter periods, before waking up again. Regarding the performance indexes, Figures 24.9 and 24.10 show the average values of loss probability and packet latency times for both the SR and the AM, respectively. The AM estimates of latency times were obtained with Eqs. 24.19 and 24.20, and loss probabilities were computed as in Eq. 24.21. The proposed model can be viewed as an interesting estimation tool, which can be adopted for controlling AR and LPI capabilities in next-generation network devices. The model can be included in the optimization procedures in order to

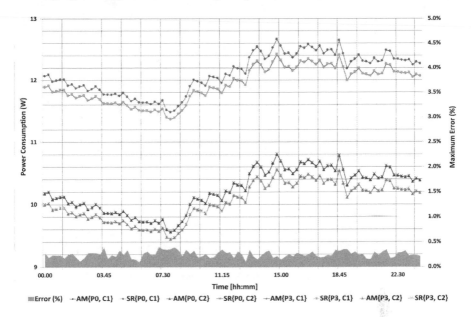

FIGURE 24.8 Energy consumption estimated by the analytical model according to various configurations of C- and P-states, and maximum estimation error of the analytical model with respect to the SW router per each time slice.

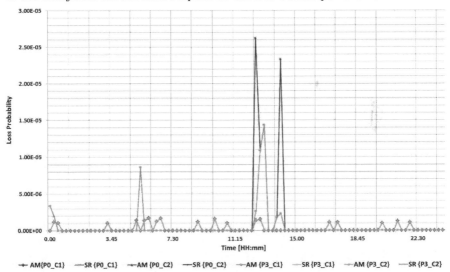

FIGURE 24.9 Packet loss probability estimated by the AM and measured on the SR with respect to different P- and C-states.

dynamically and optimally change the energy-aware device configuration with respect to the estimated traffic load and performance requirements. Such op-

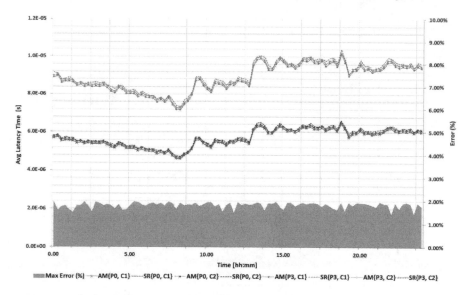

FIGURE 24.10 Average packet latency estimated by the AM and measured on the SR with respect to different P- and C-states, and maximum relative estimation error.

timization procedures have to periodically select the optimal pair of $\{C_x, P_y\}$ states, which minimizes the device's power consumption while meeting the estimated load and network performance requirements (i.e., loss rate, latency times, etc.). Thus, given the estimate of incoming traffic load, in terms of λ, β, j_{max} and ν, the optimization problem can be formalized as:

$$\begin{cases} min_{\{C_x, P_y\}} \bar{\Phi}(\lambda, \beta) \\ \bar{W} < W^* \\ P_{loss} < P_{loss}^*, \end{cases} \qquad (24.24)$$

where W^* and P_{loss}^* are the maximum admissible average values of latency and loss probability, respectively. Thus, the optimization procedure starts by considering all the $\{C_x, P_y\}$ pairs that satisfy the performance constraints in Eq. 24.24. This is performed by estimating the average latency values with Eq. 24.20, and the loss probabilities through Eq. 24.21, by numerically obtaining the first N coefficients of the Taylor series of the PGF in Eq. 24.13. For all the $\{C_x, P_y\}$ pairs satisfying the above constraints, we calculate the estimated value of the average power consumption with Eq. 24.8. The configuration guaranteeing the minimum consumption is finally selected. The optimization procedure works in every time window, by finding the best $\{C_x, P_y\}$ pair that minimizes energy consumption and satisfies the packet latency and loss constraints. The performance constraints of the model were $W^* = 10 \ \mu s$, and $P_{loss}^* = 10^{-5}$. We decided to let the optimization procedure know the exact values of time window traffic parameters (λ, β, j_{max} and ν) in advance.

This choice gives us the chance of evaluating the optimization procedure performance in the absence of errors due to incoming traffic estimation. The results in Figs. 24.11 and 24.12 only depend on the accuracy level of the AM. We decided to take the $\{C_1, P_0\}$ SR configuration as a term of comparison, since it represents the most "conservative" case for network performance. Fig. 24.11 reports the power consumption for both cases considered, and underlines how the optimization procedure allows saving about 16-17% of energy with respect to the fixed $\{C_1, P_0\}$ configuration. Regarding network performance, Fig. 24.12 shows that the latency constraints are fully satisfied in all the time windows. Moreover, we reported in Fig. 24.12 also the performance of the $\{C_2, P_3\}$ configuration, which, contrarily to the optimization procedure results, overcomes the constraint in several cases. The measured values of packet loss probability confirm the fulfilment of network performance constraints.

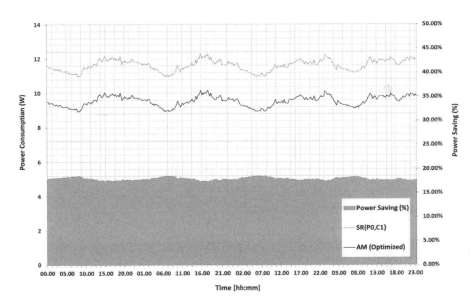

FIGURE 24.11 SR energy consumption values with both the optimization procedure, and the fixed pair $\{C_1, P_0\}$. The energy savings of the optimization procedure with respect to the $\{C_1, P_0\}$ pair are also reported.

24.5   Conclusions

In this chapter we provided the current state-of-the-art in energy efficiency for fixed telecommunication networks, both as regards improvements that can be introduced in today's networking equipment and perspectives for the Future Internet. To this aim, we have identified the main sources of power consumption in network devices and we identified the data-plane as the most energy-starving and critical element in the largest part of network device architectures. Starting from these considerations, we investigated the main issues to address in order to introduce AR and LPI capabilities in network devices. These strategies aim to modulate the capacities of network device resources (e.g., link bandwidths, computational capacities of packet processing engines, etc.) according to current traffic loads and service requirements. To configure and control the usage of AR and LPI capabilities with respect to the estimated workload, we proposed an analytical model suitable to be adopted inside optimization procedures. The model explicitly takes a large set of features and parameters into account, like, for example, the transition times to enter low-consumption sleeping states, or packet latencies and loss probabilities. The validation results were performed by using a Linux-based SR with AR and LPI primitives and real-world traffic traces, and demonstrate how the proposed model can effectively represent energy- and network-aware performance indexes. Moreover, also an optimization procedure based on the model has been proposed and experimentally evaluated. The results show that such procedure can allow saving more than 16-17% of energy with respect to a device with only LPI capabilities enabled.

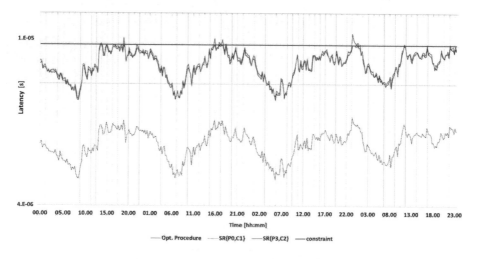

FIGURE 24.12 SR latency values measured in each time window with both the optimization procedure, and the fixed pair $\{C_1, P_0\}$.

Acknowledgements

This work has been supported by the ECONET(low Energy Consumption NETworks) project co-funded by the European Commission under the 7th Framework Programme (FP7).

Bibliography

[1] C. Bianco, F. Cucchietti, and G. Griffa, "Energy consumption trends in the next generation access network – a telco perspective," in *Proc. 29th International Telecommunications Energy Conf. (INTELEC 2007)*, Rome, Italy, Sept. 2007, pp. 737 –742.

[2] M. Webb, "SMART 2020: Enabling the low carbon economy in the information age," *The Climate Group London*, 2008. [Online]. Available: http://www.ecoc2008.org/documents/SYMPTu_Webb.pdf

[3] G.-Q. Zhang, Q.-F. Yang, S.-Q. Cheng, and T. Zhou, "Evolution of the internet and its cores," *New Journal of Physics*, vol. 10, no. 12, pp. 1–11, Dec. 2008.

[4] D. Neilson, "Photonics for switching and routing," *IEEE J. of Selected Topics in Quantum Electronics*, vol. 12, no. 4, pp. 669 –678, July-Aug. 2006.

[5] M. Bohr, "A 30 year retrospective on Dennard's MOSFET scaling paper," *IEEE Solid-State Circuits Newsletter*, vol. 12, no. 1, pp. 11 –13, Winter 2007.

[6] R. S. Tucker, R. Parthiban, J. Baliga, K. Hinton, R. W. A. Ayre, and W. V. Sorin, "Evolution of WDM optical IP networks: A cost and energy perspective," *IEEE J. Lightwave Technol.*, vol. 27, no. 3, pp. 243 –252, Feb. 2009.

[7] J. Baliga, R. Ayre, K. Hinton, W. V. Sorin, and R. S. Tucker, "Energy consumption in optical ip networks," *IEEE J. Lightwave Technol.*, vol. 27, no. 13, pp. 2391–2403, July 2009.

[8] S. Nedevschi, L. Popa, G. Iannaccone, S. Ratnasamy, and D. Wetherall, "Reducing network energy consumption via sleeping and rate-adaptation," in *Proc. 5th USENIX Symposium on Networked Systems Design and Implementation*, ser. NSDI'08, Boston, MA, June 2008, pp. 323–336. [Online]. Available: http://dl.acm.org/citation.cfm?id= 1387589.1387612

[9] R. Bolla, R. Bruschi, K. Christensen, F. Cucchietti, F. Davoli, and S. Singh, "The potential impact of green technologies in next generation wireline networks - Is there room for energy savings optimization?" *IEEE Commun. Mag.*, vol. 49, no. 8, pp. 80–86, Aug 2011.

[10] R. Bolla, R. Bruschi, F. Davoli, and F. Cucchietti, "Energy efficiency in the future internet: A survey of existing approaches and trends in energy-aware fixed network infrastructures," *IEEE Communications Surveys & Tutorials*, vol. 13, no. 2, pp. 223 –244, May 2011.

[11] "IEEE 802.3az energy efficient ethernet task force," Institute of Electrical and Electronics Engineers (IEEE). [Online]. Available: http://grouper.ieee.org/groups/802/3/az/public/index.html

[12] R. Bolla, R. Bruschi, A. Carrega, and F. Davoli, "Green network technologies and the art of trading-off," in *Proc. IEEE INFOCOM 2011 Green Communications and Networking Workshop (IEEE INFOCOM GCN)*, Shanghai, China, April 2011, pp. 301 –306.

[13] R. Bolla, R. Bruschi, and A. Ranieri, "Green support for pc-based software router: Performance evaluation and modeling," in *Proc. 2009 IEEE International Conference on Communications (IEEE ICC 2009)*, Dresden, Germany, June 2009.

[14] C. Gunaratne, K. Christensen, B. Nordman, and S. Suen, "Reducing the energy consumption of ethernet with adaptive link rate (ALR)," *IEEE Transactions on Computers*, vol. 57, no. 4, pp. 448 –461, April 2008.

[15] A. Klemm, C. Lindemann, and M. Lohmann, "Modeling IP traffic using the batch markovian arrival process," *Computer Networks*, vol. 54, no. 2, pp. 149–173, Oct 2003.

[16] G. Choudhury, "An MX/G/1 queueing system with a setup period and a vacation period," *Queueing Syst. Theory Appl.*, vol. 36, no. 1-3, pp. 23–38, Nov. 2000.

[17] B. T. Doshi, "A note on stochastic decomposition in a GI/G/1 queue with vacations or setup times," *J. of Applied Probability*, vol. 22, no. 2, pp. 419 –428, June 1985.

[18] R. Bolla and R. Bruschi, "Pc-based software routers: High performance and application service support," in *Proc. of the ACM SIGCOMM 2008 Workshop on Programmable Routers for Extensible Services of Tomorrow (PRESTO 2008)*, Seattle, WA, USA, Sept. 2008, pp. 27–32.

[19] ——, "Energy-aware load balancing for parallel packet processing engines," in *Proc. of the 1-st IEEE Conference on Green Communications (GreenCom 2011)*, Sept. 2011, pp. 105–112.

[20] MAWI Woring Group Traffic Archive, Sample Point F. [Online]. Available: http://mawi.nezu.wide.ad.jp/mawi/samplepoint-F/20080318/

Author Contact Information

Raffaele Bolla, Franco Davoli, and Paolo Lago are with DITEN-University of Genoa, Italy, and CNIT, University of Genoa Research Unit, Italy, Email: raffaele.bolla@unige.it, franco@dist.unige.it, paolo@reti.dist.unige.it. Roberto Bruschi is with CNIT, University of Genoa Research Unit, Italy, Email: roberto.bruschi@cnit.it.

25

Energy Efficient VI Planning over Converged Optical Network and IT Resources

Anna Tzanakaki

Athens Information Technology, Greece

Markos Anastasopoulos

Athens Information Technology, Greece

Konstantinos Georgakilas

Athens Information Technology, Greece

Shuping Peng

University of Essex, UK

Reja Nejabati

University of Essex, UK

Eduard Escalona

University of Essex, UK

Dimitra Simeonidou

University of Essex, UK

CONTENTS

25.1   Introduction

As the availability of high-speed Internet access is increasing at a rapid pace, distributed computing systems that are able to support a large variety of existing and upcoming applications are gaining increased popularity. Over the last decade, large-scale computer networks supporting both communication and computation were extensively employed to run distributed applications that deal with customer support, internet control processes, web content presentation, file sharing and new emerging applications such as UHD IPTV, 3D gaming, virtual worlds etc. In addition to the increased expectations from residential and business users, new requirements arising by scientific applications necessitate the deployment of high-performance data centers incorporating significant computing e.g., storage, processing power, memory, and network resources. This introduces an increase in the scale of information processing from Petabyes of Internet data to Exabytes at the end of this decade [1]. Traditionally, data center constellations used to support this type of information processing are locally installed at the customer site. However, this solution suffers from several limitations: including: a) heavy instrumentation on IT infrastructure and personnel needs, b) increased operational and maintenance costs, c) limited reliability etc. To address these issues cloud computing has been adopted as the natural evolution of traditional data centers, in which access to computing resources is provided on an on-demand basis. In this case, customers gain access to remote computing resources that they do not have to own and are charged for this elastic type of service based on the utilization level of the required resources. However, cloud computing services need to be supported by specific IT resources that may be remotely and geographically distributed, requiring connectivity through a very high capacity and increased flexibility and dynamicity network. A strong candidate to support these needs is optical networking due to its inherent abundant capacity, long reach transmission capabilities and recent technology advancements including dynamic control planes, elastic technologies etc. In this context, an infrastructure comprising converged optical network and IT resources that are jointly optimized in terms of infrastructure design and operation can be envisioned as the suitable solution to support the Future Internet.

On the other hand, in order to maximize the utilization and efficiency of infrastructures, supporting converged network and IT resources, the concept of virtualization of physical resources [2] can be additionally applied. The concept of virtual infrastructures (VIs) facilitates sharing of physical resources among various end users and virtual operators, introducing a new business model that suits well the nature and characteristics of the Future Internet and enables new exploitation opportunities for the underlying physical infrastructure. Through the adoption of VI solutions, optical network and IT resources can be deployed and managed as logical services, rather than physical resources. This results into enhanced agility, remote access to geographically distributed infrastructures and maximization of network utilization leading to reduced capital and operational costs.

An additional consideration that needs to be taken into account, in the context of Future Internet sustainability, is the energy efficient design and operation of the associated infrastructure, as ICT is responsible for about 4% of all primary energy today worldwide, and this percentage is expected to double by 2020 [3]. Specifically in VIs, energy efficiency can be effectively addressed at the VI planning phase [2]. VI planning is responsible to generate dynamically reconfigurable virtual networks satisfying the end users'- or VI provider's-driven requirements and meeting any specific needs such as energy efficiency. Through this process the least energy consuming VIs that can support the required services are identified, in terms of both topology and resources. In the optimization process involved, joined consideration of the energy consumption of the converged network and IT resources is performed. As IT resources require very high levels of power for their operation and their conventional operating window is commonly not optimized for energy efficiency, allocating IT resources in an energy-aware manner interconnected through a relatively low energy-consuming optical network can potentially offer significant energy savings.

In the context described above, resilience of the optical network and IT resources to any kind of failures is an aspect of major importance, as these infrastructures commonly support a huge amount of data that may have very tight availability requirements. It is true to say that although supporting resilience can have a significant impact on the resource requirements and hence the corresponding power consumption, to the authors' best knowledge, very limited attention has been given to the impact of resilience on the optical network power consumption [4], [5] and none to the overall power consumption of infrastructures including both network and IT resources to date.

The scope of this chapter is to address the issue of VI planning over integrated IT and optical network infrastructures from an energy efficiency perspective. To identify the least energy consuming VIs, the detailed power consumption models and figures of the underlying physical infrastructure, including joint consideration of optical network and IT resources are taken into consideration. Mapping virtual resources to physical resources and defining the energy consumption parameters of the VIs themselves is also part of the VI planning phase. Emphasis will be given to the analytical modeling of the virtual to physical infrastructure mapping process through Mixed Integer Linear Programming (MILP) problem formulations that have been also presented in [2], [6].

The rest of the chapter is organized as follows. In Section 2, a detailed description of the energy consumption models for both optical network and IT elements is given. Then, the energy-aware VI planning process, taking into account various optimization and design scenarios, is presented in Section 3. The performance of the proposed VI planning process is examined in terms of energy consumption and utilization of physical resources in Section 4. Comparisons with other similar schemes presented in the literature are also part of the performance analysis. Finally, Section 5 concludes the chapter.

25.2   Energy Consumption Models for the Physical Infrastructure.

The estimation of the energy consumption of the physical infrastructure (PI) network resources is highly sensitive to the network architecture employed and the network technology used. In Subsections 2.1 and 2.2, a detailed description of the energy consumption models for both optical network and IT elements is presented.

25.2.1   Optical Network Elements

The estimation of energy consumption of the PI network resources is very much dependent on the network architecture employed and the specific technology choices made. The current book chapter is focusing on optical network technologies based on wavelength division multiplexing (WDM) utilizing Optical Cross-Connect (OXC) nodes to perform switching and facilitate routing at the optical layer. The overall network power consumption model is based on the power-dissipating (active) elements of the network that can be classified as switching nodes (OXC nodes), and transmission line related elements. More specifically the OXCs assumed are based on the Central Switch architecture using Micro-Electrical Mechanical Systems (MEMS), while for the fiber links a model comprising a sequence of alternating single mode fiber and dispersion compensating fiber spans together with optical amplifiers to compensate for the losses is employed. The details of these models are described in [7]-[10], with the only difference being that unlike [8] the current work assumes wavelength conversion capability available at the OXC nodes.

The network is modeled as a graph, comprising a set of nodes interconnected by a set of unidirectional links. The switching nodes are OXCs supporting a number of input and output fibers, in which each fiber employs a maximum number of wavelengths.

Fig. 25.1 illustrates the OXC architecture assumed in this work. Each OXC node comprises a set of active and passive elements. The passive elements incorporated in these nodes are: the multiplexers (MUX) and de-multiplexers (DEMUX), while the active elements indicated in Fig. 25.1 in grey color include: the photonic switching matrix, one Erbium-Doped Fiber Amplifier (EDFA) per input fiber port, one Erbium-Doped Fiber Amplifier (EDFA) per output fiber port, one Optical-Electrical-Optical (OEO) transponder per output wavelength port and one transmitter (Tx) - receiver (Rx) pair per light-

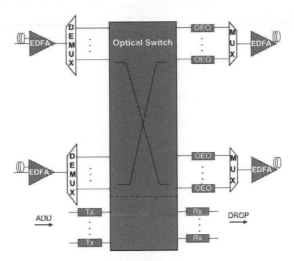

FIGURE 25.1 OXC Architecture.

TABLE 25.1 Network Equipment Power Consumption Figures

Symbol	Description	Power
$P_{port\ pair}$	Input / Output port pair of the switch fabric	0.107 mW
$P_{transponder}$	O/E/O: Line-side WDM Transponder (10G)	6W
$P_{Tx/Rx}$	E/O,O/E: Transmitter or Receiver	3.5W
P_{Edfa}	EDFA	13W

path located at the add and drop ports of the OXC. The functions supported by the aforementioned equipment are switching, optical signal amplification, signal regeneration/wavelength conversion and client signal termination respectively. It has been assumed that every network node is equipped with full wavelength conversion capability, supported by the OEO transponders available on a wavelength port basis of each OXC, lifting the wavelength continuity constraint. As the OXCs assumed in this study are symmetrical, the number of bypass (express) ports of the OXC is calculated as the product of the number of input fibers and the maximum number of wavelengths that a fiber can support. It is further assumed that the add/drop capability of the node is 50% of the bypass traffic.

The total network power consumption is determined by the power consumption of the individual OXCs and fiber links comprising the optical network. The node power consumption (P_{OXC}) depends on the four active elements comprising the OXC:

a) the switch fabric (P_{SF}),

b) the OEO transponders for transmission (P_{Tran}),

c) the wavelength converters (P_{Con}) and

d) the optical amplifiers (P_{Amp}).

Equations (25.1) to (25.4) describe mathematically the dependence of the power consumption of the node with the individual elements power consumption, whereas Table 25.1 provides a short description and typical power consumption values for the required equipment.

$$P_{SF} = ports_{total} \cdot P_{port\_pair} = (ports_{th} + ports_{a/d}) \cdot P_{port\_pair} \quad (25.1)$$

$$P_{Tran} = ports_{a/d} \cdot P_{Tx/Rx} \quad (25.2)$$

$$P_{Conv} = ports_{th} \cdot P_{Transponder} \quad (25.3)$$

$$P_{Ampl} = (f_{in} + f_{out}) \cdot P_{Edfa} \quad (25.4)$$

The power of the switch fabric in (25.1) is computed as the total port number times the power consumed by each switch path, whereas the total number of ports is the sum of bypass $ports_{tn}$ and add/drop ports $ports_{a/d}$. The transmission related power of the node in (25.2) is computed as the product of the add/drop port number and the power related to transmission equipment $(P_{Tx/Rx})$. (25.3) computes the power of the transponder equipment used for signal regeneration/wavelength conversion, as the number of bypass ports times the transponder power consumption. Finally, the power consumption of the amplifiers installed at the incident links of the node is computed by (25.4) as the product of the maximum number between incoming (f_{in}) and outgoing (f_{out}) fibers. In the case examined, we assumed a symmetric switch $M \times M$, the exact dimension of which is calculated through the network planning process.

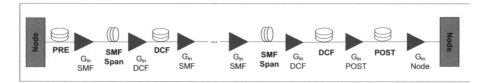

FIGURE 25.2 Link model

Fig. 25.2 illustrates the assumed fiber link model [10], where the only power consuming elements are the optical amplifiers installed per span. The span length ($span$) is assumed to be $80km$. Thus the power consumption P_ℓ of a fiber link ℓ is length-dependent and is calculated as depicted in (25.5)

$$P_\ell = \left\lfloor \frac{length(\ell)}{span} \right\rfloor P_{Edfa} \quad (25.5)$$

The total network power consumption of the physical infrastructure is computed by (25.6)

$$P_{Net} = \sum_{n \in N} P_{OXC_n} + \sum_{\ell \in L} P_\ell \quad (25.6)$$

Finally, it should be noted that 50% overhead in energy consumption is assumed addressing the cooling associated power dissipation

25.2.2 Energy Consumption Models for Computing Resources

The physical IT infrastructures in this chapter are considered to be data centers i.e. facilities used to house computer systems and associated components. Data centers are primarily used to process data (servers) and store data (storage equipment). The collection of this processing and storage equipment is referred to as the "IT equipment". The rest of the IT infrastructure facilities, such as cooling mechanisms etc. correspond to "infrastructure equipment". A large part of the energy consumption of a data center resides in the IT equipment and the cooling system [11], making infrastructure equipment (especially cooling) an important factor when trying to reduce the energy consumption in data centers.

In this chapter the analysis is based on a linear power consumption model that mainly concentrates on the power consumption associated with the CPU load of IT resources. More specifically p_s is referred to the CPU resources and E_s to the energy consumption for utilizing a portion $u_s = p_s/p_s^{max}$ of the maximum CPU capabilities, p_s^{max}, of server s. For simplicity, the following linear energy consumption model has been adopted [12], [15]

$$E_s \left(u_s \right) = P_{idle}^s + \left(P_{busy}^s - P_{idle}^s \right) u_s \qquad (25.7)$$

where P_{idle}^s and P_{busy}^s is the energy consumption of IT server s at idle state and under full load, respectively. For further details regarding the technical specifications of the IT servers the reader is referred to [12]. Note that that 50% overhead in energy consumption is assumed addressing the cooling associated power dissipation.

25.3 Energy Aware VI Planning

The VI planning is the process of using historical operating data and estimated future virtual resource (VR) requirements, to determine the optimal design of the VI. The objective of VI planning is to identify the topology and determine the virtual resources required to implement a dynamically reconfigurable VI based on both optical network and IT resources. This VI will not only meet customer's specific needs, but will also satisfy the virtual infrastructure provider's (VIP's) requirements for minimum energy consumption in our case. The Energy Aware VI planning problem is formulated through a Mixed Integer Linear Programming (MILP) model that aims at minimizing jointly the total energy that is consumed by optical network components including WDM transponders, amplifiers, switches and IT resources.

25.3.1   VI Problem Formulation

The VI planning problem is formulated using a network that is composed of one resource layer that contains the physical infrastructure (1st lower layer) and will produce as an output the virtual infrastructure layer (2nd lower layer) illustrated in Fig. 25.3.2. In the general case, randomly selected nodes in the physical infrastructure generate demands d $(d = 1, 2, ..., D)$ to be served by a set of IT servers s $(s = 1, 2, ..., S)$. The IT locations (demand destinations) at which the services will be handled, are not specified and are of no importance to the services themselves.

The objective of the current problem formulation is to minimize the total cost of the resulting network configuration. Depending on the virtualization approach, the planned network may be optimized emphasizing on network resources, on IT servers, or on their combination. These cases may be described via the following cost functions:

1. Optical Network Resources:

$$\text{Minimize } F = \sum_g C_g(u_g) \tag{25.8}$$

2. IT server Resources:

$$\text{Minimize } F = \sum_s C_s(u_s) \tag{25.9}$$

3. Combination of Network and IT:

$$\text{Minimize } F = \sum_g C_g(u_g) + \sum_s C_s(u_s) \tag{25.10}$$

where $C_g(u_g)$ denotes the total cost for installing and operating capacity u_g of link g of the PI and $C_s(u_s)$ the total cost (CapEx and OpEX) for processing demands u_s on server s. For simplicity, in the current approach C_g and C_S are linear functions of u_g and u_s respectively.

In the general case, the VI planning problem should be solved taking into account a set of constraints that guarantee the efficient and stable operation of the resulting infrastructures. The following constraints should be considered:

C1. Every demand has to be processed at a single IT server. This allocation policy reduces the complexity of implementation and increases the reliability of the resulting VI.

C2. The planned VI must have sufficient optical link capacity for all demands to be transferred to their destinations.

C3. The capacity of each link in the VI should be realized by specific PI resources.

C4. The planned VI must have adequate IT server resources such as CPU, memory, disk storage to support all requested services.

In order to formulate the basic VI planning problem, the binary variable a_{ds} is introduced to indicate whether demand d is assigned to server s or not, and it equals 1 if and only if demand d is processed on server s. Thus, constraint C1 can be expressed analytically by the following equation

$$\sum_s a_{ds} = 1, \quad d = 1, 2, ..., D \tag{25.11}$$

At the same time, for each demand d, its demand volume h_d should be realized by means of a number of lightpaths assigned to paths of the VI. Let $p = 1, 2, ..., P_{ds}$ be the candidate path list in the VI for the lightpaths required to support demand d at server s and x_{dps} the non-negative number of lightpaths allocated to path p. The following demand constraints should be satisfied in the VI [13]:

$$\sum_s \sum_p a_{ds} x_{dps} = h_d, \quad d = 1, 2, ..., D \tag{25.12}$$

Summing up the lightpaths through each link $e(e = 1, 2, ..., E)$ of the VI, the required link capacity y_e for link e as described via C2 is given by

$$\sum_d \sum_s \sum_p \delta_{edps} x_{dps} \leq y_e, \quad d = 1, 2, ..., D \tag{25.13}$$

where $\delta_{edps} x_{dps}$ is a binary variable defined as follows

$$\delta_{edps} = \begin{cases} 1, & \text{if link e of VI belongs to path p realizing demand d at server s} \\ 0, & \text{otherwise} \end{cases} \tag{25.14}$$

Using the same rationale, the capacity of each link e in the VI is allocated by identifying the required lightpaths in the PI. The resulting PI lightpath z determine the load of each link g ($g = 1, 2, ..., G$) of the PI, and hence its capacity u_g. Assuming that $q = 1, 2, ..., Q$ is used for denoting the PI's candidate path list realizing link e, then, the following demand constraint for link e should be satisfied:

$$\sum_q z_{eq} = y_e, \quad e = 1, 2, ..., E \tag{25.15}$$

where the sum is taken over all paths q on the routing list Q_e of link e. Introducing the link-path incidence coefficients for the PI·

$$\gamma_{geq} = \begin{cases} 1, & \text{if link g of PI belongs to path q realizing link e of VI} \\ 0, & \text{otherwise} \end{cases} \tag{25.16}$$

Based on (C3), the general formula specifying the PI capacity constraint can be stated as :

$$\sum_e \sum_q \gamma_{geq} z_{eq} \leq u_g, \quad g = 1, 2, ..., G \tag{25.17}$$

where G is the total number of links in the PI and the summation for each link g is taken over all lightpaths in the PI layer.

Apart from link capacity constraints (25.13) an (25.17) for the VI and PI, respectively, the total demands that are assigned to each server should not exceed its capacity p_s, $s = 1, 2, ..., S$. The latter capacity corresponds to the underlying physical resources, such as CPU, memory, disk storage etc. The inequality specifying servers' capacity constraints is given by

$$\sum_d \sum_p a_{ds} c_{ds} (x_{dps}) \leq p_s, \quad s = 1, 2, ..., S \tag{25.18}$$

where the summation is taken over all demands that arrive at server s and $c_{ds}(x_{ds})$ is a parameter specifying the computational requirements for demand d on server s. In practice, this parameter is determined by the set of relevant benchmarks for computer systems provided by the Standard Performance Evaluation Corporation (SPEC) [16].

25.3.2   VI Planning With Resilience Considerations

In this subsection, a modeling approach using MILP suitable for the energy aware planning of resilient VIs formed over an integrated IT and optical network infrastructure, is proposed and implemented, based on the analysis presented in [6]. The resilience scheme considered is that of 1:1 protection for both optical network and IT resources. Specifically, in case of a failure of the primary IT server, demands are forwarded to a secondary (protection) IT server, while in case of an optical link failure demands are routed to their destination through a secondary (protection) path.

Similarly to the basic model, the VI Planning problem with resilient consideration is formulated using a network that is composed of one resource layer that contains the physical infrastructure and will produce as an output the virtual infrastructure layer. Again, the PI is described through a sixteen-node topology corresponding to the NSFNET [18] in which randomly selected nodes generate demands d to be served by a set of IT servers s. The granularity of demands is the wavelength and the IT locations at which the services will be handled are not specified and are of no importance to the services themselves. However, uninterrupted service provisioning is of crucial importance in the deployment of transport optical networks. To this end, countermeasures against failures of the IT servers and optical links of the PI that may lead to service disruption should be taken into account during the VI planning process. In the proposed planning algorithm, a possible failure of the primary IT server, s_i, is treated by forwarding demands to a secondary IT server, s_j, with $s_i \neq s_j$. On the other hand, in case of failure of an optical link, demands are routed to their destination via alternative paths.

Similarly to the basic VI planning model, constraints (C1)-(C4) are also introduced. However, they are appropriately modified in order to take into account the resilient considerations. First of all, in order to protect the converged optical network and IT resource system from a possible failure of an IT server, equation (25.11) is modified and the binary variable a_{ds} is now used to indicate whether demand d is assigned to one of the IT servers s_i, s_j or not. This variable takes value equal to 1 if and only if demand d is processed on primary server, s_i, or in case of it's failure on the secondary, s_j. Moreover, it is assumed that each demand can be assigned only to one IT server at a time.

Another important aspect in the design of VIs is to ensure that there are adequate network resources to support demand d at the primary server s_i or the secondary, s_j. An additional consideration is that in order to protect the planned network from a possible failure of a physical layer link g, a link re-establishment mechanism is introduced. This mechanism ensures that demands will be routed via alternative paths in case of failure of link g. Assuming that $r = 1, 2, ..., R_q$ are the candidate restoration paths for link g, then the following protection capacity constraints should be satisfied [14]:

$$\sum_r z'_{gr} = u_g \quad g = 1, 2, ..., G \tag{25.19}$$

Finally, the required protection capacity for the remaining operating links ℓ of the PI is given by

$$\sum_r \beta_{\ell gr} z'_{gr} = u'_g \quad \ell, g = 1, 2, ..., G \tag{25.20}$$

where u'_g denotes the protection capacity for physical layer link g and $\beta_{\ell gr}$ is a binary variable taking value equal to 1 if link ℓ belongs to path r restoring link g; 0 otherwise.

Using the same rationale for the cost functions with the basic model, the planned network may be optimized either for minimum usage of optical network resources or for energy. In the former case the objective function is described via the following equation

$$F^{SP} = \sum_g l_g \left(u_g + u'_g\right) \tag{25.21}$$

while in the latter via

$$F^E = \sum_g k_g \left(u_g + u'_g\right) + \sum_s E_s \left[\sum_d \sum_p a_{ds} c_{ds} \left(x_{dps}\right)\right] \tag{25.22}$$

25.4   Numerical Results and Comparisons

To investigate the energy efficiency of the proposed VI design scheme, the multilayer architecture illustrated in Fig.25.3.2 is considered: the lower layer depicts the PI and the layer above depicts the VI. For the PI the NSFNET reference topology has been used in which four randomly selected nodes generate demands to be served by two IT servers located in Salt Lake City and

TABLE 25.2 Virtual to physical mapping

Virtual link	Capacity (wavelengths)	Physical Layer Paths realizing virtual links	Capacity of PI paths (wavelengths)
Y_1	25	u_3	45
Y_2	10	u_5	10
Y_3	40	$u_8 - u_9$	40
Y_4	40	$u_1 - u_4$	40
Y_5	10	u_{12}	40
Y_6	40	$u_7 - u_{15} - u_{20} - u_{23}$	40
Y_7	10	u_{17}	10
Y_8	15	$u18$	15

Pittsburgh. Furthermore, we assume a single fiber per link, 40 wavelengths per fiber, and wavelength channels of 10 Gb/s each. It is also assumed that each IT server can process up to 2 Tb/s and its power consumption ranges from 6.6 to 13.2 KW, under idle and full load, respectively [12].

An example of the optimal VI topology design for a scenario in which five nodes that are located in Seattle, San Diego, Houston, Pittsburgh and Princeton generate demands equal to 25 wavelengths each, is depicted in Fig. 25.3(a). In this scenario, the generated VI topology consists of 8 virtual links and 8 virtual nodes, while all demands are routed to the IT server in Salt Lake City. The capacity of each virtual link along with its mapping to the PI is given in Table 25.2 where e.g. it is observed that virtual link Y6 connecting Pittsburgh and Salt Lake City is realized via physical layer path u7-u15-u20-u23, with capacity 40 wavelengths. With the further increase of the generated traffic, due to the limited network capacity the IT server in Salt Lake City is switched off and the traffic demands are transferred to the IT server in Pittsburgh. The updated VI is depicted in Fig. 25.3(b).

In Fig. 25.4, the performance of the proposed energy aware VI design is compared to the demand allocation scheme presented in [17] where demands from each source node are assigned to its closest IT server. Note that "closest" refers to the shortest distance between a source node and a data center. Comparing these two schemes, it is observed that the energy aware VI design consumes significantly lower energy for serving the same amount of demands compared to the closest IT scheme, which is of the order of 30%, as in this case fewer IT servers are activated to serve the same amount of demands (Fig. 25.5). Given that the power consumption required for the operation of the IT servers is dominant in this type of networks, switching-off the unused IT resources achieves significant reduction of energy consumption. Furthermore, it is observed that in both schemes the average power consumption increases almost linearly with the number of demands. However, the relative benefit of the energy aware design decreases with the number of demands and converges to the performance of the closest IT approach when approaching full system load.

TABLE 25.3 Virtual to physical mapping with resilient considerations

Virtual link	Capacity (wavelengths)		Physical Layer Paths realizing virtual links	PI paths Capacity (wavelengths)
	P	S		
Y_1	10	0	(P) $u_1 - u_2$	10
Y_2, Y_3	0	0	-	-
Y_4	0	10	(S) $u_1 - u_{44}$	10
Y_5	20	10	(P) u_5	20
	20	10	(S) $u_2 - u_4 - u_8 - u_9$	10
Y_6	0	10	(S) $u_8 - u_9 - u_{13} - u_{14}$	10
Y_7	10	0	(P) u_{17}	10
Y_8	40	40	(P) u_{18}	20
	40	40	(P) $u_{19} - u_{22}$	20
	40	40	(S) $u_{20} - u_{21} - u_{23}$	20
	40	40	(S) $u_{19} - u_{22}$	20
Y_{10}	20	20	(P) $u12 - u_{17} - u_{18}$	20
	20	20	(S) $u_{13} - u_{14}$	20

The analysis is extended to cover the case of VI planning with resilience considerations. An example of the optimal VI topology design for a scenario in which four source nodes that are located in Seattle, San Diego, Houston and Princeton generate demands equal to 10 wavelengths each, is depicted in Fig. 25.6. In this scenario, the generated VI topology consists of 10 virtual links and 7 virtual nodes, while all demands are routed to the IT server in Pittsburgh. In case of failure, the primary IT server demands are routed to the secondary located in Princeton. Furthermore, in case of failure of a working path, additional capacity has been reserved to support a link-disjoint backup path. The working and protecting capacity of each virtual link along with its mapping to the PI is given in Table 25.3 where e.g. it is observed that virtual link Y5 connecting Houston and Pittsburgh is realized via the working physical layer path u12-u17-u18 with capacity 20 and the protecting path u13-u14.

In Fig. 25.7, the performance of two variations of the proposed energy aware VI design with and without (w/o) protection mechanisms is compared to the demand allocation scheme presented where demands from each source node are assigned to its closest IT server [17]. Comparing these two schemes, it is observed that the energy aware VI design without protection mechanism consumes significantly lower energy for serving the same amount of demands compared to the closest IT scheme: in the former approach only one IT server is activated to serve the same amount of demands. Given that the power consumption required for the operation of the IT servers is dominant in this type of networks, switching off the unnecessary IT resources achieves significant reduction of energy consumption. It is interesting to note that the energy aware scheme enhanced with protection mechanisms, achieves significantly lower power consumption compared to the closest IT approach, and for high traffic demands the energy consumption computed for the energy aware planning approach becomes equal to that obtained through the closest IT scheme without protection.

The results above clearly indicate that there is a trade-off between the utilization of optical network resources and the number of active IT servers. Specifically, since the energy cost for activating an IT server predominately affects the overall network's energy consumption, the energy aware VI planning scheme forwards traffic to a single IT server. However, in this case more optical resources are employed since data need to travel longer distances to arrive at their destination (IT server). In case where the VI is planned using the Closest IT scheme, all demands are routed to their closest IT server and therefore minimize the utilization of the optical network resources. However, this planning scheme increases the total number of active IT servers. This is illustrated in Fig. 25.8 and Fig. 25.9 where for example, in the closest IT scheme without protection it is observed that less than 20% of the total optical network resources are employed to transfer demands from the source nodes to the three active IT servers. On the other hand, the energy aware scheme routes all demands to a single IT server at the expense of artificially having to over-utilize the optical network resources.

From the above results, depending on which virtualization approach is employed the following conclusions are drawn:

i. If the planned network is optimized emphasizing on network resources (closest IT approach), demands are routed to their destinations via the shortest paths, thus, guaranteeing efficient utilization of the optical network resources. At the same time, the impact of the operation of IT resources in the infrastructure is ignored, thus leading to a suboptimal operation of IT servers with regards to IT resource utilization and operational energy consumption.

ii. If the planned network is optimized emphasizing on IT resources, demands are optimally allocated to IT servers. In this case, the impact of the proposed planning scheme on the network infrastructure and the associated resources is not optimized, thus leading to an inefficient allocation and exploitation of optical network resources.

iii. In order to optimize the integrated infrastructure including both optical network and IT resources as a whole, the infrastructure planning process has to target the optimization both for network and IT resources jointly. In this case, demands are routed to their destination via the shortest paths, keeping at the same time the number of active IT servers minimum.

25.5   Conclusions

This chapter focuses on the energy efficiency aspects of infrastructures incorporating network and IT resources in support of cloud computing services for which access to computing resources is provided on an on-demand basis. As cloud computing services need to be supported by specific IT resources that maybe remote and geographically distributed, connectivity between IT resources and end users need to be provided through a very high capacity and increased flexibility and dynamicity network. Optical networking is proposed to support the required connectivity, as it is offering abundant capacity, long reach transmission capabilities as well as dynamic control plane options through recent technology advancements.

The concept of virtualization of physical resources is applied over the integrated optical network and IT resources with the aim to maximize the utilization and efficiency of the converged network and IT infrastructure. The concept of virtual infrastructures facilitates sharing of physical resources (optical network and IT) among various users and virtual operators, introducing new business models and exploitation opportunities that are in line with the nature of the Future Internet. Through the adoption of virtualization the physical infrastructure is enhanced with increased agility, remote access to geographically distributed resources and efficiency in the utilization of resources that facilitate the reduction of capital and operational costs.

Addressing the issue of Future Internet sustainability, this chapter mainly concentrates on energy efficiency considerations in the infrastructure scenario described above. More specifically it addresses energy efficiency in the context of virtual infrastructures at the VI planning phase [2] as VI planning is responsible to generate dynamically reconfigurable virtual networks satisfying the end user requirements and meeting any specific needs such as energy efficiency. To identify the least energy consuming VIs, the detailed power consumption models and figures of the underlying physical infrastructure, including joint consideration of optical network and IT resources are taken into consideration. Mapping the virtual to physical resources and defining the energy consumption parameters of the VIs themselves is also part of the VI planning phase. Emphasis will be given to the analytical modeling of the virtual to physical infrastructures mapping process through MILP problem formulations. Our modeling results quantify significant energy savings of a complete solution jointly optimizing the allocation and provisioning of both network and IT resources. This effect is further emphasized when resilience considerations are included in the scenarios under evaluation and the associated models used to study these. This is achieved taking advantage of the act that IT resources require very high levels of power for their operation therefore allocating IT resources in an energy-aware manner interconnected through a relatively low energy-consuming optical network offers significant energy savings. However, it should be noted that this is achieved at the expense of a somewhat artificial increase in the utilization of the optical network resources.

Bibliography

[1] M. Handley, "Why the Internet only just works," *BT Technology Journal*, vol. 24, no. 3, 2006.

[2] A.Tzanakaki et al.,"Energy Efficiency in integrated IT and Optical Network Infrastructures: The GEYSERS approach," *in Proc. of IEEE IN-FOCOM 2011, Workshop on Green Communications and Networking* (2011).

[3] M. Pickavet et al., "Worldwide energy needs for ICT: The rise of power-aware networking", *in Proc. IEEE ANTS*, pp. 1-3, Dec. 2008.

[4] A. Jirattigalachote, C.Cavdar, P.Monti, L.Wosinska, A. Tzanakaki, "Dynamic Provisioning Strategies for Energy Efficient WDM Networks with Dedicated Path Protection," *Optical Switching and Networking Journal*, Elsevier, Special Issue on Green Communications and Networking, accepted for publication.

[5] A. Muhammad, P. Monti, I. Cerutti, L. Wosinska, P. Castoldi, A. Tzanakaki, "Energy-Efficient WDM Network Planning with Protection Resources in Sleep Mode," *in Proc. of Globecom 2010*, Miami, Florida, USA, 6-10 Dec. 2010.

[6] A.Tzanakaki et al., "Energy efficiency considerations in integrated IT and optical network resilient infrastructures," in Proc. of ICTON 2011, June 2011.

[7] A.Tzanakaki et.al, "Dimensioning the future Pan-European optical network with energy efficiency considerations," *JOCN* 3, 272-280 (2011).

[8] A. Tzanakaki et al., "Power Considerations towards a Sustainable Pan-European Network," JWA061, OFC2011.

[9] S. Aleksic, "Analysis of Power Consumption in Future High-Capacity Network Nodes," *Journal of Optical Communications and Networking*, vol. 1, no. 3, p. 245, 2009.

[10] K. M. Katrinis and A. Tzanakaki, "On the Dimensioning of WDM Optical NetworksWith Impairment-Aware Regeneration," IEEE/ACM Transactions on Networking, vol. 19, no. 3, pp. 735-746.

[11] Z.Davis, "Power Consumption and cooling in the data center: A survey," http://www.greenbiz.com/sites/default/files/document/Custom016C45F77410.pdf

[12] Oracle Data Sheet, "Sun Oracle DataBaseMachine," http://www.oracle.com/us/products/database/database-machine-069034.html

[13] E. Kubilinskas, P. Nilsson, M. Pioro, "Design Models for Robust Multi-Layer Next Generation Internet Core Networks Carrying Elastic Traffic," in proc. of DRCN 2003, 61-68 (2003).

[14] E. Kubilinskas, Faisal Aslam, Mateusz Dzida and Michal Pioro, "Recovery, Routing and Load Balancing Strategy for an IP/MPLS Network," *Managing Traffic Performance in Converged Networks, Lecture Notes in Computer Science*, vol. 4516/2007, 65-76, 2008.

[15] Xiaobo Fan, Wolf-Dietrich Weber, and Luiz Andre Barroso, "Power provisioning for a warehouse-sized computer," SIGARCH Comput. Archit. News 35, 2, pp. 13-23, June 2007.

[16] Standard Performance Evaluation Corporation (SPEC) (www.spec.org).

[17] K. Bouyoucef, I. Limam-Bedhiaf, and O.Cherkaoui, "Optimal Allocation Approach of Virtual Servers in Cloud Computing," in Proc. of 6th Euro-NF conference on Next Generation Internet, July 2010.

[18] J. M. Gutierrez, K. Katrinis, K Georgakilas, A. Tzanakaki, and O. B. Madsen, "Increasing the cost-constrained availability of WDM Networks with Degree-3 Structured Topologies", in Proc. ICTON 2010, 2010.

Author Contact Information

Anna Tzanakaki, Markos Anastasopoulos, and Konstantinos Georgakilas are with Athens Information Technology, Greece, Email: atza@ait.gr, manast@ait.gr, koge@ait.edu.gr. Shuping Peng, Reja Nejabati, Eduard Escalona, and Dimitra Simeonidou are with University of Essex,UK, Email: shuping.peng@gmail.com, rnejab@essex.ac.uk, eescal@essex.ac.uk, dsimeo@essex.ac.uk.

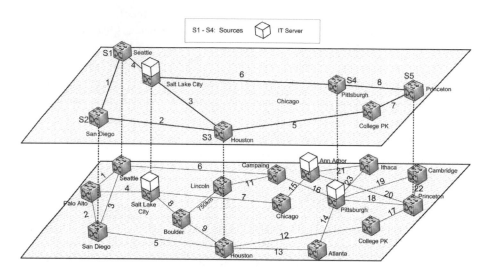

(a) Average demands/source = 25 wavelengths

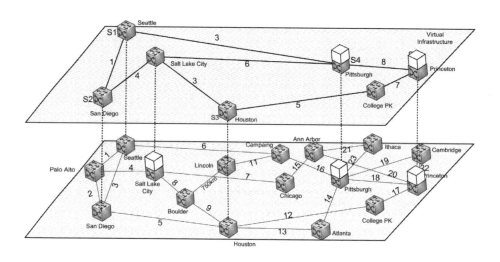

(b) Average demands/source = 40 wavelengths

FIGURE 25.3 Example of the virtualization of a physical infrastructure

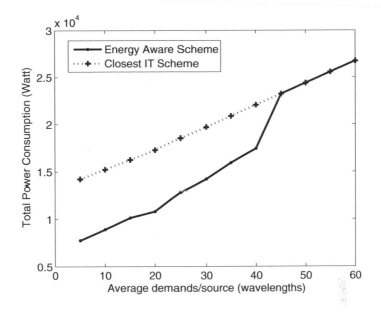

FIGURE 25.4 Comparison of the energy aware scheme with the closest IT server demand allocation scheme

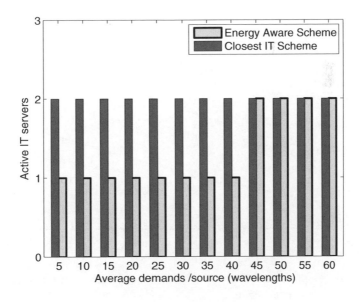

FIGURE 25.5 Number of Active IT Servers

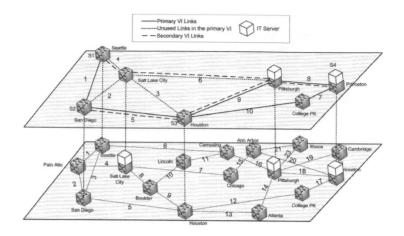

FIGURE 25.6 Resilient Virtual Infrastructure architecture over a converged optical network and IT servers

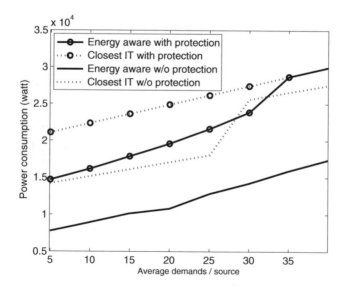

FIGURE 25.7 Comparison of the energy aware scheme with the closest IT server demand allocation scheme

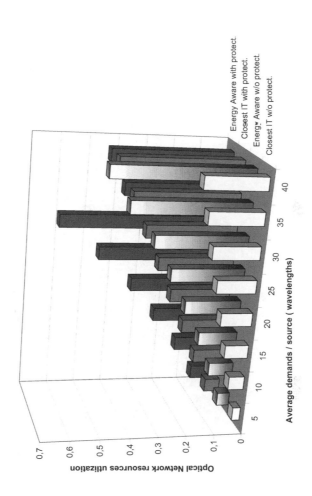

FIGURE 25.8 Utilization of optical network resources versus demands for various planning schemes.

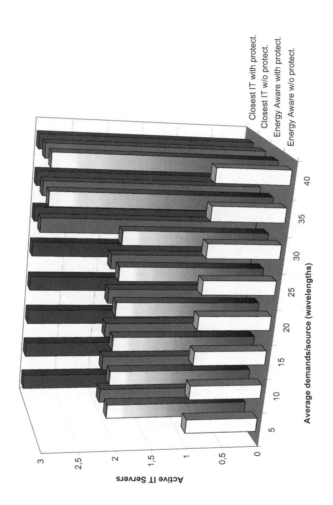

FIGURE 25.9 Number of Active IT servers versus demands for various planning schemes

26

Energy Aware Techniques In IP-Over-WDM Backbone Networks

Antonio Cianfrani

DIET - "Sapienza" University of Roma, Italy

Angelo Coiro

DIET - "Sapienza" University of Roma, Italy

Vincenzo Eramo

DIET - "Sapienza" University of Roma, Italy

Marco Listanti

DIET - "Sapienza" University of Roma, Italy

CONTENTS

26.1 Introduction

The energy consumption of the Internet is exploding due to the increase of the number of devices connected to it as well as the increase of traffic volume that is estimated in about 50% per year, fueled by mobile data and video applications. It is crucial to improve network efficiency to prevent the Internet from being throttled by an energy bottleneck; specifically, the challenge is to achieve an increase of the technological and operational efficiency faster than the rate of traffic growth.

Recent studies [1–4] explored the fundamental limitations of Internet energy consumption over the next fifteen years (2010-2025). The basic common assumptions of these studies are:

i) the network infrastructure is divided in two broad segments, namely the *access network*, typically implemented by passive optical network (PON) technology, and the *backbone network*, composed of switches, routers and transport devices, i.e. transmitters and receivers;

ii) the traffic increase in the considered time period is assumed in the range of 2 or 3 orders of magnitude in terms of aggregate bit rate;

iii) the current equipment provisioning and used network design and operational practices are maintained along the whole period;

iv) according to technological evolution, an estimated average value of decrease of the energy per bit of networking equipment in the range of 15-20% per year is considered.

As explained in [1], the main conclusions of these studies can be summarized in the following points:

- At present (2010) the access network is about two orders of magnitude higher power than the long-haul core segment.

- During the considered time interval, a shift away from a primarily access power dominated network is foreseen; the overall energy consumption of switching and transport equipment will progressively reach that of the access segment to become the dominant component after the 2020.

- All of the technologies, both in access and in backbone segments, show increases in energy efficiency over the 2010-2020 decade, corresponding to an increase in efficiency, anyway, a growing of the total power consumption is foreseen too; so, the rate of efficiency improvement is expected to be slower than the traffic growth rate.

The above mentioned results suggest some guidelines for the design of future Internet:

i) an improvement of the energy efficiency of backbone network elements and transmission systems is needed as they will be the dominant components of the overall energy consumption in the core network;

ii) rate-adaptive processing and dynamic power management in devices hardware should be implemented to enable traffic load proportional processing; moreover, the use of optical bypass can reduce the number of routers required, thus reducing network energy consumption;

iii) the network elements operate should operate at the maximum possible utilization, so significant improvements in the overall network control procedures should be achieved.

In the rest of this chapter, we consider a typical network architecture composed of two different layers: the IP layer and the Wavelength Division Multiplexing (WDM) optical layer (see Figure 26.2). The former performs the switching operation in the electronic domain on a packet basis whilst the latter processes aggregated traffic flows directly in the optical domain, providing the IP layer with optical circuits, named *lightpaths*. The IP logical topology is mapped on the WDM network, composed of Optical Cross-Connects (OXCs) and fiber links. Each IP link, considered as unidirectional, corresponds to a dedicated WDM lightpath. A GMPLS control plane [5] is considered; in case of failures, GMPLS is able to reconfigure the optical layer, by establishing/deleting lightpaths. The OXC in the optical layer originates/terminates lightpaths, or, if the node is a transit node for a lightpath, switches it between the relevant incoming and outgoing optical interfaces. A specific IP interface (NIC: Network Interface Card) is associated to each logical IP link originated or terminated by the node; a NIC interacts with Optical Layer by means of an E/O converter.

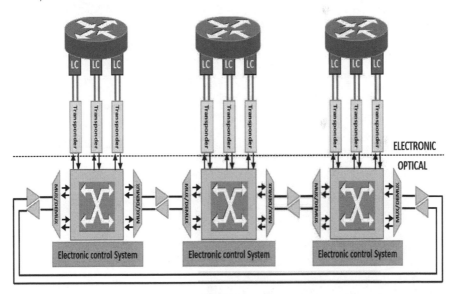

FIGURE 26.1 Architecture of an IP over WDM network.

As further observation, it is worth remembering that, at present, networks are dimensioned statically, based on a peak traffic load plus a reserve; this peak capacity is directly connected with the network peak energy consumption. Moreover the traffic presents significant periodic fluctuations over day/month time basis, e.g., during night hours, as shown in Figure 26.2 where the traffic trend over an Italian ISP backbone link [6] is reported. Unfortunately this traffic drop is not followed by a corresponding decrease in the amount of energy consumed by the network. Moreover, the traffic fluctuation is not taken also into account by the network topology design policies, i.e. current networks are generally designed with the target of minimizing the capital expenditure (CAPEX) without taking into account their energy consumption that represents instead the major contributor of operational expenditure (OPEX).

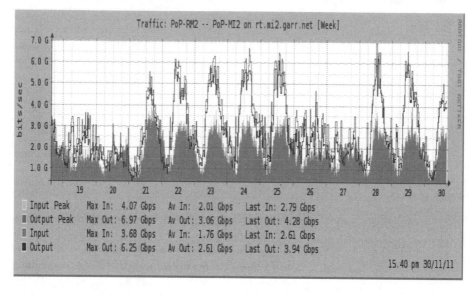

FIGURE 26.2 **Traffic statistics measured over the backbone link of the Italian GARR network [6].**

The previous considerations lead to two wide complementary families of strategies to improve network energy efficiency:

- *Power Aware Network Topology Design* (PA-NTD)

- *Energy Aware Traffic Engineering* (EA-TE).

The strategies belonging to the first family take as input a traffic matrix, representing the traffic demands among the higher layer network nodes (typically the IP layer) and the lower layer network topology, generally referred to as physical topology, implemented by the WDM optical layer, and provides the higher layer topology (the virtual topology) and the traffic routing, exploiting proper traffic grooming strategy aiming at minimizing the total power consumed by the network.

Although PA-NTD is able to drastically improve the energy efficiency of the network, it does not consider the variation experienced by traffic flows in time. In fact, since the traffic changes during a day following a typical day-night trend, it is possible to further reduce the network energy consumption putting in idle some network devices when they are not strictly needed acting on the routing of traffic flows. This is the goal of the EA-TE strategies. They implement dynamic mechanisms to efficiently switch off link and/or nodes when the traffic demand slows down in order to make the power consumption of the network proportional to the carried traffic. Besides the need of defining efficient heuristics to decide the set of devices to put in idle, such mechanisms should also consider several issues related to their practical applicability, such as routing instabilities, packet loss or out-of-order delivering, network congestion, etc., that may be experienced during the switch-off of network devices.

PA-NTD and EA-TE strategies will be dealt with in the next two sections.

26.2   Power Aware Network Topology Design

Considering the above described network scenario, the Power Aware Network Topology Design (PA-NTD) problem can be divided in two different subproblems:

i) the *Power Aware Virtual Topology Design* (PA-VTD) problem which deals with finding the set of lightpaths, i.e. the Virtual Topology (VT), and routing the traffic matrix on the VT;

ii) the *Power Aware Routing and Wavelength Assignment* (PA-RWA) problem which deals with routing the lightpaths on the physical topology and assigning them a wavelength.

The VTD problem aims at minimizing the power required at the IP layer that is mainly related to IP routers equipment, such as line cards and switching matrices. Instead, the goal of the RWA problem is the minimization of the power required at the optical layer; this is mainly related to optical amplifiers deployed along WDM links and is much less than what required by electronic devices. In [7] it is evaluated that about 90% of power is consumed by the IP layer and 10% by optical WDM links. line plane.

26.2.1   Power Aware Virtual Topology Design (PA-VTD)

The PA-VTD problem is also known as *traffic grooming* problem since it deals with grooming the smaller traffic flows of the higher network layer into the higher bit-rate traffic flows established at the lower layer. In IP over WDM networks this means aggregating the IP layer traffic into the lightpaths established in the optical network that directly interconnect IP routers. In doing this two opposite strategies can be distinguished: the link-by-link and the end-to-end grooming.

In case of link-by-link, the grooming operation is performed at each intermediate node for any traffic flow. In this case, lightpaths are terminated at each intermediate node and all traffic is switched at the IP layer. Such a strategy is also referred to as non-bypass strategy since traffic is not allowed to optically bypass a node. On the opposite, the end-to-end grooming strategy consists of aggregating the whole traffic request between a node pair and routing it on a single lightpath, which directly interconnects the two end-nodes. Such a strategy is also known as single-hop lightpath bypass since each IP traffic flow is allowed to be routed on a single hop on the VT, i.e. on the IP topology. The resulting VT is the same as the physical topology, in the case of non-bypass, whilst, in the case of single-hop bypass, is composed of a link for each traffic request in the traffic matrix.

Although, in many cases, these two opposite strategies do not lead to the optimal solution of the VTD problem, since they constrains the VT to a predetermined topology, if we compare them in terms of energy consumption, a general indication can be inferred about how to construct efficient heuristics where both the non-bypass and the single-hop constraints are removed.

This kind of study has been performed in [8,9], where results demonstrate that, as the total traffic offered to the network grows, the end-to-end strategy leads to a significant saving with respect to the link-by-link case. Such a result can be explained observing that, when the offered traffic is low, and in particular when each traffic relation is lower than the capacity of each lightpath, which is equal to the bit-rate of a line card, the end-to-end strategy leads to a considerable bandwidth waste since each lightpath carries only a small amount of traffic with respect to its capacity. Thus, line cards are underutilized leading to a considerable energy waste. Instead, as the offered traffic grows and each traffic relation becomes comparable with the line card capacity, the end-to-end strategy allows to fully use the capacity provided by each lightpath reducing the total consumption with respect to the link-by-link case. In particular, the higher the average bandwidth requested by a traffic relation is, the better the end-to-end strategy performs.

An alternative strategy is to allow each IP traffic flow to be routed on a multi-hop path in the VT, removing any constraint on the VT itself. The problem has been faced in [7,10]. In both papers the problem is formalized as a Mixed Integer Linear Programming (MILP) problem and heuristic solutions are provided to solve the problem in a reasonable time.

In [7] each IP router is assumed to consume an amount of power given by the number of line cards deployed on it and each line card is assumed to consume a fixed amount of power regardless of the actual carried traffic. Thus, the goal is to minimize the total number of line cards installed network routers that is equivalent to minimize the total number of lightpaths. The proposed solution works in a greedy way. The first step is to order traffic relations in decreasing order of requested bandwidth; then each relation is processed to be routed on the current VT. The VT is constructed step-by-step starting from an empty topology, i.e. with no links. At each step the current traffic relation is routed on the shortest path with enough capacity, if any, otherwise a direct link is added in the VT connecting the two end-nodes and the traffic relation is routed on that link. The followed principle is to route traffic using the spare capacity as much as possible. Concerning the traffic flow ordering criterion, it has the objective of constructing direct links for higher traffic relations, trying to route smaller flows on multi-hop paths, trying to minimize the total network load. The proposed solution is compared with the direct-bypass and the non-bypass strategy and with the optimal solution given by the MILP formulation. Results are showed in Figure 26.3. They are expressed as percentage saving with respect to the non-bypass strategy versus the average traffic between each node pair; the line card bit-rate is assumed to be 40 Gbits/s. Such results confirm that when the average bandwidth requested by each traffic relation is close to the capacity of each line card, the direct bypass strategy performs close to the optimal solution. At low load, instead, the proposed solution allows to save more energy than the direct-bypass strategy; however about a further 15% is needed to achieve the optimal solution.

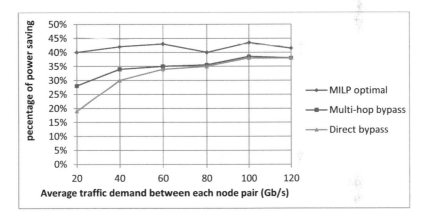

FIGURE 26.3 Percentage of power saving with respect to the non-bypass strategy for the 15-nodes 21-links NSFNET network; the line card bit-rate is equal to 40 Gbits/s.

The study in [10] considers a different power consumption model. It is composed of two different terms.

- The term $P_{TX}(B_{TX})$ is related to the power required by a Transmitter-Receiver (TX-RX); it does not depend on the actual load but only on the bit-rate B_{TX}, i.e. a fixed amount of power is required for each TX-RX;

- The term $P_{SW}(\lambda)$ is related to the power required to electronically switch the traffic; it is assumed to linearly grow with the amount λ of switched traffic.

Regardless of the specific functionality which the two terms of power consumption are related with, the key point is that each lightpath established in the network can be assumed to require a fixed part of power plus a variable part proportional to the carried traffic. Clearly, the ratio between the two terms, indicated as ν^0, has an important impact on the optimal VT.

In this study two different solutions are proposed. The first one is an iterative greedy algorithm very similar to the one proposed in [7]. The only difference is that, when a traffic flow is processed to be routed, the power needed to route it on the shortest path with enough capacity is evaluated and compared with the power needed to add a new direct link and route the flow on it; then the least energy alternative is chosen. This makes sense because, due to the variable term of power consumption, a traffic flow is assumed to consume power even if it is routed on lightpaths that are already established. Authors also propose a solution based on a genetic algorithm. Results showed in this study are interesting since they are carried out as the ratio ν^0 grows. The parameter ν^0 is defined as follows:

$$\nu^0 = \frac{P_{SW}(\lambda = B_{TX})}{P_{TX}(B_{TX})}. \tag{26.1}$$

Figure 26.4 shows the average number of transmitters which each node is equipped with, which represents the average nodal degree, versus the ratio ν^0, for the optimal and the heuristic solutions. Results demonstrate that, as ν^0 grows, the optimal VT becomes more and more meshed. In fact, when the fixed term becomes negligible with respect to the variable one, the main goal is to minimize the total network load and the optimal solution tends to a full-meshed VT. For high values of ν^0 the solution can be easily found and the two proposed solution perform very close to the optimal one (see Figure 26.5; instead, for low values of ν^0, the genetic algorithm outperforms the greedy algorithm.

It has to be outlined that current electronic devices, such as line cards, switching matrices, etc., have a power consumption profile that is almost independent of the load; this correspond to a very low ν^0. In this case the goal of the PA-VTD problem is to minimize the number of lightpaths trying to aggregate traffic as much as possible. In such a situation finding a good VT is a hard task especially when traffic relations are small compared with the capacity of a lightpath (B_{TX}). On the other hand, thanks to the introduction of mechanism such as Dynamic Voltage Scaling, Dynamic Frequency Scaling, Low Idle Power, etc., current electronic devices are moving towards more load-proportional power profiles; this will clearly improve the energy efficiency of the network and simplify the PA-VTD problem. However, since a full proportionality between the power consumption and the load is unlikely to be achieved also in the future, the PA-VTD problem remains a key point to improve the energy efficiency of the network. Another important indication is that as electronic devices will move towards more load-proportional power profiles, the optimal VT will require an increasing number of lightpaths for

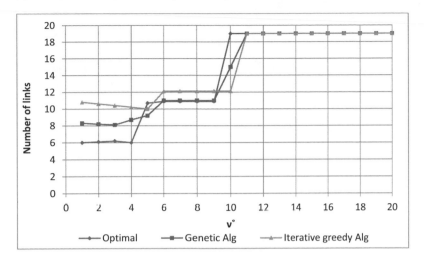

FIGURE 26.4 Average number of transmitters per node in the US backbone 24-nodes network.

the same amount of traffic; this will lead to a growth of the share of power required by the optical network layer. Thus, improving the energy efficiency of the optical layer will become more and more important.

26.2.2 Power Aware Routing and Wavelength Assignment (PA-RWA)

In this section three different solutions aiming at improving the energy-efficiency of the optical network layer will be examined. The general problem takes in input the physical topology and a traffic matrix representing the set of lightpaths, i.e. the VT, that must be established. Considering a transparent optical network, the main goal is to use the least possible number of optical amplifiers that are the main contributors of power consumption [11].

Optical amplifiers are placed along optical fibers, typically with a spacing of about 100 Km, and amplify all wavelengths carried in a fiber. They generally consume a fixed amount of power regardless of the number of wavelengths that are actually carrying traffic. Thus, in order to save energy, the aim is to keep completely unused as many optical fibers as possible in order to put optical amplifiers deployed along them in idle state.

In [12] authors propose two different heuristic algorithms to solve the PA-RWA problem, namely *Most Used Path* (MUP) and *Ordered Lightpath-MUP* (OL-MUP); both of them belong to the general family of least cost path algorithms; this means that the cost of a path is given by the sum of the costs of each link belonging to it and the path with the minimum cost is chosen. The basic idea of MUP is to assign to each fiber link a cost proportional to the power that it requires and dependent on its state. Specifically, if a link is

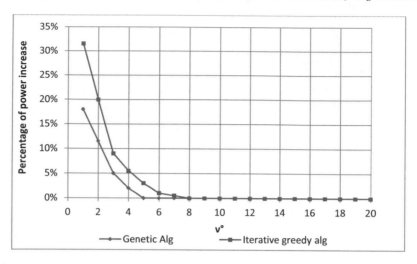

FIGURE 26.5 **Relative power consumption normalized to the power consumption of the optimal VT in the US backbone 24-nodes network.**

already used, it gets a null cost since using it does not require any additional power; instead, if a link is unused, it gets a cost equal to the total power required by optical amplifiers deployed along it.

$$LC_{i,j} = LF(L_{i,j}) \cdot PC_{i,j} \tag{26.2}$$

$$LF(L_{i,j}) = \begin{cases} 0 \; if \; L_{i,j} > 0 \\ 1 \; otherwise \end{cases} \tag{26.3}$$

In (26.2) the cost $LC_{i,j}$ of a link (i,j) is given by two terms: i) $PC_{i,j}$ is the power consumption of the link (i,j) that is due to optical amplifiers deployed along it; it is a static term that does not depends on the link load; ii) $LF(L_{i,j})$ is a variable term that depends on the load $L_{i,j}$ of the link (i,j); the load, i.e. the variable $L_{i,j}$, is the number of used wavelengths.

OL-MUP differs from MUP since it orders lightpaths before to route them; specifically it selects the proper lightpath to be routed according to a particular criterion. For such a reason OL-MUP can only be applied to the Static Lightpath Establishment (SLE) scenario since it requires that the traffic matrix is known in advance; on the contrary MUP can be used in both SLE and Dynamic Lightpath Establishment (DLE) scenarios. As outlined in [11], a drawback of the MUP algorithm is that, by assigning a null cost to used fibers, the average path length increases very much; this leads to a bandwidth waste and, consequently, to an increase of the total number of needed fibers.

For such a reason, a different RWA algorithm, named Load Based Cost (LBC), is proposed in [11]. It is still a least cost path algorithm but link costs are computed in a different way with respect to MUP. Specifically, as in MUP, the cost of each fiber link is given by the product of a fixed term, related to the power required by the fiber, and a load dependent term. However, the load dependent term can assume a number of different values, not just zero or one. Specifically, it is computed according to the number of used wavelengths by a cost function composed of two branches: let W be the number of wavelengths of a WDM link, the first branch decreases in the interval $[0, W/2]$ whilst the second one increases in the interval $[W/2, W]$. The reason for the increasing branch is that, when fiber loads are high, the probability to find an available wavelength along the whole path rapidly decreases due to the wavelength continuity constraint; this could determine the need to use a new fiber, on some links of the chosen path, during the wavelength assignment phase, or could lead to a block. The decreasing branch, instead, aims at avoiding using slightly loaded fibers. This makes sense in the DLE scenario where each lightpath has a finite lifetime; thus, less loaded fibers are likely to be powered off when lightpaths established on them will end. In [11] both LBC and MUP are evaluated and compared with the shortest path algorithm and the optimal solution of the RWA problem. Results reported in Figure 26.6 show the energy efficiency of a given algorithm, defined as ratio between the power consumption in case of the optimal solution and the power consumption given by the specific algorithm, versus the average traffic load.

A different approach to improve the energy efficiency of the optical network can be found in [13]. The proposed solution is based on an iterative algorithm that tries to switch-off optical fiber links when the traffic load slows down. The solution does not act on the RWA algorithm, that is assumed to be a classical power-unaware RWA algorithm, but tries to switch off as many fibers as possible by rerouting lightpaths established on them in the rest of the network. Specifically, fibers are ordered according to a given criterion and at each step a fiber is selected to be switched off; if all lightpaths established on it can be rerouted on the rest of the network using the classical RWA algorithm, the fiber is switched off, otherwise the fiber remains active. A number of criteria are proposed to properly order fibers, taking into account power consumption metrics and topological considerations. Such a solution allows a significant energy saving, up to 70% when the traffic load scales to 20% of the peak load, and leads to a power consumption that is almost proportional to the offered load. A possible drawback of such an approach concerns the complexity associated with its practical implementation; in fact, a practical solution should avoid service interruption during the rerouting operation of lightpaths and consider QoS requirements.

26.3 Energy Aware Traffic Engineering

Energy-Aware Traffic Engineering strategies aim at bringing the network energy consumption nearly proportional to the amount of carried traffic. As well

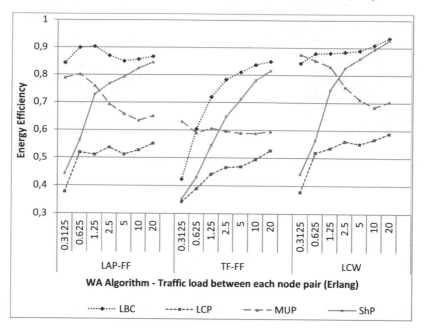

FIGURE 26.6 **Energy efficiency of the Load Based Cost (LBC), Most Used Path (MUP), Shortest Path (SP) and Least Congested Path (LCP) algorithms versus the average traffic load between each node pair Three different wavelength assignment algorithms are considered: LAP-FF (Least Additional Power-First Fit), TF-FF (Two Phases-First Fit) and LCW (Least Cost Wavelength).**

known, Internet design follows two main principles: redundancy and resource over-provisioning; the combined effect of these two principles enable Internet to support traffic variations and possible faults avoiding QoS degradations. Taking in mind the traffic fluctuations, it results that network resources are underutilized for most of the time and their consumption is independent of the load.

EA-TE strategies generally aims at aggregating traffic flows over a subset (possibly the minimum one) of the network elements, allowing other links and interconnection devices to be put in power saving mode. These solutions should satisfy the traffic demand variations in time, preserve connectivity and QoS, for instance by limiting the maximum utilization over any link, and ensure a minimum level of path redundancy.

A generic block diagram of a EA-TE policy is shown in Figure 26.7 [14]. In the off-line plane, a representation of the network topology, a power model of the network devices, and a traffic matrix estimation (if needed) are used to pre-compute the set of paths, so as to minimize the network energy consumption and the complementary set of links to be put in low power mode. The decisions taken by the computation block are transferred to the network elements to be installed in their routing/forwarding tables. The traffic information are collected on-line and transferred to the processing functional blocks residing in the off line plane.

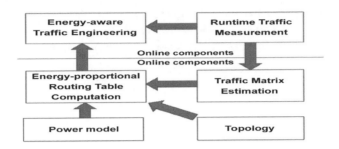

FIGURE 26.7 General block diagram of an Energy Aware Traffic Engineering (EA-TE) mechanism (Source: [14]).

EA-TE strategies differentiate for the different ways adopted to implement the functional blocks of the generic scheme shown in fig. 2. Different choices concern:

a) the centralized or distributed implementation of the energy-saving path computation function;

b) the algorithm used to determine the energy-saving path set;

c) the type of information utilized to compute the energy-saving paths;

d) the type of routing information and the mechanism used to exchange them between the network elements;

e) the path updating period.

These criteria are utilized to define the EA-TE methods presented in the next sections.

As far as the application of EA-TE strategies is concerned, an interesting problem consisting in determining when the EA-TE strategies can be usefully applied and their expected potential benefits. In other words, it would be useful to have a rough preliminary estimate of the possible savings that the adoption of smart energy saving policies may entail and in which cases these advantages can be achieved. Some studies face this problem, e.g. [15].

These studies use a very simple network model, mainly based on purely topological properties, and consider, together with the benefit due to the link/node power-off, the average effect of traffic load over network links due to path rerouting. The energy consumption of a link interface is modeled by a constant part, that includes the fraction of the node energy cost due to the link, and a variable part that is proportional to the traffic loading the link.

Although the approach is oversimplified, it is able to capture some interesting properties on the field of applicability of EA-TE strategies.

It is to be observed that, the variable part considered in the energy consumption model means that power management capabilities provided by the HW level be implemented in link interfaces. These techniques aim at modulating capacities of network device (e.g., link bandwidths, computational capacities of packet processing engines, etc.) according to current traffic loads. An overview of these techniques is given in [16–18].

Lessons learned by the application of these prediction models are:

i) the advantage of application of sleeping techniques increases as the network size and the number of alternative paths increase i.e., normally, EA-TE methods are the more convenient the more the network is large and the number of devices grows;

ii) EA-TE strategies are not convenient in small networks, where computation of energy-saving path sets is limited by the low value of path redundancy;

iii) there is a trade-off between EA-TE techniques and power management capabilities, if the device consumption is independent of the load, or lightly variable, the use of sleep modes is very effective; on the other hand, if efficient power management techniques are implemented in the network devices, the effectiveness of sleep mode approaches decreases.

A further aspect that has to be considered for EA-TE strategy application is that, unfortunately, putting interfaces of switches or routers to sleep can have serious drawbacks due to the network protocols operations. As discussed in [19], the main elements to be considered are the possible interactions of a specific EA-TE strategy with:

i) *Layer 2 control protocols*: it is needed to consider, on the one hand, the interaction between the EA-TE algorithms and the adaptive learning algorithms, used by switches to populate their forwarding tables and, on the other hand, the effect of the topology changes on the procedures of the spanning tree computation periodically performed by switches; the risk is to have a continuous spanning tree recomputation activity with significant loss of throughput.

ii) *Layer 3 control protocols*: a specific attention must be devoted to the reaction of routing protocols at a sleeping event of a network interface; as a matter of example, if a router put an interface to sleep, OSPF (Open Shortest Path First) would generate a flood of LSA (Link-State Advertisement) packets indicating the that link is down, followed by a recomputation of the Shortest Path First (SPF) algorithm by all routers. Analogously, Border Gateway Protocol (BGP) may suffer from route oscillations and, occasionally, forwarding loops.

Summarizing, it appears that EA-TE strategies are promising, but they should take into account the network protocol operation in order to avoid negative interactions, or, alternatively, their application will require some changes to the current protocol specifications [18].

Starting from this consideration, EA-TE strategies can be classified in two broad classes on the basis of the interaction with the existing IP routing protocols:

- *EA-TE Overlay strategies*;

- *EA-TE Integrated strategies.*

The overlay strategies works independently of the IP routing protocols: the links to be switched off are computed, by means of a specific algorithm, neglecting the impact of the topological changes on the IP routing protocol. On the contrary, in the integrated solutions, the power saving algorithm works together with the IP routing protocol and the selection of the links to be switched off are performed trying to avoid, as much as possible, the negative effects of the IP topology reconfiguration procedures.

26.3.1   EA-TE Overlay Strategies

The scope of a generic EA-TE Overlay strategy is to define network paths so that a set of routers and/or links can pass in standby state. The classical formulation of this problem is given in [20]. The input parameters of the problem are:

- the network topology, represented by a graph $G(V, L)$ composed by N nodes and L links;

- the links capacity, represented by values C_{ij} for every link (i, j) ;

- the traffic matrix, represented by values t^{sd} for traffic from node s to node d;

- the power consumption of each link and node, represented by P_i and P_{ij} respectively.

The problem output is the set of routers and/or links that can be put in standby state so that the total power consumption is minimized, subject to flow maximum link utilization constraints.

The Integer Linear Programming (ILP) formalization of the problem is here reported. In the formulation, the binary variables x_i and x_{ij} represents the power states of nodes and links i.e., $x_i = 0$ if node i is powered down, $x_i = 1$ otherwise.

$$\min \; [\sum_{i=1}^{N} \sum_{j=1}^{L} x_{ij} P_{ij} + \sum_{i=1}^{N} x_i P_i] \qquad (26.4)$$

s.t.:

$$\sum_{j=1}^{N} f_{ij}^{sd} - \sum_{j=1}^{N} f_{ji}^{sd} = \begin{cases} t^{sd} \ \forall s, d, i = s \\ -t^{sd} \ \forall s, d, i = d \\ 0 \ \forall s, d, is, \neq d \end{cases} \tag{26.5}$$

$$f_{ij} \leq \alpha C_{ij} x_{ij} \tag{26.6}$$

$$\sum_{i=1}^{N} x_{ij} + \sum_{j=1}^{N} x_{ij} \leq M x_i \tag{26.7}$$

The objective function (26.4) is the total power absorbed by the network. Equation (26.5) represents the flow conservation condition, where the variable f_{ij}^{sd} represents the amount of traffic from s to d crossing link (i, j). Equation (26.6) introduces a constraint about the maximum link utilization α. The last Equation (26.7), where $M \geq 2^N$, simply states that a node can be powered down only if all its incoming and outgoing links are powered down: .

This problem is a typical multi-commodity minimum flow problem with capacity constraints and it is known to be NP-hard, so heuristics have to be defined to solve this problem in practical cases. The idea proposed in [20] is to rank the network elements, i.e. a link or a router, and to proceed step by step. At each step a network element is selected and it is checked if it can be powered off. A network element can be powered off if it is possible to find a routing solution on the residual network, i.e. the network without the availability of the selected element.

The heuristics proposed in [20] differ on the basis of the strategy used to rank the network elements, whereas the used path computation algorithm is always the classical shortest paths one. Hierarchical topologies composed of three levels of nodes (core, edge and aggregation) are considered. In particular the authors show that randomly generating topologies with 160 nodes (10 core nodes, 30 edge nodes, and 120 aggregation nodes), it is possible to save about up to 30% of router and 50% of links when traffic is 20% with respect to peak hours.

Other strategies [21–23], that can be considered as extensions of the previous one, have been successively proposed. As an example, the general problem formulation is enhanced in [21] by considering:

- the introduction of line-cards into the router model: the energy saving increases considering the ability to put in standby state the whole line-card if all its ports are in standby;

- the availability of a set of precomputed paths for each source destination pair: instead of searching the solution among all the possible paths, a set of candidate paths, the k-shortest ones, is considered;

- the introduction of an extra constraint on packet delay, obtained limiting the length of the candidate paths.

In this way an EA-TE solution able to minimize the energy consumption distributing the traffic among the set of precomputed paths is found. In this work also a possible implementation in an MPLS network with OSPF routing protocol is proposed.

In [22] the model of network interfaces is further enhanced considering their ability to adapt the transmission rate. The authors assume that new techniques, such as the frequency change and the Dynamic Voltage Scaling (DVS), that have been successfully applied to general purpose processors will be available for network devices. Two different algorithms, referred to as EATe, are defined.

The first algorithm assumes low idle power consumption and good ability to save energy varying link rate. In this scheme, the algorithm shifts links to lower energy regions while being careful not to move the corresponding links on the alternative paths to higher energy levels.

The second algorithm assumes that rate adaptation provides modest benefits and the idle consumption of the network element is relatively high. In this case the algorithm strives to remove traffic from as many links as possible to let them enter a sleep state. In both cases a distributed on-line implementation of EATe algorithms is proposed.

In [23] the problem formulation is modified considering that in core networks physical links are usually realized by multiple physical cables i.e., a logical bundled link. In this way, the energy saving problem become the problem of finding the minimum set of physical cables needed to satisfy the traffic requirement. This problem is NP-hard and an heuristic is proposed. The heuristic works in two phases.

- In the first phase an initial path computation is performed by solving a classical minimum flow problem i.e, the total flow summed over all links is minimized; after that all cables with no traffic are powered down.

- In the second phase a mechanism to scan links is defined and an iterative procedure is executed: at each step the link with the greatest spare capacity is selected and one of its cable is evaluated for removal. If the cable can be removed the minimum flow paths computation procedure is performed.

The weak point of the EA-TE Overlay approaches is the coexistence with an IP routing protocol. EA-TE Overlay solutions determine links and/or router to put to sleep without considering how to perform this operation in the real IP infrastructure. In other words, to avoid the sleeping links to be used to route traffic, the overlay strategies have to force the removal of the powered off links from the IP logical topology. Unfortunately, this procedure leads to a new virtual IP topology and the IP routing protocol begins a reconfiguration phase that can have transient disruptions. In fact, if a link is switched off, it is seen by an IP routing protocol, typically OSPF, as a topological modification causing protocol restoration procedure; in particular:

- OSPF generates a control message i.e., a Link State Advertisement (LSA), flooding round to update the router topology databases;

- all network routers starts a paths re-computation procedure.

Considering that a great number of network links can be powered down, the reconfiguration phase can highly impact network performance. A further problem is the worsening of response time to failure events. When router interfaces are powered down, the routing protocol adjacencies among neighbor routers are lost; if a failure happens, a re-establishment of adjacencies is needed, this procedure could require a remarkable time, even tens of seconds. Finally, EA-TE Overlay strategies are all based on the knowledge of the traffic matrix, that represent an open issue especially in a classical IP network.

26.3.2 EA-TE Integrated Strategies

The EA-TE Integrated strategies are based on the interaction between the energy saving algorithm and the IP routing protocol. In other words, the algorithm used to detect the network links and/or router to put to sleep takes into account the routing protocol operations. In this way, solutions having low-impact over control plane procedures could be obtained.

A first class of EA-TE Integrated strategies is presented in [24]. In this work, the classical problem of computing OSPF links weights to perform Traffic Engineering [25] has been slightly modified to take into account energy saving. The new problem allows the evaluation of OSPF weights that minimize network links involved in packet routing. Maximum values of link utilization and shortest path routing are assumed as constraints. A simplified version of this problem is here reported:

$$\min \ \sum_{l \in L} \alpha_l y_l \tag{26.8}$$

s.t.:

$$\sum_{p \in P_W} \sum_{l \in L} w_l x_p \delta_{pl} \leq \sum_{l \in L} w_l \delta_{pl} \quad \forall q \in P_k, k \tag{26.9}$$

The decision variables of the problem are

- x_p, representing the shortest paths used to transmit packets; $x_p = 1$ only if p is the shortest path:

- y_l, representing the active links; $y_l = 1$ if link l is selected to be powered down;

- the link weights w_l.

The objective function reported in (26.8), where α_l is the power consumption of an active link l, represents the global energy consumption of network links. Together with the classical flow and link load constraints (analogous to expressions (26.5) and (26.6) of formulation related to EA-TE Overlay Strategies), this formulation introduce the constraint (26.9). This equation forces the path selected to route traffic i.e., $p \in P_W$, to be the shortest one among all network paths. In (26.9) $\delta_{pl} = 1$ only if path p includes link l, P_W represent the set of shortest paths and k represent a source destination pair.

The previous formulation, being problem a classical multi commodity flow problem, is NP-hard. So in [24] heuristics are proposed and their performance are evaluated with reference to the USA network topology reported in Figure 26.8, the percentage of network links that can be powered off is computed. It is shown that the different heuristics are able to power off about the 50% of network links when traffic decreases above 40% of its peak value.

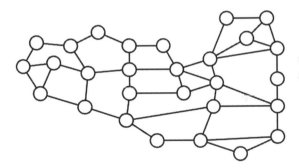

FIGURE 26.8 Reference USA network topology (28 nodes and 43 links).

The implementation of such a strategy requires the recomputation and consequent change of the OSPF link weights, when traffic decreases. OSPF protocol handles a single link weight modification as a topological change and so a new LSA is generated. In this way, the same problems of EA-TE Overlay Strategies, due to protocol restoration procedure, arise.

An EA-TE Integrated Strategies trying to overcome this problem is defined in [26]. This strategy, named Energy Aware Routing (EAR), is fully integrated with the OSPF routing protocol:

- it is able to reduce the number of IP router interfaces used to forward packets with no IP topology changes;

- it can be implemented by using the same OSPF data structures;

- it does not interfere with OSPF normal operations.

The basic idea of the path computation strategy adopted in EAR is that the IP links used to route traffic can be reduced if a subset of Shortest Path Trees (SPTs) is used instead of the full set (one for each router) as in case of classical OSPF. In this way, the IP interfaces not more used can be switched in a sleep mode. Normally, a router only computes its own SPT via the Dijkstra's algorithm; diversely, the EAR operation is based on the concept of "SPT exportation": a set of routers, called "exporters", forces the use of their SPTs to another set of routers, called "importers". The new Path Tree of the importer is referred to Modified Path Tree (MPT). The exportation mechanism is possible because all the IP routers have a consistent view of the IP network topology in their OSPF databases.

The effect of an exportation is the reduction of active links i.e., links used to route traffic, and the modification of network paths. It is important to define some rules that guarantee the correctness of the new paths i.e., loops avoidance, when multiple exportations are executed. In other words simple rules, to establish if more moves can be performed at the same time, must be defined. This relationship among exportation is defined compatibility. The authors demonstrates that new routing paths are loop-free if the following rules is respected:

- an SPT associated to an exporter node cannot be modified by any exportation, i.e. the routing path trees of exporter nodes must remain the shortest ones.

The problem of saving energy reducing the number of active links becomes a new problem: finding the set of exportations assuring the highest energy saving. To better describe this new problem, referred to as EAR problem, some notations are introduced. The IP network is represented by a directed graph $G(V, E)$. Starting from G and considering the exportation mechanism an undirected graph $H = (M, C)$ is defined. Each node $m \in M$ represents an exportation and the edges $c \in C$ indicate the compatibility relationships between the edges: if there exists an edge c between the nodes m_1 and m_2, then m_1 and m_2 are compatible. Moreover, each node m is characterized by a weight $w(m)$ which represents the number of IP links that the exportation m allows to put to sleep. In $H = (M, C)$, a set of compatible exportations is represented by a clique $K_H(n)$ where n is the number of nodes of the clique. A clique is a complete subgraph of H in which every pair of nodes is adjacent.

Each clique $K_H(n)$ of H is characterized by its weight $W(K_H)$, given by the sum of the weights of all the clique nodes. In EAR case, since two compatible exportation cannot switch off the same IP links, $W(K_H)$ is the number of IP links that the set of compatible exportations represented by $K_H(n)$ allows to put to sleep. So the EAR problem is equivalent to find the complete subgraph of H characterized by the maximum weight. A simplified version of this problem, obtained if every node has the same weight, is known in literature as the maximum clique problem of a graph, and it is *NP-hard*; as a consequence EAR problem is *NP-hard*, too.

The authors propose an heuristic, called Max_Compatibility heuristic, that aims at finding a set of compatible moves. In Figure 26.9 performance of the proposed method are shown. The used reference network is similar to the one reported in Figure 26.9. The figure of merit used is the η_e parameter that represents the percentage of links that pass in standby with respect to the maximum number of links that can pass in standby state i.e, the total number of network links except links needed to guarantee network connectivity. The η_e parameter is reported as a function of the scaling factor α, representing the traffic decreases with respect to peak hours, and ρ_{max}, representing the maximum admitted link load.

Results show that about the 30% of network links, that correspond to $\eta_e = 50\%$ can be put in standby state when $\alpha \leq 50\%$. The authors also show that the impact of such procedure on network performance is limited: about the 80% of network paths are not modified by EAR procedure; moreover it is demonstrated that, in the case of equal OSPF weight links, the maximum paths length increase is equal to 2 hops.

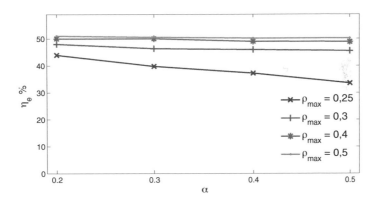

FIGURE 26.9 EAR performance as a function of traffic decreases α and link maximum utilization ρmax.

The most attracting feature of the EAR strategy is that it can be implemented into OSPF routing protocol implementation. The authors propose a centralized implementation with the following features:

- a network router is elected as the EAR coordinator;

- the EAR coordinator computes the routers roles applying the Max_Compatibility heuristic;

- the EAR coordinator sends a specific OSPF message to each importer router specifying the associated exporter router.

- the importers computes their MPTs and put in standby state a subset of network links

This procedure is completely transparent to non-importer routers and, if an advanced standby state for router interfaces is introduced, avoids OSPF convergence procedures.

Bibliography

[1] R. Tucker, "Green optical communications - Part I: Energy limitations in transport," *IEEE J. Selected Topics in Quantum Electronics*, vol. 17, no. 2, pp. 245–260, Mar. 2011.

[2] ——, "Green optical communications - Part II: Energy limitations in networks," *IEEE J. Selected Topics in Quantum Electronics*, vol. 17, no. 2, pp. 261–274, Mar. 2011.

[3] C. Lange, D. Kosiankowski, R. Weidmann, and A. Gladisch, "Energy consumption of telecommunication networks and related improvement options," *IEEE J. Selected Topics in Quantum Electronics*, vol. 17, no. 2, pp. 285–295, Mar. 2011.

[4] R. S. Tucker, J. Baliga, R. Ayre, K. Hinton, and W. V. Sorin, "Energy consumption in ip networks," in *the 34th European Conference on Optical Communication 2008, ECOC 2008*, Sept. 2008.

[5] E. Mannie, "Generalized Multi-Protocol Label Switching (GMPLS) Architecture," RFC 3945, 2004. [Online]. Available: http://www.ietf. org/rfc/rfc3945.txt

[6] GARR Network. [Online]. Available: http://www.garr.it/rete/ statistiche/backbone

[7] G. Shen and R. Tucker, "Energy-minimized design for IP over WDM networks," *IEEE/OSA J. Optical Communications and Networking*, vol. 1, no. 1, pp. 176–186, June 2009.

[8] W. V. Heddeghem, M. D. Groote, W. Vereecken, D. Colle, M. Pickavet, and P. Demeester, "Energy-efficiency in telecommunications networks: Link-by-link versus end-to-end grooming," in *2010 14th Int. Conf. Optical Network Design and Modeling (ONDM)*, Feb. 2010, pp. 1–6.

[9] E. Yetginer and G. Rouskas, "Power efficient traffic grooming in optical wdm networks," in *Proc. IEEE Global Telecommunications Conference, 2009, GLOBECOM 2009*, Dec. 2009, pp. 1–6.

[10] A. Ahmad, A. Bianco, E. Bonetto, D. Cuda, G. Castillo, and F. Neri, "Power-aware logical topology design heuristics in wavelength-routing networks," in *2011 15th Int. Conf. Optical Network Design and Modeling (ONDM)*, Feb. 2011, pp. 1–6.

[11] A. Coiro, M. Listanti, A. Valenti, and F. Matera, "Power-aware routing and wavelength assignment in multi-fiber optical networks," *IEEE/OSA J. Optical Commun. Networking*, vol. PP, no. 99, pp. 816–829, 2011.

[12] Y. Wu, L. Chiaraviglio, M. Mellia, and F. Neri, "Power-aware routing and wavelength assignment in optical networks," in *the 35th European Conference on Optical Communication 2009. ECOC '09*, Sept. 2009, pp. 1–2.

[13] A. Coiro, M. Listanti, A. Valenti, and F. Matera, "Reducing power consumption in wavelength routed networks by selective switch off of optical links," *IEEE J. Selected Topics in Quantum Electronics*, vol. 17, no. 2, pp. 428–436, Mar. 2011.

[14] N. Vasi, P. Bhurat, D. Novakovi, M. Canini, S. Shekhar, and D. Kosti, "Responsive, energy-proportional networks," EPFL, Tech. Rep., July 2010.

[15] L. Chiaraviglio, D. Ciullo, M. Mellia, and M. Meo, "Modeling sleep modes gains with random graphs," in *Proc. 2011 IEEE Conf. on Computer Communications Workshops (INFOCOM WKSHPS)*, Apr. 2011, pp. 355–360.

[16] A. Wierman, L. Andrew, and A. Tang, "Proc. power-aware speed scaling in processor sharing systems," in *Proc. IEEE INFOCOM 2009*, Apr. 2009, pp. 2007–2015.

[17] K. J. Christensen, C. Gunaratne, B. Nordman, and A. D. George, "The next frontier for communications networks: power management," *Computer Communications*, vol. 27, no. 18, pp. 1758–1770, 2004.

[18] A. P. Bianzino, C. Chaudet, D. Rossi, and J.-L. Rougier, "A survey of green networking research," 2010. [Online]. Available: http://arxiv.org/abs/1010.3880

[19] M. Gupta and S. Singh, "Greening of the internet," in *Proc. the 2003 Conference on Applications, technologies, architectures, and protocols for computer communications*, ser. SIGCOMM '03. New York, NY, USA: ACM, 2003, pp. 19–26.

[20] L. Chiaraviglio, M. Mellia, and F. Neri, "Reducing power consumption in backbone networks," in *Proc. IEEE Int. Conf. Commun. 2009. ICC'09*, June 2009, pp. 1–6.

[21] M. Zhang, C. Yi, B. Liu, and B. Zhang, "GreenTE: Power-aware traffic engineering," in *2010 18th IEEE Int. Conf. Network Protocols (ICNP)*, Oct. 2010, pp. 21–30.

[22] N. Vasić and D. Kostić, "Energy-aware traffic engineering," in *Proc. the 1st Int. Conf. on Energy-Efficient Computing and Networking*, ser. e-Energy '10. New York, NY, USA: ACM, 2010, pp. 169–178.

[23] W. Fisher, M. Suchara, and J. Rexford, "Greening backbone networks: reducing energy consumption by shutting off cables in bundled links," in *Proc. the first ACM SIGCOMM workshop on Green networking*, ser. Green Networking '10, 2010, pp. 29–34.

[24] S. S. Lee, P.-K. Tseng, and A. Chen, "Link weight assignment and loop-free routing table update for link state routing protocols in energy-aware internet," *Future Generation Computer Systems*, vol. 28, no. 2, pp. 437 – 445, 2012.

[25] B. Fortz and M. Thorup, "Internet traffic engineering by optimizing ospf weights," in *IEEE INFOCOM 2000*, vol. 2, 2000, pp. 519–528.

[26] A. Cianfrani, V. Eramo, M. Listanti, and M. Polverini, "An OSPF enhancement for energy saving in IP networks," in *Proc. 2011 IEEE Conf. Computer Communications Workshops (INFOCOM WKSHPS)*, Apr. 2011, pp. 325–330.

Author contact information

Antonio Cianfrani, Angelo Coiro, Vincenzo Eramo, and Marco Listanti are with DIET - "Sapienza" University of Roma, Italy, Email: cianfrani@diet.uniroma1.itcoiro@infocom.uniroma1.itvincenzo.eramo@uniroma1.it marco@infocom.uniroma1.it.

27

Energy-Aware Network Management and Content Distribution

Luca Chiaraviglio

Politecnico di Torino, Torino, Italy

Ibrahim Matta

Boston University, USA

CONTENTS

27.1   Introduction

Energy-efficient communication has become a challenging problem in the last few years. Current estimates [1] show that the Information and Communication Technology sector (ICT) consumes between 2% and 10% of the worldwide energy consumption, and this trend is expected to grow even more in the future due to the proliferation of both networked and networking devices. Internet Service Providers (ISP), are becoming sensitive to reducing the power consumption of their infrastructure, due to increasing energy costs and new business opportunities that can be realized by "going green". At the same time, Content Providers (CP) are faced with a constant increase in the number of users coupled with the need to reducing the energy consumption of both server farms and cooling systems. Therefore, both ISPs and CPs could potentially realize great benefits if energy-efficient techniques would be fully developed for network devices [2, 3] and servers [4, 5].

Starting from the seminal work of [6], several approaches have been proposed in order to reduce power consumption of ICT. For example, in [7] the authors consider the minimization of the power consumed by an ISP network. Moreover, new recent approaches like [8] aim at reducing the power consumption of big CPs considering the variation in electricity prices. Nevertheless, none of these previous studies considers the minimization of the *total* power consumption of both ISP and CP, so that great power savings can be achieved by considering a mutual objective.

In [9] the authors solve jointly the traffic engineering and content distribution problem, showing that great improvements in Quality of Service (QoS) can be obtained if CP and ISP pursue the same objective. However, this solution does not consider power consumption at all and consequently the power waste can be huge.

In this chapter, we propose a new approach to reducing power consumption for ISPs *and* CPs. In particular, we solve a multi-objective problem in which a CP and an ISP *cooperate* to reduce overall power consumption. We assume that the ISP is the owner of a network infrastructure. Additionally, we represent the CP infrastructure as a set of servers placed in different cities. We assume that users request contents from the CPs. In particular, we assume that the CP manages aggregate user requests, so that an aggregate traffic request is measured or estimated toward each egress node of the ISP. Then, we aim at controlling the whole system composed of the ISP and the CP in order to find the minimal set of network resources and servers that minimize the total power consumption while satisfying the current content requests. Traditionally, ISPs and the CPs are not willing to share sensitive information such as the network topology and the servers' load. Therefore, our approach in this chapter is distributed between the ISP and the CP to limit the amount of exchanged information. In particular, we develop different algorithms based on two techniques: the dual decomposition and the Benders decomposition. We test our model on real ISP topologies and considering realistic power figures. Our results show that there is ample room to pursue a cooperative green approach, so that large power savings can be easily achieved.

27.2   Problem Formulation

The main goal of our approach is to minimize the power consumption jointly between the CP and the ISP. In particular, we assume that the ISP is the owner of the network infrastructure, so that it manages a physical topology, i.e. a set of nodes and links. The CP instead is composed of a number of servers connected to the ISP. When a user (terminal) asks for a CP's resource, we assume that the resource is replicated over the CP infrastructure, so that the user can be potentially served by *any* of the servers of the CP.

An informal description of the problem we consider is the following: **Given** i) a physical ISP topology comprising routers and links, in which links have a known capacity, ii) the knowledge of the average amount of traffic demands from users, iii) the power consumption of each link and node, and iv) the power consumption of each CP's server; **Minimize** the total power consumption of the ISP network and the CP infrastructure; **Subject to** maximum link utilization, maximum admissible delay, and maximum server utilization. More formally, we represent the ISP topology as a di-graph $G = (V, E)$, where V is the set of vertices and E is the set of edges. Vertices represent network nodes, while edges represent network links. We denote by $N = |V|$ and $L = |E|$ the total number of nodes and links, respectively. Let C_l be the capacity of link l, and let $U_l^{MAX} \in [0, 1]$ be the maximum link utilization that can be tolerated.[1] S is the set of servers of the content provider. Denote by W_s the maximum load allowed on server $s \in S$. Let R_t be the traffic demand between terminal $t \in T$ and the content provider S. Moreover, let x^{st} be continuous variables representing the amount of traffic between a source node s and a terminal t. We divide x^{st} into x_m^{st} and x_b^{st} to denote the amount of traffic originating from the content provider under consideration and from other content providers, respectively. Actually, x_b^{st} are constants, i.e. the considered CP can not modify these traffic demands. On the other hand, we assume that x_m^{st} are continuous variables so that a traffic demand R_t from terminal t can be served by anyone of the $s \in S$ CP servers, while satisfying load and delay constraints.

In order to describe the network topology, we can either choose between a *node-link* formulation or a *path-link* formulation [10]. In particular, in the node-link formulation each node enforces the flow conservation law, while in the path-link formulation the flow conservation is done on the entire path. We

---

[1]Link utilization is normally kept below 100% due to QoS requirements.

decided to choose the path-link formulation since: i) it requires a lower number of variables than the node-link formulation ($O(N^2)$ instead of $O(N^3)$); ii) ISP can easily control the paths since they are pre-computed prior to launching the problem; iii) additional constraints, such as a minimum number of disjoint paths for each source and destination pair, can be easily enforced.

More formally, let δ_{lp}^{st} be constants which take the values of 1 if link l belongs to path p carrying demand from s to t, 0 otherwise. Let z_p^{st} and q_p^{st} be continuous variables representing the amount of traffic from s to t on path p for the considered CP and for other CPs, respectively. Let $\mathcal{P}(s,t)$ be the set of pre-computed paths from s to t. Additionally, let f_l be the total amount of flow on link l, split into f_l^{cp} and f_l^{bg} for the considered CP and for other CPs, respectively. Let D_l be the delay on link l, which can be modeled as $D_l = 1/(C_l - f_l) + D_l^p$, where $1/(C_l - f_l)$ and D_l^p denote the queueing delay and the propagation delay of link l, respectively.

Finally, we consider the power consumption of devices. Let y_n and y_s be binary variables which take the value of 1 if node n and server s are powered on, respectively. $P_n^c y_n$ and $P_s^c y_s$ represent the static amount of power (startup energy cost) consumed by node n and server s when powered on. Moreover, let $P_l^d(f_l)$, $P_n^d(\sum_{l \in \mathcal{L}(n)} f_l)$, $P_s^d(\sum_{t \in T} x_{cp}^{st})$ be monotonically increasing convex functions representing the dynamic power consumption of link l, node n and server s, respectively. $\mathcal{L}(n)$ denotes the set of links incident to node n.

Given the previous notations, we recall the classic formulation of [9] and then present a new green formulation.

27.2.1   Classical Design

The objective of the classic design problem presented in [9] is the minimization of the overall delay experienced by all users in the network. The problem is formalized as follows:

$$\mathbf{C}: \quad \min \sum_l f_l^{cp} D_l \quad \text{subject to:} \tag{27.1}$$

$$\sum_{s \in S} x_m^{st} = R_t \quad \forall t \in T \tag{27.2}$$

$$\sum_{t \in T} x_m^{st} \leq W_s \quad \forall s \in S \tag{27.3}$$

$$\sum_{p \in \mathcal{P}(s,t)} q_p^{st} = x_b^{st} \quad \forall s,t \tag{27.4}$$

$$x_m^{st} = \sum_{p \in \mathcal{P}(s,t)} z_p^{st} \quad \forall s,t \tag{27.5}$$

$$f_l^{cp} = \sum_{s,t,p \in \mathcal{P}(s,t)} \delta_{lp}^{st} z_p^{st} \quad \forall l \in E \tag{27.6}$$

$$f_l^{bg} = \sum_{s,t,p \in \mathcal{P}(s,t)} \delta_{lp}^{st} q_p^{st} \quad \forall l \in E \tag{27.7}$$

$$f_l = f_l^{bg} + f_l^{cp} \leq C_l U_l^{MAX} \quad \forall l \in E \tag{27.8}$$

Control variables: $z_p^{st} \geq 0$, $q_p^{st} \geq 0$.

More precisely, Eq.(27.2) ensures that the total traffic demand R_t from terminal t is served by the CP, Eq.(27.3) limits the total traffic on each server by the maximum load, Eq.(27.4) splits the traffic demand x_b^{st} among the different paths, and Eq.(27.5) computes the total amount of traffic from each server to each terminal of the CP. Finally, Eq.(27.6)-(27.8) compute the total flow on each link and impose the maximum link utilization constraint. This formulation does not take into account power consumption at all: we therefore compute the power consumption of devices as a post-processing phase.

The C problem falls into the class of convex optimization problems, for which finding a local optimum is equivalent to finding the global optimum.

27.2.2 Centralized Green Model

In order to consider the power consumption we propose a novel approach in which CP and ISP share information to minimize the global power consumption. Starting from our previous work [11] under the power consumption model as in Table 27.1, we first define a centralized model by introducing the estimated demands $\widetilde{x}_m^{st}$ and the estimated delay $\widetilde{d}_a$ as additional variables. We then define the green centralized problem (G) as follows:

$$\mathbf{G}: \quad \min\left(P_{TOT} = P_{CP} + P_{ISP}\right) \quad \text{s.t.:}$$

$$P_{ISP} = \sum_{l \in E} P_l^d(f_l) + \sum_{n \in V} \left[P_n^d \left(f_{l \in \mathcal{L}(n)} \right) + P_n^c y_n \right] \qquad (27.9)$$

$$\sum_{s \in S} \widetilde{x}_m^{st} = \widetilde{R}_t \quad \forall t \in T \qquad (27.10)$$

$$\sum_{p \in \mathcal{P}(s,t)} q_p^{st} = x_b^{st} \quad \forall s, t \qquad (27.11)$$

$$\widetilde{x}_m^{st} = \sum_{p \in \mathcal{P}(s,t)} z_p^{st} \quad \forall s, t \qquad (27.12)$$

$$f_l = \sum_{s,t,p \in \mathcal{P}(s,t)} \left[\delta_{lp}^{st} z_p^{st} + \delta_{lp}^{st} q_p^{st} \right] \leq C_l U_l^{MAX} \quad \forall l \qquad (27.13)$$

$$d_l \geq a_i f_l + b_i \quad \forall l, i \in I \qquad (27.14)$$

$$d_a = \frac{\sum_l d_l}{|T|} \qquad (27.15)$$

$$\sum_{l \in \mathcal{L}(n)} f_l \leq M_n y_n \quad \forall n \in V \qquad (27.16)$$

$$P_{CP} = \sum_{s \in S} \left[P_s^d \left(x_m^{st} \right) + P_s^c y_s \right] \qquad (27.17)$$

$$\sum_{s \in S} x_m^{st} = R_t \quad \forall t \in T \qquad (27.18)$$

$$\widetilde{d}_a \leq D^{MAX} \qquad (27.19)$$

$$\sum_{t \in T} x_m^{st} \leq W_s \quad \forall s \in S \qquad (27.20)$$

$$\sum_{t \in T} x_m^{st} \leq M_s y_s \quad \forall s \in S \qquad (27.21)$$

$$\widetilde{d}_a = d_a \qquad (27.22)$$

$$\widetilde{x}_m^{st} = x_m^{st} \quad \forall S \times T \qquad (27.23)$$

Control variables: $z_p^{st} \geq 0$, $q_p^{st} \geq 0$, $y_n \in \{0,1\}$, $y_s \in \{0,1\}$.

The objective function of the G problem is the minimization of the total power consumption of the ISP and the CP, adopting as control variables the amount of traffic between every server-terminal pair, i.e. $z_p^{st} \geq 0$ and $q_p^{st} \geq 0$.

Considering the ISP, Eq.(27.9) computes its total power consumption. Eq.(27.10) imposes that estimated CP traffic is equal to estimated terminal demand. Notice that here we assume that $\widetilde{R}_t$ is the ISP estimation of total traffic R_t from client t. Routing constraints are specified by Eq.(27.11) and (27.12). The total flow on each link is computed and constrained by Eq.(27.13). Then, Eq.(27.14) computes the total delay for each link, using the additional variables $d_l \geq 0$. The delay function is approximated by I linear segments as in [9]. Finally, the average network delay is computed by Eq.(27.15).

Considering the CP, Eq.(27.17) computes its total power consumption. Eq.(27.18) guarantees the traffic demand constraint. Eq.(27.19) bounds the average delay of users. Eq.(27.20) limits the maximum load on each server.

Finally, Eq.(27.16) and Eq.(27.21) impose powering-on a network node and a server, respectively, if their incoming/outgoing flows are larger than zero, adopting a big-M method, i.e. $M_s \geq W_s$ and $M_n \geq \sum_{l \in \mathcal{L}(n)} C_l$ Moreover, Eq.(27.22) and Eq.(27.23) act as consistency constraints, guaranteeing that the estimated values are always equal to the real ones.

The equivalent model G belongs to the class of mixed-integer problems, that can be solved using standard optimization programs.

27.3 Distributed Algorithms

Two considerations hold for the G model: (i) the problem can be completely split between the ISP and the CP using a decomposition technique, (ii) after the problem is split the amount of information shared by ISP and CP is limited. In the following, we describe the algorithms we propose to efficiently solve the distributed problem.

27.3.1 Dual Green Algorithm

We first apply a dual decomposition technique to derive a distributed algorithm, following a well-known procedure in the literature [12], and whose main ideas where presented in our previous work [13]. After the decomposition is applied, the ISP uses an estimation of the traffic demands, while the CP uses an estimation of the users' delay. In particular, we first introduce the Lagrange multipliers λ^{st} and μ_a associated with the consistency constraints of Eq.(27.22) and Eq.(27.23). The Lagrange multipliers are then shared between the ISP and the CP. We define the ISP problem as follows:

D-GreenISP: $\quad \min \left(P_{ISP} - \sum_{st} \lambda^{st} \widetilde{x}_m^{st} + \mu_a d_a \right) \quad$ s.t.: (27.9)-(27.16)

Control variables: $z_p^{st} \geq 0$, $q_p^{st} \geq 0$, $y_n \in \{0,1\}$.

The CP instead solves the following problem:

D-GreenCP: $\quad \min \left(P_{CP} + \sum_{st} \lambda^{st} x_m^{st} - \mu_a \widetilde{d}_a \right) \quad$ s.t.: (27.17)-(27.21)

with control variables: $x_m^{st} \geq 0$, $\widetilde{d}_a \in \mathcal{R}^+$, $y_s \in \{0,1\}$.

In order to get an optimal solution, the **D-GreenISP** and the **D-GreenCP** are solved using an iterative method that involves the Lagrange multipliers. In particular, at each iteration k the Lagrange multipliers are updated using a subgradient method:

$$\lambda^{st}(k+1) \quad = \quad \lambda^{st}(k) - \alpha_k \left[\tilde{x}_m^{st}(k) - x_m^{st}(k) \right] \quad \forall s,t \qquad (27.24)$$

$$\mu_a(k+1) \quad = \quad \mu_a(k) - \alpha_k \left[\tilde{d}_a(k) - d_a(k) \right] \qquad (27.25)$$

with α_k small or diminishing step size. The intuition is that the Lagrange multipliers act as penalty/reward for the objective functions. For example, when $\tilde{x}_m^{st}(k) - x_m^{st}(k) > 0$ the associated multiplier $\lambda^{st}(k+1)$ is decreased. When $\lambda^{st}(k+1)$ is positive, it acts as a reward for the ISP and a penalty for the CP. In our example, at iteration $k+1$ the ISP will decrease $\tilde{x}_m^{st}(k+1)$ since the associated reward $\lambda^{st}(k+1)$ is decreased, and the CP will increase $x_m^{st}(k+1)$ since the associated penalty $\lambda^{st}(k+1)$ is decreased. Note that at equilibrium, i.e. when Eq.(27.22) and (27.23) hold, the solution of the distributed algorithm is optimal.

Since the Lagrange multipliers' update needs the demands and the delays from both the CP and the ISP, we propose the adoption of a trusted third-party server (TS) to delegate the manipulation of the Lagrange multipliers. The TS can be controlled by a trusted authority that ensures that both ISPs and CPs are actively cooperating in reducing power consumption. The trusted authority incentives the cooperation between ISP and CP.[2] Fig. 27.3(left) illustrates the exchanged parameters between ISP, CP and TS. The dotted line indicates that the ISP use estimated or measured values for R_t.

The dual algorithm then works as follows: the Lagrange multipliers are initialized by the TS, then the **D-GreenISP** and the **D-GreenCP** are solved in parallel by the ISP and the CP, respectively, using the current Lagrange multipliers. At the end of each iteration the TS updates λ^{st} and μ_a using Eq.(27.24) and (27.25). The distributed problems are iteratively solved until a maximum number of iterations k_{MAX} is reached.[3]

Let P_{TOT}^G be the optimal total power consumption obtained from the centralized G model. Let $P_{TOT}^{D-G}(k)$ be the total power consumption at iteration k obtained from the dual algorithm. Since the subgradient method adopted in Eq.(27.24)-(27.25) is not a descent method, we keep track of the best distributed solution found so far:

$$P_{TOT}^{D-G}(k_{Best}) = \min_{i=1,...,k} P_{TOT}^{D-G}(i) \qquad (27.26)$$

where k is the current iteration. We also define the precision error for the current solution as:

$$e_P(k) = \left| P_{TOT}^{D-G}(k) - P_{TOT}^G \right| \qquad (27.27)$$

---

[2]Normally multiple ISPs and CPs are present. Nevertheless, our solution can be adopted also in this case. For example, a federation of trusted servers can be deployed.

[3]Another stopping criterion might be to test that $|\lambda^{st}(k+1) - \lambda^{st}(k)|$ and $|\mu_a(k+1) - \mu_a(k)|$ are very small.

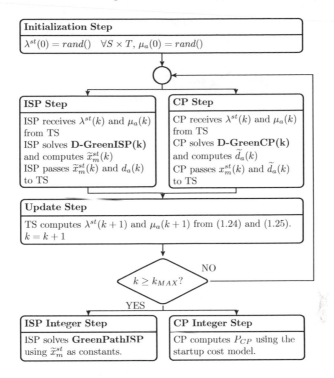

FIGURE 27.1 Dual Green algorithm (D-G) ([13], © 2011 IEEE).

Similarly, we define $e_P(k_{Best})$ as the precision error considering the best distributed solution found so far. As reported in [14] $e_P(k_{Best})$ depends mainly on α_k. In particular, if the associated Lagrangian is a continuous function and a diminishing step size rule is adopted for α_k, then $e_P(k_{Best}) \to 0$ as $k \to \infty$, i.e. the distributed solution converges to optimum. However, in our case the Lagrangian associated with the G model is not continuous, due to the presence of the integer variables y_n and y_s. Therefore, the resulting distributed problem does not converge to an optimal solution [15]. Additionally, the distributed problem does not even converge to an equilibrium point, since the consistency constraints are not assured by the distributed approach. This impacts negatively the QoS of users, because traffic demands and delays are not properly estimated.

To overcome these problems, we propose to solve the distributed solution using only the continuous part of the power functions, then we add the integer variables locally at a second step. We name this algorithm Dual Green (D-G). In particular, the ISP and the CP initially solve the distributed step of D-G, representing the power consumption as a convex function, i.e. $y_n = 0 \; \forall n$, $y_s = 0 \; \forall s$. After few iterations, the problem converges to an optimal solution, for which both $\widetilde{x}_m^{st}$ and $\widetilde{d}_a$ are correctly estimated. Interestingly, at the end of this step both the ISP and the CP have agreed on a possible feasible solution.

Then as a second step, the ISP optimizes the power consumption using the integer variables y_n and the estimated traffic demand $\widetilde{x}_m^{st}$ computed in the first step, as follows:

$$\textbf{GreenPathISP} \quad \min\,(P_{ISP}) \quad \text{s.t.:}$$

$$(27.9), (27.11) - (27.16) \tag{27.28}$$

Control variables: $z_p^{st} \geq 0$, $q_p^{st} \geq 0$, $y_n \in \{0, 1\}$.

Notice that the **GreenPathISP** problem optimizes the power consumption over the set of paths taking as inputs the traffic demands $\widetilde{x}_m^{st}$.

In parallel, the CP computes its power consumption from the demands x_m^{st} using the integer variables y_s. Fig. 27.1 shows a schematic description of the D-G algorithm.

27.3.2   Benders Green Algorithm

The second approach to obtain a distributed solution relies on the Benders decomposition. This technique is well known in transportation systems [16] when the problem structure allows to separate some variables from the others. The intuition behind this technique is to individuate the variables that prevent from splitting the original problem into a set of new small problems. Such variables are named complicating variables. In particular, with the Benders decomposition two new problems are defined: the subproblem and the master problem. The subproblem uses parameterized values of the complicating variables. The master problem can instead modify the complicating variables, but at each iteration it adds new constraints in order to take into account the solution obtained by the subproblem. The added constraints are called Benders cuts. A detailed description of the Benders method can be found in [15].

In our case, the complicating variables are the traffic demands x_m^{st}: intuitively, once they are fixed to constant values, the original problem can be split between the ISP and the CP. In particular, the ISP solves the following subproblem:

$$\textbf{B-GreenISP} \quad \min\,(P_{ISP}) \quad \text{s.t.:}$$

$$(27.9), (27.11) - (27.16) \tag{27.29}$$

$$d_a \quad \leq \quad D^{MAX} \tag{27.30}$$

$$\widetilde{x}_m^{st} \quad = \quad x_m^{st}(k) \quad \forall s, t \tag{27.31}$$

Control variables: $z_p^{st} \geq 0$, $q_p^{st} \geq 0$, $y_n \in \{0, 1\}$.

Eq.(27.30) bounds the average delay of users. Eq.(27.31) guarantees that estimated CP traffic is equal to real CP traffic. Notice that $x_m^{st}(k)$ are the parameterized traffic demands passed by the CP at iteration k. Let us introduce $\theta^{st}(k)$ as the dual variables associated with Eq.(27.31). Notice also

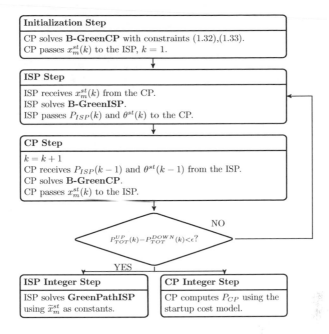

FIGURE 27.2 Benders Green algorithm (B-G).

that the possible problem infeasibility can be easily overcome by adding additional variables to the **B-GreenISP** problem using the reformulation reported in [15]. In brief, Eq.(27.12) is relaxed and additional variables are introduced to formulate an always-feasible subproblem.

The CP solves instead the following master problem:

$$\textbf{B-GreenCP} \quad \min\left(P_{CP} + \gamma\right) \quad \text{s.t.:}$$

$$(27.17) - (27.18), (27.20) - (27.21) \quad (27.32)$$

$$\gamma \geq \widetilde{P}_{ISP}^{min} \quad (27.33)$$

$$\sum_{s,t} \theta^{st}(\nu)\left[x_m^{st} - x_m^{st}(\nu)\right] \leq \gamma - P_{ISP}(\nu) \quad \nu = 1...k-1 \quad (27.34)$$

Control variables: $x_m^{st} \in \mathcal{R}^+$, $\gamma \in \mathcal{R}^+$, $y_s \in \{0,1\}$.

Two considerations hold for the CP master problem: (i) a new penalty variable γ is introduced to take into account the ISP power consumption, (ii) Eq.(27.34) are the Benders cuts that bound γ from below. Intuitively, γ is a lower bound on the ISP power consumption. $\widetilde{P}_{ISP}^{min}$ is an estimation of the ISP power consumption used to bound γ. Notice that when $\gamma = P_{ISP}$ the solution of the master problem is optimal.

Finally, an upper and lower bound on the total power consumption are

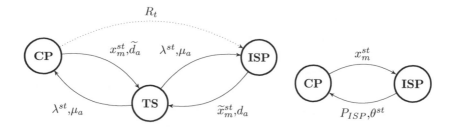

FIGURE 27.3 Exchanged parameters for the D-G algorithm ([13], © 2011 IEEE) (left) and B-G algorithm (right).

TABLE 27.1 Power Consumption Model [W] [11], © 2010 Association for Computing Machinery, Inc. Reprinted by permission.)

P_n^c	$P_l^d(f_l)$	$P_n^d\left(\sum_{l\in\mathcal{L}(n)}f_l\right)$	P_s^c	$P_s^d\left(\sum_{t\in T}x_{cp}^{st}\right)$
100	$20f_lA_l$	$20\sum_{l\in\mathcal{L}(n)}f_l$	200 ± 100	$(40\pm20)\sum_{t\in T}x_{cp}^{st}$

computed:

$$P_{TOT}^{UP}(k) = P_{CP}(k) + P_{ISP}(k) \qquad (27.35)$$
$$P_{TOT}^{DOWN}(k) = P_{CP}(k) + \gamma(k) \qquad (27.36)$$

When the difference between the upper and lower bound is below a given threshold ϵ, the algorithm ends and a near-optimal solution is returned.

Fig. 27.2 shows a schematic description of the proposed Benders green algorithm (B-G). The CP first solves the **B-GreenCP** problem, then the ISP uses the parameterized demand x_m^{st} to solve the **B-GreenISP** subproblem. Then the CP solves the **B-GreenCP** using P_{ISP} and θ^{st} passed from the ISP. The procedure is iterated until the power bounds are sufficiently close. Finally, both ISP and CP solve the integer step as the D-G algorithm. Fig. 27.3 (right) illustrates the exchanged parameters for the B-G algorithm. Interestingly, differently from the D-G algorithm, in this case all the parameters are directly exchanged between the ISP and the CP.

27.4   Performance Evaluation

We test the effectiveness of the model proposed using ISP backbone topologies obtained from RocketFuel [17]. The topologies are first pre-processed using a simple shortest path algorithm to obtain the set of paths, using their measured link weights as link costs. In particular, for each (s,t) we compute up to two completely disjoint paths. This reflects normal behavior of ISP which guarantees alternate paths for failure protection. Moreover, the capacity C_l

is set to 10 Gbps for each link, since the topologies considered are those of tier-1 ISPs. Links are utilized up to 50% of their capacity, i.e. $U_l^{MAX} = 0.5 \quad \forall l \in E$, to avoid congestion and to guarantee QoS. We assume that nodes are connected by optical links, in which the optical carrier is regenerated by amplifiers. For each link we randomly assign a number of amplifiers A_l uniformly distributed between 1 and 5.

Considering the CP, the maximum load on each server is set to the network capacity offered at that node, i.e. $W_s = \sum_{l \in \mathcal{L}(s)} C_l U_l^{MAX}$. We consider the case in which the CP infrastructure is composed of 15 servers, adopting different strategies for server placement over the ISP topology. The CP traffic demand R_t is modeled according to a Pareto distribution, with a variable lower bound R_t^{min} and a constant upper bound R_t^{MAX} given by the total capacity offered at that node, i.e. $R_t^{MAX} = \sum_{l \in \mathcal{L}(t)} C_l U_l^{MAX}$. Unless otherwise specified, $D^{MAX} = 300$ ms and $R_t^{min} = 100$ Mbps.

Tab.27.1 describes the model used to evaluate the power consumption. Here we are assuming next-generation devices able to adapt their power with traffic flow. Considering the ISP, the power consumption of nodes is composed of a constant term P_n^c due to the chassis static power plus an additional term P_n^d which scales linearly with traffic flow. The constant values are extracted by interpolating the power measurements of real devices under high load [18]. Moreover, the power consumption of a link P_l^d depends linearly on both the load and the number of amplifiers A_l between nodes, as reported in [19].

Focusing on CP, the server power consumption is also modeled by a static term P_s^c and a dynamic term P_s^d: in this case instead the slope is higher due to the presence of backup elements and power supplies, which actually double the server power consumption. Moreover, an additional random variation of 50% in the server power is introduced to model energy price fluctuation as reported in [8]. For the sake of simplicity we do not consider any additional background traffic of other CPs, since our goal is mainly to assess the maximum power savings achievable by the whole system composed of the ISP and the considered CP. Finally, 50% of randomly chosen nodes are selected as terminals t.

Unless otherwise specified, we assign the servers of the considered CP to nodes using a preferential degree placement. ISP nodes are first grouped according to the city in which they are located, and then the groups are sorted by decreasing number of links with other cities. Finally, the CP servers are assigned to the cities with the highest connection degree, one CP server per city. A random node inside each selected city is chosen as server location.

27.4.1 Centralized Model

We start analyzing the performance of the green centralized G model against the classic formulation C. We first evaluate how the degree of cooperation impacts the total power consumption, then we consider the variation of QoS constraints and finally we investigate the impact of CP servers' location.

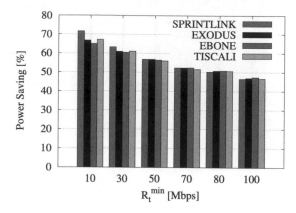

FIGURE 27.4 Average power saving against traffic variation for different topologies ([11], © 2010 Association for Computing Machinery, Inc. Reprinted by permission.).

27.4.1.1   Impact of Cooperation

We solve optimally both C and G problems, considering different scenarios. Fig.27.4 details the power savings against increasing R_t^{min}, for different topologies. Values are averaged over 10 runs, where for each run we chose a different set of terminal nodes. Astonishingly, G is able to save from 47% to more than 65% of power relative to C, for all the topologies considered, with a maximum power saving of 71% for the SprintLink topology. This is due to the fact that the green model explicitly takes into account the power consumption of devices, powering-on only the minimal amount of resources needed to satisfy the traffic demand. Moreover, the savings are decreasing as the traffic increases, suggesting that a larger number devices need to be powered on to meet the requirements.

To give more insight, we add a parameter α to weigh differently the CP and the ISP power, so that the objective function of G becomes $\alpha P_{CP} + (1 - \alpha)P_{ISP}$ and $\alpha \in [0,1]$. With $\alpha = 0$ just the minimization of power within the ISP infrastructure is pursued, while with $\alpha = 1$ only the optimization of the content provider is considered. Intuitively, a CP-only power optimization may result in choosing a more energy-efficient server but so much more distant that more power is consumed over the network. Conversely, an ISP-only power optimization may yield a server that is closer to the terminal but whose power cost is high. Thus, a joint power optimization should provide the right balance and yield higher power savings.

Fig.27.5 details the total power consumption as α varies, considering the different topologies. In this case the best savings can be obtained only when both the CP and ISP power are taken into account. For example, the total power of the SprintLink topology is more than 41kW with $\alpha = 1$, decreasing to 23kW with $\alpha = 0.5$. In this case the largest part of power consumption is due to ISP, so that the power due to CP does not increase significantly as α

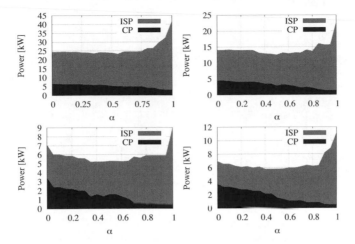

FIGURE 27.5 Power consumption components versus α. From left to right: **SPRINTLINK, EXODUS, EBONE and TISCALI topologies** ([11], © 2010 Association for Computing Machinery, Inc. Reprinted by permission.)

decreases further. Instead, if we consider for example the EBone topology, the total power is clearly minimized only when $\alpha = 0.5$. The intuition suggests that a great amount of power is wasted if the ISP and CP individually optimize their own power consumption, while less power is required if both of them jointly pursue the minimization of the total power of the entire system, i.e. $\alpha \approx 0.5$.

27.4.1.2 Impact of QoS Constraints

In order to assess the impact of QoS constraints, we consider the variation in the maximum delay D^{MAX}. Fig.27.6 reports the power consumption for both C and G for different D^{MAX} values, considering the SprintLink topology.[4] Values are averaged over 10 different runs, while error bars report minimum and maximum values. Interestingly, with C the power grows notably with decreasing D^{MAX}, while G consumes around the same amount of power even for small D^{MAX}. The intuition suggests that G can more wisely adapt to stricter QoS constraints, leading to higher power savings (around 52%) even for small D^{MAX}.

27.4.1.3 Impact of CP Location

In this setup we adopt a completely random node selection for placing the servers. In particular, Fig.27.7 reports a comparison of the power consumed under the random and the preferential assignment. Results are obtained for the SprintLink topology, using the G formulation. The minimum amount of traffic varies between 10 Mbps and 100 Mbps. The power saving is computed

---

[4]Here we have enforced the maximum admissible delay constraint also for the C model.

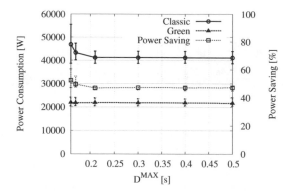

FIGURE 27.6 Maximum admissible delay for the SPRINTLINK topology ([11], © 2010 Association for Computing Machinery, Inc. Reprinted by permission.).

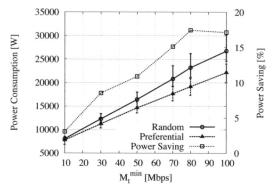

FIGURE 27.7 Preferential and random placement comparison.

as the relative difference between the two assignments. As expected, the system consumes a consistent amount of additional power if servers are placed randomly, and the power consumption is also characterized by a greater variability. In particular, considering $R_t^{min} = 100$ Mbps the solution with random placement consumes more than 26kW on average while the preferential assignment consumes just 22kW, corresponding to a power saving of more than 17%.

27.4.2   Distributed Algorithms

In this section we analyze the performance of the distributed algorithms. We first consider the performance of D-G and then we consider the B-G algorithm.

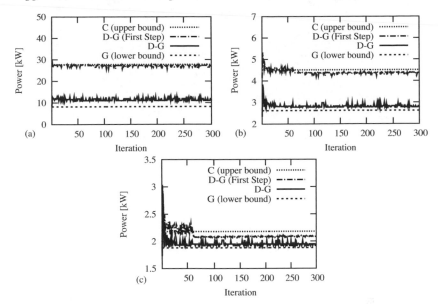

FIGURE 27.8 Power consumption of the C, G and D-G algorithms: (a) 100-200 power model, (b) 10-20 power model, (c) 1-2 power model ([13], © 2011 IEEE).

27.4.2.1 Comparison

We start by running the D-G algorithm over the SprintLink topology, since it is one of the largest topologies of RocketFuel in terms of nodes and links. Unless otherwise specified, we assume that the ISP knows exactly the total traffic of each client, i.e. $\widetilde{R}_t = R_t$.[5] Moreover, we use the power model presented in Tab.27.1. We name this power model as "100-200". Finally, we set a step size rule $\alpha_k = 1000/k$ for updating the Lagrange multipliers.

Fig. 27.8-(a) reports the power consumption variation of the D-G algorithm versus the number of iterations, considering the 100-200 model. Notice that here, differently from the original scheme of Fig. 27.1, we perform the ISP and CP Integer Step at each iteration to better assess the dynamic behavior of the algorithm. We report also a lower bound, i.e. the power consumption of the G algorithm. The figure reports also the upper bound, namely the classic (C) centralized solution. Finally, the figure reports the power consumption of the D-G algorithm at the end of the distributed step, before **GreenPathISP** is solved. Several considerations hold in this case: (i) the solution of D-G is always close to the lower bound even after few iterations, so that the maximum power loss is less than 17%,[6] (ii) the power consumption of the first step of

---

[5]$\widetilde{R}_t$ is measured or computed from previous estimations of the traffic demands.

[6]The power loss is computed as the difference between D-G and G in terms of power savings, i.e. $P_{loss} = \dfrac{P_{TOT}^{D-G} - P_{TOT}^{G}}{P_{TOT}^{C}}$.

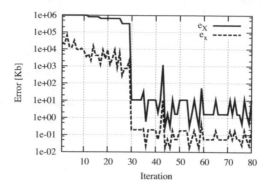

FIGURE 27.9 e_x and e_X for the SprintLink topology [13] © 2011 IEEE.

D-G is instead close to the upper bound, (iii) the optimization performed by **GreenPathISP** is essential to obtain large savings, since P_{TOT} drops from more than 26 kW in the first step to less than 13 kW at the end of the algorithm.

We then investigate how the startup cost impacts the total power consumption. Fig. 27.8-(b) shows the results for the 10-20 model, i.e. $P_n^c = 10$ W and $P_s^c = 20 \pm 10$ W. As expected, the power consumption of the G algorithm is lower in this case, and the bounds are closer too. Interestingly, the D-G is even closer to the lower bound, since the linear part of the power function that is optimized in the distributed solution step becomes predominant. These phenomena are even more evident with the 1-2 model (Fig. 27.8-(c)): in this case the upper and lower bound are even closer, suggesting that with small power step sizes the solution of the D-G algorithm approaches that of the G one.

We define the mean error of the traffic demands at each iteration:

$$e_x(k) = \frac{\sum_{s,t} |\widetilde{x}_m^{st}(k) - x_m^{st}(k)|}{|S \times T|} \tag{27.37}$$

In a similar way we define the maximum error at each iteration:

$$e_X(k) = \max_{s,t} |\widetilde{x}_m^{st}(k) - x_m^{st}(k)| \tag{27.38}$$

Notice that when the algorithm converges $e_X \approx 0$, i.e. $\widetilde{x}_m^{st} \approx x_m^{st} \quad \forall s, t$.

Fig. 27.9 reports both e_x and e_X at each iteration. Interestingly, e_x falls below 100 Kb after 30 iterations, while e_X is bounded below 100 Kb after 45 iterations. This means that only few iterations are sufficient to guarantee QoS for users, since the estimated demands $\widetilde{x}_m^{st}$ are close to the real ones x_m^{st}.

We then run the B-G algorithm over the SprintLink topology with the same set of parameters and the 100-200 model. In particular, we set $\widetilde{P}_{ISP}^{min} = P_{ISP}^G(1 - \Delta_P)$, where P_{ISP}^G is the optimal power consumption of the G algorithm and $\Delta_P \in (0, 1]$. Moreover, we run the ISP and CP Integer Step at each iteration. Fig.27.10 reports the power consumption of the B-G algorithm for

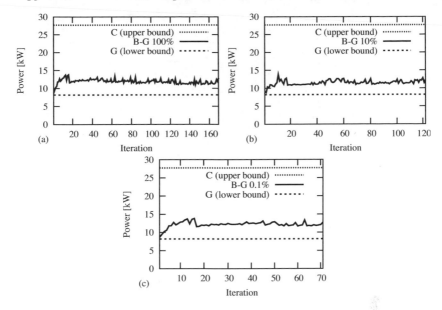

FIGURE 27.10 Power consumption of B-G: (a) $\Delta_P = 100\%$, (b) $\Delta_P = 10\%$, (c) $\Delta_P = 0.1\%$.

different values of Δ_P. The figure reports also the bounds as the D-G case. Also in this case, the distributed solution is quite close to the lower bound, with a maximum power efficiency loss of 18%, for all the cases.

27.4.2.2 Sensitivity Analysis

We then consider the impact of parameters on the performance of the D-G algorithm. In particular, we start considering the case in which the power consumption is strictly proportional with the current load. This case can be representative of future energy-aware devices, able to completely adapt the power consumption to the current load [2]. Fig. 27.11 (left) reports the precision error $e_P(k_{Best})$ for $k \in [1, 300]$, considering different diminishing step size rules for α_k. We set $k_{max} = 300$ to limit the convergence time. Small step sizes lead to very slow convergence, since the Lagrange multipliers change very slowly. For example, with $\alpha_k = 10/k$ the error is always higher than 9%, meaning that the distributed solution is quite far from the centralized one. However, also large values tend to be inaccurate since large oscillations are induced. By choosing instead the intermediate value of $1000/k$, the D-G algorithm converges to the optimal solution in less than 50 iterations with a precision of less than 0.0001%.

Fig. 27.11 (center and right) show the power consumption of the ISP and CP, respectively. Interestingly, all the step sizes are able to reach at least a near-optimal power consumption for the ISP, $1000/k$ and $1000/\sqrt{k}$ the noisiest ones due to the large steps used. If we consider instead the CP power

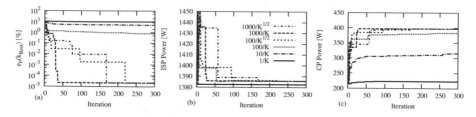

FIGURE 27.11 Impact of different decreasing α_k considering the SprintLink topology: (a) $e_P(k_{Best})$, (b) ISP power consumption, (c) CP power consumption ([13], © 2011 IEEE).

consumption, then only when α is greater than $100/\sqrt{k}$, the CP converges to the optimal power allocation, while all the other values are quite far from the optimal solution.

To better assess the computational time of D-G, we compute the CPU time required to solve the problem at each iteration. In particular, since the ISP Step and the CP Step can be processed in parallel, we take the maximum between the CPU times: $ctime(k) = \max(ctime_{ISP}(k), ctime_{CP}(k))$. We then compute the total cost of running the algorithm at iteration k as $cctime(k) = cctime(k-1) + ctime(k)$, where $cctime(1) = ctime(1)$. We assume that the CPU times required to perform the Initialization and the Update steps of D-G are negligible.

Fig. 27.12(left) reports $ctime(k)$ and $cctime(k)$ for the SprintLink topology and $1000/k$ step size rule. All the times have been measured by running D-G on the NEOS server [20] to obtain a reliable measurement on a widely known system. Interestingly, $ctime(k)$ is nearly constant, so that 30 iterations require 52 minutes to be completed. Clearly, a tradeoff emerges between solution precision and admissible computational time.

Finally, we introduce a precision error for $\widetilde{R}_t$, so that $\widetilde{R}_t = R_t(1 + \Delta_R)$. This reflects the case in which R_t is over-estimated by the ISP, for example by measurements.[7] Fig. 27.12(right) reports $e_P(k_{Best})$ for different Δ_R. For $\Delta_R = 0\%$ and $\Delta_R = 1\%$ the algorithm converges to the optimal power consumption and $e_P(k_{Best})$ falls below 0.01% in less than 90 iterations in both cases. For $\Delta_R = 10\%$ instead the D-G algorithm requires 0.46% of additional power than the G algorithm, which rises to more than 16% with $\Delta_R = 50\%$ even after 300 iterations. Therefore, the solution produced by the D-G algorithm involves a small amount of additional power only when the precision error in the estimated demand is reasonably small.

We now consider the impact of parameters on the B-G algorithm, assuming first that power is strictly proportional to load. We first evaluate $e_P(k_{Best})$ for different values of $\widetilde{P}_{ISP}^{min}$ considering the SprintLink topology. In particular, we set $\widetilde{P}_{ISP}^{min} = P_{ISP}^G(1 - \Delta_P)$, where P_{ISP}^G is the optimal power consumption of

---

[7]We do not consider the under-estimated case since it introduces packet-loss and consequently QoS violation for users.

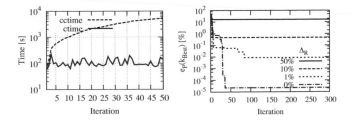

FIGURE 27.12 (left) $ctime(k)$ and $cctime(k)$ of D-G considering the SprintLink topology and $1000/k$ step size rule. (right) $e_P(k_{Best})$ for different precision values Δ_R [13] © 2011 IEEE.

the G algorithm and $\Delta_P \in (0, 1]$. Unless otherwise specified, we set $\epsilon = 0.01$ as stopping criterion. Let us stress that $\widetilde{P}_{ISP}^{min}$ acts as lower bound for the variable γ. Fig. 27.13-(a) reports the optimality gap $e_P(k_{Best})$ for different values of Δ_P. $\Delta_P = 100\%$ corresponds to the case when the CP can not estimate the power consumption of the ISP, hence $\widetilde{P}_{ISP}^{min} = 0$. This yields a slow convergence time, since the B-G algorithm takes more than 160 iterations to optimally converge. On the contrary, when $\widetilde{P}_{ISP}^{min}$ is close to P_{ISP}^G, i.e. $\Delta_P = 0.1\%$, the B-G algorithm converges in less than 80 iterations with an error below 0.001%. Finally, when $\widetilde{P}_{ISP}^{min}$ is known with a moderate error, the algorithm converges with a number of iterations between the extremes.

To give more insight, Fig. 27.13-(b) reports the upper and lower bounds on the total power for $\Delta_P = 10\%$. As reported by [15], the lower bound is an increasing function. In this case the gap between the bounds is lower than 30W after 20 iterations, reflecting the fact that the upper bound is already close to the optimal solution. In fact, differently from the D-G algorithm, the performance of the B-G algorithm in terms of distance from the optimal solution can be easily evaluated from the gap between the algorithm bounds.

Finally, we consider the computational times of the B-G algorithm. Fig. 27.13-(c) reports $ctime(k)$ and $cctime(k)$ for $\Delta_P = 10\%$. Notice that $ctime(k) = ctime_{ISP}(k) + ctime_{CP}(k)$ since the **B-GreenISP** and the **B-GreenCP** problems are solved sequentially. The CPU time is measured with NEOS as for the D-G algorithm. Interestingly, in this case $ctime(k)$ is linearly increasing with time, since the **B-GreenCP** problem adds a new Benders cut constraint at each iteration. Nevertheless, $ctime(k)$ is one order of magnitude less than the D-G algorithm up to 40 iterations, yet it constantly increases as the number of iterations grows.

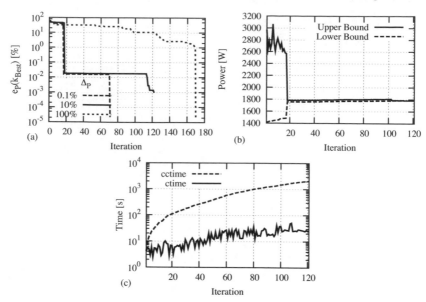

FIGURE 27.13 B-G algorithm results: (a) Optimality gap of the best solution with different values for Δ_P. (b) Upper and lower bound for $\Delta_P = 10\%$. (c) Computational time for $\Delta_P = 10\%$.

27.5   Discussion

In light of the obtained results, in this section we discuss the characteristics of the presented algorithms. In particular, we consider the following aspects for the centralized algorithm and the distributed ones.

Power Saving: The centralized G algorithm finds the optimal solution that minimizes the power consumption. All the distributed solutions are suboptimal, but the gap is limited thanks to the improvements introduced. However, when energy-proportional devices without initial step of power are considered, the solution from the distributed algorithm is optimal.

Shared Information: The green centralized algorithm G requires the maximum amount of shared information, including: network topology, servers' load, link flow, traffic demand, source-destination traffic, power consumption of routers, links and servers. Considering the Benders-based algorithm, the ISP and CP share a moderate amount of information, including: source-destination traffic and total power consumption of ISP. The dual-based algorithm instead requires the smallest amount of shared information, requiring only the exchange of the Lagrange multipliers and optionally the traffic demands.

Computational Time: The computational time of each iteration is constant for the dual-based algorithm. On the contrary, the computational time of the Benders-based algorithm linearly increases at each iteration. Moreover, the ISP and CP can parallelize the distributed step of the dual-based algorithm. The Benders approach instead is completely sequential.

Tuning: Regarding how the parameters are set and tuned for the distributed algorithms, the Benders-based algorithm is easy to tune: the only parameter to set is $\widetilde{P}_{ISP}^{min}$, that can be estimated for example by measurements. Then, the algorithm precision is simply controlled by the threshold ϵ. Instead the dual-based algorithm is harder to tune, since the chosen step size α_k greatly influences the precision error.

Termination Rule: The centralized algorithms terminate when the problem is solved. In the dual-based approach instead the TS communicates the termination to the ISP and the CP. Finally, for the Benders decomposition case the termination rule is governed by the CP.

27.6 Related Work

Several approaches have been proposed in order to reduce the power consumption of ISPs and CPs. In [7] the authors formulate the problem of minimizing the total power consumption of ISPs, grounding the need of adopting energy-saving approaches with real measurements. In particular, they solve the problem optimally for small networks, showing that great savings can be achieved while guaranteeing QoS for users. In [21] the authors solve the problem also for large networks, showing that even simple energy-efficient heuristics can save a considerable amount of ISP power. More recently, in [22] and [23] the authors propose both approximate and optimal algorithms to minimize the ISP power consumption, considering also the case in which the power of the device is proportional to its current load. Nevertheless, all these approaches assume a fixed traffic matrix, so that the optimization is performed only by the ISP. In our work instead, we act also on the traffic matrix to find the optimal configuration in terms of power consumption, so that large savings are realized for both the ISP and the CP.

At the same time, studies have considered how to reduce globally the power consumption of large data centers and CPs. For example, new recent approaches like [8] and [24] aim at reducing the power consumption of big Content Providers (CP) considering the variation in electricity prices at different locations. In [25] the authors consider also renewable electricity to reduce energy consumption produced by carbon-intensive sources. Nevertheless, all of these previous studies focus only on the CP, thus completely ignoring the impact on the ISP in terms of power consumption. Considering jointly ISP and CP power consumption seems to be a viable solution to globally reduce power consumption.

In [9] the authors solve jointly the traffic engineering and content distribution problem, showing that great improvements in QoS can be obtained if

CP and ISP pursue the joint minimization of users' delay. However, as we have shown in this chapter, such approach can waste a considerable amount of power. Pursuing energy-efficiency while guaranteeing QoS for users seems essential for both ISPs and CPs.

Finally, in [26] the authors show that caching content in "content" routers saves energy by reducing the number of hops traversed when content is served from the cache to the client. A similar approach has been previously investigated by [27] in the context of multi-hop wireless networks, proving a substantial energy reduction when lost packets are retransmitted from an intermediate caching node rather than from the original source. Our model is different from these works since it formulates a joint energy optimization problem wherein the CSP and the ISP are separate entities that exchange minimal information to solve the corresponding distributed optimization algorithm.

27.7   Conclusions

In this chapter we have proposed a new approach in which both CP and ISP cooperate to reduce the overall power consumption. We have first proposed a completely centralized model that minimizes the overall power consumption. Then, we have proposed two distributed algorithms that limit the amount of shared information between ISP and CP, such as the network topology or the servers' load.

Simulations performed on realistic topologies show that large power savings are achievable, up to 71% relative to a classic formulation of the joint problem minimizing the users' delay. Moreover, we have shown that a preferential server placement saves an additional 15% over the overall power consumption with respect to a random one. Finally, our results indicate that the distributed solutions are near-optimal for all considered scenarios, with a maximum power efficiency loss of 17%.

As future work, we will consider energy optimization models where content is cached/replicated inside the network as in [26, 27] and so content and network management are integrated.

Acknowledgment

The research leading to these results has received funding from the European Union Seventh Framework Programme (FP7/2007-2013) under grant agreement n. 257740 (Network of Excellence "TREND". The work of I. Matta was also in part supported by NSF grant CNS-0963974.

Bibliography

[1] M. Webb, "Smart 2020: Enabling the low carbon economy in the information age," *The Climate Group London*, 2008.

[2] R. Bolla, R. Bruschi, K. Christensen, F. Cucchietti, F. Davoli, and S. Singh, "The potential impact of green technologies in next-generation wireline networks–is there room for energy saving optimization?" *IEEE Communication Magazine (COMMAG)*, 2011.

[3] F. Idzikowski, S. Orlowski, C. Raack, H. Woesner, and A. Wolisz, "Saving energy in ip-over-wdm networks by switching off line cards in low-demand scenarios," in *Optical Network Design and Modeling (ONDM), 2010 14th Conference on.* IEEE, 2010, pp. 1–6.

[4] D. Meisner, B. Gold, and T. Wenisch, "Powernap: eliminating server idle power," *ACM SIGPLAN Notices*, vol. 44, no. 3, pp. 205–216, 2009.

[5] L. Barroso and U. Holzle, "The case for energy-proportional computing," *Computer*, vol. 40, no. 12, pp. 33–37, 2007.

[6] S. S. M. Gupta, "Greening of the internet," in *Proc. ACM SIGCOMM*, Karlsruhe, Germany, Aug. 2003.

[7] J. Chabarek, J. Sommers, P. Barford, C. Estan, D. Tsiang, and S. Wright, "Power awareness in network design and routing," in *Proc. IEEE INFO-COM 2008*, Phoenix, USA, April 2008.

[8] A. Qureshi, R. Weber, H. Balakrishnan, J. Guttag, and B. Maggs, "Cutting the electric bill for internet-scale systems," in *Proc. the ACM SIG-COMM 2009 conference on Data communication.* ACM, 2009, pp. 123–134.

[9] W. Jiang, R. Zhang-Shen, J. Rexford, and M. Chiang, "Cooperative content distribution and traffic engineering in an isp network," in *Proc. the eleventh international joint conference on Measurement and modeling of computer systems.* ACM, 2009, pp. 239–250.

[10] M. Pioro and D. Medhi, "Routing, flow, and capacity design in communication and computer networks," 2004.

[11] L. Chiaraviglio and I. Matta, "Greencoop: cooperative green routing with energy-efficient servers," in *Proceedings of the 1st International Conference on Energy-Efficient Computing and Networking.* ACM, 2010, pp. 191–194.

[12] S. Boyd and L. Vandenberghe, *Convex optimization.* Cambridge Univ Pr, 2004.

[13] L. Chiaraviglio and I. Matta, "An energy-aware distributed approach for content and network management," in *IEEE INFOCOM 2011 Workshop on Green Communications and Networking*, Shanghai, China, April 2011.

[14] L. Xiao, M. Johansson, and S. Boyd, "Simultaneous routing and resource allocation via dual decomposition," *IEEE Trans Communications*, vol. 52, no. 7, pp. 1136–1144, 2004.

[15] A. Conejo, E. Castillo, R. Minguez, and R. Garcia-Bertrand, *Decomposition techniques in mathematical programming: engineering and science applications.* Springer, 2006.

[16] J. Cordeau, G. Stojkovic, F. Soumis, and J. Desrosiers, "Benders decomposition for simultaneous aircraft routing and crew scheduling," *Transportation science*, vol. 35, no. 4, p. 375, 2001.

[17] N. Spring, R. Mahajan, and D. Wetherall, "Measuring isp topologies with rocketfuel," in *ACM SIGCOMM Computer Communication Review*, vol. 32, no. 4.   ACM, 2002, pp. 133–145.

[18] E. Bonetto, L. Chiaraviglio, D. Cuda, G. Gavilanes Castillo, and F. Neri, "Optical technologies can improve the energy efficiency of networks," in *Proc. 35th European Conference on Optical Communication, 2009. ECOC'09.*   IEEE, 2009, pp. 1–4.

[19] L. Chiaraviglio, M. Mellia, and F. Neri, "Energy-aware backbone networks: a case study," in *Proc. the First Int. Workshop on Green Communications (GreenComm 2009)*, Dresden, Germany, June 2009.

[20] J. Czyzyk, M. Mesnier, and J. Moré, "The neos server," *Computational Science & Engineering, IEEE*, vol. 5, no. 3, pp. 68–75, 1998.

[21] L. Chiaraviglio, M. Mellia, and F. Neri, "Reducing power consumption in backbone networks," in *Proc. IEEE ICC'09*, Dresden, Germany, June 2009.

[22] M. Andrews, A. Anta, L. Zhang, and W. Zhao, "Routing for energy minimization in the speed scaling model," in *Proc. IEEE INFOCOM, 2010.*   IEEE, 2010, pp. 1–9.

[23] ——, "Routing and scheduling for energy and delay minimization in the powerdown model," in *INFOCOM, 2010 Proceedings IEEE*.   IEEE, 2010, pp. 1–5.

[24] L. Rao, X. Liu, and W. Liu, "Minimizing electricity cost: Optimization of distributed internet data centers in a multi-electricity-market environment," in *Proc. IEEE INFOCOM, 2010.*   IEEE, 2010, pp. 1–9.

[25] K. Le, R. Bianchini, M. Martonosi, and T. Nguyen, "Cost-and energy-aware load distribution across data centers," *Proceedings of HotPower*, 2009.

[26] U. Lee, I. Rimac, and V. Hilt, "Greening the internet with content-centric networking," in *Proc. the 1st International Conference on Energy-Efficient Computing and Networking*.   ACM, 2010, pp. 179–182.

[27] N. Riga, I. Matta, A. Medina, C. Partridge, and J. Redi, "An energy-conscious transport protocol for multi-hop wireless networks," in *Proc. the 2007 ACM CoNEXT conference*.   ACM, 2007, p. 22.

Author Contact Information

Luca Chiaraviglio is with Electronics Department, Politecnico di Torino, Corso Duca degli Abruzzi 24, Torino, Italy, Email: luca.chiaraviglio@gmail.com. Ibrahim Matta is with Computer Science Department, Boston University, 111 Cummington Street, Boston, USA, Email: matta@bu.edu.

28

Energy Saving Strategies in Fixed Access Networks

Björn Skubic, Alexander Lindström, Einar In de Betou, Ioanna Pappa

Ericsson Research, Ericsson AB, Sweden

CONTENTS

This chapter provides an overview of energy saving strategies in fixed access networks. Different energy saving strategies are analyzed, including technology aspects, architectural aspects and the energy saving potential of dynamic power management. A comparison of energy consumption for state-of-the-art fixed access architectures (based on Ethernet PtP, GPON, XG-PON, ADSL2+, VDSL2, etc) is provided. As a basis for the comparison, models for power consumption of the different technologies are presented.

28.1 Introduction

Energy efficiency has become an increasingly important aspect of network design, both due to the increasing operational costs related to energy consumption as well as due to the increasing awareness of global warming and climate change. This has spurred an intense research effort within the ICT sector on strategies for reducing power consumption. With continued focus on climate change and increasing volatility in energy costs, energy efficiency will likely be an increasingly important aspect in future network design.

In many ways the information and communications technologies (ICT) sector can be seen as part of the solution for reducing green house gas emissions in other sectors of society. New services enabled by emerging communication technologies will offer carbon lean substitutes to traditional services [1]. However, with continued growth of the ICT sector and the networked society, energy efficiency considerations are becoming increasingly important in order to cope with the expected traffic growth. Communication networks are experiencing a growing energy consumption, with network traffic growing quicker than energy efficiency improvements provided by evolution of technology. Efficiency improvements in technology, such as energy efficiency improvements in CMOS, are unable to compensate for growing network capacity [2]. In the future, networks may in fact be constricted in evolution by power consumption rather than capacity. As a result, new energy saving strategies and concepts will be important for the evolution of future communication networks.

In today's networks, energy consumption is dominated by access [2], due to the shear amount of distributed network elements (Fig. 28.1). The fixed access is here defined as the wired "last mile". As we move further out in the network towards the customer, power consumption of individual network elements tend to decrease while the number of elements increases rapidly. There is also a tendency for an increasing proportion of power consumption to be associated with interfaces and transmission rather than switching and routing. As a function of capacity, the power associated with interfaces and transmission is more flat compared to switching and routing. For this reason, with continued traffic growth and increased capacity requirements, power consumption in the metro and backbone is expected to gain significance.

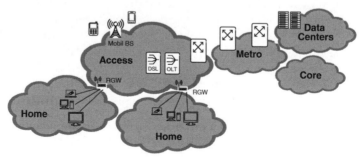

FIGURE 28.1 Network overview

Considering the access part of the network, which is particularly important in terms of energy consumption, there are some basic strategies that can be employed for reducing network power. Strategies relevant for the access are shown in Table 28.2. Beyond general technology evolution, strategies include increasing the sharing of network resources and increasing the load adaptive behavior of the network. Ultimately, new technologies and system concepts could also provide a basis for more energy efficient communications. Increased resource sharing can be achieved on a component level through e.g. more integrated components, on an equipment level through increased sharing of internal equipment elements or on a network level through increased sharing of network nodes. Similar considerations apply for load adaptive behavior which can be achieved on a component level, equipment level or network level. Hence, the strategies presented in Fig. 28.2 should be seen as general. The strategies themselves are not necessarily independent nor non-contradictory. As an example there could be a conflict between increased network sharing with more integrated equipment design, and load adaptive behavior with more modular design. This presents a trade-off that needs to be understood. Different energy saving concepts will be discussed in the remaining of this chapter. In section 28.2 we discuss different system concepts for the access and compare power consumption from a technology point of view. In section 28.3 we consider architectural aspects of power consumption and in section 28.4 we look at dynamic power management for achieving more load adaptive power consumption behavior.

Energy efficient components / subsystems	• General technology evolution
Increased sharing of network resources	• Integrated components • Integrated equipment design • Node-consolidation
Load adaptive power consumption	• Components with load adaptive power • Node and link power management functionality (sleep, doze, etc.) • Network-wide power management (traffic steering, energy-awareness)
New system concepts	• New technologies, modulation formats, etc.

FIGURE 28.2 Strategies for reducing energy consumption in the fixed access.

28.2 Technology

One of the fundamental determinants of power consumption in the access is the technology choice. Different technologies offer different power-per-line potential that can be directly compared. Concerning choice of technology there are however other aspects that need to be considered beyond power, where different technologies offer different merits. One important factor for choice of technology is the existing infrastructure and the extent to which existing infrastructure can be reused with a certain technology. Migration aspects not only constricts the evolution of the network from a cost point of view, but also from a sustainability perspective. Re-use of infrastructure is an important aspect, as the deployment phase typically presents a non-negligible contribution to the carbon footprint of the complete life-cycle of a system [3].

In terms of comparing power for different technologies, there is a span in power consumption depending on system design choices. Despite this variation it is still possible to provide representative models for the different technologies that can be used for technology comparisons. Models for power consumption are preferably split into system common parts which are modeled in a similar way for all systems and system specific parts containing the technology dependent contributions which are modeled individually for each system. When modeling common contributions we can differentiate between different types of equipment such as: the customer premisses equipment (CPE), shelfs/racks placed at the central offices (CO) and monolithic nodes (e.g. pizza boxes) typically placed at remote nodes (e.g. cabinets, buildings). These different equipment types all provide different static contributions to the technology dependent per port power consumption. Data presented in this chapter is to large extent a consolidation of data from various sources such as [4–6]. Data should be taken as for state-of-the-art equipment (year 2012) in the active state. It should be noted that different technologies have reached different degrees of maturity which may affect the future potential of further reducing power.

28.2.1   Equipment Models

Table 28.1 presents model power parameters for basic CPE components, common shelf/rack components as well as common parts of a monolithic node. Note that there is a significant difference in power-per-port depending on port density of the equipment. The model shelf includes backplane, power supply (redundant), management, and L2 switching. Here we assume a shelf capacity of 16 slots of which typically 2 slots would be reserved for uplink. The required switching capacity in a shelf is modeled separately and is based on an assumed guaranteed sustainable data rate per customer which varies from case to case. Each slot card has a baseline power in addition to the system specific contributions. To the CPE parts, which are modeled in greater detail, one should also add an additional loss (e.g. 20 %) associated with DC-DC conversion.

28.2.2   System Specific Contributions

For system specific parts (Table 28.2 and Table 28.3) we identify the main elements and associated power consumption. These are here categorized into

TABLE 28.1
Baseline power parameters for the CPE, shelf and monolithic node.

Component/subsystem	Power [W]
CPE SoC baseline ($\sim$100Mb/s)	0.40
CPE SoC baseline (1Gb/s)	0.80
CPE External memory	0.12
CPE Misc	0.30
CPE Additional residential gateway features	0.50
CPE GE port	0.80
CPE FE port	0.30
CPE FXS	1.10
Shelf baseline (16 slots, 2 reserved for uplink)	100
Line card baseline	5.00
Monolithic node baseline ($\leq$ 12 ports)	5.00
Monolithic node baseline ($>$ 12 ports)	10.00
Shelf/monolithic node L2 switching capacity (per Gb/s)	1.00

the TRx, TRx electronics and processing. For a DSL port the basic elements are the line driver (TRx), analogue front-end (TRx electronics) and the digital signal processing/network processer (processing). An optical port typically consists of a laser diode, line driver, photo diode, limiting amplifier, SERDES and CDR. Different configurations are possible with implications on power consumption, where for example the SERDES is either included in the system on chip (SoC) or as an independent component. For each port type reasonable assumptions have been made where power is split into the categories TRx, TRx electronics and processing contributions. For the shelf and monolithic node model an assumption on number of ports per slot card or element is also required for the different technologies. We assume the following number of ports per slot card: Ethernet PtP (16), GPON/EPON (8), 10G PON (4), 10/10 GE-PON (2), and DSL (32). Based on the presented parameters various baseline configurations can be established which are compared in the subsequent section. The per-line power consumption in the central office is obtained by calculating the power for a fully occupied shelf and dividing by the number of associated downlink lines. For PONs we assume a splitting ratio of 1:32 for number of lines per PON port. The shelf power consumption is obtained by summing the shelf baseline power, power associated with switching, slot card baseline contributions, downlink port contributions and uplink port contributions. The switching power is calculated based on an estimate of the total required aggregate downlink sustainable data rate for the shelf. The CPE power consumption is calculated by summing appropriate baseline contributions with the power of the CPE uplink port.

TABLE 28.2
Parameters for port specific power consumption of uplink ports (mainly CPE).

Component/subsystem	Power [W]			
	TRx	TRx electr.	Processing	Total
ADSL2+	0.30	0.30	0.40	1.00
VDSL2(8, 12a, 17a)	0.50	0.50	0.80	1.80
VDSL2(30a)	0.70	0.70	1.00	2.40
PtP electrical (100M)	0.20	-	0.10	0.30
PtP electrical (1G)	0.50	-	0.30	0.80
PtP optical (100M)	0.40	0.25	0.20	0.85
PtP optical (1G)	0.70	0.40	0.50	1.60
GPON	1.20	0.50	0.60	2.30
XG-PON	1.80	0.60	0.80	3.20
EPON	1.00	0.40	0.50	1.90
GE-PON (10G/1G)	1.80	0.60	0.70	3.10
GE-PON (10G/10G)	2.50	0.90	1.00	4.40

TABLE 28.3
Parameters for port specific power consumption of downlink ports (mainly shelf and monolithic node).

Component/subsystem	Power [W]			
	TRx	TRx electr.	Processing	Total
ADSL2+	0.30	0.30	0.05	0.65
VDSL2(8, 12a, 17a)	0.50	0.50	0.05	1.05
VDSL2(30a)	0.60	0.60	0.05	1.25
PtP electrical (100M)	0.20	-	0.10	0.30
PtP electrical (1G)	0.50	-	0.30	0.75
PtP optical (100M)	0.40	0.25	0.20	0.85
PtP optical (1G)	0.70	0.40	0.50	1.60
GPON	2.00	0.60	2.50	5.10
XG-PON	3.00	0.90	4.20	8.10
EPON	2.00	0.60	2.20	4.80
GE-PON (10G/1G)	3.00	0.90	4.50	8.40
GE-PON (10G/10G)	3.00	0.90	5.60	9.50

28.2.3   Per-Line Comparison

Based on the presented models, power-per-line comparisons for different systems and architectures can be made. For fixed access the most interesting metric for comparing energy efficiency of different technologies/solutions is power-per-line, where power is compared for different system alternatives, grouped into categories with equivalent service capabilities. The measure bits-per-Joule, obtained by dividing the bit-rate (bits/s) by power (W), is more troublesome in the access as it depends on the bit-rate definition (peak-rate, sustainable data rate, etc.) and the degree of over subscription in the access. In addition, since the most interesting comparisons concern solutions that provide similar services, a power-per-line comparison is adequate.

A comparison of conventional technologies is presented in Fig. 28.3. A large part of the CPE power is associated with baseline contributions which limits the relative importance of the choice of link technology. Considering only the central office site we find however that there are significant differences in power between the different technologies. Note that the comparison here is purely power-per-line for the last mile system (assuming maximum system utilization). The comparison does not consider the different reach capabilities of the different systems which may lead to further differentiation in power consumption as discussed in section 28.3. Advantages of fiber are lower power-per-line in the central office, longer reach and higher bit-rates. The main advantage of copper is the re-use of existing infrastructure. Power saving modes have not been considered in the power-per-line comparisons and may affect the different systems differently. In particular we can expect power saving modes to have larger effect on central office equipment for PtP and DSL compared to PON. This is due to the lower average utilization (and larger opportunity to sleep) of the single client ports in PtP and DSL compared to the shared ports in PON.

Beyond today's systems there is a range of system candidates that can cater for the increasing bandwidth demands in the access. There is currently no consensus concerning next-generation technology for fixed access. Several different system concepts have been proposed for next-generation access such as e.g. wave-length division multiplexing (WDM)-PON, stacked time division multiplexing (TDM)-PON, hybrid WDM/TDM-PON, orthogonal frequency division multiplexing (OFDM)-PON and coherent ultra-dense (UD)-WDM-PON. These different technologies all present different trade-offs in terms of cost, power and performance that need to be understood. An important consideration is that infrastructure deployed for today's systems to large extent needs to be re-used in these next-generation systems. Hence, architectures that are deployed today will constrict systems which are relevant for the future. Detailed models of next generation systems can be found in Ref. [7]. Results show that power-per-line is similar for several of the considered systems but with slight differences in the distribution between the customer, remote node and central office equipment. In a pure power-per-line comparison of the different

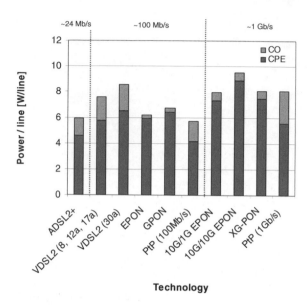

FIGURE 28.3 Power comparison of different technologies. The different technologies have been grouped into three basic service categories.

systems, WDM-PON and PtP systems offer slightly lower power consumption compared to alternatives. However, there are other considerations than just pure system power-per-line that must be considered as will be discussed in the next section.

28.3   Architecture

It is the power-per-line of the last mile system which ultimately terminates at the customer premises which presents the dominating contribution to power consumption in the access network. Beyond the power-per-line consumption of the last mile system, various architectural aspects can affect the total network power-per-line consumption. One avenue for decreasing network power is to increase the sharing of network resources. The following aspects can be exploited in the access:

- Lower power per port in network elements with larger port counts

- Higher degree of utilization of network elements in larger nodes

- Fewer aggregation points and links by consolidating nodes

In order to understand the potentials we first discuss ongoing trends in the access. Near-term evolution of the fixed access is driven by increasing bandwidth demands, resulting in deeper fiber penetration (Fig. 28.4) as customers are migrating from traditional copper technologies (e.g. ADSL2+) to higher speed technologies. This trend is a result of the decreasing reach for higher bit-rate copper technologies pushing active DSL network equipment closer towards the customers. Eventually, increasing bandwidth demands will result in a fiber-to-the-home (FTTH) scenario. This trend is illustrated in Fig. 28.4.

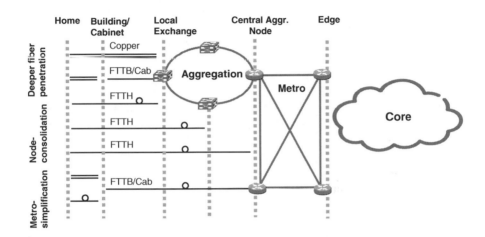

FIGURE 28.4 Fixed access architectures, illustrating the evolution towards deeper fiber penetration, node-consolidation and metro-simplification.

Beyond increasing fiber penetration, several operators are considering node-consolidation, where local exchange offices in traditional service areas are consolidated to central aggregation nodes covering larger service areas, as a means of reducing total cost of ownership (TCO). This vision is driven by a desire to reduce operational costs associated with having a plenitude of local exchange offices. Node-consolidation requires long reach fixed access systems. In order to achieve a reduction in central office sites by a factor 10, systems providing reach up to 60-90 km could be required [8]. Conventional technologies have certain drawbacks. Long reach TDM-PON systems require active amplifiers (i.e. reach extenders) in the field, while point-to-point (PtP) systems lead to extreme fiber and port counts that need to be catered for in the ducts and at operator sites.

Node-consolidation also potentially offers power saving opportunities. The most frequently discussed variant of node-consolidation involves increasing the passive reach of the last mile system in order to reduce the number of operator sites and increase the number customers per site. Alternatively, node consolidation could also be beneficial higher up in the network structure by consolidating network nodes and simplifying the metro network. In the first case, large gain is seen already for small consolidation factors. One reason is the fact that power-per-port is reduced as port counts go from tens to hundreds. This is also visible in the models in section 28.2, where we see a lower power-per-line for a port in a shelf configuration compared to a monolithic node. Another reason for the large gain already at small consolidation factors is the reduced overbuilding of the network. This is an effect of the granularity offered by system equipment and the degree of over dimensioning of the network in order to cater for future customers. Higher utilization factors of equipment can be achieved by centralizing equipment at fewer sites. As consolidation reaches beyond a certain factor the continued gain in network power-per-line occurs at a reduced rate. Links and aggregation points are reduced but savings are split over a large number of clients. In conclusion the main gain of node-consolidation in terms of power is already seen for quite small degrees of consolidation.

An alternative to conventional node-consolidation is metro-simplification (Fig. 28.4), which is based on the same principle as node-consolidation, but applied higher up in the network. Metro-simplification allows for power savings primarily through fewer aggregation points and links. WDM-PON systems with long reach and high sustainable bandwidths is one enabling technology for metro-simplification.

One aspect that must be considered when discussing node-consolidation is the increasing power typically associated with long reach systems. If a technology change is required to support larger service areas, there is a step increase in power-per-line as the consolidation factor is increased. Alternatively, amplifiers or more powerful optics could be used only for those particular lines where long reach is required. In any case, optimal degree of node-consolidation is the result of a trade-off between energy savings (of node consolidation) and increased power of long reach systems due to alternative technology, amplifiers or other types of reach extenders. The actual savings due to node-consolidation based on long reach systems still need to be quantified.

28.4   Dynamic Power Management

One of the most promising avenues for reducing power in the access is through load adaptive techniques. This is due to the combination of large contribution to network power and low average network utilization in the access compared to other parts of the network. Large power savings can in principle be achieved by reducing power consumption at low load or simply powering off elements that are not in use. Beyond the inherent load adaptive behavior in network elements (which typically is small), dynamic power management can be used to actively increase the load adaptive behavior (Fig. 28.5). There are two main avenues for achieving this, i.e. sleep mode and rate adaptation (Fig. 28.6). Which avenue is most effective depends on the network element, subsystem or component. For optical transceivers there is little gain in power by reducing the rate of operation and sleep mode presents a more promising avenue. For integrated circuits, power consumption scales with the square of the applied voltage and dynamic voltage scaling is effectively used to conserve energy. Inherent to dynamic power management is the trade-off of energy for performance. Ideally the power management is tuned to provide just the required performance.

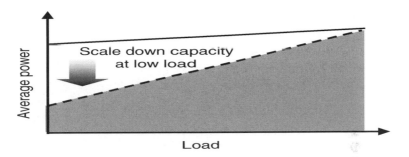

FIGURE 28.5 The inherent load adaptive behaviour of power for network nodes, subsystems and components is typically small but the effect can be increased by means of different techniques.

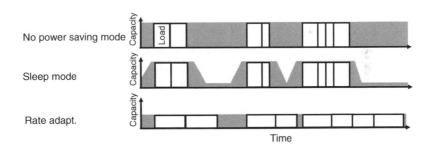

FIGURE 28.6 Illustration of sleep mode and rate adaptation. Squares represent load while the shading represents the operating rate of the equipment.

In general, resource management can be an important enabler for dynamic power management for minimizing power consumption. Resource management is used to re-allocate network resources in such a way that the potential for sleep mode and rate adaptation are maximized. Due to the typical tree shaped structure of the access network, opportunities for shifting resources across different links are limited. However, to the extent that resources can be shifted internally within a network element, resource management is important for minimizing the internal power consumption of a network element.

28.4.1   Sleep Mode

There has been an intense research effort recently on various sleep modes for fixed access systems and in particular for the CPE. The objective of sleep mode is to minimize the power of the CPE whilst still complying with service level requirements. One important differentiator between different approaches is the intended time scale of sleep, e.g. standby mode where sleep follows the daily variations in usage or fast sleep where subsystems/components sleep on a micro/millisecond time scale. Fig. 28.7 shows the potential savings of a standby mode for a simple G-PON optical network terminal (ONT) where all elements that can be put into idle mode autonomously by the ONT have been put into idle mode. As seen in Fig. 28.7, for such a simple style ONT there is a power saving potential of ~30% at times when the ONT is not utilized. In the standby state, further reductions of the standby power may be possible by also exploiting fast sleep for various interfaces such as the uplink TRx. However, this introduces additional complexity and requires coordination across the link with periodic TRx wake-up in order to catch potential incoming wake-up signals. Periodic TRx wake-up needs to be arranged frequently enough so that the system can wake-up quickly in response to activity. Ideally the power of the standby state could be further brought down to 25% of the full ONT power by exploiting TRx sleep in the standby state. An alternative to powering off the full TRx is to limit sleep mode to powering off the Tx. This limits the savings during standby (Tx contribution to TRx power is ~30%) but reduces requirements for periodic wake-up.

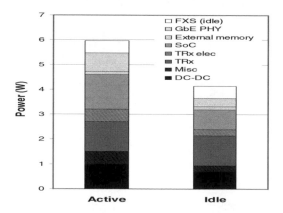

FIGURE 28.7 Power consumption of a simple style G-PON ONT in the active state and in an idle state when everything that can be put to idle mode autonomously has been put to idle mode.

Fast sleep is where components/subsystems are put to sleep on a milli/micro second time scale during low load traffic in order to increase load adaptive power behavior. The potential for fast sleep is governed by component/subsystem wake-up time, transitional power consumption associated with wake-up, service level requirements and implications on component/subsystem stress by the cyclic powering. In particular, for PON there have been proposals that involve cyclic sleep of the TRx. In the active state, power saving potential due to fast sleep are likely to be around 10-20% (Fig. 28.7). Schemes involving powering off the ONT transceivers require coordination between the ONT and optical line terminal (OLT) and introduce additional complexity to the access system. The required coordination between the ONT and OLT requires control signals to be defined between the two units. Challenges for the implementation of low power modes involving powering off the transceiver, include defining appropriate control signals and associated control intelligence for scheduling the low power states and developing fast wake-up circuitry and hardware for quickly transitioning between the different power states. Several different signaling schemes have been proposed both within the context of 10G Ethernet-PON (10GE-PON) and 10-Gigabit-capable (XG-PON). Ref. [9] describes a signaling scheme for dynamically scheduling synchronized sleep periods of all ONTs participating in the cyclic sleep mode. Ref. [10] defines a sleep and periodic wake-up (SPW) method and associated control messages for scheduling ONT sleep. Ref. [11] describes a sleep signaling scheme for 10GE-PON referred to as adaptive lock step (ALS) scheduling and also describes fast wake-up circuitry for an ONT evaluated using the ALS scheme. In the XG-PON standard, control messages have been defined for supporting two modes of low power operations referred to as *cyclic sleep* and *doze mode* [12]. One variant of fast sleep is cyclic sleep which is a state consisting of timed transitions between sleep and awake. One of the more interesting parameters for the operation of cyclic sleep of an ONT is the length of the cyclic sleep period. The length of this period is governed by QoS requirements. For stringent QoS requirements, opportunities for sleep are however limited. Service-awareness of the sleep mode control could then in principle enhance savings by adjusting the cyclic sleep period based on requirements of active traffic.

28.5   Conclusions

The access network represents a significant contribution to network power consumption in today's networks. There are three main aspects of importance for understanding future trends and bottlenecks associated with power consumption and these are technology/system design, architecture and dynamic power management. In this chapter we presented models for power consumption of state-of-the-art technology showing the power-per-line potential of different systems. A critical question for the future is to understand which are the low power future system alternatives that can support future bandwidth requirements and how these concepts can support savings achievable on an architectural level (by increased sharing of network resources) as well as by dynamic power management. Concerning architecture we noted that there are significant savings for node-consolidation already for very small degrees of consolidation, while the additional gain eventually decreases with increasing consolidation factor. Hence, for node-consolidation there is an expected sweet spot between savings on an architectural level and savings in pure system power consumption. Finally we discussed significant savings that may come from dynamical power management. In the access, large savings are expected already for quite simple standby modes where CPE power is reduced at times of idleness. More aggressive power saving modes are also being explored that can both reduce the power of the standby state and provide additional savings also during low load traffic.

28.6 Glossary

ADSL2+: Asymmetric digital subscriber line.

CDR: Clock and data recovery.

CPE: Customer premisses equipment.

CO: Central office.

DSL: Digital subscriber line.

EPON: Ethernet-PON.

FE: Fast Ethernet.

FTTH: Fiber-to-the-home.

FXS: Foreign exchange subscriber.

G-PON: Gigabit-capable PON.

GE: Gigabit Ethernet.

GE-PON: Gigabit Ethernet PON.

ICT: Information and communication technologies.

OFDM: Orthogonal frequency division multiplexing.

OLT: Optical line terminal.

ONT: Optical network terminal.

ONU: Optical network unit.

PON: Passive optical network.

PtP: Point-to-point.

Rx: Receiver.

SERDES: Serializer/deserializer.

SoC: System on chip.

TCO: Total cost of ownership.

TDM: Time division multiplexing.

TDMA: Time division multiple access.

TRx: Transceiver.

Tx: Transmitter.

UD-WDM: Ultra-dense WDM.

VDSL2: Very high-rate digital subscriber line.

WDM: Wave-length division multiplexing.

XG-PON: 10-gigabit capable PON.

Bibliography

[1] T. C. Group, "Smart 2020: Enabling the low carbon economy in the information age," Tech. Rep., 2008.

[2] R. Bolla, R. Bruschi, F. Davoli, and F. Cucchietti, "Energy efficiency in the future internet: A survey of existing approaches and trends in energy-aware fixed network infrastructures," *IEEE Communications Surveys & Tutorials*, vol. 13, pp. 223–244, May 2011.

[3] Ecobilian, "Developing a generic approach for ftth solutions using lca methodology," FTTH Council Europe, Genval Belgium, Tech. Rep., 2007.

[4] K. Grobe, M. Roppelt, A. Autenrieth, J.-P. Elbers, and M. Eiselt, "Cost and energy consumption analysis of advanced wdm-pons," *IEEE Communications Magazine*, vol. 49, pp. s25–s32, Feb. 2011.

[5] *GPON power conservation*, ITU-T Std. ITU-T G-series Recommendations - Supplement 45, 2009.

[6] "Code of conduct on energy consumption of broadband equipment (version 4)," European Commission, Ispra, Tech. Rep., 2011.

[7] "D4.2.1 - technical assessment and comparison of next-generation optical access system concepts," EU FP7 project OASE, Tech. Rep., 2011.

[8] "D2.1 - requirements for next-generation optical access," EU FP7 project OASE, Tech. Rep., 2010.

[9] O. Haran, L. Khermosh, and V. Vaisleib, "Methods and devices for reducing power consumption in a passive optical network while maintaining service continuity," Israel Patent 0 263 127 A1, Oct. 22, 2009.

[10] R. Kubo, J. Kani, Y. Fujimoto, N. Yoshimoto, and K. Kumozaki, "Proposal and performance analysis of a power-saving mechanism for 10 gigabit class passive optical network systems," in *NOC 2009*, Valladolid, Spain, Jun. 2009.

[11] S.-W. Wong, S.-H. Yen, P. Afshar, S. Yamashita, and L. G. Kazovsky, "Demonstration of energy conserving tdm-pon with sleep mode onu using fast clock recovery circuit," in *OFC 2010*, San Diego, CA, Mar. 2010.

[12] *10-Gigabit-capable passive optical network (XG-PON) systems*, ITU-T Std. ITU-T G-series Recommendations - G987.x, 2010.

Author Contact Information

Björn Skubic, Alexander Lindström, Einar In de Betou, and Ioanna Pappa are with Ericsson Research, Ericsson AB, Sweden, Email: bjorn.skubic@ERICSSON.COM.

Index